Automotive Simulation World Congress 2016 and Electronics Simulation Expo 2016

Munich, Germany
7 - 8 June 2016

ISBN: 978-1-5108-4104-8

Printed from e-media with permission by:

Curran Associates, Inc.
57 Morehouse Lane
Red Hook, NY 12571

Some format issues inherent in the e-media version may also appear in this print version.

Printed by Curran Associates, Inc. (2017)

ANSYS Inc.
Southpointe
2600 ANSYS Drive
Canonsburg, PA 15317
USA

Phone: 844.Go.ANSYS
844.462.6797
Fax: 724.514.9494

www.ansys.com

Additional copies of this publication are available from:

Curran Associates, Inc.
57 Morehouse Lane
Red Hook, NY 12571 USA
Phone: 845-758-0400
Fax: 845-758-2633
Email: curran@proceedings.com
Web: www.proceedings.com

TABLE OF CONTENTS

POWERTRAIN: COOLING & LUBRICATION

POWERTRAIN: LIGHTWEIGHTING

BODY & CHASSIS: BRAKES & PD PUMPS

BODY & CHASSIS: AERO OPTIMIZATION

BODY & CHASSIS: WIND NOISE

EMOBILITY: MOTORS & GENERATORS

EMOBILITY: ELECTRIC DRIVES

BODY AND CHASSIS: MANUFACTURING & STRUCTURAL INTEGRITY

CLOSING PLENARY SESSION

OPENING PLENARY SESSION

ANSYS

Simulating Smart, Efficient Vehicles

Welcome Address
Automotive Simulation World Congress & ANSYS Electronics Simulation Expo

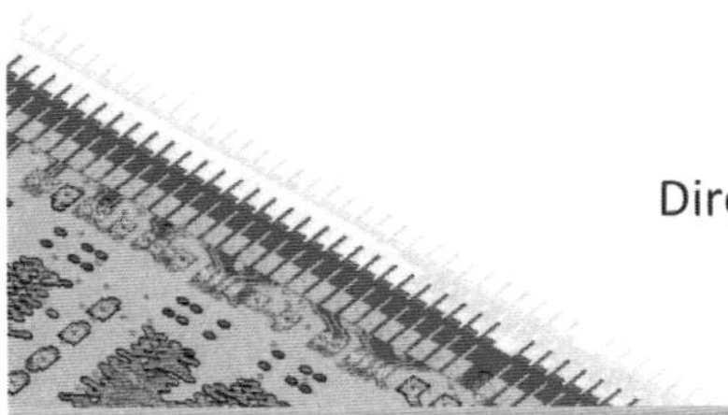

Sandeep Sovani
Director, Global Automotive Industry
ANSYS, Detroit

Automotive and High-Tech Fusion

ANSYS

Major investments in autonomous vehicle test facilities

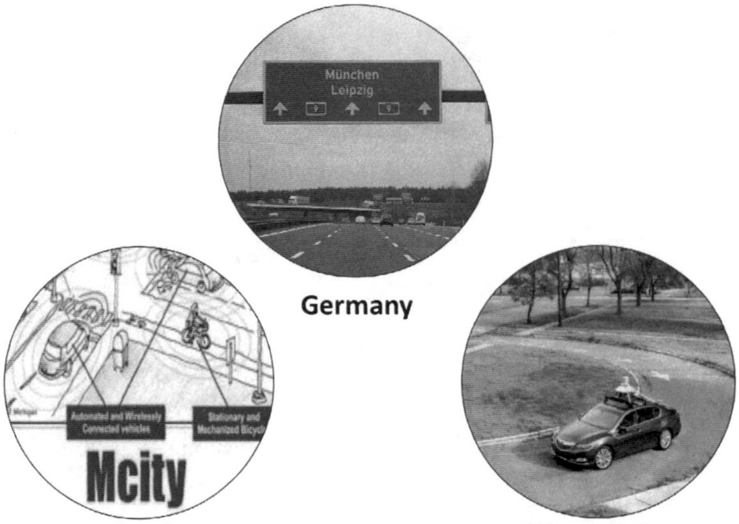

Germany

Detroit

Silicon Valley

ANSYS

Role of simulation is greater than ever

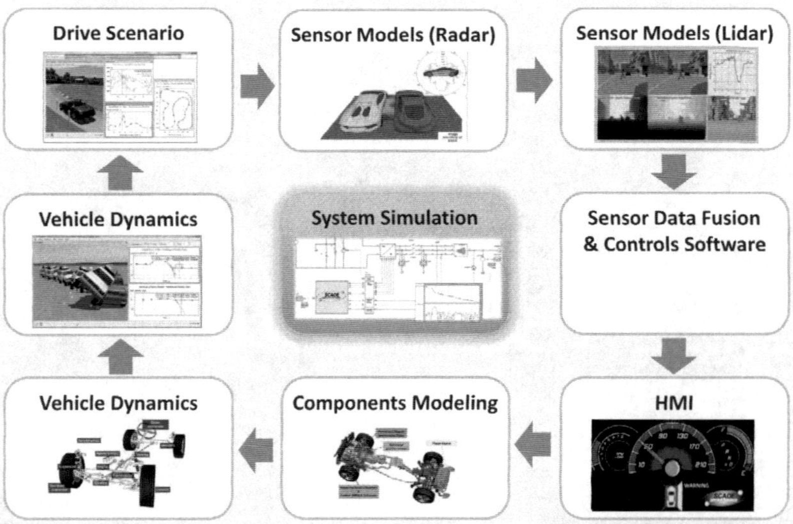

ADAS and Autonomous Driving – Control Loop

ANSYS

Role of simulation is greater than ever

Chip-Package Systems

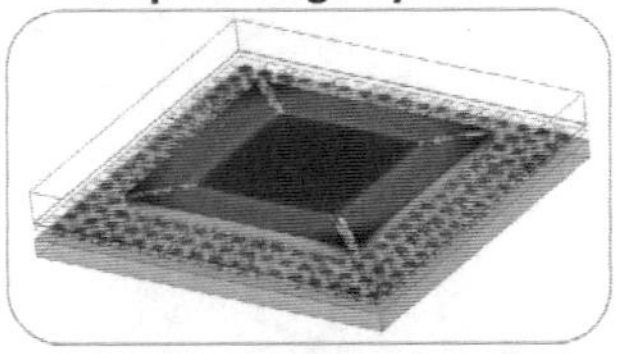

Antennas

Human-Machine Interfaces

Embedded Software

ANSYS

Role of simulation is greater than ever

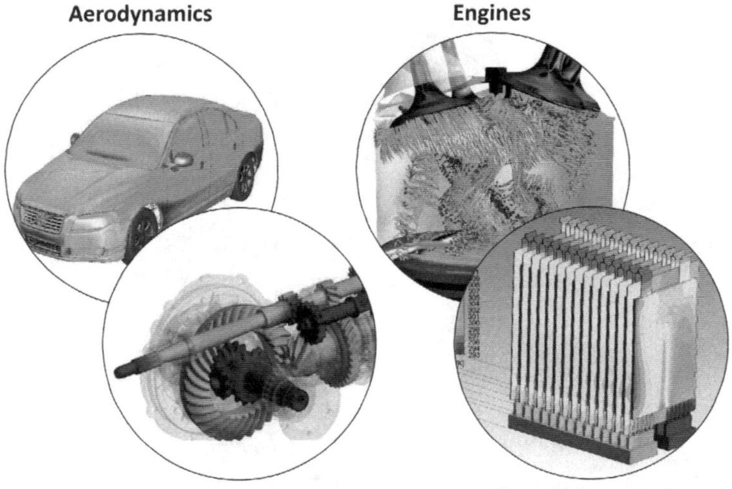

ANSYS

Automotive Simulation World Congress – Journey

2012 Detroit | 2013 Frankfurt | 2014 Tokyo | 2015 Detroit

2012 +
2013 +
2014 +
2015

250 presentations

From 15+ major automotive countries

1400+ attendees

From 450+ organizations

ANSYS

ANSYS Electronics Simulation Expo

Canada, China, France, Germany, India, Israel, Japan, Korea, Sweden, United Kingdom

2014 +
2015 +

180 presentations

From 10 major High Tech countries

1700+ attendees

From 300+ organizations

ANSYS

2016 Conference Overview

ANSYS

2017

Automotive Simulation World Congress

2017 Asia

2017 Europe

ANSYS Electronics Simulation Expo

Stay Tuned On:

www.ansys.com/aswc
www.ansys.com/aese

ANSYS

Awards

Automotive Simulation World Congress

1. Best Engine Simulation
2. Best Powertrain Simulation
3. Best Body & Chassis Simulation
4. Best eMobility & Autonomous Driving Simulation
5. Best Systems, HPC & Cloud Simulation
6. Best Electronics Simulation
7. Best Automotive Simulation

ANSYS Electronics Simulation Expo

1. Best Electronics Simulation

ANSYS

Conference Sponsors – Thanks for your support!

Platinum Sponsors

amazon web services · Hewlett Packard Enterprise · intel

Gold Sponsors

CRAY · Lenovo · sgi

Silver Sponsors

BETA CAE Systems SA · CEI · FUJITSU · mode FRONTIER · noesis Empower your experience · sharc · OpenIT

Media Sponsor

ATZ

ANSYS

Conference Organization Team – Thank you!

ANSYS Customer Excellence Team | ANSYS Account Managers Team

ANSYS

#aswc2016

ANSYS

ANSYS

Engineering the Internet of Things

Jim Cashman

President & CEO
ANSYS Inc

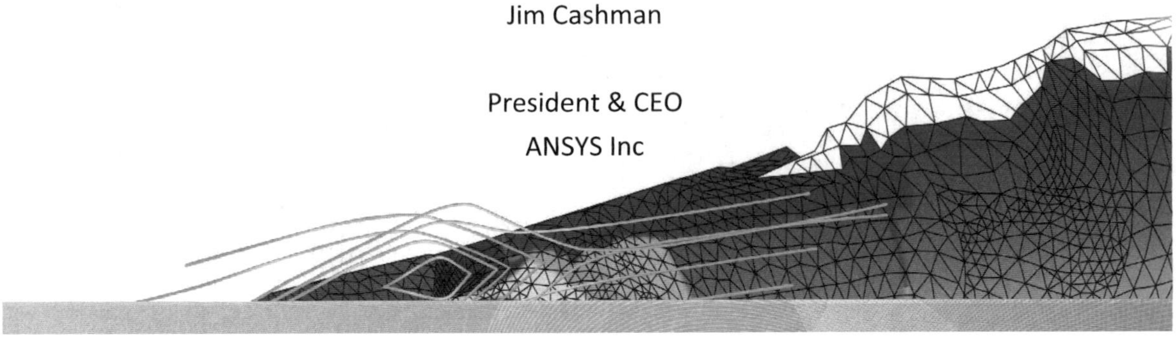

Congratulations to BMW on Centennial: 1916-2016

Industry 4.0 – The 4th industrial revolution has started, enabled by the Internet of Things

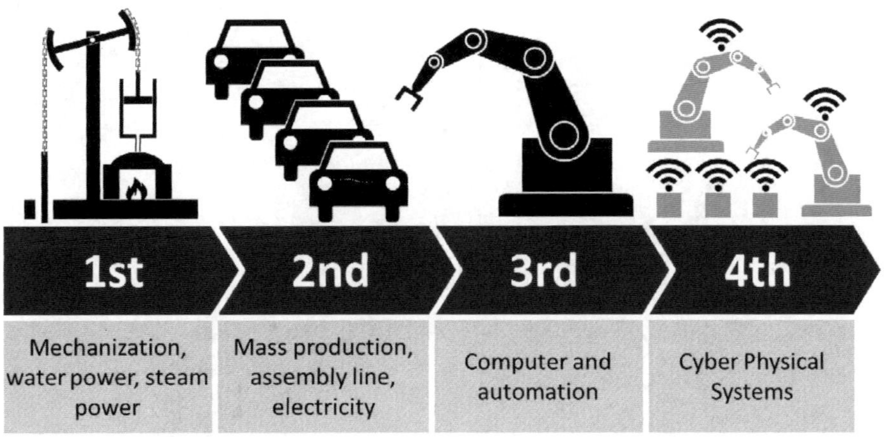

Image Credit: *Christoph Roser at AllAboutLean.com ; Wikimedia Commons*

A confluence of technology advances is driving the IoT revolution

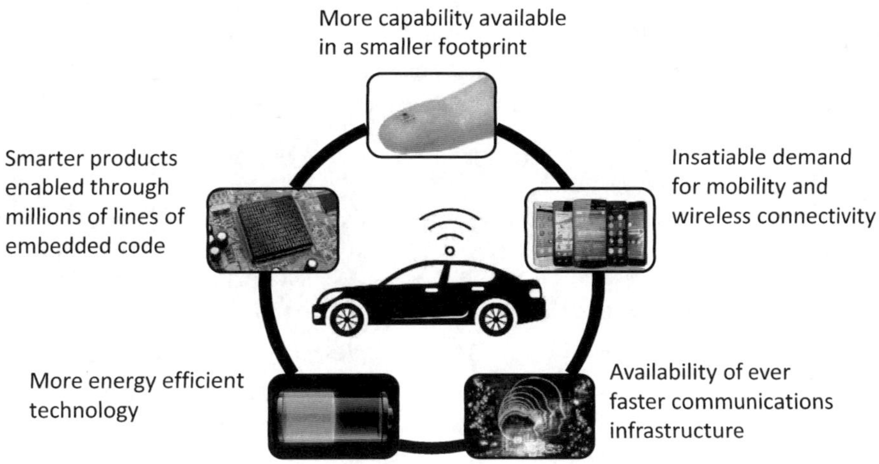

The Internet of Things is making pervasive changes across many industries

ANSYS

Jim Fowler, CIO General Electric

"GE is undergoing the largest and most important transformation in our 140 year history. We are starting to combine our physical world of machines with the logical and digital world of software"

Jeffery Immelt, CEO General Electric

"If you went to bed last night as an industrial company, you are going to wake up in the morning as a software and analytics company"

Sources: General Electric https://www.youtube.com/watch?v=UUw4SfXzMrw
Amazon Web Services https://www.youtube.com/watch?v=i1yW6vWCpgk

ANSYS

THE WALL STREET JOURNAL.

Home World U.S. Politics Economy Business Tech Markets Opinion Arts Life Real Estate

Bosch CEO Remakes Company Around 'Internet of Things'

German manufacturer envisions smart everything, embraces Silicon Valley

'Everything we do in this networked world has to be open,' says Volkmar Denner, chief executive of private conglomerate Bosch. PHOTO: FLORIAN PELJAK FOR THE WALL STREET JOURNAL

" . . . aims to connect everyday wares and devices over the Internet, fueling the rise of 'smart' homes and cars"

Reference: http://www.wsj.com/articles/bosch-ceo-remakes-company-around-internet-of-things-1431336782

ANSYS

The connected, autonomous car is a key hub in the Internet of Things

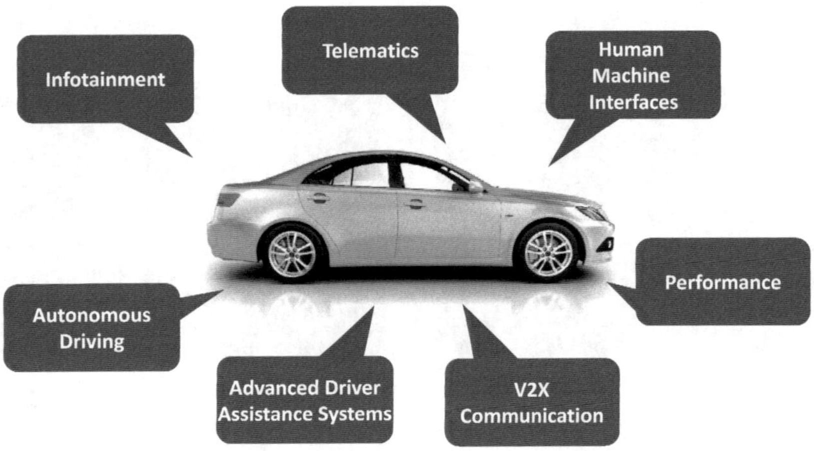

ANSYS

Underlying IOT technologies need to be engineered thoroughly for reliability and safety

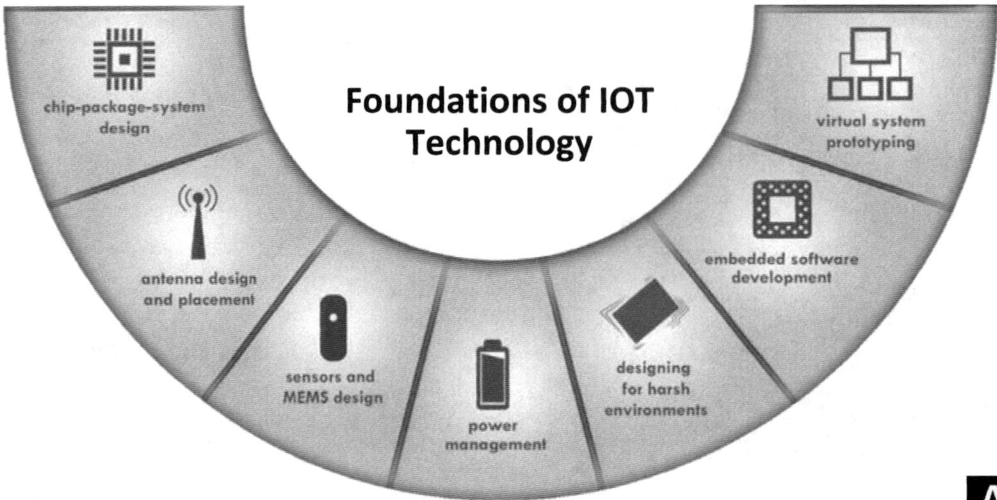

While connected car features are a delight for consumers, they are a nightmare for automotive engineers

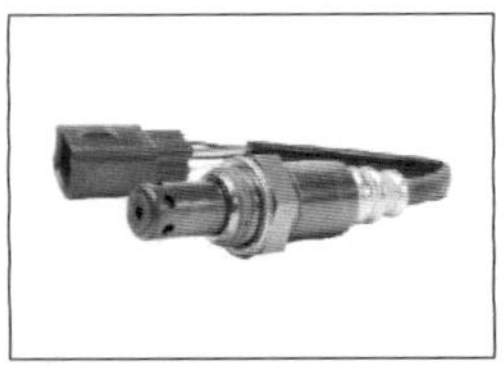

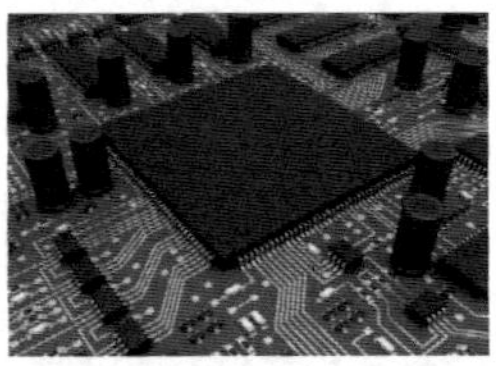

100s of sensors X 30 – 100 ECUs X 20 – 100 Million Lines of Code X Dozens of antennas

= Extreme complexity

Complexity invites failure

Something left unaddressed somewhere in the millions of combinations of design parameters and operating conditions, can cause catastrophe

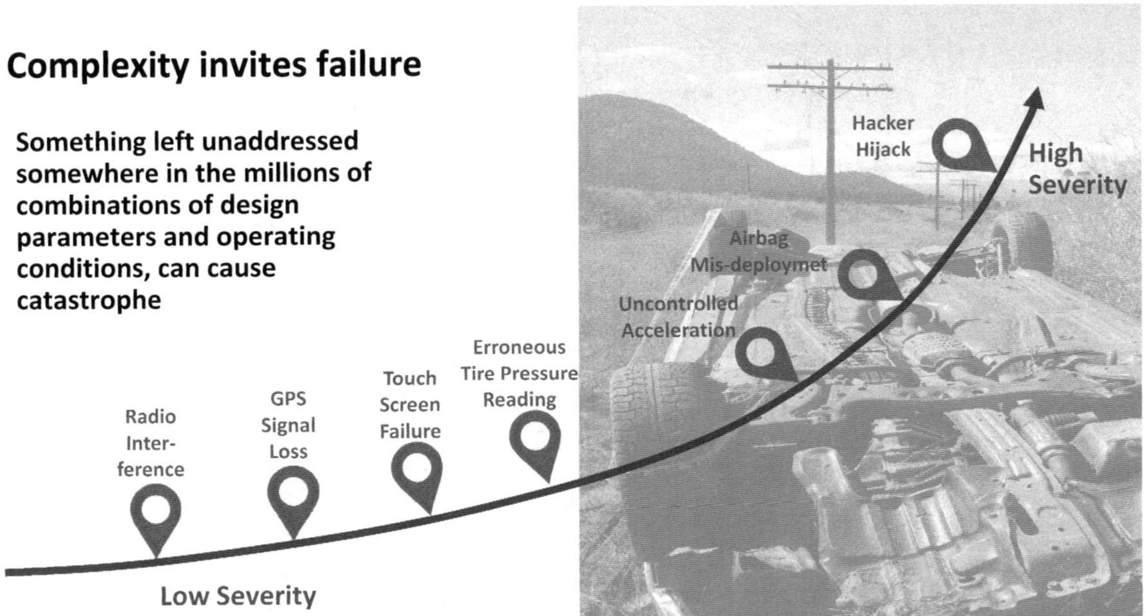

Case Study: Unintended Acceleration

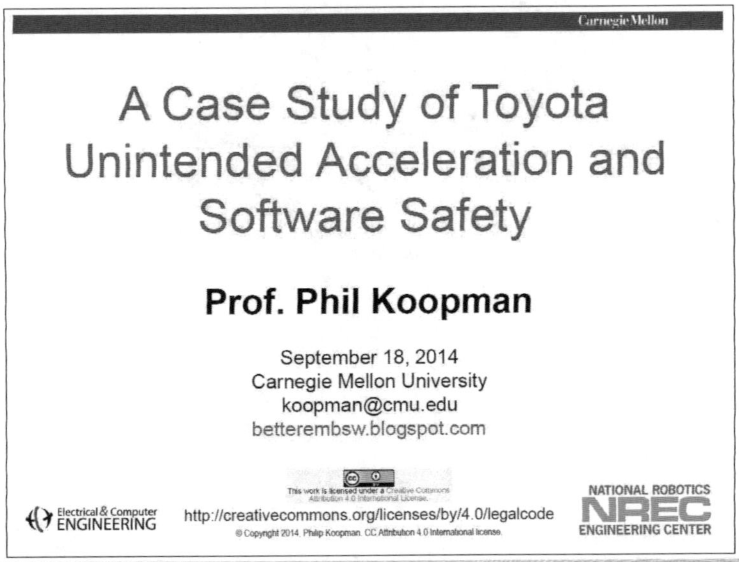
Carnegie Mellon

A Case Study of Toyota Unintended Acceleration and Software Safety

Prof. Phil Koopman

September 18, 2014
Carnegie Mellon University
koopman@cmu.edu
betterembsw.blogspot.com

This work is licensed under a Creative Commons Attribution 4.0 International License.
http://creativecommons.org/licenses/by/4.0/legalcode
© Copyright 2014. Philip Koopman. CC Attribution 4.0 International license.

Electrical & Computer ENGINEERING

NATIONAL ROBOTICS NREC ENGINEERING CENTER

ANSYS

Case Study: Unintended Acceleration

Carnegie Mellon

Didn't Vehicle Testing Make It Safe?

- Vehicle level testing is useful and important
  - Can find unexpected component interactions
- But, it is impracticable to test everything at the vehicle level
  - Too many possible operating conditions, timing sequences
  - Too many possible faults, which might be intermittent
    - Combinations of component failures + memory corruption patterns
    - Multiple software defects activated by a sequence of operations

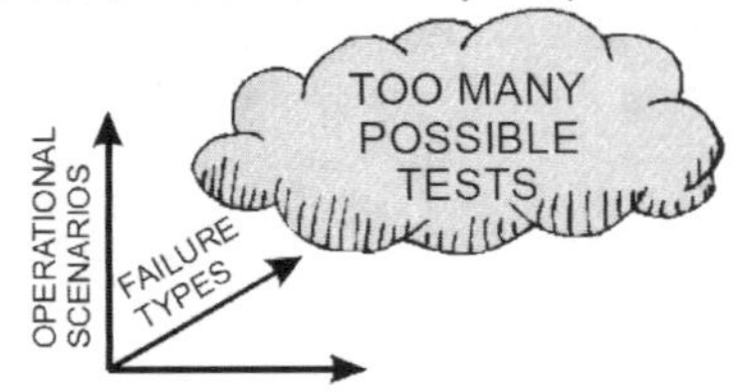

Electrical & Computer ENGINEERING

19

ANSYS

Simulation is a key enabler in engineering the IOT

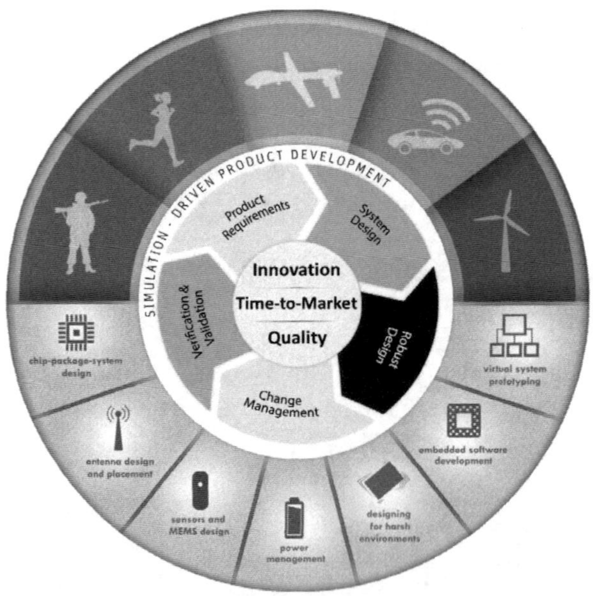

Simulation makes it possible to test 1000s of design scenarios to optimize performance and reduce malfunction

ANSYS

Simulation example: Identify and mitigate EMI in upfront design

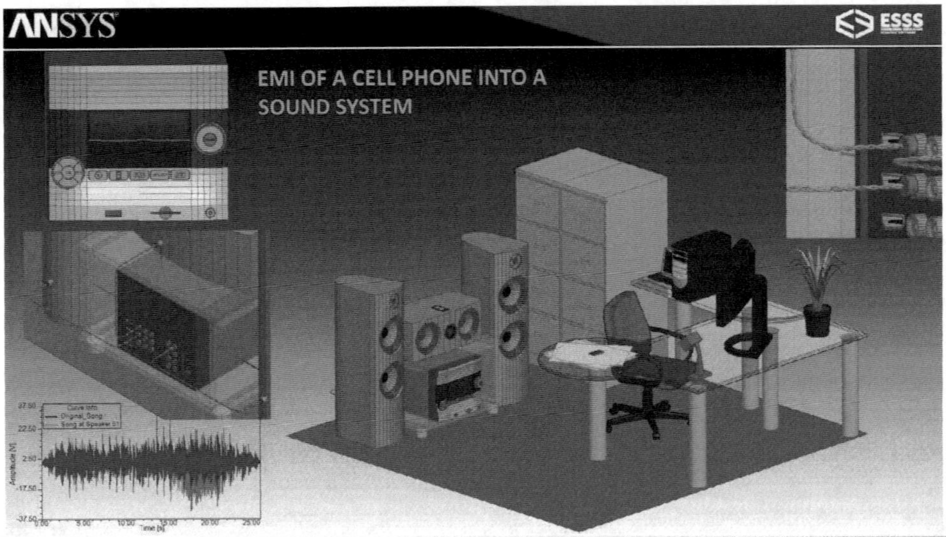

Simulation example: Virtual development of autonomous radar

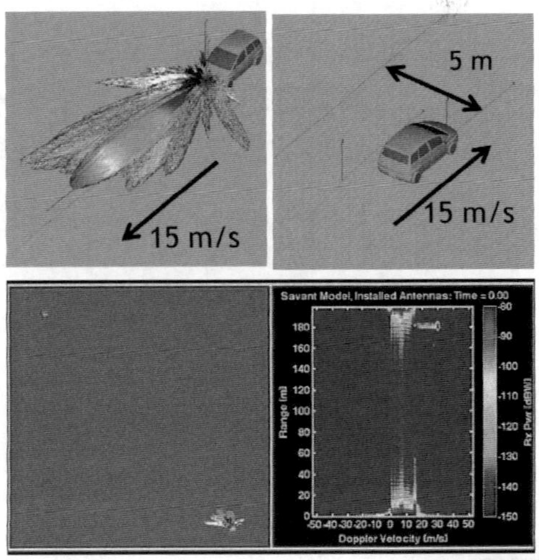

Benefits of engineering simulation cannot be overlooked

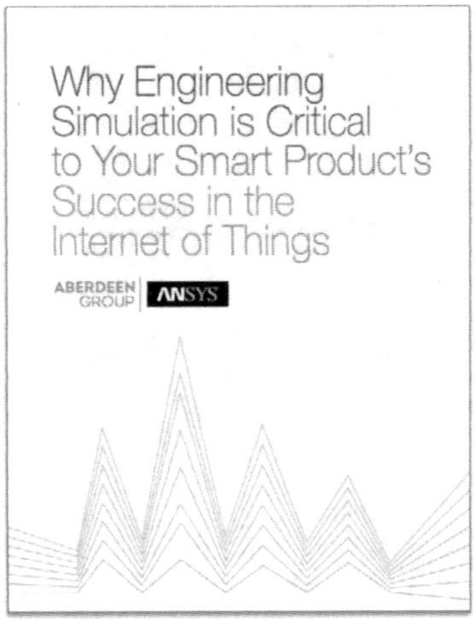

Download at: www.ansys.com/iot

The Benefits Cannot be Overlooked

SIMULATION vs NO SIMULATION
Simulated Environments Experience:

Length of Development Time	▼	**9x** reduction
Overall Product Cost	▼	**4x** reduction
Warranty Costs	▼	**89%** more likely to decrease
Number of Change Orders After Release to Manufacturing	▼	**2.5x** more likely to decrease
Percent Successful New Product Introduction (NPI) Rate	▲	**67%** more likely to achieve

Source: The Aberdeen Group

ANSYS

ANSYS simulation is enabling companies to expedite IOT innovation

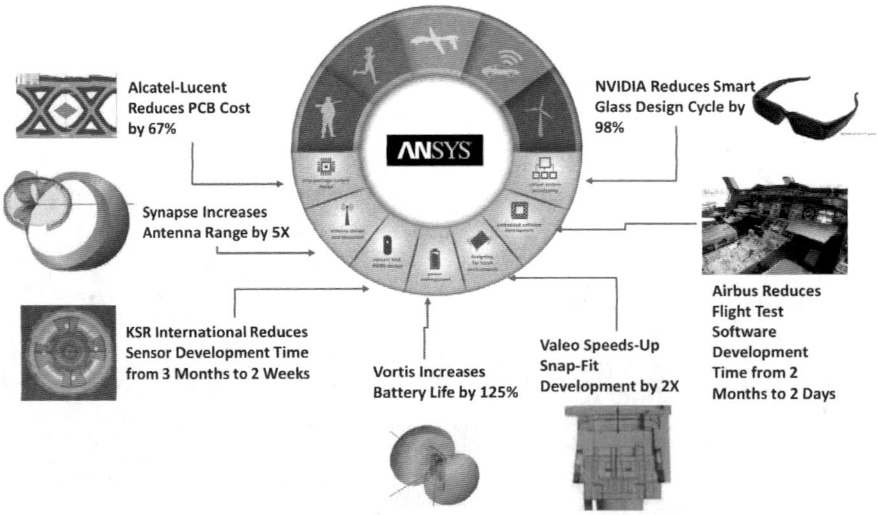

ANSYS

Best-in-class companies take a holistic approach to product development, and need a consolidated simulation platform

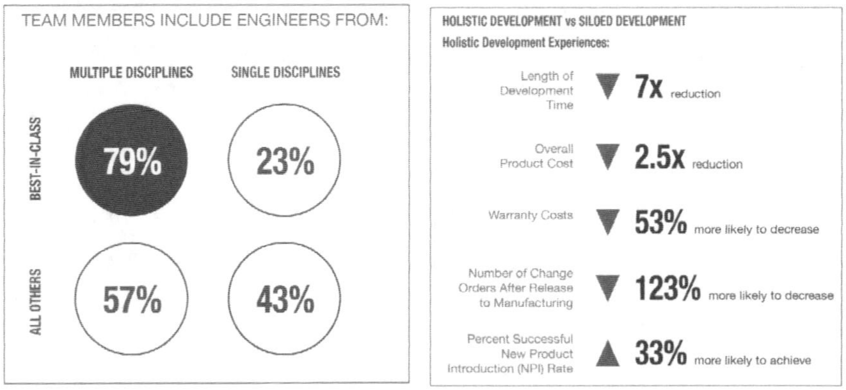

ABERDEEN GROUP | **39%** of respondents indicated that lack of collaboration is the top product development challenge they face everyday

ANSYS

ANSYS has consistently lead the way in providing a consolidated simulation platform

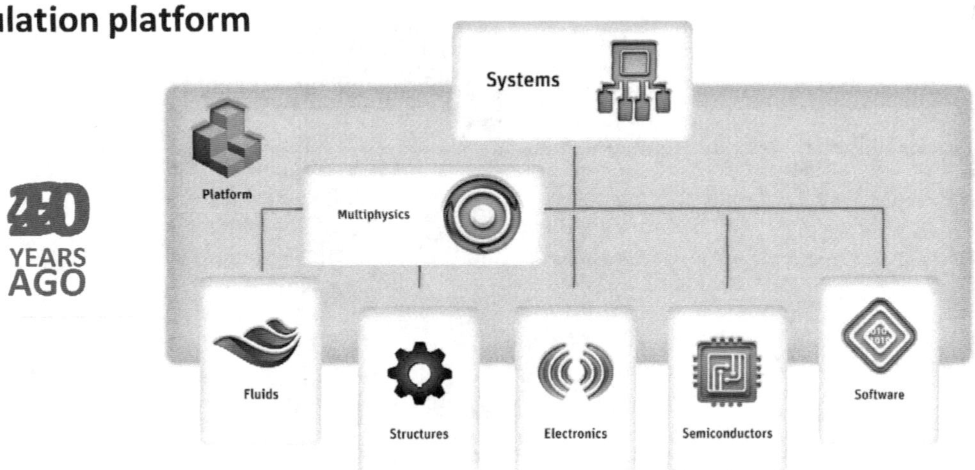

ANSYS

Consistent investment in our vision

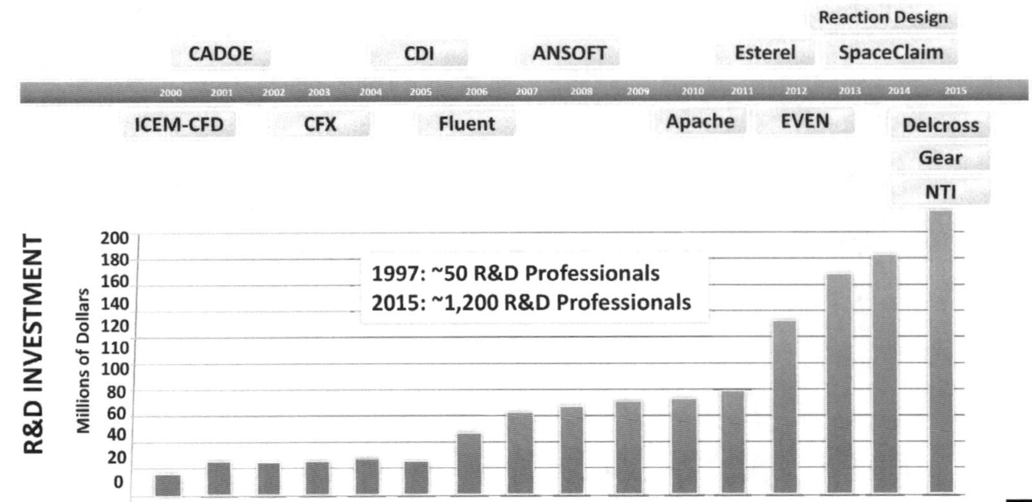

ANSYS 17.0

Investing in a proven strategy

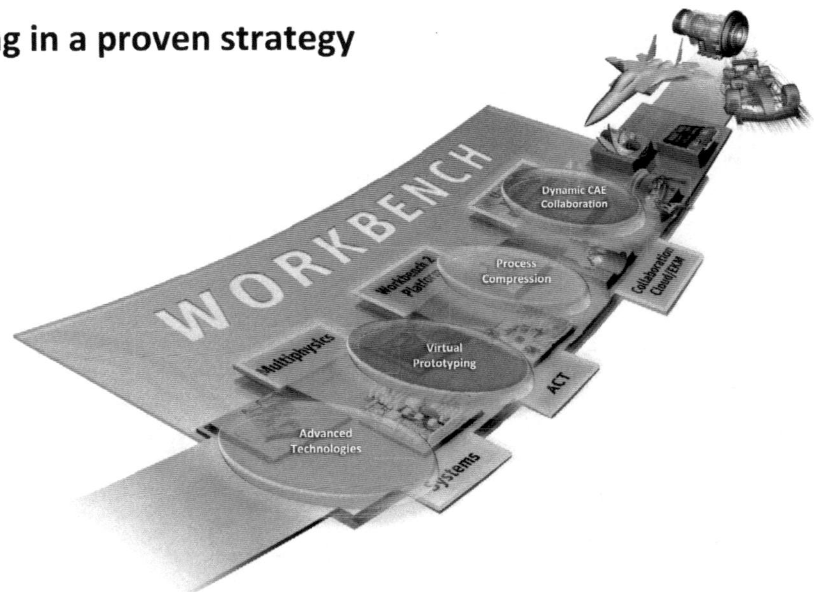

Advanced Technology: Example - 10x faster transient solutions

17 Advanced Technologies

Maxwell 3D R16	Maxwell 3D R17
50,000 Tets	50,000 Tets
75 sec per time step	3.6 sec per time step
About 8 hrs total	About 22.5 mins total

Advanced Technologies

Advanced Technology: Example - 10x faster meshing

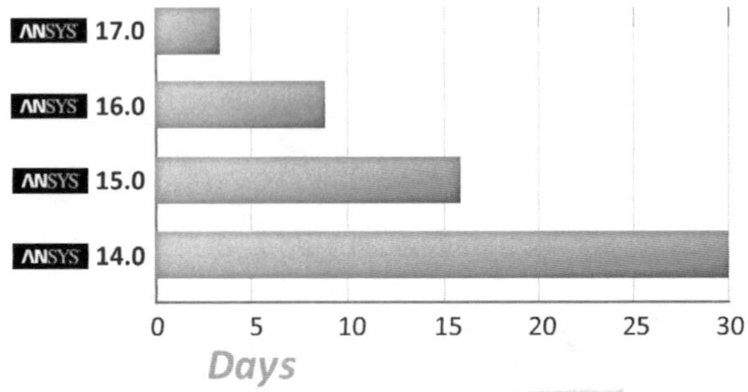

Volvo S80
Courtesy of Volvo

Virtual Prototyping

Virtual Prototypes: Example -10x productivity for system verification

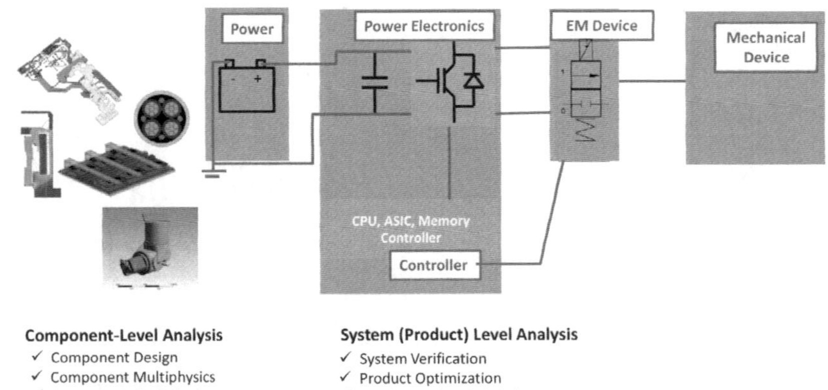

Component-Level Analysis
- ✓ Component Design
- ✓ Component Multiphysics
- ✓ Component Optimization

System (Product) Level Analysis
- ✓ System Verification
- ✓ Product Optimization
- ✓ Test for Product Reliability

Process Compression: Example
- 10x faster structural insight (HPC)

Process Compression

- Order of magnitude improvement in recent releases
- Clear industry leadership

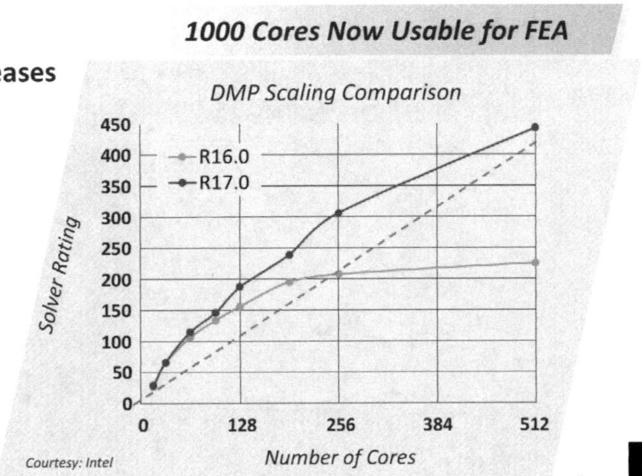

ANSYS

Dynamic CAE collaboration: Example
– 10x more complete system

Dynamic CAE Collaboration

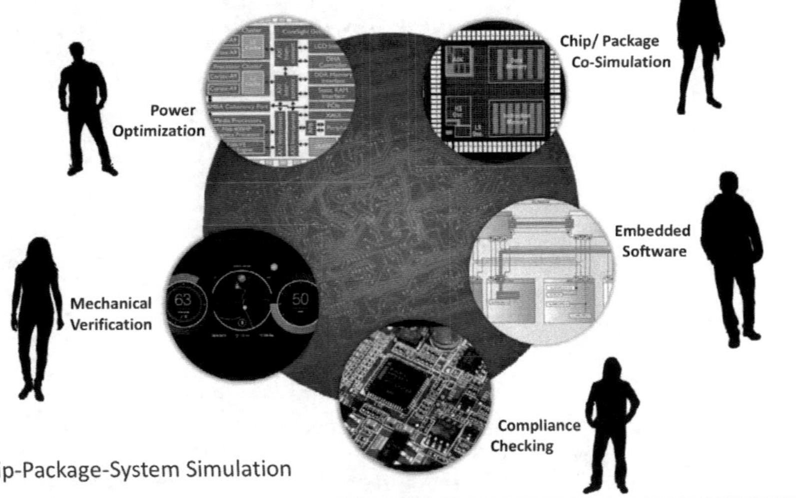

End-to-End Chip-Package-System Simulation

ANSYS

The evolution of simulation

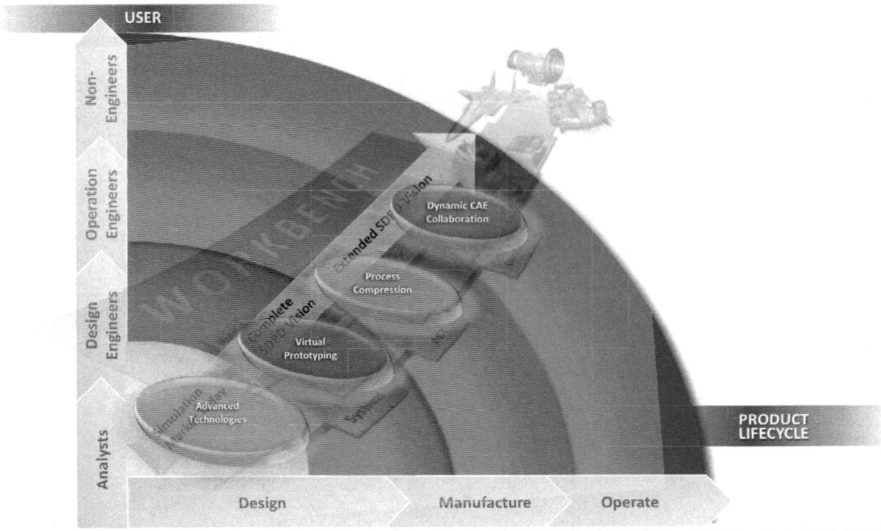

ANSYS

ANSYS and GE partner to create smarter machines

PRODUCT LIFECYCLE

ANSYS And GE Work Together To Enable A World Of Smarter Industrial Machines

Released : 29 Sep 2015

PITTSBURGH, Sept. 29, 2015 /PRNewswire/ -- ANSYS (NASDAQ: ANSS) announced today that it joined GE's (NYSE: GE) Digital Predix early adopter program. This move combines the power of the leading engineering simulation platform with the first and only industrial cloud offering for industrial data and analytics, GE's Predix cloud platform. As part of the Predix early adopter program, the two companies will demonstrate a pilot integration of ANSYS flagship simulation platform with GE's Predix industrial internet platform today at GE's Minds + Machines event in San Francisco.

Cloud Platform for the Industrial Internet

ANSYS

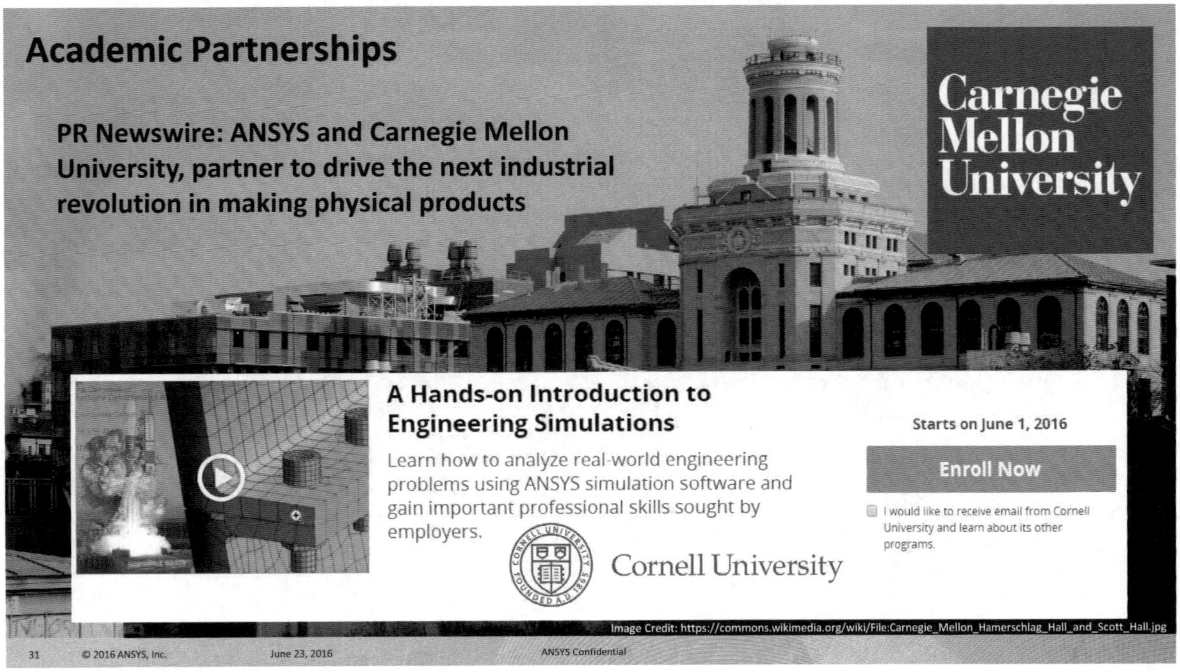
Academic Partnerships
PR Newswire: ANSYS and Carnegie Mellon University, partner to drive the next industrial revolution in making physical products
Carnegie Mellon University
A Hands-on Introduction to Engineering Simulations
Learn how to analyze real-world engineering problems using ANSYS simulation software and gain important professional skills sought by employers.
Cornell University
Starts on June 1, 2016
Enroll Now
I would like to receive email from Cornell University and learn about its other programs.
Image Credit: https://commons.wikimedia.org/wiki/File:Carnegie_Mellon_Hamerschlag_Hall_and_Scott_Hall.jpg
31
© 2016 ANSYS, Inc.
June 23, 2016
ANSYS Confidential

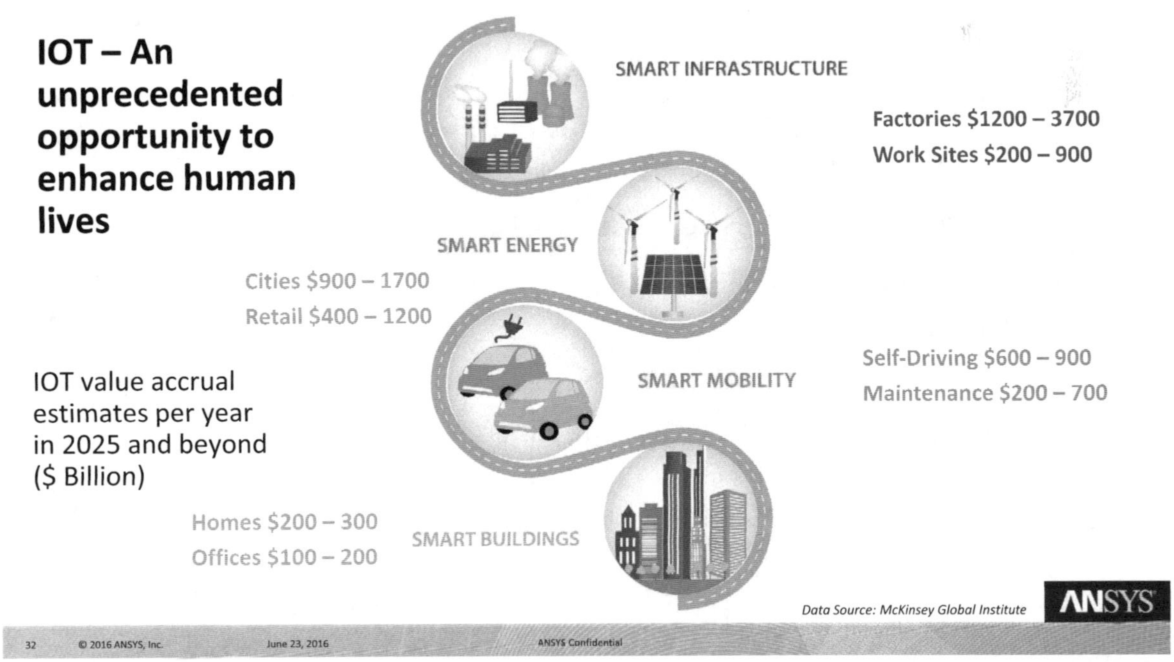
IOT – An unprecedented opportunity to enhance human lives
IOT value accrual estimates per year in 2025 and beyond ($ Billion)
SMART INFRASTRUCTURE
Factories $1200 – 3700
Work Sites $200 – 900
SMART ENERGY
Cities $900 – 1700
Retail $400 – 1200
SMART MOBILITY
Self-Driving $600 – 900
Maintenance $200 – 700
SMART BUILDINGS
Homes $200 – 300
Offices $100 – 200
Data Source: McKinsey Global Institute
ANSYS
32
© 2016 ANSYS, Inc.
June 23, 2016
ANSYS Confidential

Automotive Simulation WORLD CONGRESS

June 7-8, 2016 Munich

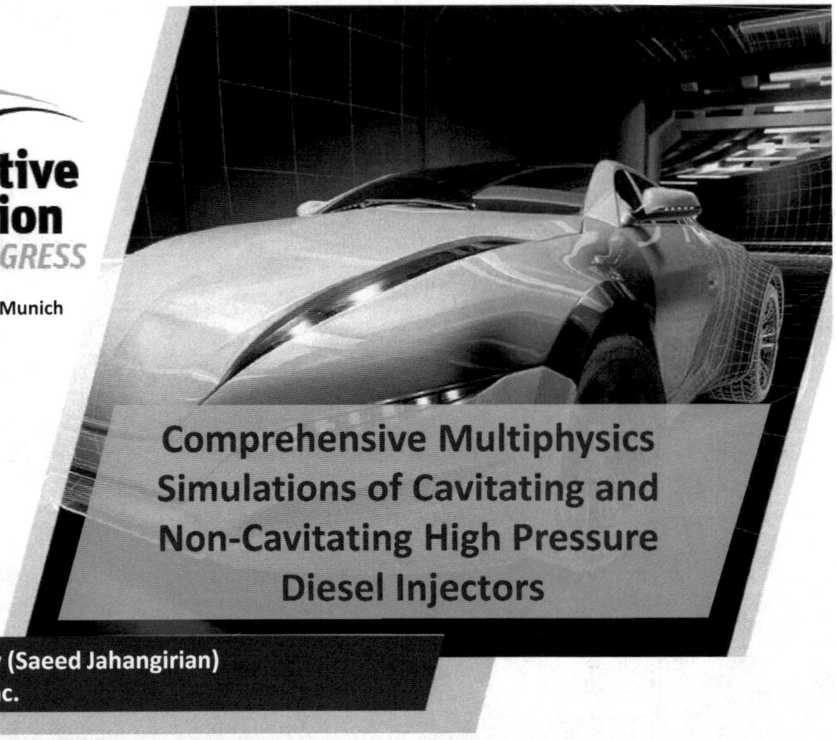

Comprehensive Multiphysics Simulations of Cavitating and Non-Cavitating High Pressure Diesel Injectors

Laz Foley (Saeed Jahangirian)
ANSYS Inc.

Outline

- **High Fidelity CFD Simulations**
  - ECN Spray C and D
- **Multiphysics and System Solutions**
  - Fluid Structure Interaction (FSI)
  - CFD/EM Co-Simulation
  - GT-Suite/Maxwell Co-Simulation
  - Simplorer/Modelon System Modeling

Background

- Engine Combustion Network (ECN) has initiated comprehensive injector studies using diesel Spray A, but has recently extended research to diesel Sprays C and D and gasoline Spray G.
- Spray C and Spray D are single-orifice diesel injectors (similar to Spray A, but more symmetrical and less affected by manufacturing irregularities).
- The nozzle in Spray C is larger than Spray A (nominal dia. 0.20 mm vs. 0.090 mm). Spray D, and particularly Spray C, are expected to be more prone to cavitation than Spray A.
- Hydraulic characterization of Sprays C and D has been done at CMT [8].

Objective

- Characterize cavitation region, if observed, in simulations.
- Analyze influences of modeling assumptions (e.g., compressibility, viscous heating, physical property implementation) on results.

Spray C Geometry & Details

Specifications for Spray C injectors of the Engine Combustion Network (Single Hole)	
Common rail fuel injector	Bosch 3-22
Fuel injector nominal diameter	0.20 mm
Nozzle K factor (Cylindrical)	K = 0
Nozzle shaping	5% hydroerosion
Flow with 10 MPa pressure drop	200 cc/min

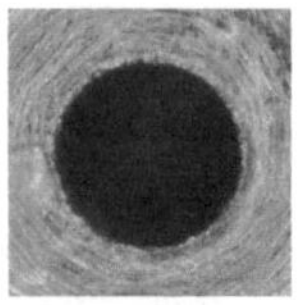

Orifice Exit

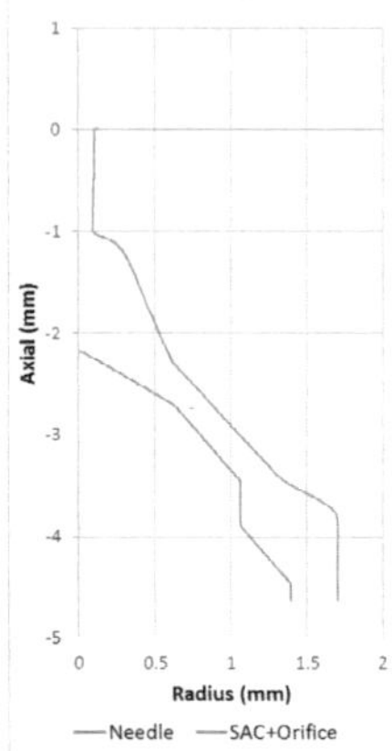

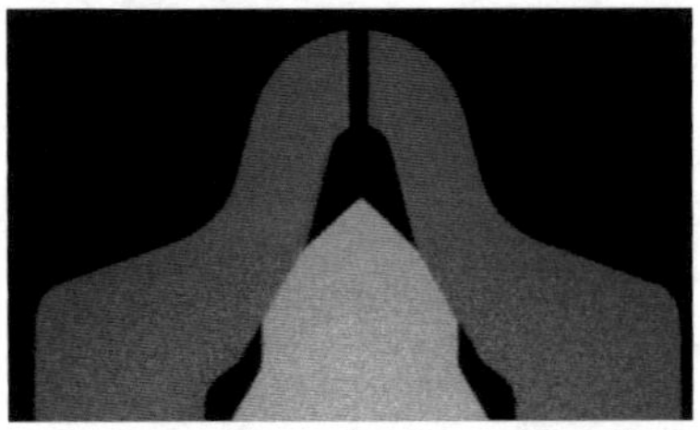

X-Ray Tomography Cross-Section

Ref: http://www.sandia.gov/ecn/cvdata/targetCondition/SpCNozGeo.php

Spray D Geometry & Details

Specifications for Spray D injectors of the Engine Combustion Network (Single Hole)	
Common rail fuel injector	Bosch 3-22
Fuel injector nominal diameter	0.186 mm
Nozzle K factor (Convergent)	K = 1.5
Nozzle shaping	hydroerosion to Cd = 0.86
Flow with 10 MPa pressure drop	228 cc/min

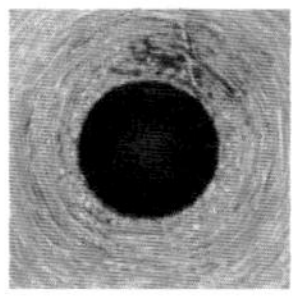

Orifice Exit

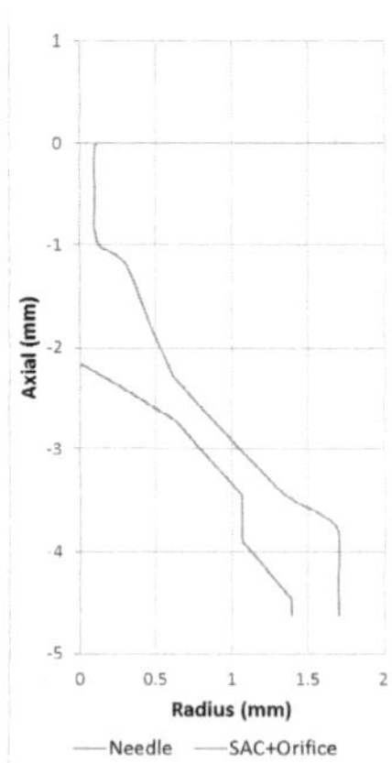

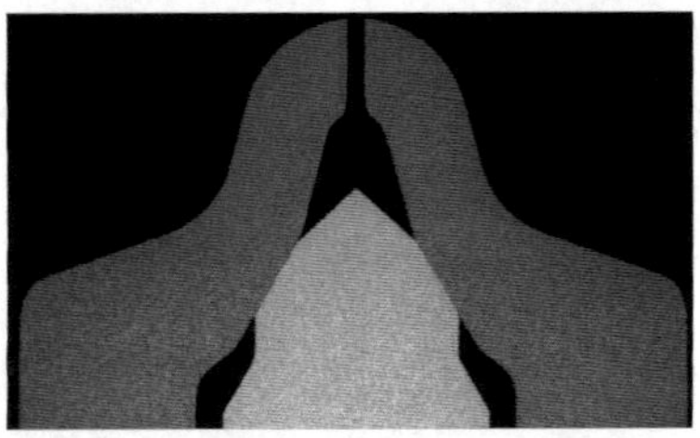

X-Ray Tomography Cross-Section

Ref: http://www.sandia.gov/ecn/cvdata/targetCondition/SpDNozGeo.php

Comparison of Spray C and D Geometries

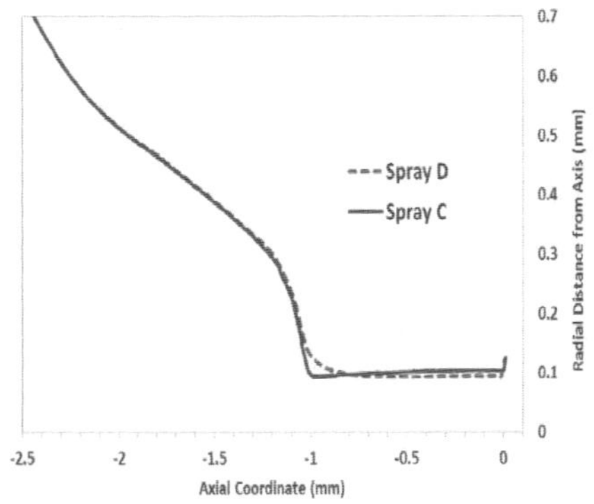

- The nozzle inlet in Spray C has a sharper and less smooth profile than Spray D.
- The nozzle profile of Spray C is slightly diverging and nozzle diameter is slightly larger than Spray D.
- Note the sharp corner at outlet of Spray D has been made smooth in the mesh.

Automotive Simulation WORLD CONGRESS

Simulation Boundary Conditions

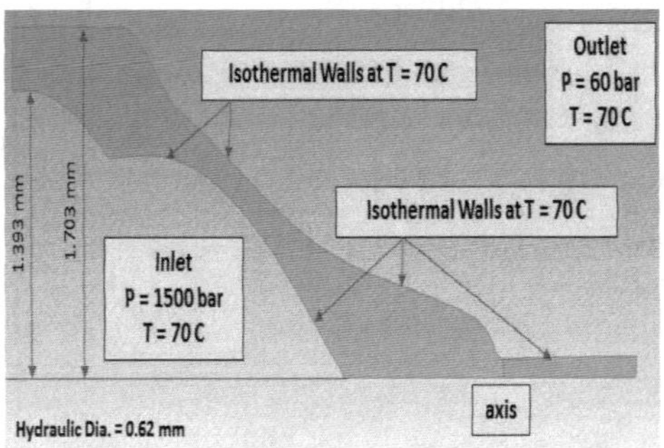

- Configuration of the nozzle is fixed ("steady-state" opening). The needle has already been "unseated"; no "transient needle motion" in the simulations.
- Fuel is n-dodecane. Injection pressure is 150 MPa.
- Fuel reservoir temperature is 70°C. Injector walls assumed to be at 70°C (warm water is in contact with the injector).

Automotive Simulation WORLD CONGRESS

ANSYS Fluent Solution Details

ANSYS Fluent 17.0 is utilized for simulations.	
Liquid fuel	n-Dodecane
Solver	Steady-State Pressure Based Coupled Solver
Equation of State	Non-linear function of *p,T* for liquid, ideal gas for vapor (see following slides)
Cavitation Model	Zwart-Gerber-Belamri [1,2]
Turbulence	RANS SST k-ω (with Compressibility/Viscous Energy Effects)
Spatial Discretization	2nd order for Momentum, TKE, TDE, Energy, QUICK for Volume Fraction (void)
Cell Type	2D Quad mesh with 10 boundary layers (First Layer Thickness = 1 μm)

Automotive Simulation WORLD CONGRESS

Physical Properties CRITICAL to Solution Accuracy!

- **Multiple scenarios studied (for example for Spray C)**
  - **(1) Incompressible liquid and vapor flow; the mass flow rate is 11.269 kg/s.**
  - **(2) Compressible liquid only; the mass flow rate is 11.179 kg/s.**
  - **(3) Compressible liquid and vapor; the mass flow rate is ~10.90 kg/s.**

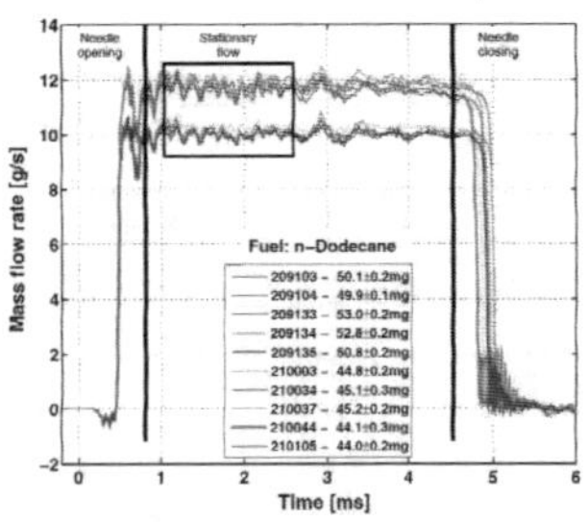

Measurements at CMT [8]

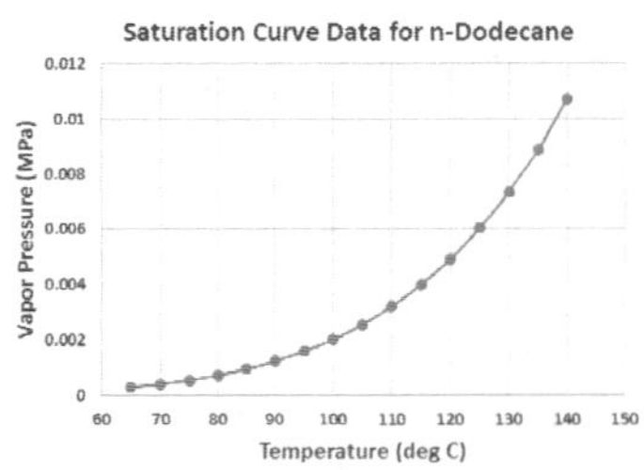

NIST Data

Automotive Simulation WORLD CONGRESS

Detailed Liquid Properties

Property	Type	Reference	Range of Validity	Implementation
Density	Variable, non-linear function of p, T	Caudwell et al. [3]	298.15-473.15 K Up to 200 MPa	UDF
Viscosity	Variable, non-linear function of p, T	Caudwell et al. [3]	298.15-473.15 K Up to 200 MPa	UDF
Specific Heat	Table	NIST Chemistry Webbook [4]	65-140 °C	Piecewise-Linear
Conductivity	0.157 at 150 MPa and 70 °C	Arienti & Sussman [5]	-	Constant
Standard State Enthalpy	-6.6355e+07 at 70 °C	NIST Chemistry Webbook [4]	-	Constant
Speed of Sound	Variable, non-linear function of p, T	Derived from eq. $1.06 * \sqrt{\frac{\partial P}{\partial \rho}}$ using ρ in [3] to match data in Khasanshin [6]	Validity Range of Density in [3]	UDF

Detailed Vapor Properties

Property	Type	Reference	Range of Validity	Implementation
Density	Variable	Ideal Gas*	-	Ideal Gas
Viscosity	Table	NIST Chemistry Webbook [3]	70-140 °C	Piecewise-Linear
Specific Heat	Table	NIST Chemistry Webbook [3]	65-140 °C	Piecewise-Linear
Conductivity	0.011543 at 70 °C	NIST Chemistry Webbook [3]	-	Constant
Standard State Enthalpy	-9049463 at 70 °C	NIST Chemistry Webbook [3]	-	Constant
Vapor Pressure	Table	NIST Chemistry Webbook [3]	65-140 °C	Piecewise-Linear

* Similar variations to the table data for density using the saturation vapor phase data from NIST tables.

Spray C Results

- **Calculated results for mass flow rate, spray momentum and effective velocity are compared against measured data at CMT [8].**

Mass Flow Rate Calculated [g/s]	Mass Flow Rate Measured for Nozzle #210003 [g/s]
10.83	*10.06±0.12**
Momentum Calculated [N]**	**Momentum Measured for Nozzle #210003 [N]**
6.49	*5.65±0.06*
Effective Velocity* Calculated [m/s]**	**Effective Velocity Measured [m/s]**
599.3	*561.8±3.6*

* Standard deviation of 5 nozzle measurements
** Momentum is calculated as "Flow Rate" of axial velocity at outlet of nozzle
*** Effective Velocity = (Momentum Flux)/(Mass Flow Rate)

Spray C Results [1]

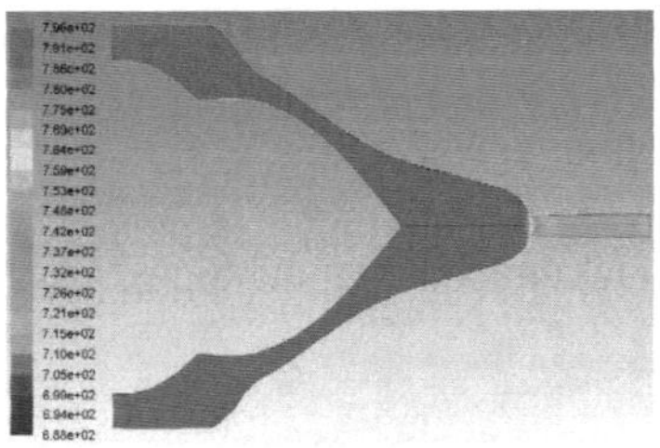
Liquid Density [kg/m^3]

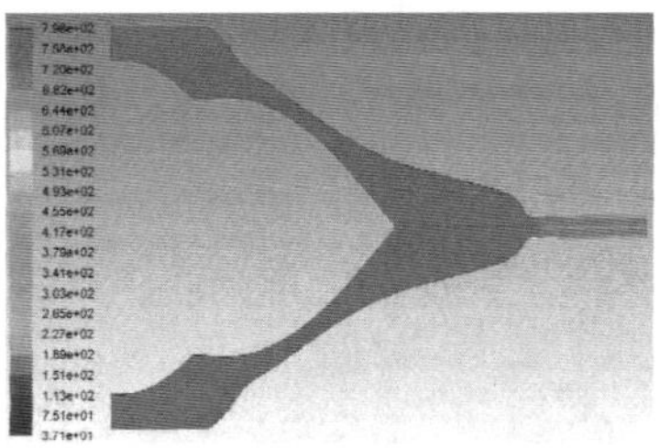
Mixture Density [kg/m^3]

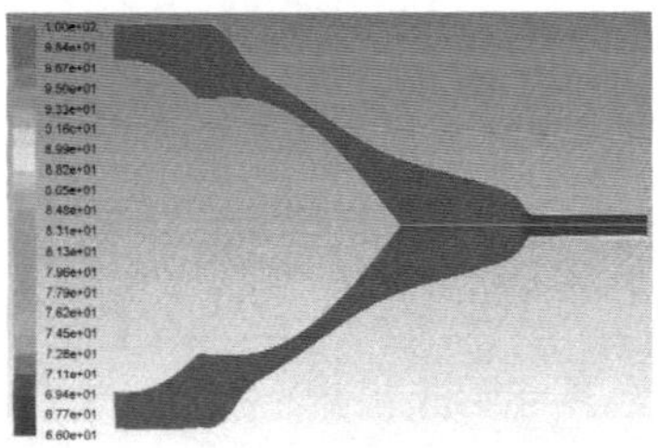
Mixture Temperature [°C]

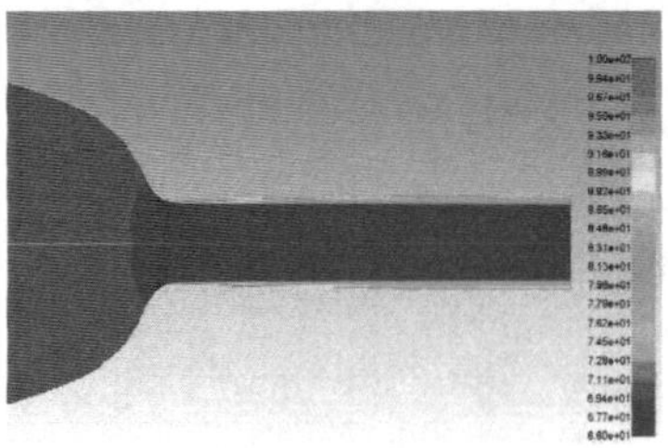
Mixture Temperature [°C]

Spray C Results [2]

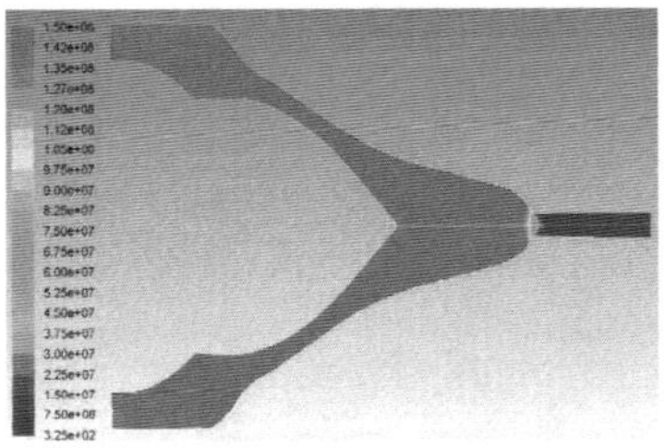

Mixture Static Pressure [Pa]

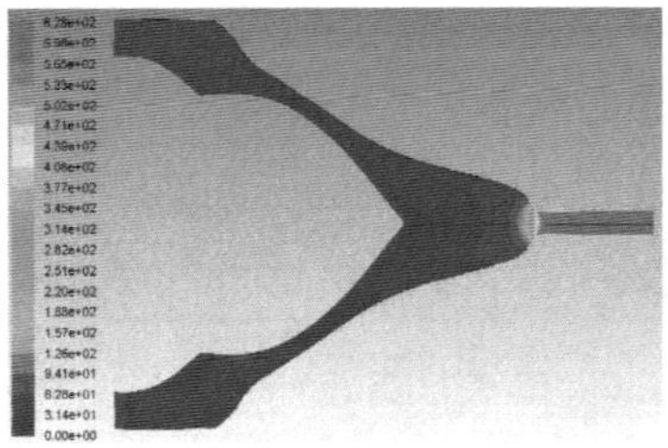

Mixture Velocity Magnitude [m/s]

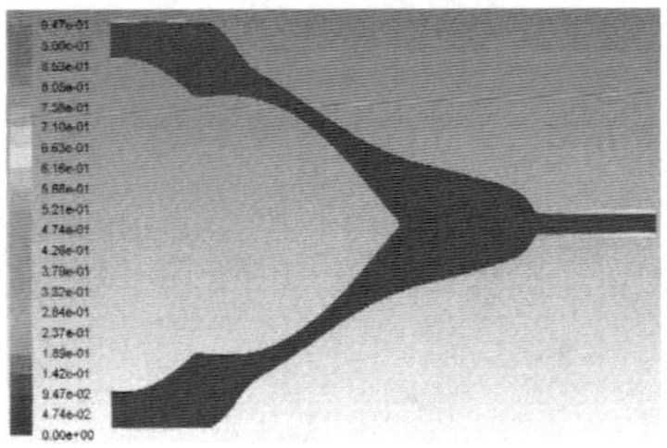

Vapor Volume Fraction

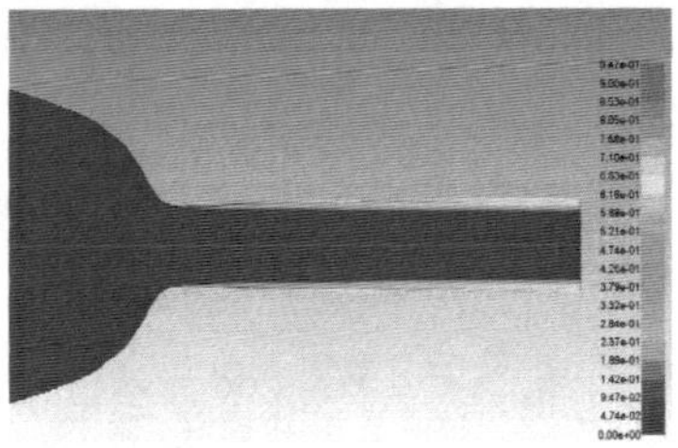

Vapor Volume Fraction

Automotive Simulation WORLD CONGRESS

Spray D Results

- **Calculated results for mass flow rate, spray momentum and effective velocity are compared against measured data at CMT [8].**

Mass Flow Rate Calculated [g/s]	Mass Flow Rate Measured for Nozzle #210003 [g/s]
11.26	*11.71±0.18**
Momentum Calculated [N]**	**Momentum Measured for Nozzle #210003 [N]**
6.62	*6.82±0.12*
Effective Velocity* Calculated [m/s]**	**Effective Velocity Measured [m/s]**
587.9	*582.6±7.1*

* Standard deviation of 5 nozzle measurements
** Momentum is calculated as "Flow Rate" of axial velocity at outlet of nozzle
*** Effective Velocity = (Momentum Flux)/(Mass Flow Rate)

Automotive Simulation WORLD CONGRESS

Spray D Results

* No cavitation region (void formation) is observed in Spray D.

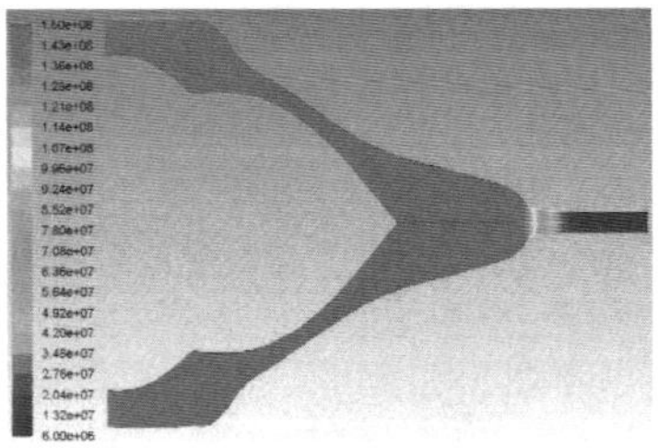

Mixture Static Pressure [Pa]

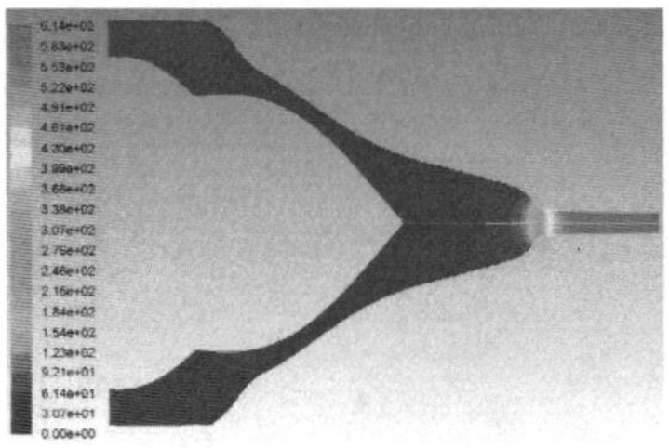

Mixture Velocity Magnitude [m/s]

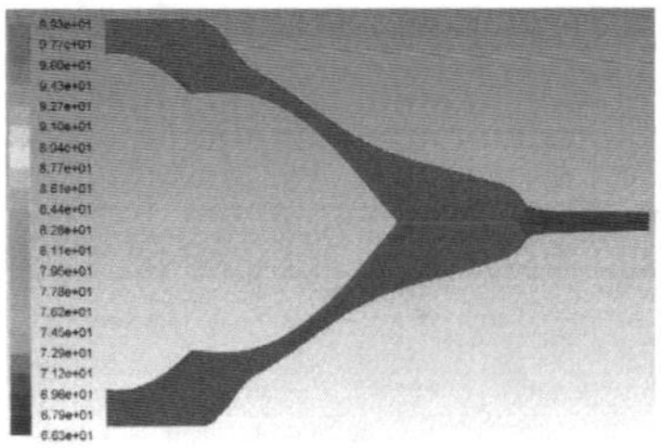

Mixture Temperature [°C]

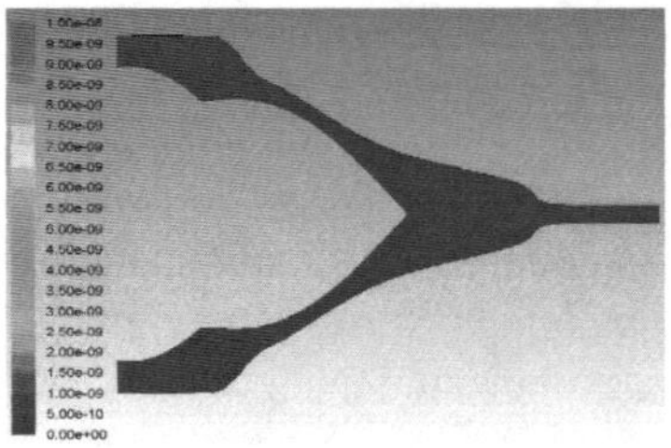

Vapor Volume Fraction*

Mesh Sensitivity: Spray C

- A mesh sensitivity study was performed by adapting the mesh by a factor of 2 in both directions (4x larger mesh).
- No major changes observed in important quantities; minor changes in maximum TKE*, volume fraction and temperature.

	Mesh Count	Momentum – Calculated [N]	Momentum – Measured for Nozzle #210003 [N]	Mass Flow Rate – Calculated [g/s]	Mass Flow Rate – Measured for Nozzle #210003 [g/s]	Average Outlet Axial Velocity (m/s)
Baseline	22919	6.52	***5.65±0.06***	10.89	***10.06±0.12***	572
Adapted	91676	6.49	***5.65±0.06***	10.83	***10.06±0.12***	571

* Maximum TKE is ~12% lower and max turbulent viscosity is ~20% lower in the finer mesh; however, distribution is very similar to the base mesh.

Comparison of Spray C & Spray D at Injector Nozzle Outlet

- **Despite larger diameter, Spray C has a smaller injection flow rate than Spray D in measurements [8]. This is explained by cavitation in Spray C.**
- **The experimental trends for decreased mass flow rate and momentum in Spray C are captured well by the model.**

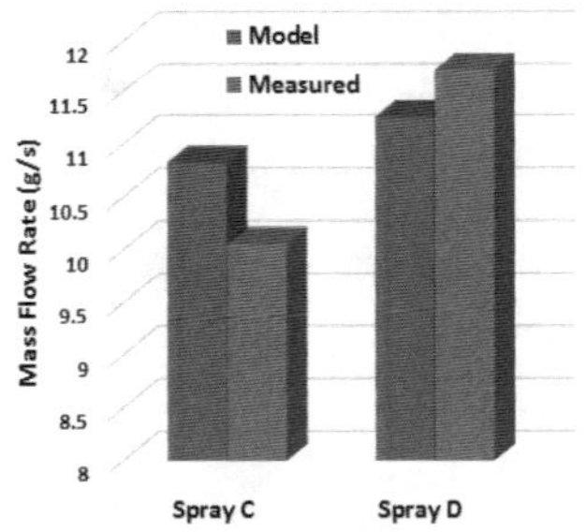

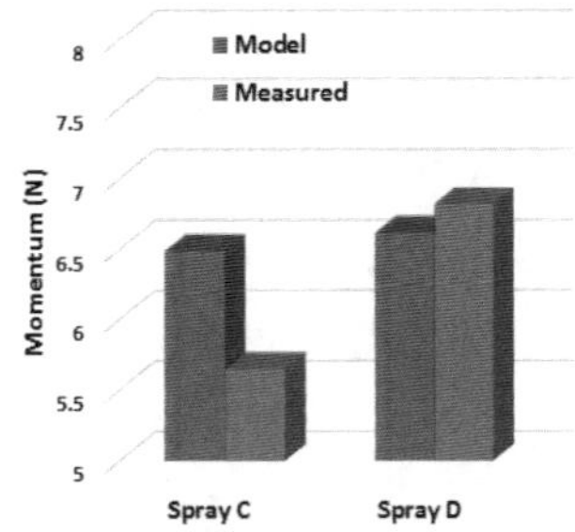

Comparison of Spray C & Spray D at Injector Nozzle Outlet

- **Minor differences in axial outlet velocity and significant differences in density and outlet mass fluxes are observed.**
- **The differences explain different spray cone angles and penetration lengths observed in experiments of Westlye, et al. [9].**

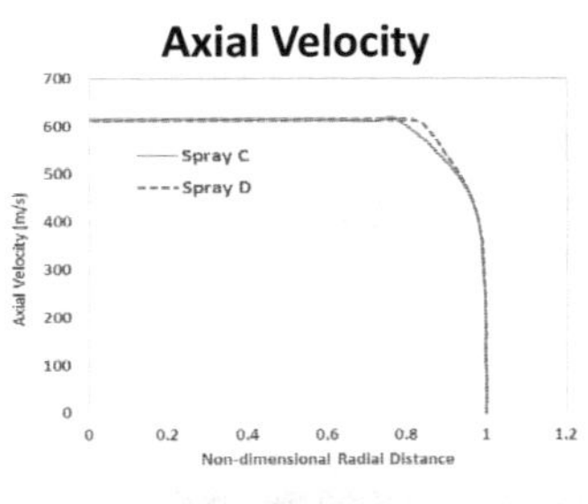

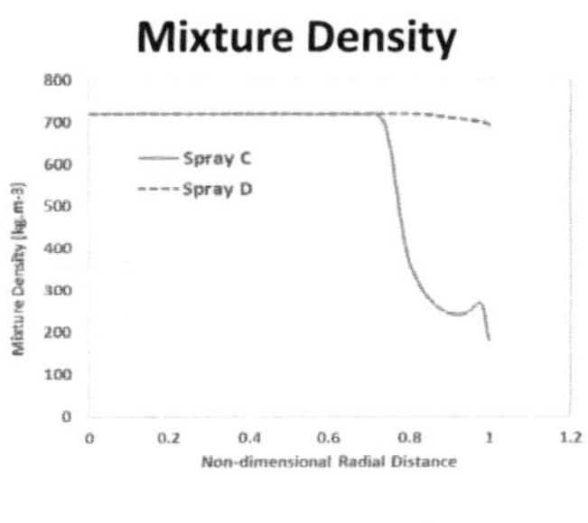

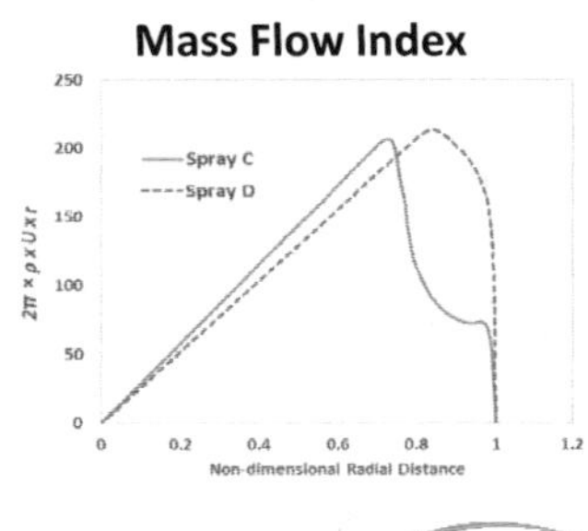

Conclusions and Future Work [1]

- ECN Sprays C and D results show good agreement with the measured mass flow rate, momentum and effective velocity.
- Spray C is cavitating, but Spray D does not show any cavitation region.
- It was found important for the solution accuracy to include compressibility of both liquid and vapor.
- Inclusion of "turbulent viscous heat generation" and "pressure work" is important. It will raise the temperature and affect liquid/vapor properties specifically in the cavitating zone close to the nozzle wall.
- The 2D wireframe was not smooth resulting fluctuations in pressure and vapor formation along the nozzle wall (particularly for Spray C).

Automotive Simulation WORLD CONGRESS

Conclusions and Future Work [2]

- Accuracy of properties for injector simulations is critical. Whenever data was available, variable T- or P- dependent properties were incorporated.
- It appears that the community still needs more validated databases for fuel physical properties, particularly, real fuels.
- Future Work:
  - Repeat the simulations when new liquid/vapor properties are available.
  - Consider 3D simulations on a more realistic representation of geometry.
  - Consider an Eulerian Ω-Y model (ELSA) to predict liquid penetration length and extend domain to include spray chamber.

Automotive Simulation WORLD CONGRESS

FSI: Fuel Injector Leakage

- **2-way thermal-structural FSI**
  - Viscous work, pressure loads and external thermal loads affect shape of the leakage path and the temperature distribution
  - Force/displacement and thermal coupling
- **Transferred quantities**
  - Force, HTC, Near Wall Temperature (from CFD)
  - Deformation, Wall Temperature (from FEA)
  - Coupled field elements utilized

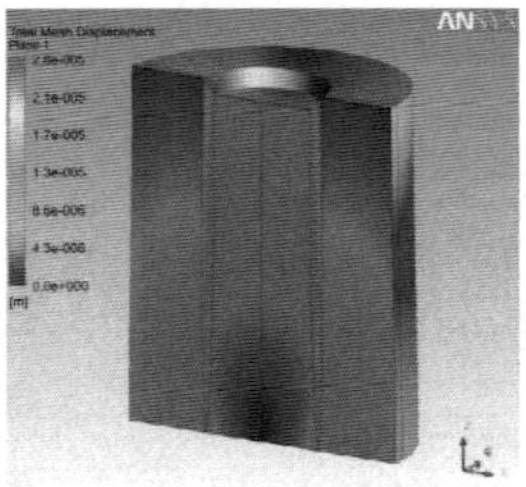

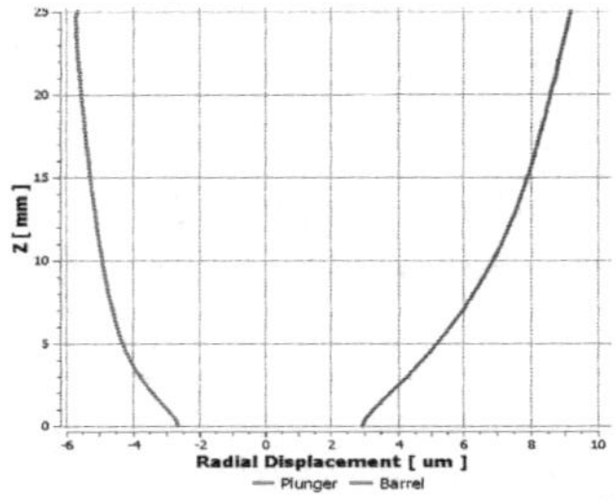

FSI: 1 DOF Approaches

- **In some simulations, needle movement is not pre-defined**
  - Force balance acting on needle defines its velocity and location
- **One-Degree of Freedom (1DOF) force balancing with a user function**
- **1DOF can be used for**
  - Transient run with force calculation OR
  - Steady state run with force calculation to calculate equilibrium position of the needle

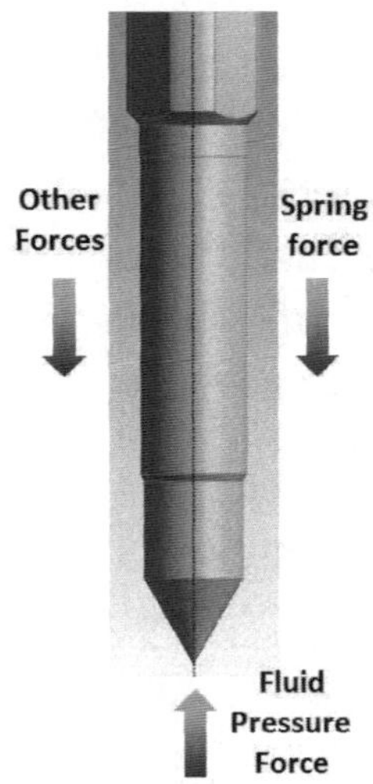

Coupling Maxwell-Fluent For Solenoids

- **Electromagnetic – CFD Coupling via Maxwell-Fluent coupling can be used for**
  - Thermal analysis of solenoids & actuators where heat loss passed to Fluent and Temperature received by Maxwell
  - Force calculation on needle where force passed to Fluent and displacement received by Maxwell
- **Example**
  - Two-way coupling - Maxwell 2D RZ magnetostatic and Fluent 2D steady state

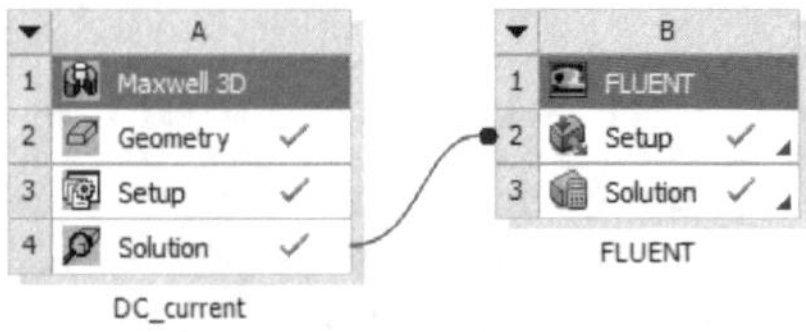

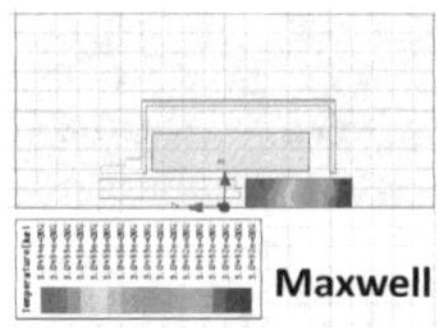

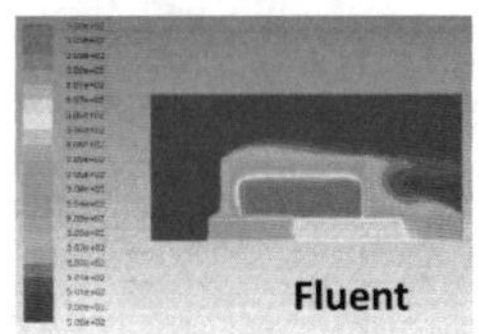

GT-Suite Maxwell Co-Simulation [1]

- **Dynamic magnetic behavior, including eddy current activity, has an effect on motion and the electronic circuit, which in turn both affect the magnetic behavior.**
- **GT-Suite provides hydraulic forces, mechanical stops, electric current, and voltage**
- **Maxwell returns magnetic force and back EMF to GT-Suite**

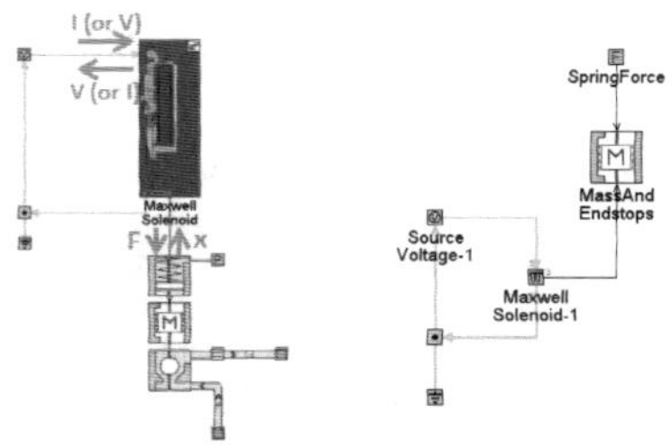

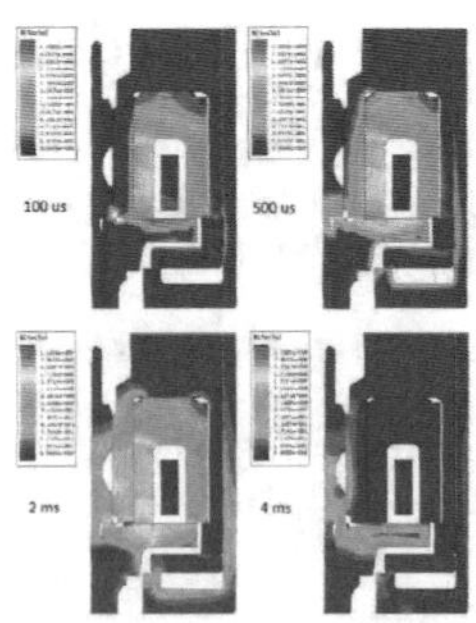

GT-Suite Maxwell Co-Simulation [2]

XPI Injector

Pilot Valve

Valve Stroke 47 um
Preload 114 N
Over-travel force 10 N
Pull-In current 18 A
Hold current 12 A
Boost voltage 50 V

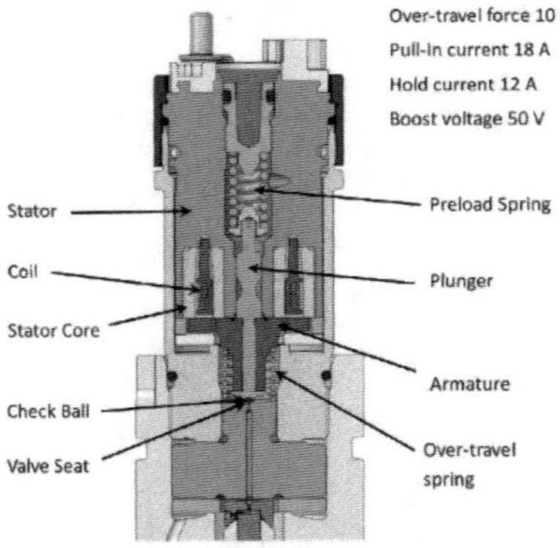

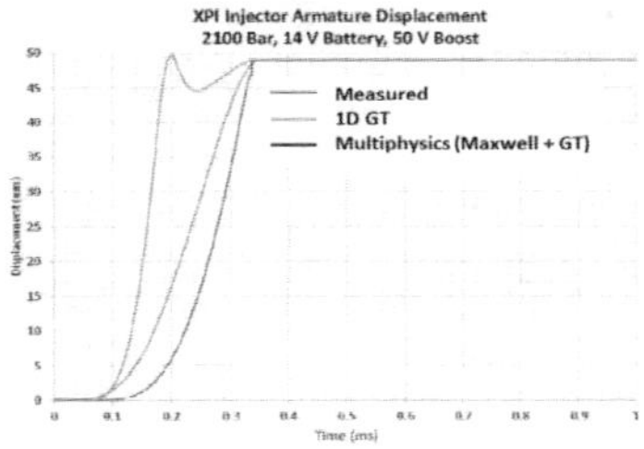

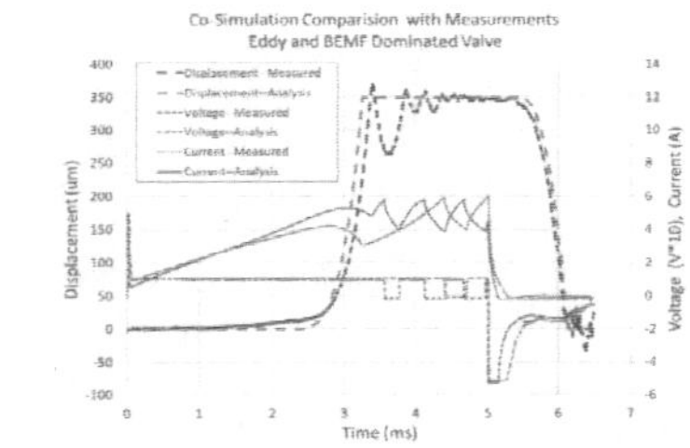

GTI User Conference 2015

Magnetic Actuator FEA Co-Simulation: Ansys Maxwell and GT-SUITE

Ben Manring and Ahmad Sabri **Cummins Fuel Systems**

Iakovos Papadimitriou and Shawn Harnish **Gamma Technologies**

Mark Solveson and Xiao Hu **ANSYS**

Automotive Simulation WORLD CONGRESS

Simplorer: Modelica Library Support

Modelon

MODELICA

Modelica Standard Library

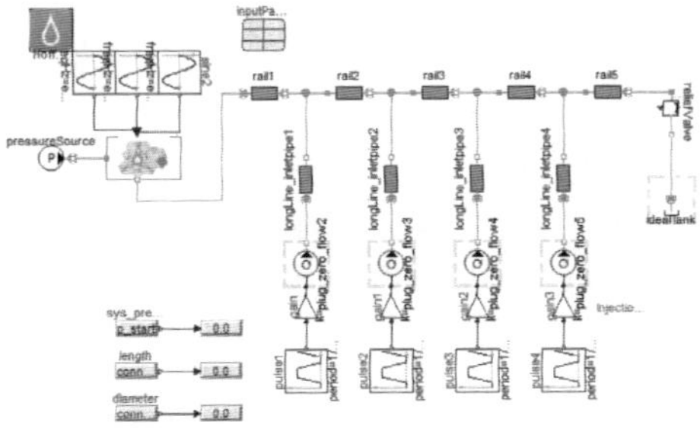

Information	Information
Interfaces	Interfaces for hydraulic components
Templates	Templates for hydraulic components
Examples	Demonstration examples of the components of this package
Cylinders	Models of cylinders
DirectionalControl	Danfoss valves
Elements	Library of elements for design of hydraulic components
HydraulicResistance	Models of hydraulic resistance
Lines	Models of long lines
Pumps	Models of hydraulic pumps
Motors	Models of hydraulic motors
Sources	Models of hydraulic sources
Restrictions	Models for restrictions in hydraulic components
Sensors	Sensor for hydraulic quantities
Utilities	Utilities
Valves	Models of hydraulic valves
Visualizers	Classes for result visualization
Volumes	Models of volumes
Fluids	Fluid media models
Units	Package containing units and related types and functions

Automotive Simulation WORLD CONGRESS

Automotive Simulation
WORLD CONGRESS

June 7-8, 2016 Munich

Questions

References [1]

[1] P. J. Zwart, A. G. Gerber, and T. Belamri. "A Two-Phase Flow Model for Predicting Cavitation Dynamics“, Fifth International Conference on Multiphase Flow, Yokohama, Japan. 2004; also, see Cavitation Models in ANSYS FLUENT Theory Guide.

[2] Li, H., Vasquez, S. A., “Numerical Simulation of Steady and Unsteady Compressible Multiphase Flows”, Proceedings of the ASME International Mechanical Engineering Congress and Exposition, 2012, Houston, Texas, USA, Paper No. IMECE2012-87928.

[3] Caudwell, D.R., Trusler, J.P.M., Vesovic, V. and Wakeham, W.A., “The Viscosity and Density of n-Dodecane and n-Octadecane at Pressures up to 200 MPa and Temperatures up to 473 K.” International Journal of Thermophysics, Vol. 25, No. 5, 2004.

[4] http://webbook.nist.gov/chemistry/fluid/

References [2]

[5] Arienti, M., Sussman, M., "A High-Fidelity Study of High-Pressure Diesel Injection", SAE Paper 2015-01-1853.

[6] Khasanshin, T. S., Shchamialiou, A. P., Poddubskij, O. G., "Thermodynamic Properties of Heavy n-Alkanes in the Liquid State: n-Dodecane." International Journal of Thermophysics, Vol. 24, No. 5, 2003.

[7] Padilla-Victoria, H., Iglesias-Silva, G. A., Ramos-Estrada, M., Hall K. R., "A Correlation to Predict Speed of Sound in Liquids: 1. n-Alkanes (≥C5) and their Mixtures at High Pressures", Fluid Phase Equilibria 338 119-127, 2013.

[8] R. Payri, J. Gimeno, J. Cuisano, J. Arco, " Hydraulic characterization of diesel engine single-hole injectors, Fuel, 180 (2016) 357–366. (DOI: 10.1016/j.fuel.2016.03.083).

[9] Westlye, et al., " Penetration and Combustion Characterization of Cavitating and Non-Cavitating Fuel Injectors under Diesel Engine Conditions", SAE Paper 2016-01-0860 .

Flow and cavitation in Diesel fuel injector nozzles

Dr Foivos Koukouvinis & Prof Manolis Gavaises

School of Computer Science, Mathematics & Engineering Sciences
City University London, UK

1

Introduction & Motivation | Theoretical details | Cases examined | Indicative results | Concluding Remarks

Presentation Outline

- **Introduction and motivation**
  Why cavitation/bubble dynamics?
- **Theoretical details**
  FLUENT capabilities on cavitation/bubble dynamics
- **Cases examined**
  City's cavitation test rig
  Caterpillar injectors
  Immensed Boundary method
- **Indicative results**
- **Conclusions**
  Including potential improvements and additions

The 3000bar potential

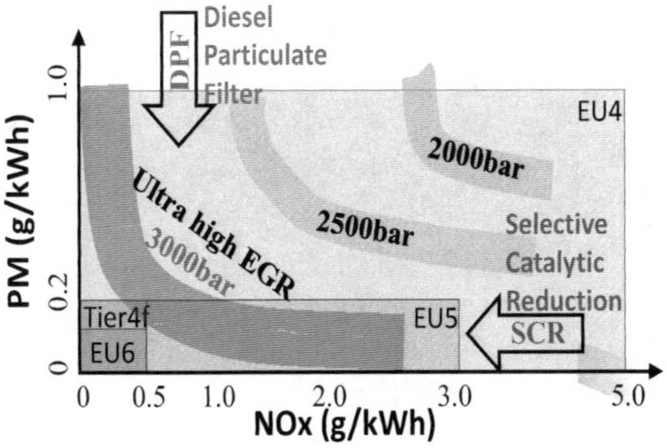

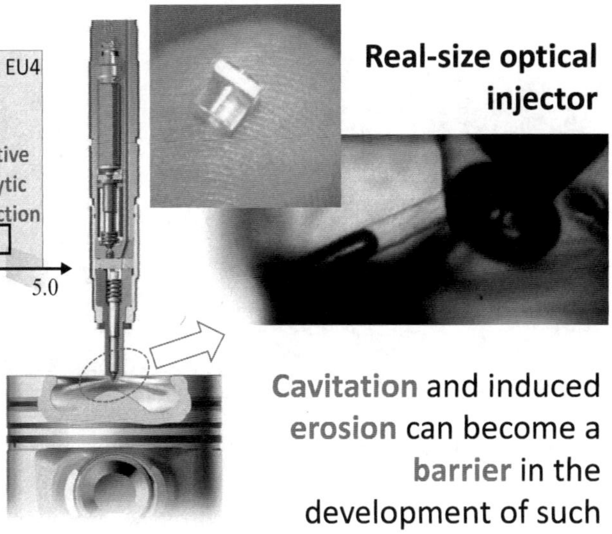

Real-size optical injector

Emission regulations can be met without the need for an expensive after-treatment device with **high fuel pressurisation**

Cavitation and induced erosion can become a barrier in the development of such systems

Cavitation erosion pictures

1-10m

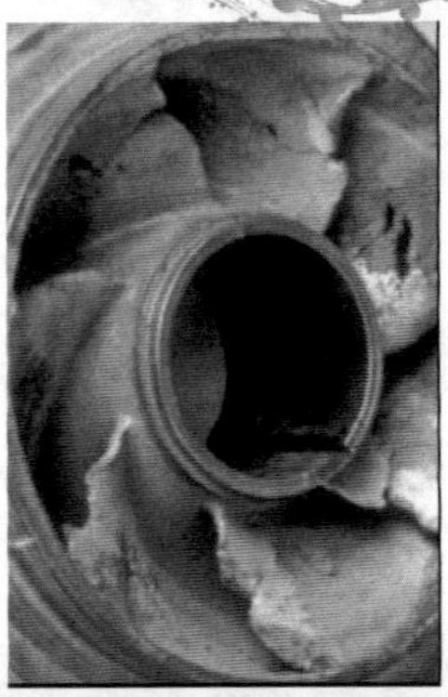

Hydraulic turbines

cm to m

Marine propeller blades

~0.1 m

IC engines piston rings

μm to mm

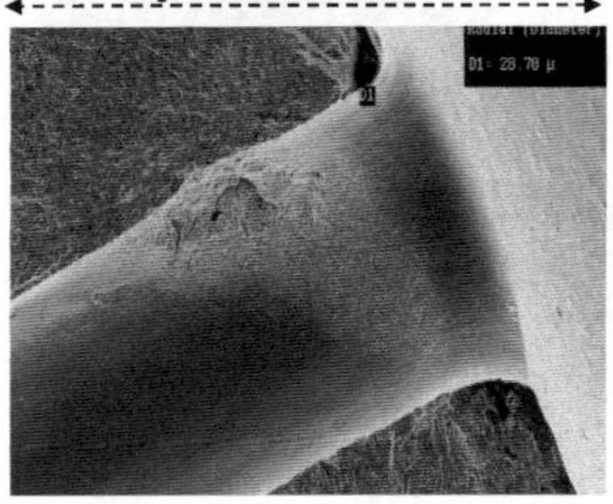

Fuel injection equipment

What causes erosion?

Time scale: 10^{-15}s

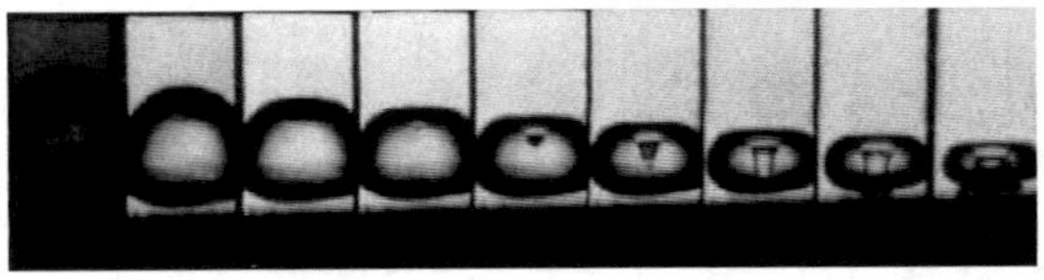

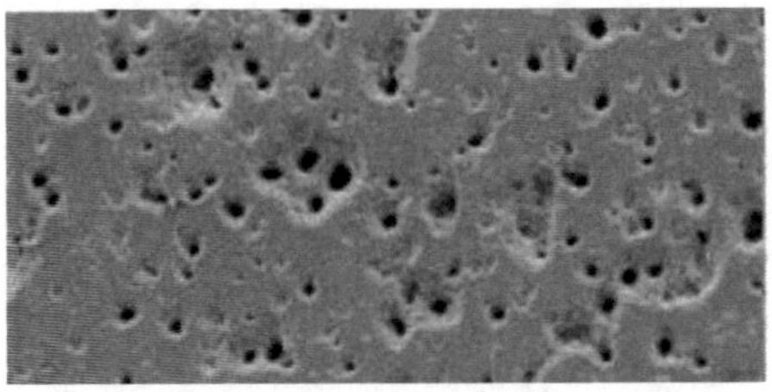

$$P_{acoustic} = \frac{\rho_L}{4\pi\ell}\frac{d^2V}{dt^2}$$

V: Bubble volume
t: time
l: distance from wall

Sudden compression during collapse may give pressures ~Gpa

Solver types:	Pressure-based	Density based – explicit	Density based – implicit
	Originally developed for incompressible flows. Solve for entropic variable set [p, u, v, w,...].	Originally developed for compressible flows. Solve for conservative variable set [ρ, ρu, ρv, ρw,...].	
Advantages	Consistent Mach number formulation. Inherently implicit, thus do not have time-step constrains for stability.	1. Highly efficient for very violent flows, with compressibility effects. 2. No need for convergence! Simple time-marching. 3. Embarrassingly parallel; easy to parallelize high speed gain.	1. No time step constraints for time stepping. 2. Overcomes the problems of the explicit variant with Mach -> 0.
Disadvantages	1. Problematic in cases with sudden/large changes of speed of sound (e.g. cavitation). 2. Problematic for HEM; in saturation there are infinite states with the same pressure.	Very small time steps to ensure stability. Problematic when	1. Needs acceleration techniques (Preconditioning) and Dual time stepping to be consistent in time. 2. These remove the

Cavitation modelling	Eulerian: Immiscible model (VOF, LS, etc..)	Eulerian Lagrangian: Full RP equation	Eulerian: Mass transfer models	Eulerian: Homogenous Equilibrium models (HEM) e.g. barotropic
Advantages	Capable of accurately resolving all bubble dynamics phenomena, inclusion of surface tension, vapour mass transfer (what mass transfer?) and gas content, provided exact nuclei content is known (!)	Correct inclusion of the Rayleigh – Plesset equation, with surface tension, gas content..	1. Mass transfer is a generic way to represent cavitation (simplicity) – link to compressibility. 2. Inherently describes cavitation as a non-equilibrium thermodynamic effect. 3. Can include mechanical and thermal non-equilibrium.	Simplifies the system of PDEs, assuming instantaneous mass transfer.
Disadvantages	Immense computational cost for practical applications, Too difficult to determine nuclei content which will be used as input.	1. Coupling between the phases can be problematic/diffusive. 2. The Rayleigh-Plesset equation is not always valid (axis-symmetry) 3. Difficult to include bubble interactions.	Since bubbles are under-resolved, surface tension cannot be included.	1. Inherently the fluid is in thermodynamic equilibrium. 2. Omits slip velocity of the vapour phase, or any other non-equilibrium effects. 3. No surface tension

FLUENT capabilities on cavitation/bubble dynamics

- Capable of simulating VOF bubbles (vaporous and/or gaseous) with interface tracking & compressible solver

Collapse of air bubble near a free surface

Collapse of air bubble in water. R_0=10μm, P_{g0}=6800Pa, P_∞=1bar

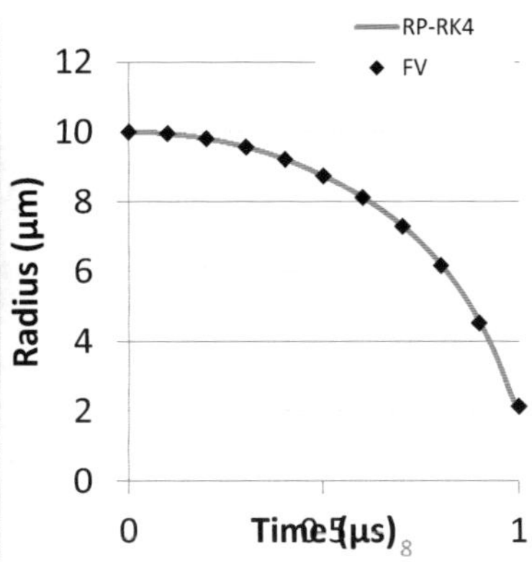

Introduction & Motivation	Theoretical details	Cases examined	Indicative results	Concluding Remarks

UDFs for nozzle flows

- Capable of simulating real fluid materials (e.g. Stiffened Gas, Tait, arbitrary material EOS), provided a UDRGM UDF with relevant properties is given
- Example, diesel injector flow with custom EOS for diesel liquid. Phase change/cavitation is included

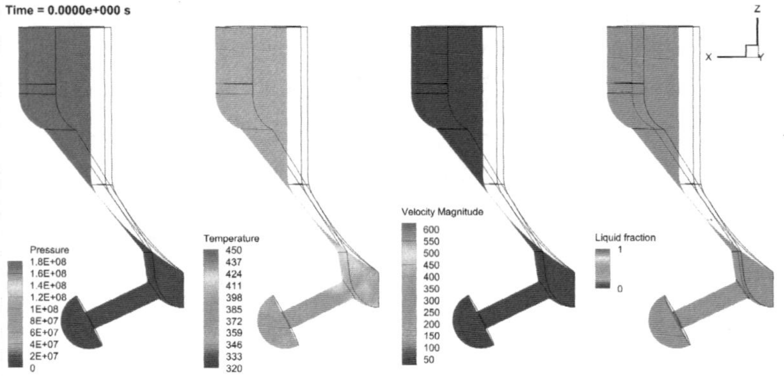

Limitation: only one compressible phase can be involved (in Fluent 17 combination of Real Gas and Ideal gas is supported).

9

Introduction & Motivation	Theoretical details	Cases examined	Indicative results	Concluding Remarks

Presentation Outline

- **Introduction and motivation**
  Why cavitation/bubble dynamics?
- **Theoretical details**
  FLUENT capabilities on cavitation/bubble dynamics
- **Cases examined**
  City's cavitation test rig
  Caterpillar injectors
  Immersed Boundary method
- **Indicative results**
- **Conclusions**
  Including potential improvements and additions

City's cavitation rig

-Inlet pressure ~45-55bar
- Operating temperature 40°C
- Outlet pressure ~17bar
- Needle lift 1mm

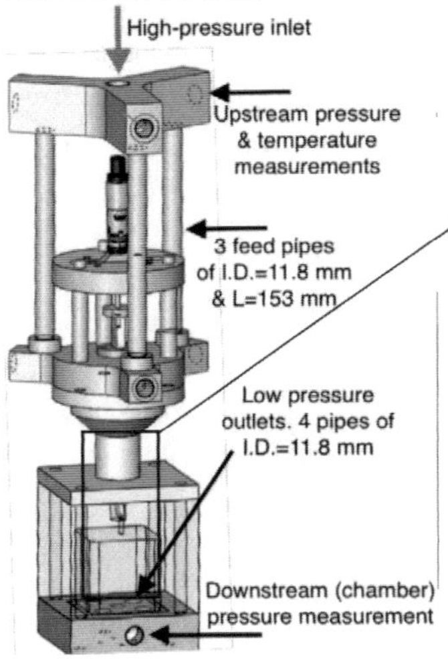

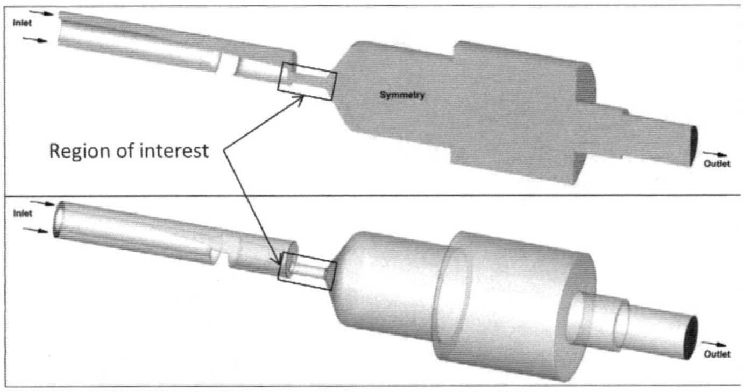

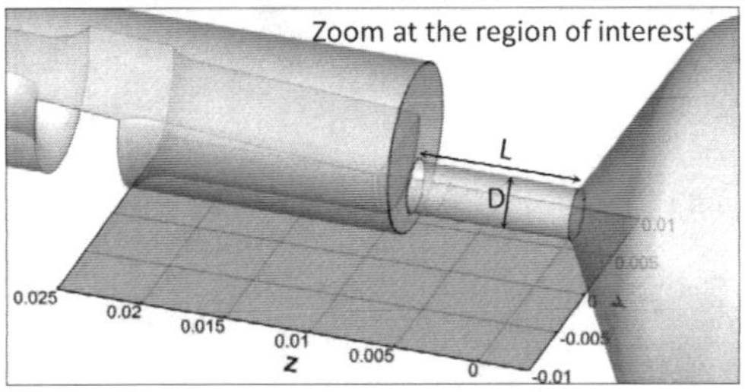

L~10mm
D~3mm

11

Cavitating Channel – Test Section

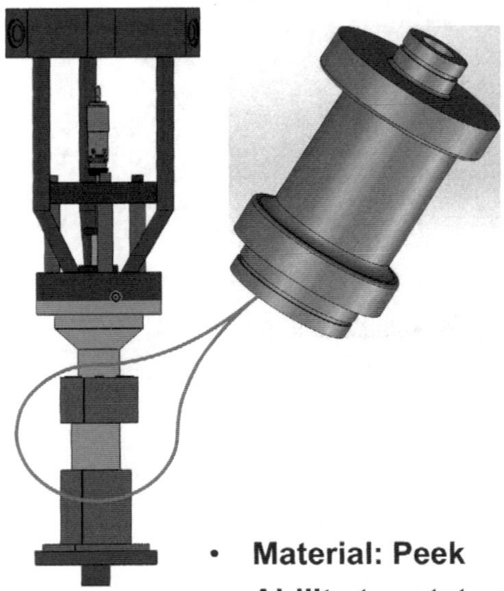

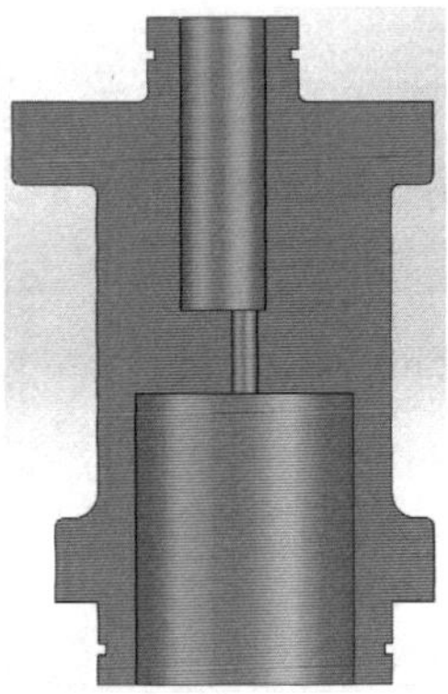

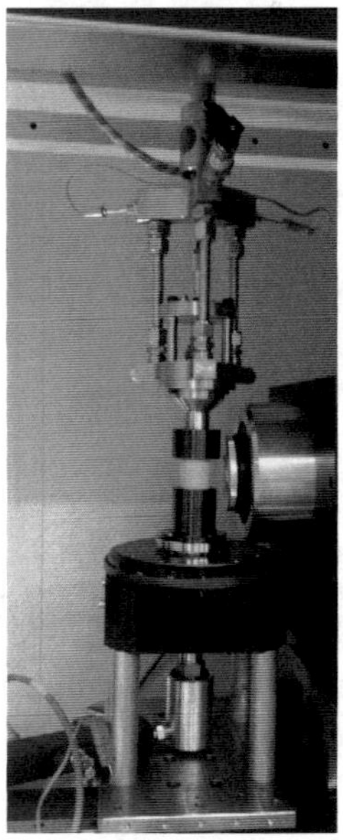

- **Material: Peek**
- **Ability to rotate around hole axis**
- **Micro Computed Tomography**

CITY UNIVERSITY LONDON

X-ray mCT measurements and visualisation

Shadowgraphy

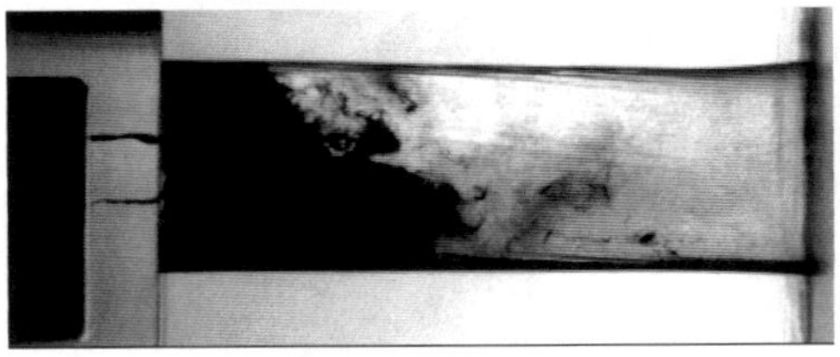

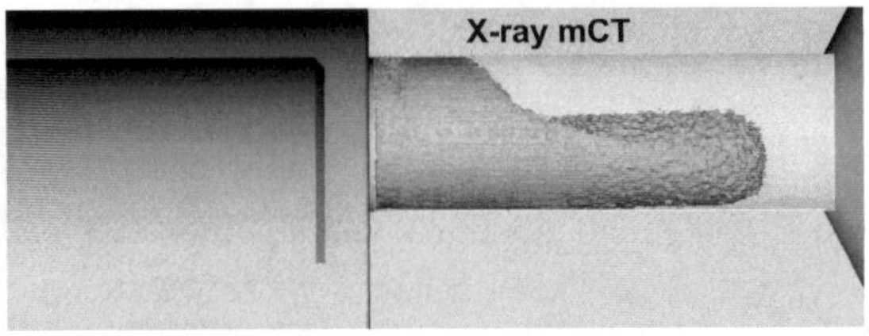

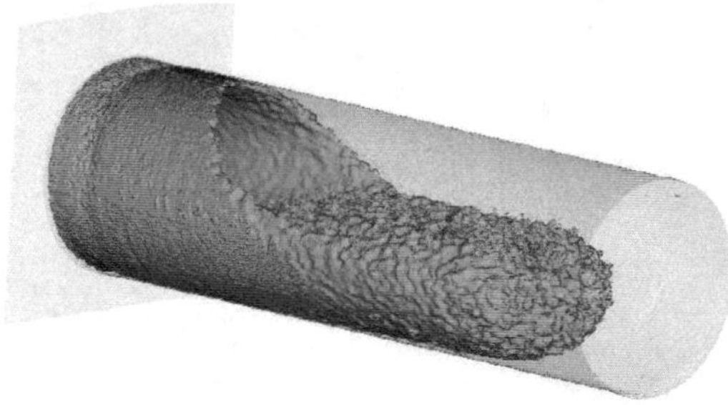

CITY UNIVERSITY LONDON

Introduction & Motivation | Theoretical details | Cases examined | Indicative results | Concluding Remarks

Accurate surface reconstruction

The actual geometry of the constriction is used:
- CT scan of the surface
- High resolution spline interpolation of the surface
-Geometry generation
- Mesh

Idealized geometry results to significant under prediction of the flow rate (error ~10%)

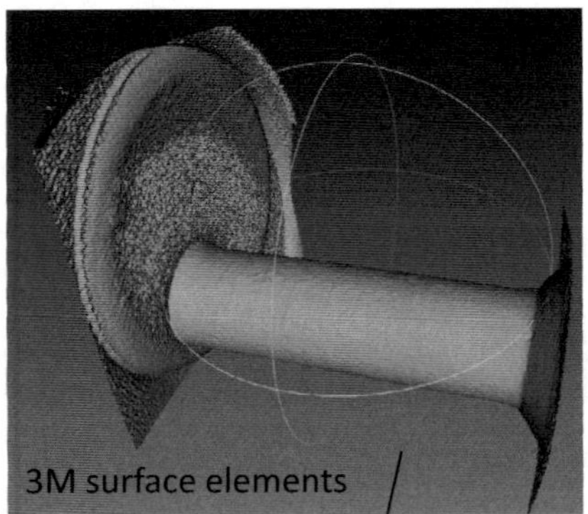

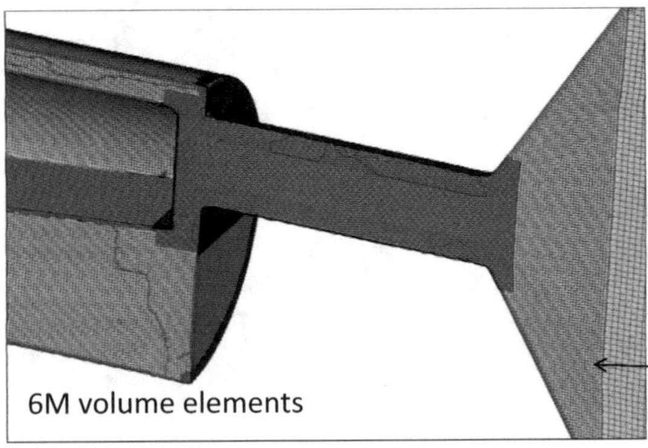

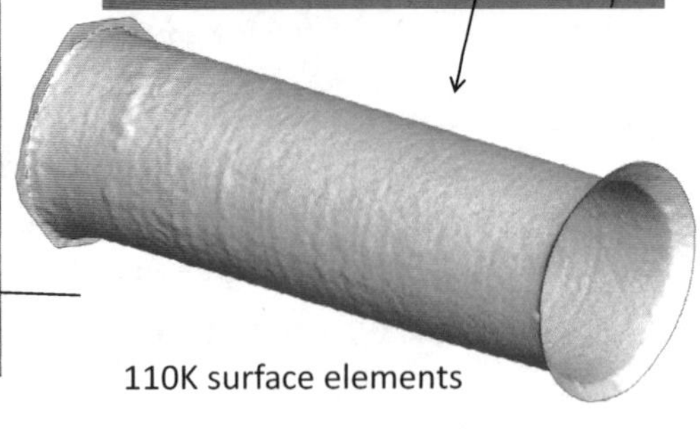

Operating conditions

Conditions tested:

	P_{up} (bar)	P_{down} (bar)	*Cn* (-)	*σ* (-)	*Re* (-)	u_{av} (m/s)	m_{flux} (kg/s)
1	43.05	17.30	1.50	0.67	56119	56.12	0.33
2	55.00	17.45	2.18	0.46	65000	65.08	0.38

➢Properties estimated at 40°C:

Property	*Liquid – Tait equation*	*Vapour*
ρ_0 (kg/m³)	830	1.123
B (Pa)	172013986	
n (-)	7.15	
c_0 (m/s)	1217.3	
p_{ref} (= p_{vap}, Pa)	17200	
μ (Pa.s)	2.16E-03	8.00E-06

$$p = B\left[\left(\frac{\rho}{\rho_0}\right)^n - 1\right] + p_{ref}$$

$$B = \frac{\rho_0 c_0^{\ 2}}{n}$$

15

LES – CN2.2

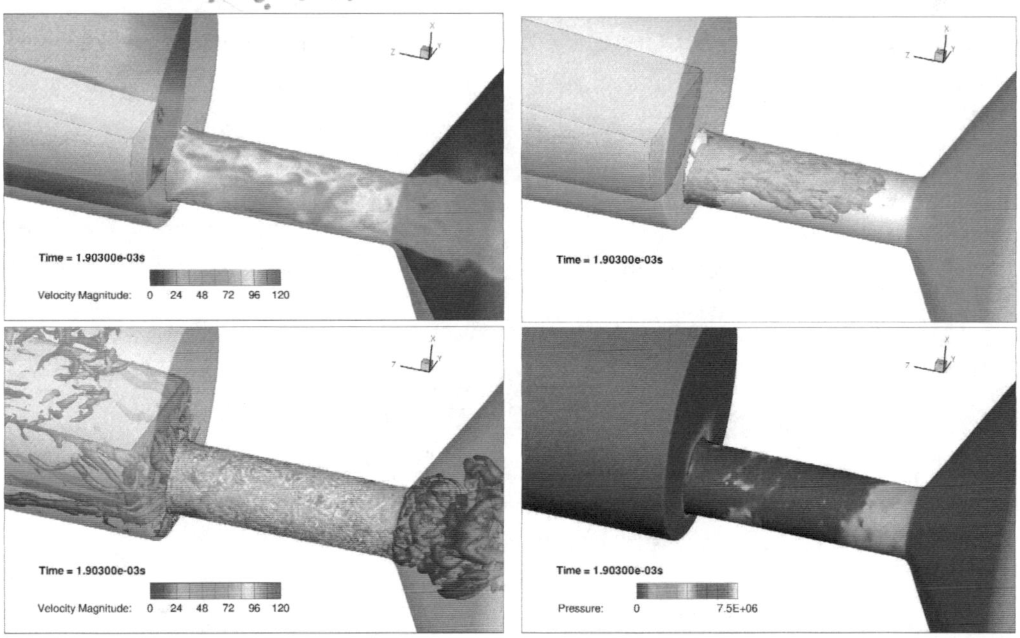

Validation – flow rate

➢Test rig – mass flow rate comparison.

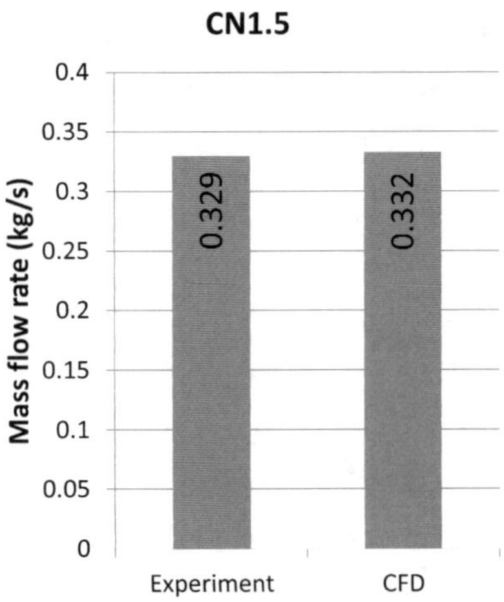

Error ~1%

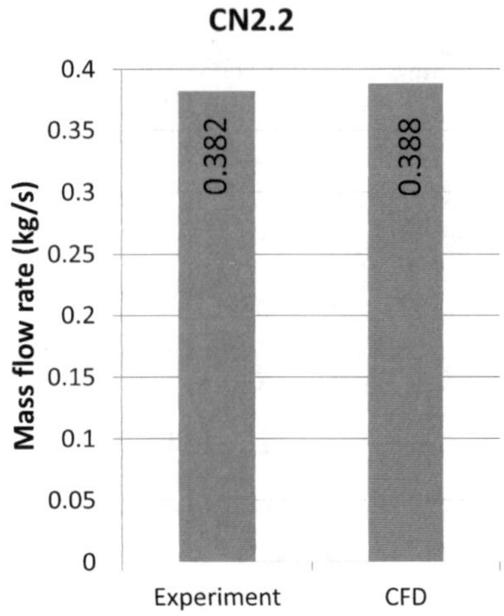

Error ~1.5%

Validation – Volume fraction

➢Test rig – average volume fraction comparison inside the channel

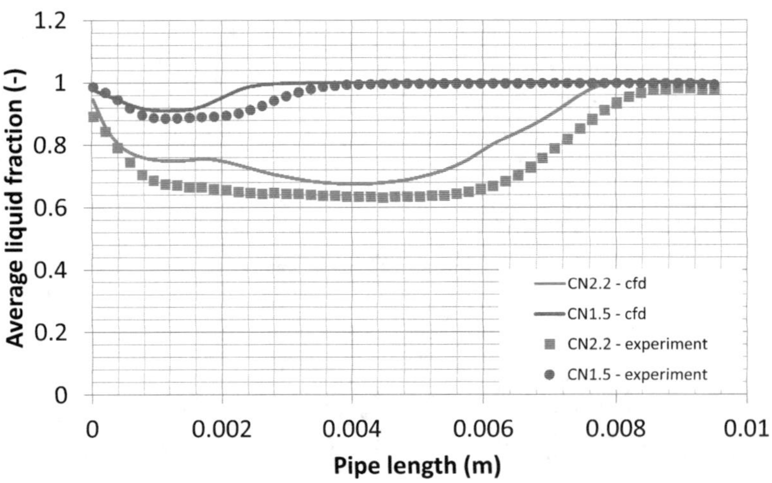

Validation – Volume fraction STD

➢Test rig – standard deviation of volume fraction across the channel

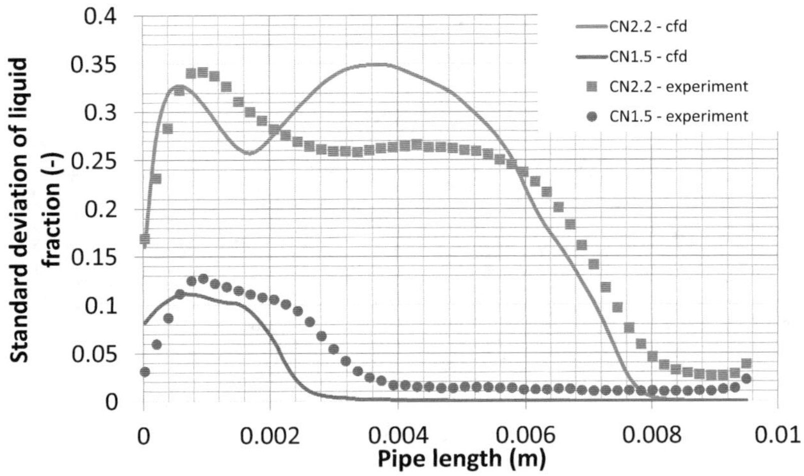

Introduction & Motivation | Theoretical details | Cases examined | Indicative results | Concluding Remarks

Erosion

➢Test rig, CT scan – erosion comparison

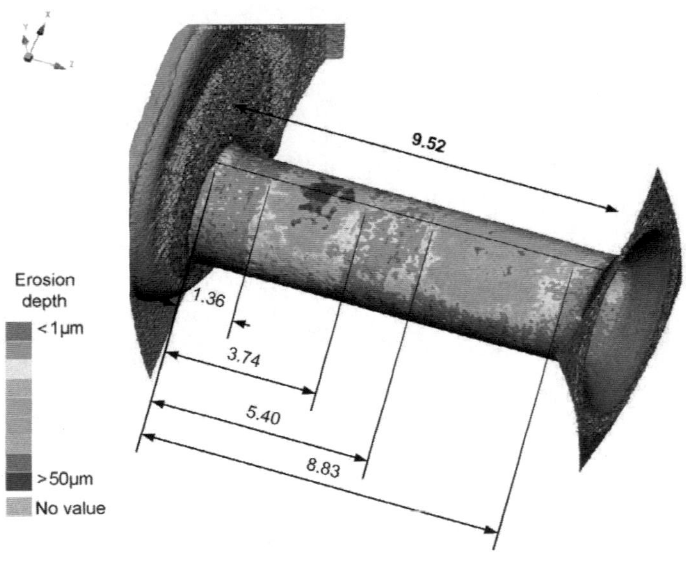

Erosion is located in the area of intense cavitation, in two separated zones:

- Zone 1: from ~1.36 to ~3.74mm downstream the orifice entrance.

- Zone 2: from 5.4 to 8.83mm downstream the orifice entrance.

Introduction & Motivation | Theoretical details | Cases examined | **Indicative results** | Concluding Remarks

Validation - Erosion

➤Test rig, CFD results – erosion location comparison

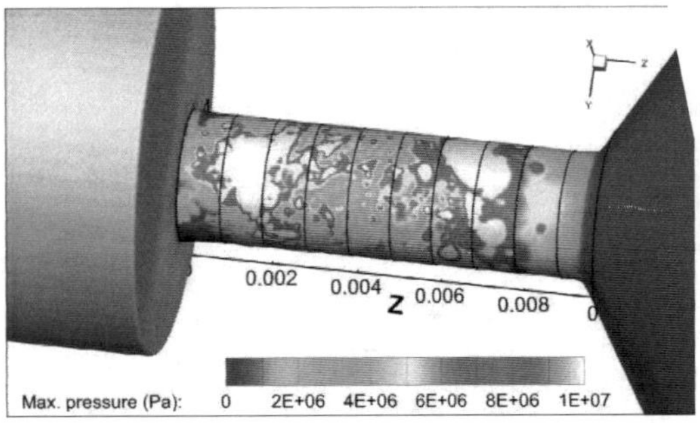

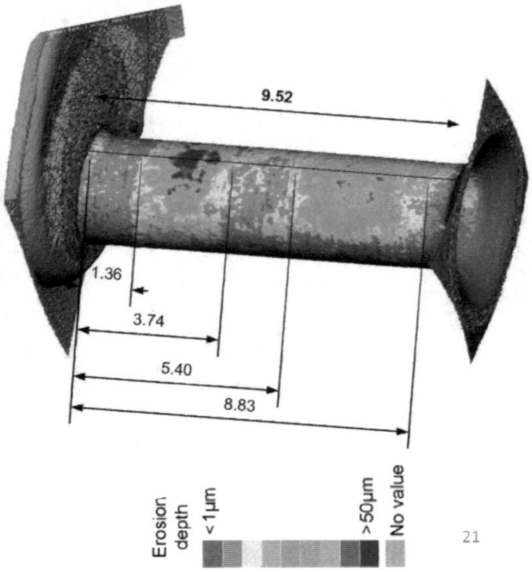

21

Introduction & Motivation | Theoretical details | Cases examined | Indicative results | Concluding Remarks

Presentation Outline

➤ Introduction and motivation
    Why cavitation/bubble dynamics?

➤ Theoretical details
    FLUENT capabilities on cavitation/bubble dynamics

➤**Cases examined**
    City's cavitation test rig
    Caterpillar injectors
    Immersed Boundary method

➤**Indicative results**

➤ Conclusions
    Including potential improvements and additions

Injectors examined

-Inlet pressure ~1800bar (pulse)
- Inlet temperature 353K
- Outlet pressure 50bar
- Needle motion governed by a lift law
- Adiabatic walls
- CN = 39.5 or σ (=1/CN) = 0.025
- Re ~ 30000

Both pressure pulse and needle motion are estimated with 1D CFD techniques

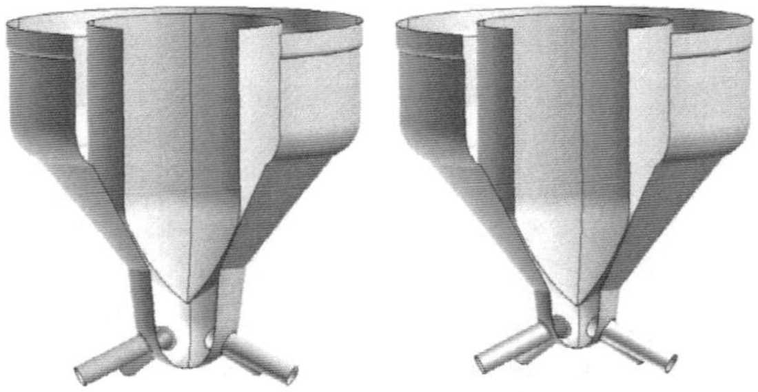

Design A Design B

Geometries provided by Caterpillar Fuel Systems

23

Meshing

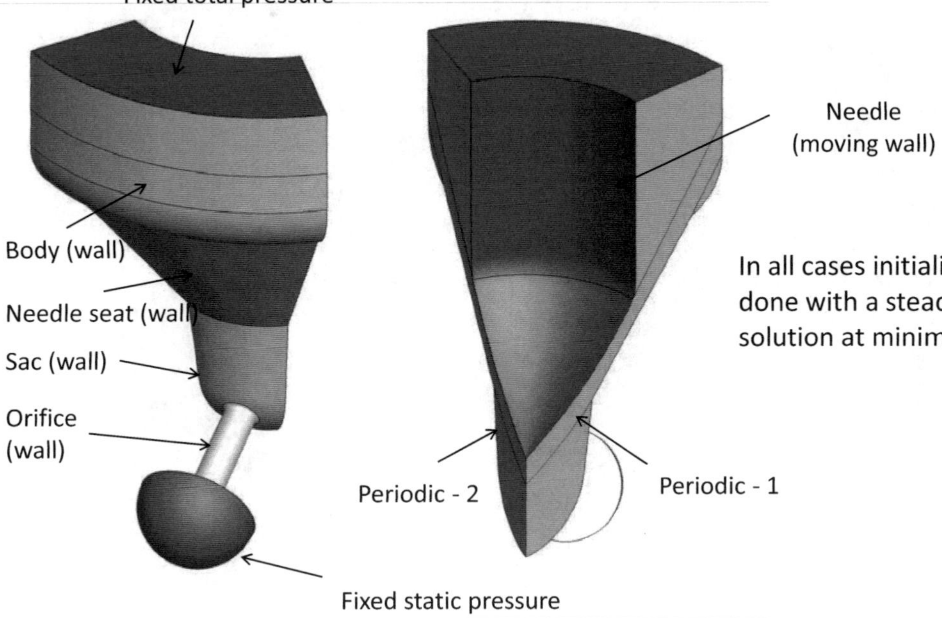

In all cases initialization is done with a steady state solution at minimum lift.

24

Meshing

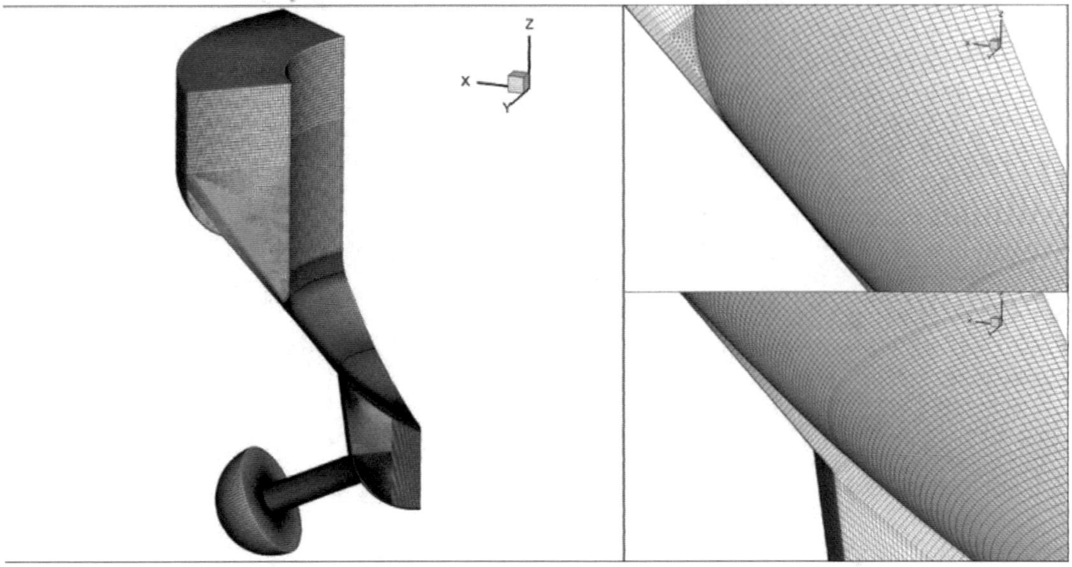

➢ At low lifts mesh stretching. At higher lifts mesh layer addition/removal

Validation - Erosion

➢Accumulated pressure peaks over time – Design A

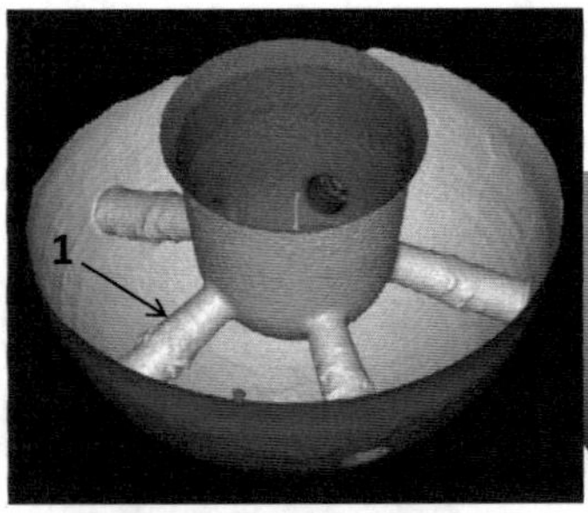

->No signs of high pressures on needle
-> No signs of high pressures on case
-> High pressures at the whole periphery of the orifice (1)

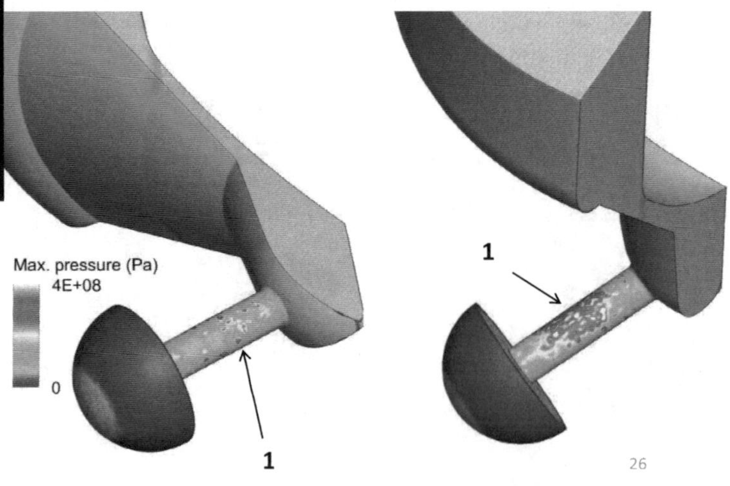

Validation - Erosion

➢Accumulated pressure peaks over time – Design B

->No signs of high pressures at the lower hole surface
-> High pressures on upper hole surface (1)
-> High pressures on needle at R~0.6mm (2)
-> Sporadic high pressures on sac (3)

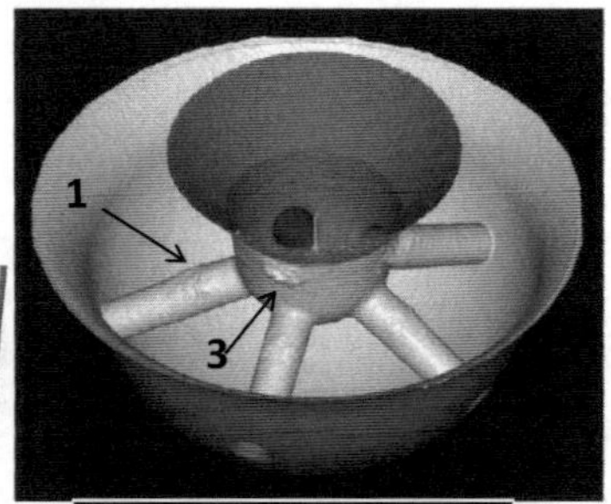

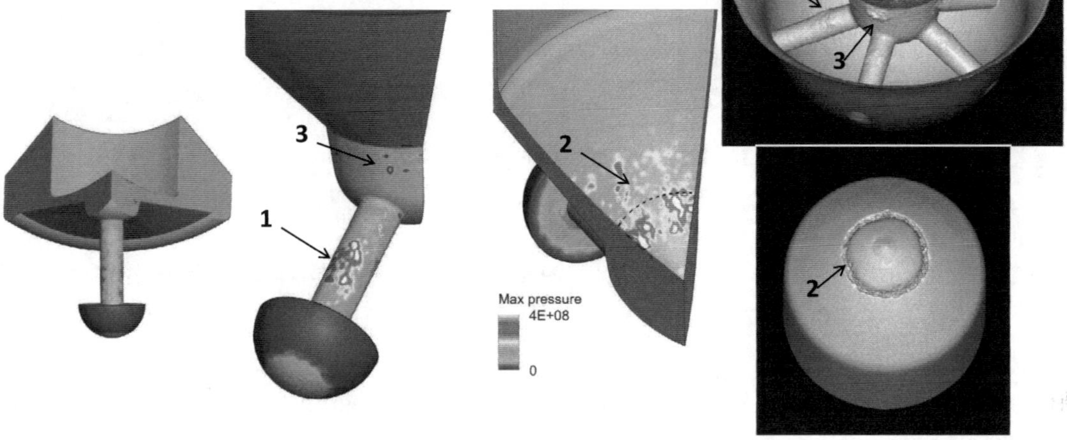

Presentation Outline

➢ Introduction and motivation
Why cavitation/bubble dynamics?

➢ Theoretical details
FLUENT capabilities on cavitation/bubble dynamics

➢**Cases examined**
City's cavitation test rig
Caterpillar injectors
Immersed Boundary method

➢**Indicative results**

➢ Conclusions
Including potential improvements and additions

Conclusions

- Many capabilities are offered in FLUENT as is; the software has a broad scope of applicability in complicated physics.

- Complex nozzle geometries have been simulated successfully using the pressure-based compressible flow solver, which is required for capturing bubble collapse effects; this is a pre-requisite for indicating locations of cavitation erosion. Results are in good agreement with experimental data

- While the density-based solver can be an attractive alternative (and more physical) for many cases, involving phase change (large variations of density), compressibility effects (including large variations of the speed of sound), in its current state it is somewhat limited.

- Potentially, a density-based solver solving the conservative variable set would have an advantage in cavitating/ compressible multiphase cases, including subcritical to critical transitions.

Simulation Methodology for High Pressure Diesel Injection & Combustion Development

Junmei Shi
Simulation Team Leader
Advanced Injection and Combustion Center (AICC)

with a collective contribution of
N. Guerrassi, G. Dober, P. Lopez Aguado, K. Krapic, E. Gomez Santos

ASWC June 07-08 2016, Munich, Germany

DELPHI
Innovation for the Real World

Outline

- AICC mission
- Simulation methodology development roadmap
- Key Diesel injection parameters
- Link between nozzle flow and spray
- Advanced Simulation Methodologies:
  - Real-Gas Fluid fuel property model
  - 1D (FIE system) – CFD (nozzle flow) coupling for injection diagnostic
  - LES + x-ray investigation of jet breakup
- Conclusions & Outlook

ASWC – Automotive Simulation World Congress, June 07-08, 2016, Munich, Germany

DELPHI

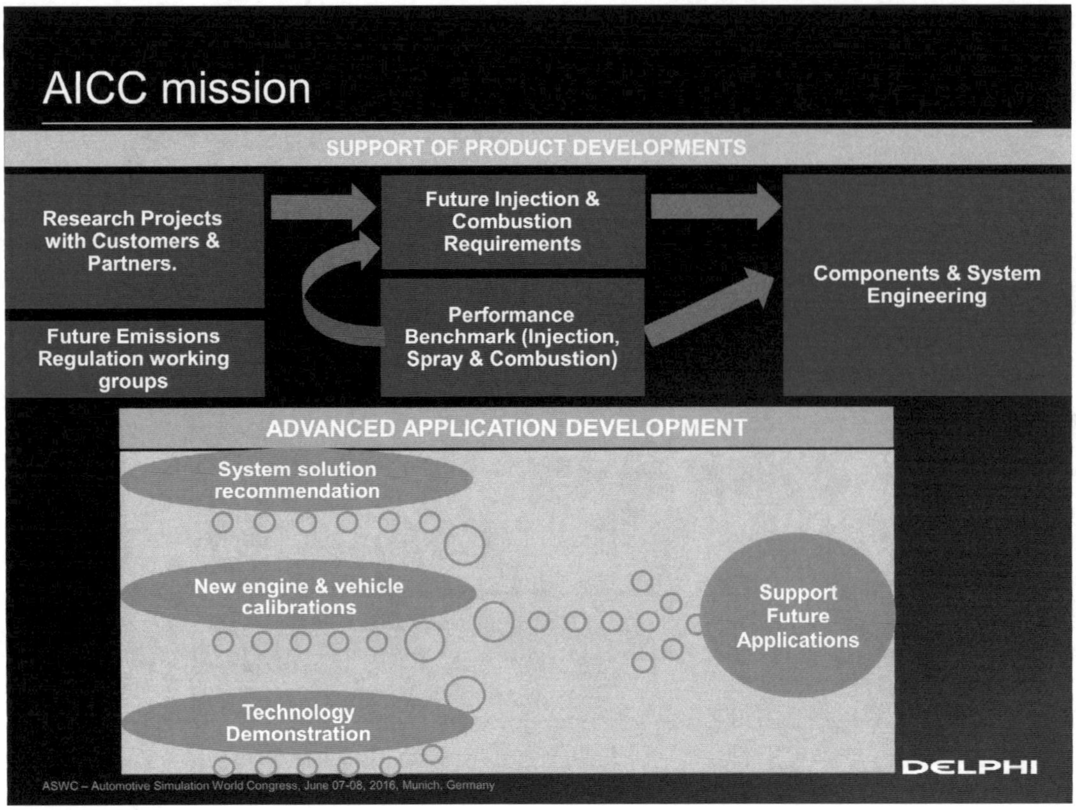
AICC mission
SUPPORT OF PRODUCT DEVELOPMENTS
Research Projects with Customers & Partners.
Future Emissions Regulation working groups
Future Injection & Combustion Requirements
Performance Benchmark (Injection, Spray & Combustion)
Components & System Engineering
ADVANCED APPLICATION DEVELOPMENT
System solution recommendation
New engine & vehicle calibrations
Technology Demonstration
Support Future Applications
ASWC – Automotive Simulation World Congress, June 07-08, 2016, Munich, Germany
DELPHI

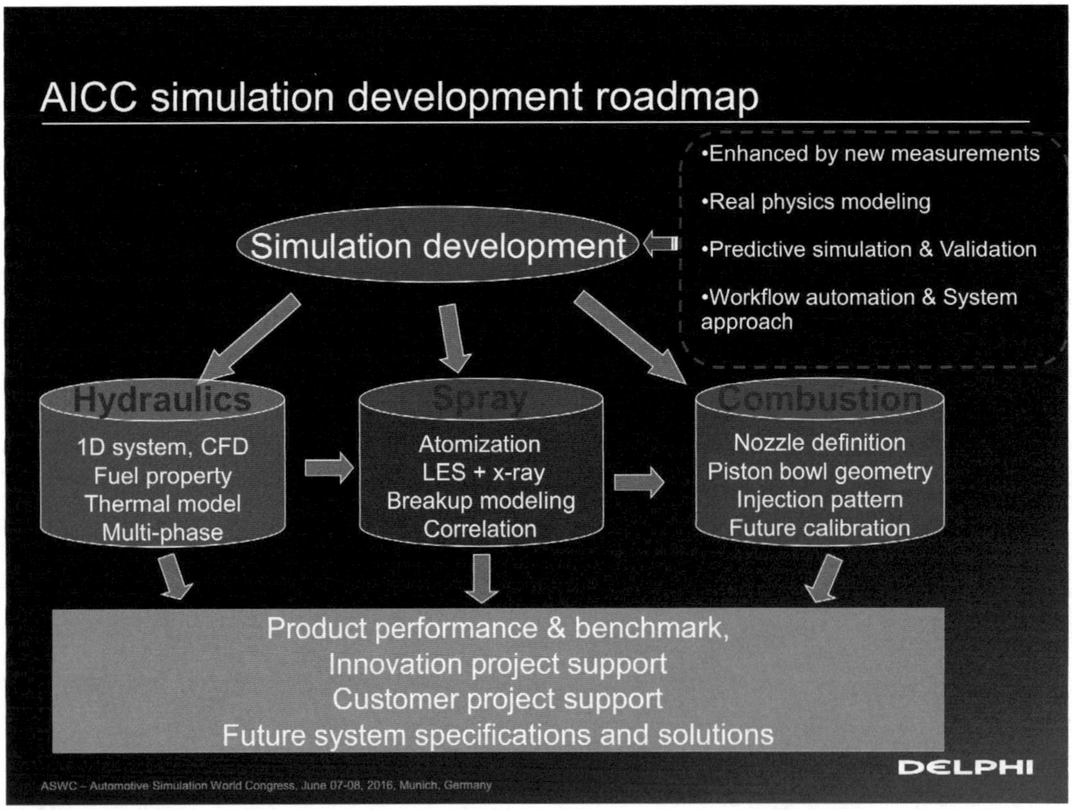
AICC simulation development roadmap
•Enhanced by new measurements
•Real physics modeling
•Predictive simulation & Validation
•Workflow automation & System approach
Simulation development
Hydraulics
1D system, CFD
Fuel property
Thermal model
Multi-phase
Spray
Atomization
LES + x-ray
Breakup modeling
Correlation
Combustion
Nozzle definition
Piston bowl geometry
Injection pattern
Future calibration
Product performance & benchmark,
Innovation project support
Customer project support
Future system specifications and solutions
ASWC – Automotive Simulation World Congress, June 07-08, 2016, Munich, Germany
DELPHI

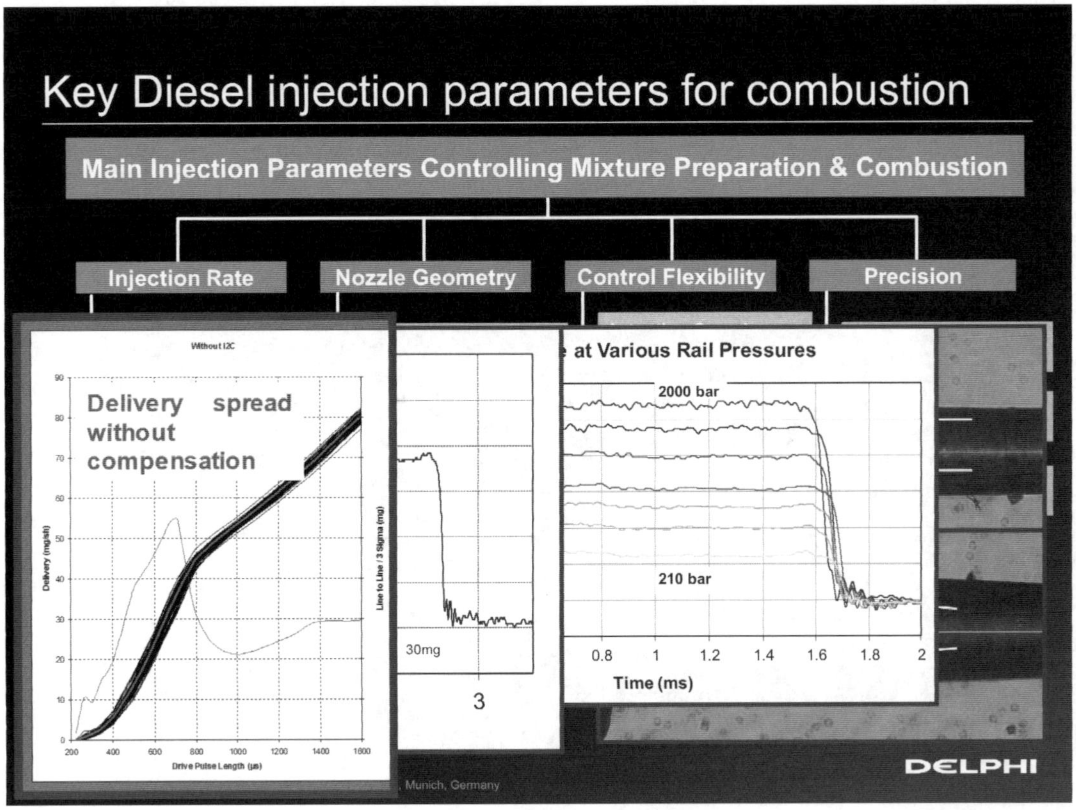
Key Diesel injection parameters for combustion
Main Injection Parameters Controlling Mixture Preparation & Combustion
Injection Rate
Nozzle Geometry
Control Flexibility
Precision
Without I2C
Delivery spread without compensation
Delivery (mg/st)
Drive Pulse Length (µs)
Line to Line / 3 Sigma (mg)
30mg
3
at Various Rail Pressures
2000 bar
210 bar
Time (ms)
Munich, Germany
DELPHI

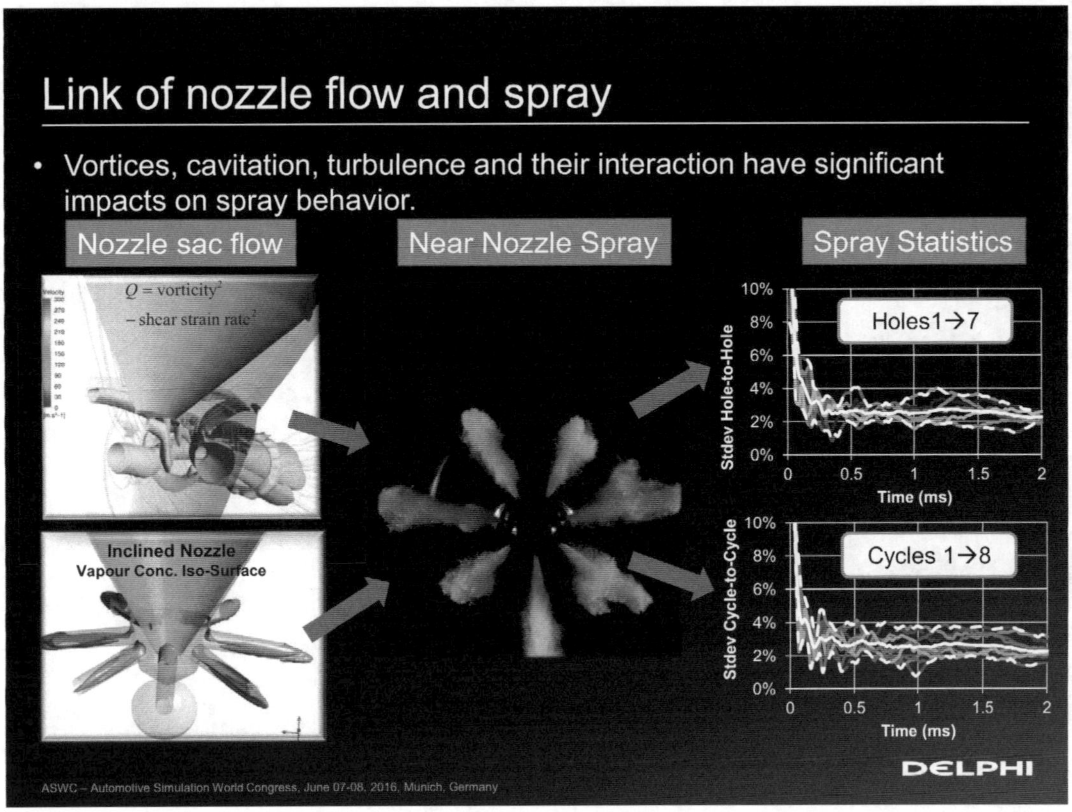
Link of nozzle flow and spray
• Vortices, cavitation, turbulence and their interaction have significant impacts on spray behavior.
Nozzle sac flow
Near Nozzle Spray
Spray Statistics
Q = vorticity² – shear strain rate²
Inclined Nozzle
Vapour Conc. Iso-Surface
Holes1→7
Stdev Hole-to-Hole
Time (ms)
Cycles 1→8
Stdev Cycle-to-Cycle
Time (ms)
ASWC – Automotive Simulation World Congress, June 07-08, 2016, Munich, Germany
DELPHI

Advanced simulation development enhanced by new measurement techniques

New Experiments
Hydraulics, Sprays, and Combustion

- Instrumented hydraulic rig (ROI, pressure wave, Temp)
- Time & field resolved high pressure momentum rig
- High contrast X-ray imaging
- Micro-visualisation rig (cavitation, near spray)
- Macro-optical spray characterization
- Transparent engine
- Single cylinder engine

New Modelling
Fuel Injection System, Sprays & Combustion

- 3-phase model
- RGFM fuel property model
- Coupled simulation: FIE system (1D) and Nozzle flow (CFD)
- Scale Resolved Simulation
- Breakup energy model
- Coupled nozzle flow and spray/combustion
- Detailed chemistry / Tabulate chemistry

Targeted outcomes

- Improved understanding of the physical processes
- Nozzle flow dynamics
- Atomization
- Correlations of nozzle design, spray & combustion
- Combustion system optimisation

ASWC – Automotive Simulation World Congress, June 07-08, 2016, Munich, Germany

Real-Gas-Fluid fuel property model (1)

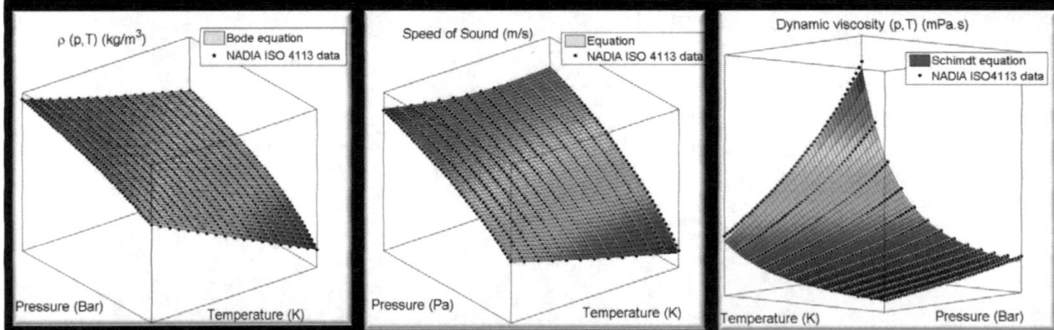

Real-Gas Fluid enthalpy equation requires consistency between ρ and c_P

$$c_P = \left.\frac{\partial h}{\partial T}\right|_P = c_P(P_0, T) + \int_{P_0}^{P} \left\{ \left[\left(\frac{\partial^2 \rho}{\partial T^2}\right)_P - \frac{2}{\rho}\left(\frac{\partial \rho}{\partial T}\right)_P^2 \right] \frac{T}{\rho^2} \right\} dP$$

$$c_T = \left.\frac{\partial h}{\partial p}\right|_T = \frac{1}{\rho}\left(1 + \frac{T}{\rho}\left.\frac{\partial \rho}{\partial T}\right|_p\right)$$

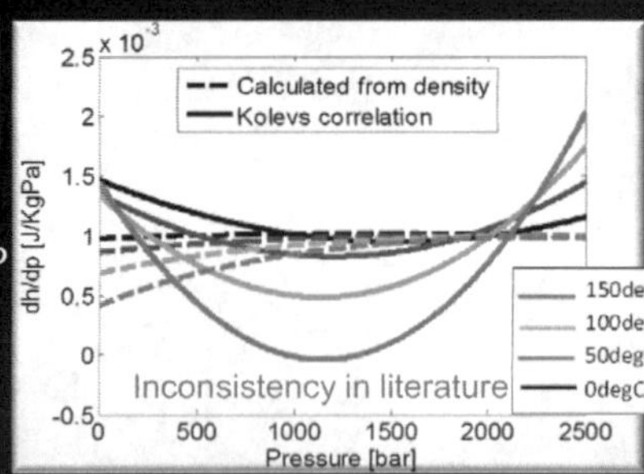

ASWC – Automotive Simulation World Congress, June 07-08, 2016, Munich, Germany

Real-Gas-Fluid fuel property model (2)

- Variable fuel property and thermal hydraulic modelling is essential for the sharp-edged inlet design.
- Inlet Reynolds number is the characteristic number for this design.

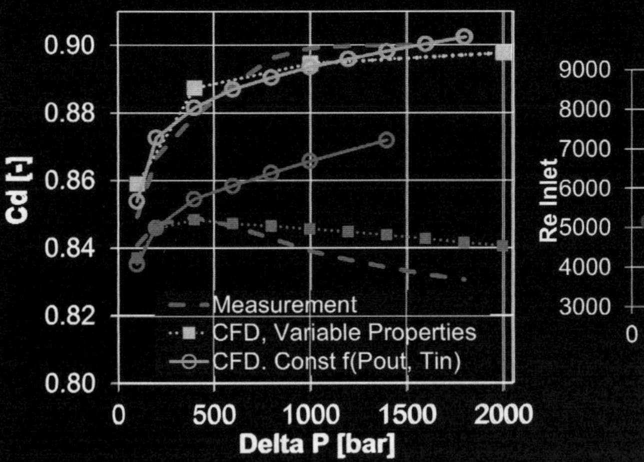

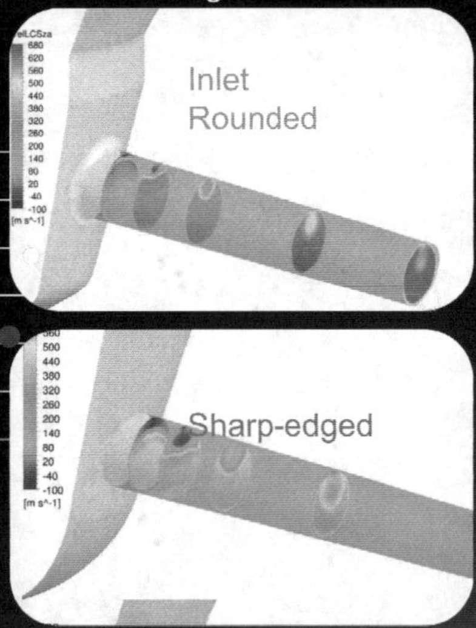

ASWC – Automotive Simulation World Congress, June 07-08, 2016, Munich, Germany

DELPHI

Coupled simulation approaches FIE system – Nozzle flow – Spray

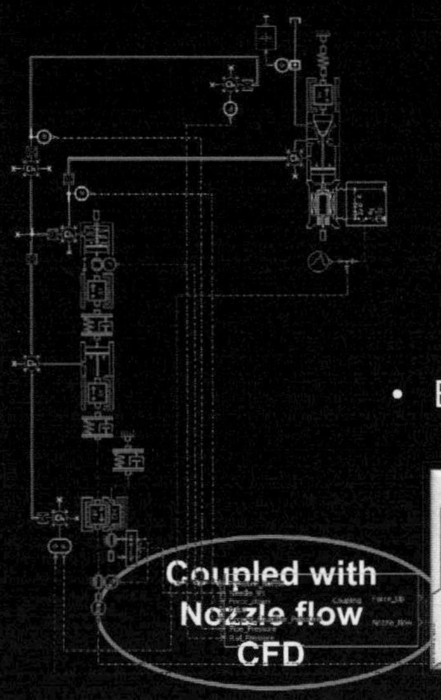

- Coupling 1D (FIE system) – CFD (nozzle flow)
  - Injector performance analysis
    - Small quantity injection diagnostic
    - Multiple injection diagnostic
  - Injector instrumentation support
  - Precise input data for combustion simulation
- Explicit coupling Nozzle flow – Spray & Combustion

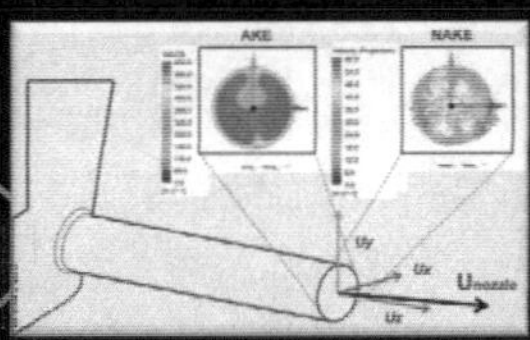

Eulerian nozzle flow

Nozzle file:
velocity
volume fraction
turbulence

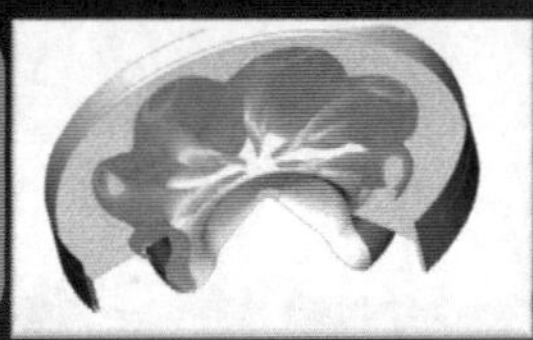

Lagrangian spray model

ASWC – Automotive Simulation World Congress, June 07-08, 2016, Munich, Germany

DELPHI

Numerical sensors for flow and spray diagnostic

Hydraulic diagnostic

- Mass flow / Cd
- Needle force
- P in the sac
- T in the injection hole
- Flow structure (vortex, turbulence)
- Cavitation in seat, sac, hole

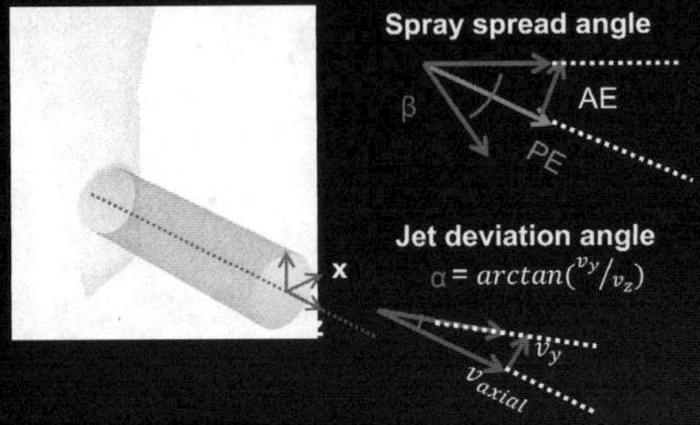

$$AKE = \frac{1}{2}v_z^2, \quad NAKE = \frac{1}{2}(v_x^2 + v_y^2), \quad TKE = \frac{1}{2}\overline{u_i'u_i'}$$

Spray diagnostic

- Axial velocity at injection hole exit, Jet momentum
- Vertical velocity at injection hole exit, jet deviation angle
- Penetration energy (PE): Axial kinetic energy
- Atomisation energy (AE): Turbulent kinetic energy, Non-axial kinetic energy, Cavitation energy

ASWC – Automotive Simulation World Congress, June 07-08, 2016, Munich, Germany

DELPHI

Small quantity injection diagnostic Validation

Hydraulic measurement – 2phase BP=50bar

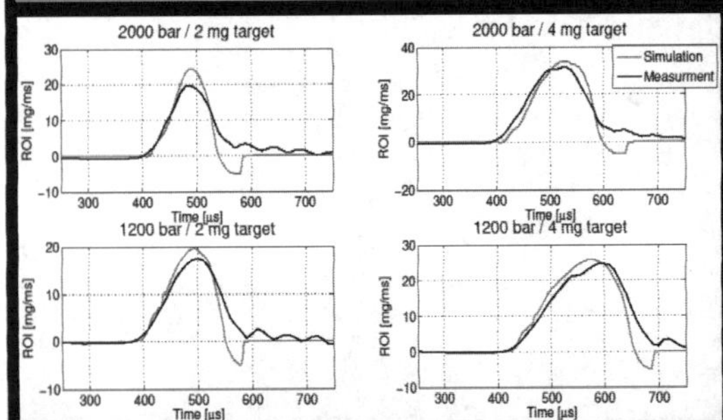

Momentum measurement – 3phase BP=1bar

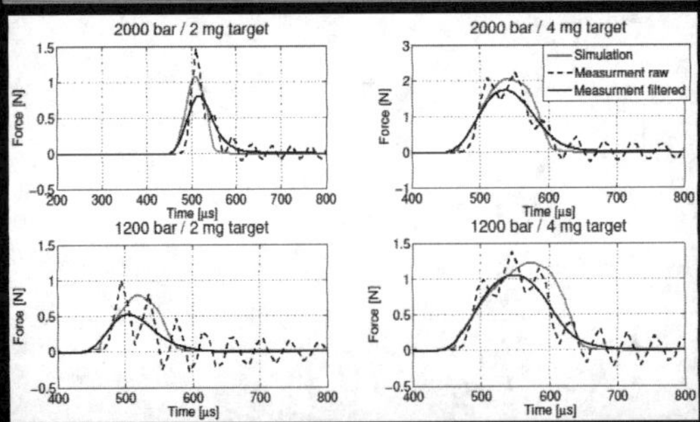

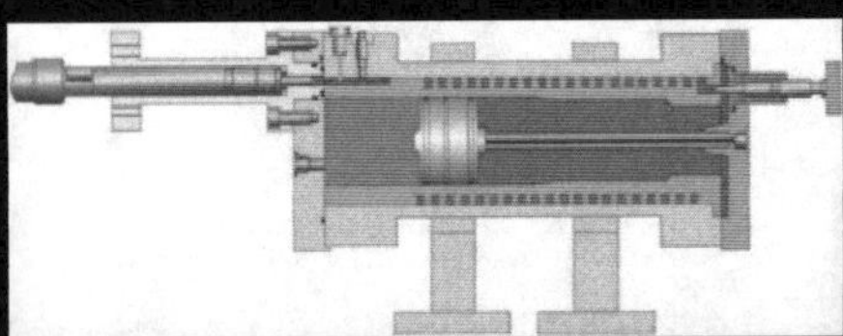

DELPHI

ASWC – Automotive Simulation World Congress, June 07-08, 2016, Munich, Germany

Small quantity injection diagnostic
Injection velocity

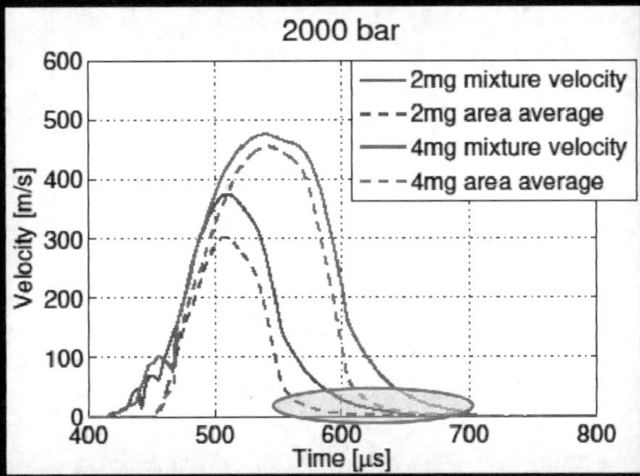

- Pilot injection also with high jet velocity
- The area averaged velocity is significantly lower than the real liquid jet velocity, especially during the early opening and late closing phase.

ASWC – Automotive Simulation World Congress, June 07-08, 2016, Munich, Germany

DELPHI

LES + x-ray investigation of primary breakup
Rounded vs. sharp edged hole inlet

SHARP EDGED | ROUNDED

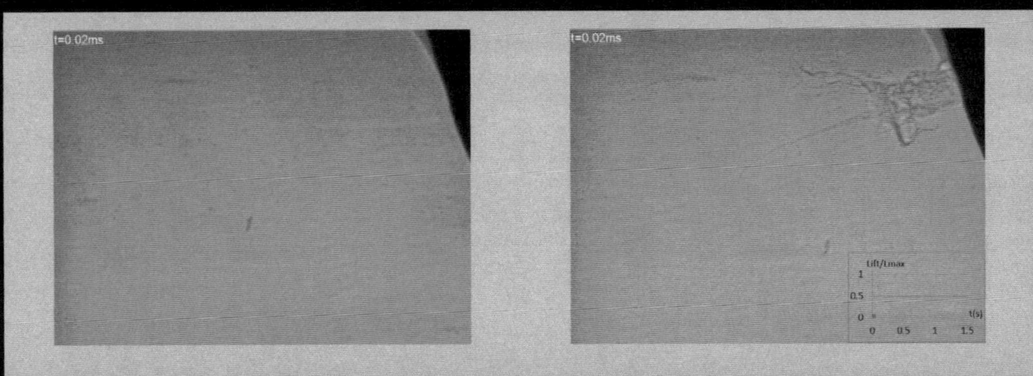

ASWC – Automotive Simulation World Congress, June 07-08, 2016, Munich, Germany

DELPHI

LES + x-ray investigation of primary breakup Rounded vs Sharp edged hole inlet

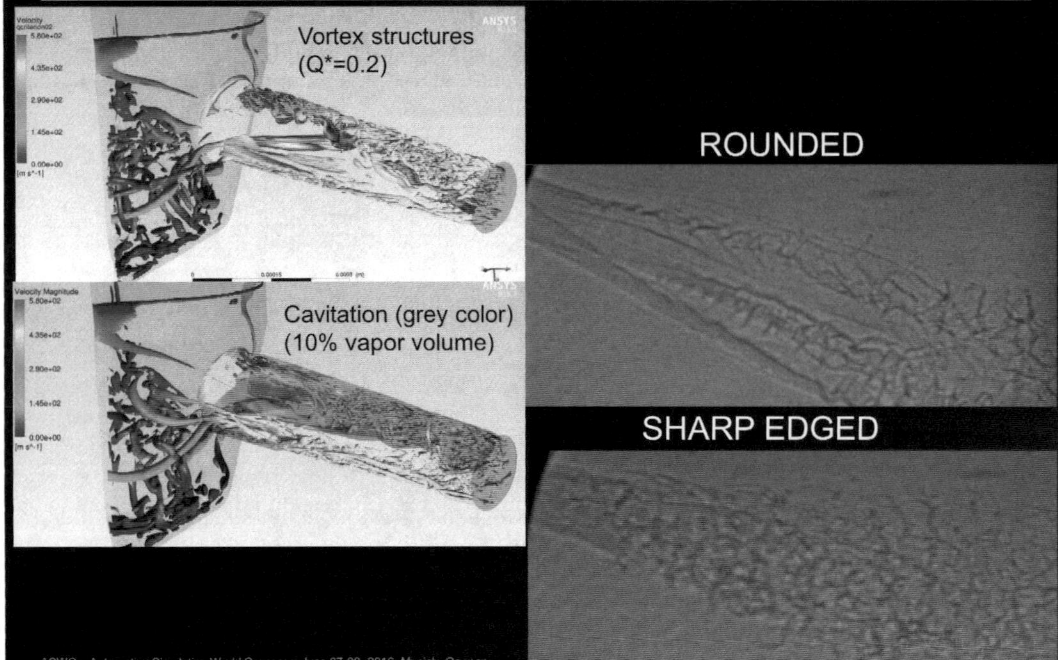

ASWC – Automotive Simulation World Congress, June 07-08, 2016, Munich, Germany

LES + x-ray investigation of primary breakup Rounded vs Sharp edged hole inlet

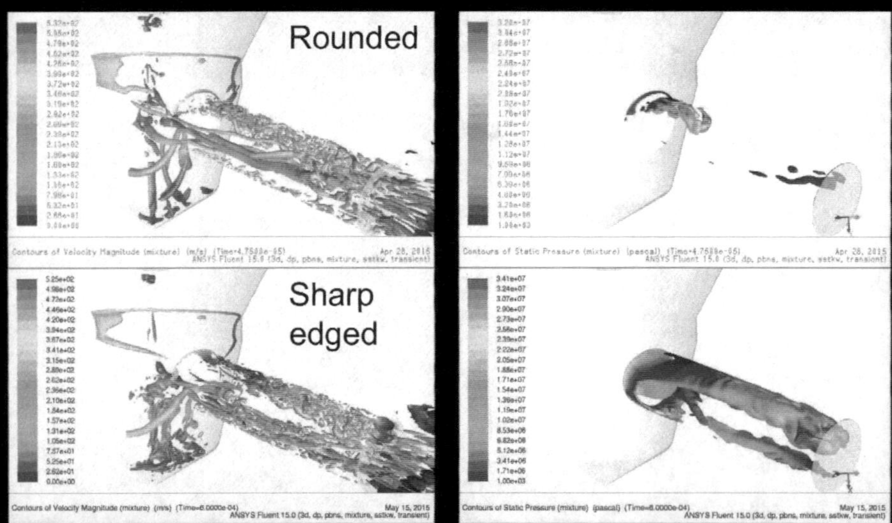

Sharp edged

Turbulent structures (Normalized Q-criterion 0.2 iso-surfaces, left) and vapor iso-surfaces (10% vapor, right)

DELPHI

ASWC – Automotive Simulation World Congress, June 07-08, 2016, Munich, Germany

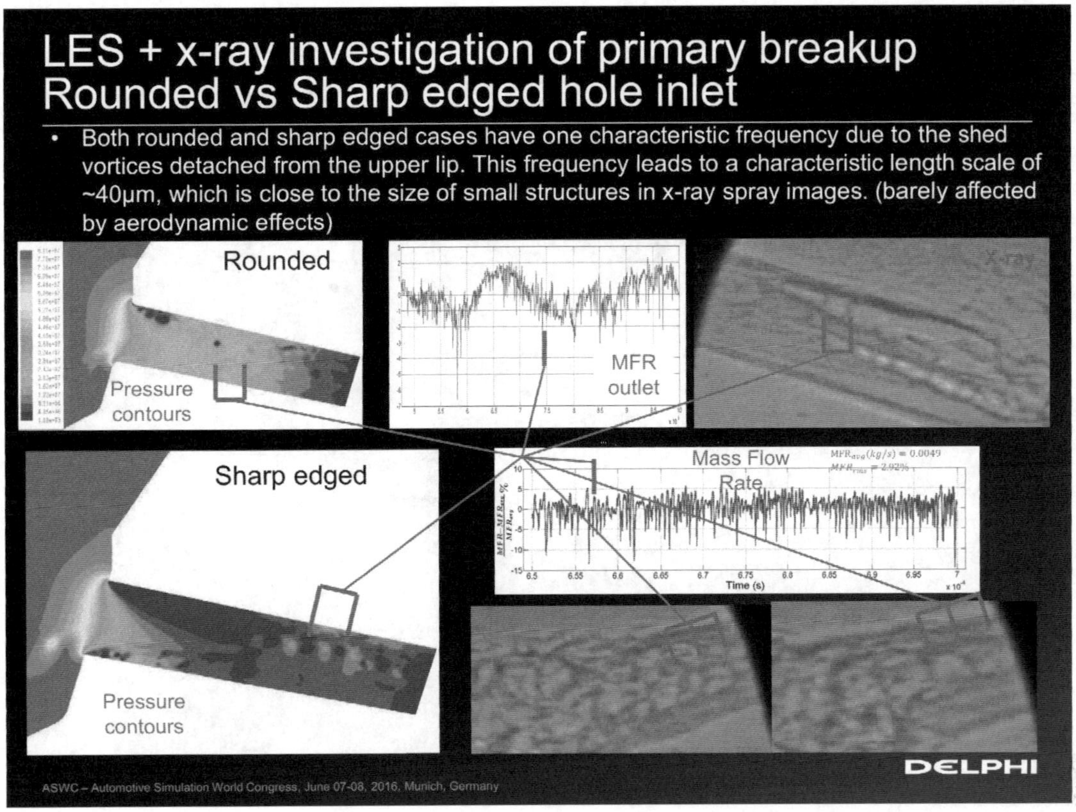
LES + x-ray investigation of primary breakup
Rounded vs Sharp edged hole inlet
• Both rounded and sharp edged cases have one characteristic frequency due to the shed vortices detached from the upper lip. This frequency leads to a characteristic length scale of ~40μm, which is close to the size of small structures in x-ray spray images. (barely affected by aerodynamic effects)
Rounded
Pressure contours
MFR outlet
X-ray
Sharp edged
Mass Flow Rate
Time (s)
Pressure contours
ASWC – Automotive Simulation World Congress, June 07-08, 2016, Munich, Germany
DELPHI

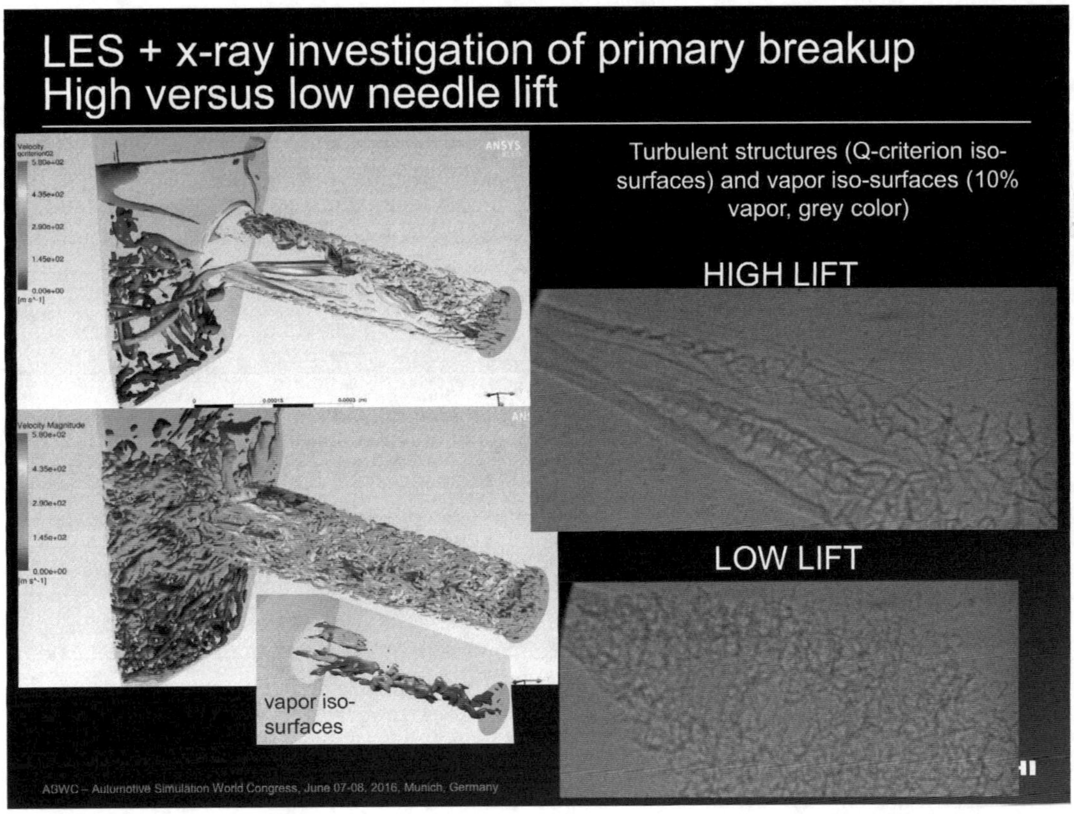
LES + x-ray investigation of primary breakup
High versus low needle lift
Turbulent structures (Q-criterion iso-surfaces) and vapor iso-surfaces (10% vapor, grey color)
HIGH LIFT
LOW LIFT
vapor iso-surfaces
ASWC – Automotive Simulation World Congress, June 07-08, 2016, Munich, Germany

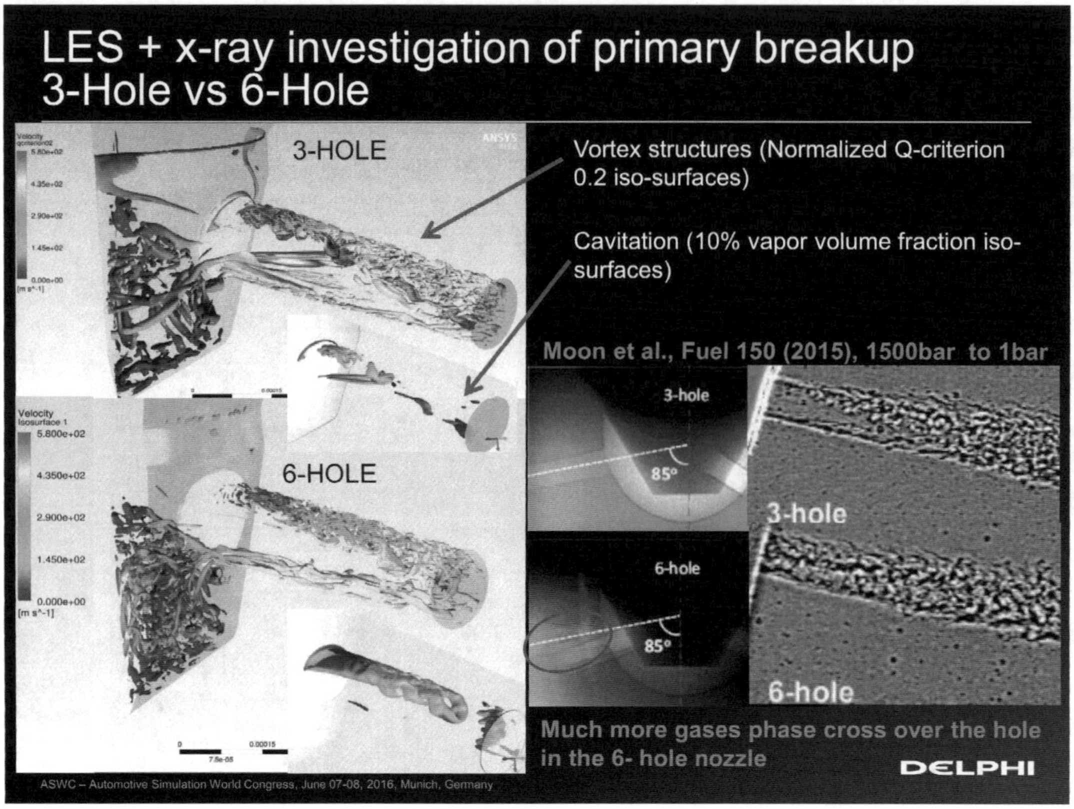
LES + x-ray investigation of primary breakup
3-Hole vs 6-Hole
3-HOLE
6-HOLE
Vortex structures (Normalized Q-criterion 0.2 iso-surfaces)
Cavitation (10% vapor volume fraction iso-surfaces)
Moon et al., Fuel 150 (2015), 1500bar to 1bar
3-hole
6-hole
85°
85°
3-hole
6-hole
Much more gases phase cross over the hole in the 6- hole nozzle
ASWC – Automotive Simulation World Congress, June 07-08, 2016, Munich, Germany
DELPHI

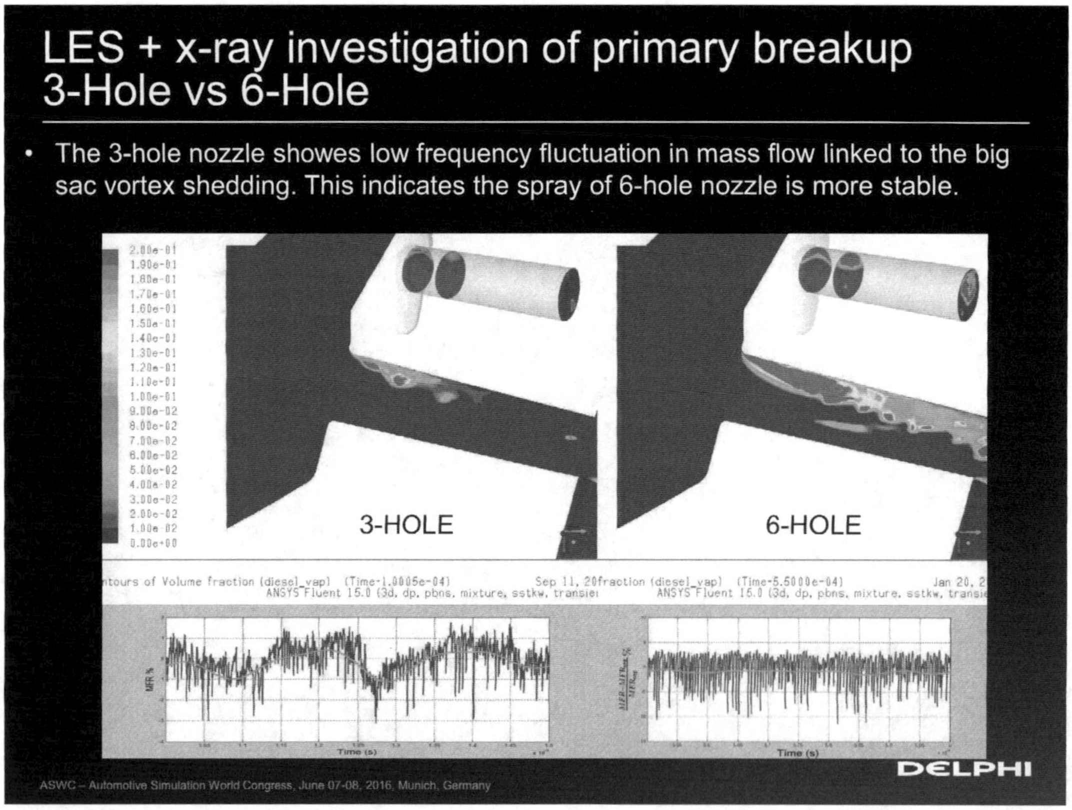
LES + x-ray investigation of primary breakup
3-Hole vs 6-Hole
• The 3-hole nozzle showes low frequency fluctuation in mass flow linked to the big sac vortex shedding. This indicates the spray of 6-hole nozzle is more stable.
3-HOLE
6-HOLE
Time (s)
Time (s)
ASWC – Automotive Simulation World Congress, June 07-08, 2016, Munich, Germany
DELPHI

Conclusions

- Complex physics modelling and new simulation methodologies have been developed.
- These new developments have provided complementary diagnostic functionalities for high pressure diesel injection and combustion analysis.
- In combination with new experiments the new simulation methodologies have produced new understandings on the physics involved in FIE products

ASWC – Automotive Simulation World Congress, June 07-08, 2016, Munich, Germany

DELPHI

Acknowledgement

- The following supports are gratefully acknowledged:
  - The Fonds National de la Recherche (FNR) funding to the PhD project of Pablo Aguado Lopez (Grant Number 588638)
  - The EU project funding, Holistic Approach of Spray Injection through a generalized multi-phase framework (HAoS) to the PhD project of Eduardo Gomez Santos (Grant number 675676)
  - The ANSYS – Delphi partnership program and technical support

ASWC – Automotive Simulation World Congress, June 07-08, 2016, Munich, Germany

DELPHI

DELPHI

Innovation for the Real World

Automotive Simulation
WORLD CONGRESS

June 7-8, 2016 Munich

Internal combustion engine simulation strategies

Ellen Meeks, Director of Reacting Flow
ANSYS, Inc.

Outline

- **Automotive Engine Simulation at ANSYS**
  - Overview of tools and specialization
- **Focus on Engine Combustion simulation**
  - Current capabilities
  - Validation studies
  - Application examples

The full power of ANSYS Workbench enables Optimization and Fluid-Structures Interaction

Design Optimization:
- Optimize a geometry
- Optimize operating conditions

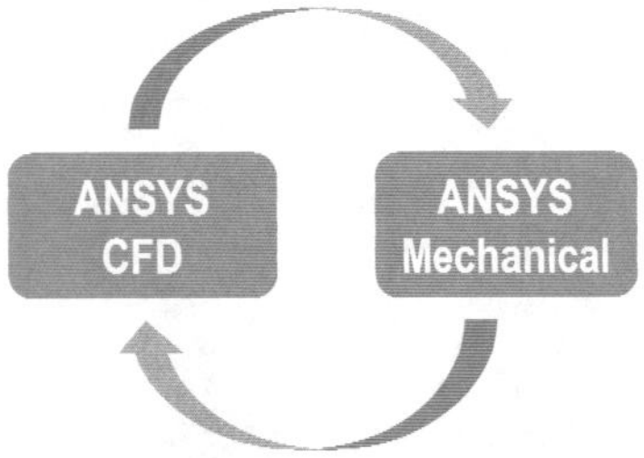

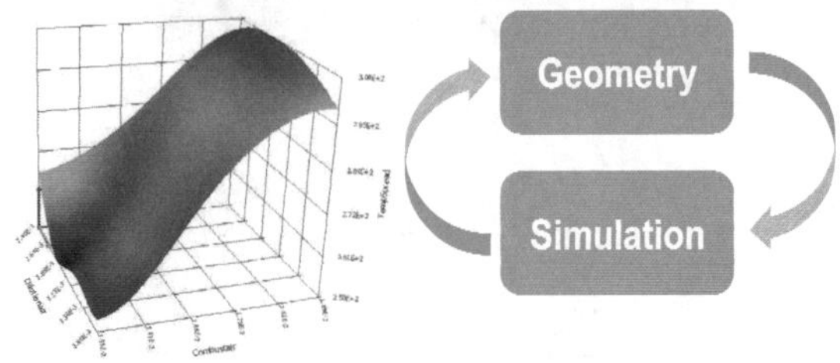

Fluid Structure Interaction (FSI)
- CFD, Thermal, Structural
- 1- and 2-way FSI Coupling

Typical Simulation Activities in IC Engine Design

- **Flow-bench studies**
- **Injector/Spray characterization**
- **Engine mapping**
- **Design exploration**
- **Design optimization**
- **Engine in the system: multicomponent & multi-physics design considerations**
  - **overall fuel efficiency, heat/work balance, fatigue/stress, reliability, controllability**

ANSYS CFD serves all of these applications

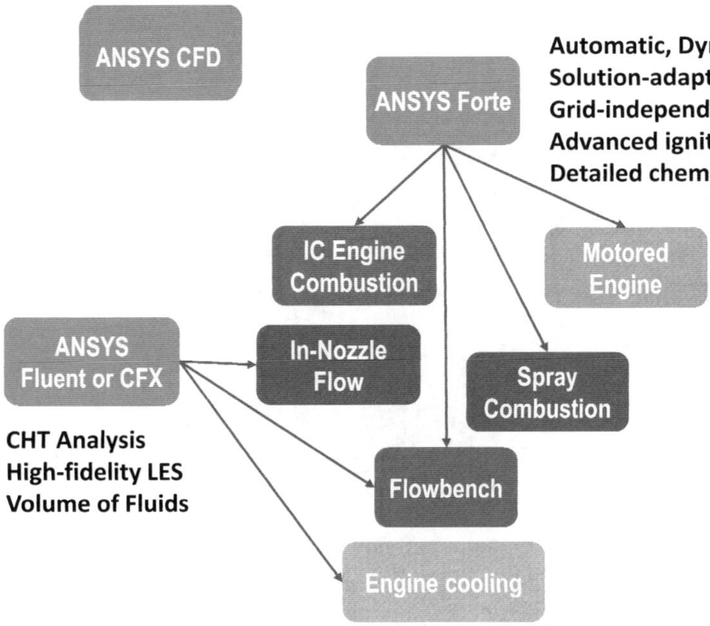

Automatic, Dynamic Mesh Generation
Solution-adaptive mesh refinement
Grid-independent spray models
Advanced ignition & flame propagation
Detailed chemistry & fuel models

CHT Analysis
High-fidelity LES
Volume of Fluids

IC engine simulation with ANSYS CFD: ANSYS Forte

- We are investing in the ANSYS Forte solution for ICE:
  - Developing new capabilities
  - Integrated with ANSYS WB, Licensing, Installation
  - Coupling with ANSYS Fluent for conjugate heat transfer (CHT)
  - Continuing advances of fuel models & fuel modeling tools
  - Performing extensive verification and validation tests
  - Continuing performance and scalability improvements

Key capabilities of ANSYS Forte

- **Set up tailored to IC Engines**
- Automatic mesh generation
  - On-the-fly, dynamic mesh
- **Well validated fuel-combustion kinetics with fast solver**
  - **Predict ignition, emissions, fuel effects**
  - **Predict flame-front location and knocking in SI engines**
- **State-of-the-art spray and spray-wall models**
  - **Capture physics accurately and robustly with mesh insensitivity**

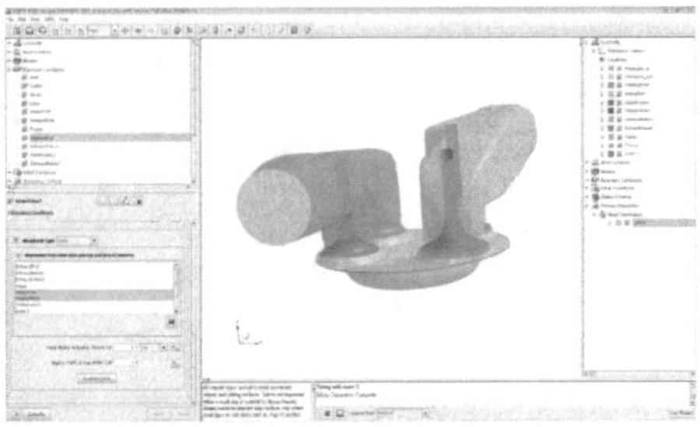

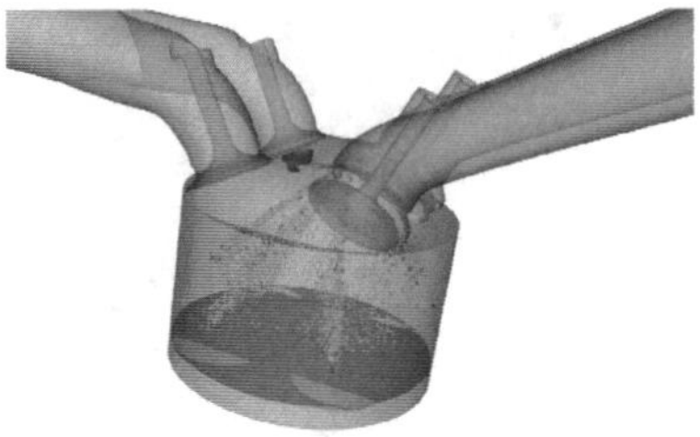

Automotive Simulation WORLD CONGRESS

Workbench Integration: Quick and easy CAD => Simulate

- IC Engine (Forte) in WB

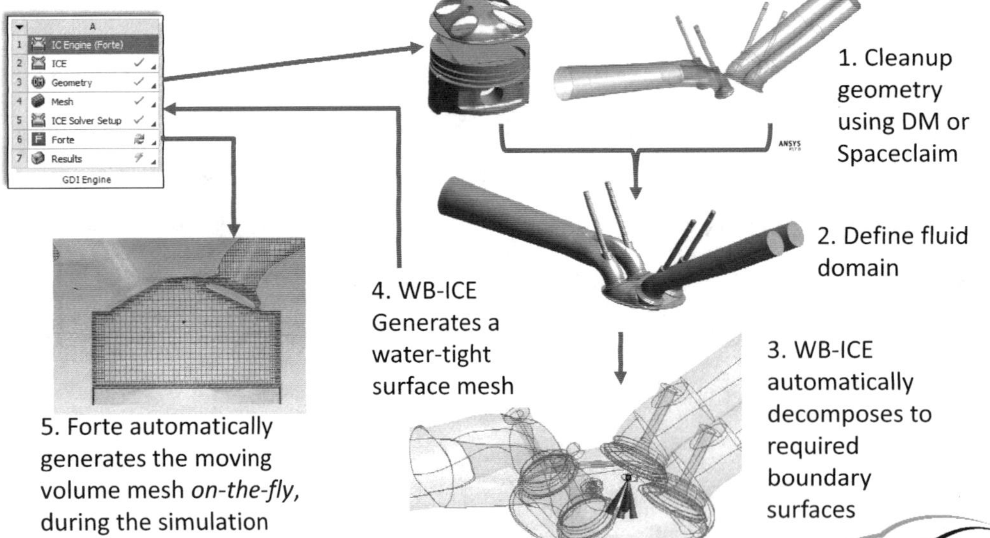

Automotive Simulation WORLD CONGRESS

Mesh-refinement controls for Forte automatic meshing

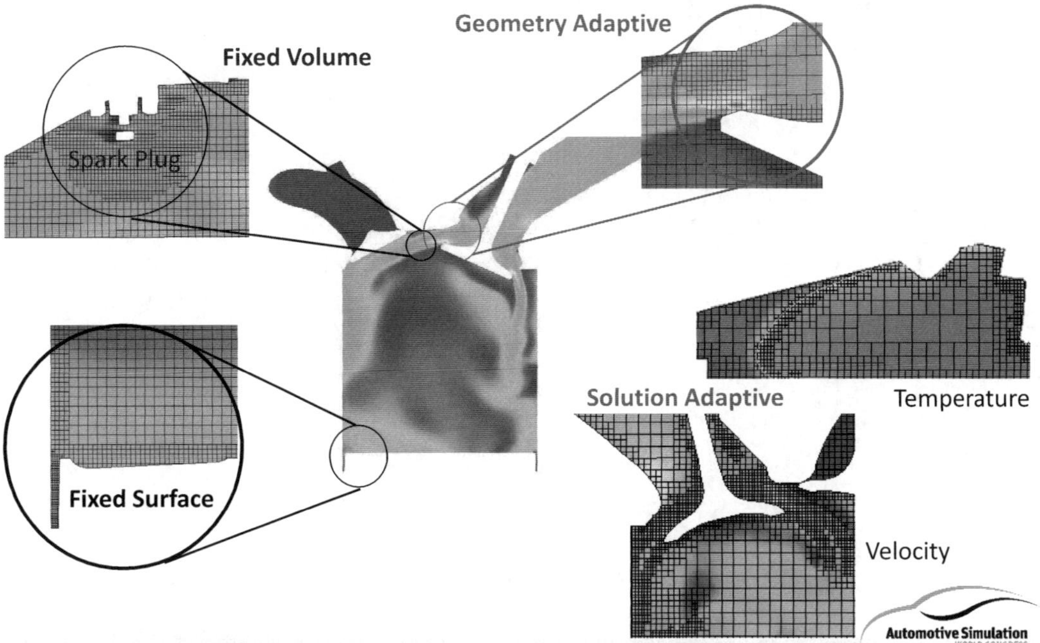

CFD-Post visualization of Forte results

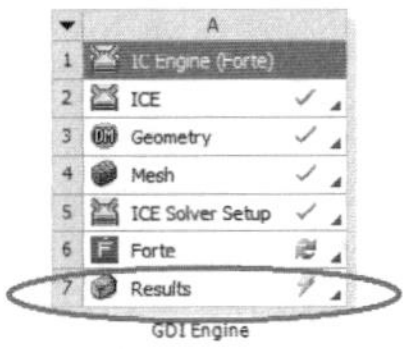

- **Including automatic reporting through WB-ICE**
- **Also fully supported: Ensight®, Fieldview®**

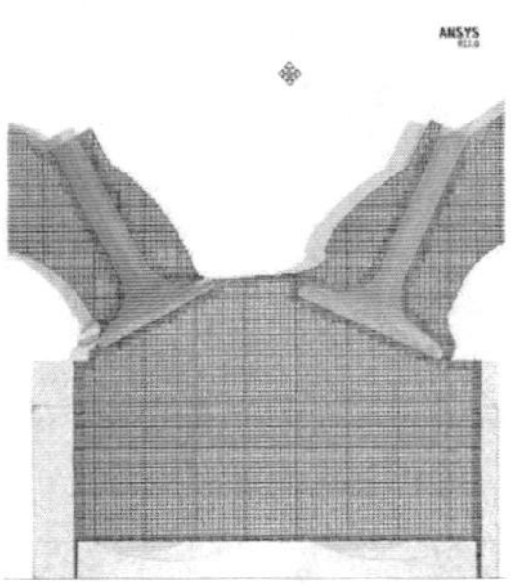

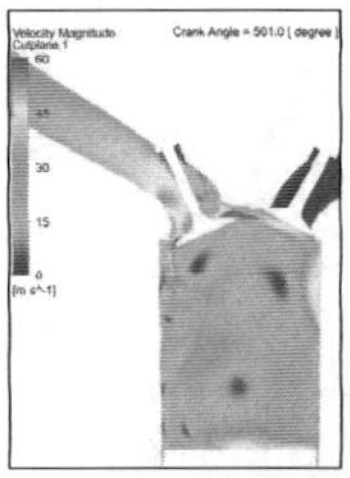

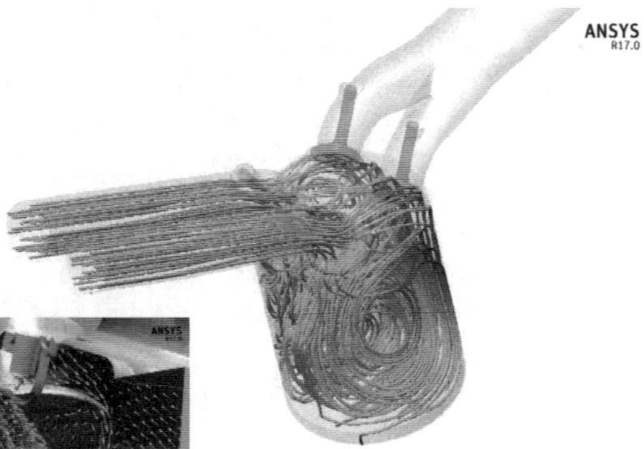

Advanced solution techniques allow practical turn-around with chemistry

A good chemistry mechanism is key to predicting engine combustion behavior:
- ignition, emissions, flame speeds, knocking, particles

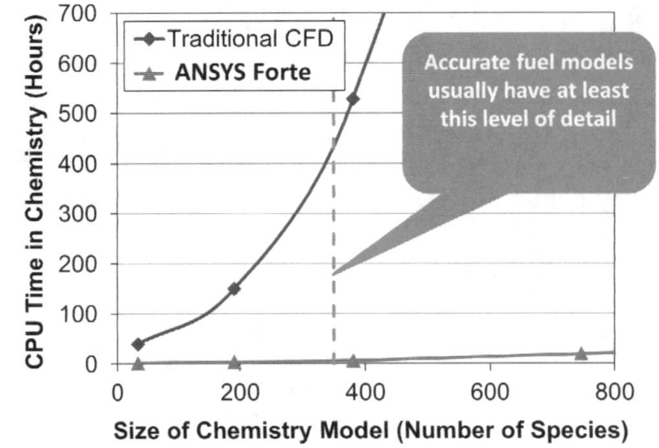

ANSYS Model Fuels Library is a source of well validated fuel-combustion mechanisms

~ 65 Validated Fuel Components

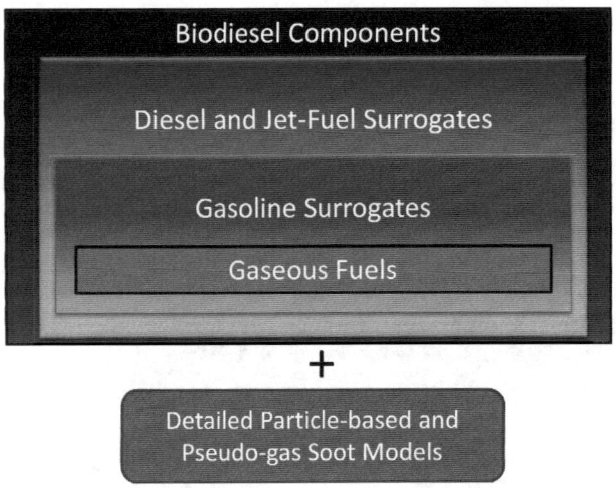

- Common core chemistry
- Consistent rules for reaction rates
- Validated fuel-blending behavior

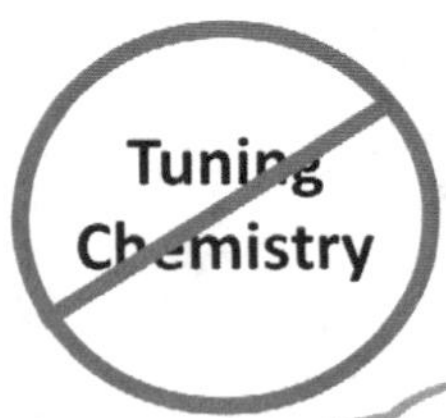

Spray: consistent results on coarse to fine meshes

- **Both liquid and vapor penetration show mesh insensitivity**

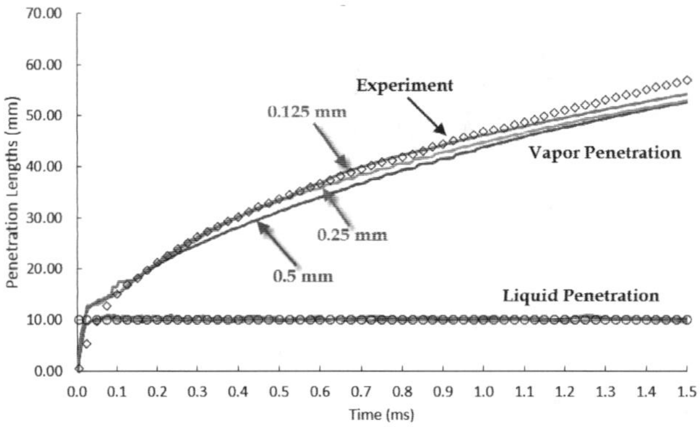

Velocity magnitude

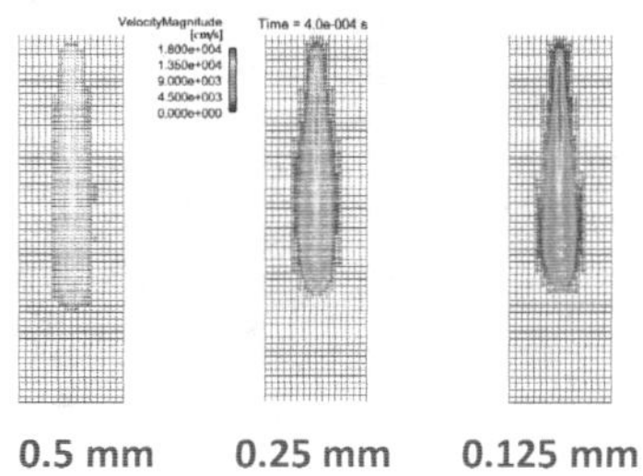

Results from:
L. Liang, et al., 2016
IMEM, April 2016

Forte includes comprehensive spray-wall interactions

- **Includes:**
  - **Droplet stick**
  - **Droplet rebound**
  - **Droplet splash**
  - **Film spread**
  - **Film separation (stripping)**
  - **Flash boiling**
  - **Heat transfer**

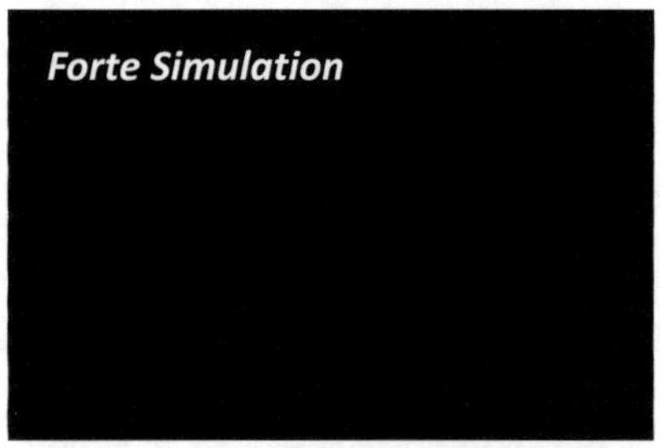

Validation for varying impingement angles:

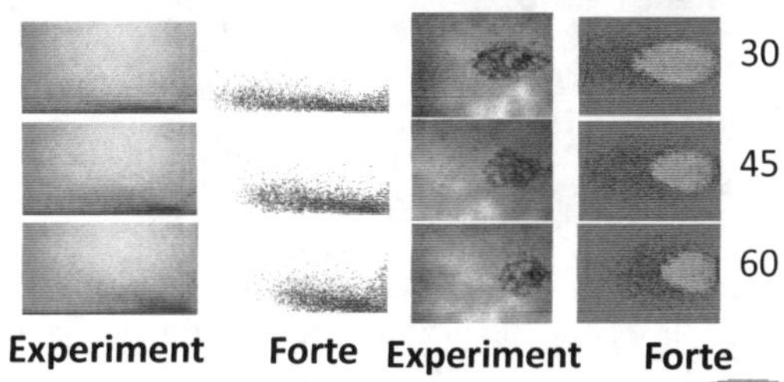

Spark ignition and flame propagation use well proven flame-front-tracking methods

The spark ignites a kernel of gas that is very small at first

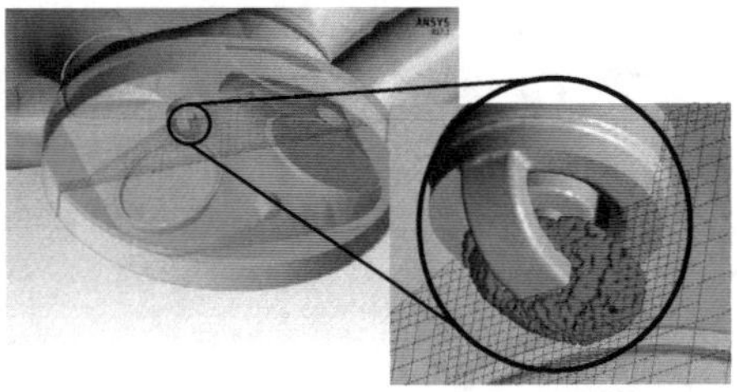

Particles are used to track the initial kernel growth and flame location

The flame (typically <~0.01 mm thick) propagates through the unburned gas

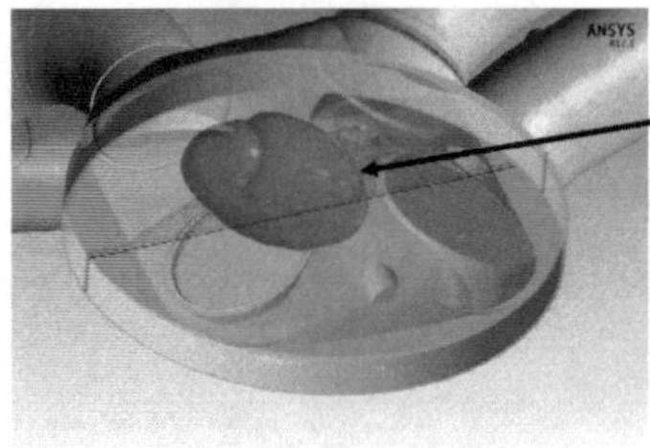

Afterwards, the level-set (G-eq.) method takes over in tracking the flame front

The flame propagation responds to local turbulence conditions and chemistry is solved throughout

SAE 2003-01-0722, 2003-01-1843, 2006-01-0243, 2007-01-0165

Built-in flame tables provide multi-component fuel effects

- **50 Fuel components included in Flame-speed Library**
- **Accounts for all factors affecting laminar flame-speed**
  - **Stratification in vapor composition**
  - **Cross-dependence of:**
    - **Pressure**
    - **Equivalence Ratio**
    - **Temperature**
    - **Dilution**
    - **Fuel structure**

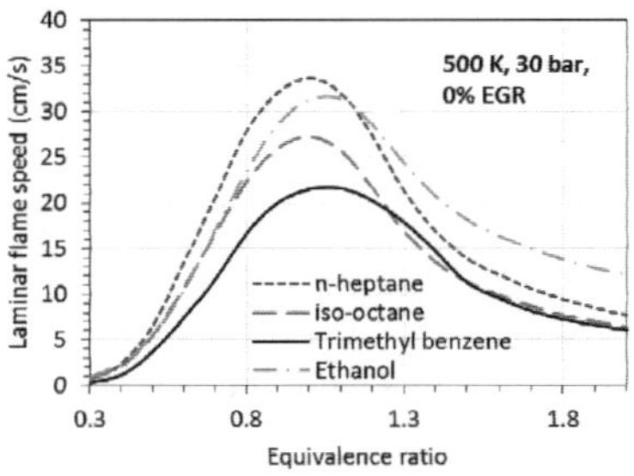

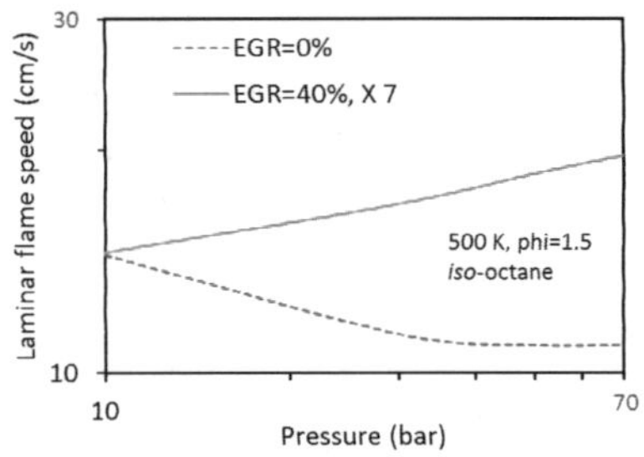

Example: Dependence of S_L on Pressure varies with EGR

Power-law correlations *cannot capture this effect*

SAE Technical Paper 2016-01-0597

Difference between a detailed vs. simple fuel model

- **Gasoline IC engine simulation: combustion phasing**
  - **ANSYS CFD simulations show good agreement with data for detailed model**

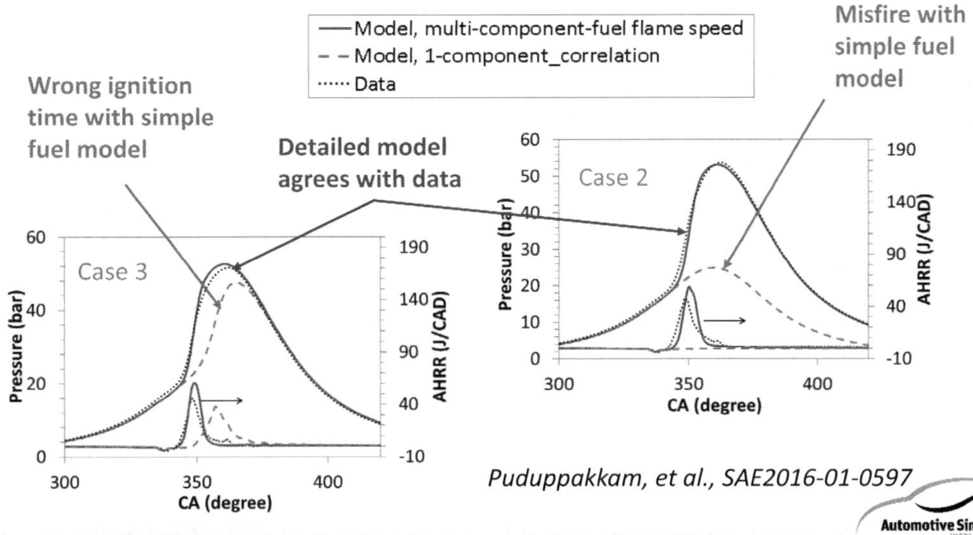

Puduppakkam, et al., SAE2016-01-0597

Automotive Simulation WORLD CONGRESS

Validation studies show predictive capabilities

- **Gasoline Direct Injected (GDI) Engines**
  - **Same model settings over broad range of conditions**
  - **Mesh insensitivity to trends**
  - **Combustion phasing, spray behavior**
- **Diesel engines**
  - **Same model setup over broad range of conditions**
  - **Ignition, IMEP, Soot and other emissions predictions**

Automotive Simulation WORLD CONGRESS

Extensive validation uses rich set of data for GDI Engines

- **Motored cases, varying:**
  - Intake temperature
  - Coolant temperature
  - Intake MAP
  - Engine rpm
- **8 Port-fuel Injection (PFI) cases, varying:**
  - Intake pressure
  - Equivalence ratio
  - Spark timing
- **9 Gasoline Direct Injection (GDI) cases, varying:**
  - Global equivalence ratio
  - Spark timing

SAE2016-01-0591

Motored engine study shows good mesh convergence

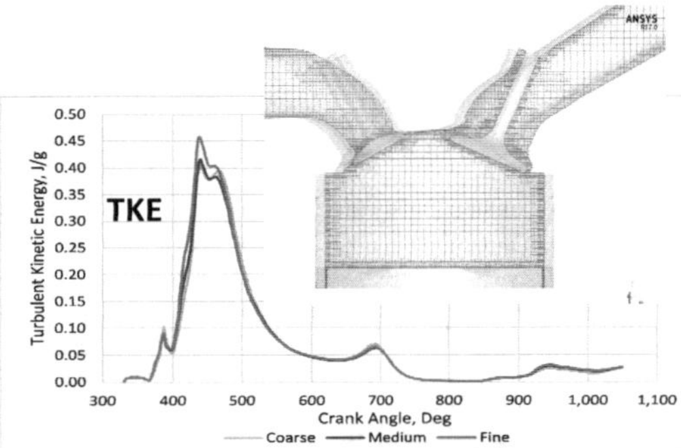

Global cell size varied from 3 mm to 1 mm

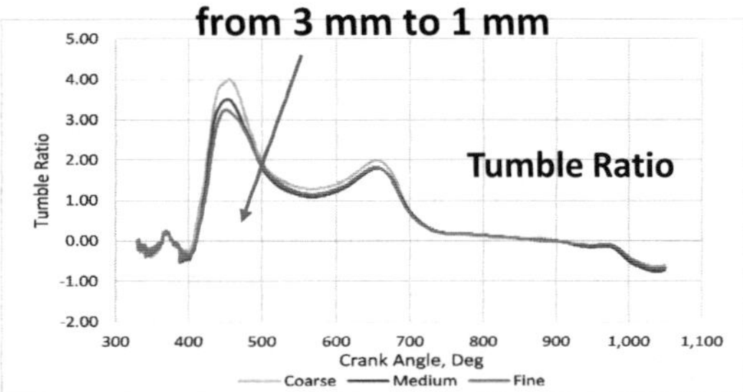

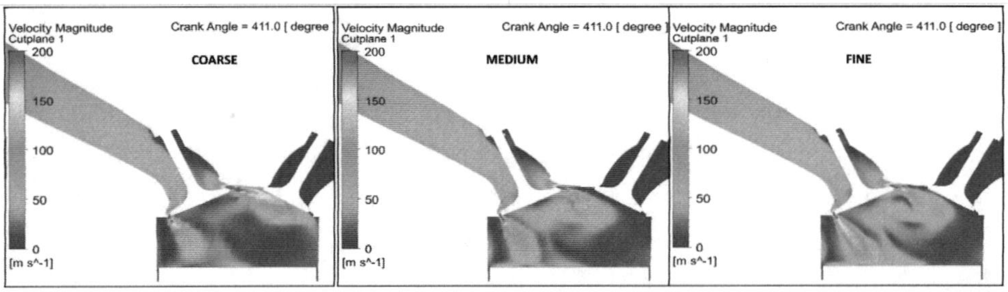

SAE2016-01-0591

Motored Engine study: validation vs. data

- **Same setup captures subtle changes in experiments**
- No calibration of inputs
  - Use measured Intake T, P
  - Use measured b.c. T

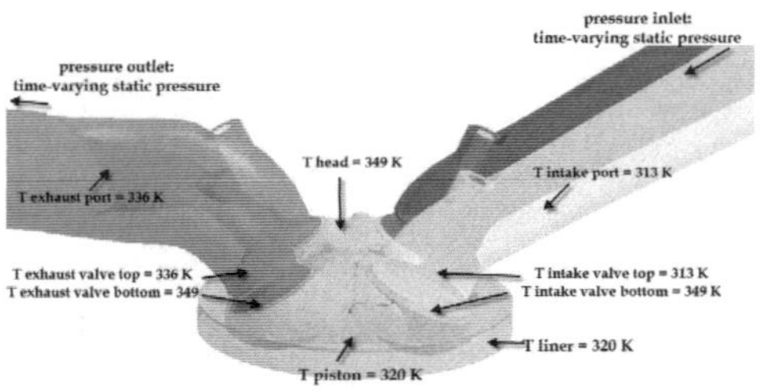

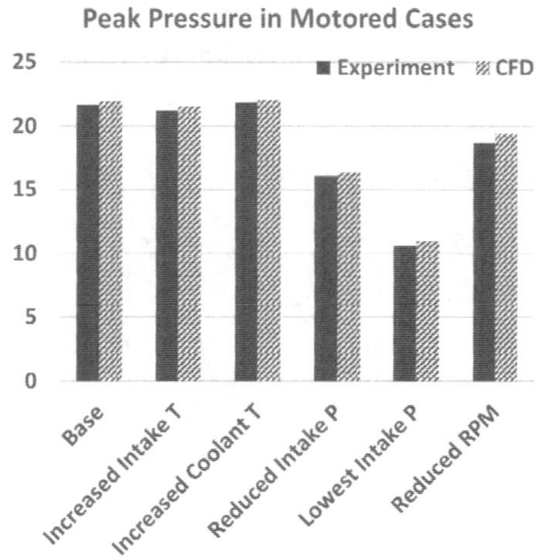

SAE2016-01-0591

PFI Study: same setup, with detailed fuel model

- **6-component fuel model matches experimental fuel properties**
  - **230 species; from ANSYS Model Fuels Library**
- All model parameters kept constant
- No calibration of boundary conditions or spark timing

Property Name	Gasoline Fuel	Surrogate Fuel
Research Octane Number	96.6	95.47
Motored Octane Number	88.6	89.39
Molar H/C Ratio	1.843	1.8429
Liquid Density [g/cm3]	0.742	0.735
L.H.V. [MJ/kg]	43	43.41

Black line = data

Red line = CFD

CoV for GDI cases=~4%

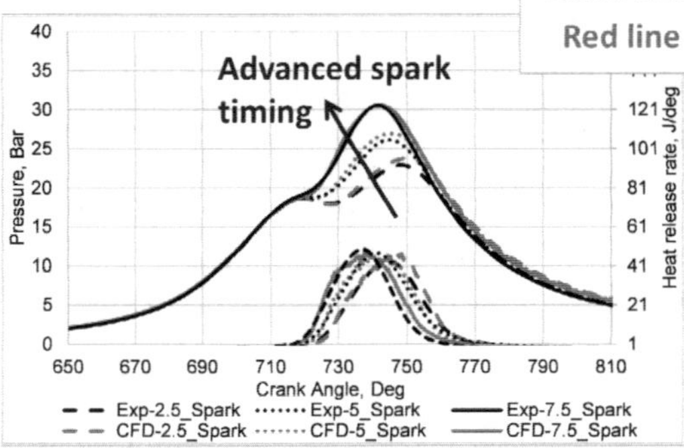

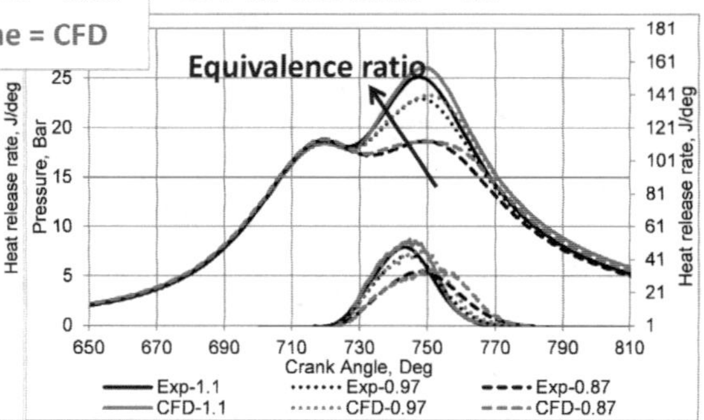

SAE2016-01-0591

Spray chamber study: establish spray settings

- **Same fuel model as in PFI cases**
- **Mesh and time-step insensitivity**
- **Visual and quantitative comparisons**

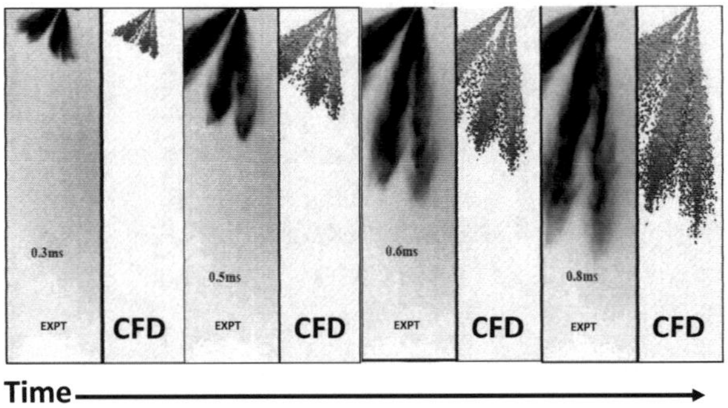

U. Wisconsin ERC Experiments

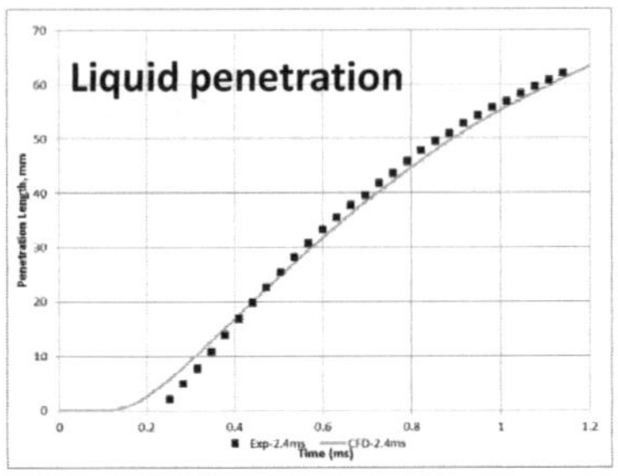

SAE2016-01-0579

Spray validation: droplet size and velocities

- **The model captures the droplet size distribution well**

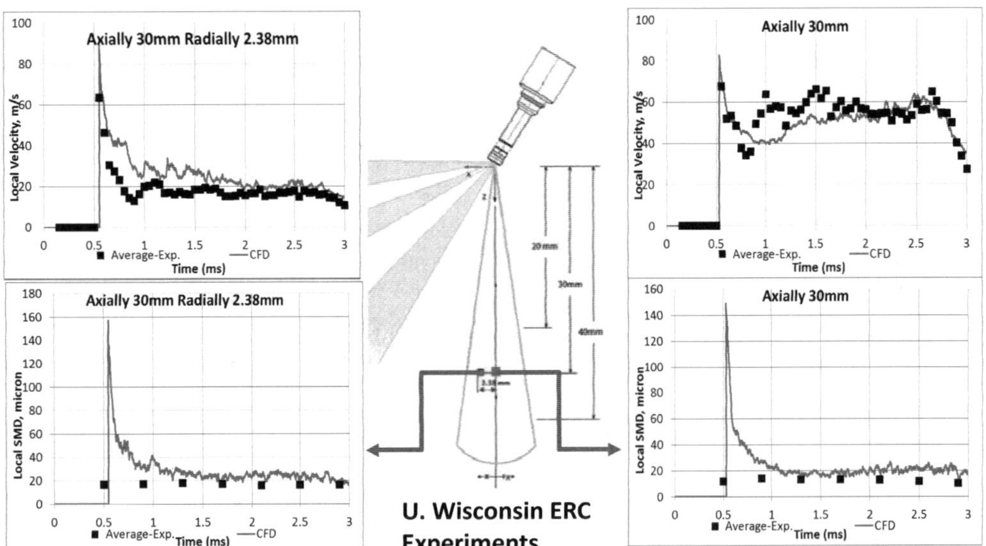

U. Wisconsin ERC Experiments

SAE2016-01-0579

GDI Engine Mode: using same model setup

- Same fuel model
- Same mesh as motored & PFI
- Same spray-model settings
- No calibration of b.c.'s or spark timing

Equivalence ratio contours show stratified charge

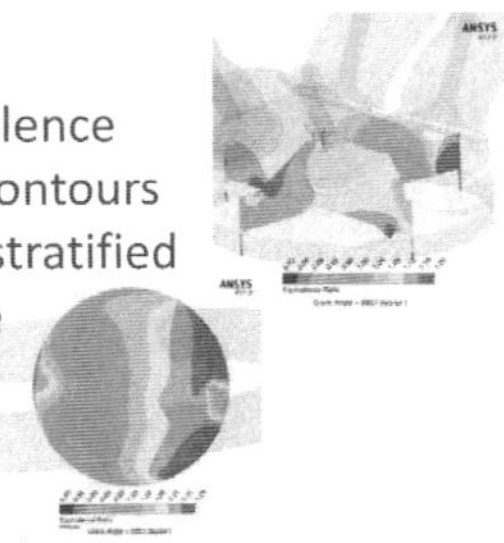

Black line = data

Red line = CFD

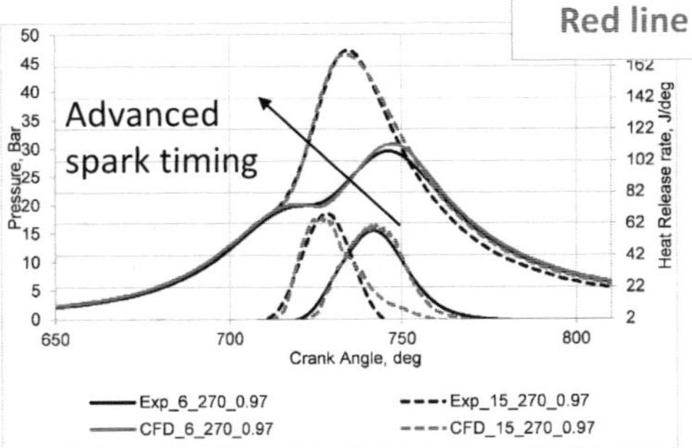

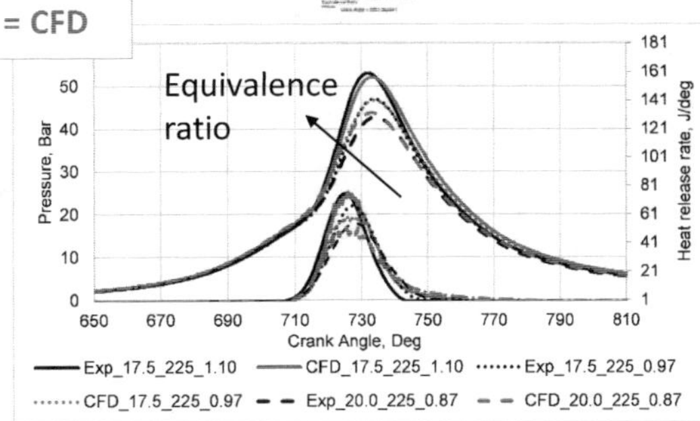

SAE2016-01-0579

CoV for GDI cases=1-3%

Predictive capabilities have also been shown for Diesel engines

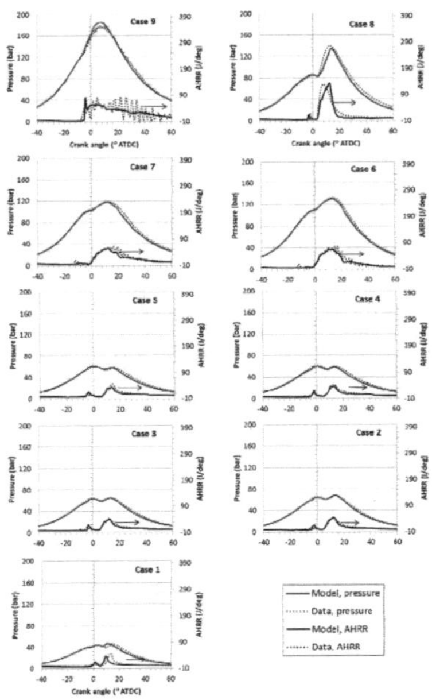

- Detailed (7-component fuel model), 500 gas-phase species, multi-step soot model
- Combustion and emissions trends captured with no change to model or model approach

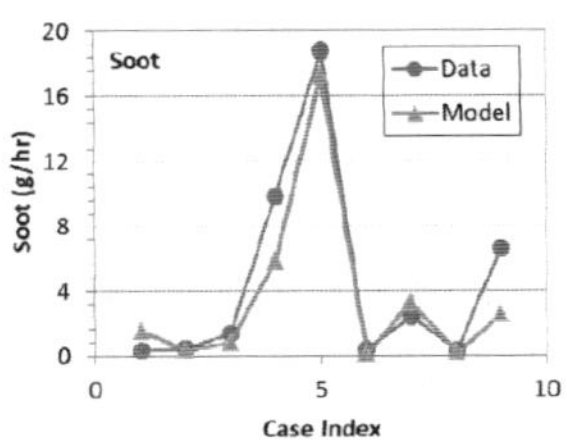

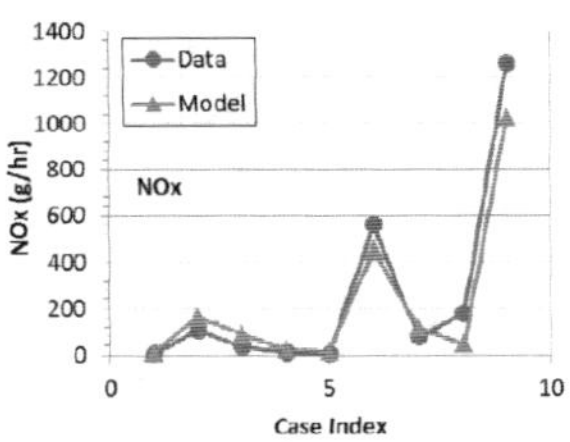

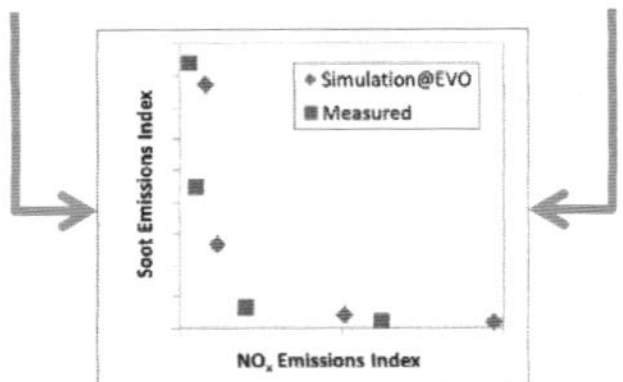

Puduppakkam, et al. SAE Powertrain Fuels and Lubes, 2014

ANSYS MFL soot models provide unprecedented capability

Ability to see where, when, how many and what size soot particles are produced during combustion

Where?

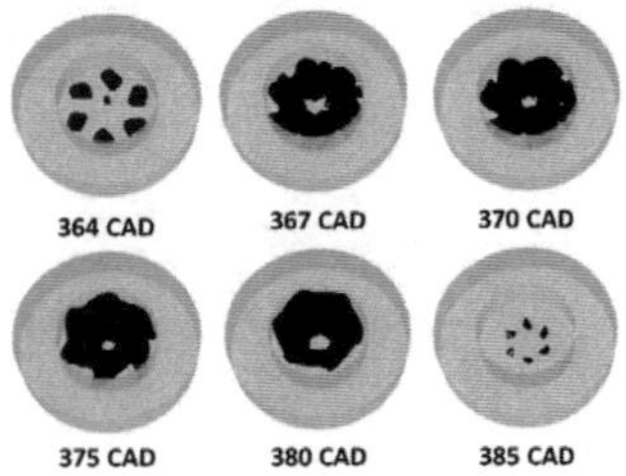

Using: ANSYS Forte CFD with Model Fuels Library

When?

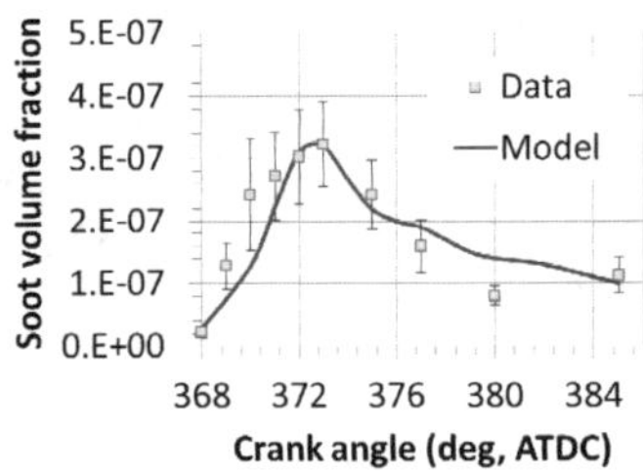

How many of What size?

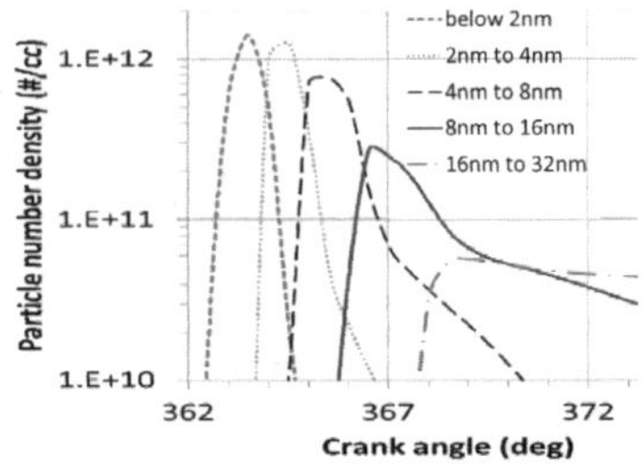

Ref: K. Puduppakkam, et al. IMEM workshop 2015

Automotive Simulation WORLD CONGRESS

Summary

- ANSYS is committed to advancing IC Engine Simulation
- Forte is the core of the ANSYS IC engine strategy
  - Significant new advances
  - Continuous improvement on accuracy, performance and scalability
  - Integration with other ANSYS tools
  - Leveraging ANSYS Model Fuels Library
- Extensive validation studies show predictive nature of models and the power of detailed chemistry
  - Spray and spray-wall interactions
  - Spark ignition and flame propagation
  - Fuel effects and emissions

Automotive Simulation WORLD CONGRESS

Automotive Simulation WORLD CONGRESS

June 7-8, 2016 Munich

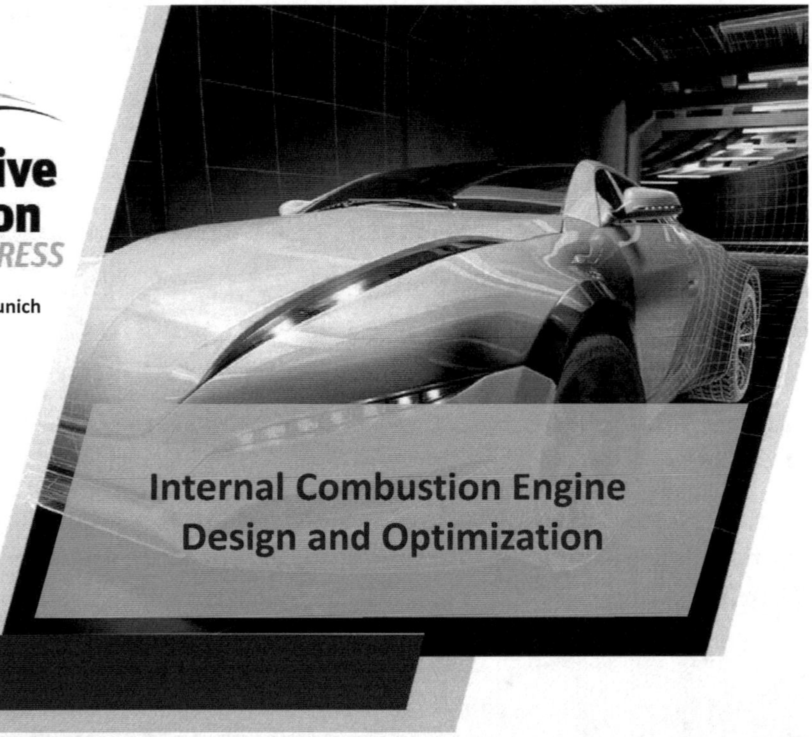

Internal Combustion Engine Design and Optimization

Laz Foley
ANSYS Inc.

Outline

- Beyond Core Combustion CFD
- 01-1D Design Assessment
  - Chemkin-Pro SI Model + GT-Power Coupling
- 3D Design Exploration and Topology Based Optimization
  - Parametrization, Meshing Morphing, Adjoint Methods
- Work Flow Automation and Utilities
- ANSYS Customization Toolkit (ACT)
- Combustion System Design and Optimization
  - Use Case 1: Injection Rate Shape Study
  - Use Case 2: Parametric Diesel Bowl Study

Beyond Core Combustion

- Combustion system design is much more than combustion CFD
- Solution paradigm requires various levels of solution fidelity and extensive design space exploration
- Solutions need to span physics disciplines including multiphysics and co-simulation; i.e. CHT, design for durability, predict failure modes etc.
- An integrated solution platform is an important element for solution execution, adoption and deployment in an engineering IT environment

0D Model For Preliminary Design Assessment

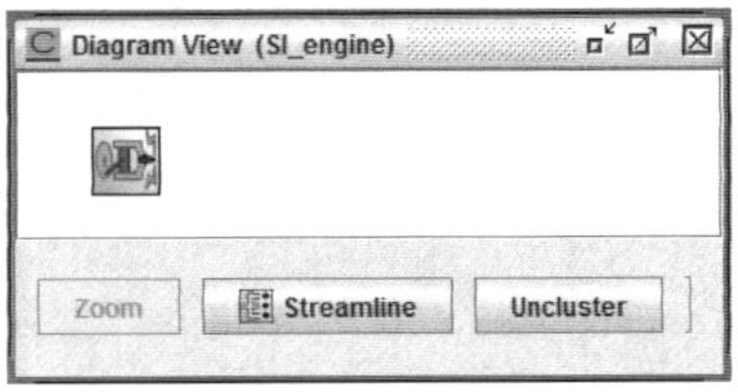

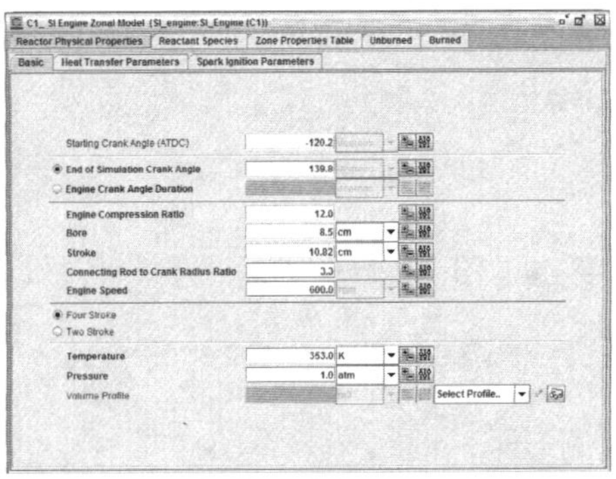

θ_{50}	θ_{10}	θ_{90}
-1.0	-10.0	7.0
1.0	-8.0	9.0
3.0	-6.0	11.0
5.0	-4.0	13.0
7.0	-2.0	15.0
9.0	0.0	17.0
11.0	2.0	19.0
13.0	4.0	21.0
15.0	6.0	23.0

Parametric combustion phasing sweep

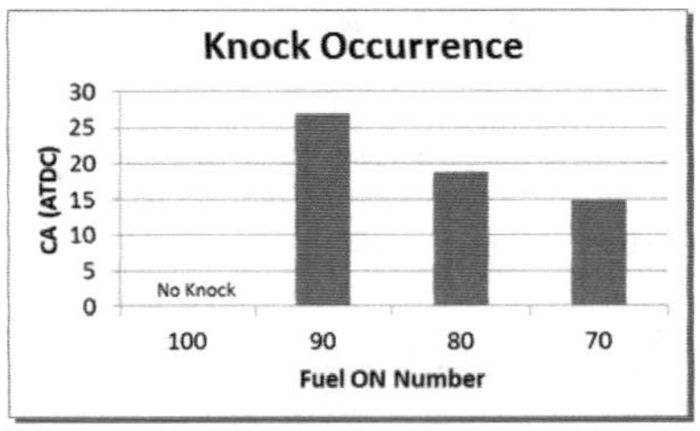

Simulations run in a few minutes with detailed kinetics and fuel models

Chemkin-Pro/GT-Power Co-Simulation

- **Planned to be a release capability in 17.2**
  - Uses Chemkin-Pro and MFL chemistry for closed cycle combustion via a user combustion model in GT-Suite

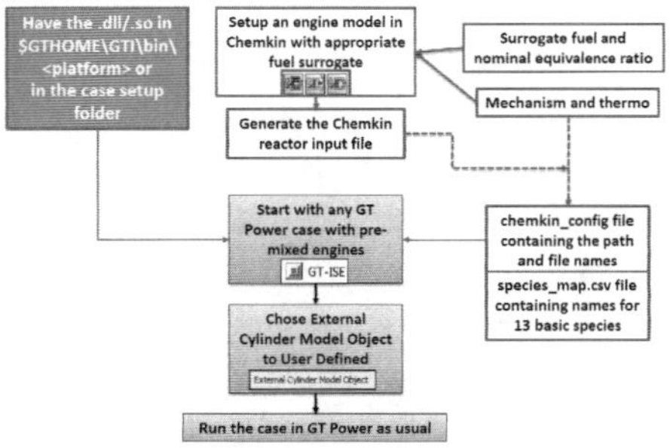

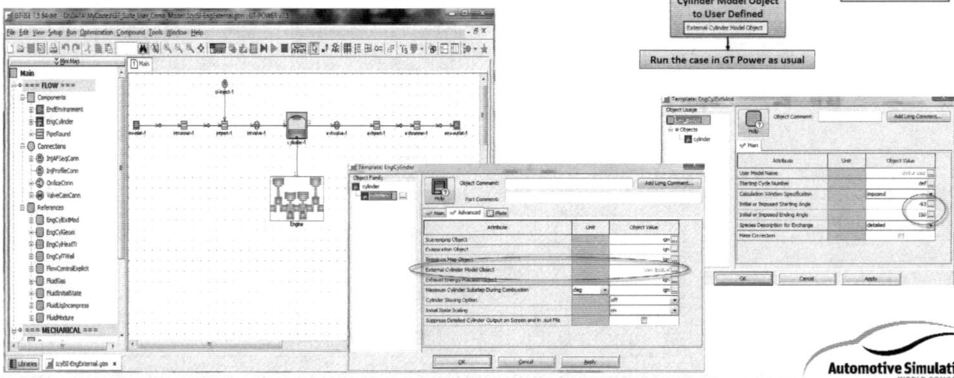

Knock Assessment: Typical Output

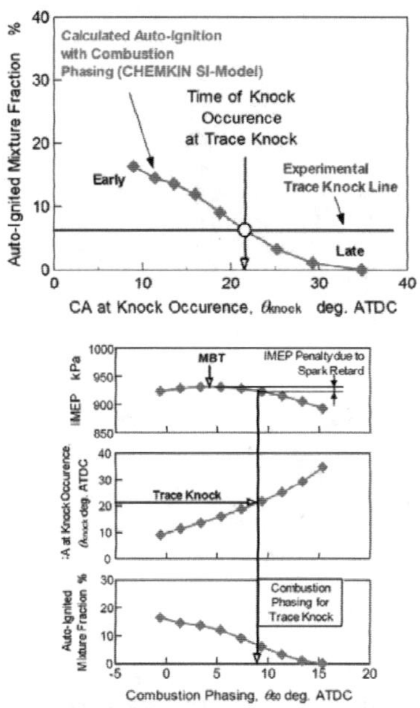

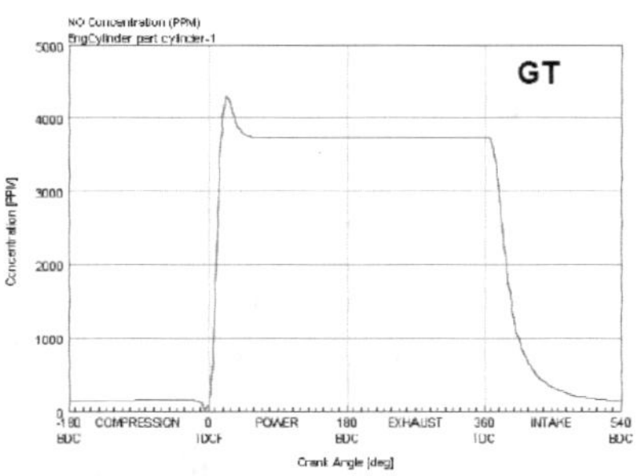

- **NO at EVO**
  - 3737 ppm in Chemkin-Pro
  - 3717 ppm in GT
    - Minor differences due to lumping of unburned species in Chemkin-Pro to fuel in GT

3D Conceptual Port Design with Optimization

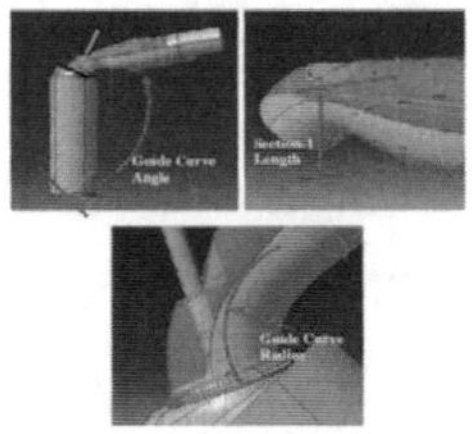

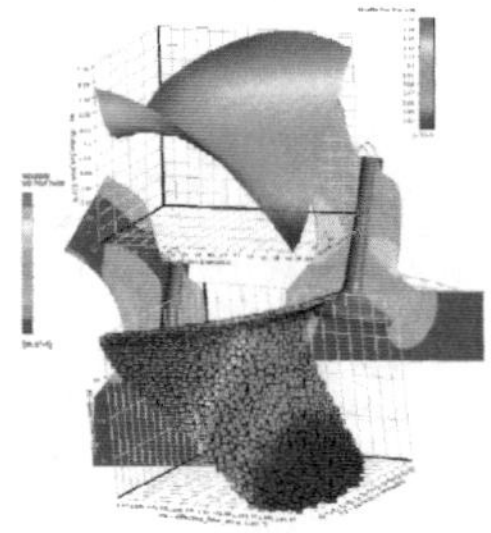

- **Parametric CAD Model**
  - Bi-directional connectivity to enable CAD to PPT "push-button" process
  - Automated and persistent work flows
  - Easily extended to design exploration/optimization
- **Non-Parametric CAD Model**
  - Topology Optimization via mesh morphing and/or adjoint methods

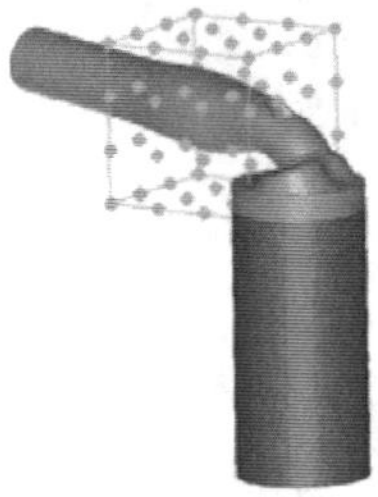

Mesh Morphing Control Volume

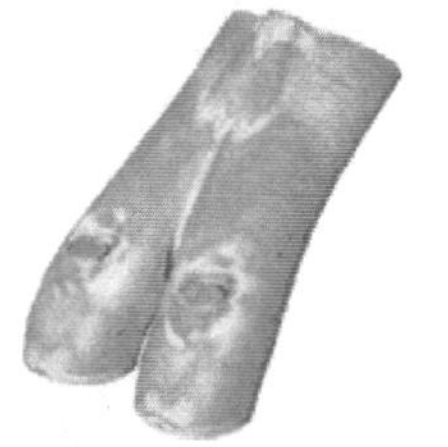

Adjoint Shape Sensitivity Contours

Automated CAD to Solution

- **Dynamic Mesh from CAD**
  - CAD to STL, ANSYS pre-processing portfolio including geometry/meshing tools
- **Generates the mesh on-the-fly**
  - With automated refinement around valve gaps and small features
  - Mesh preview in simulation environment
- **Fully automatic Cartesian meshing**
- **Body-Fitted Sector Mesh option**
  - For quick diesel design studies
  - Sector mesh generation from bowl profile and engine parameters
- **Solution adaptive meshing (SAM)**

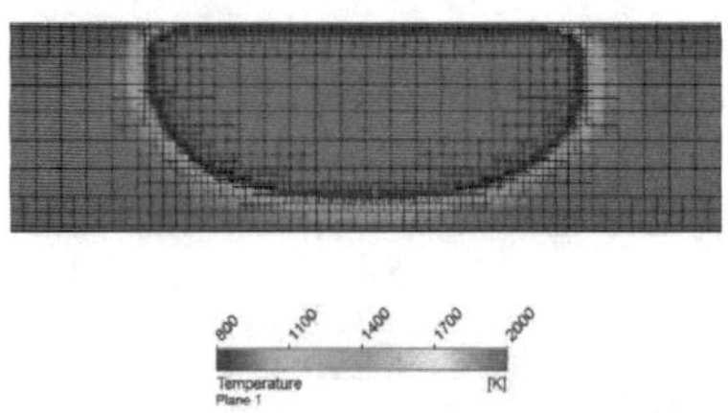

Streamlined Pre-Processing

- **Easy geometry preparation**
  - Full Engine, Sector, Symmetry, De-Featuring etc.
  - Access to all ANSYS pre-processing technologies
  - Automated and persistent work flows
- **CAD connectivity with full modeling and parametric capabilities**
  - Extendable to design points (i.e. parametric studies), DOE and optimization
- **Support for Fluent mesh file format**
  - All user named zones in mesh file preserved for re-usable and persistent work flow

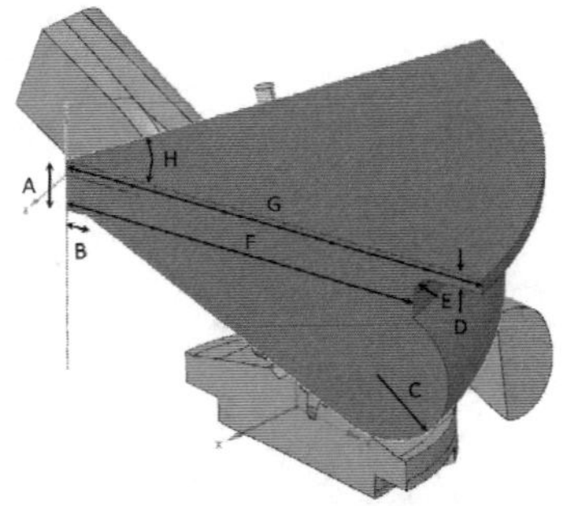

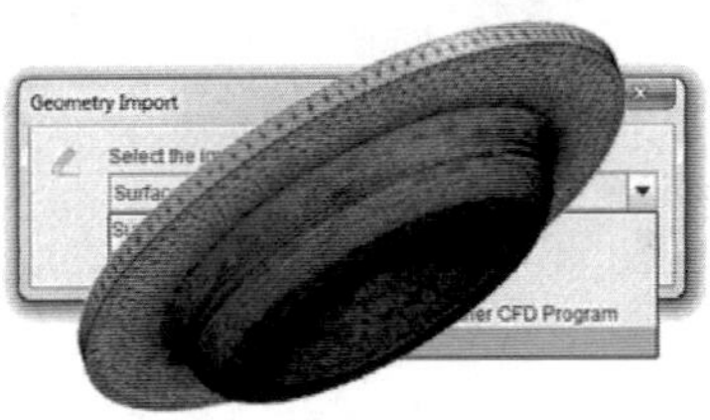

Example: Parametric Piston Bowl

- **Geometrical parameters**
  - Bowl radius, bowl depth, throat radius, lip radius, lip angle etc.
- **Combustion CFD inputs**
  - All numerical values, booleans, tables and profiles can be used for parametric studies
- **Combustion CFD outputs**
  - All performance outputs standard (PCP, burn angles/durations, GISFC, IMEP, NOx, Soot etc.)

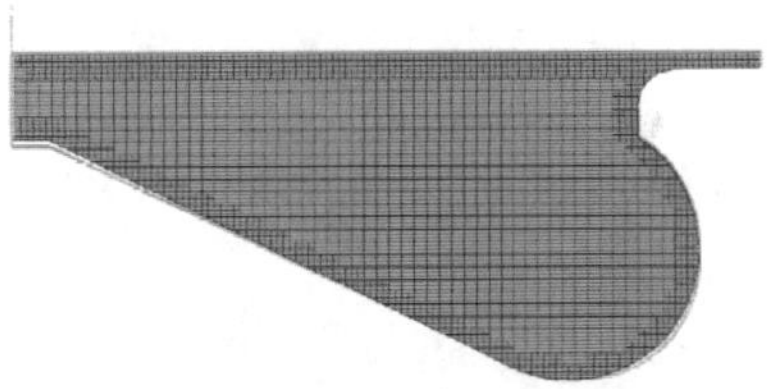

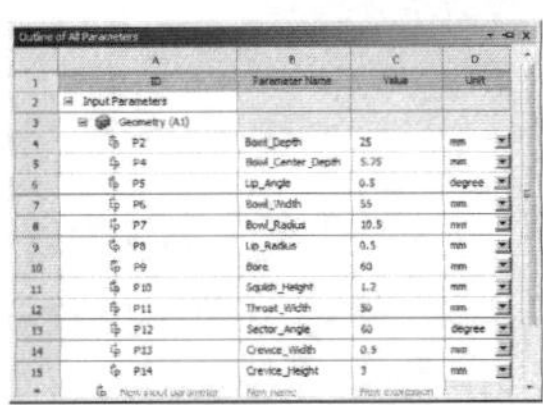

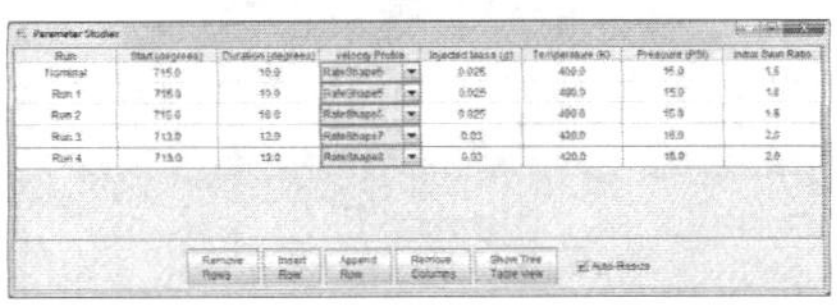

Parametric Studies

Workbench Platform and ICE

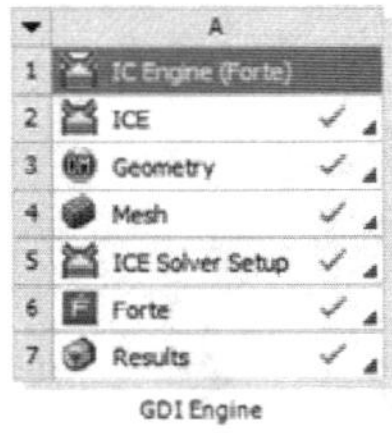

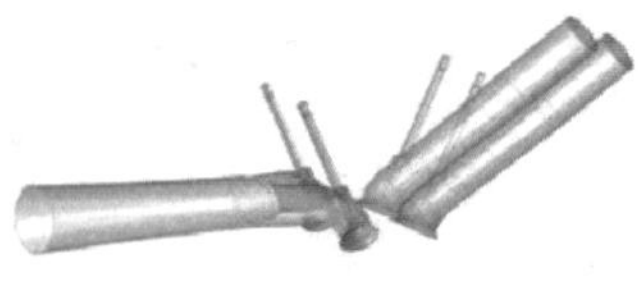

Prepare CAD using ANSYS geometry suite

Create volume defining fluid domain

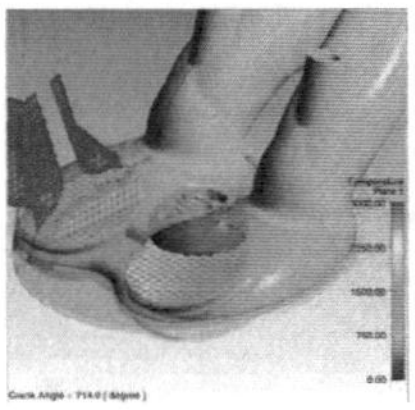

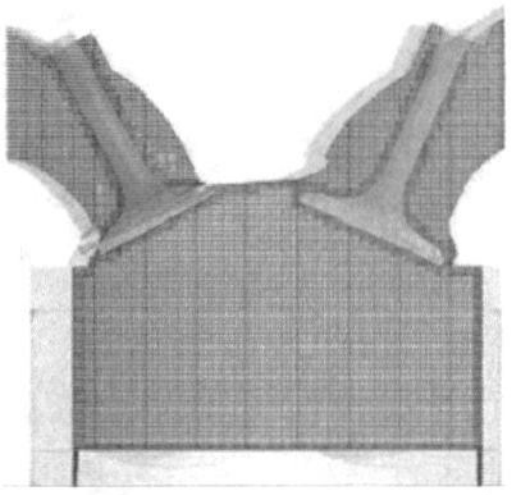

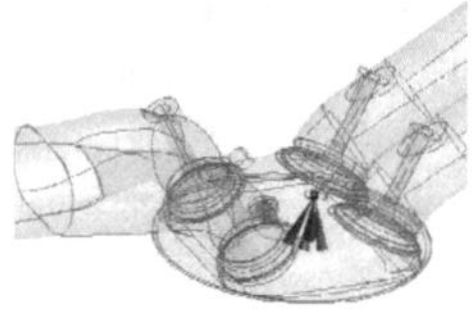

Automated on-the-fly volume meshing

Automatic decomposition of boundary surfaces and surface mesh creation

Automated pre-processing and report creation

Utilities: Gas Composition Calculator

- **Calculates gas compositions for**
  - IVC conditions
  - Initial conditions (Ports, Cylinder)
  - Intake conditions (e.g., with EGR)
- **Accounts for fuel composition**
- **Manage dual-fuel conditions**
  - Liquid-Liquid, Gas-Liquid or Gas-Gas
- **Estimates internal EGR rate**
  - Based on CR and/or T, P ratios
  - Eliminates errors that are often introduced by manual calculations

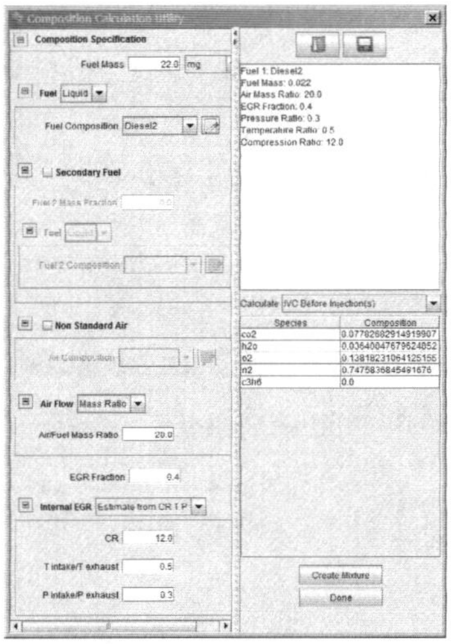

What is ACT?

- **ACT is the tool for customization of all ANSYS applications**
- **ACT enables application-specific and multi-physics needs**
  - Automate a process with application specific customized interfaces
  - Add new pre- and post- features
  - Integrate external applications into Workbench
  - Offers both third-party and ANSYS product connections
- **API's are documented (i.e. ACT Developer's Guide) and forward compatible**
  - Python and XML syntax

ACT App Store

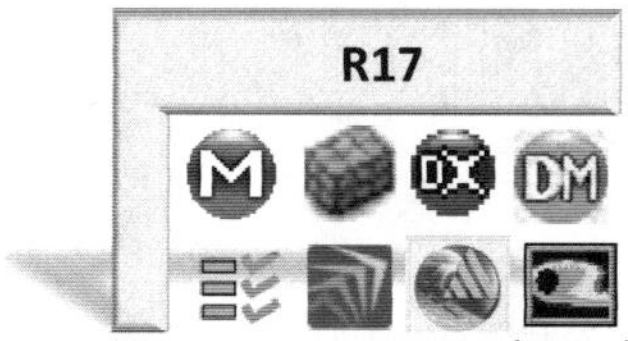

(Beta)

Automotive Simulation WORLD CONGRESS

Injection Rate Shape Study

- **Sandia Heavy Duty Diesel Engine**
- **Low-Temperature Combustion (LTC) Mode**
  - Common Rail Fuel Injector
  - Diesel #2 Fuel
  - 8 Nozzle Holes (Sector Geometry)
  - Nozzle Diameter 0.196mm
  - Engine Speed 1200 RPM
  - Total Mass of Injected Fuel 55mg
- **Baseline injection setup**
  - SOI: -10°btdc
  - Duration: 10°
  - Swirl Ratio: 0.5
  - 0 EGR

Sandia/Cummins N14 Engine Specifications	
Base	Cummins N-14, DI Diesel
Bore x Stroke	139.7mm x 152.4mm
Connecting rod length	304.8mm
Piston bowl profile	Cylindrical
Bowl width, depth	97.8mm, 15.5mm
Squish Distance	7.33mm
Compression ratio	11.2:1
Number of cylinders	1
IVC	540° APTDC
EVO	845° APTDC

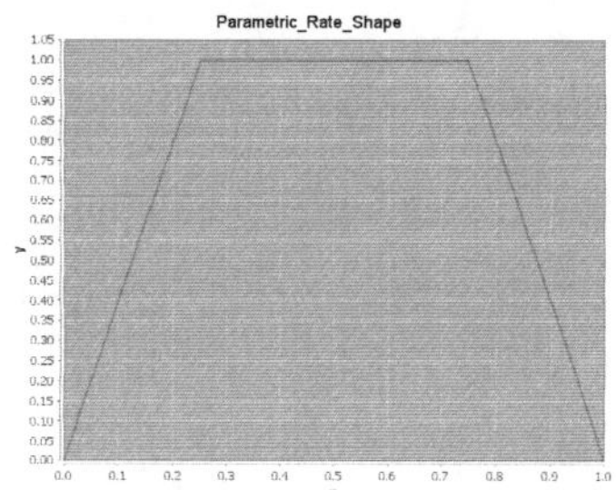

Automotive Simulation WORLD CONGRESS

Setup Summary

- **Body-fitted sector mesh with maximum cell count <10K cells**
  - n-Heptane 36 species chemistry
  - Two equation soot model
- **Total CPU time ~ 190 CPU hours**
  - Run time per case ~20 mins (4 CPU's)

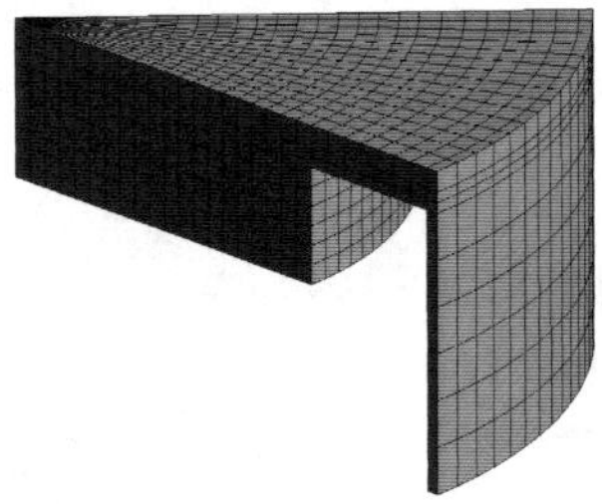

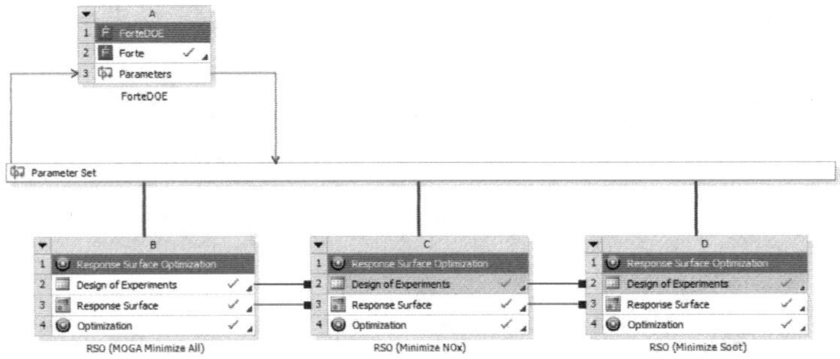

Automotive Simulation WORLD CONGRESS

June 7-8, 2016 Munich

Parametric Rate Shape Parameters

- **Single Injection with non-dimensional input parameters for "Top Hat" and "Boot"**
  - A = Ramp Up/Down Ratio
  - B = Ramp Length Ratio
  - C = Ramp Height Ratio
- **Initial conditions calculated for varying EGR rates using Gas Composition Calculator**
- **Extendable to multiple/split parametric injection shapes and fuel quantities**

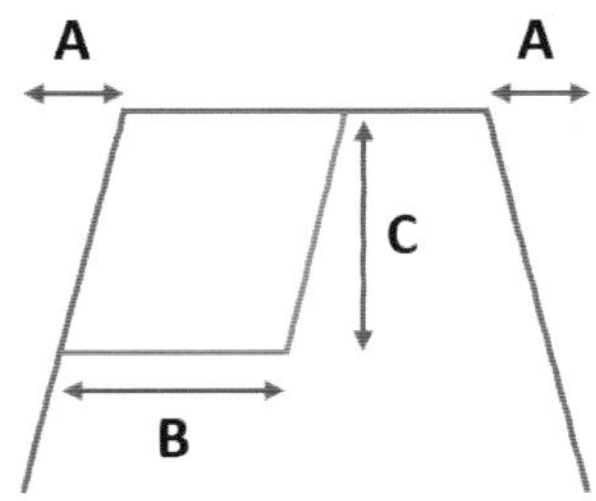

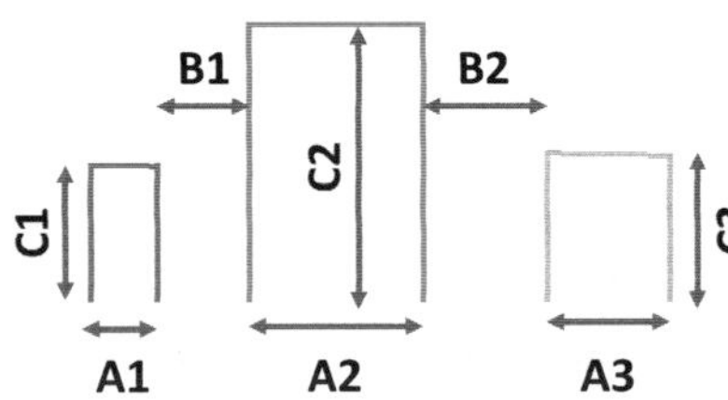

Automotive Simulation WORLD CONGRESS

June 7-8, 2016 Munich

DOE Setup and Results

- **140 Design Points in DOE Space**
  - **EGR: 0/10/20/30/40**
  - **SOI: -25/-18.75/-12.5/-6.25/0**
  - **Injection Duration: 5/7.5/10/12.5/15**
  - **Swirl Ratio: 0.5/0.625/0.75/0.875/1**
  - **Ramp Up-Down Ratio: 0/0.025/0.05/0.075/0.1**
  - **Ramp Length Ratio: 0/0.125/0.25/0.375/0.5**
  - **Ramp Height Ratio: 0/0.225/0.45/0.675/0.9**

Setup File	
Forte FTSIM	C:\temp\Sandia_Engine_LTC.ftsim
Number of Processors	4
Inputs	
Fuel Quantity (mg/cycle)	55
EGR (%)	0
Start of Injection (deg)	-10
Injection Duration (deg)	10
Ramp Up/Down Ratio	0.25
Ramp Length Ratio	0
Ramp Height Ratio	0
Swirl Ratio	0.5
Outputs	
ISFC [g/kW-h]	189.96
NOx [g/kg-fuel]	126.56
Soot [g/kg-fuel]	0.019587

Table of Schematic B2: Design of Experiments (Custom)

	A	B	C	D	E	
1	Name	P4 - EGR	P5 - StartOfInjection	P6 - RampUpDownRatio	P7 - RampLengthRatio	P8 - Ramp
120	120	30	-6.25	0.025	0.125	0.675
121	121	0	-25	0.1	0	0.9
122	122	10	-18.75	0.075	0.125	0.675
123	123	40	-25	0.1	0	0.9
124	124	30	-18.75	0.075	0.125	0.675
125	125	0	0	0.1	0	0.9
126	126	10	-6.25	0.075	0.125	0.675
127	127	30	-6.25	0.075	0.125	0.675
128	128	0	-25	0	0.5	0.9
129	129	10	-18.75	0.025	0.375	0.675
130	130	40	-25	0	0.5	0.9
131	131	30	-18.75	0.025	0.375	0.675
132	132	0	0	0	0.5	0.9
133	133	10	-6.25	0.025	0.375	0.675
134	134	30	-6.25	0.025	0.375	0.675
135	135	0	-25	0.1	0.5	0.9
136	136	10	-18.75	0.075	0.375	0.675
137	137	40	-25	0.1	0.5	0.9
138	138	30	-18.75	0.075	0.375	0.675
139	139	0	0	0.1	0.5	0.9
140	140	10	-6.25	0.075	0.375	0.675
141	141	30	-6.25	0.075	0.375	0.675
*	New Design Point					

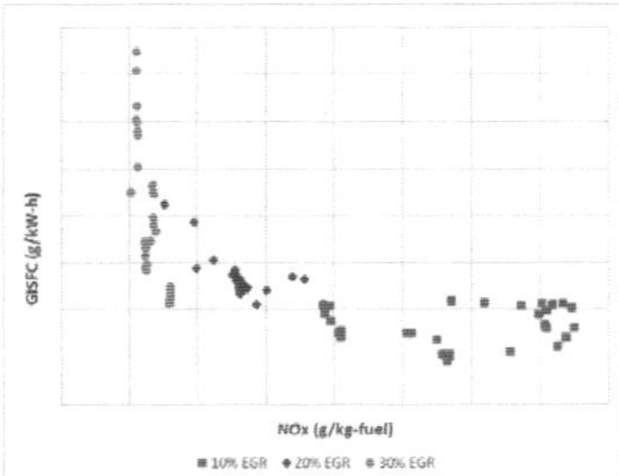

Automotive Simulation WORLD CONGRESS

Sensitivity Analysis

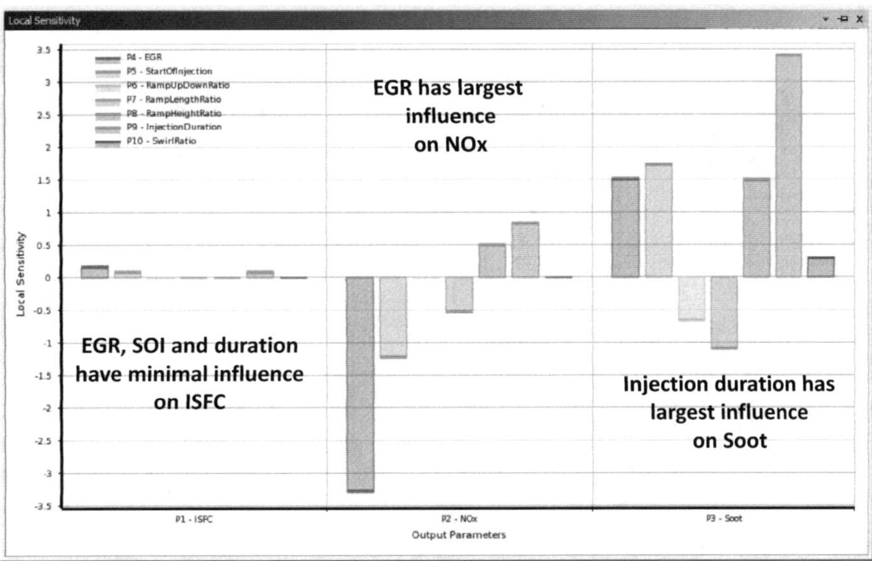

Automotive Simulation WORLD CONGRESS

Response Surface Analysis

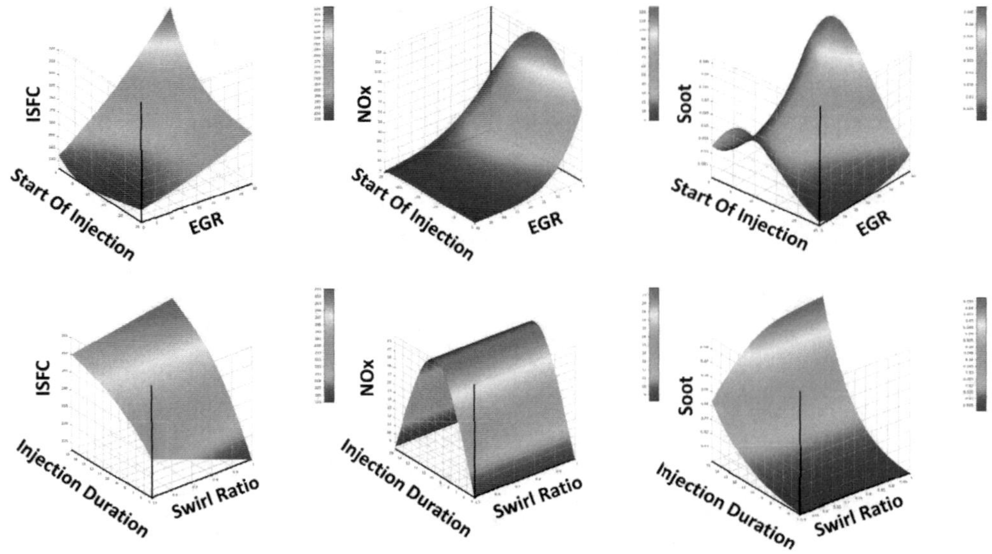

Optimization Study Injection Details

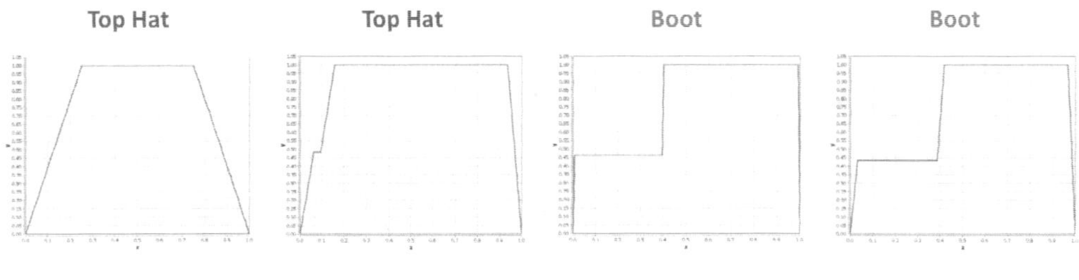

Baseline

EGR (%)	0
SOI (°btdc)	10
Duration (°)	10
Ramp Up Ratio	0.25
Ramp Down Ratio	0.25
Ramp Length Ratio	0
Ramp Height Ratio	0
Swirl Ratio	0.5

Minimize ISFC, NOx, Soot

EGR (%)	30
SOI (°btdc)	-13.5
Duration (°)	5
Ramp Up Ratio	0.06
Ramp Down Ratio	0.06
Ramp Length Ratio	0.03
Ramp Height Ratio	0.48
Swirl Ratio	0.56

Minimize NOx

EGR (%)	30
SOI (°btdc)	2.5
Duration (°)	10
Ramp Up Ratio	0.008
Ramp Down Ratio	0.008
Ramp Length Ratio	0.39
Ramp Height Ratio	0.46
Swirl Ratio	0.8

Minimize Soot

EGR (%)	20
SOI (°btdc)	8.2
Duration (°)	9.3
Ramp Up Ratio	0.03
Ramp Down Ratio	0.03
Ramp Length Ratio	0.35
Ramp Height Ratio	0.43
Swirl Ratio	0.7

Optimization Results

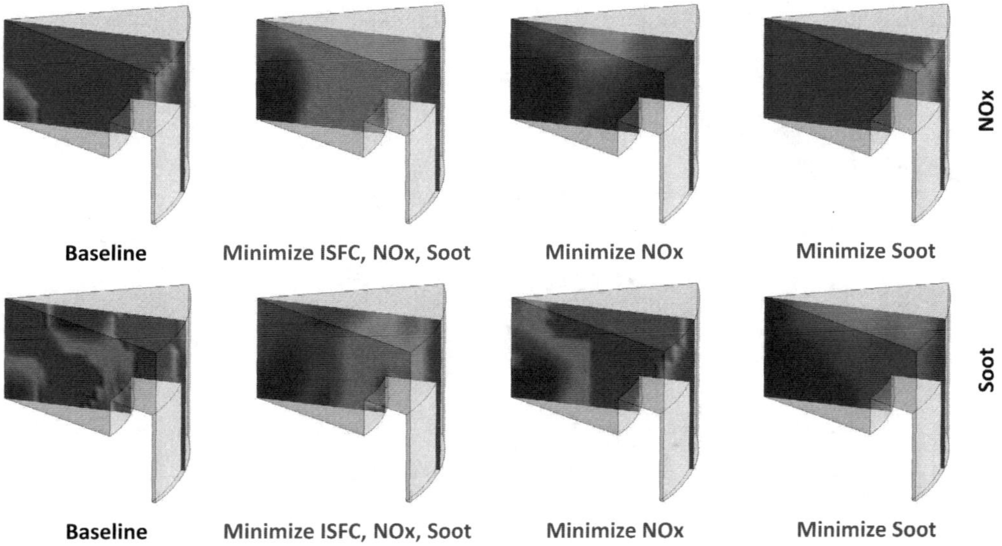

Parametric Bowl Study

- **Medium Duty Diesel Engine**
- **Same Parametric Rate Shape inputs as before**
  - Common Rail Fuel Injector
  - Diesel #2 Fuel
  - 8 Nozzle Holes
  - Nozzle Diameter 0.17mm
  - Engine Speed 1500 RPM
- **Baseline injection setup**
  - SOI: -10°btdc
  - Duration: 10°
  - Swirl Ratio: 0.5
  - 0 EGR
- **Total CPU time ~ 7200 CPU hours**
  - Run time per case ~2.5h (4 CPU's)

Bore (cm)	13
Stroke (cm)	13.5
Con-Rod Length (cm)	18.5
IVC (°btdc)	165
EVO (°atdc)	125
Bowl Volume (cc)	104
Compression Ratio	17.2
Squish Height (cm)	0.078-0.143

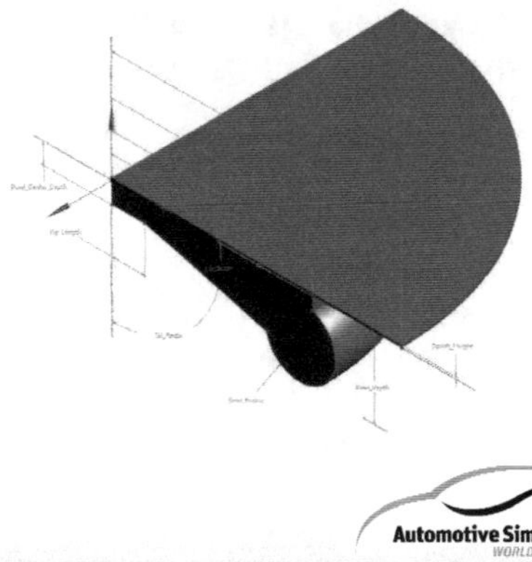

Bowls (Compression Ratio = 17.2)

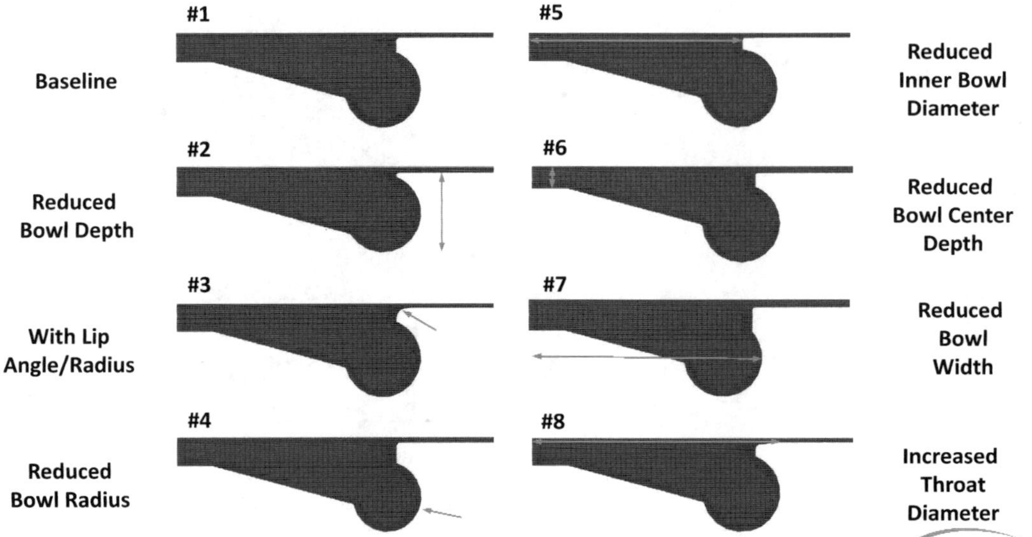

Automotive Simulation WORLD CONGRESS

DOE Setup Summary

- **728 Design Points in DOE Space**
  - EGR: 10/20/30
  - SOI: -25/-18.75/-12.5/-6.25/0
  - Injection Duration: 5/7.5/10/12.5/15
  - Swirl Ratio: 0.5/0.625/0.75/0.875/1
  - Ramp Up-Down Ratio: 0/0.025/0.05/0.075/0.1
  - Ramp Length Ratio: 0/0.125/0.25/0.375/0.5
  - Ramp Height Ratio: 0/0.225/0.45/0.675/0.9

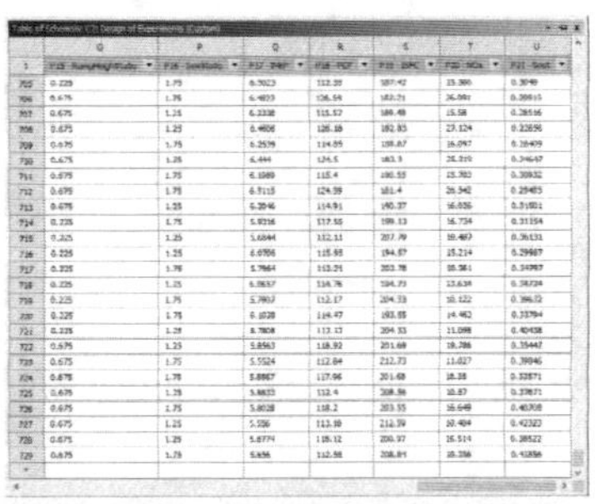

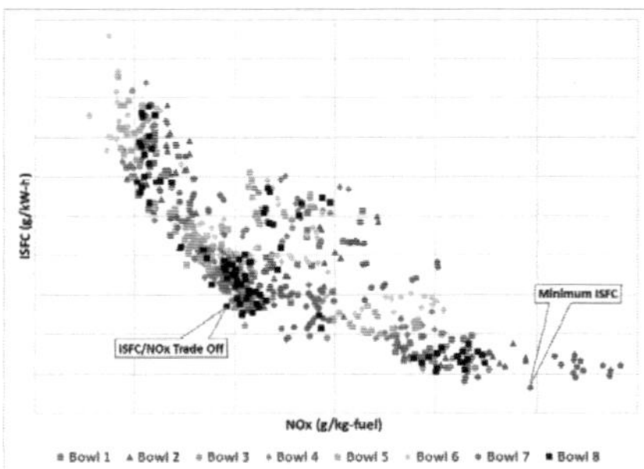

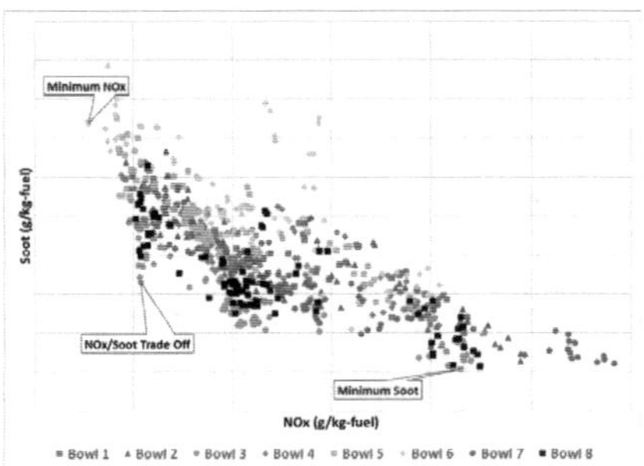

Automotive Simulation WORLD CONGRESS

Sensitivity Analysis

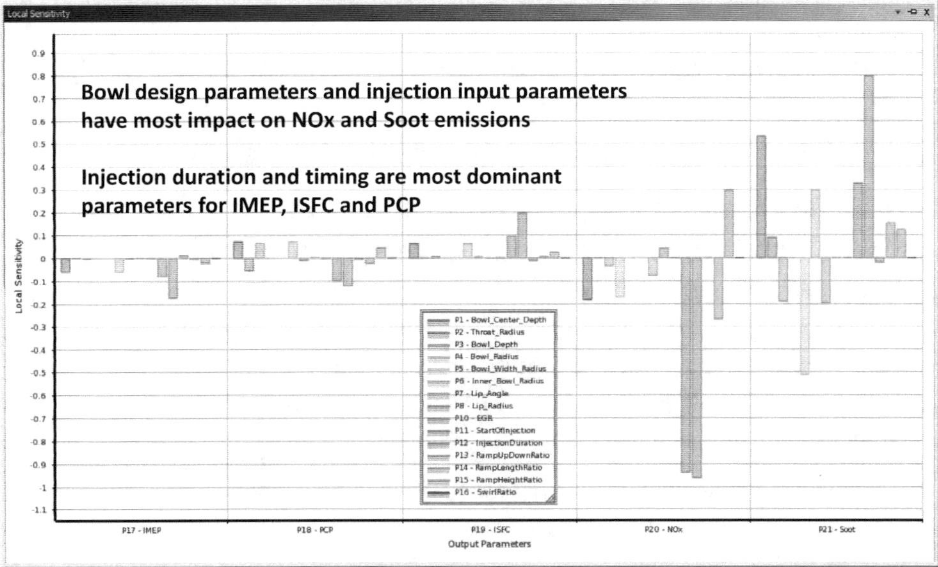

Automotive Simulation WORLD CONGRESS

DOE Observations

- **Bowl Design Parameters**
  - Increasing bowl center depth or inner bowl radius increases Soot
  - Increasing bowl width reduces Soot
  - Increasing bowl center depth or bowl depth reduces NOx
  - Increasing bowl radius reduces NOx
  - Increasing lip angle reduces NOx and Soot. Reduces Soot more.
- **Injection Parameters**
  - Increasing injection duration or retarding injection timing reduces NOx
  - Increasing injection duration increases Soot more than retarding injection timing
  - Increasing injection duration and retarding injection timing reduces IMEP. Opposite increases ISFC.
  - Increasing ramp length ratio reduces NOx but increases Soot

Automotive Simulation WORLD CONGRESS

Summary

- **0D Chemkin SI Engine model extension to perform co-simulation with GT-Power on the horizon**
  - Use detailed chemistry with 1D system code for early design assessment
- **Combustion system design exploration and optimization with an automated and integrated platform allows 100's of design scenarios to be studied**
  - Relatively easy to generate large number of designs/simulations whereas can be more time consuming to extract relevant data to drive design decisions
  - Design trade-offs can be identified given competing conflicts between performance, fuel consumption and pollutant formation
- **ACT extensions allow for customization and re-use of combustion system design scenarios**
  - Use case shown here confined to a single bowl family with constraints to preserve compression ratio

Automotive Simulation
WORLD CONGRESS
June 7-8, 2016 Munich

Questions

Automotive Simulation
WORLD CONGRESS

June 7-8, 2016 Munich

An Integrated Design Process for Turbochargers

Marc Kainz
ANSYS Germany GmbH

Contents

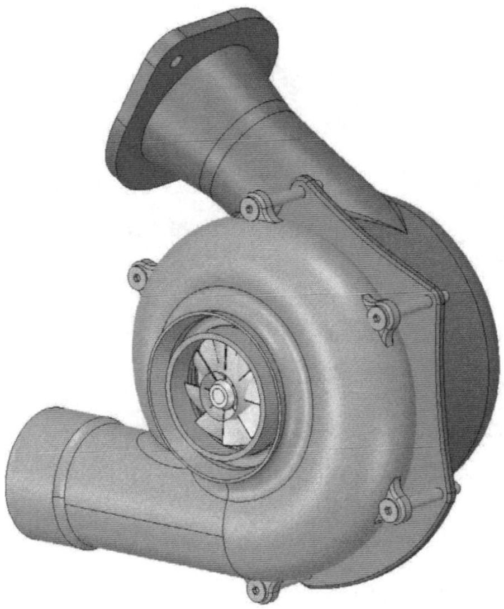

Challenges

Environmental

CO2 emissions ↓

Fuel burn ↓

Noise emissions ↓

Number of turbochargers ↑

Business

Product costs ↓

Time to market ↓

Innovation ↑

Number of simulations ↑

Automotive Simulation WORLD CONGRESS

3 June 7-8, 2016 Munich

Simulations Challenges & Trends

Complex geometries

Scallops

Waste gate

Variable turbine guide vanes

Unsteady simulations

Rotor/stator interaction

Off design performance

Inlet distortion

Acoustics

Courtesy of Cummins Turbo Technologies

Automotive Simulation WORLD CONGRESS

4 June 7-8, 2016 Munich

Simulations Challenges & Trends

- Multidisciplinary
  - Conjugate heat transfer
  - Aeromechanics
- Optimization & robust design
- Complete system simulation
- Bearings

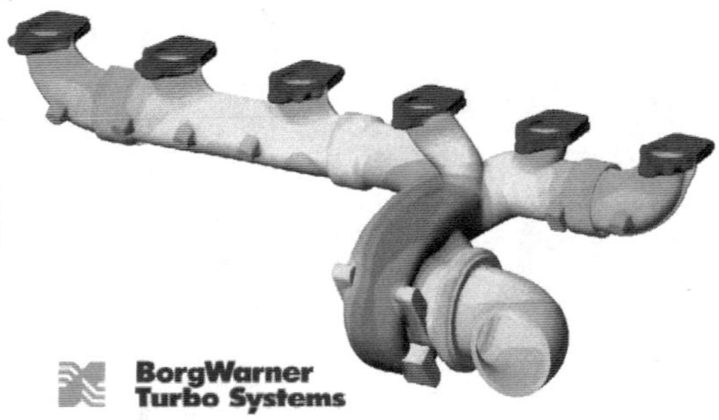

Turbocharger Compressor Design Workflow

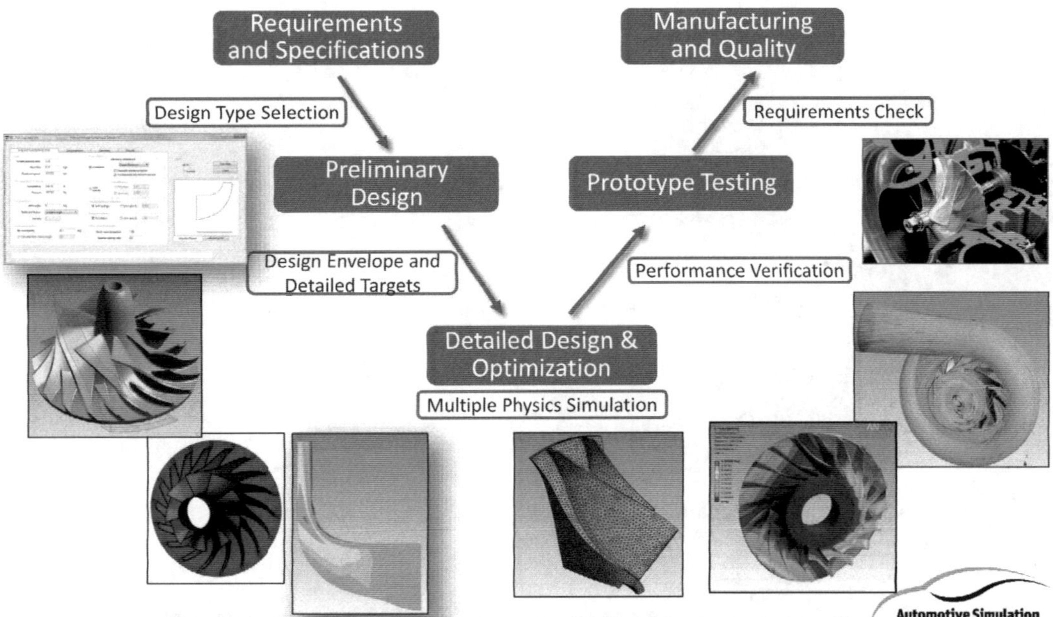

ANSYS Software & Turbo Tools

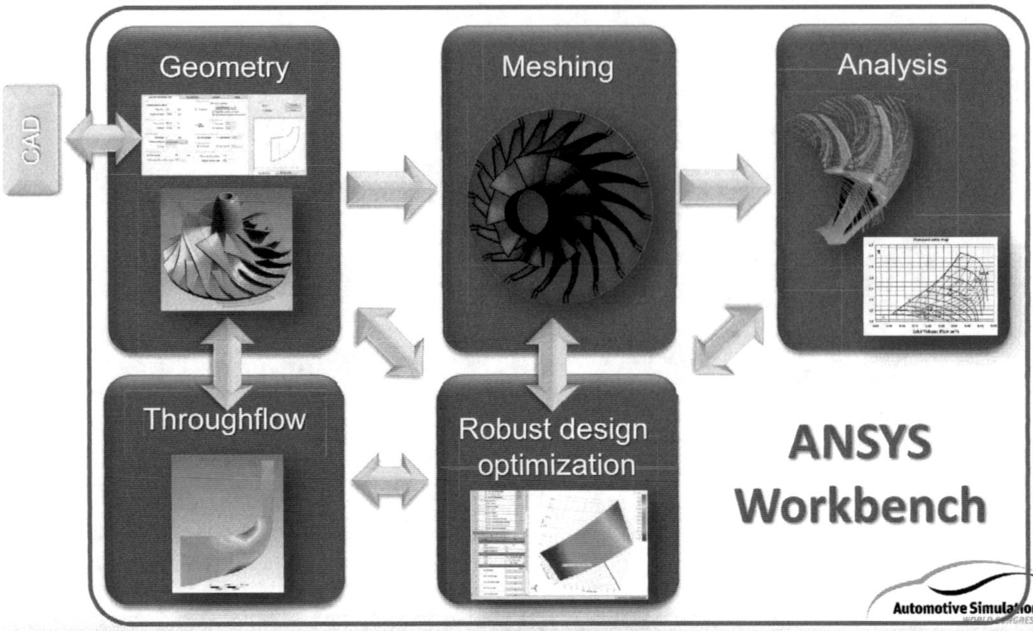

Automotive Simulation WORLD CONGRESS

Integration in ANSYS Workbench

- Easys & clear setup
- Complete multi-physics workflows
- Simply combine tools via drag & drop
- Parametrization

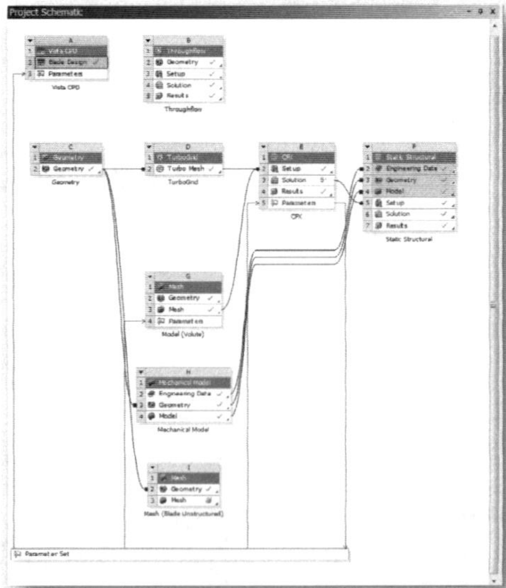

Automotive Simulation WORLD CONGRESS

Preliminary Design – Base Line Design

PCA ENGINEERS

Vista Tools

1D “mean line” design

Initial blade design based on correlations

Vista CCD for ‚Centrifugal Compressor Design‘

Vista RTD for ‚Radial Turbine Design‘

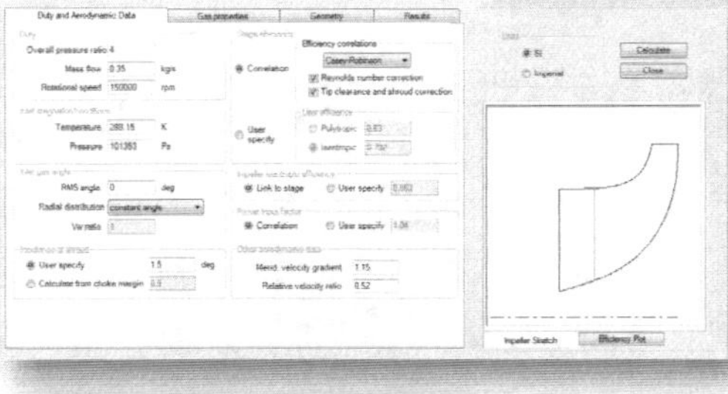

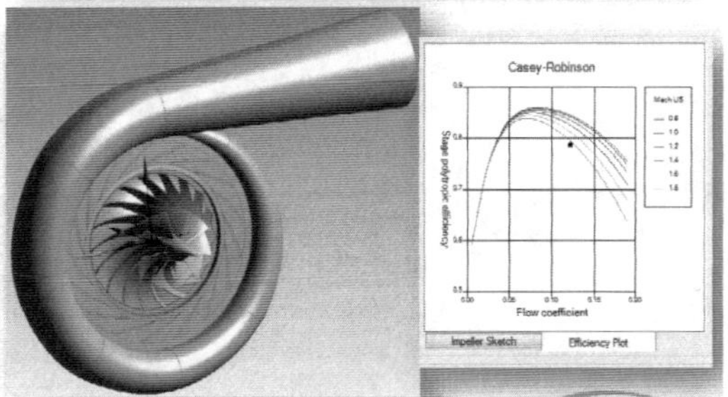

Geometry provided by PCA Engineers

Preliminary Design – Throughflow Solver

PCA ENGINEERS

Vista Throughflow

Meridional flow solver

Loss correlations

Flow solution in seconds

Initial design optimization

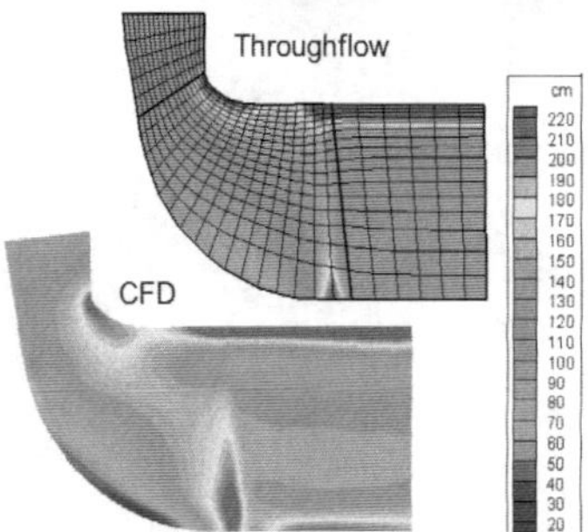

Courtesy of PCA Engineers

3D Geometry – Blade Design

- BladeGen & BladeEditor
- Blade creation in meridional view
- Blade modification via angle & thickness distributions
- Parametrization
- CAD to parametric geometry

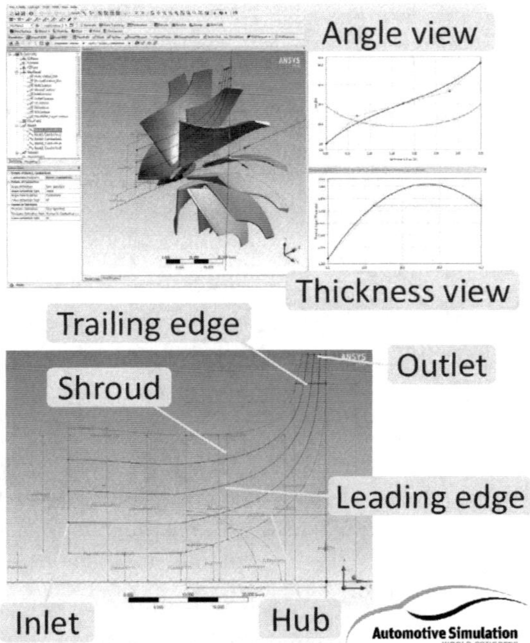

3D Geometry – Blade Design

Optimize existing CAD impeller model

Camberline and thickness distribution

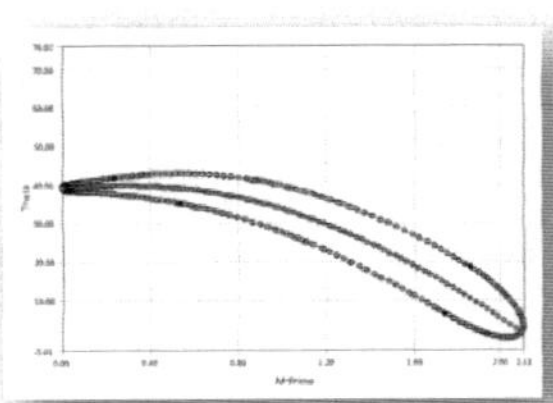

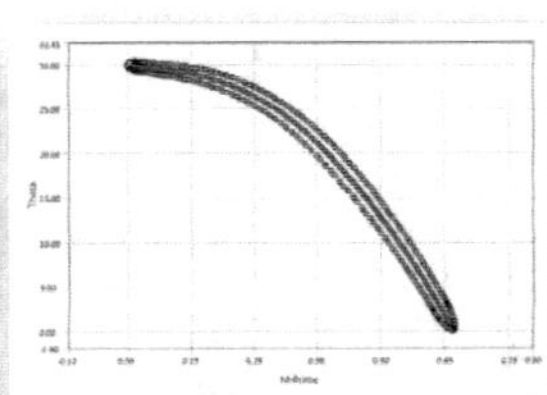

CAD model

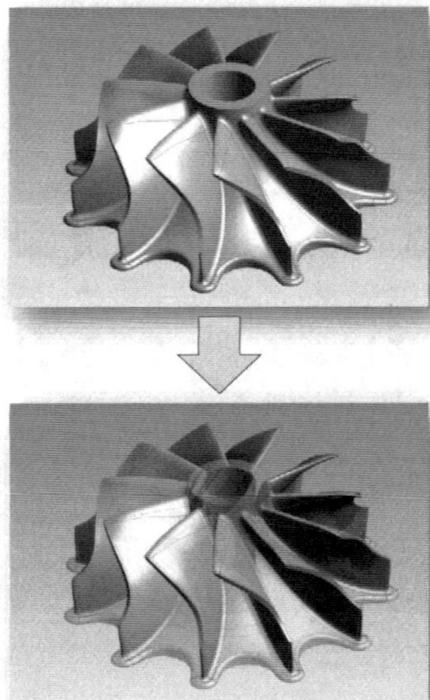

Volute Parameterisation

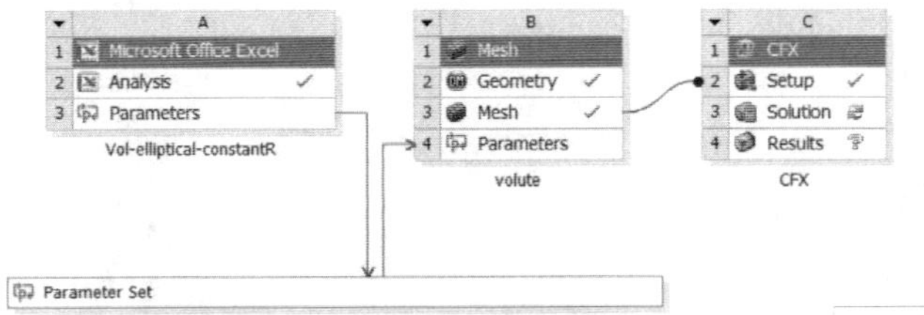

Outline of Schematic A3: Parameters

	A	B	C	D
1	ID	Parameter Name	Value	Unit
2	Input Parameters			
*	New input parameter	New name	New expression	
4	Output Parameters			
5	Vol-elliptical-constantR (A1)			
6	P1	a.0	0.37125	
7	P2	a.135	2.0147	
8	P3	a.180	2.3264	
9	P4	a.225	2.601	
10	P5	a.270	2.8492	
11	P6	a.315	3.0775	
12	P7	a.360	3.29	

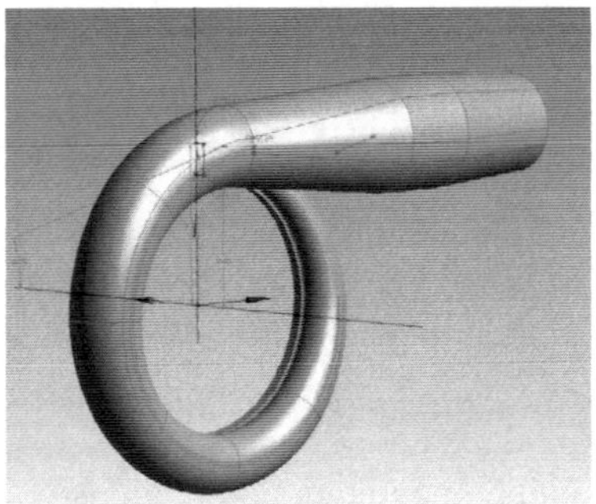

Courtesy of PCA Engineers

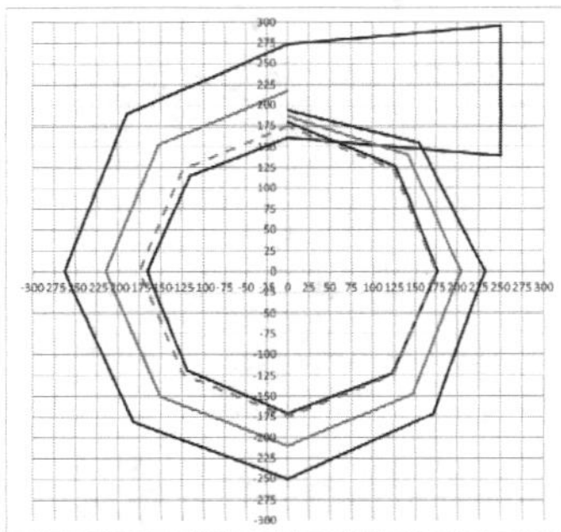

Courtesy of PCA Engineers

Automotive Simulation WORLD CONGRESS

3D Geometry – Other Parts

- Geometry creation
- Repair & prepare
- De-featuring
- Extraction of fluid volume
- Parametrization
- Bidirectional interface to CAD
- STL tool

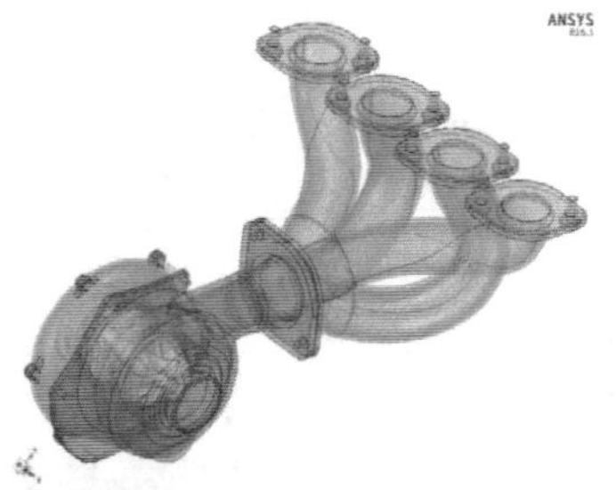

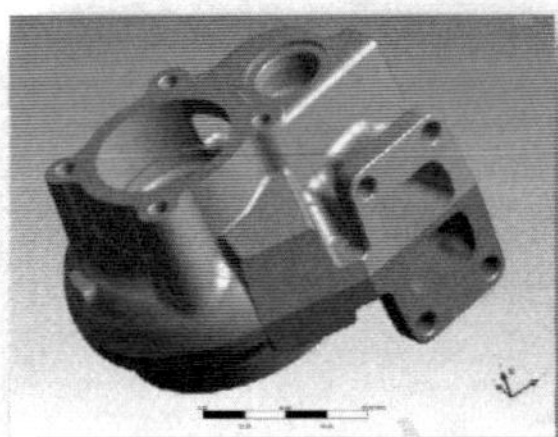

Courtesy of Voith Turbo

Courtesy of Voith Turbo

Mesh Generation – Blade Rows

- Automated mesh generation
- High quality hexahedral meshes
- Repeatable
- Scalable

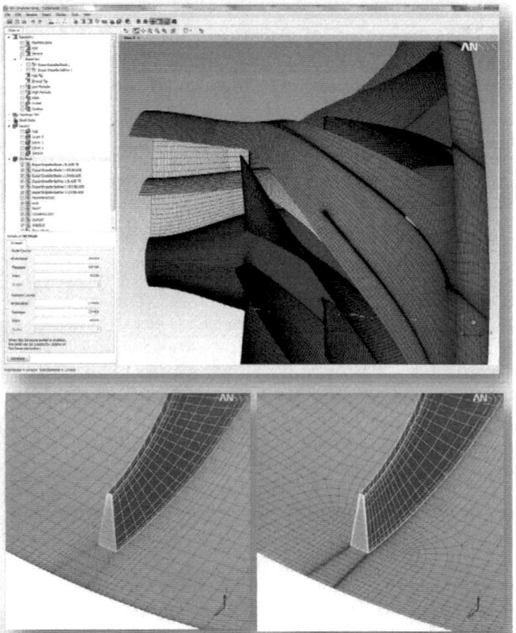

Mesh Generation – Example Compressor

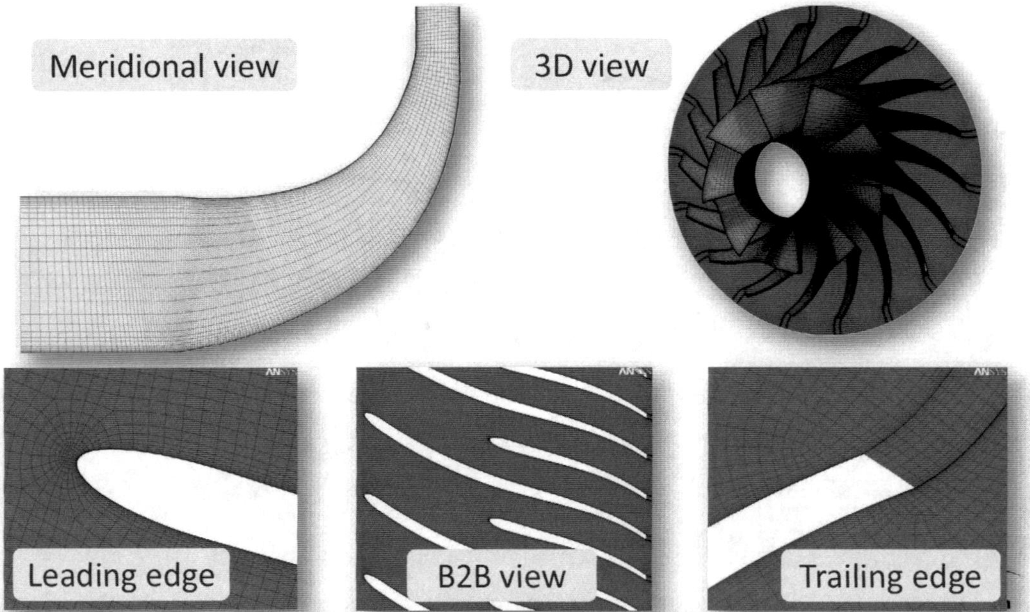

Meshing – Other Parts

- Automated meshing
- Tet-Prism meshes
- Hexahedral meshes
- Wrapping

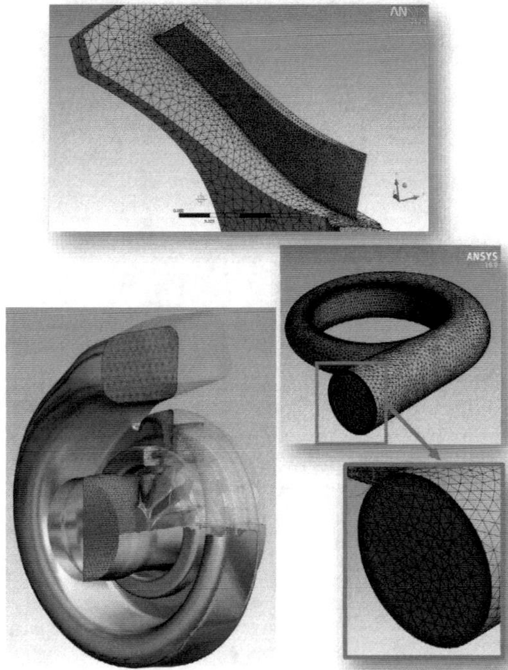

Courtesy of Cummins Turbo Technologies

Mesh Generation – Example Volute

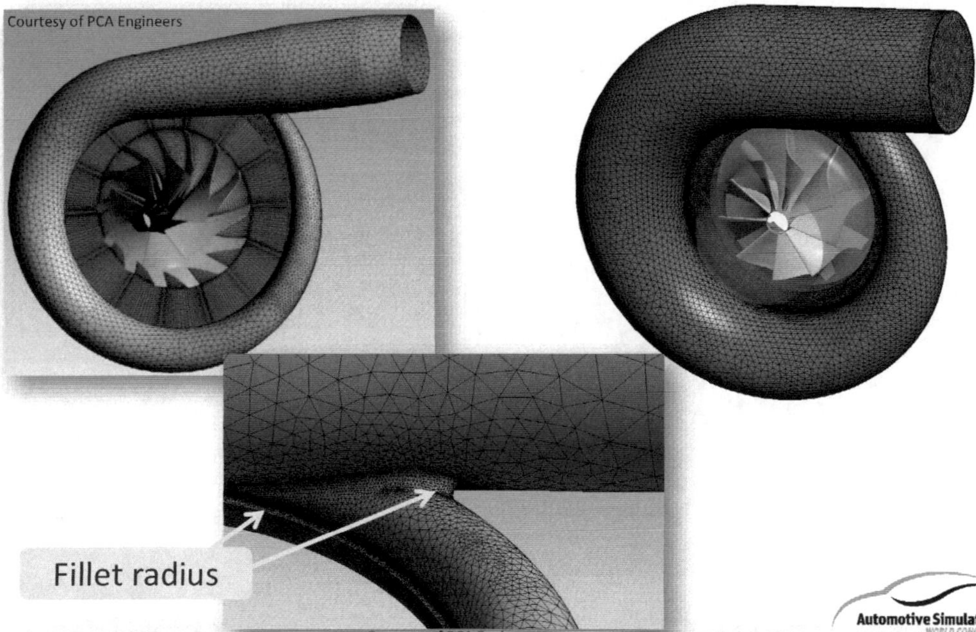

CFD – Solver

- Finite volume solver
- Fast & scalable
- Accurate & robust
- Steady & unsteady state
- Rotor/stator interfaces
- Turbulence & heat transfer
- Multi-phase & real gases

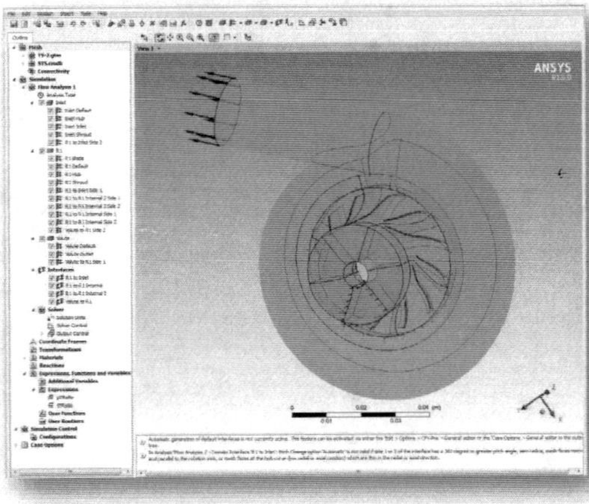

Exit Corrected Mass Flow Boundary Condition

One boundary condition for the whole operating range

$$\dot{m}_{cor} = \dot{m}\frac{\sqrt{\tilde{T}_{o1}/T_{ref}}}{\tilde{P}_{o1}/P_{ref}}$$

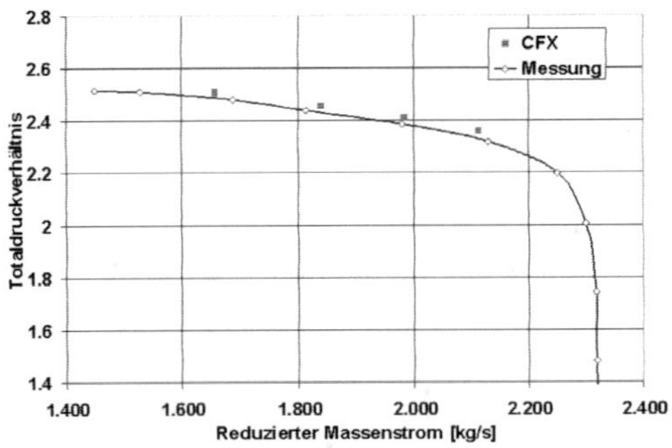

CFD – Post Processing

- Turbo mode
- 0D to 4D post processing
- Integral values
- Locations & contours
- Vectors & streamlines
- Pictures & animations
- Scripting, macros & reports

Compressor map

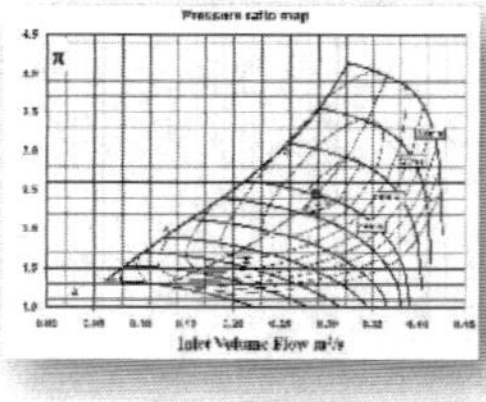

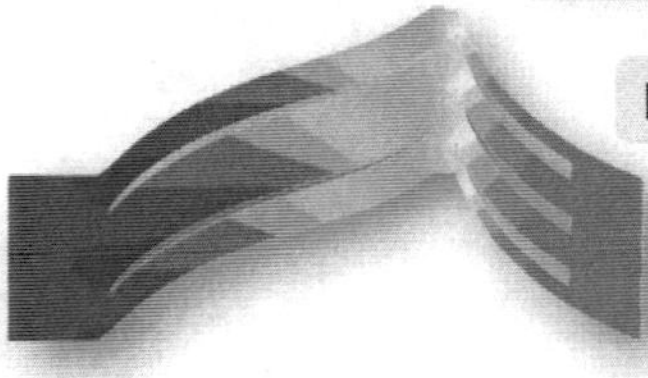

B2B view

Meridional view

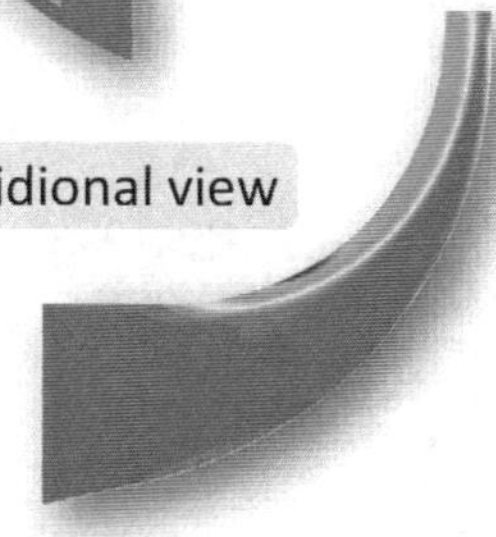

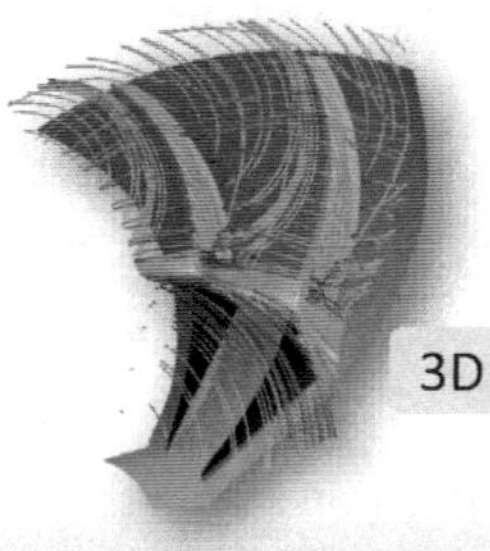

3D Streamlines

Automotive Simulation
WORLD CONGRESS

CFD – Parametrization & Optimization

- Parametrization of all ANSYS tools
- Sensitivity analysis
- Design optimization
- Robustness evaluation

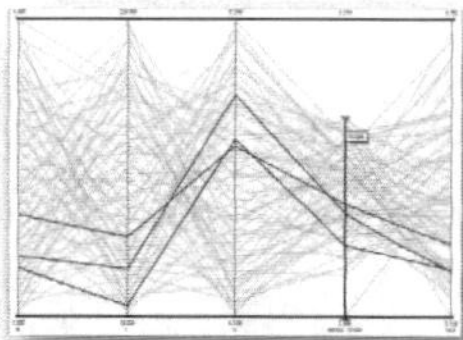

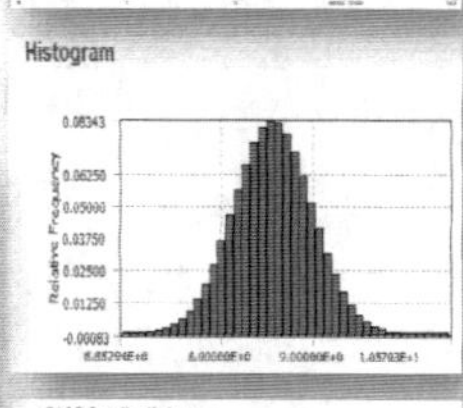

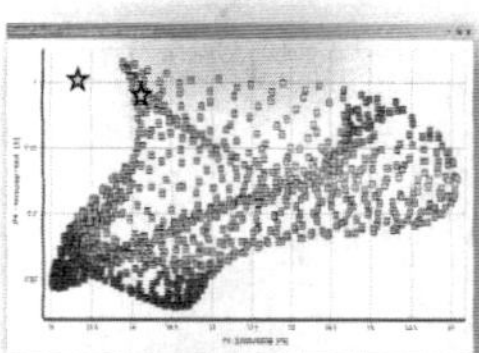

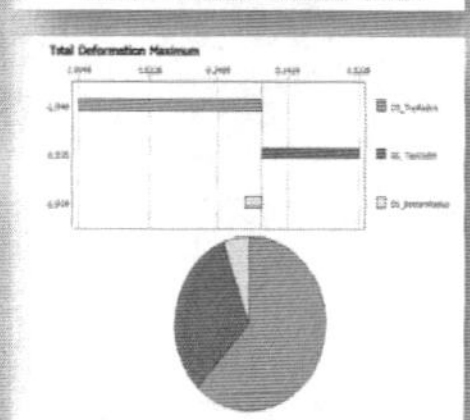

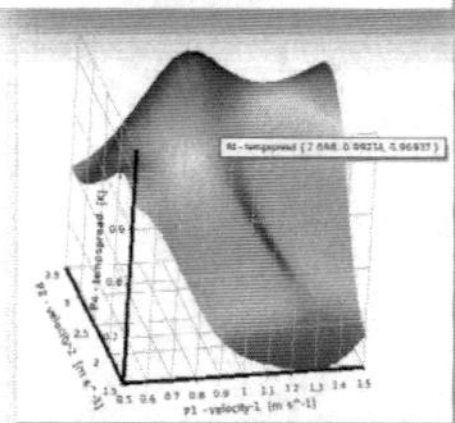

Automotive Simulation
WORLD CONGRESS

ANSYS Blade Row Analysis Methods

Steady pitch change

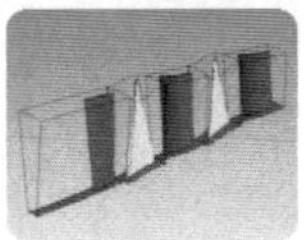

Transient pitch change

Transient full-annulus

Time domain

- Profile Transformation
- Fourier Transformation
- Time Transformation

Frequency domain

- Harmonic Transformation

Fourier Transformation (FT)

Blade pitch change via shape correction of He (Fourier based)

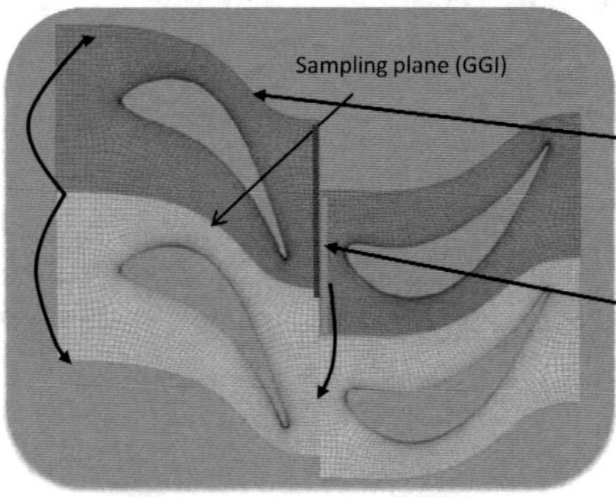

$$\phi(t) = \sum_{k=-N}^{N} A_k e^{-j(k\omega t)}$$

$$\phi(t,\theta) = \sum_{l=-M}^{M} \sum_{k=-N}^{N} A_{k,l} e^{-j(k\omega t + l\theta)}$$

- No limits on pitch ratio
- For all flow speeds (compressible or incompressible)
- Frequencies a priori given (BPF and harmonics)

Example – Radial Turbine

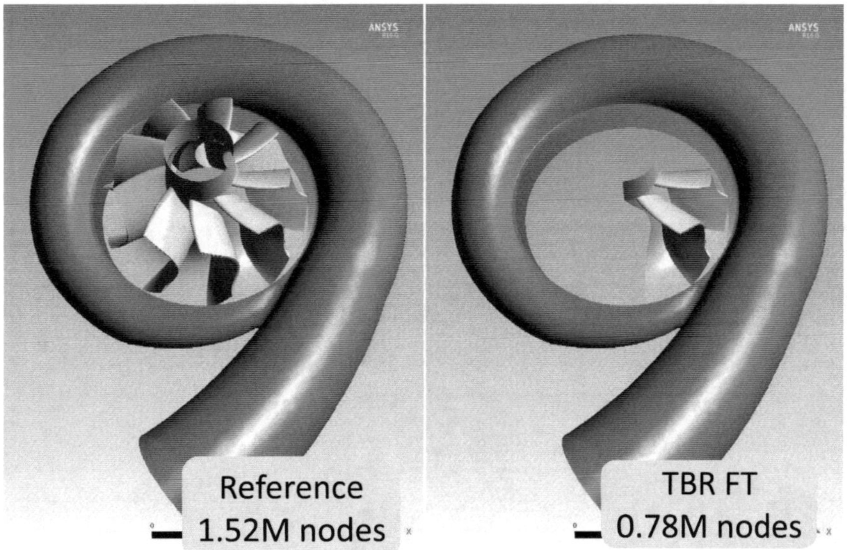

Automotive Simulation WORLD CONGRESS

Example – Radial Turbine

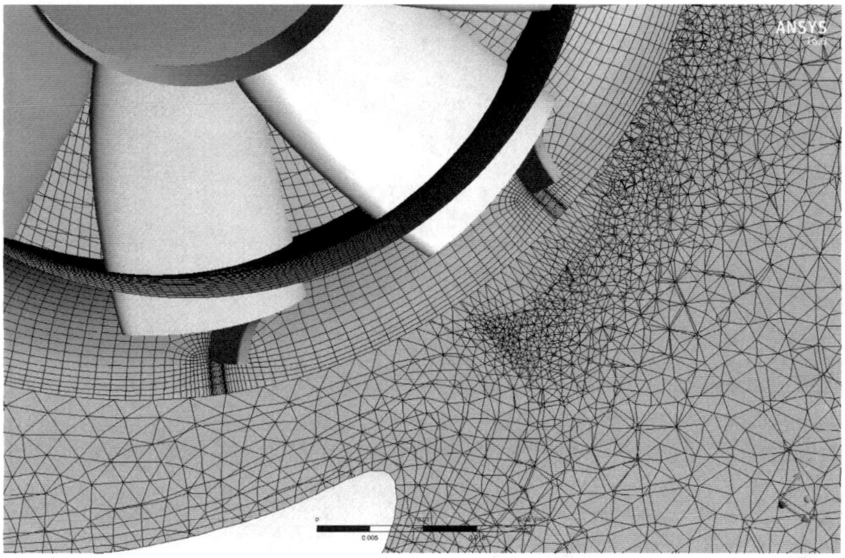

Automotive Simulation WORLD CONGRESS

Pressure @ 2.89 revs

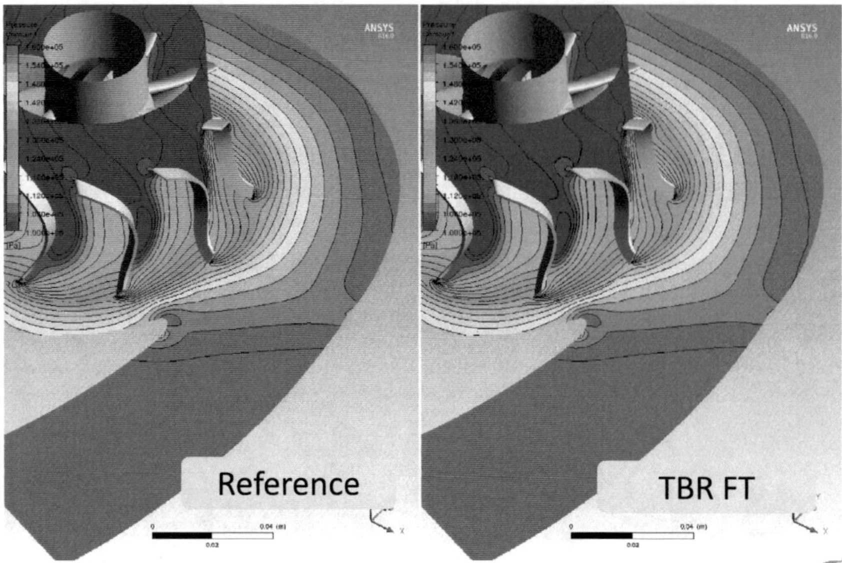

Pressure @ 2.89 revs

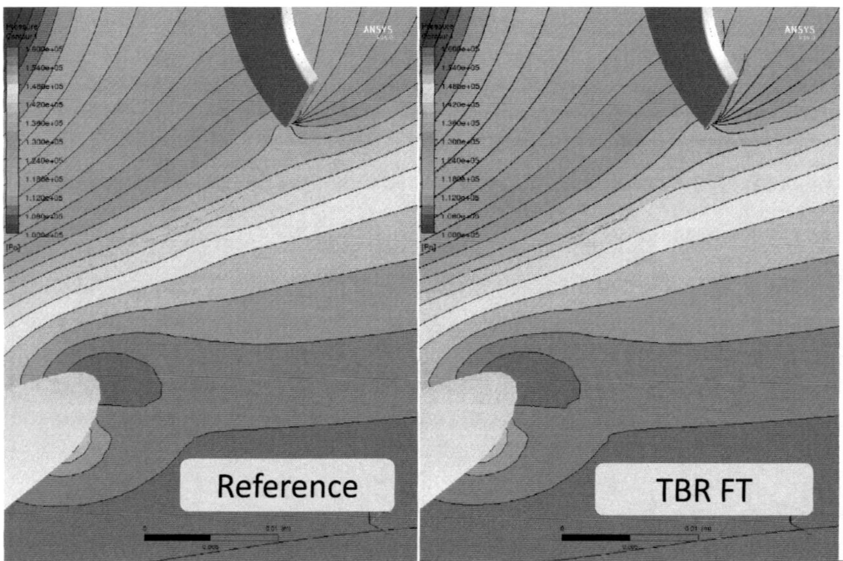

TBR Harmonic Transformation

Based on Harmonic balance / Time spectral method of Hall

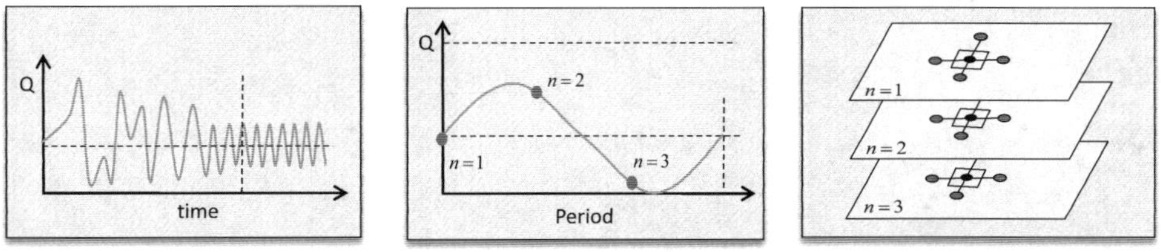

- Use harmonics to reconstruct transient solution
- Simultaneous coupled solution on several time instances
- Solved like steady-state flow simulations
- Fully accounts for all flow non-linear effects
- Arbitrary pitch changes and all flow speeds

Harmonic Transformation

- Equal Pitch inlet disturbance
- Comparing true transient solution to Harmonic Transformation (HT) solution
- Assess method <u>stability</u> and <u>convergence behavior</u> for range of passing frequencies and modes (number of time-levels)
- Example showing: disturbance (hot streaks) passing frequency 100 Hz

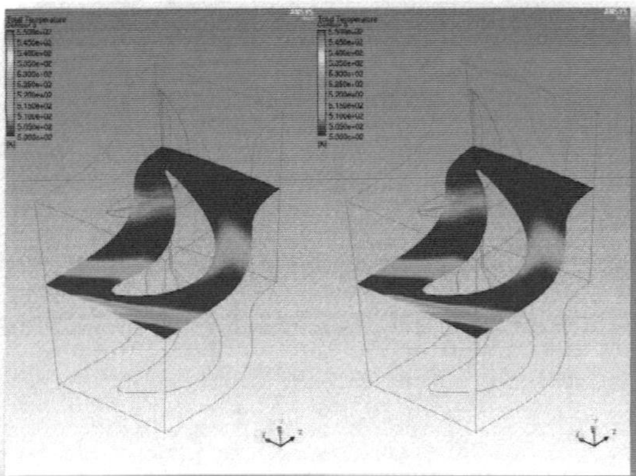

Total Temperature values

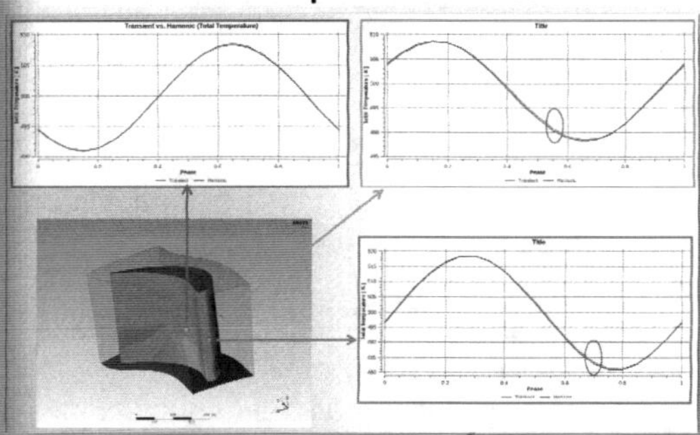

Harmonic Transformation

- Upstream blade row modeled as frozen gust (inlet disturbance) problem.
- 14 hot streaks / 21 blades → Pitch ratio = 1.5
- Pitch-change via FT method
- HA (7 modes) converged in 200 iter
- Transient FT in 4 blade passing

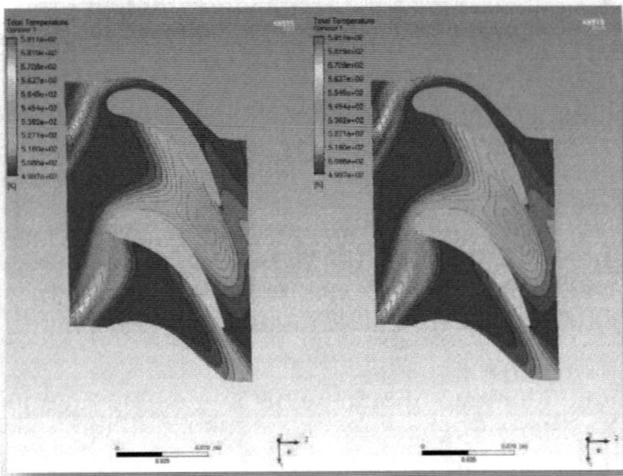

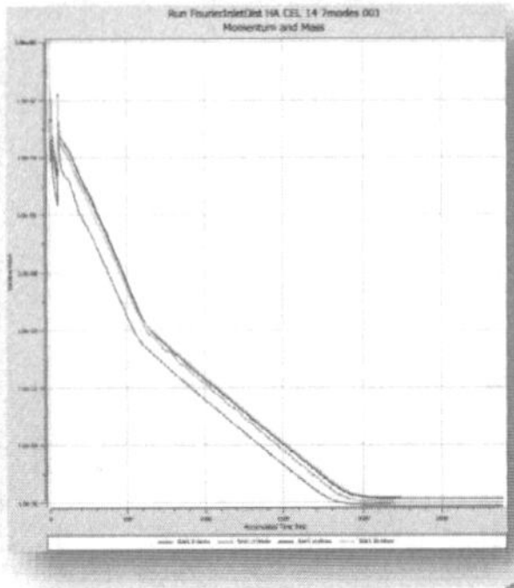

Automotive Simulation WORLD CONGRESS

Transient Blade Row with CHT

Altered solid thermal response to combine CHT with TBR

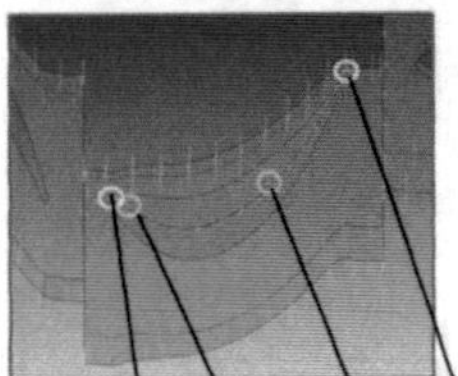

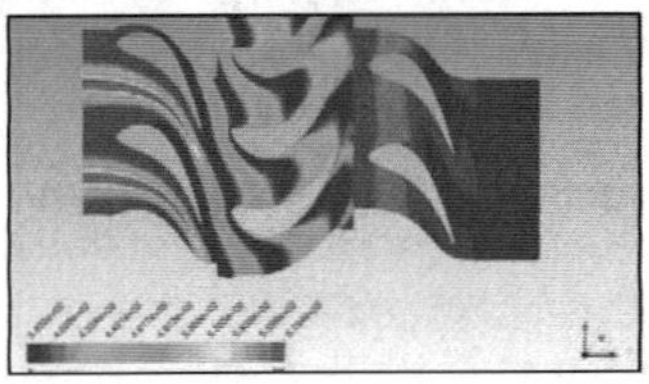

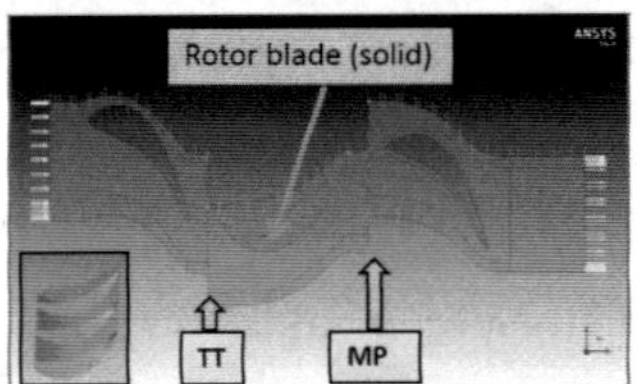

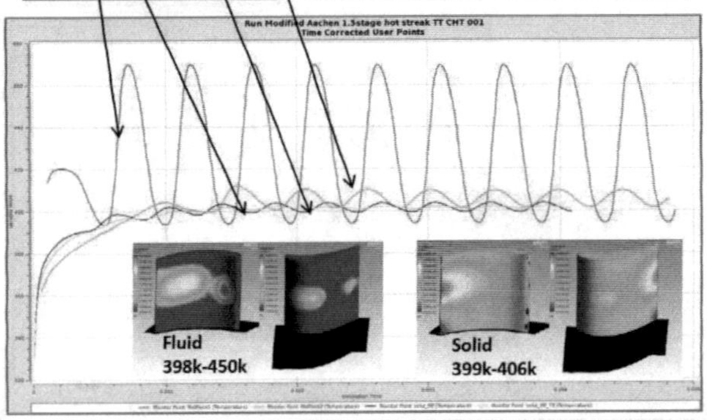

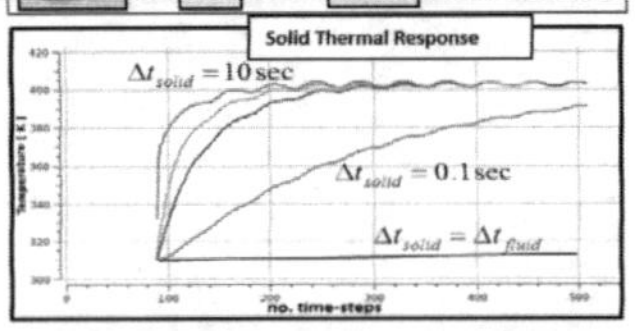

Simulation timestep = 9.718×10^{-6} s

Solid time scale $= \frac{L^2_{scale}}{Solid\ thermal\ diffusivity}$
Ave : 10.6 s

Timescale Factor $= \frac{Solid\ time\ Scale}{Simulation\ timestep}$
$= 1.0 \times 10^6$

ulation WORLD CONGRESS

Transient Blade Row – Convergence Check (β)

Cyclic and polar plots in solver monitors (β)

Overlay multiple common blade passing periods

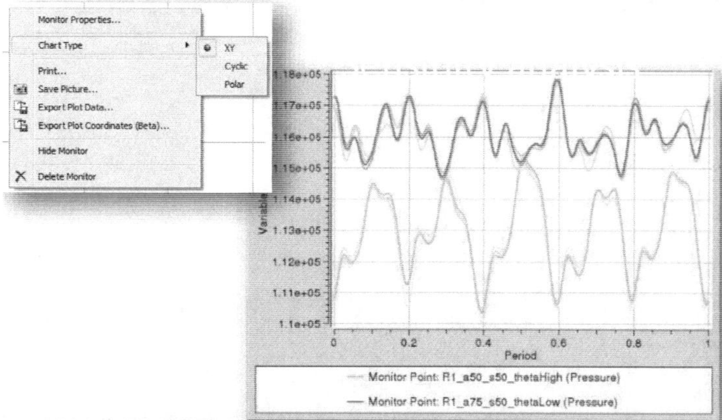

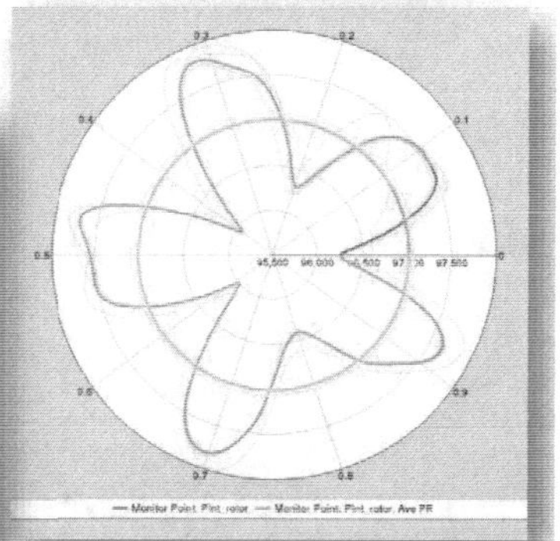

Structural Mechanics Analysis

- Mechanical deformation
- Thermal stresses
- Modal analysis
- Blade flutter
- Forced response

Static analysis

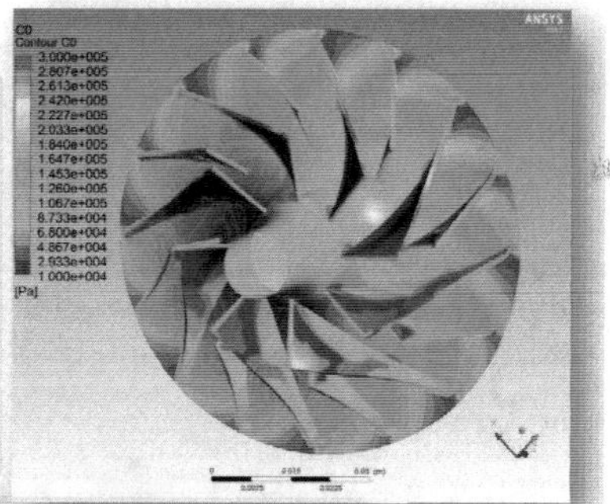

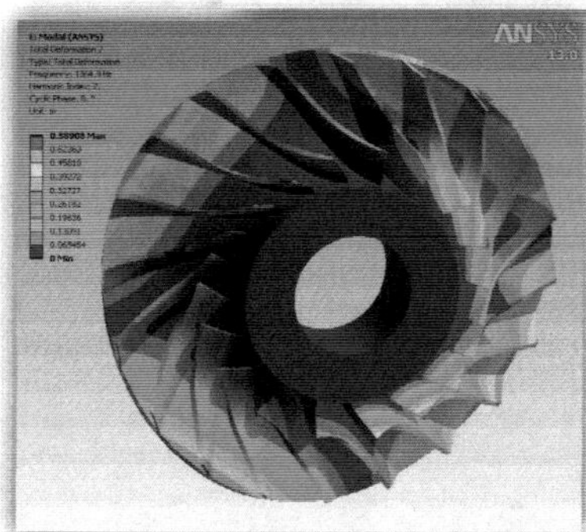

Modal analysis

Summary

ANSYS delivers sophisticated software tools for the complete turbocharger design process

Various tools can easily be combined in ANSYS Workbench

Parametrization and optimization in one user environment

EFFICIENT FSI SIMULATION FOR TURBOCHARGERS.

AUTOMOTIVE SIMULATION WORLD CONGRESS, JUNE 7TH-8TH 2016.

R. Wildhofer, W. Brack, Dr. B. Hußmann (BMW Group).

Prof. Tekn. Dr. D. Vogt (University of Stuttgart).

BMW GROUP THE NEXT 100 YEARS

FLUID-STRUCTURE INTERACTION IN TURBOCHARGERS. MOTIVATION AND BACKGROUND.

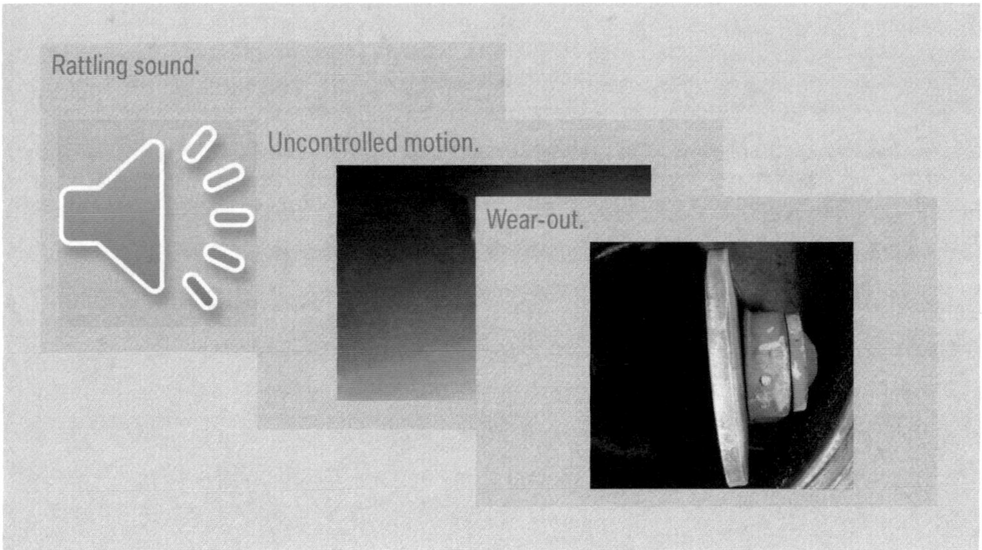

FLUID-STRUCTURE INTERACTION IN TURBOCHARGERS. MOTIVATION AND BACKGROUND.

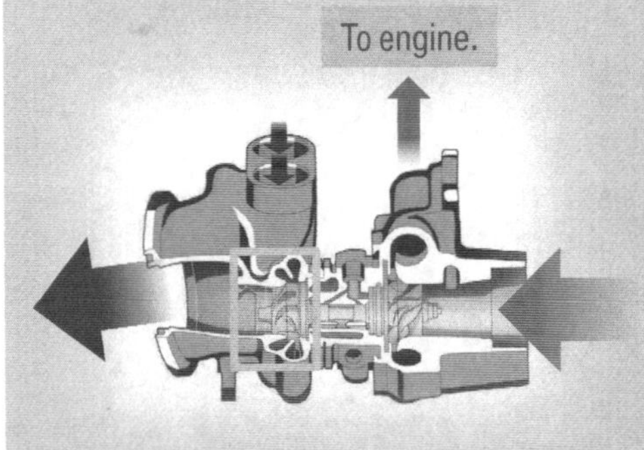

Turbochargers have been established as a mainstream technology for increasing power and efficiency of internal combustion engines.

Due to the increased amount of air, the **specific engine power** rises.

For application to various operating points, the turbine needs a **control unit**, e.g.:

- Variable guide vanes.
- Bypass valve.

FLUID-STRUCTURE INTERACTION IN TURBOCHARGERS. MOTIVATION AND BACKGROUND.

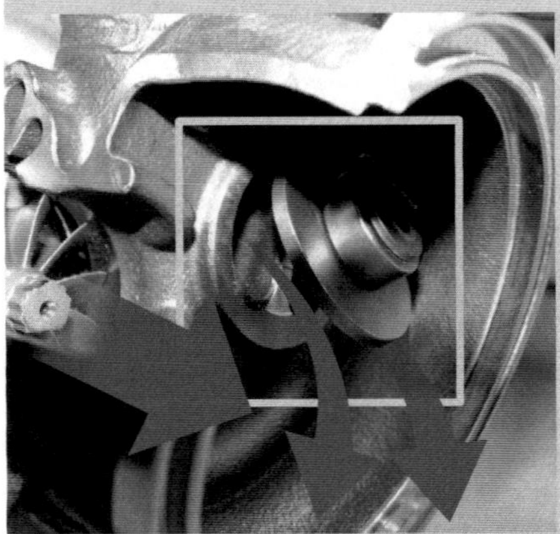

Wastegate guides surplus exhaust gas past the turbine → bypass valve.

FLUID-STRUCTURE INTERACTION IN TURBOCHARGERS. MOTIVATION AND BACKGROUND.

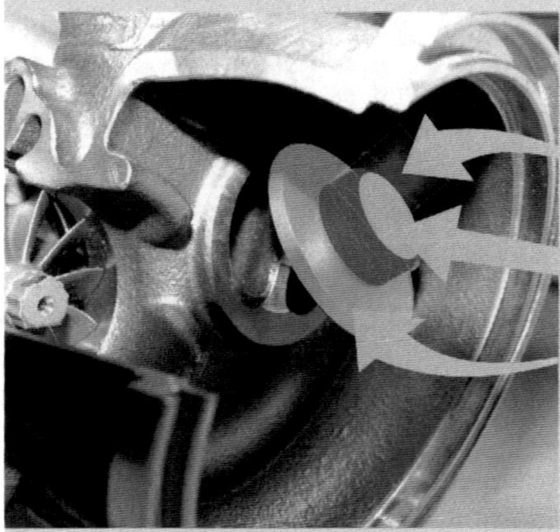

Wastegate guides surplus exhaust gas past the turbine → bypass valve.

Most common type: Threepart wastegate.

- Lever Arm.
- Rivet.
- Plate.

FLUID-STRUCTURE INTERACTION IN TURBOCHARGERS. MOTIVATION AND BACKGROUND.

Wastegate guides surplus exhaust gas past the turbine → bypass valve.

Most common type: Threepart wastegate.

Loose fitting ensures perfect sealing at low-end torque. Problem: **Rattling noise, wear-out**.

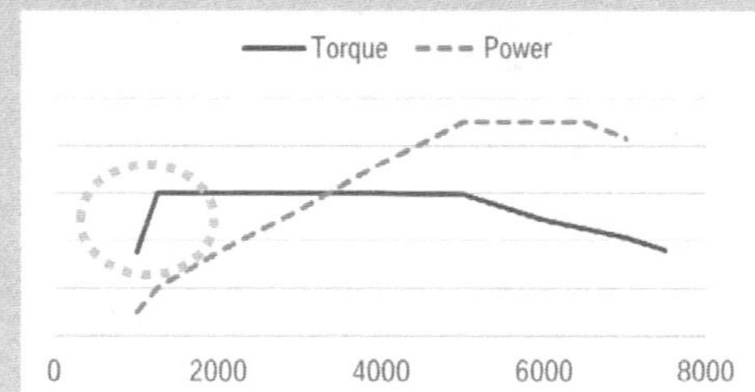

FLUID-STRUCTURE INTERACTION IN TURBOCHARGERS. KEY QUESTIONS.

1. What do we want to simulate?

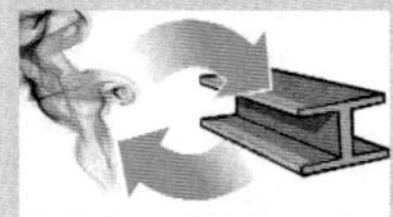

2. Do we need a two-way coupling?

3. How can we speed up the simulations?

4. Can we use synergies with existing workflows?

FLUID-STRUCTURE INTERACTION IN TURBOCHARGERS. SIMULATION OBJECTIVES.

1. What do we want to simulate?

FLUID-STRUCTURE INTERACTION IN TURBOCHARGERS. SIMULATION OBJECTIVES.

Predict

Correct body motion of wastegate.

Efficient FSI Simulation for Turbochargers. BMW Group, June 8th 2016.

Page 9

FLUID-STRUCTURE INTERACTION IN TURBOCHARGERS. SIMULATION OBJECTIVES.

Predict

Structure-born sound.

→ From impact **accelerations**.

Wear-out behaviour.

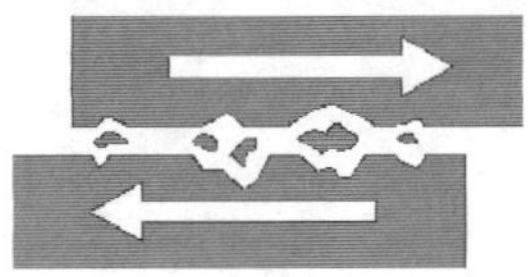

→ From body **velocities**, impact **accelerations**.

Efficient FSI Simulation for Turbochargers. BMW Group, June 8th 2016.

Page 10

FLUID-STRUCTURE INTERACTION IN TURBOCHARGERS. SIMULATION OBJECTIVES.

Predict

For different geometries:	For different operating points:
• Valve type. • Clearance. • Lever arm position.	• Load. • Engine speed.

FLUID-STRUCTURE INTERACTION IN TURBOCHARGERS. KEY QUESTIONS.

1. What do we want to simulate?

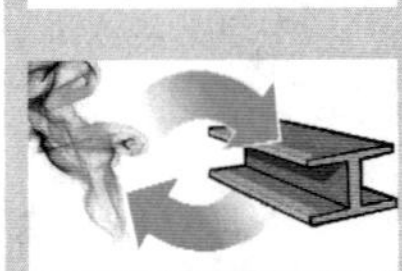

2. Do we need a two-way coupling?

3. How can we speed up the simulations?

4. Can we use synergies with existing workflows?

FLUID-STRUCTURE INTERACTION IN TURBOCHARGERS. COUPLING METHOD.

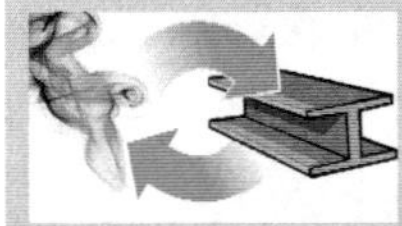

2. Do we need a two-way coupling?

Efficient FSI Simulation for Turbochargers. BMW Group, June 8th 2016. Page 13

FLUID-STRUCTURE INTERACTION IN TURBOCHARGERS. COUPLING METHOD.

ONE-WAY COUPLING.

CFD Solver
CFX

Fluid forces.

Forces →

FEM Solver
ANSYS Mechanical

Contact mechanics etc.

→ **Insufficient** for some relevant quantities (e.g. rotation).
→ Two-way coupling is needed.

Efficient FSI Simulation for Turbochargers. BMW Group, June 8th 2016. Page 14

FLUID-STRUCTURE INTERACTION IN TURBOCHARGERS. COUPLING METHOD.

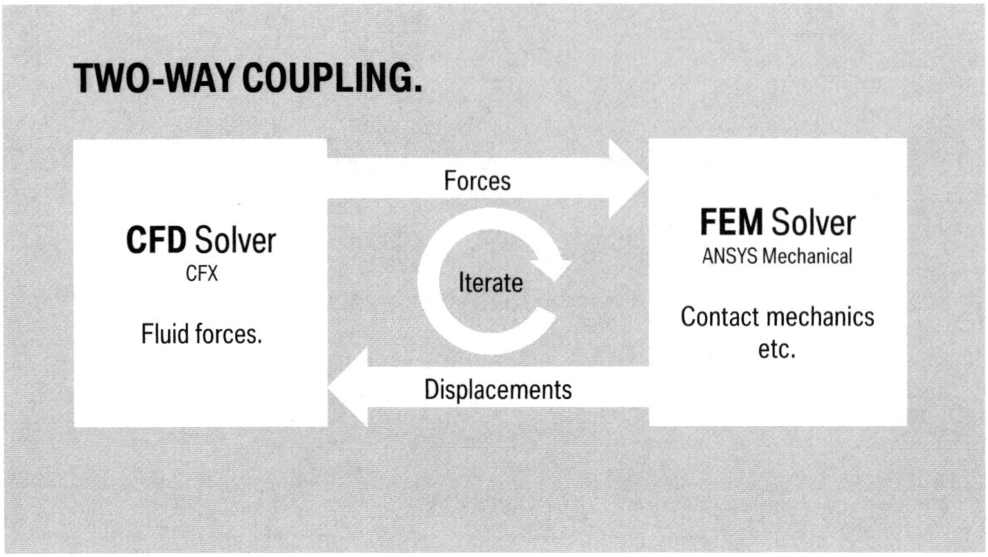

FLUID-STRUCTURE INTERACTION IN TURBOCHARGERS. COUPLING METHOD.

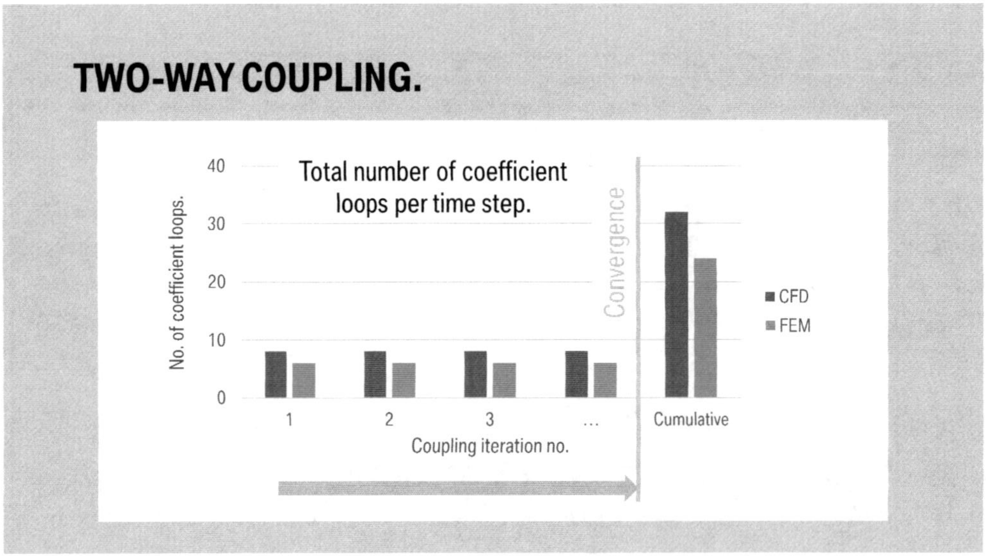

FLUID-STRUCTURE INTERACTION IN TURBOCHARGERS. KEY QUESTIONS.

1. What do we want to simulate?

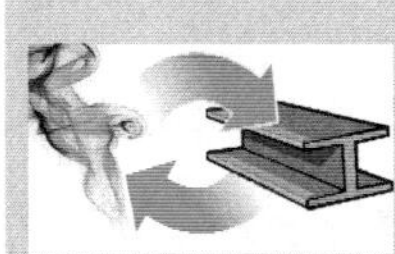

2. Do we need a two-way coupling?

3. How can we speed up the simulations?

4. Can we use synergies with existing workflows?

FLUID-STRUCTURE INTERACTION IN TURBOCHARGERS. SIMULATION SPEEDUP.

3. How can we speed up the simulations?

FLUID-STRUCTURE INTERACTION IN TURBOCHARGERS. SIMULATION SPEEDUP.

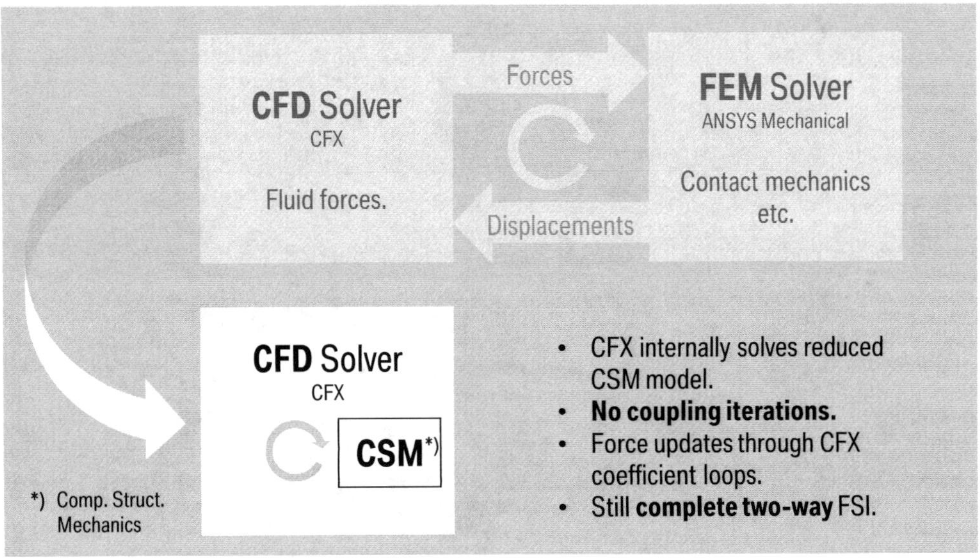

FLUID-STRUCTURE INTERACTION IN TURBOCHARGERS. SIMULATION SPEEDUP.

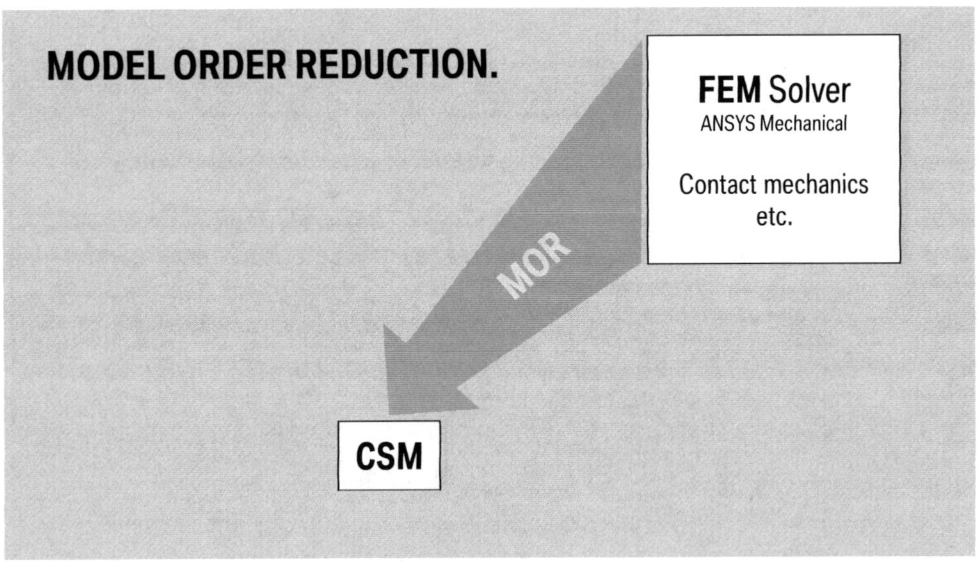

FLUID-STRUCTURE INTERACTION IN TURBOCHARGERS. SIMULATION SPEEDUP.

MODEL ORDER REDUCTION.

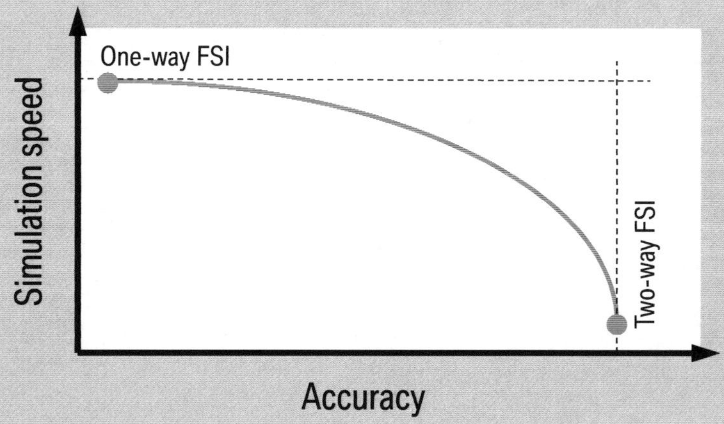

Efficient FSI Simulation for Turbochargers. BMW Group, June 8th 2016.

Page 21

FLUID-STRUCTURE INTERACTION IN TURBOCHARGERS. SIMULATION SPEEDUP.

1. RIGID BODY METHOD.

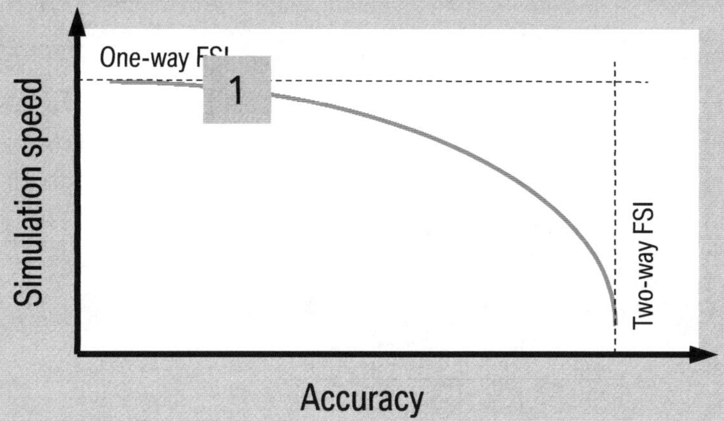

Efficient FSI Simulation for Turbochargers. BMW Group, June 8th 2016.

Page 22

FLUID-STRUCTURE INTERACTION IN TURBOCHARGERS. SIMULATION SPEEDUP.

1. RIGID BODY METHOD.

Strategy outline:

- Use CFX Rigid Body solver.
- Augment by contact model.
    - Implemented in User Fortran.
    - Different contact models (Penalty, Augmented Lagrange).

Efficient FSI Simulation for Turbochargers. BMW Group, June 8th 2016. Page 23

FLUID-STRUCTURE INTERACTION IN TURBOCHARGERS. SIMULATION SPEEDUP.

1. RIGID BODY METHOD.

Efficient FSI Simulation for Turbochargers. BMW Group, June 8th 2016. Page 24

FLUID-STRUCTURE INTERACTION IN TURBOCHARGERS. SIMULATION SPEEDUP.

2. MODAL SUPERPOSITION.

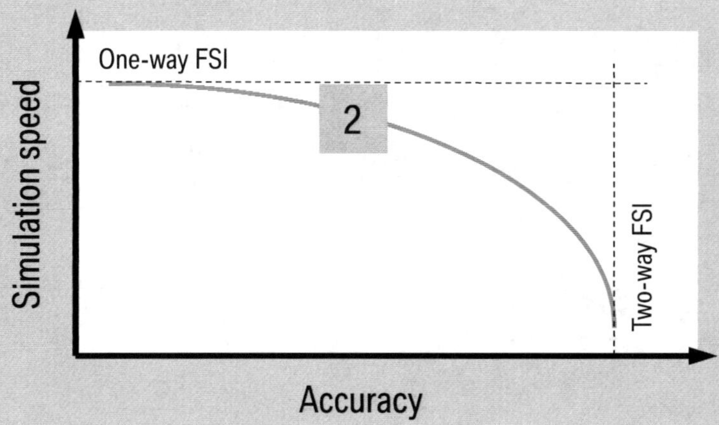

FLUID-STRUCTURE INTERACTION IN TURBOCHARGERS. SIMULATION SPEEDUP.

2. MODAL SUPERPOSITION.

Strategy outline:

- **"Linearize what can be linearized".**
  - Translatory rigid body motion.
  - Small rigid body rotations.
  - Elastic deformations.

  } Using rigid and elastic eigenmodes.
- Treat nonlinearities explicitly in **contact model**.
  - Same as in rigid body model.

FLUID-STRUCTURE INTERACTION IN TURBOCHARGERS. SIMULATION SPEEDUP.

2. MODAL SUPERPOSITION.

Efficient FSI Simulation for Turbochargers. BMW Group, June 8th 2016. Page 27

FLUID-STRUCTURE INTERACTION IN TURBOCHARGERS. SIMULATION SPEEDUP.

3. NONLINEAR MODEL ORDER REDUCTION.

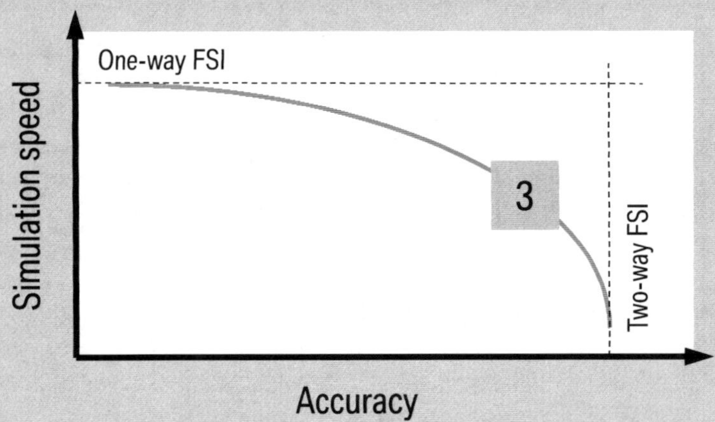

Efficient FSI Simulation for Turbochargers. BMW Group, June 8th 2016. Page 28

FLUID-STRUCTURE INTERACTION IN TURBOCHARGERS. SIMULATION SPEEDUP.

3. NONLINEAR MODEL ORDER REDUCTION.

Strategy outline:

- Incorporate all nonlinearities in the reduced order model.
- Reduction method:
  - o Create restricted solution subspace:
    - – Spanned by **snapshots** from training procedure (offline).
    - – Generic, pure FEM simulations.
  - o Reduced basis via Proper Orthogonal Decomposition.
  - o Project system matrices onto new basis.
  - o Solve reduced system (User Fortran).

FLUID-STRUCTURE INTERACTION IN TURBOCHARGERS. SIMULATION SPEEDUP.

3. NONLINEAR MODEL ORDER REDUCTION.

Efficiency of the method:

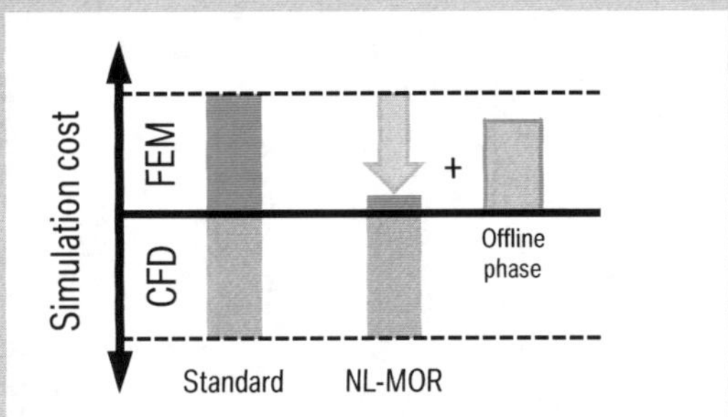

FLUID-STRUCTURE INTERACTION IN TURBOCHARGERS. KEY QUESTIONS.

1. What do we want to simulate?

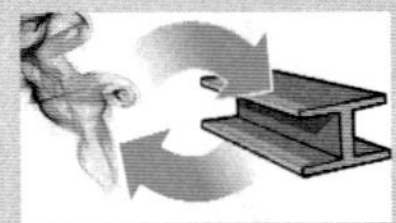

2. Do we need a two-way coupling?

3. How can we speed up the simulations?

4. Can we use synergies with existing workflows?

FLUID-STRUCTURE INTERACTION IN TURBOCHARGERS. USING SYNERGIES.

4. Can we use synergies with existing workflows?

FLUID-STRUCTURE INTERACTION IN TURBOCHARGERS. USING SYNERGIES.

The **mechanics part** can be easily extended to other problem settings:

- Exhaust valve flutter.
- Tumble flap motion.
- ...

FLUID-STRUCTURE INTERACTION IN TURBOCHARGERS. USING SYNERGIES.

The **"parent" CFX** process can also be any of the existing (transient) workflows:

- Catalytic converter heating.
- Heat transfer calculation.
- In-cylinder flow.

FLUID-STRUCTURE INTERACTION IN TURBOCHARGERS. CONCLUSIONS.

Efficient simulation methodology for fluid-structure interaction:

- Two-way coupling.
- Fast.
- Integration of nonlinearities.
- Modular and extendible.

Efficient FSI Simulation for Turbochargers. BMW Group, June 8th 2016. Page 35

R. Wildhofer, W. Brack, Dr. B. Hußmann (BMW Group).
Prof. Tekn. Dr. D. Vogt (University of Stuttgart).

BMW GROUP THE NEXT 100 YEARS

FLUID-STRUCTURE INTERACTION IN TURBOCHARGERS. PICTURE CREDITS.

Page 7: revamo.de
thinglink.com
play.google.com
clipartpanda.com

Page 9: github.com

Automotive Simulation WORLD CONGRESS

June 7-8, 2016 Munich

Highly Efficient Design Optimization for Automotive Applications

Alberto Clarich, Engineering Manager
ESTECO SpA

Summary

- **Introduction to modeFRONTIER**

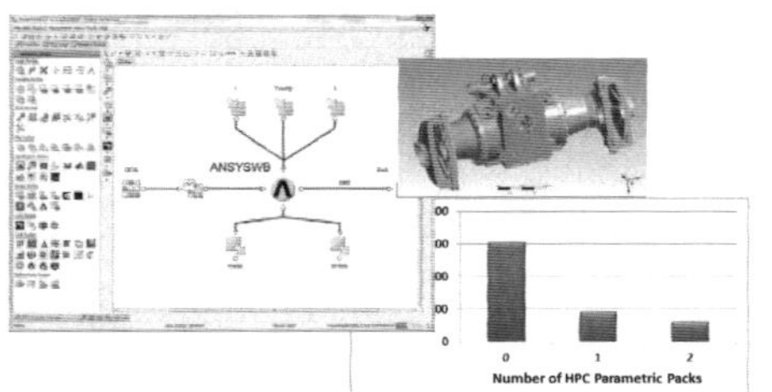

- **Introduction to ANSYS HPC**
- **Real-world application**
  >> Optimization of Heat Exchangers
- **HPC Performance**
- **Results**

About ESTECO

ESTECO is an independent technology provider that delivers first-class software solutions aimed at perfecting the simulation-driven design process. With more than 15 years' experience, we support engineers and companies in designing better, more efficient products.

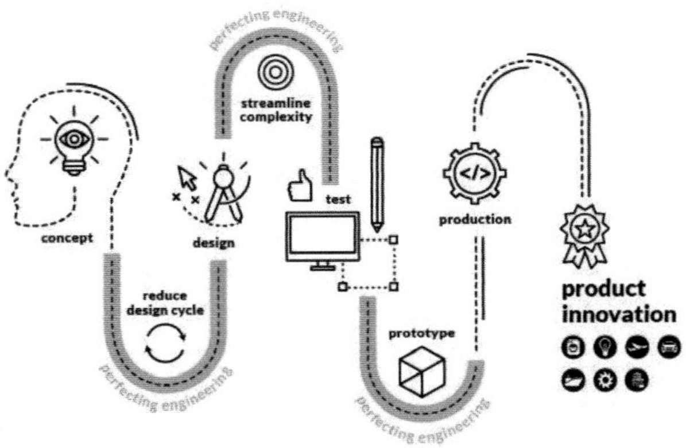

We specialize in customer-focused solutions for numerical optimization, CAE integration, process automation and simulation data management

Where we are

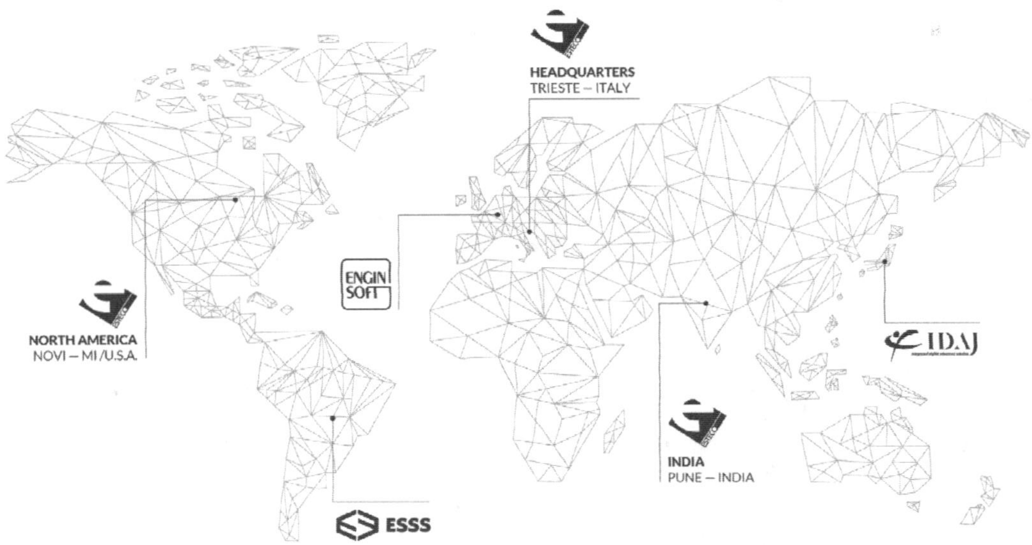

ESTECO clients

Over 300 international clients have relied on our software to design better and more efficient products across a wide spectrum of industrial sectors.

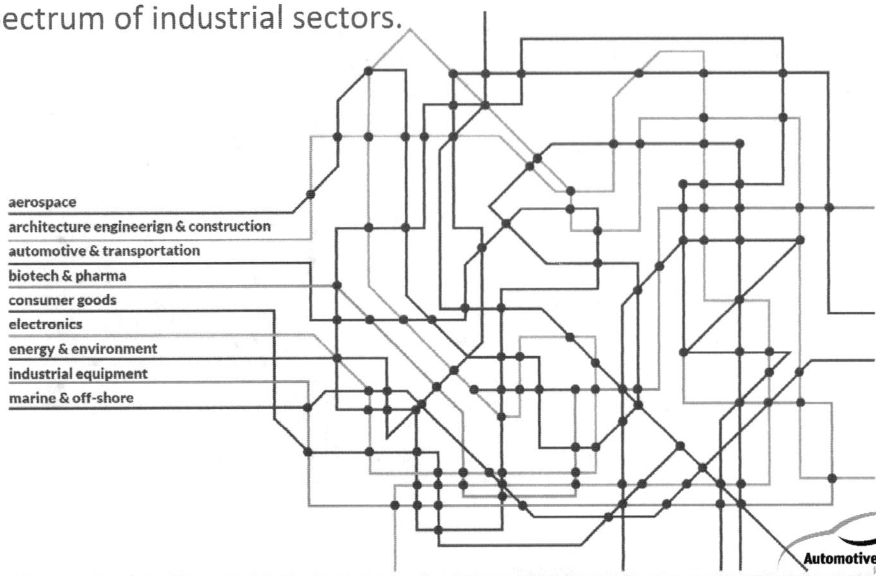

Automotive Simulation WORLD CONGRESS

ESTECO Technology

Our aim is to **decrease the tedium** in engineering analysis and **increase creativity** by developing and maintaining **cutting-edge software** that enables integration, optimization and advanced data analysis.

Increases efficiency of design simulation tools

inspires decision making

accelerates product innovation

Our first-class technology

Automotive Simulation WORLD CONGRESS

What can you do with modeFRONTIER

INTEGRATION AND PROCESS AUTOMATION

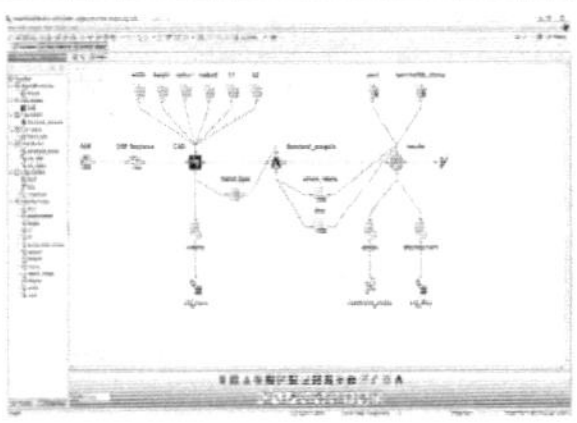

OPTIMIZATION AND RESPONSE SURFACES

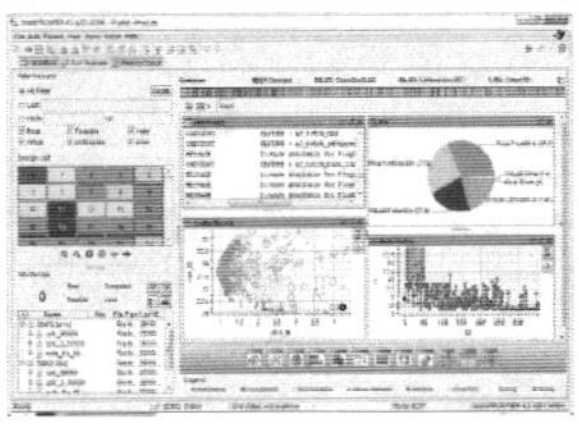

ROBUST DESIGN AND RELIABILITY

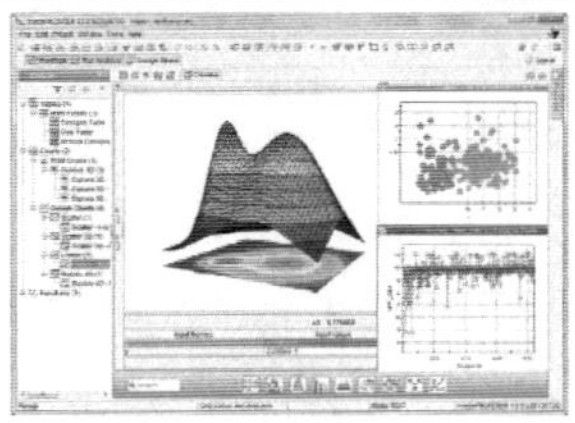

DESIGN SPACE EXPLORATION

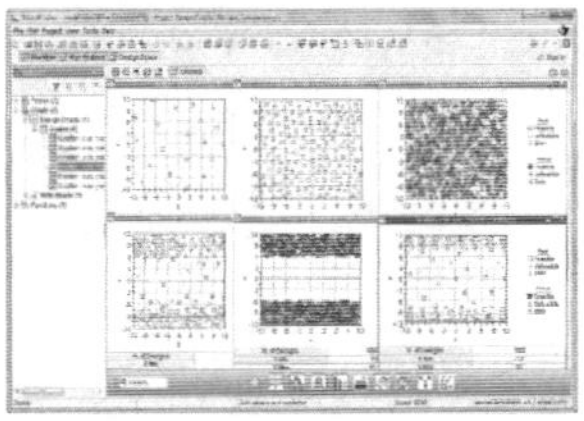

ANALYTICS AND VISUALIZATION

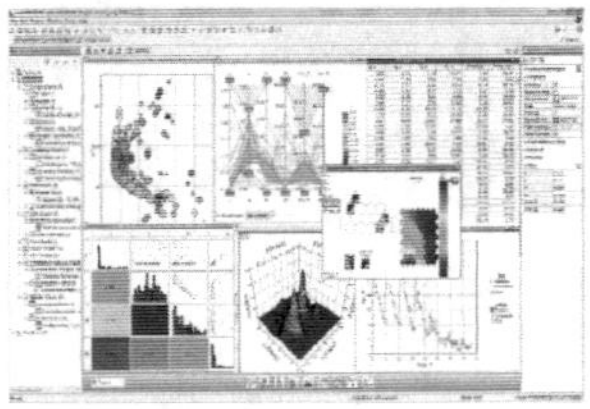

DECISION MAKING

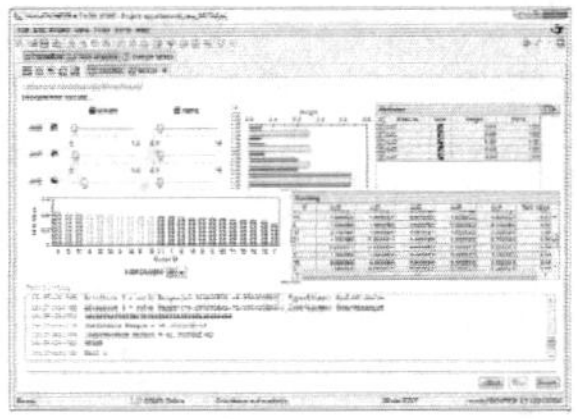

....Optimization can be an expensive task

Performing trade studies and optimization analysis means **running a huge number of calculations** depending on the problem complexity in terms of

model fidelity / size	number of disciplines involved	number of parameters	robust analysis

Scaling up with modeFRONTIER algorithms

modeFRONTIER optimization algorithms benefit from concurrent executions

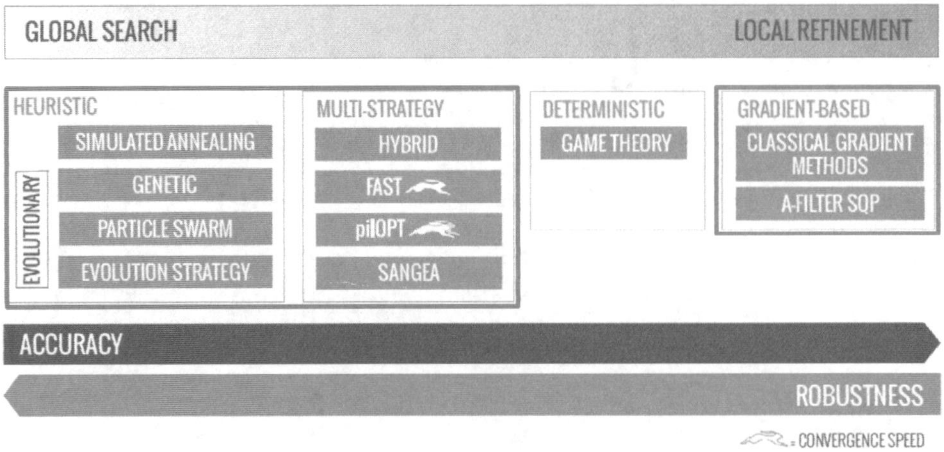

- Can evaluate up to **n** (= <u>number of variables</u>) simultaneous designs
- Can evaluate up to **m** (= <u>population size</u>) simultaneous designs

Automotive Simulation WORLD CONGRESS

New in mF2016: ANSYS HPC Parametric Pack

Support for ANSYS HPC Parametric Pack license

modeFRONTIER is going to support the very effective HPC license consumption scheme from ANSYS Workbench

modeFRONTIER users will be able to launch multiple concurrent design evaluations without checking out additional Workbench licenses, taking advantage of their own ANSYS HPC Parametric Pack

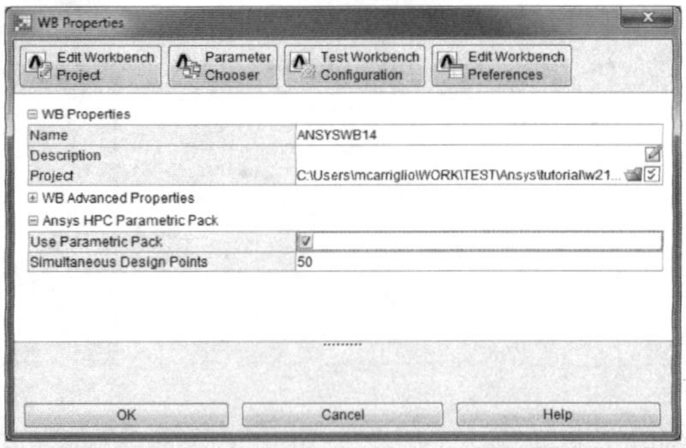

License Usage During Simultaneous Designs Update

Simultaneous solve is a great way to reduce time to ROI!!!

But!

- ANSYS products require licenses as each software component is executed
- To update n design points simultaneously you need *n* licenses.

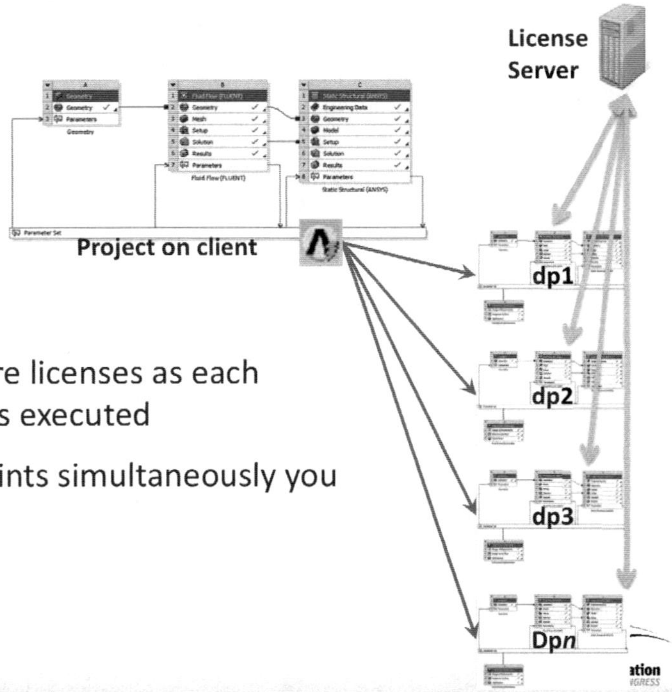

Licensing solution: ANSYS HPC Parametric Pack

- Allow users to run *n* design points simultaneously, amplifying a single set of "base" license(s)
  - All the keys in the project, multiple solvers, HPC Packs, etc.
- Scalable, similar to ANSYS HPC Pack
- Enabled by ANSYS Workbench, Design Points Updates, Reserved Licensing, and RSM.

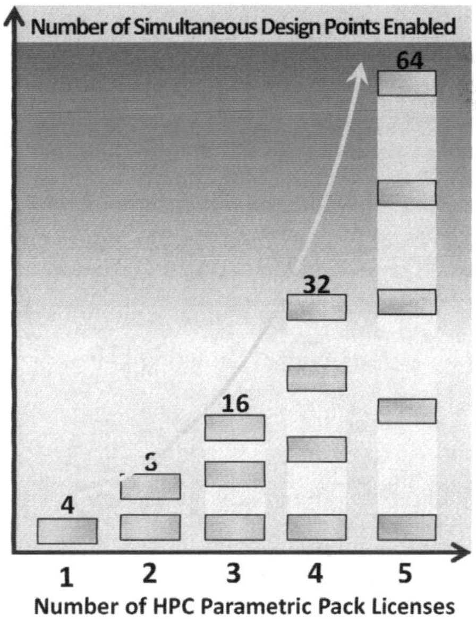

ANSYS direct interface in modeFRONTIER

mF workflow can be automatically created using direct ANSYS WB interface

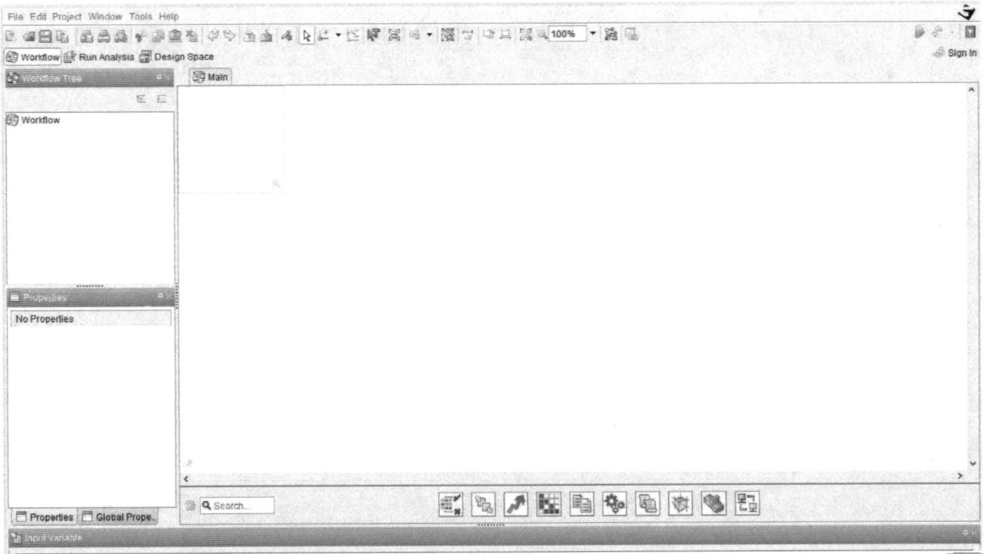

Application example: OptimHEX Project

Funded by Eurostars, powered by EUREKA and the European Community

eurostars™

GOAL

Developing a new methodology for Heat Exchanger design

HYDAC SUPSI ESTECO UNIVERSITÀ DEGLI STUDI DI TRIESTE

OptimHEX Project: Heat Exchanger

Channel length [mm]	400
Cooler height [mm]	400
Cooler depth [mm]	94
Inner Medium	Water/Glycol 50/50
Inner medium flow [l/min]	100
Air flow [kg/s]	1.70 (6120 kg/h)
Cooling power range @ 30°C ETD [kW]	[25-30]

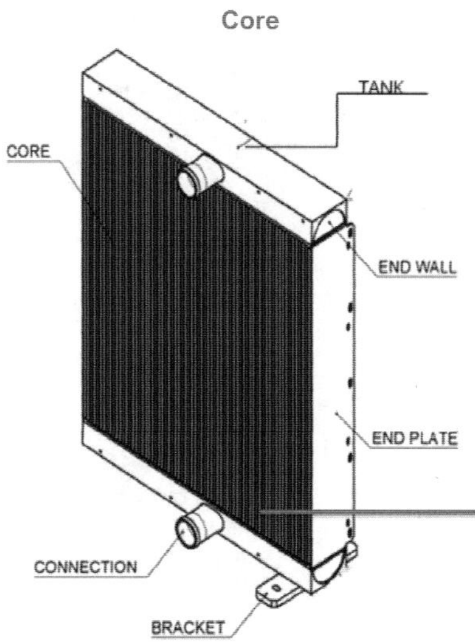

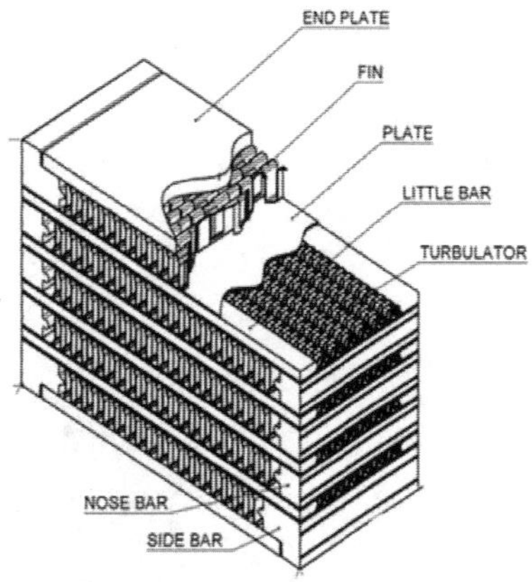

Automotive Simulation WORLD CONGRESS

Parameters selection

Type		Variable	Description	Symbol	Range / Value	Unit
Geometrical Parameters	Air	X1	Air fin thickness	t_{air}	[0.12 – 0.4]	mm
		X2	Air fin pitch	p_{air}	[2 - 8]	mm
		X3	Air channel height	h_{air}	[6 – 16]	mm
		X4	Air fin length	l_{air}	[2, 3, 5, 8, 12, 20, 30]	mm
		X5	Air fin angle	α_{air}	[0 – 20]	°
		X6	Air fin width ratio (top / bottom)	wr_{air}	[0.15 – 1]	–
		X7	Air fin offset coefficient	Ko_{air}	[0 - 1]	–
		X8	Air channel length	lc_{air}	[45, 63, 82, 94, 113, 140, 160]	mm
	Liquid	X9	Liquid fin thickness	t_{liq}	[0.12 – 0.4]	mm
		X10	Liquid fin pitch	p_{liq}	[4 – 10]	mm
		X11	Liquid channel height	h_{liq}	[1.5 – 5]	mm
		X12	Liquid fin length	l_{liq}	[1.5, 2, 4, 8, 16, 30]	mm
		X13	Liquid fin angle	α_{liq}	[0 - 20]	°
		X14	Liquid fin width ratio (top / bottom)	wr_{liq}	[0.15 – 1]	–
		X15	Liquid fin offset coefficient	Ko_{liq}	[0 – 1]	–
		X16	Liquid channel length	lc_{liq}	[200 - 1650]	mm
		X17	Dividing plate thickness	t_p	[0.3 – 1]	mm
Operational parameters	Air	X18	Approaching air mass flow density	mfd_{air}	[0.5 – 20]	$\frac{kg}{s \cdot m^2}$
		X19	Air inlet temperature	T_{air}	298.15	K
	Liquid	X20	Liquid type	liq_{type}	1 = VG68 2 = VG320 3 = WG 40 %	–
		X21	Approaching liquid mass flow density	mfd_{liq}	[7 – 700]	$\frac{kg}{s \cdot m^2}$
		X22	Liquid inlet temperature	T_{liq}	338.15 for oils 328.15 for WG	K

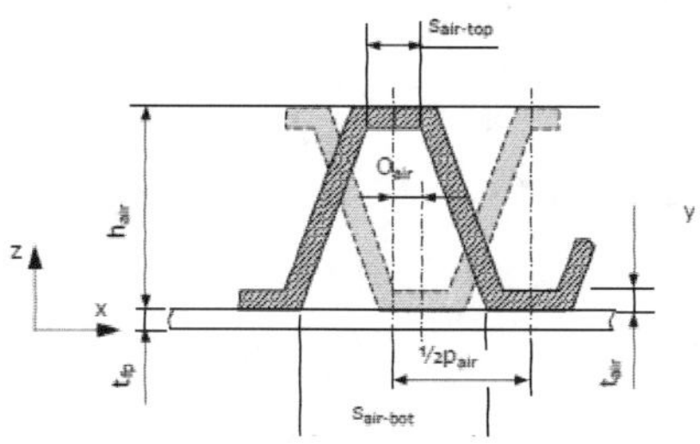

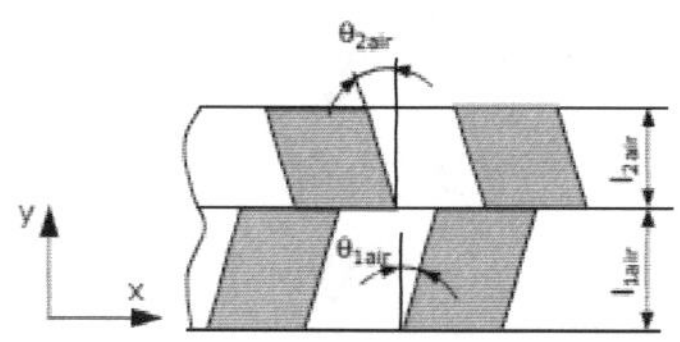

Automotive Simulation WORLD CONGRESS

CAD Parameterization

- Parametric CAD model exported to ANSYS WB

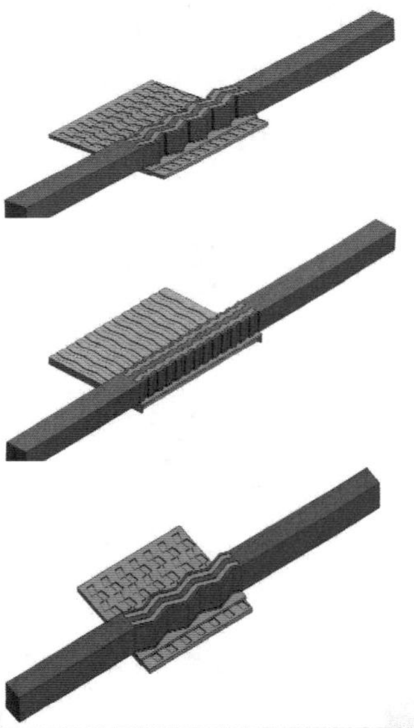

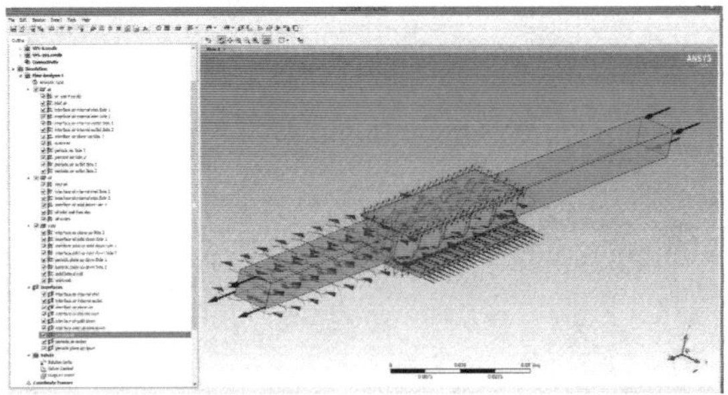

- Originally the HECE model was defined by a full length channel
- After the comparisons of results, it has been decided to reduce the model to a single unit

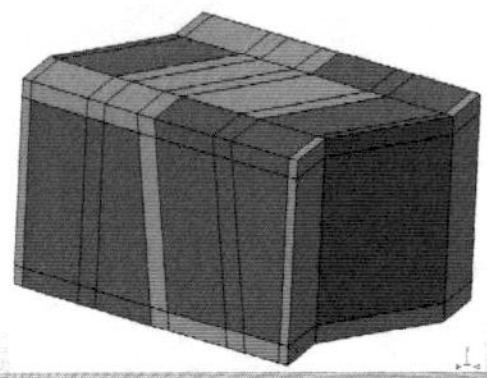

Automotive Simulation WORLD CONGRESS

HECE Numerical model

Periodic inlet/outlet with imposed mass flow and constant Heat Flux

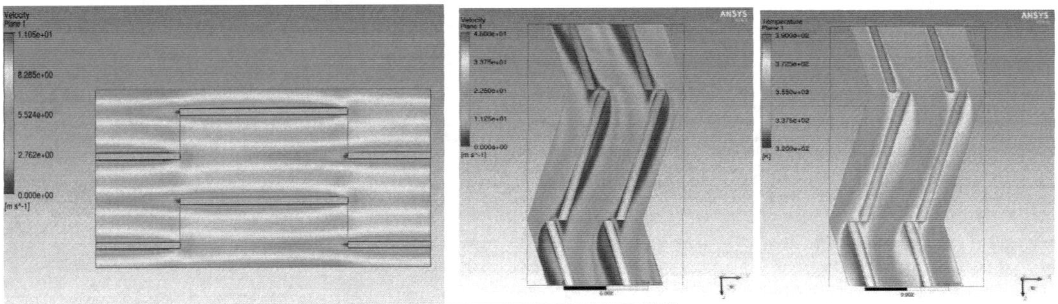

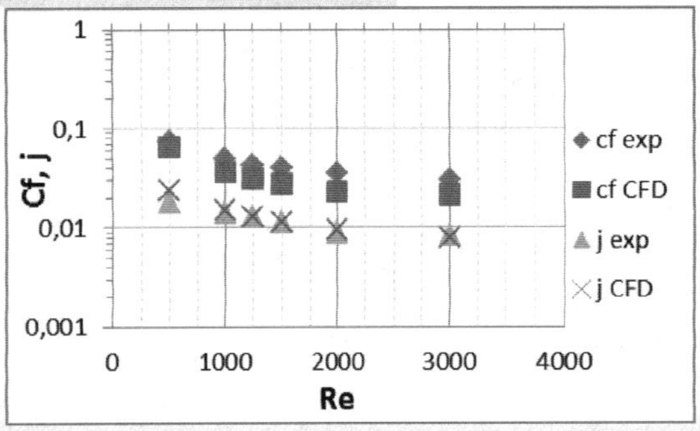

- Good agreement with experimental data (*Kays & London**)

**W.M. Kays and A.L. London, Compact heat exchangers Repr. ed. 1998 with corrections.*

Automotive Simulation WORLD CONGRESS

Hexa meshing - macro

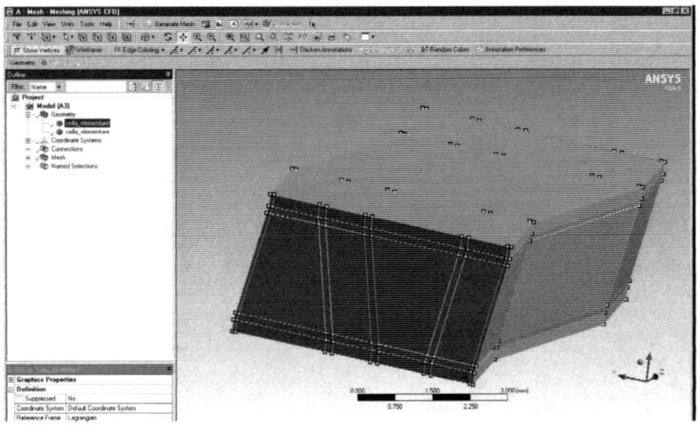

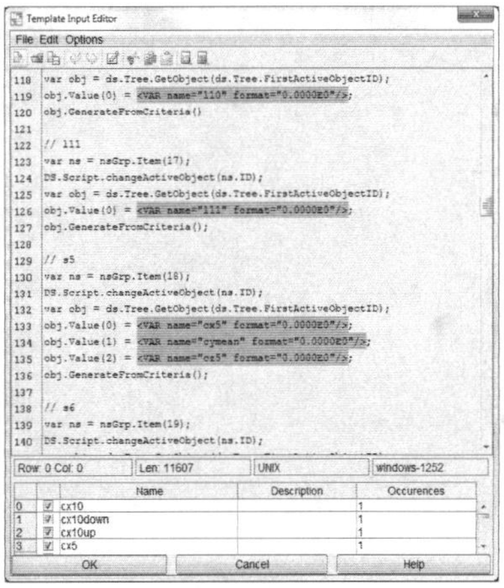

- ANSYS WB model with meshing **blocks** – **macro** defined in mF for automation
- All relevant faces are assigned to a different named selection via **VBS macro** then meshing and boundary condition definition is performend in ANSYS CFX

Automotive Simulation WORLD CONGRESS

Automation of HECE simulation in modeFRONTIER

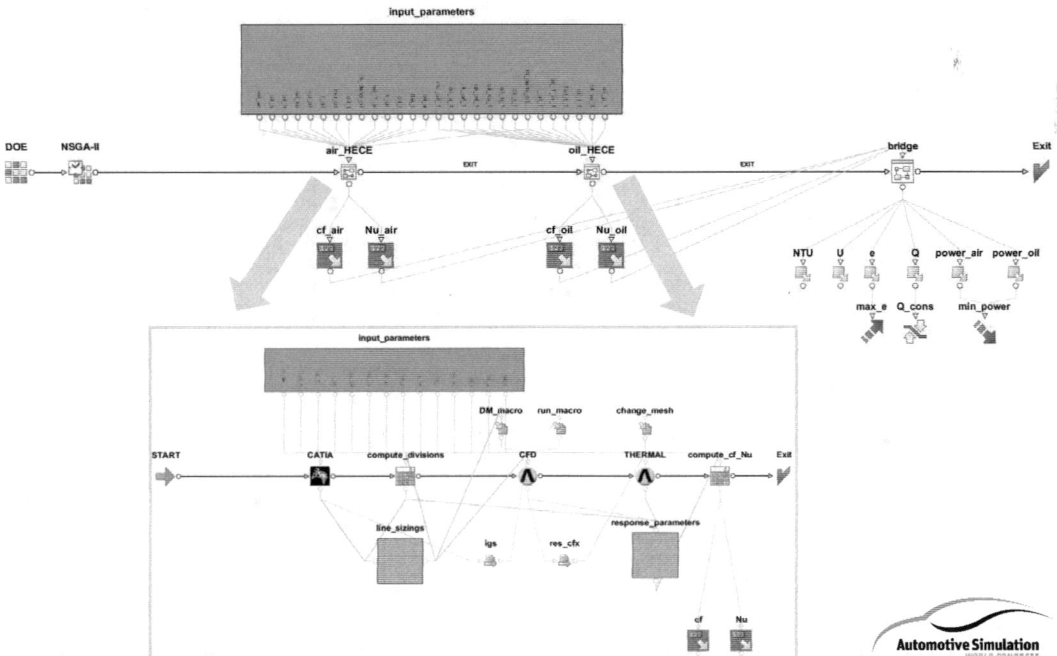

Automotive Simulation WORLD CONGRESS

Bridge Model: ε-NTU method

- e: - HE effectiveness
  - function of NTU, the ratio of the enthalpy flows and the flow arrangement
- NTU: number of heat transfer units
- The relation between e and NTU is known for certain flow arrangements

E.g., crossflow heat exchanger with both fluids unmixed, approximate relation:

$$\varepsilon = 1 - \exp\left[\frac{NTU^{0.22}}{C_r}\left[\exp\left(-C_r NTU^{0.78}\right) - 1\right]\right]$$

$$C_r = \frac{C_{min}}{C_{max}} = \frac{(\dot{m}c_p)_{min}}{(\dot{m}c_p)_{max}}$$

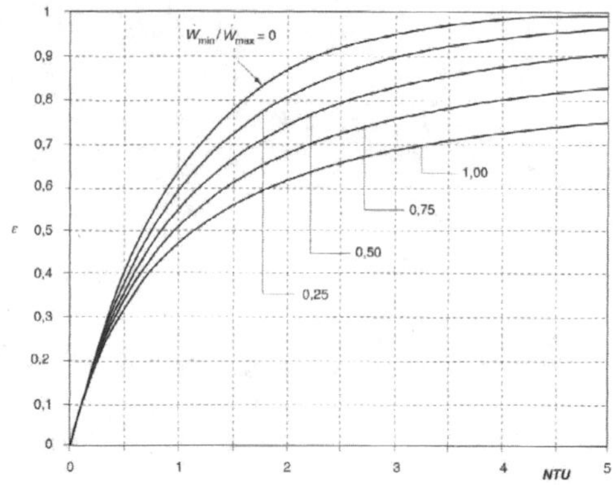

Source: Walter Wagner, Wärmeaustauscher

Automotive Simulation WORLD CONGRESS

Optimization example: results

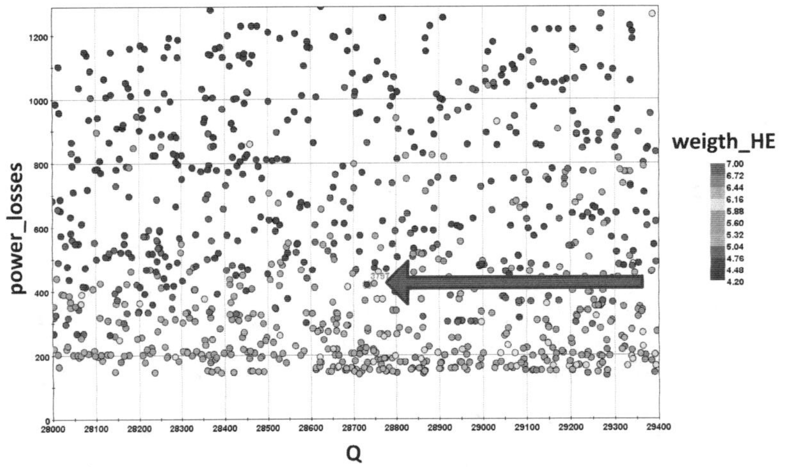

>> NSGAII (genetic algorithm) 20 generations x 48 designs for each population

Automotive Simulation WORLD CONGRESS

Optimization example: results

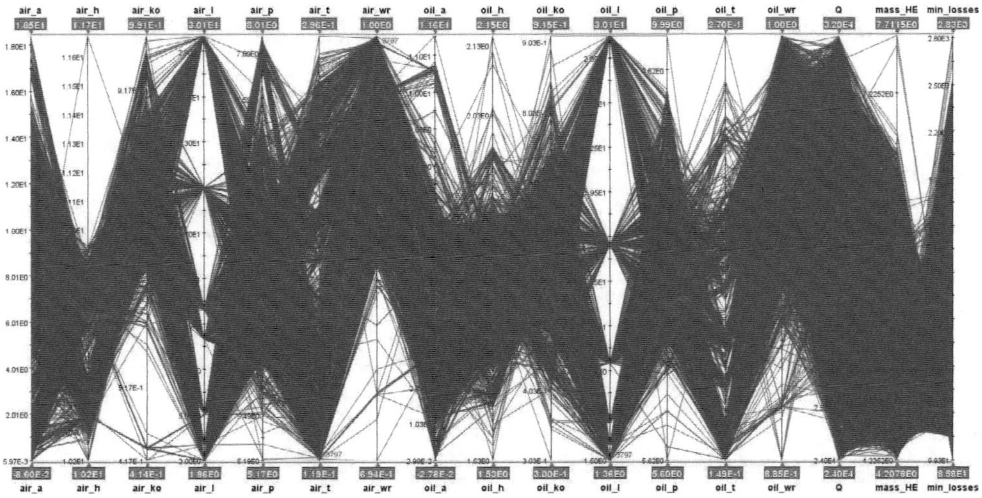

>> Optimal design (ID 3797) chosen accordingly to best trade-off between desired Q, minimum losses and minimum mass

Automotive Simulation WORLD CONGRESS

Optimization example: results

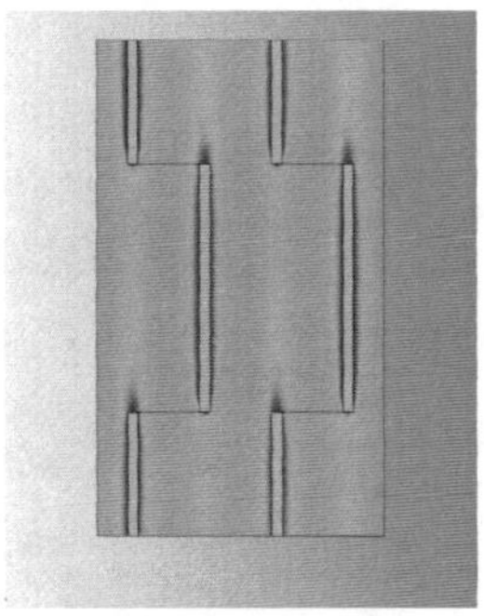

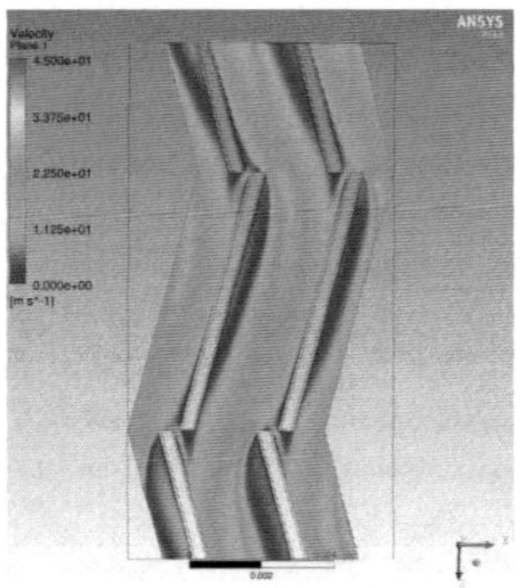

Variable	Baseline configuration	Optimized configuration
Core weight [kg]	6.18	4.65 (-25%)
Cooling power [kW]	28.55	28.73 (+1%)
Power losses [W]	464	420 (-9%)

Automotive Simulation WORLD CONGRESS

Time for simulation:
- Standard, sequential execution

modeFRONTIER submits design evaluations directly **on the workstation**:

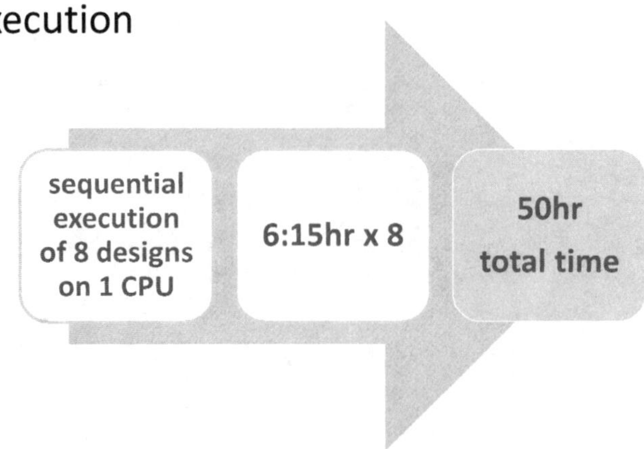

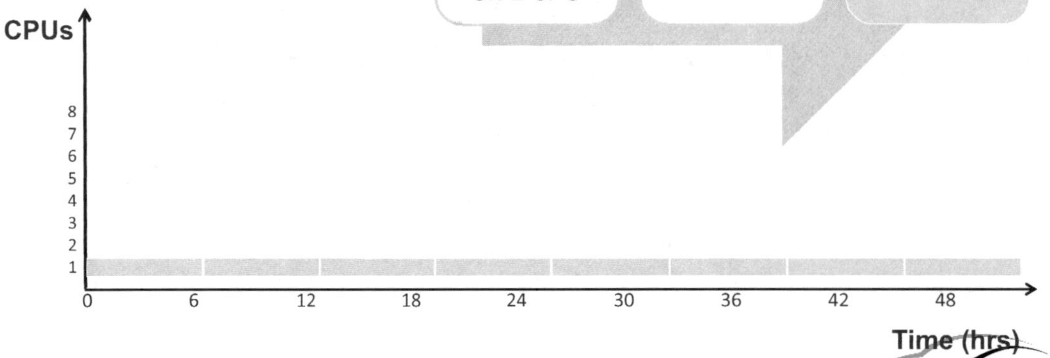

Time for simulation:
- Sequential execution supported by HPC

modeFRONTIER submit design evaluations **on a cluster via ANSYS RSM**:

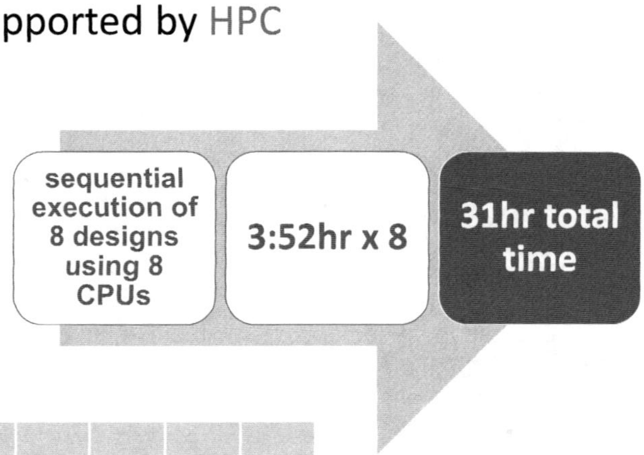

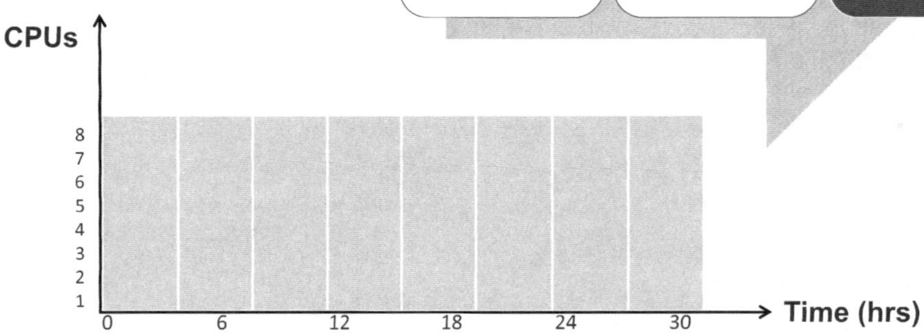

Time for simulation:

- Simultaneous execution supported by HPC Parametric Pack

modeFRONTIER submitting designs through **2 ANSYS HPC Parametric Packs:**

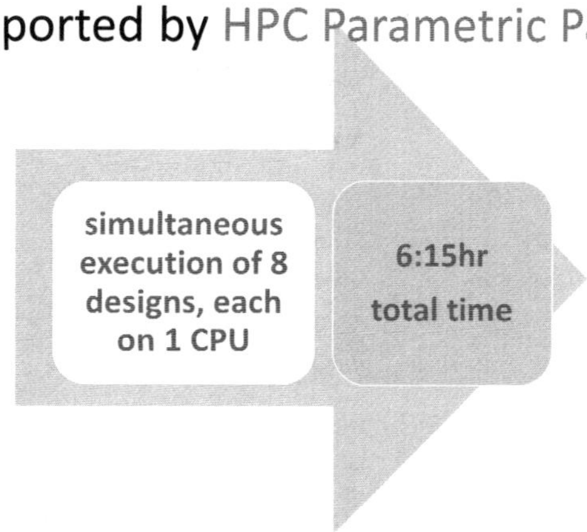

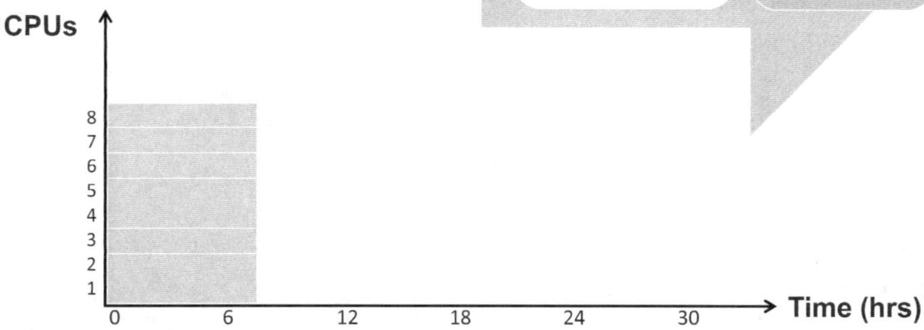

Conclusions and benefits

The new features of the **ANSYS direct integration node** enable users to **fully exploit the HPC Parametric Pack** to:

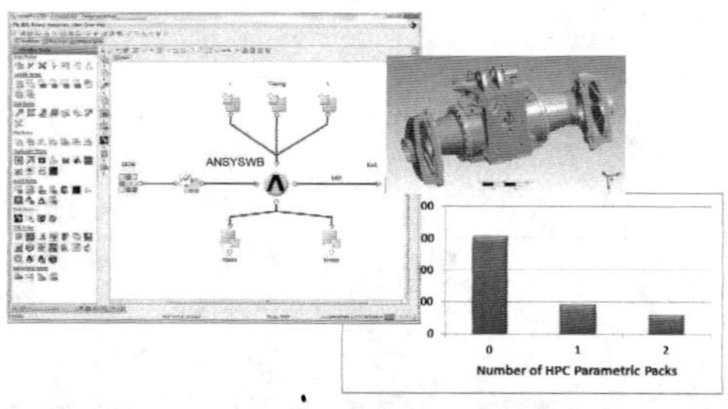

>> get **optimization results before** and dedicate more time to post processing and data analysis

>> fully exploit the **process automation** capabilities of modeFRONTIER in combination with the accuracy of the ANSYS multiphysics solvers

>> **remove computational limits** and engage in true MDO analysis

>> **save on HPC setup costs**

Thanks for the attention!
ESTECO
EXPLORE DESIGN PERFECTION
esteco.com

Automotive Simulation WORLD CONGRESS

June 7-8, 2016 Munich

CAE simulations in the cloud, in novel software containers

Wolfgang Gentzsch, President
UberCloud

UberCloud

- July 2012: Building user community with 'cloud experiments', 190 so far, feedback, measuring traction, cloud roadblocks
- March 2013: Starting HPC container development => Apps are easy packageable, portable, accessible, usable, and scalable
- November 2014: Opening online marketplace with currently 35 stores & established provider ecosystem (appl. software resources, software, expertise)
- April 2016: CAE software containers:, ANSYS, CD-adapco, CFD Support, CFturbo, Dacolt, Flow Science, Friendship Systems, LS-Dyna, NICE DCV, Numeca, OpenFOAM, Red Cedar, Simulia, . . .
- On Clouds Resources from Advania, Amazon AWS, Microsoft Azure, Bull extreme factory, CPU 24/7, OzenCloud . . .

Your major computing tools:

workstations , servers, and clouds

Let's analyze the **benefits & challenges** for 20 million engineers & scientists today

Automotive Simulation WORLD CONGRESS

3 June 7-8, 2016 Munich

But, workstations have limited capacity

57 % of users are **dissatisfied** with their desktop computing capacity*

- **Computing: too slow**
- **Memory: too small**

* Source: US Council of Competitiveness: http://www.compete.org/

Automotive Simulation WORLD CONGRESS

4 June 7-8, 2016 Munich

Benefits of servers

- More compute power and memory
- Higher quality design and products
- Reducing product failure early
- Shorten time to market

=> Strengthen your company's competitiveness.

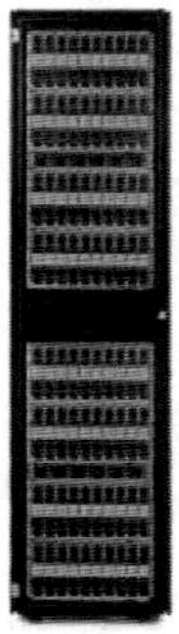

However:
Servers are expensive and complex

Typical Three-Year Server TCO

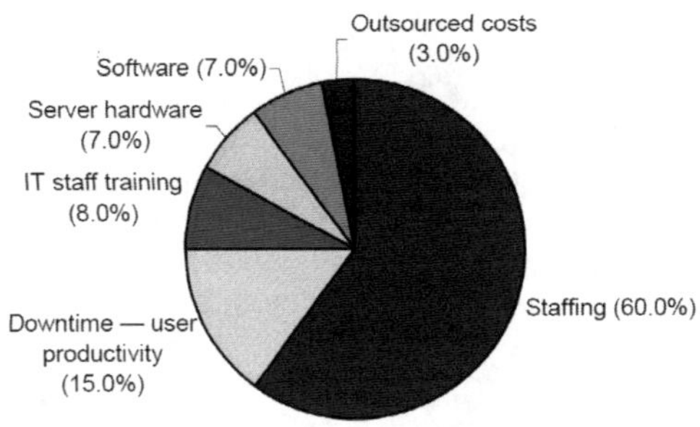

Note: The figure is based on over 300 interviews conducted across numerous platforms, presented in composite form.

Source: IDC, 2007

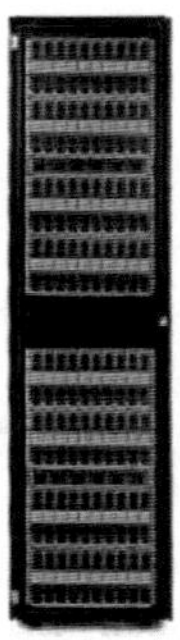

$70,000 server => $500K – $1M cost over 3 years

More challenges with in-house servers

Our 200+ UberCloud simulation experiments exposed the following pain points:

- Long procurement cycles for buying h/w and s/w
- High Total Cost of Ownership (TCO)
- Complex technical software and infrastructure
- Bursty usage
- Rapid technology refresh cycles
- Need for trained server experts.

Benefits of clouds

- More (infinite) computing
- On demand, pay per use, at your fingertips
- Scaling resources dynamically, up and down
- Choice, with multiple providers
- Result: better, faster, cheaper.

But: Cloud Challenges

- Security / your intellectual property
- Traditional software licensing
- Data transfer
- Cloud expertise is needed
- Loosing control over your jobs and data
- First access to clouds is often cumbersome

You want to learn more about Cloud Computing ?

And how it benefits YOUR business ?

Start with ANSYS in a free experiment

or a professional personal hands-on training !

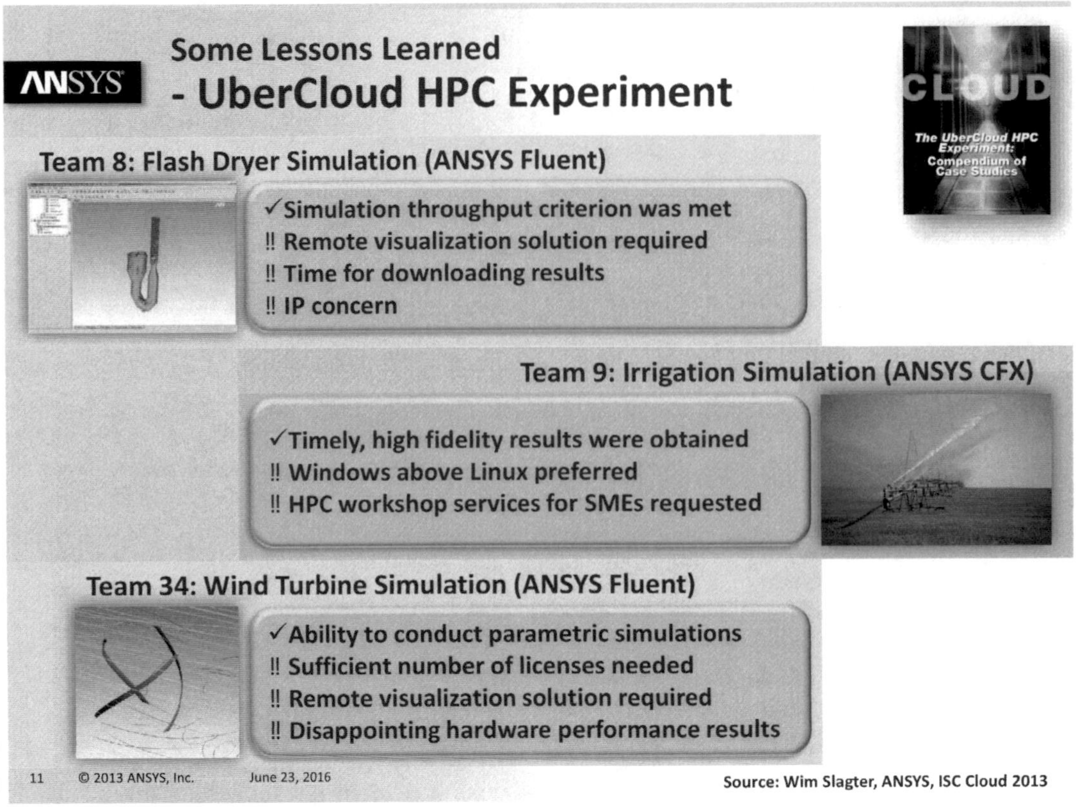
ANSYS
Some Lessons Learned
- UberCloud HPC Experiment
CLOUD
The UberCloud HPC Experiment: Compendium of Case Studies
Team 8: Flash Dryer Simulation (ANSYS Fluent)
✓Simulation throughput criterion was met
!! Remote visualization solution required
!! Time for downloading results
!! IP concern
Team 9: Irrigation Simulation (ANSYS CFX)
✓Timely, high fidelity results were obtained
!! Windows above Linux preferred
!! HPC workshop services for SMEs requested
Team 34: Wind Turbine Simulation (ANSYS Fluent)
✓Ability to conduct parametric simulations
!! Sufficient number of licenses needed
!! Remote visualization solution required
!! Disappointing hardware performance results
11 © 2013 ANSYS, Inc. June 23, 2016
Source: Wim Slagter, ANSYS, ISC Cloud 2013

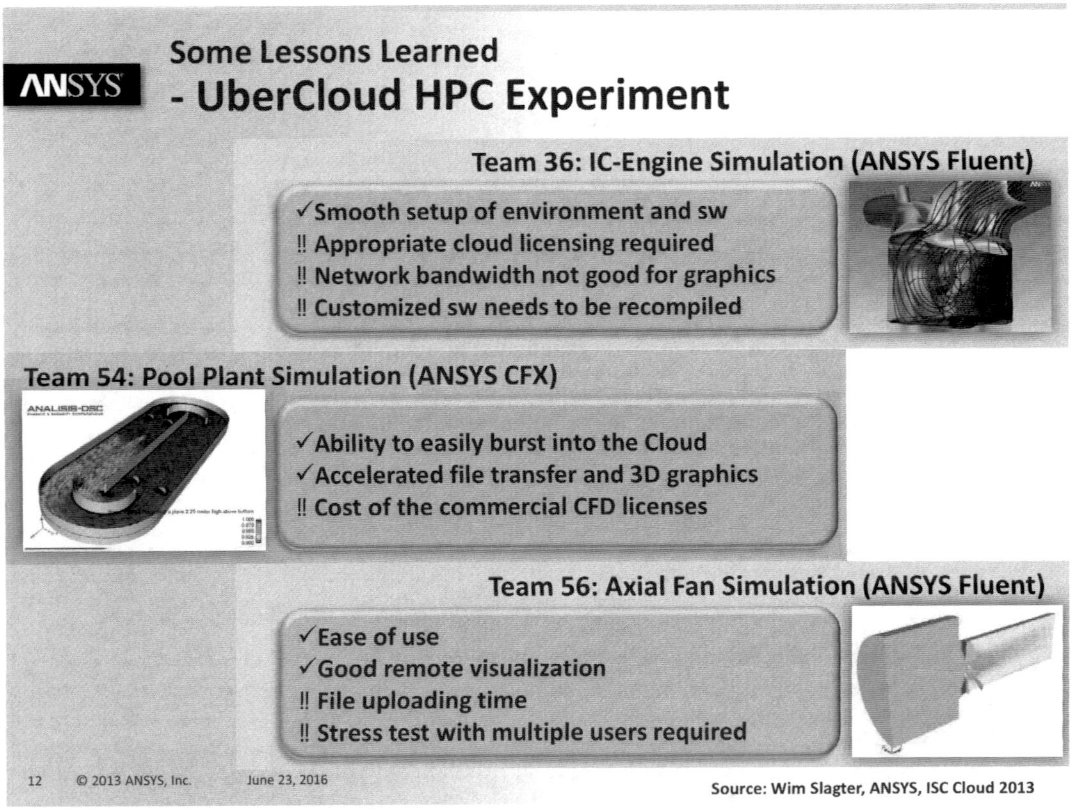
ANSYS
Some Lessons Learned
- UberCloud HPC Experiment
Team 36: IC-Engine Simulation (ANSYS Fluent)
✓Smooth setup of environment and sw
!! Appropriate cloud licensing required
!! Network bandwidth not good for graphics
!! Customized sw needs to be recompiled
Team 54: Pool Plant Simulation (ANSYS CFX)
✓Ability to easily burst into the Cloud
✓Accelerated file transfer and 3D graphics
!! Cost of the commercial CFD licenses
Team 56: Axial Fan Simulation (ANSYS Fluent)
✓Ease of use
✓Good remote visualization
!! File uploading time
!! Stress test with multiple users required
12 © 2013 ANSYS, Inc. June 23, 2016
Source: Wim Slagter, ANSYS, ISC Cloud 2013

Removing CAE Cloud Challenges:

Linux CAE Containers

UberCloud application containers

- Based on Docker, enhanced for engineering & scientific application software like ANSYS
- Software packages designed to deliver the tools that an engineer needs
- Ready to execute, in an instant. No need to install software, deal with complex OS commands, or configure.
- Software is pre-installed, configured, and tested, and running on bare metal, without loss of performance.

Containers: Build once, run anywhere

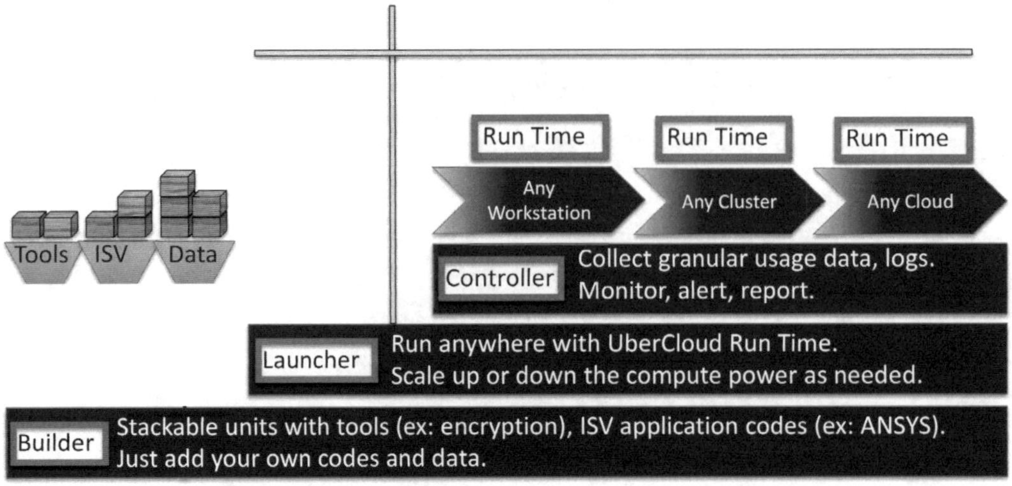

Containers: Build once, run anywhere

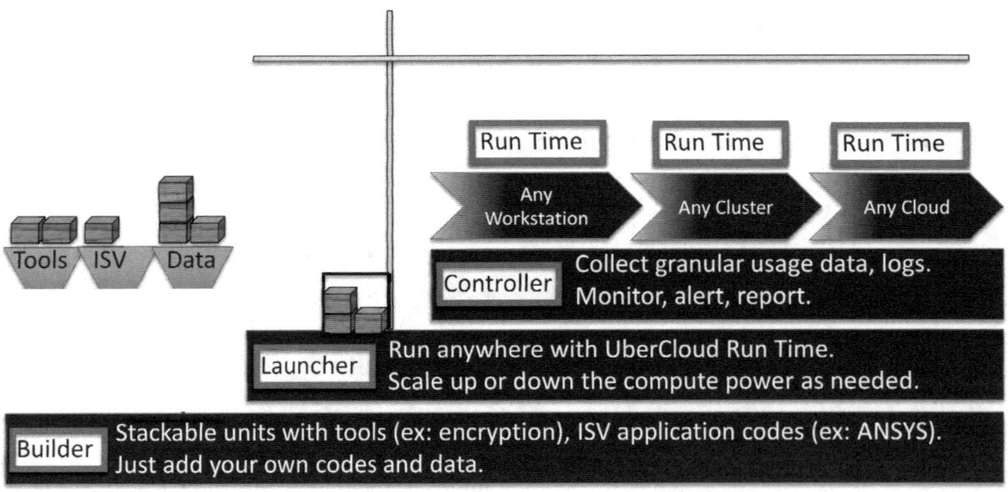

Containers: Build once, run anywhere

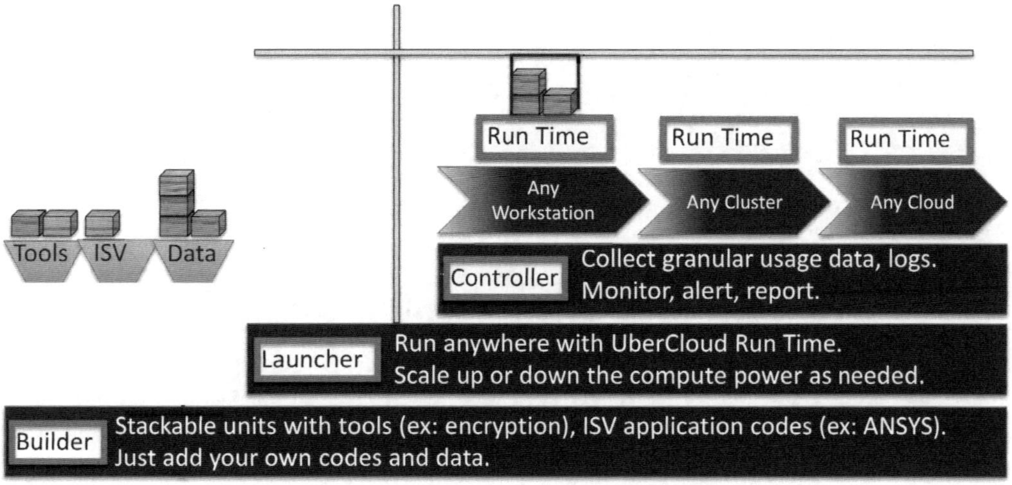

Containers: Build once, run anywhere

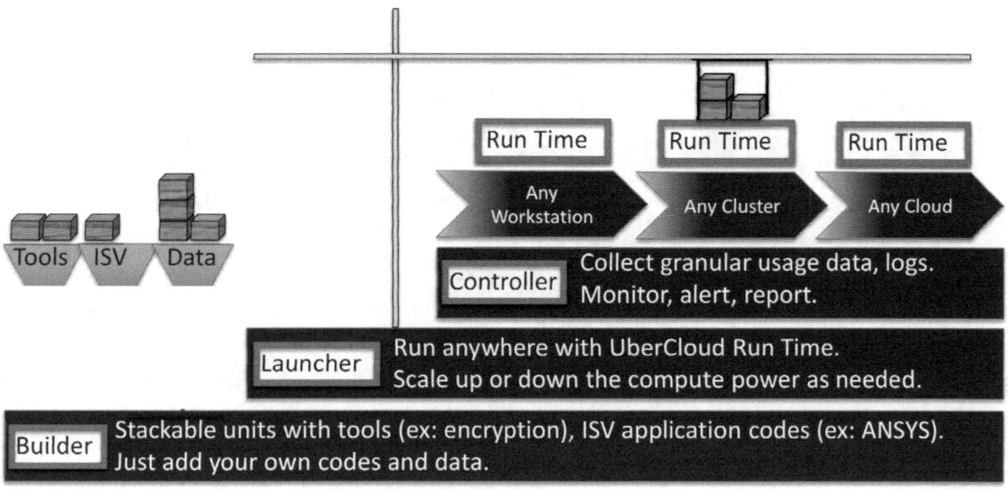

UberCloud Containers: Reducing / Removing Cloud Challenges

CAE Cloud Challenges	UberCloud
Security	√
Portability	√
Compliance	√
Data Transfer	√
Standardization	√
Software Licenses	√
Resource Availability	√
Transparency of Market	√
Cost & ROI Transparency	√
No Cloud Expertise Needed	√

UberCloud Marketplace

Ready for Computing as a Service, for your professional engineering and scientific simulation projects? Need software licenses, computing resources, storage capacity and expert consulting, packaged into one solution stack, available at your fingertips? The UberCloud Marketplace is your one-stop shopping mall to get access to these resources, on-demand, when you need them. And, stay tuned, currently we are working on opening 20+ more marketplace stores; and one could be yours.

Search for products | Search

Software Providers:

ANSYS
ANSYS has pioneered the development and application of simulation methods to solve the most challenging engineering problems. Show 4 products

ANSYS on OzenCloud
Ozen Engineering - the ANSYS Partner & Distributor in California - provides software, training, technical support and consulting services. Show 4 products

Autodesk
A360 project collaboration software helps design, engineering, and project teams work together efficiently on a centralized platform. Show 3 products

CADFEM
The CADFEM Engineering Simulation Cloud provides an easily accessible high-performance CAE environment for ANSYS Workbench. Show 8 products

CAE Fidesys
CAE Fidesys is an easy-to-use and effective next-generation CAE tool for performing a full cycle of engineering-strength analysis. Show 9 products

CD-adapco
STAR-CCM+ is an entirely integrated engineering process for solving problems involving flow (of fluids or solids), heat transfer and stress. Show 5 products

COMSOL - Coming Soon
COMSOL is the developer of COMSOL Multiphysics, an interactive environment for engineering & scientific modeling and simulation. Show 2 products

CST - Computer Simulation Technology
CST offers accurate, efficient computational solutions for 3D electromagnetic design & analysis, operating in a wide range of frequencies. Show 8 products

DYNAmore - Coming Soon
Dynamore supports engineers to solve non-linear mechanical problems with FE solver LS-DYNA and LS-OPT for optimization. Show 2 products

ESI OpenFOAM - Demo Store
World-leading provider in Virtual Product Engineering, leveraging the physics of materials and eliminating the need for physical prototypes. Show 3 products

Next step:
1-day personal hands-on ANSYS Cloud Training

Run your own application and data with ANSYS in the cloud:

- **24 hours, on 32 cores, with 8 hours mentoring and professional services**
- **Mentoring: introduction, training, Q&A, ...**
- **Professional services: support for ANSYS, cloud, and container handling**
- **Case study with challenges, benefits, lessons learned**
- **TCO: Total Cost of Ownership (TCO) analysis comparing in-house server versus cloud usage**

Thank You !

Wolfgang Gentzsch, The UberCloud

Uber Cloud

Automotive Simulation
WORLD CONGRESS

June 7-8, 2016 Munich

Multiphase CFD-Analysis for Filling of Automotive SCR Tank Systems as Basis for Design Optimization

Sandy Meyer, Engineer
Fahrzeug – Entwicklung Sachsen GmbH

Outline

- Introduction
- CFD – Model and Boundary Conditions
- CFD Solver Settings
- Results
- Conclusion

Introduction: The SCR – Technology

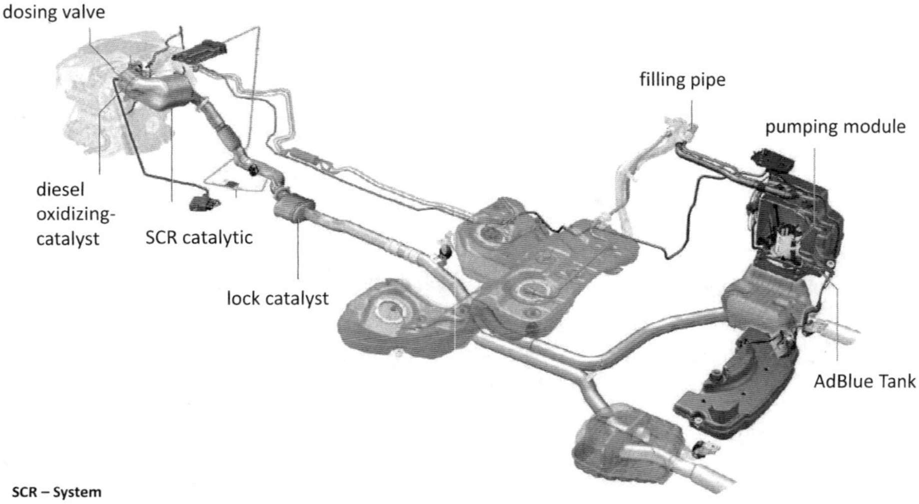

SCR – System

From: Selbststudienprogramm 622 Audi clean Diesel 2.Generation, 2013

Introduction: The SCR – Technology

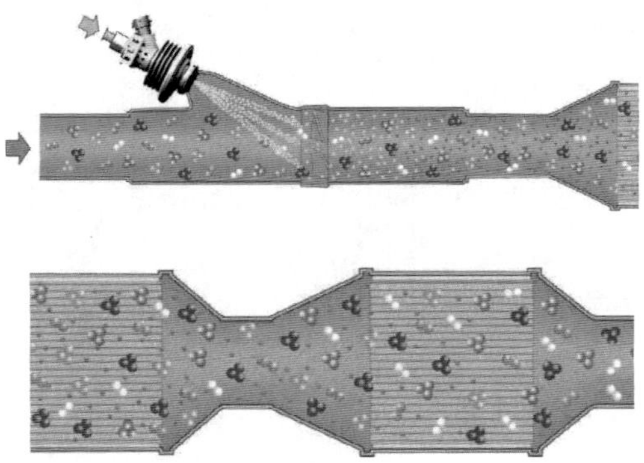

- injection of AdBlue
- reaction to ammonia and carbondioxid
- ammonia reacts with the nitrogen oxides to nitrogen and water in the scr catalytic at high temperature
- the emission, especially the nitrogen oxides, can be reduced

From: Selbststudienprogramm 446 Die 2,5l-TDI-EURO-V-Motoren mit SCR-System im Crafter, 2009

Introduction: Technical Standards

NEFZ		On the road
0.8 l per 1.000 km Service-intervall: 15.000 or 30.000 km **Refill beyond service-intervall possible (theoretical)**	→	1.5 l per 1.000 km Service-intervall: 15.000 or 30.000 km **Refill after 8.000 or 16.000 km necessary**

Introduction: Technical Standards

Current

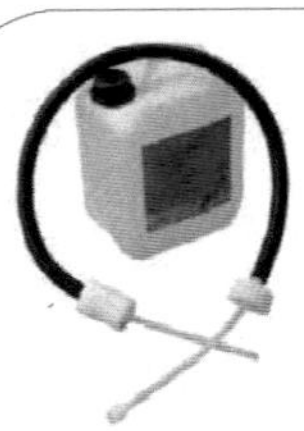

Refillcan

Kruse-bottle

Refill time SCR tank: > 8 min
(13 bottles, 38 s per bottle)

Volume flow rate: 3l/min

→

Future

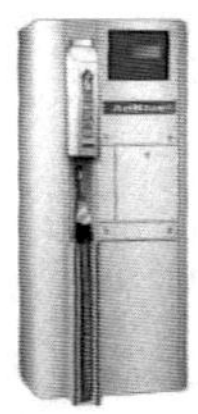

pump nozzle (truck)

Refill time SCR tank: 37 s

Volume flow rate: max. 40l/min

CFD – Model and Boundary Conditions

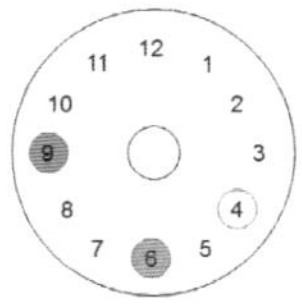

- ANSYS Fluent R16.1
- SCR – Tank with capacity of 25 l
- Volume flow rate: max. of 40 l/min (extrem conditions)
- diffent positions of pump nozzle
- comparison to test

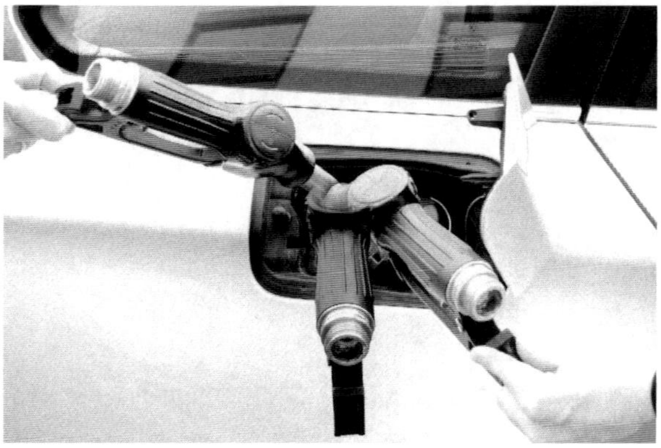

CFD – Model and Boundary Conditions

- polyhedron-mesh
- 3-Layer-Model of the pump nozzle, filling pipe and vent pipe
- different element's edge length (0.4 – 6 mm)
- 150.000 shell elements
- 700.000 finite volumes

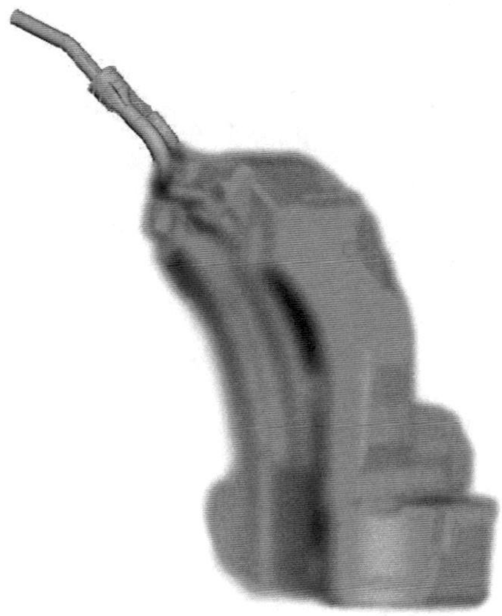

CFD – Model and Boundary Conditions

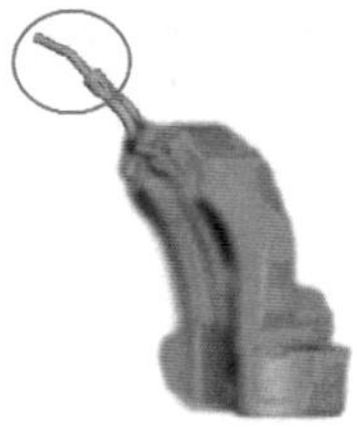

mass-flow-inlet: 0.7267 kg/s (volume flow rate 40 l/min, AdBlue)

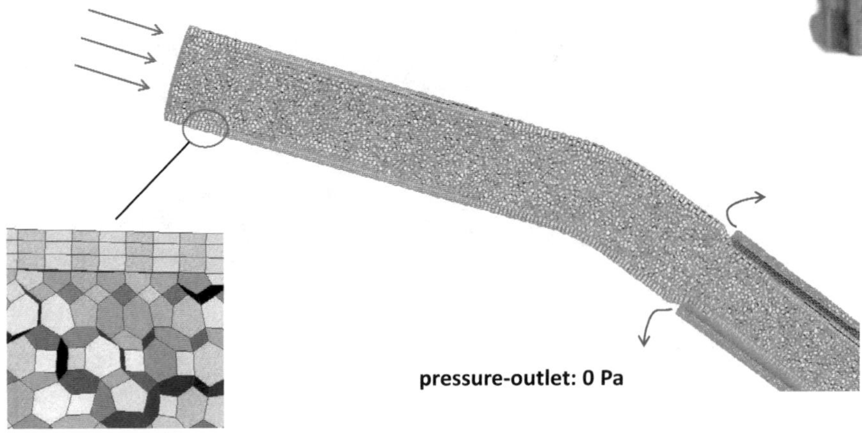

CFD Solver Settings

- transient
- pressure-based solver
- Volume-of-Fluid (VOF)
- primary phase: AdBlue
- secondary phase: Air
- RLZ-k-ε-Model
- implicit

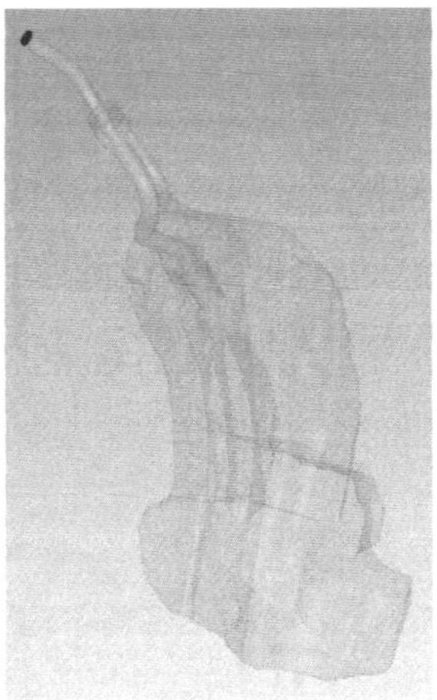

- initialization: standard with air volume fraction 1
- SIMPLE – algorithm
- time step: 10E-3 s
- maximum of 5 iterations per time step

CFD Solver Settings

Explicit Solution	Implicit Solution

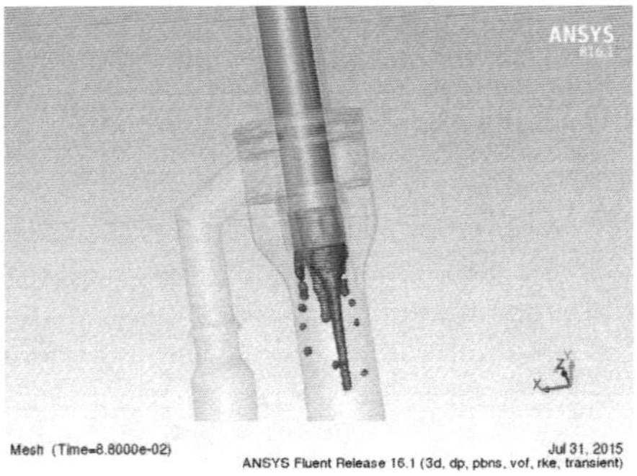

- divergence at further calculation
→ variable time step or adapt mesh

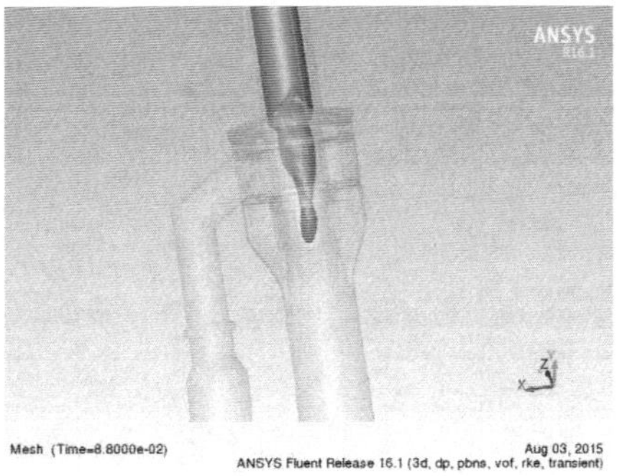

- convergence
→ fixed time step and adequately computing time

Results

- volume flow rate: 40 l/min
- mass flow fraction > 5%
- simulation time: 36 s
- pump nozzle in 6h position
- computing time: 36 h with 32 CPUs

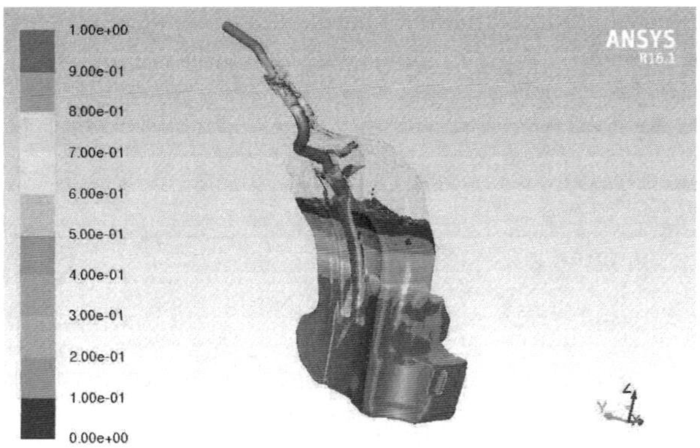

Contours of Volume fraction (adblue) (Time=3.0000e+01) Oct 26, 2015
ANSYS Fluent Release 16.1 (3d, dp, pbns, vof, rke, transient)

Results

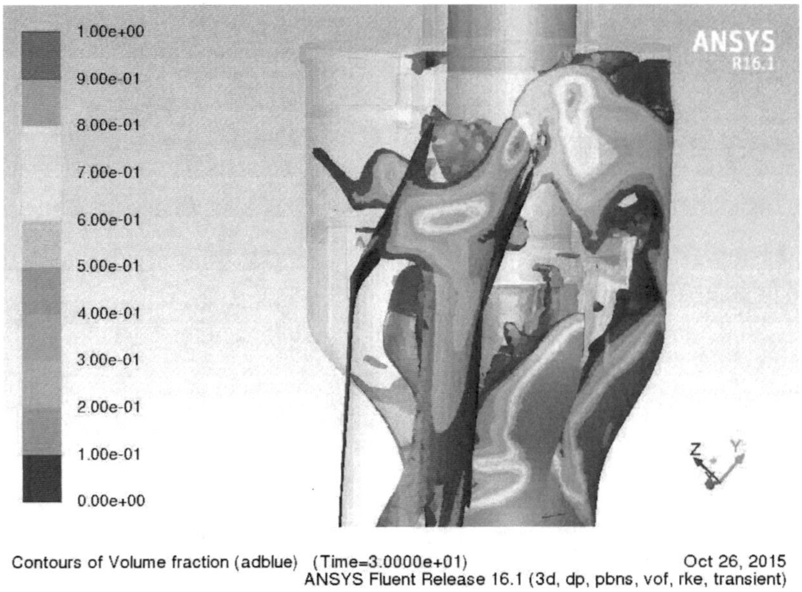

Results

- **Results of pump nozzle in 6h position with different volume flow rates**

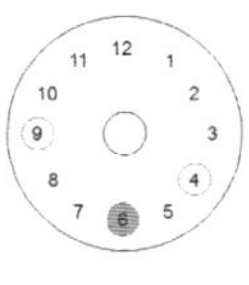

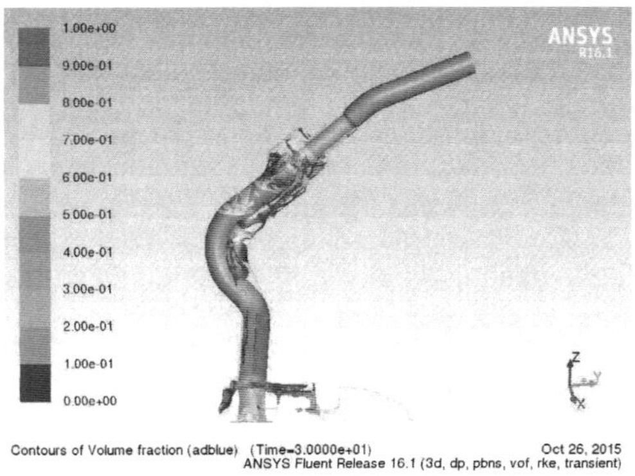

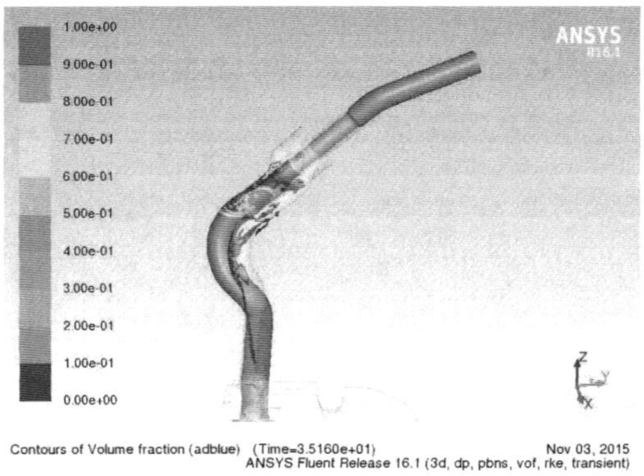

volume flow rate: 40 l/min **volume flow rate: 35 l/min**

Results

- **Analysis of Pre – Shot Off**

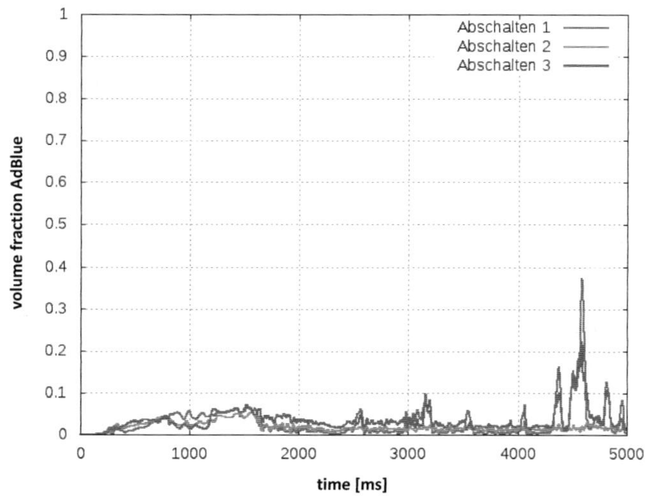

- Pre – Shot Off: 4.500 ms (= 3l)

Results

- **Validation**

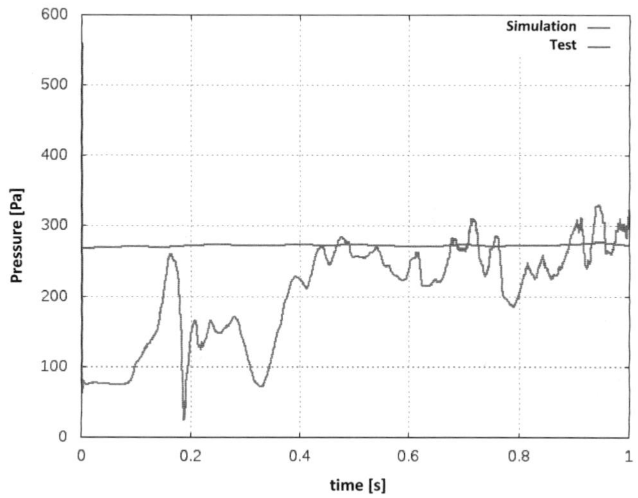

- measuring interval simulation: 1 ms
- measuring interval test: 40 ms
- differences due to the number of data points

Conclusion

- cooperation of test – simulation – construction
- deduction of suitable steps
- implementation of design changes
- reduction of costs for the production of STL-parts and for tests

Automotive Simulation
WORLD CONGRESS

Automotive Simulation

Thanks for your Attention!

Automotive Simulation
WORLD CONGRESS

June 7-8, 2016 Munich

Sealing for pumps and turbines
- simulation-driven development of next-generation mechanical seals -

Hideyuki Inoue, Head of Research Department
EAGLE INDUSTRY CO., LTD.

TOPICS

- About EAGLE INDUSTRY
- Our Products
- About Mechanical Seals
  - Conventional Mechanical Seal
  - Next-Generation Mechanical Seal
- Calculation and Results
  - Motivation
  - Mesh Generation
  - Steady Analysis
  - Transient Analysis
  - Results
- Future Plan

About EAGLE INDUSTRY

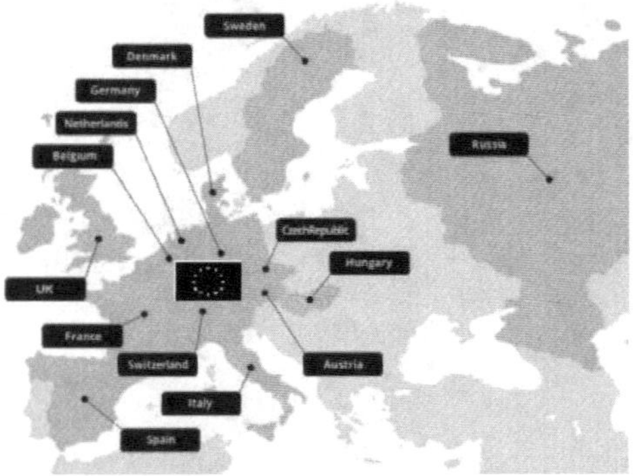

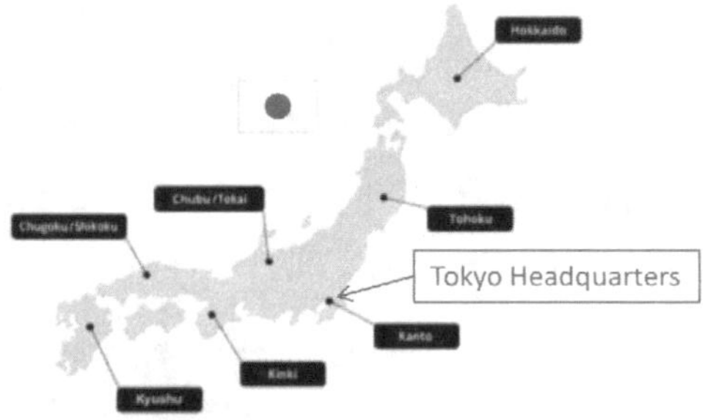

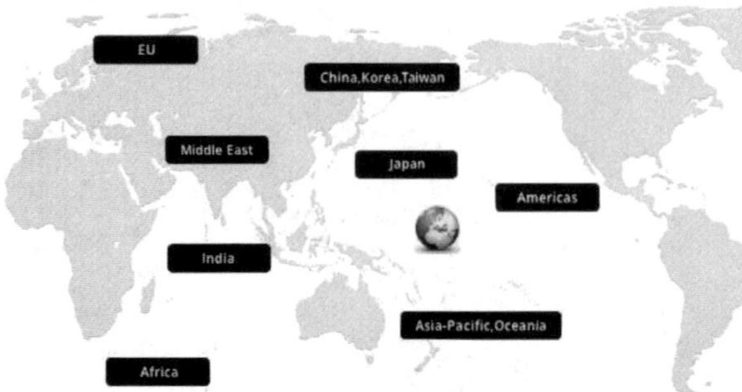

About EAGLE INDUSTRY

EKK R&D Center, Saitama Plant

About EAGLE INDUSTRY

Sales

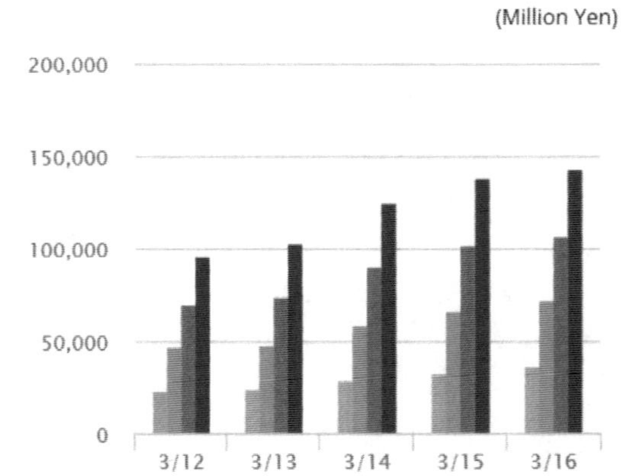

Business Segment Information

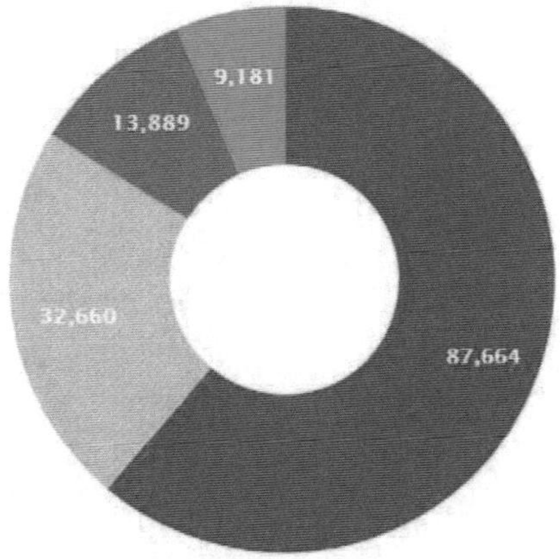

■ For Automotive and Construction Machinery Industries ■ For General Industrial Machinery Industry
■ For Marine Industry ■ For Aerospace and Optoelectronics Industries

Automotive Simulation WORLD CONGRESS

Products

① Petrochemical, Steel, Pharmaceutical, Water, Pulp & Paper, Food

Dry gas seals

② Semiconductor

Magnetic fluid seals

③ Automobile, Construction Machinery

Lip seals for automotive air-conditioner compressors

④ Housing Equipment

Valves for spray toilets

⑤ Aerospace

Aircraft engine seals

⑥ Energy

Non-contact seals

⑦ Sensing & Imaging

Sensing and imaging products

⑧ Marine

Oil lubricated stern tube seals

Automotive Simulation WORLD CONGRESS

Conventional Mechanical Seal

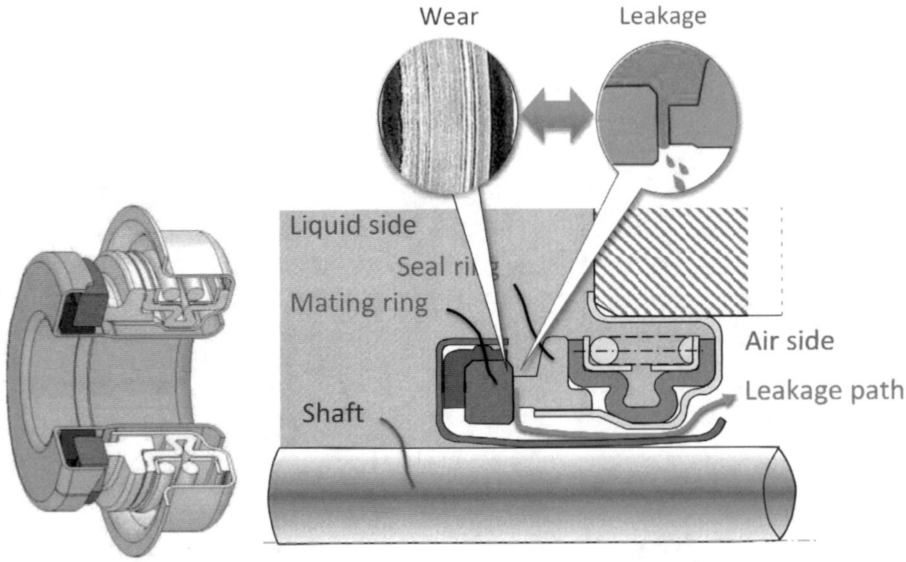

- Both sealing and lubricating functions need to be realized.

Conventional Mechanical Seal

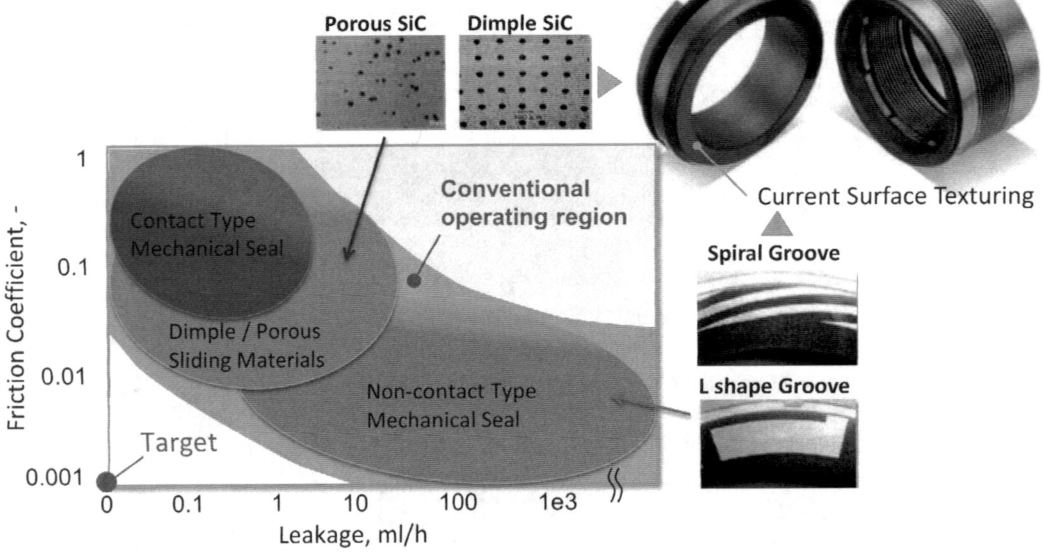

The market wants "low-friction" and "low-leakage" mechanical seal.
⇒We proposed a new surface structure of mechanical seal in order to maximum both the functions.

Next-Generation Mechanical Seals

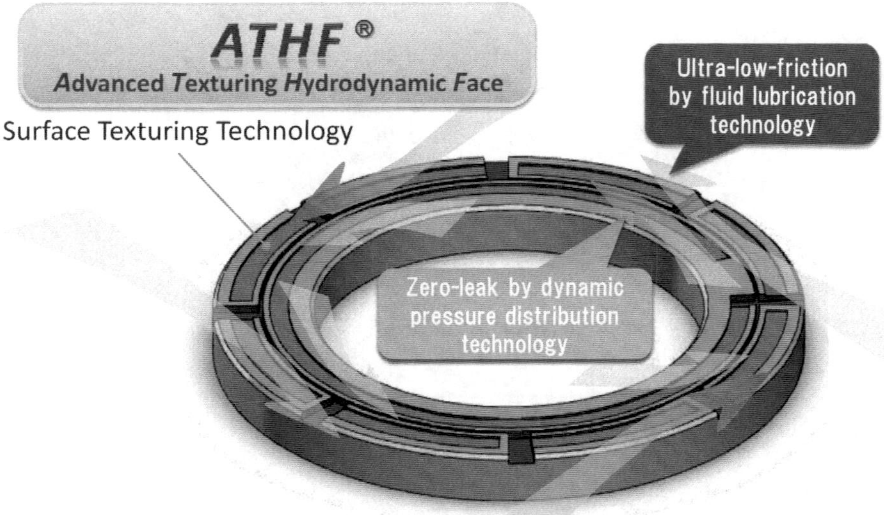

- It's available ultra-low-friction reduced 98% to existing face by a fluid lubrication technology.
- It's available both functions of sealing and lubricating by dynamic pressure distribution control technology.

Next-Generation Mechanical Seals

- Fundamental equation of seal leakage

$$q = \frac{r\omega_1^2 h^3}{40\nu} - \frac{h^3}{12\eta}\frac{\partial p}{\partial r}\bigg|_{r=r_i} + \frac{V_{\theta 1}}{2}h$$

Centrifugal force term — Pressure flow term — Shear flow term

h: Film thickness
$\partial p/\partial r$: Pressure gradient at radial direction
η: Viscosity

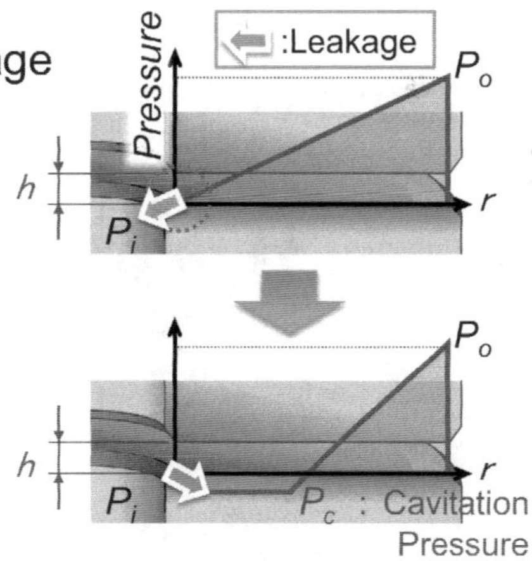

- If we can control the $\partial p/\partial r$ to be zero or negative, the seal leakage could be zero or even realize the pumping effect.

➡ The key factor is the "controlling of pressure distribution" by surface texturing.

Next-Generation Mechanical Seals

- Fundamental structure of sliding surface

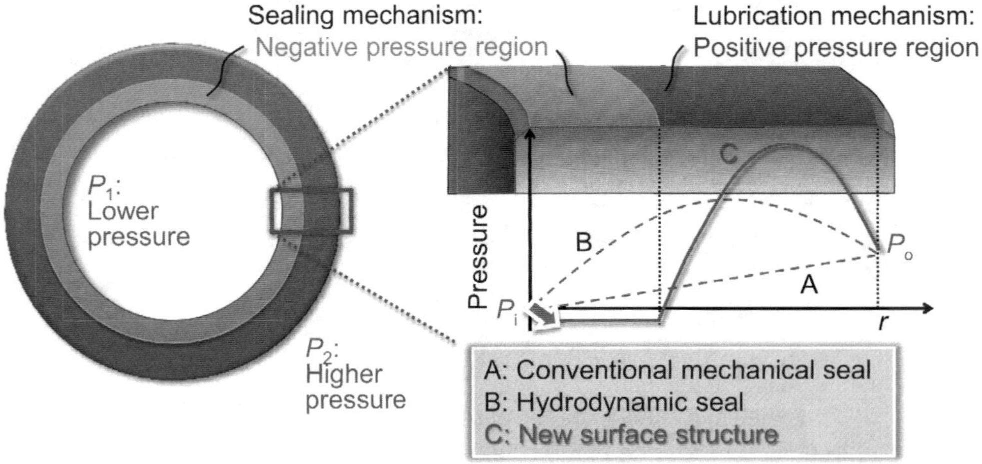

Next-Generation Mechanical Seals

- Fundamental structure of sliding surface

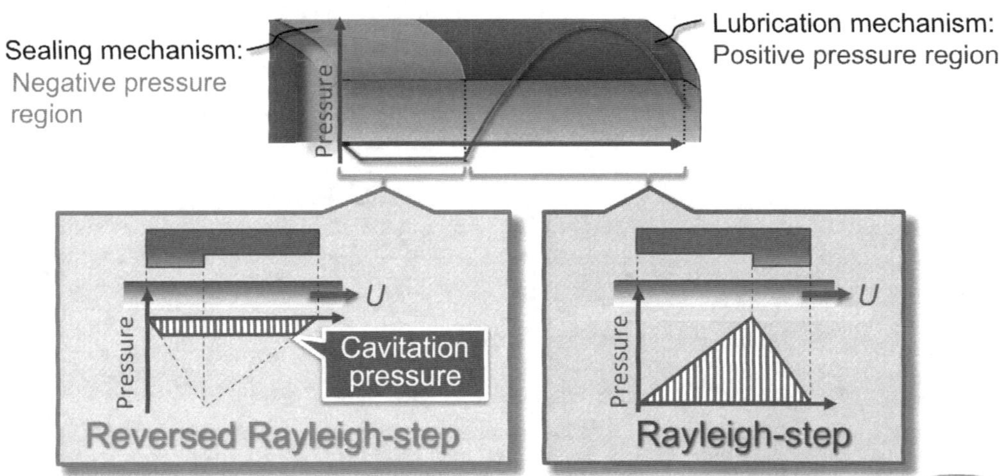

Next-Generation Mechanical Seals

- Experimental conditions

Experimental conditions	
Material	SiC x Glass
Load	28 N
Inner/outer dir.	19.5mm/27mm
Sliding velocity	1.8m/s
Sealed fluid	Distilled water
Fluid temp.	Room temp.
Pressure	0.15MPaG

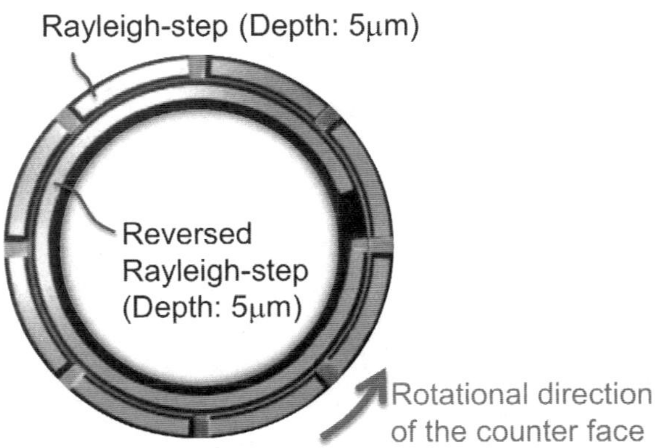

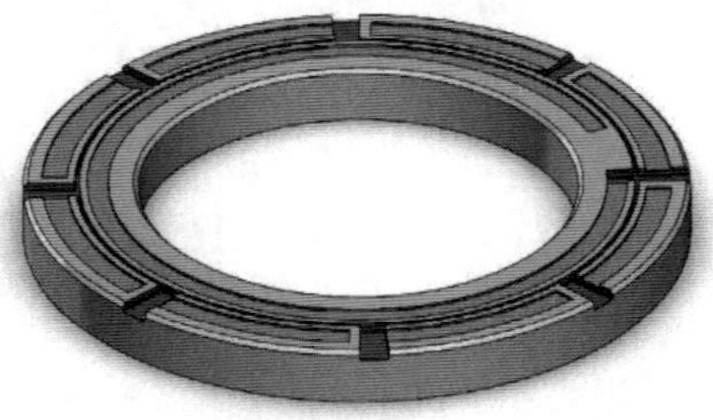

Next-Generation Mechanical Seals

- Cavitation and Pumping effect

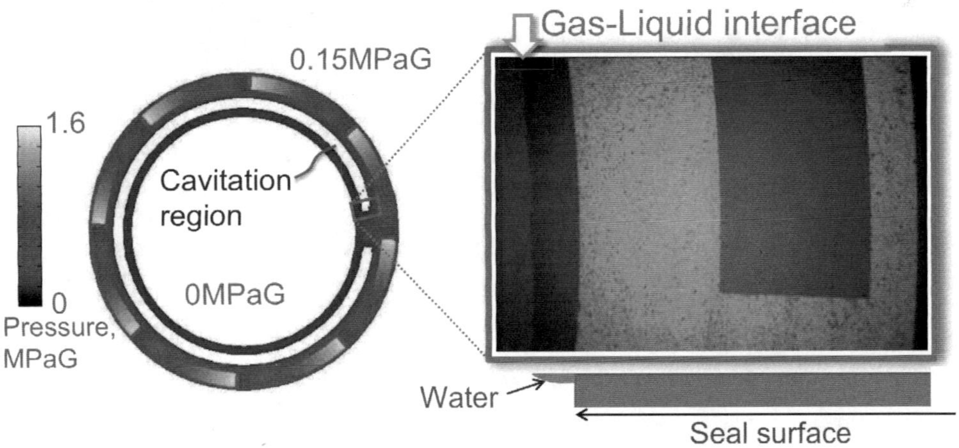

- The pumping effect have been observed.

Next-Generation Mechanical Seals

▶ Friction coefficient

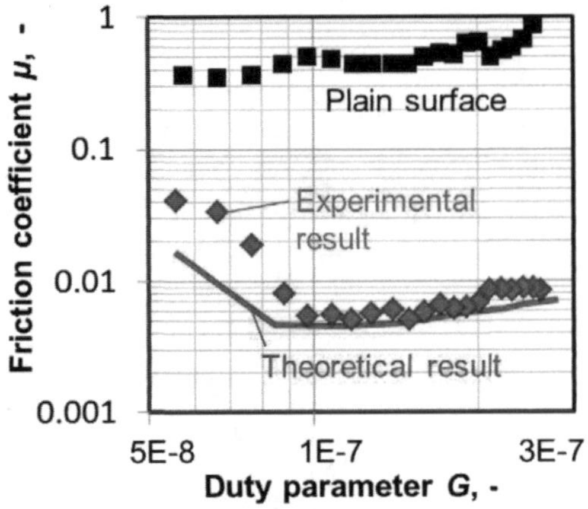

- The minimum friction coefficient of the seal shows 0.005.
- Almost 99% down in friction coefficient compared with the plain surface.

Calculation and Results

▶ Motivation

- Confirm Zero Leakage
- Optimize Groove Depth

Calculation and Results

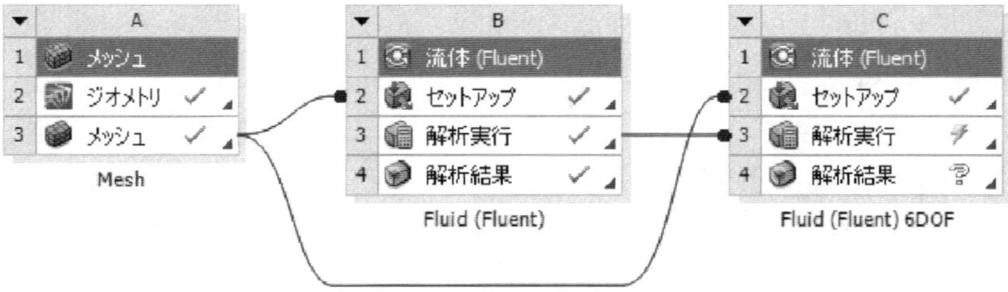

Mesh Generation — Space Claim — ANSYS Mesh

Steady Analysis — Fluent

Transient Analysis — Fluent — Dynamic Mesh (6DOF)

Calculation and Results

Mesh Generation

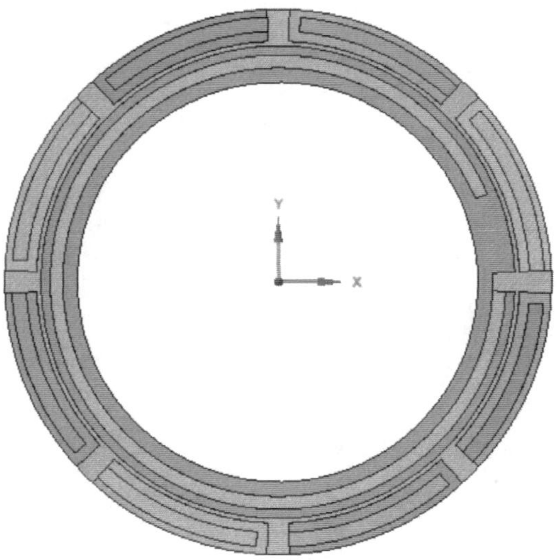

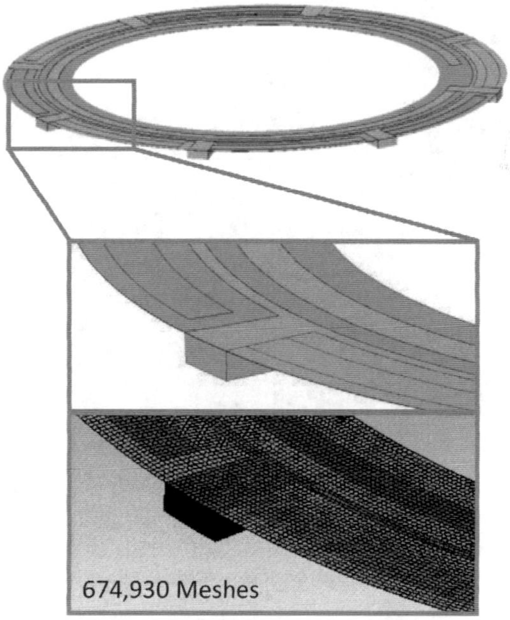

Calculation and Results

Steady Analysis

Analysis Conditions	
Filem Thickness	10µm
Inner/Outer Dia.	22mm/30mm
Sliding Velocity	5,250rad/s
Fluid	Water/Vapor
Inlet Pressure	0.15MPaG
Outlet Pressure	0.0MPaG

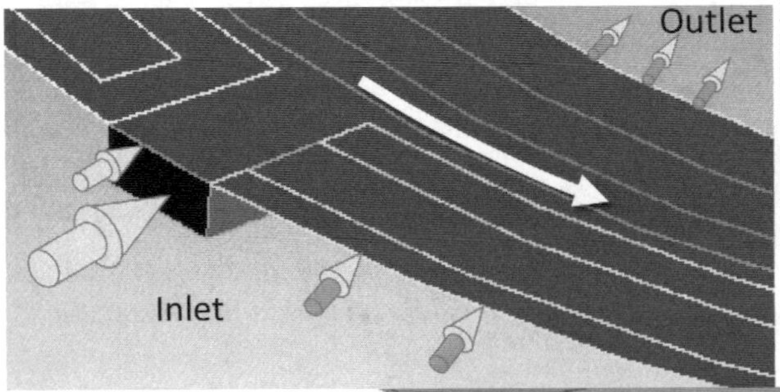

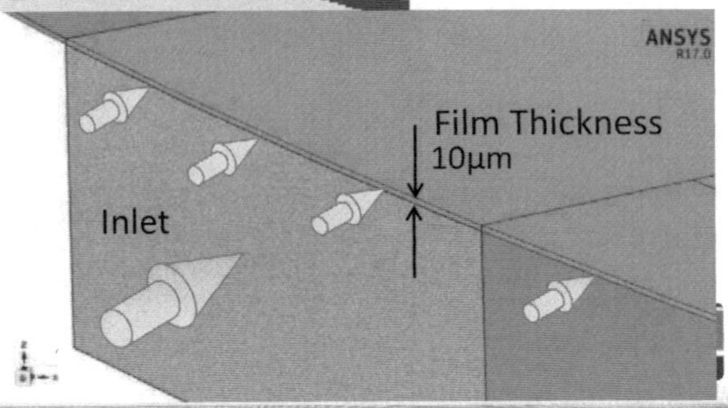

Calculation and Results

Transient Analysis (6DOF)

Analysis Conditions	
Load	50N, 72N, 90N, 120N
Inner/Outer Dia.	22mm/30mm
Sliding Velocity	5,250rad/s
Fluid	Water/Vapor
Inlet Pressure	0.15MPaG
Outlet Pressure	0.0MPaG

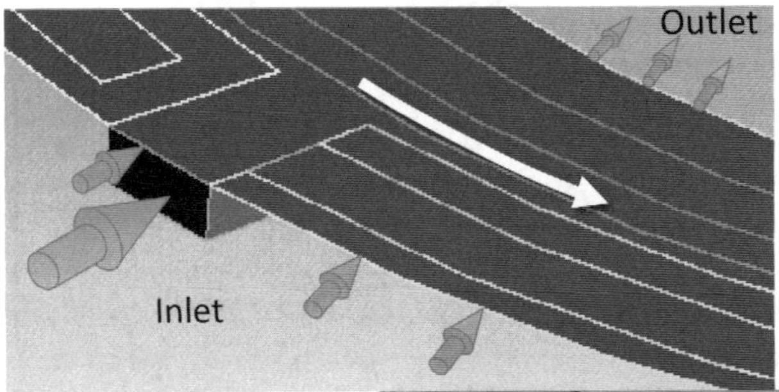

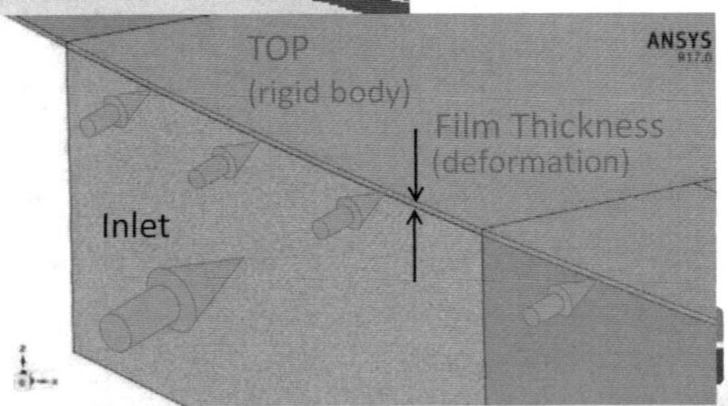

Calculation and Results (Transient Analysis)

Pressure Distribution

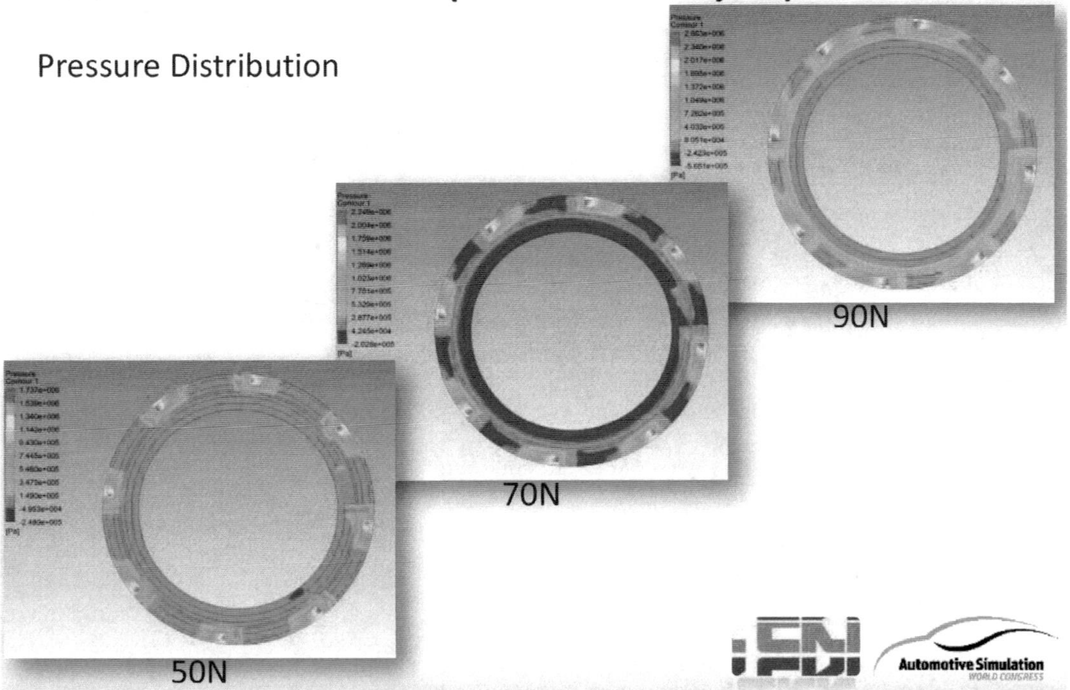

Automotive Simulation
WORLD CONGRESS

21 June 7-8, 2016 Munich

Calculation and Results (Transient Analysis)

Phase Fraction

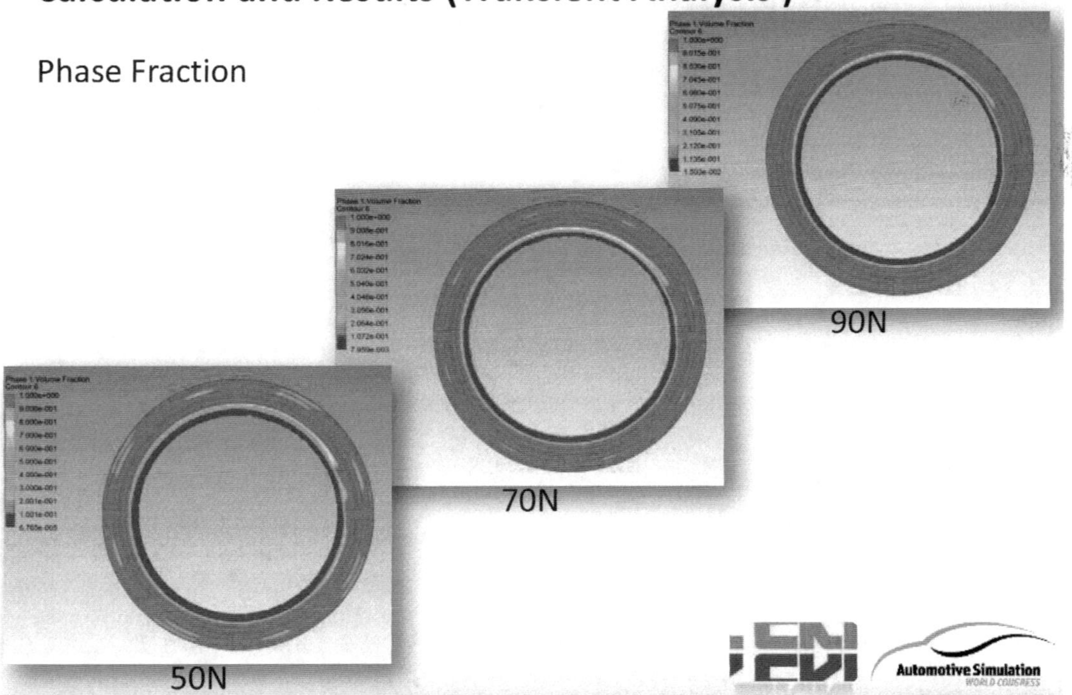

Automotive Simulation
WORLD CONGRESS

22 June 7-8, 2016 Munich

Calculation and Results (Transient Analysis)

Fluid Velocity

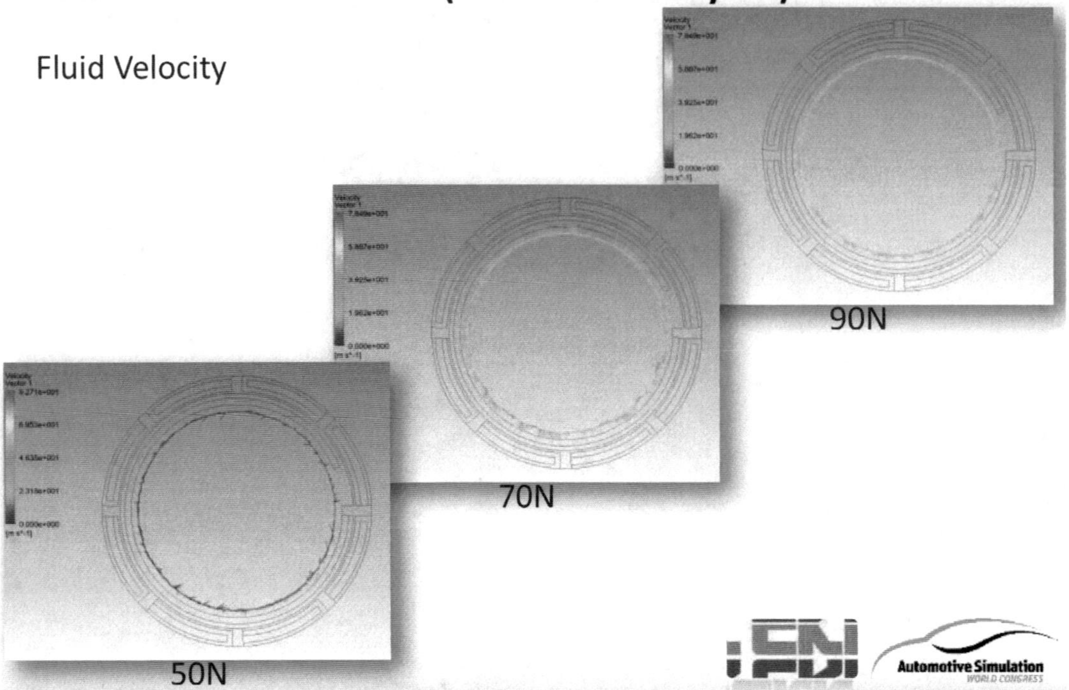

Calculation and Results (Transient Analysis)

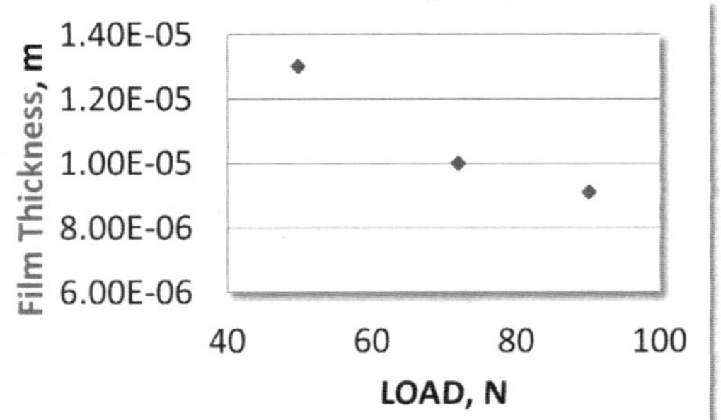

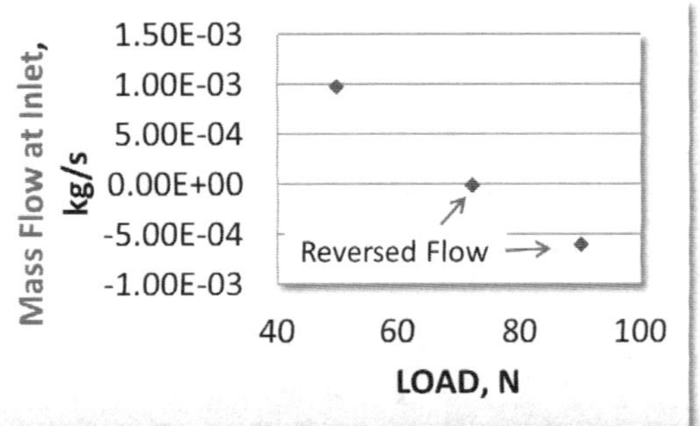

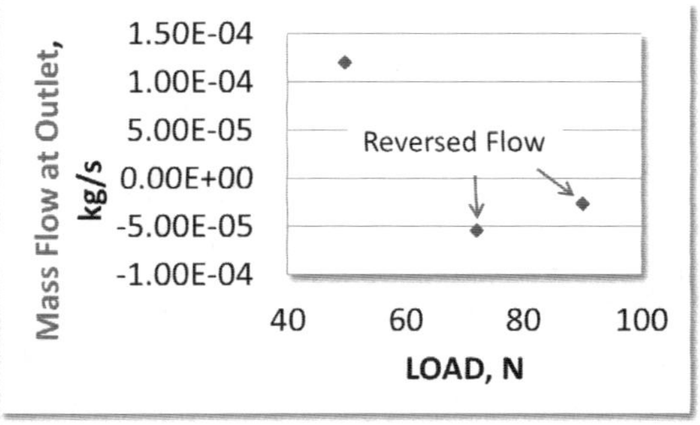

We achieved zero-leakage!

Future Plan

Optimization

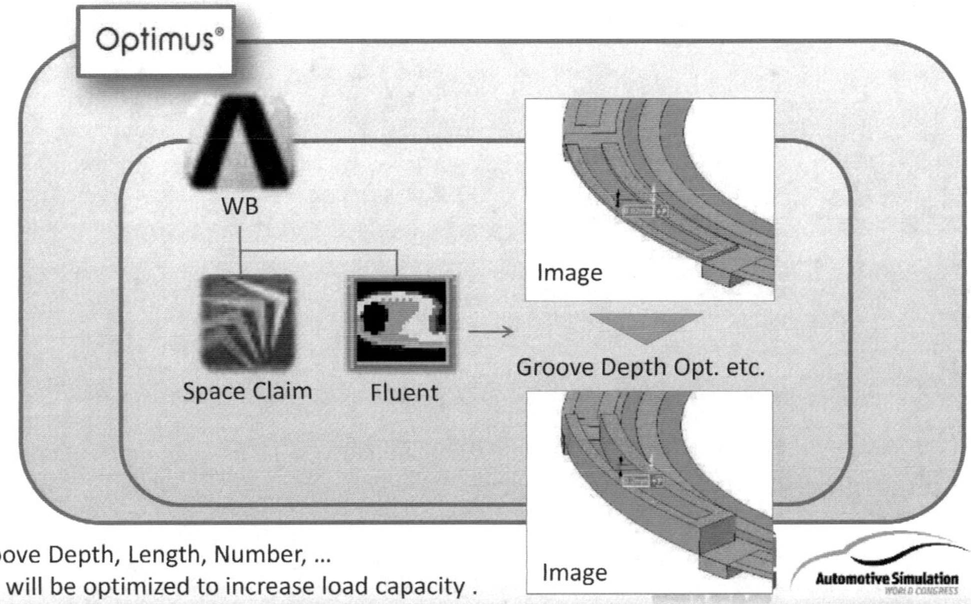

Groove Depth, Length, Number, ...
will be optimized to increase load capacity .

THANK YOU FOR YOUR ATTENTION

Efficient simulation of boiling effects at engine cooling jackets

by Andreas Boemer and Philipp Morsch, DEUTZ AG

The engine company.

Automotive Simulation WORLD CONGRESS

June 7-8, 2016 Munich

DEUTZ

Efficient simulation of boiling effects at engine cooling jackets
Boiling

The engine company. DEUTZ

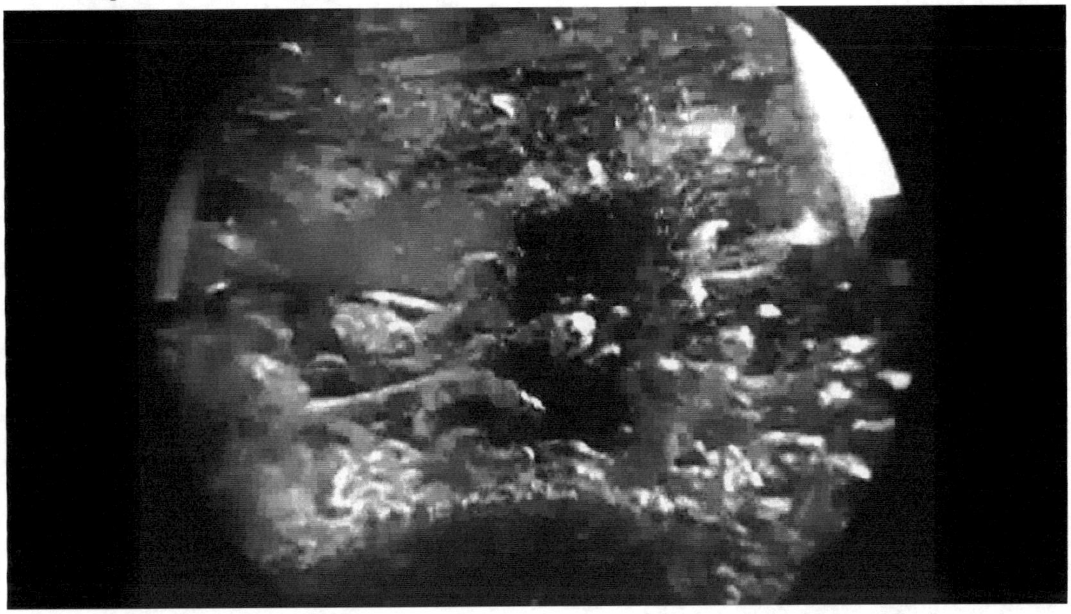

Source: www.youtube.com/watch?v=KgykbVBnxKk

Efficient simulation of boiling effects at engine cooling jackets
Engine cooling jacket

The engine company. DEUTZ

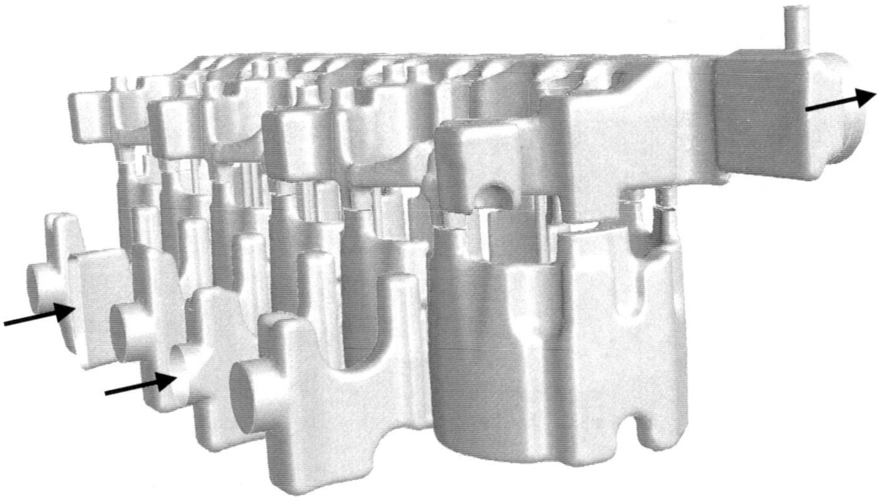

VE-TT3 · Boemer & Morsch 04.05.2016

3

Efficient simulation of boiling effects at engine cooling jackets
Engine cooling jacket

The engine company. DEUTZ

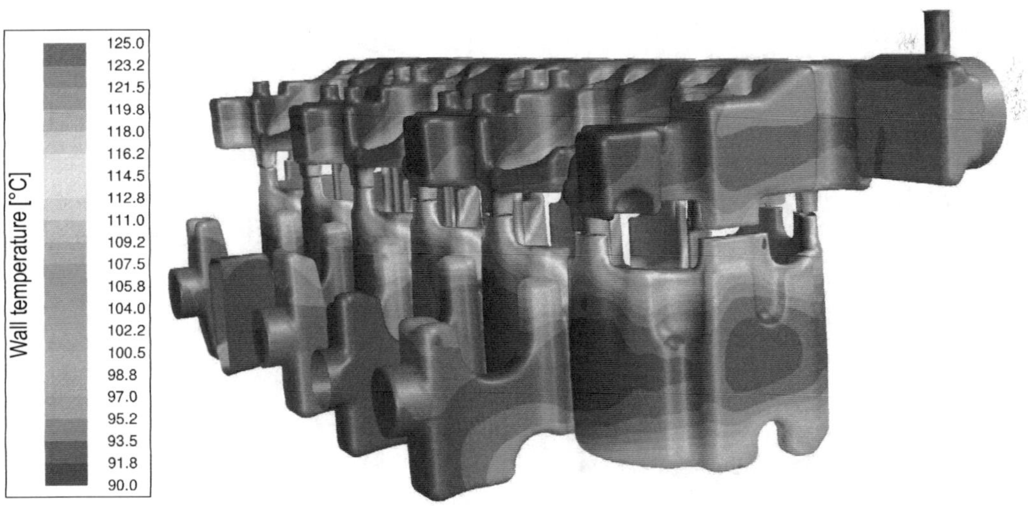

VE-TT3 · Boemer & Morsch 04.05.2016

4

Efficient simulation of boiling effects at engine cooling jackets
Engine cooling jacket CFD simulation – no pain, no gain

The engine company. DEUTZ

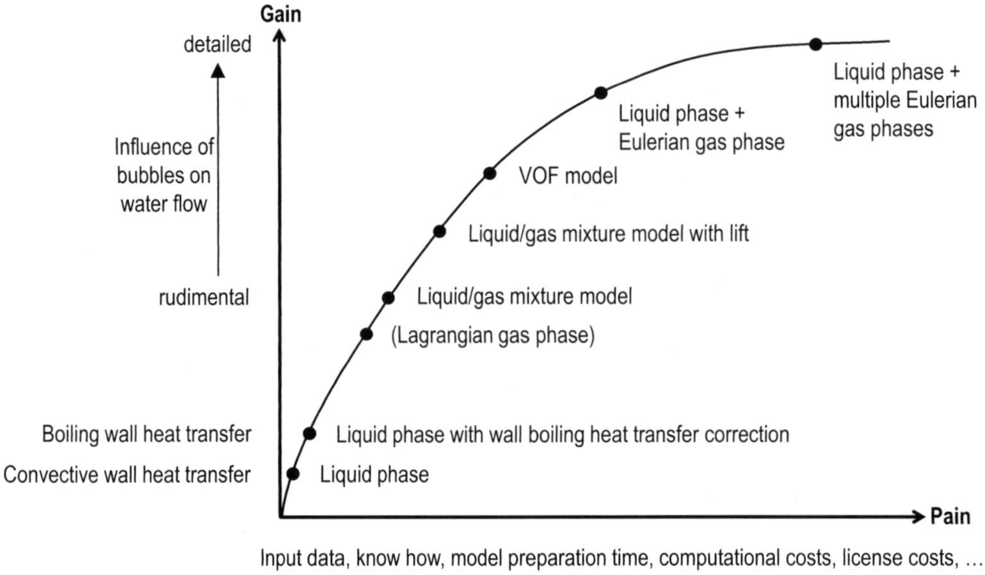

VE-TT3 · Boemer & Morsch 04.05.2016

5

Efficient simulation of boiling effects at engine cooling jackets
Engine cooling jacket CFD simulation – no pain, no gain

The engine company. DEUTZ

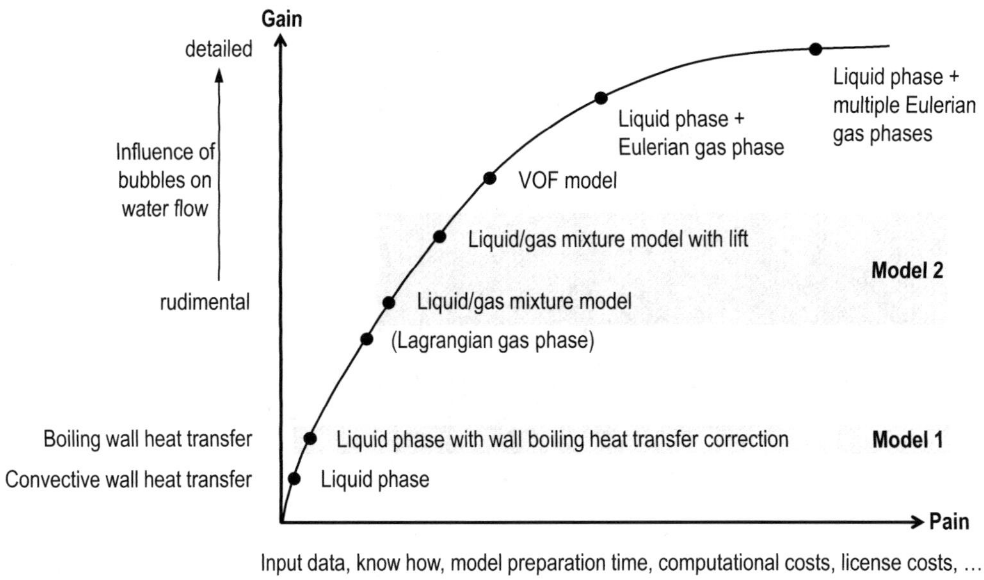

VE-TT3 · Boemer & Morsch 04.05.2016

6

Efficient simulation of boiling effects at engine cooling jackets
Model 1 – Single liquid phase

The engine company. DEUTZ

Wall heat transfer coefficient

$\alpha = \alpha_{conv} + \alpha_{boil}$

from standard FLUENT simulation

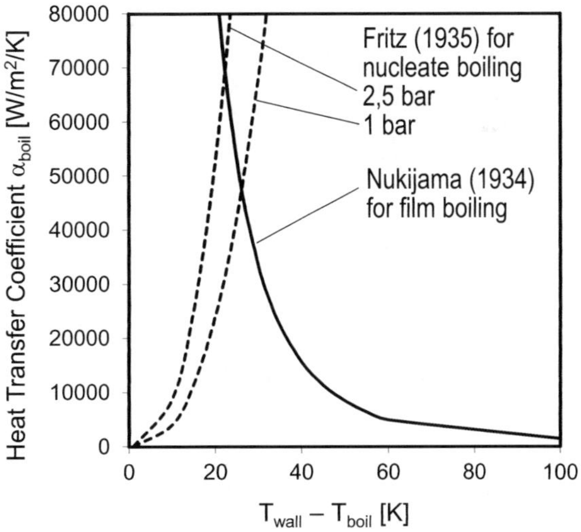

Efficient simulation of boiling effects at engine cooling jackets
Model 1 – Single liquid phase

The engine company. DEUTZ

Wall heat transfer coefficient

$\alpha = \alpha_{conv} + \alpha_{boil}$

from standard FLUENT simulation

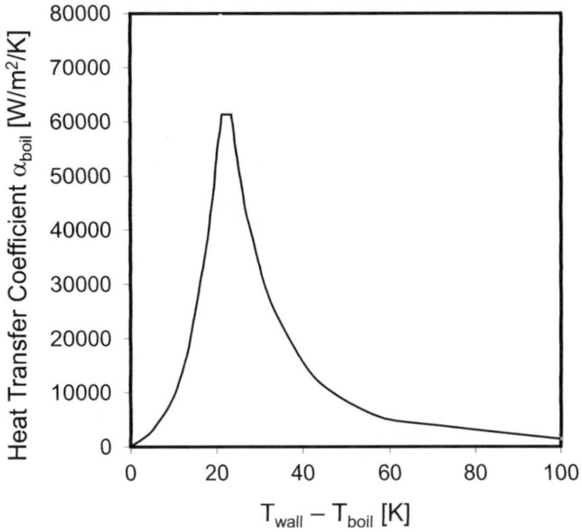

Efficient simulation of boiling effects at engine cooling jackets
Model 2 – Liquid/gas mixture

The engine company. DEUTZ

- **Standard FLUENT Mixture Model with calculation of**
  - gas volume fraction, drag and lift
  - evaporation, condensation in fluid cells
  - convective wall heat transfer
- **Wall boiling model Chen (1966)**
  - enhancement of convective wall heat transfer by bubble movement
  - additional boiling wall heat transfer
  - evaporation at hot walls

Efficient simulation of boiling effects at engine cooling jackets
Model 2 – Liquid/gas mixture

The engine company. DEUTZ

- **Standard FLUENT Mixture Model with calculation of**
  - gas volume fraction, drag and lift
  - evaporation, condensation in fluid cells
  - convective wall heat transfer
- **Wall boiling model Chen (1966)**
  - enhancement of convective wall heat transfer by bubble movement
  - additional boiling wall heat transfer
  - evaporation at hot walls

„Semi Mechanistic Boiling Model"
Punekar and Das (2013)
UDF provided by ANSYS, Inc.

Efficient simulation of boiling effects at engine cooling jackets
Model 2 – Liquid/gas mixture

The engine company. DEUTZ

- **Standard FLUENT Mixture Model with calculation of**
  - gas volume fraction, drag and lift
  - evaporation, condensation in fluid cells
  - convective wall heat transfer
- **Wall boiling model Chen (1966)**
  - enhancement of convective wall heat transfer by bubble movement
  - additional boiling wall heat transfer
  - evaporation at hot walls, distribution of steam mass to multiple boundary layers:

„Semi Mechanistic Boiling Model“
Punekar and Das (2013)
UDF provided by ANSYS, Inc.

DEUTZ/RWTH Aachen
Bachelor Project
Morsch (2016)

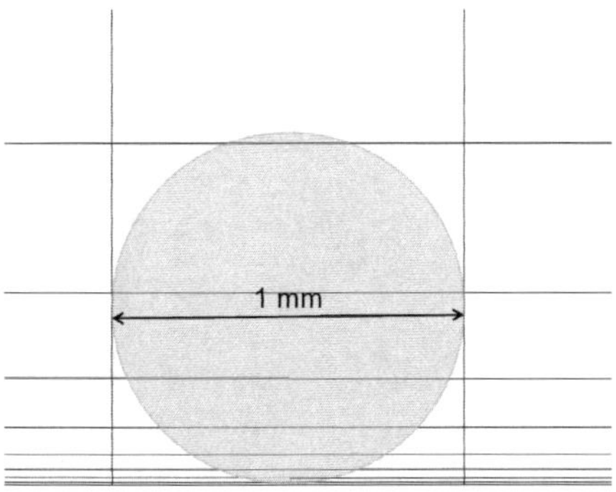

D_{bubble} ~ 1 mm
(Krepper et al., 2007)

VE-TT3 · Boemer & Morsch 04.05.2016

11

Efficient simulation of boiling effects at engine cooling jackets
Model 2 – model parameters to adjust

The engine company. DEUTZ

- **Evaporation/Condensation**
  - at walls

$$\dot{q}_{wall} = AF \cdot \dot{q}_{conv} + \dot{q}_{boil}$$

↗ steam generation $HTV \cdot \dot{q}_{boil}$

↘ water temperature rise $(1 - HTV) \cdot \dot{q}_{boil}$

$$\dot{m}_{evap,wall} = \frac{HTV \cdot \dot{q}_{boil}}{h_{fg}}$$

AF „Augmentation Factor“

HTV „Heat To Vapour“

  - in fluid cells

$$\dot{m}_{evap,fluid} = C_{Evap} X_f \rho_f \frac{T_f - T_{boil}}{T_{boil}}$$

$$\dot{m}_{cond,fluid} = C_{Cond} X_g \rho_g \frac{T_{boil} - T_g}{T_{boil}}$$

VE-TT3 · Boemer & Morsch 04.05.2016

12

Efficient simulation of boiling effects at engine cooling jackets
Model 2 – model parameters to adjust

The engine company. DEUTZ

- **Evaporation/Condensation**
  - at walls

  $HTV = 0.1$

  - in fluid cells

  $C_{Evap} = 1$
  $C_{Cond} = 1$

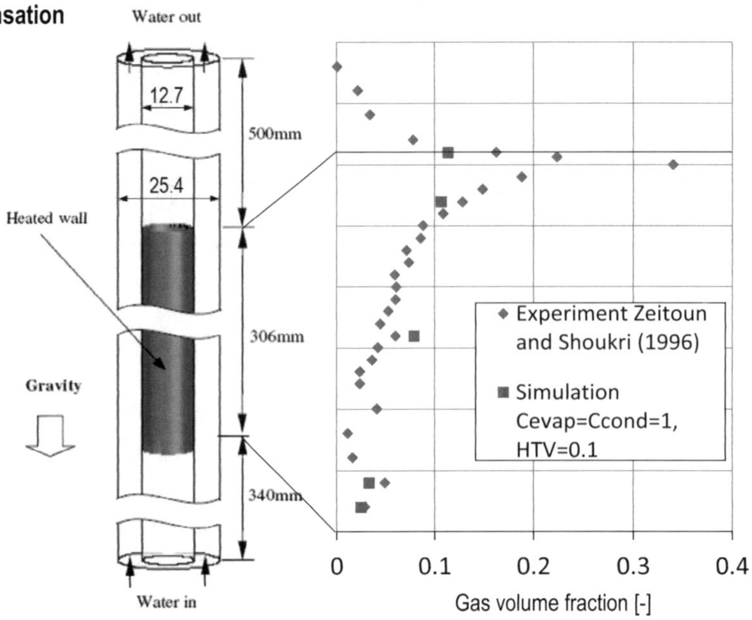

Efficient simulation of boiling effects at engine cooling jackets
Model 2 – model parameters to adjust

The engine company. DEUTZ

- **Lift**
  Data from Robinson (2001)
  => Characteristic bubble diameter = 0.1 mm

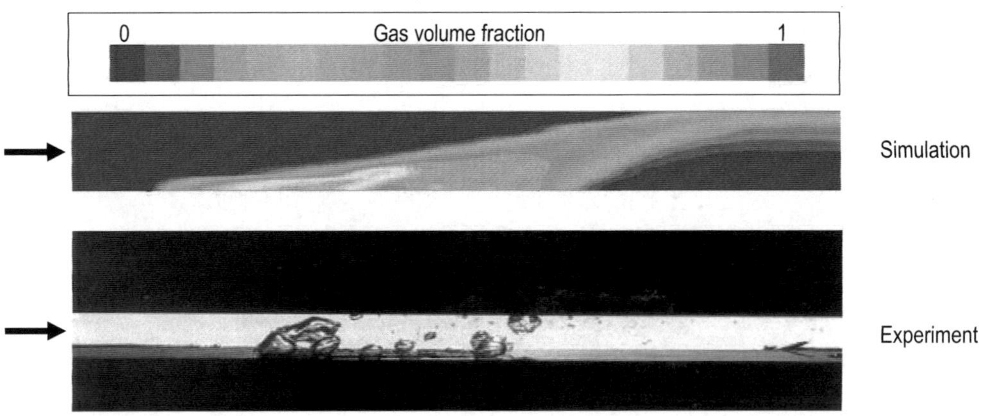

Efficient simulation of boiling effects at engine cooling jackets
CFD parameters

The engine company. DEUTZ

- **Software**
  FLUENT 16.2
- **Surface roughness**
  - hydraulic smooth
  - hydraulic rough
- **Turbulence**
  kω-sst, Y^+-insensitive wall treatment
- **Grid**
  - standard (wall-function approach, Y^+ ~30)
  - fine (resolved near-wall region, Y^+ ~1)

 15

Efficient simulation of boiling effects at engine cooling jackets
Test case with hydraulic developed flow

The engine company. DEUTZ

Test case by Steiner et al. (2004)

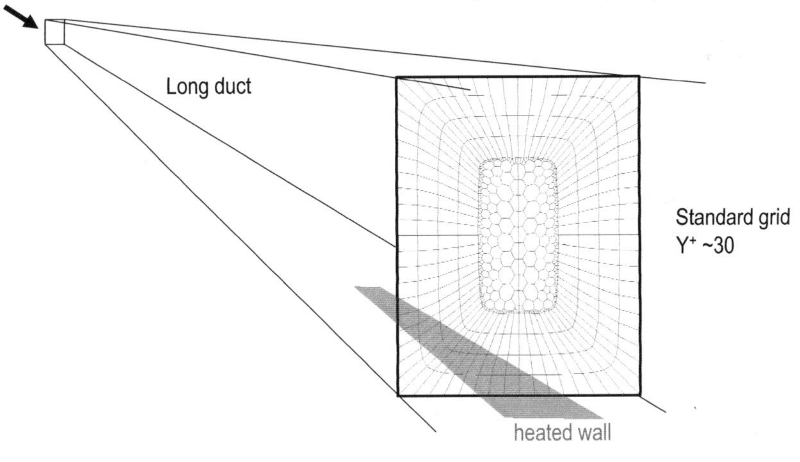

 16

Efficient simulation of boiling effects at engine cooling jackets
Test case with hydraulic developed flow

The engine company. DEUTZ

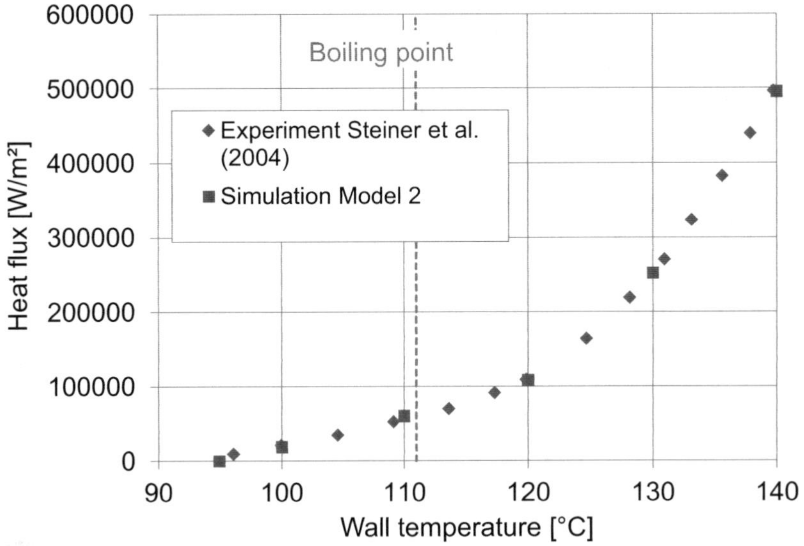

$AF = 1$

Efficient simulation of boiling effects at engine cooling jackets
Test case with hydraulic non-developed flow

The engine company. DEUTZ

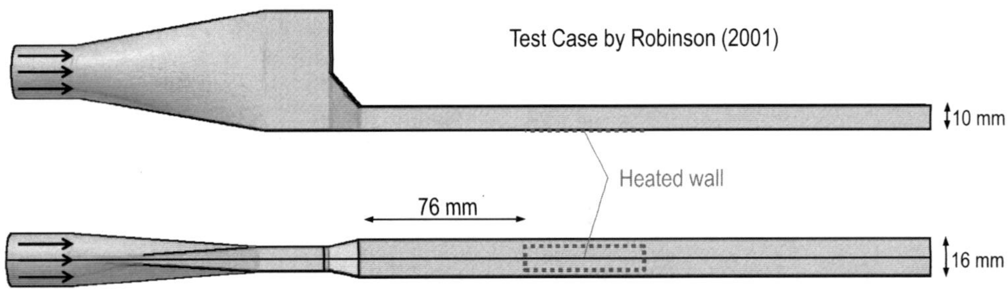

Standard grid

first boundary layer 0.5 mm

Y^+~30

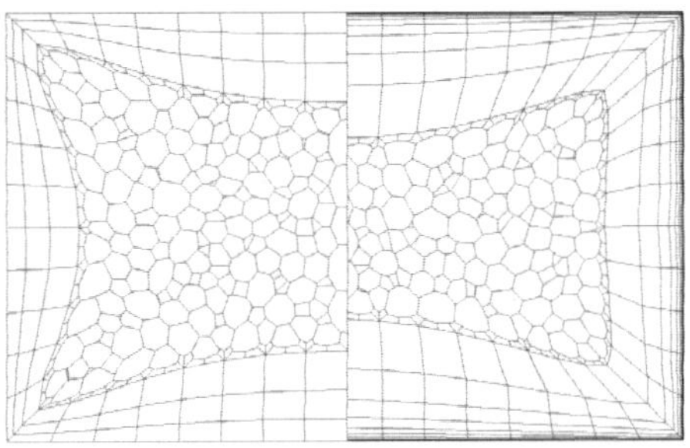

Fine grid

first boundary layer 0.01 mm

Y^+~1

Efficient simulation of boiling effects at engine cooling jackets

Test case with non-developed flow

The engine company. DEUTZ

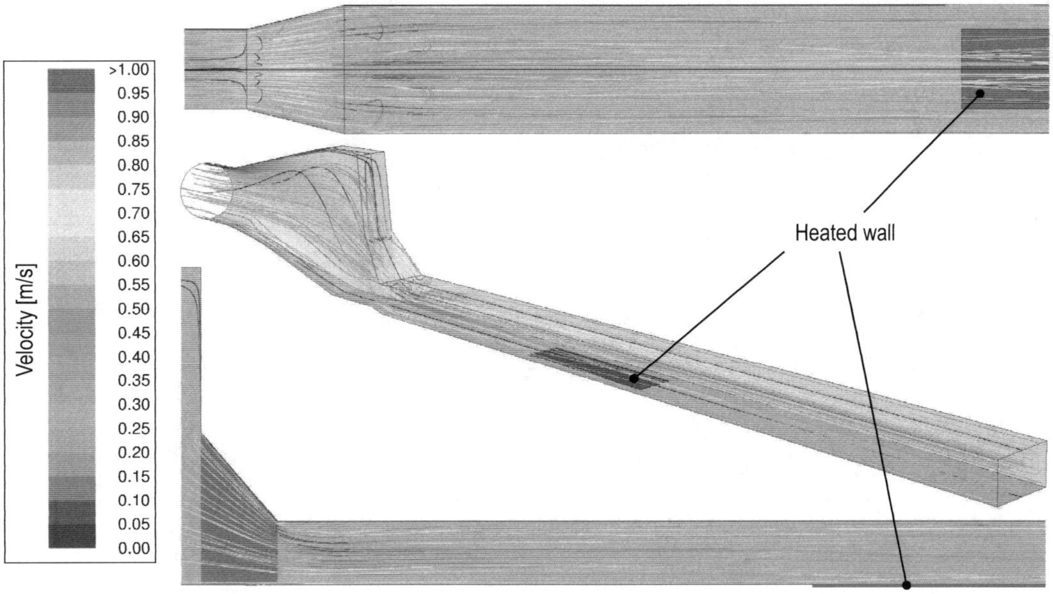

Efficient simulation of boiling effects at engine cooling jackets

Test case with hydraulic non-developed flow

The engine company. DEUTZ

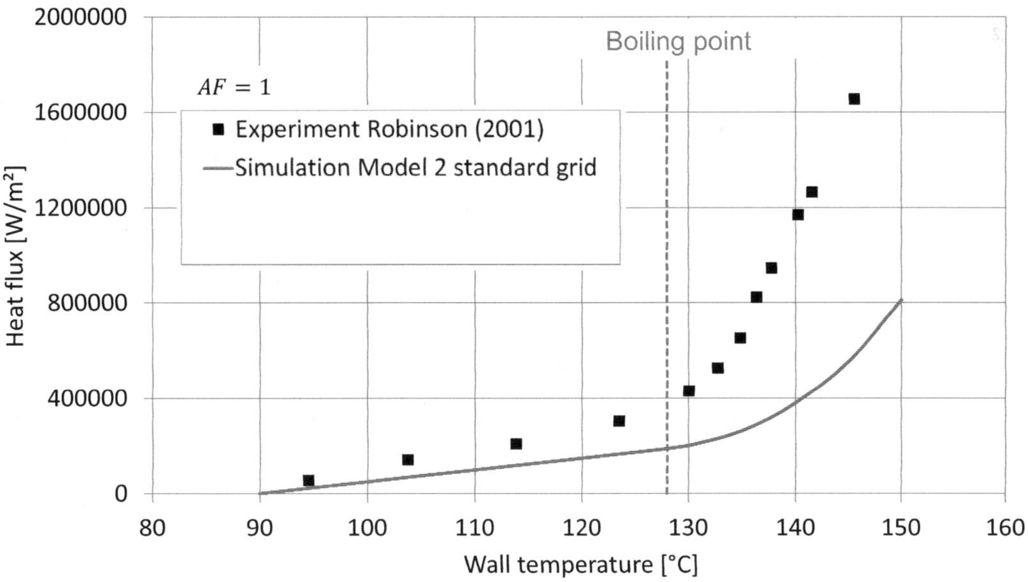

Efficient simulation of boiling effects at engine cooling jackets
Test case with hydraulic non-developed flow

The engine company. DEUTZ

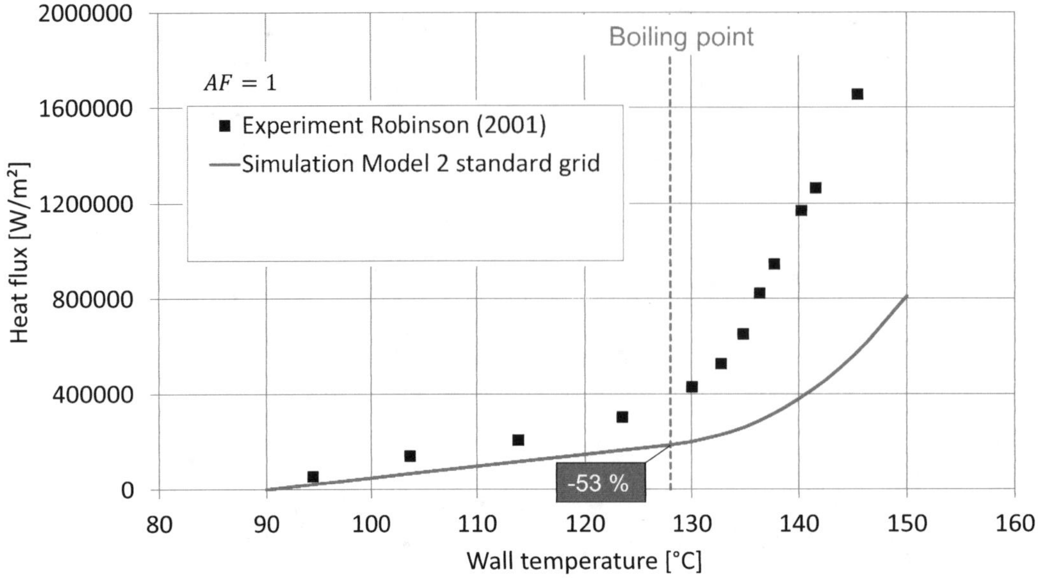

Efficient simulation of boiling effects at engine cooling jackets
Test case with hydraulic non-developed flow

The engine company. DEUTZ

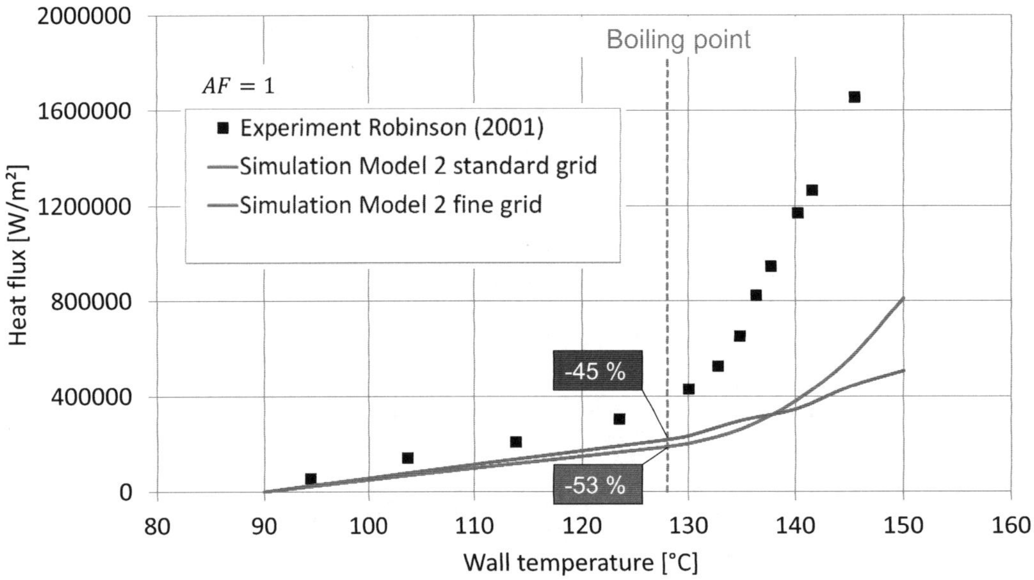

Efficient simulation of boiling effects at engine cooling jackets

Test case with hydraulic non-developed flow

The engine company. DEUTZ

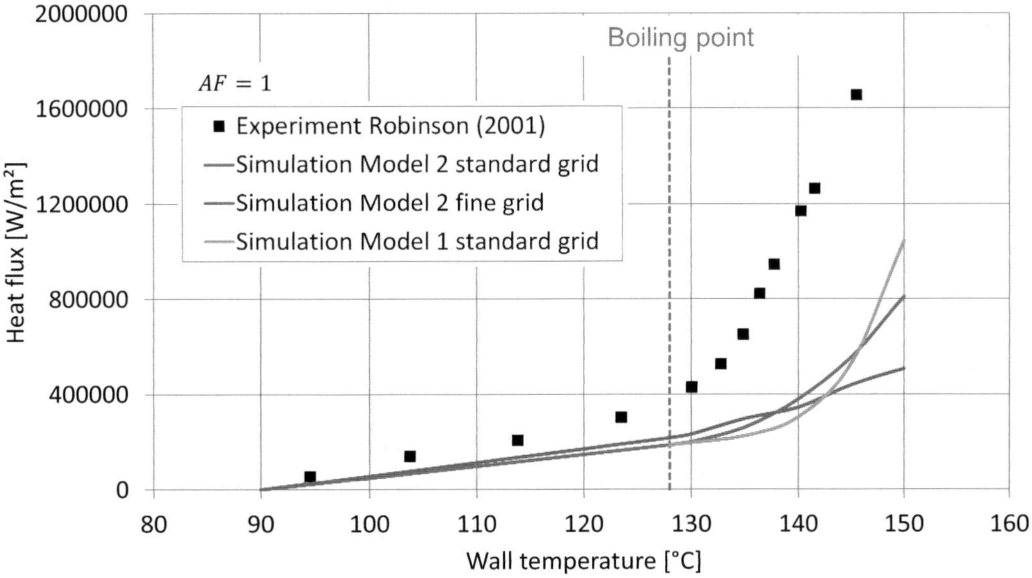

VE-TT3 · Boemer & Morsch 04.05.2016

23

Efficient simulation of boiling effects at engine cooling jackets

Test case with hydraulic non-developed flow

The engine company. DEUTZ

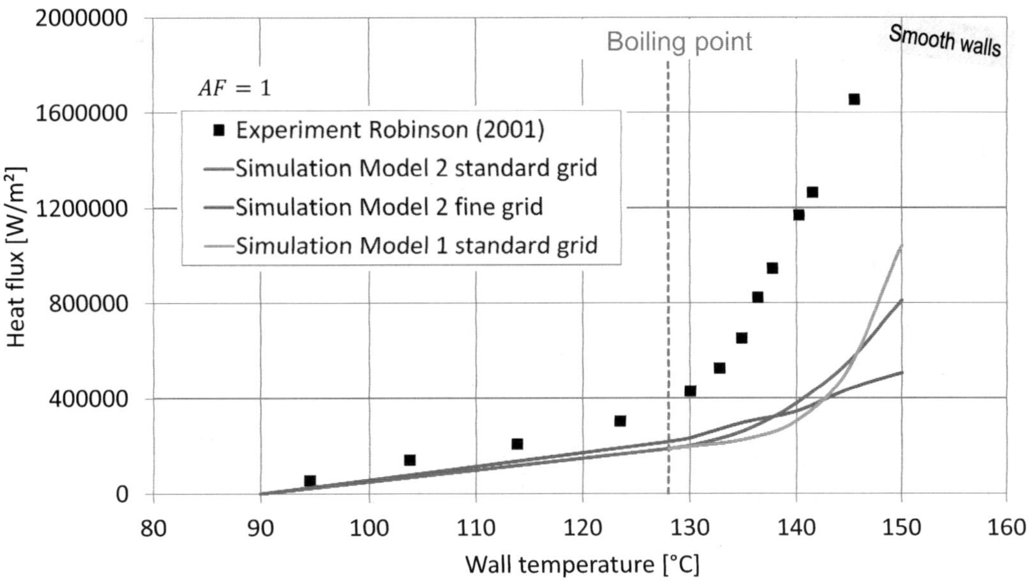

VE TT3 · Boemer & Morsch 04.05.2016

24

Efficient simulation of boiling effects at engine cooling jackets
Test case with hydraulic non-developed flow

The engine company. DEUTZ

Standard grid

first boundary layer 0.5 mm

Fine grid

first boundary layer 0.01 mm

Rough walls

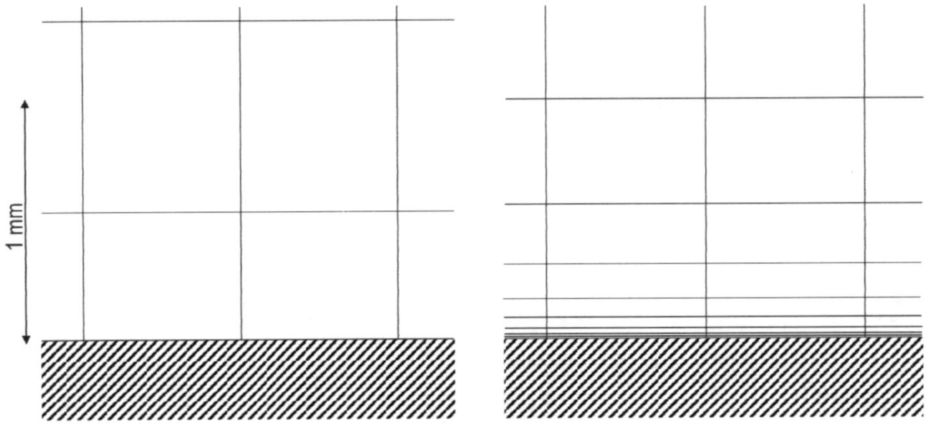

VE-TT3 · Boemer & Morsch 04.05.2016

25

Efficient simulation of boiling effects at engine cooling jackets
Test case with hydraulic non-developed flow

The engine company. DEUTZ

Standard grid

first boundary layer 0.5 mm

Fine grid

first boundary layer 0.01 mm

Rough walls

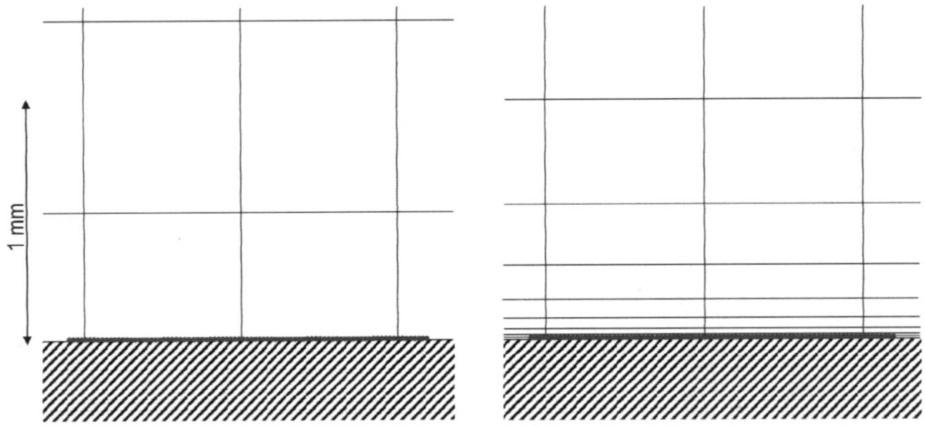

Ra = 0.017 mm

Experiment Robinson (2001)

VE-TT3 · Boemer & Morsch 04.05.2016

26

Efficient simulation of boiling effects at engine cooling jackets
Test case with hydraulic non-developed flow

The engine company. DEUTZ

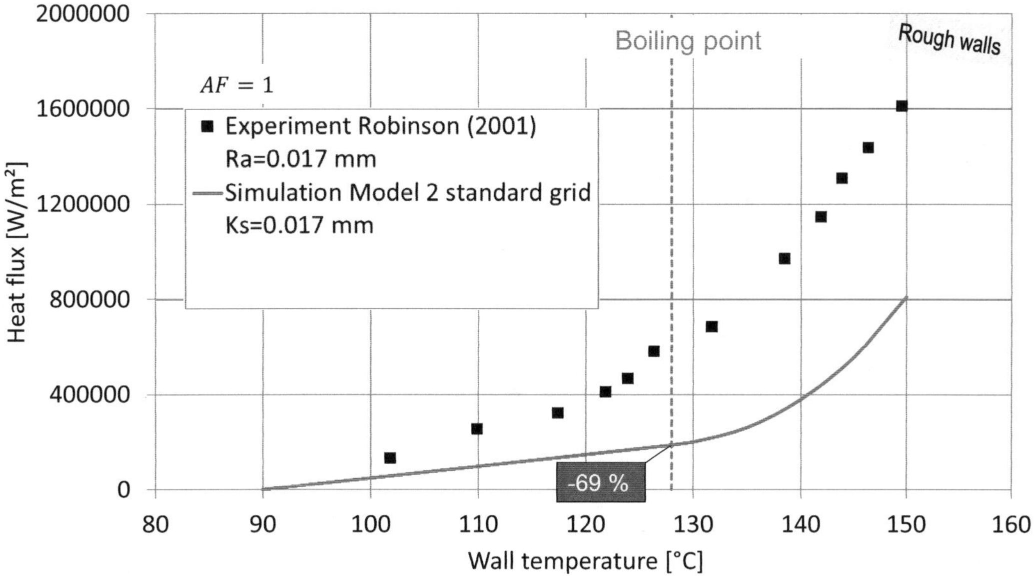

Efficient simulation of boiling effects at engine cooling jackets
Test case with hydraulic non-developed flow

The engine company. DEUTZ

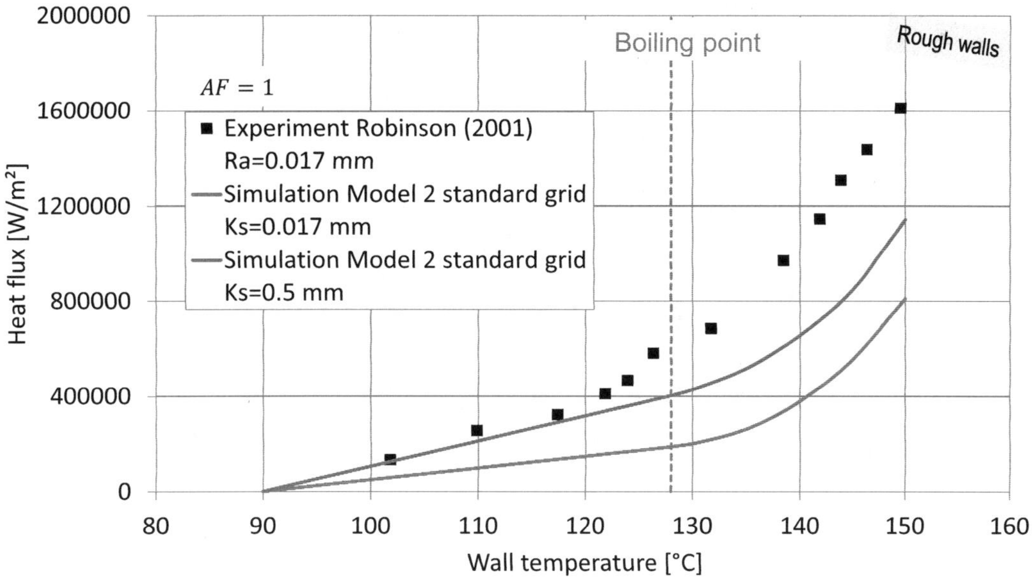

Efficient simulation of boiling effects at engine cooling jackets
Conclusions

The engine company. DEUTZ

- **Standard FLUENT**

  - Convective heat transfer simulation at non-developed flows needs improvement.

  - Lift in the Mixture Model is an issue.

- **Boiling extension**

  - Model 1 is limited, but efficient.

  - Model 2 (Semi Mechanistic Boiling Model) has higher potential, but needs improvement:

    - Application of AF is no general solution

    - Distribute $\dot{m}_{evap,wall}$ to multiple wall layers (Morsch, 2016)

    - Distribute $(1 - HTV) \cdot \dot{q}_{boil}$ to multiple wall layers, ensure $T_{cell} \leq T_{boil}$

    - Need a model for HTV

    - Need heat flux boundary condition

  - Adaption of Model 2 to the VOF Model?

VE-TT3 · Boemer & Morsch 04.05.2016

29

Efficient simulation of boiling effects at engine cooling jackets
Thanks

The engine company. DEUTZ

Thanks to the FLUENT support

- Hemant Punekar
- Holger Grotjans
- Lars Göllnitz
- Marco Oswald
- Peter Payandeh
- Ulrich Postl

VE-TT3 · Boemer & Morsch 04.05.2016

30

Efficient simulation of boiling effects at engine cooling jackets
Literature

The engine company. DEUTZ

Fritz, W.: Berechnung des Maximalvolumens von Dampfblasen, Physik. Z. 36, 379/84 (1935)

CHEN, J. C.: A Correlation for Boiling Heat Transfer of Saturated Fluids in Convective Flow, Ind. Eng. Chem. Process Des. Dev., Vol. 5, No. 3, pp. 322-329 (1966)

Krepper, E.; Koncar, B., Egorov, Y.: CFD modelling of subcooled boiling – Concept, validation and application to fuel assembly design, Journal of Nuclear Engineering and Design, Elsevier B.V., Amsterdam, The Netherlands (2007)

Morsch, P.: CFD-Simulation von konvektivem und evaporativem Wärmeübergang in Wassermänteln, Bachelor Thesis, RWTH Aachen, Germany (2016)

Nukijama, S.: The Maximum and Minimum Values of the Heat Q Transmitted from Metal to Boiling Water under Atmospheric Pressure, Journal of the Japan Society of Mechanical Engineers, Vol. 37, no. 206, pp. 367-374 (1934)

Punekar, H., Das, S.: Numerical Simulation of Subcooled Nucleate Boiling in Cooling Jacket of IC Engine, SAE 2013-01-1651 (2013)

Robinson, K.: IC Engine Coolant Heat Transfer Studies, PhD Thesis, University of Bath, UK (2001)

Steiner, H., Kobor, A., Gebhard, L.: A wall heat transfer model for subcooled boiling flow, Proceedings of The ASME – ZSIS International Thermal Science Seminar II, Bled, Slovenia (2004)

Combining CFD, FEA and scripting for efficient oil vane pump integrity assessment.

Christopher Lynch

simulate to innovate

WHO ARE WE?

simulate to innovate

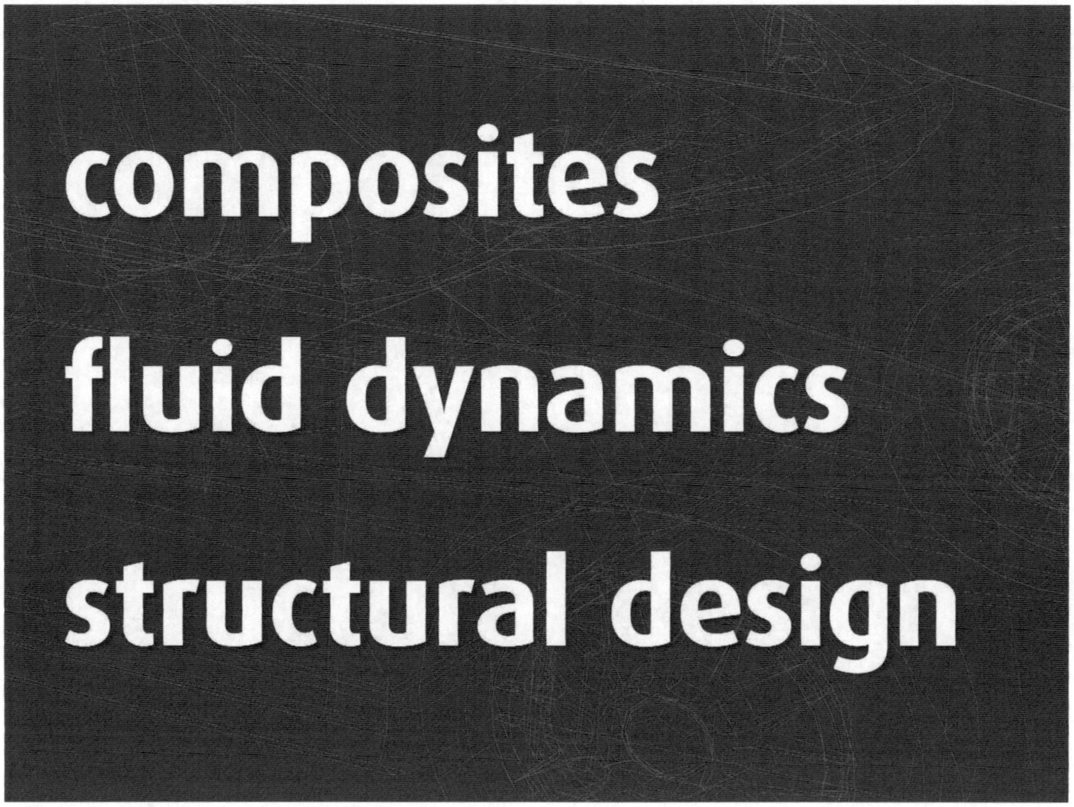
composites
fluid dynamics
structural design

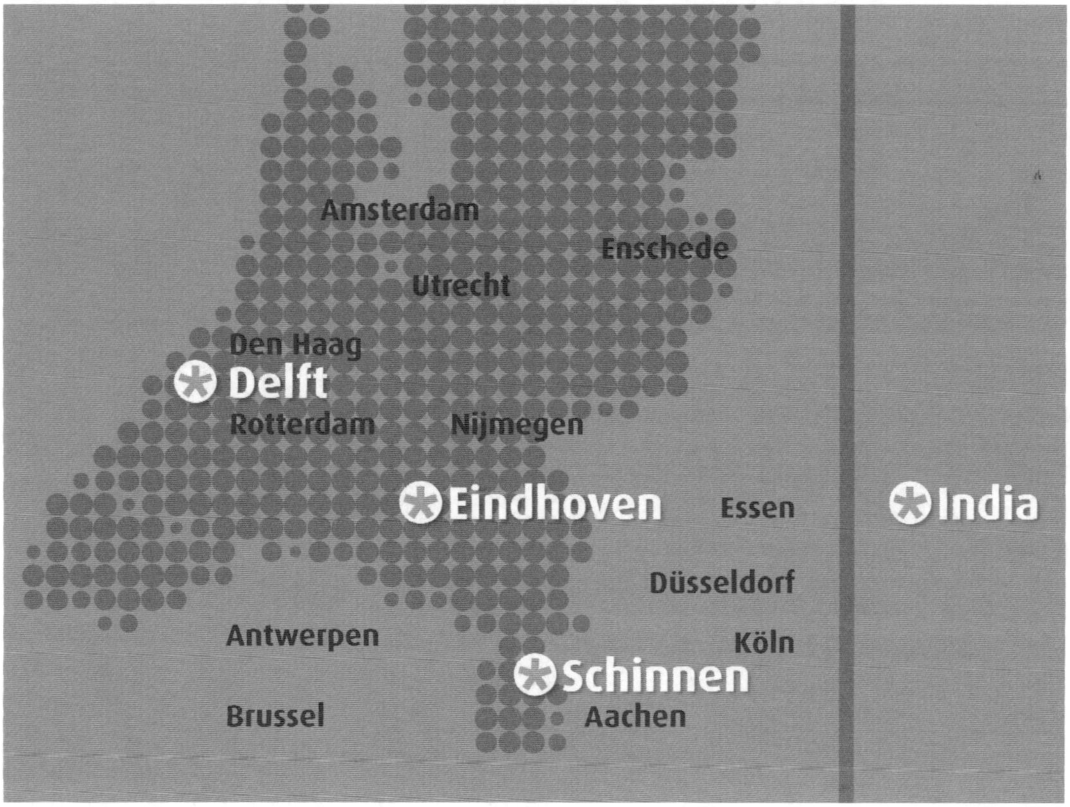
Amsterdam
Enschede
Utrecht
Den Haag
Delft
Rotterdam
Nijmegen
Eindhoven
Essen
India
Düsseldorf
Antwerpen
Köln
Schinnen
Brussel
Aachen

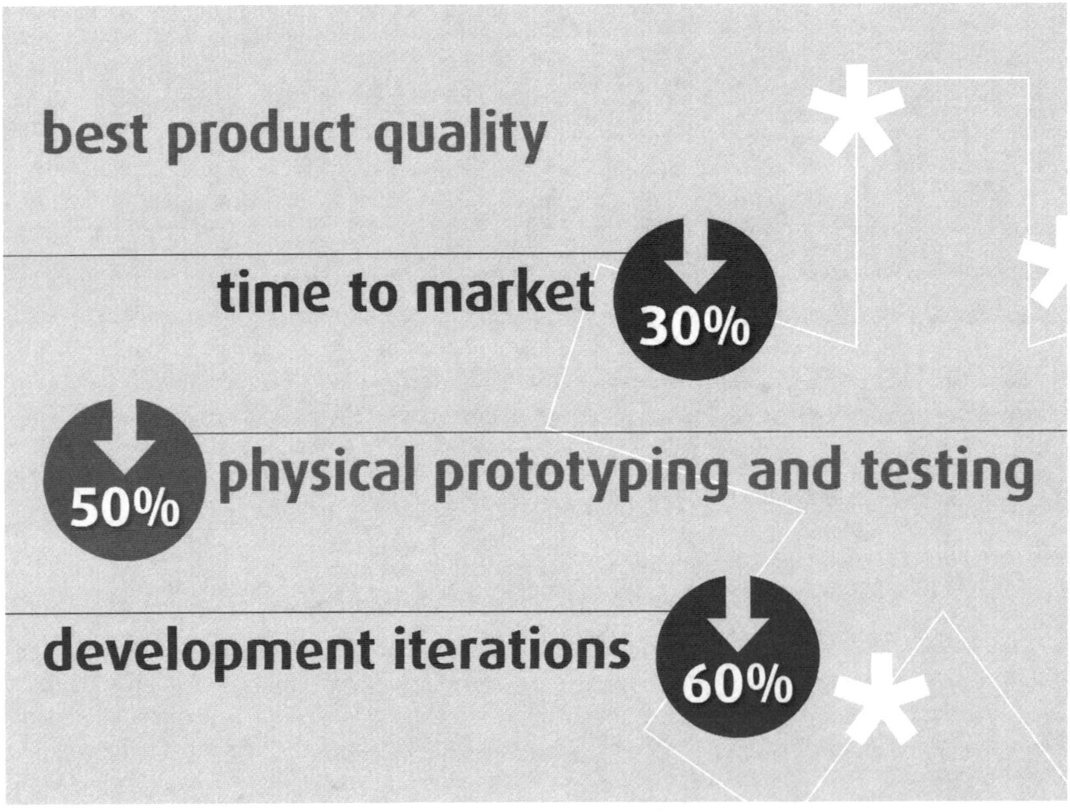
best product quality
time to market
30%
50%
physical prototyping and testing
development iterations
60%

"My product has to meet high standards regarding safety"

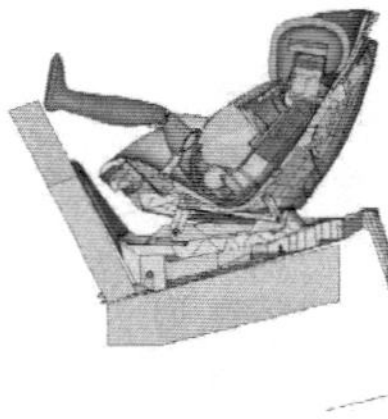

"My product design fails, but I don't know when, why and where?"

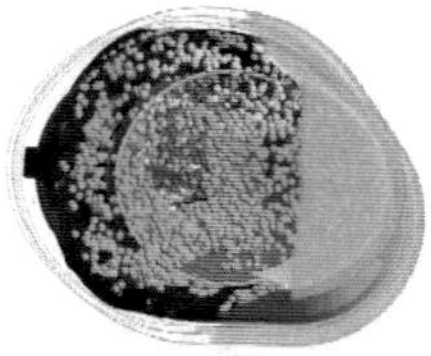

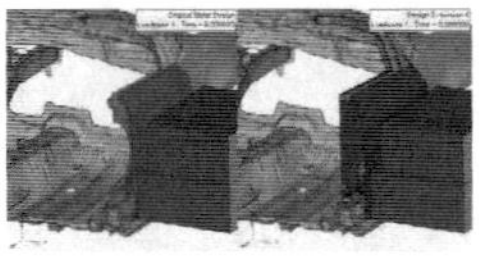

"I want to redesign my product and improve the quality and performance."

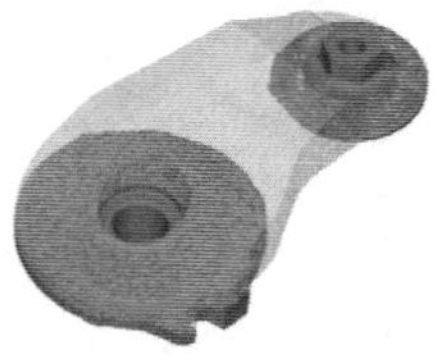

"How much weight can be reduced by using topology optimization?"

APPLICATION BACKGROUND

simulate to innovate

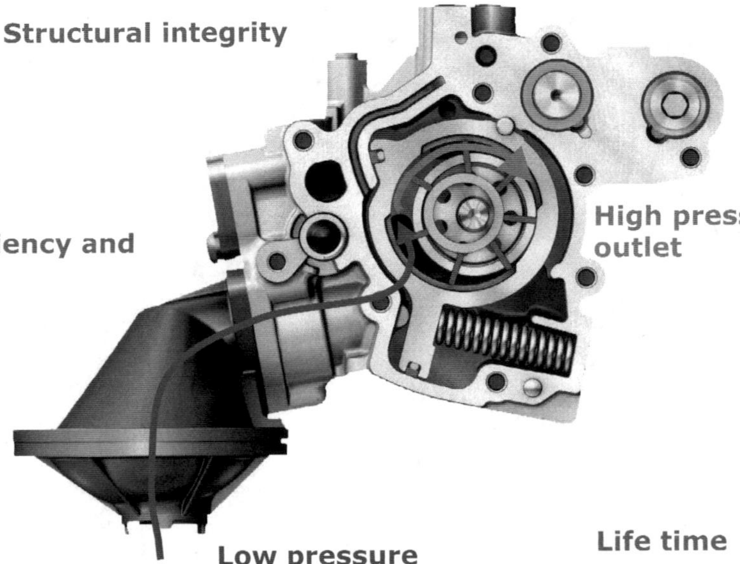

Source: http://www.audi-technology-portal.de/en/drivetrain/engine-efficiency-technologies/demand-controlled-oil-pump

Challenges

- Increased volume flow & higher pressure
- Reduced packaging space
- No reliability concessions
- "Delivery must have been yesterday"

Old working practice does not suffice

Experiments

Experience

GOAL

simulate to innovate

Provide insight in:

Flow behavior

Deformation and dynamics

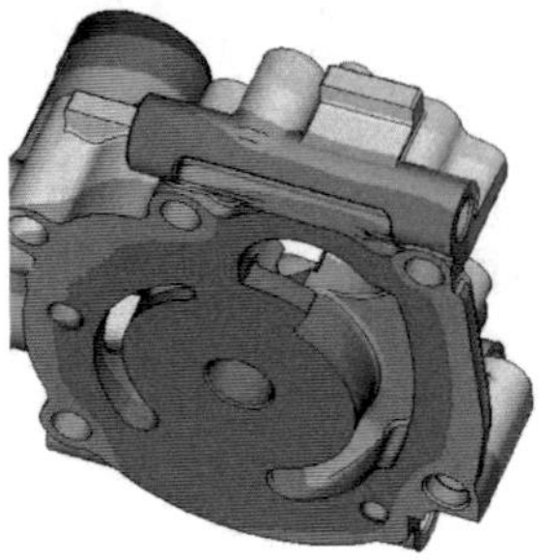

Stresses

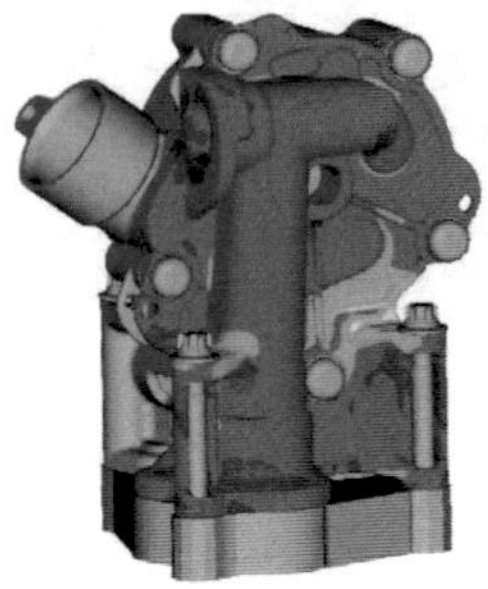

PROJECT PROCEDURE

simulate to innovate

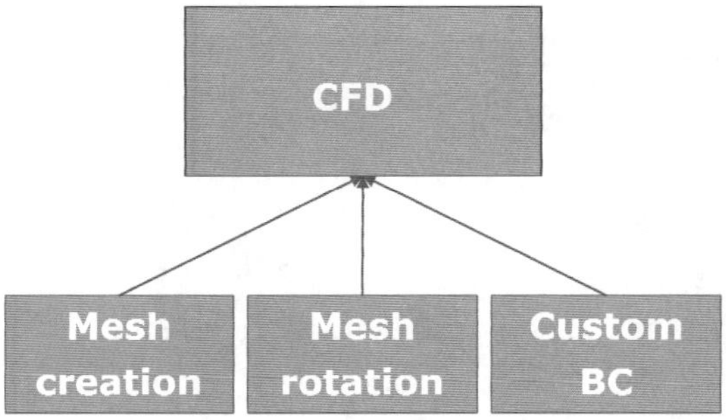

CFD model

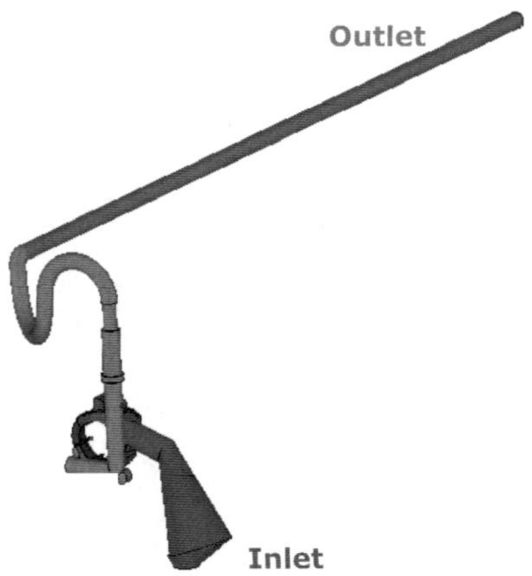

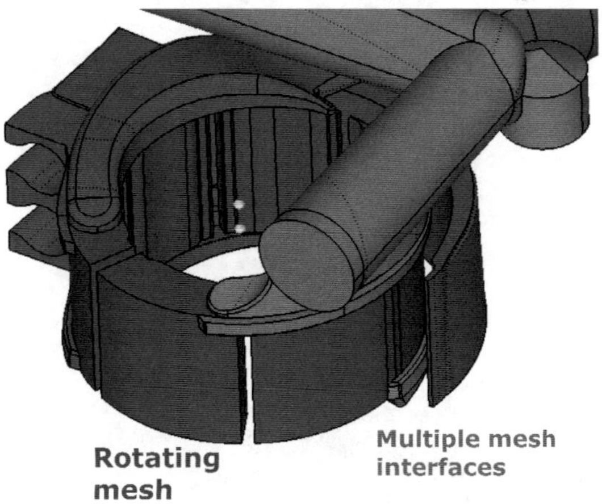

- Transient simulation
- Tait equation for compressibility
- k-ω-SST turbulence model

Mesh creation (Scripted)

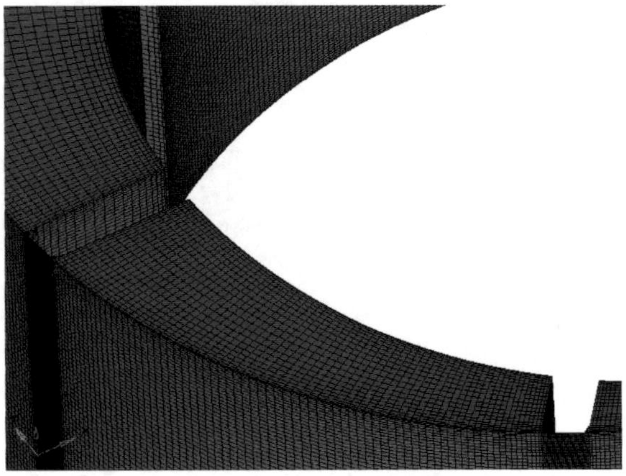

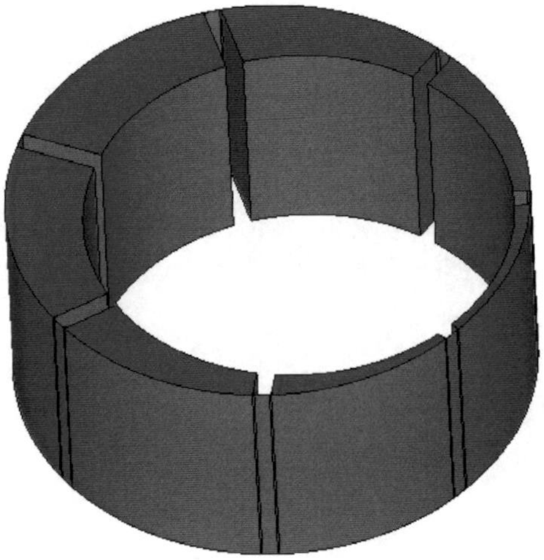

Mesh rotation (Scripted)

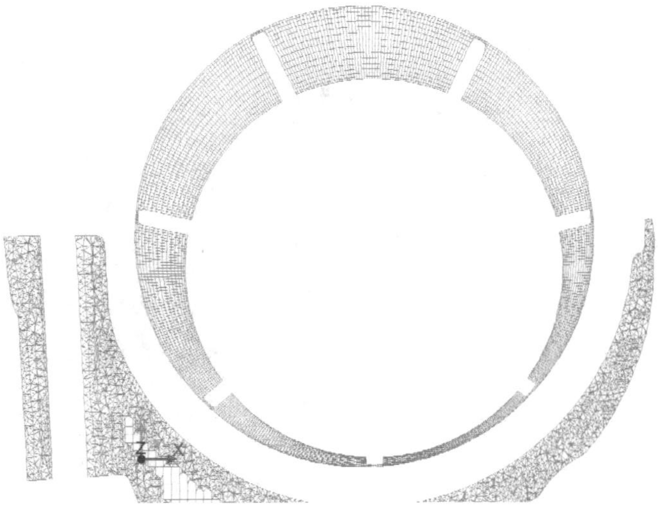

Boundary reflection (Scripted)

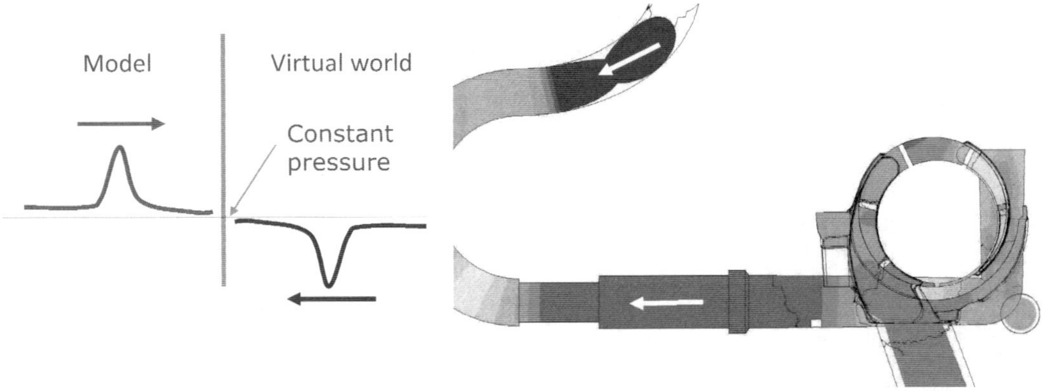

Boundary reflection (Scripted)

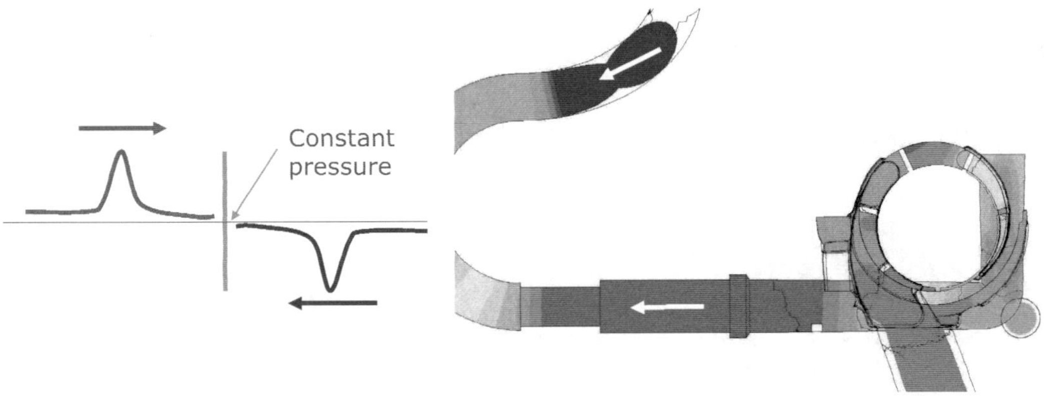

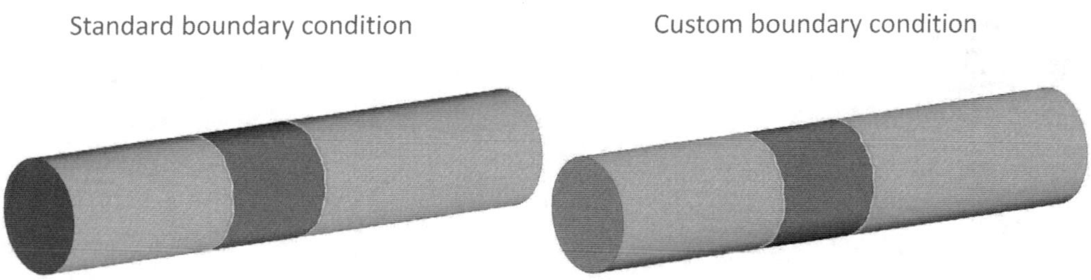

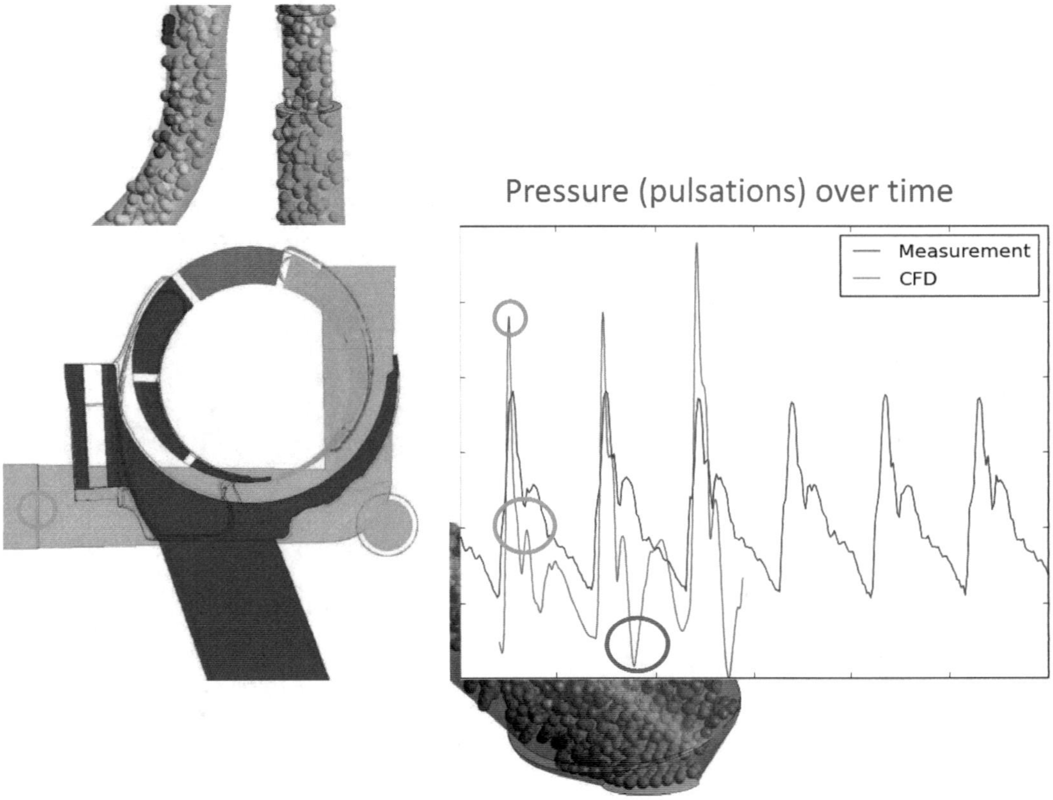
Pressure (pulsations) over time
Measurement
CFD

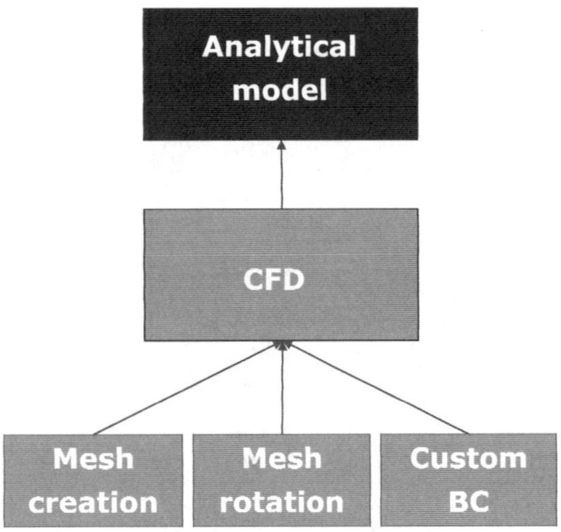
Analytical model
CFD
Mesh creation
Mesh rotation
Custom BC

The analytical model

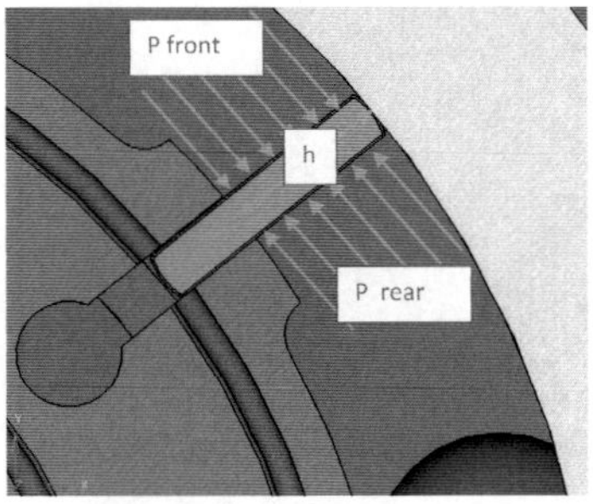

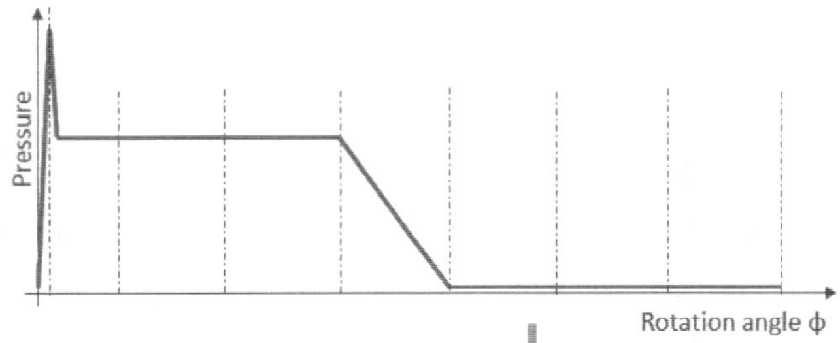

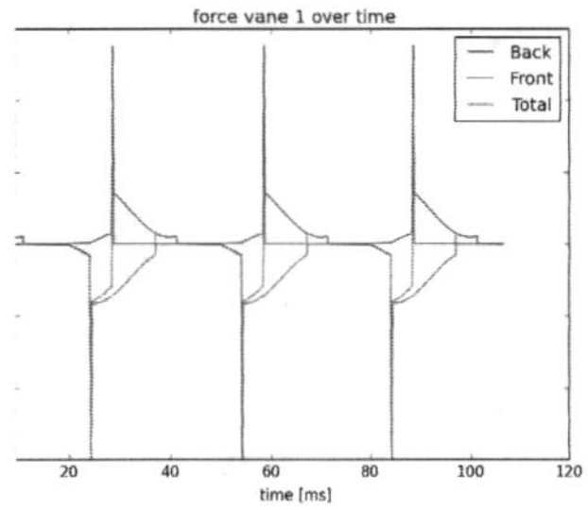


```
146.875000, -2.992388
146.880000, -2.993168
146.885000, -2.993939
146.890000, -2.994700
146.895000, -2.995451
146.900000, -2.996192
146.905000, -2.996924
146.910000, -2.997646
146.915000, -2.998358
146.920000, -2.999060
146.925000, -2.999752
*AMPLITUDE, NAME =Force_Vane1_Front, DEFINITION = TABULAR
0.000000, 0.000000
0.005000, 0.003721
0.010000, 0.007461
0.015000, 0.011219
0.020000, 0.014995
0.025000, 0.018790
0.030000, 0.022604
0.035000, 0.026435
0.040000, 0.030286
```

CFD

Force vs time

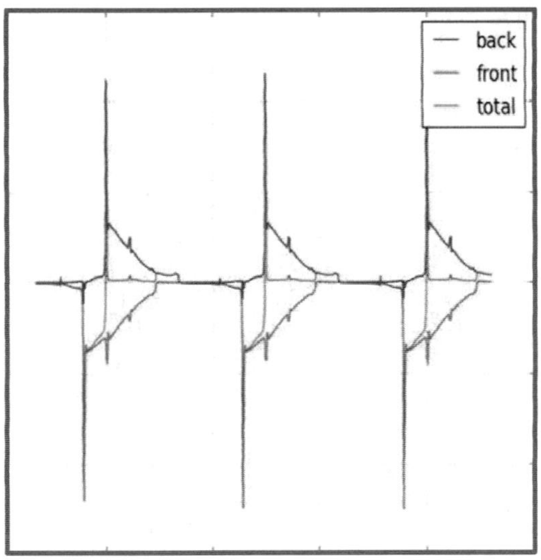

Analytical model

Force vs time

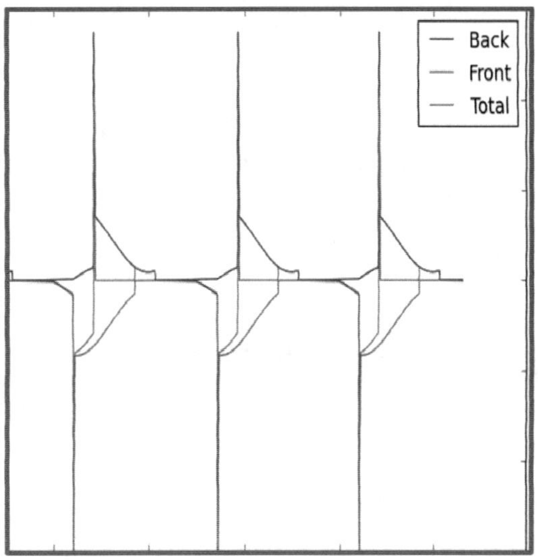

Validity

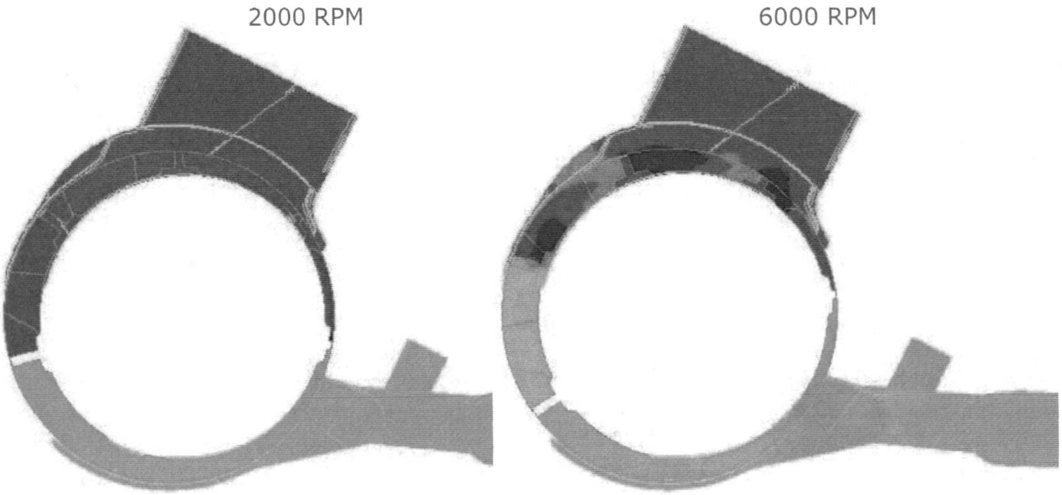
2000 RPM
6000 RPM

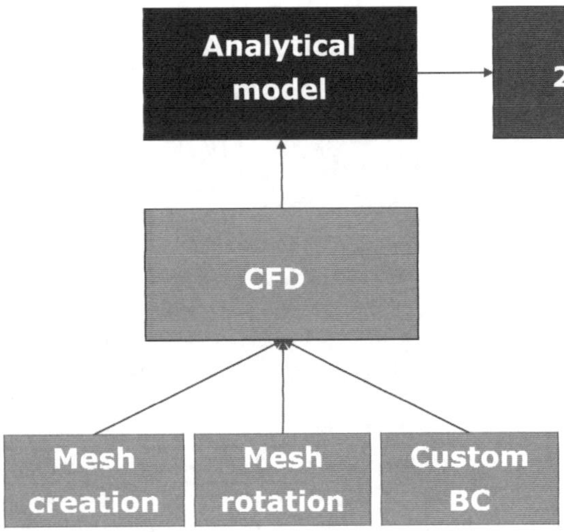
Analytical model
2D/3D FEA
CFD
Mesh creation
Mesh rotation
Custom BC

Transient FEA

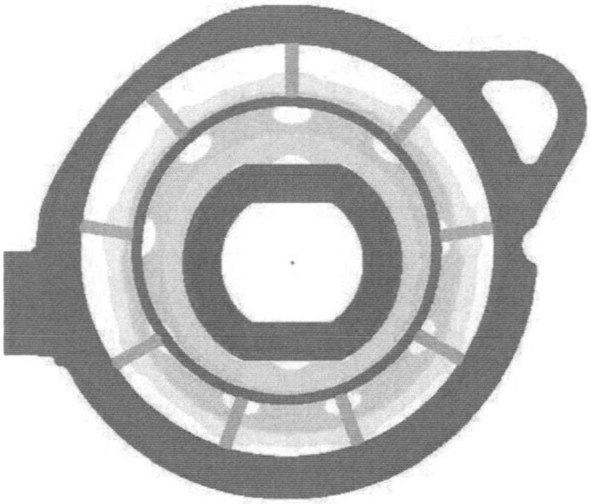

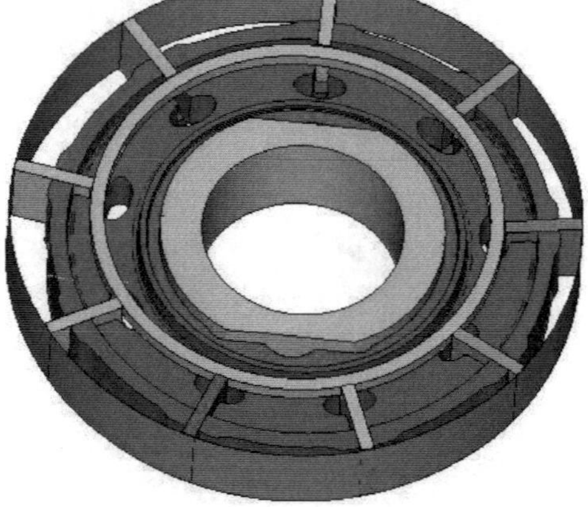

External influences

- Forces
- Friction
- Speed variation
- Impact

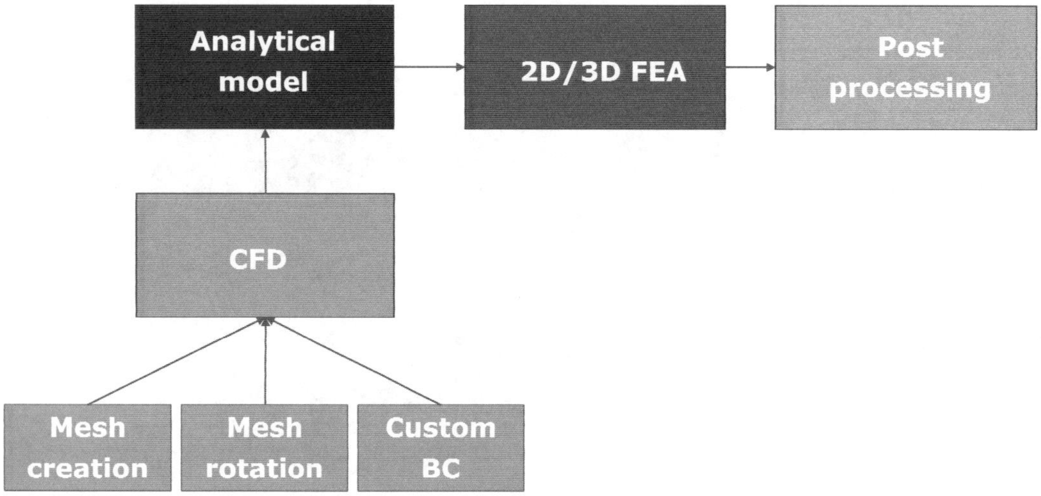
Analytical model
2D/3D FEA
Post processing
CFD
Mesh creation
Mesh rotation
Custom BC

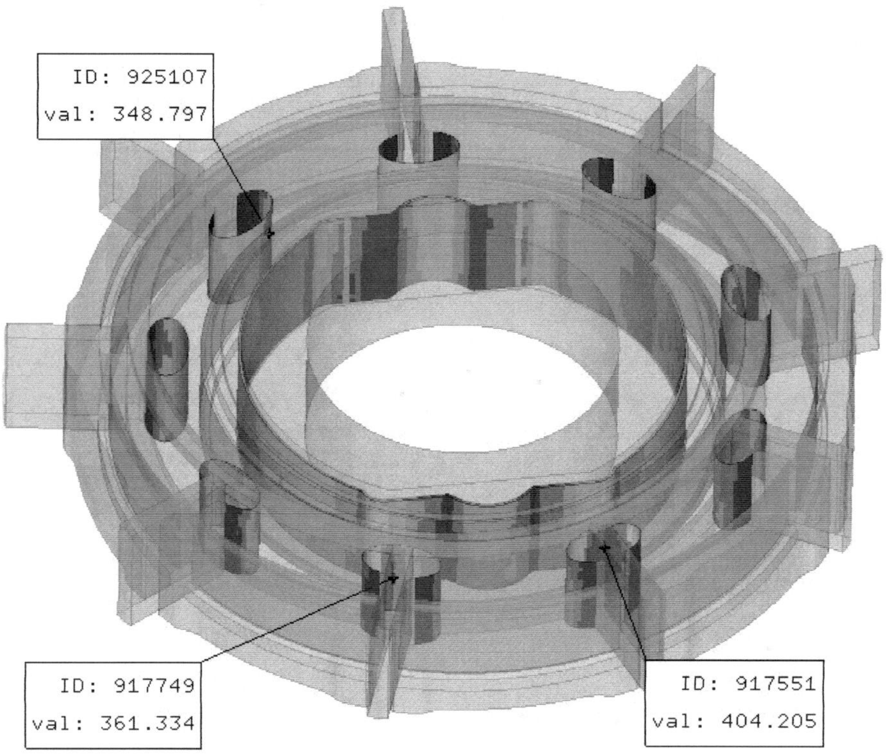
ID: 925107
val: 348.797
ID: 917749
val: 361.334
ID: 917551
val: 404.205

Safety factors (Scripted)

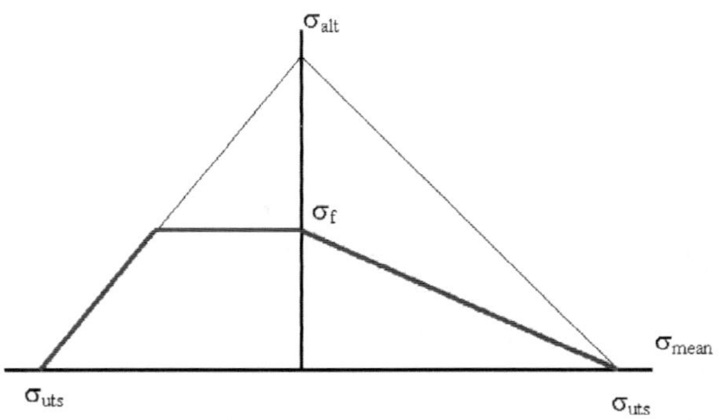

Provide insight in:

Flow behavior	Dynamics	Stresses

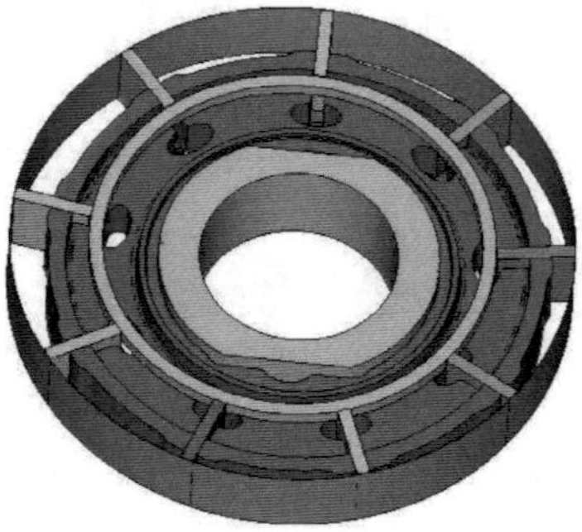

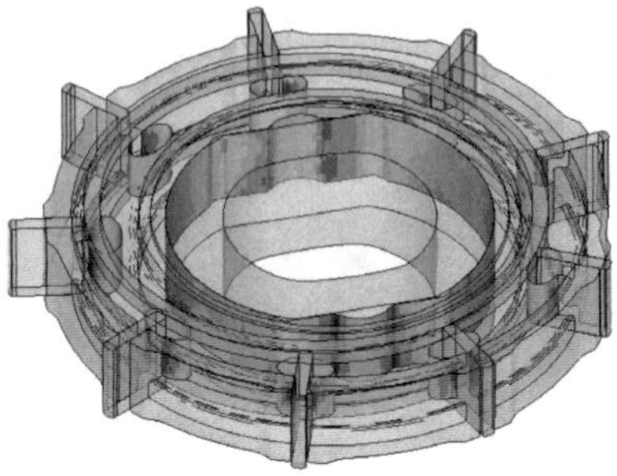

WWW.CODE-PS.COM

FOLLOW US ON LINKEDIN

Schinnen
Breinder 26
6365 CX Schinnen

+31 (0)88 115 1900
info@code-ps.com

Delft
Rotterdamseweg 183c
2629 HD Delft

+31 (0)88 115 1920
info@code-ps.com

Eindhoven
Torenallee 26-06
5617 DB Eindhoven

+31 (0)88 115 1940
info@code-ps.com

India
No.136, Arcot Road
Saligramam, Chennai

+91 (0)44 6686 4416
vijay.nath@code-ps.com

ADVANCED SIMULATION METHODS FOR LIGHT-WEIGHTING WITHIN A PASSENGER CAR POWERTRAIN

ANSYS Automotive Simulation World Congress
8th June 2016
David Vickerman

www.grm-consulting.co.uk

Contents

- Introduction to GRM
- Introduction to GTAM/GSAM
- ULTRAN Project Case Study
- Summary

www.grm-consulting.co.uk

BACKGROUND OF GRM CONSULTING

- UK based Finite Element Analysis (FEA) consultancy with a key focus on optimisation and CAE led design.
- UK & EU distributors of Vanderplaats Research & Development (VR&D) design optimisation software.
- Working in many industries including:
  - Automotive
  - Medical
  - Defence
  - Rail
  - Motorsport
  - Aerospace
  - Marine

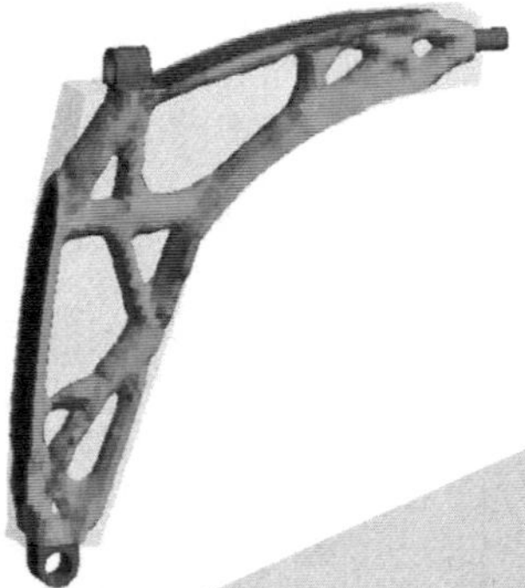

www.grm-consulting.co.uk

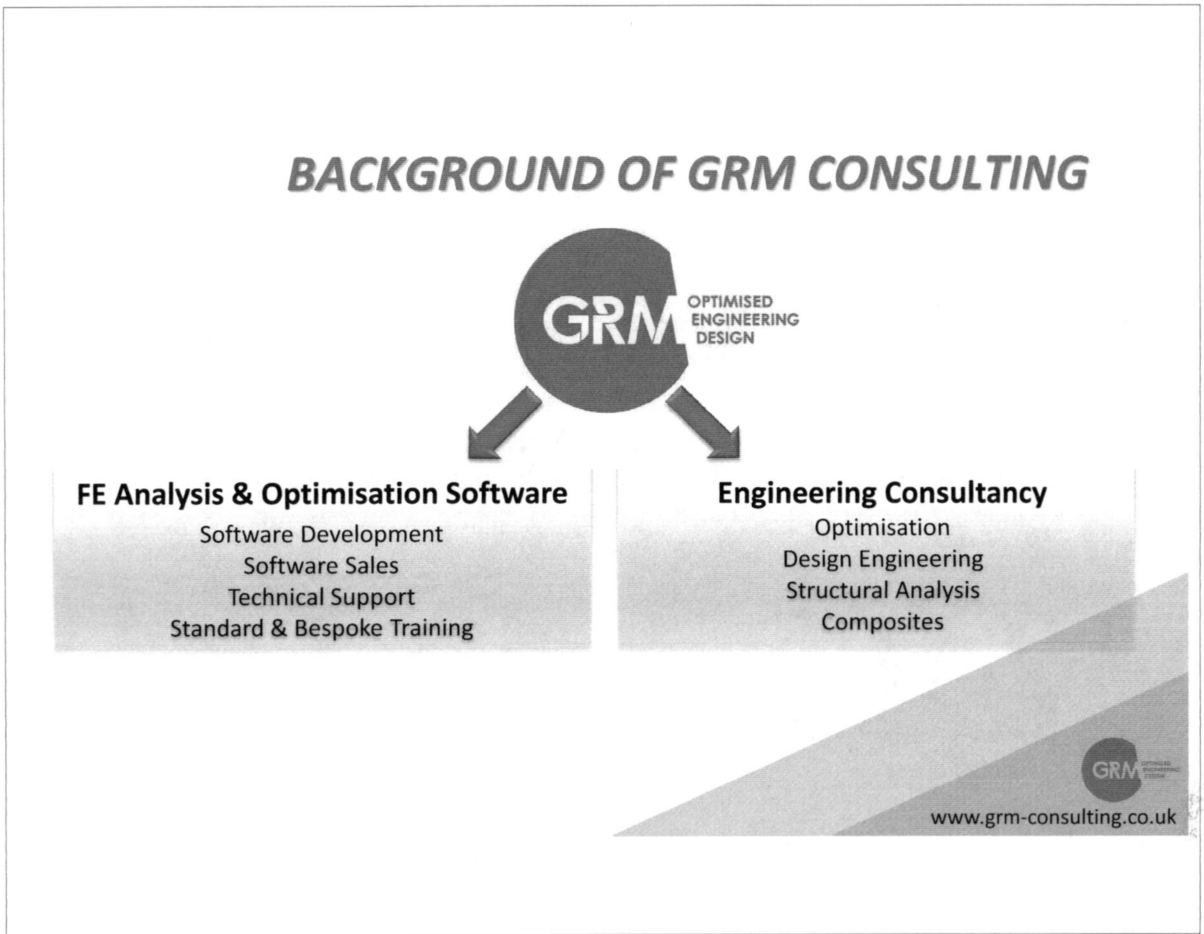
BACKGROUND OF GRM CONSULTING
GRM
OPTIMISED ENGINEERING DESIGN
FE Analysis & Optimisation Software
Software Development
Software Sales
Technical Support
Standard & Bespoke Training
Engineering Consultancy
Optimisation
Design Engineering
Structural Analysis
Composites
www.grm-consulting.co.uk

EXPERIENCE
Automotive
• Crash
• Interior
• Powertrain
Heavy Industry
• Waste Vehicle
• Plant Machinery
Medical Equipment
• MRI Scanners
• Drug Dispenser
Rail
• Seat design and simulation
• Crashworthiness
Aerospace
• Wing Intercostals
• Engine Mount Frames
• Seating Systems
Civil
• Street Lamp Abuse Testing
Marine
• America's Cup
Defence
• Blast mitigation
• Ballistic protection
Consumer Goods
• Boiler Systems
• Packaging
Formula 1
• Composite Chassis Design
• Suspension Design
www.grm-consulting.co.uk

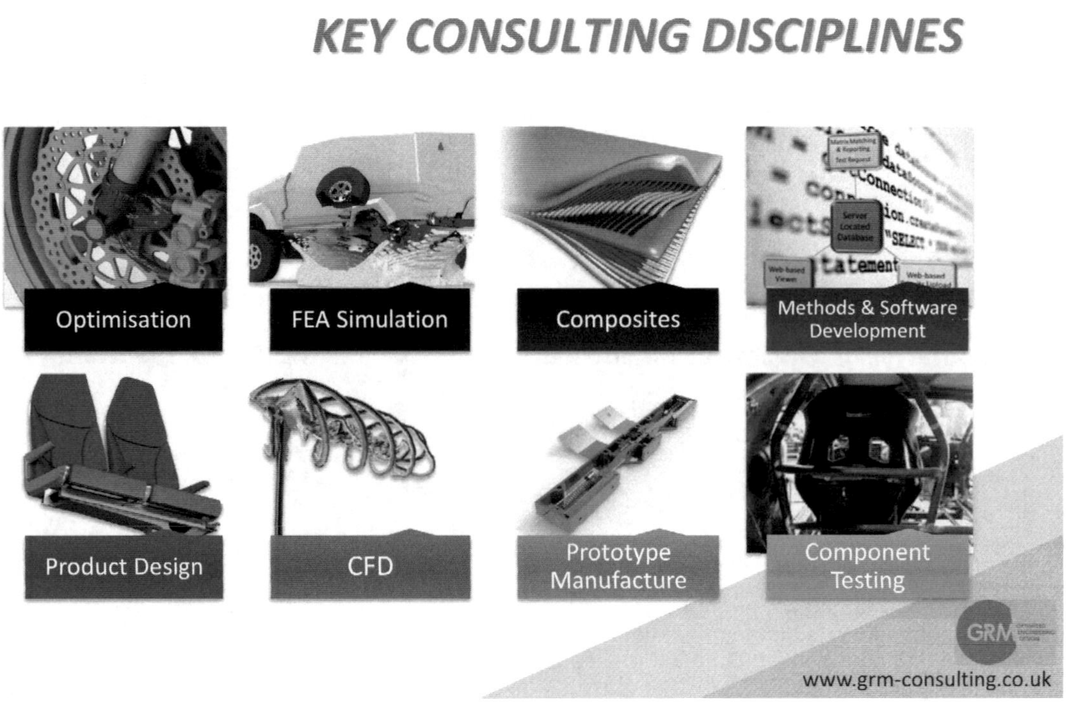
KEY CONSULTING DISCIPLINES
Optimisation
FEA Simulation
Composites
Methods & Software Development
Product Design
CFD
Prototype Manufacture
Component Testing
GRM
www.grm-consulting.co.uk

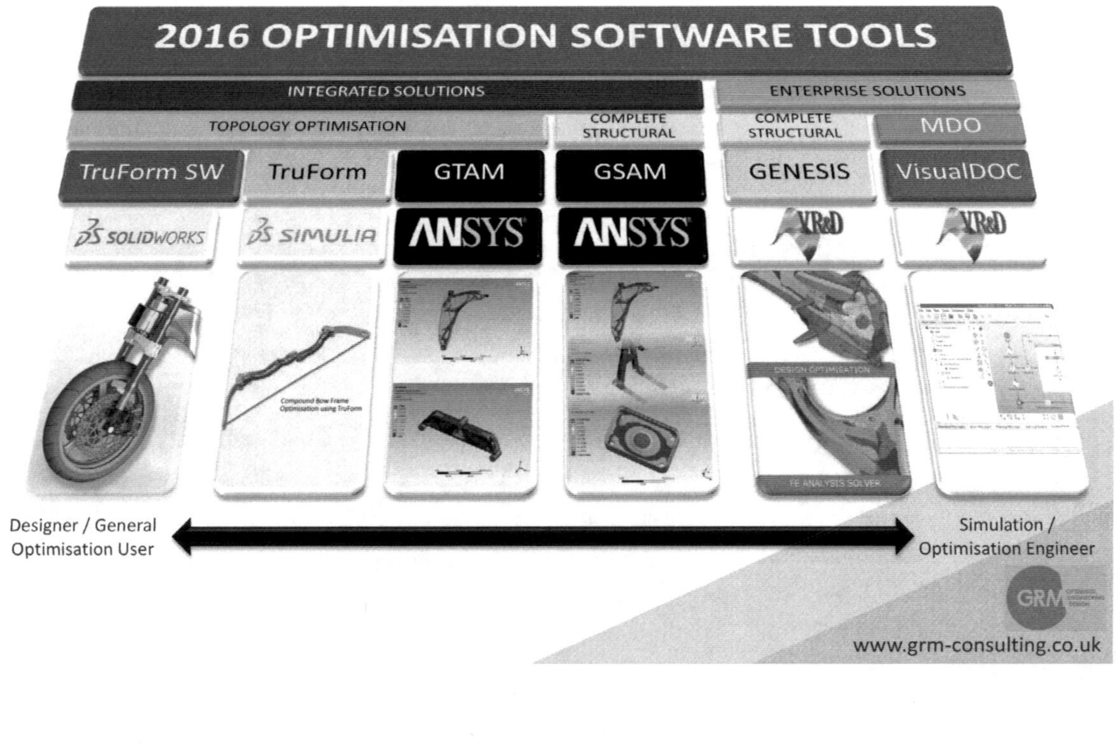
2016 OPTIMISATION SOFTWARE TOOLS
INTEGRATED SOLUTIONS
ENTERPRISE SOLUTIONS
TOPOLOGY OPTIMISATION
COMPLETE STRUCTURAL
COMPLETE STRUCTURAL
MDO
TruForm SW
TruForm
GTAM
GSAM
GENESIS
VisualDOC
SOLIDWORKS
SIMULIA
ANSYS
ANSYS
VR&D
VR&D
DESIGN OPTIMISATION
FE ANALYSIS SOLVER
Designer / General Optimisation User
Simulation / Optimisation Engineer
GRM
www.grm-consulting.co.uk

GRM's MISSION

GRM'S MISSION

Through its own, and its partner's, products and services be the first choice for optimised engineering design solutions for organisations all over the world.

www.grm-consulting.co.uk

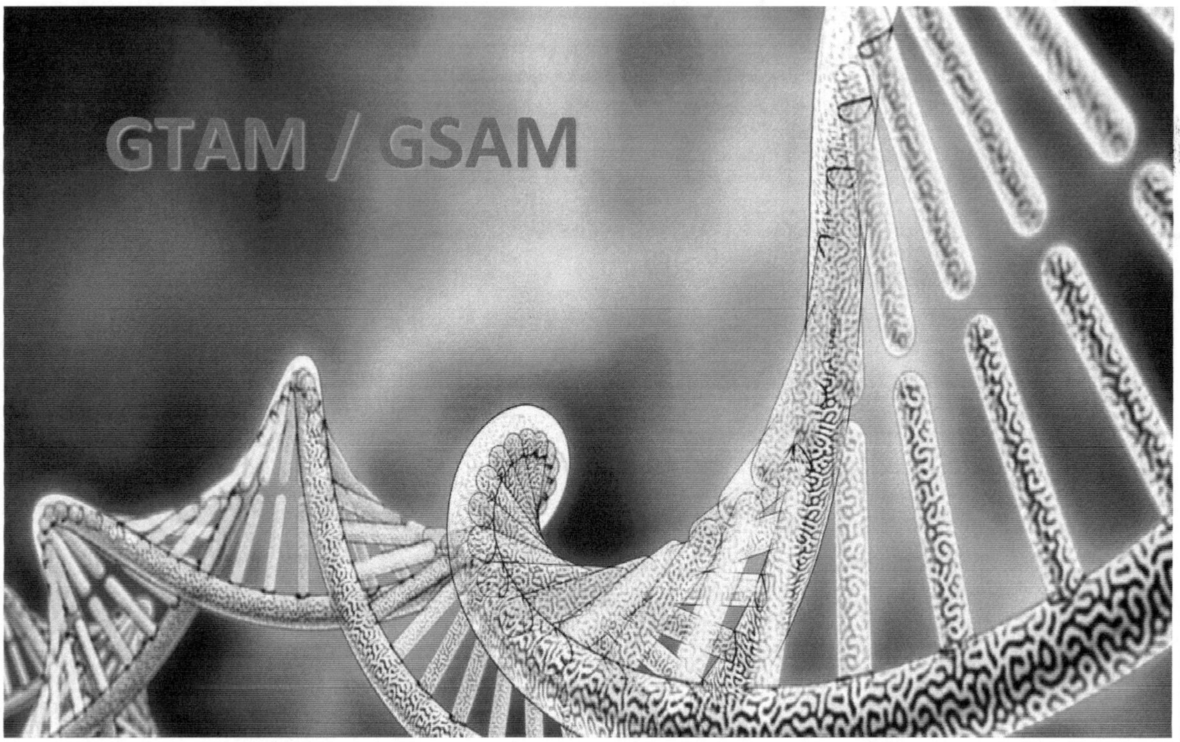

WHAT IS TOPOLOGY OPTIMISATION?

- **Topology optimisation** is a mathematical approach that defines the optimum material layout within a given package envelope, for a given set of loads and boundary conditions, such that the resulting layout meets a prescribed set of performance targets.

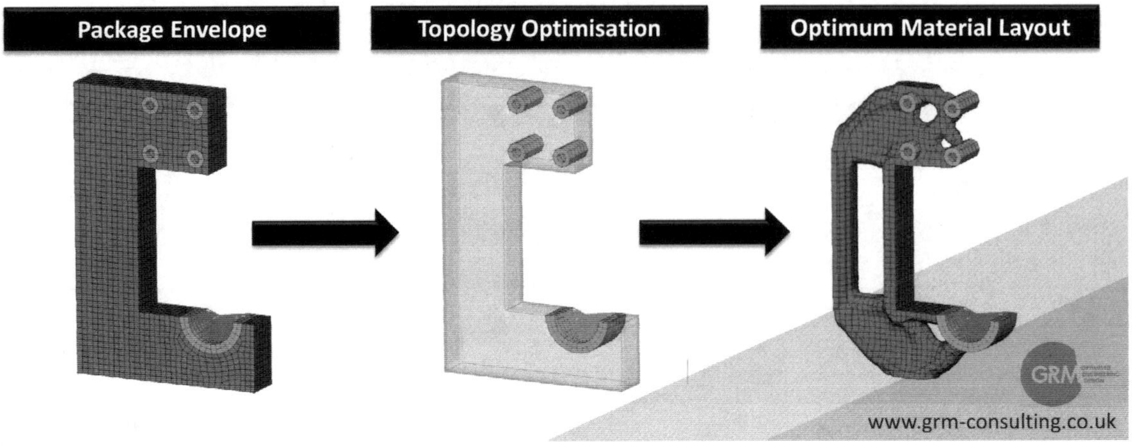

VR&D Genesis

- **VR&D Genesis** provides the most complete suite of optimisation capabilities.
- Each method is available for all analysis methods within **VR&D GENESIS** and can efficiently consider combination of multiple loading requirements.

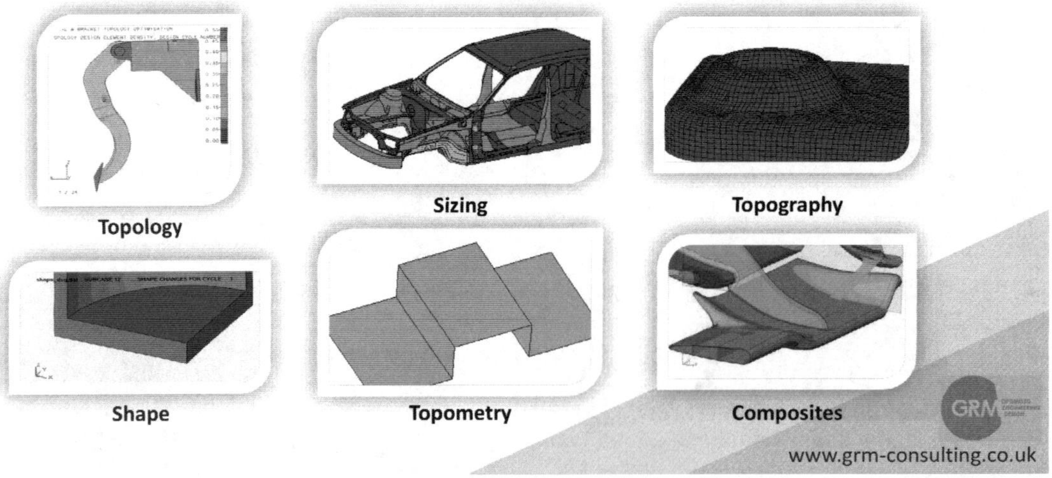

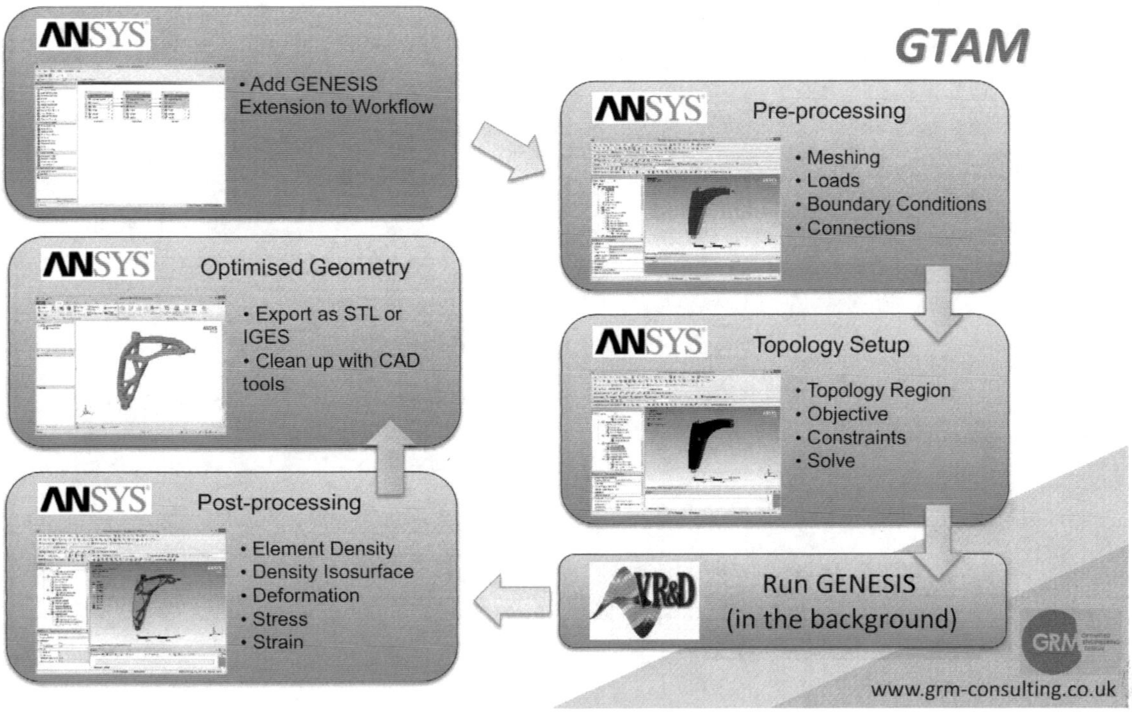

VR&D GENESIS

- **VR&D Genesis** provides the most complete suite of optimisation capabilities.
- Each method is available for all analysis methods within **VR&D GENESIS** and can efficiently consider combination of multiple loading requirements.

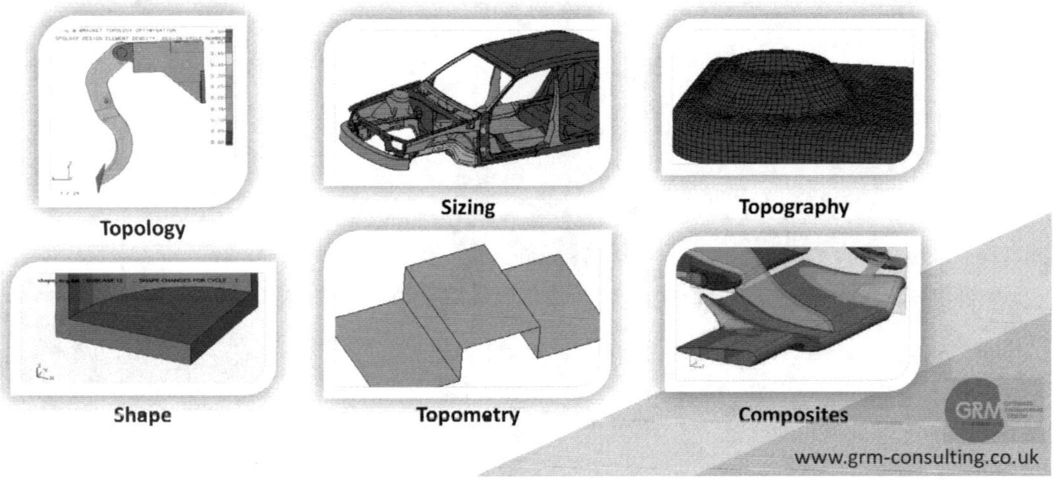

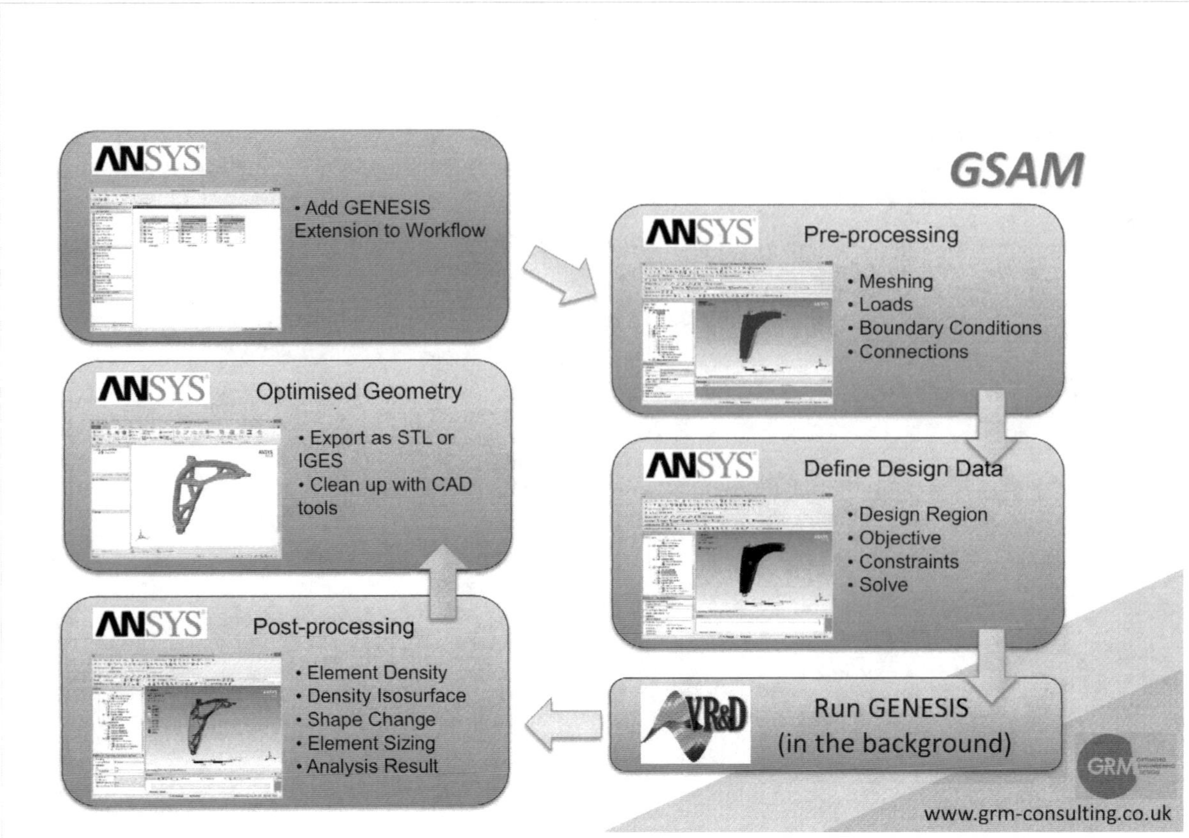
GSAM
ANSYS
• Add GENESIS Extension to Workflow
ANSYS Pre-processing
• Meshing
• Loads
• Boundary Conditions
• Connections
ANSYS Define Design Data
• Design Region
• Objective
• Constraints
• Solve
VR&D
Run GENESIS (in the background)
ANSYS Post-processing
• Element Density
• Density Isosurface
• Shape Change
• Element Sizing
• Analysis Result
ANSYS Optimised Geometry
• Export as STL or IGES
• Clean up with CAD tools
www.grm-consulting.co.uk

ULTRAN PROJECT
CASE STUDY
www.grm-consulting.co.uk

WHAT IS ULTRAN?

- 3 year collaborative research project.
- Aims to provide a step change in powertrain weight.
- Target saving across powertrain of 25% while meeting strict stiffness, strength and NVH requirements.
- The ULTRAN project is partially funded by Innovate UK as part of the IDP9 Technology Challenge in Low Carbon Vehicles programme.

American Axle & Manufacturing GRM Newcastle University Jaguar Land Rover Ricardo WMG The University of Warwick Tata Steel University of Southampton

Project Partners

Innovate UK
Technology Strategy Board

www.grm-consulting.co.uk

GRM'S ROLE IN ULTRAN

Explore new Computer-Aided Engineering (CAE) techniques which would allow multi-disciplinary optimisation of components within a passenger car powertrain.

www.grm-consulting.co.uk

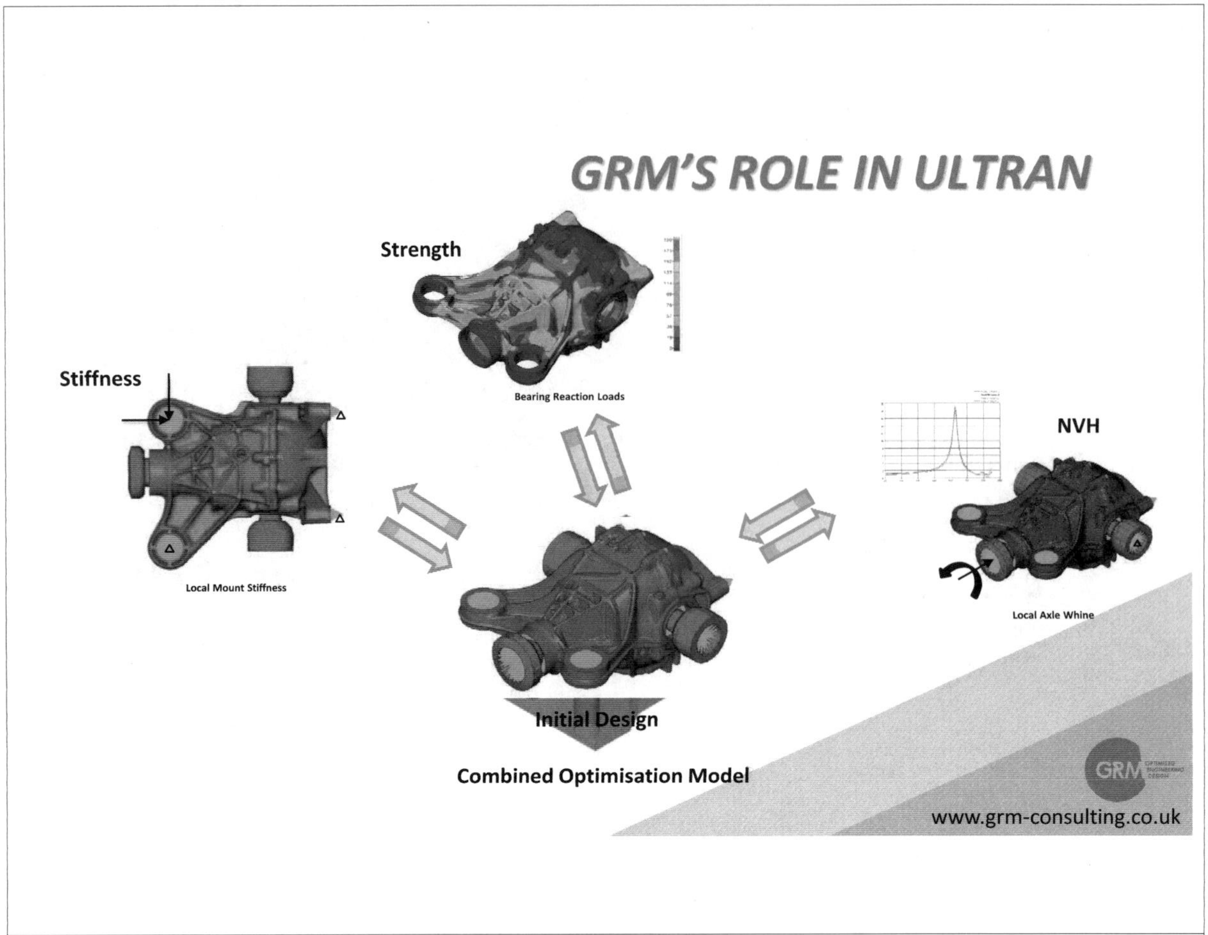
GRM'S ROLE IN ULTRAN
Strength
Bearing Reaction Loads
Stiffness
Local Mount Stiffness
NVH
Local Axle Whine
Initial Design
Combined Optimisation Model
GRM
www.grm-consulting.co.uk

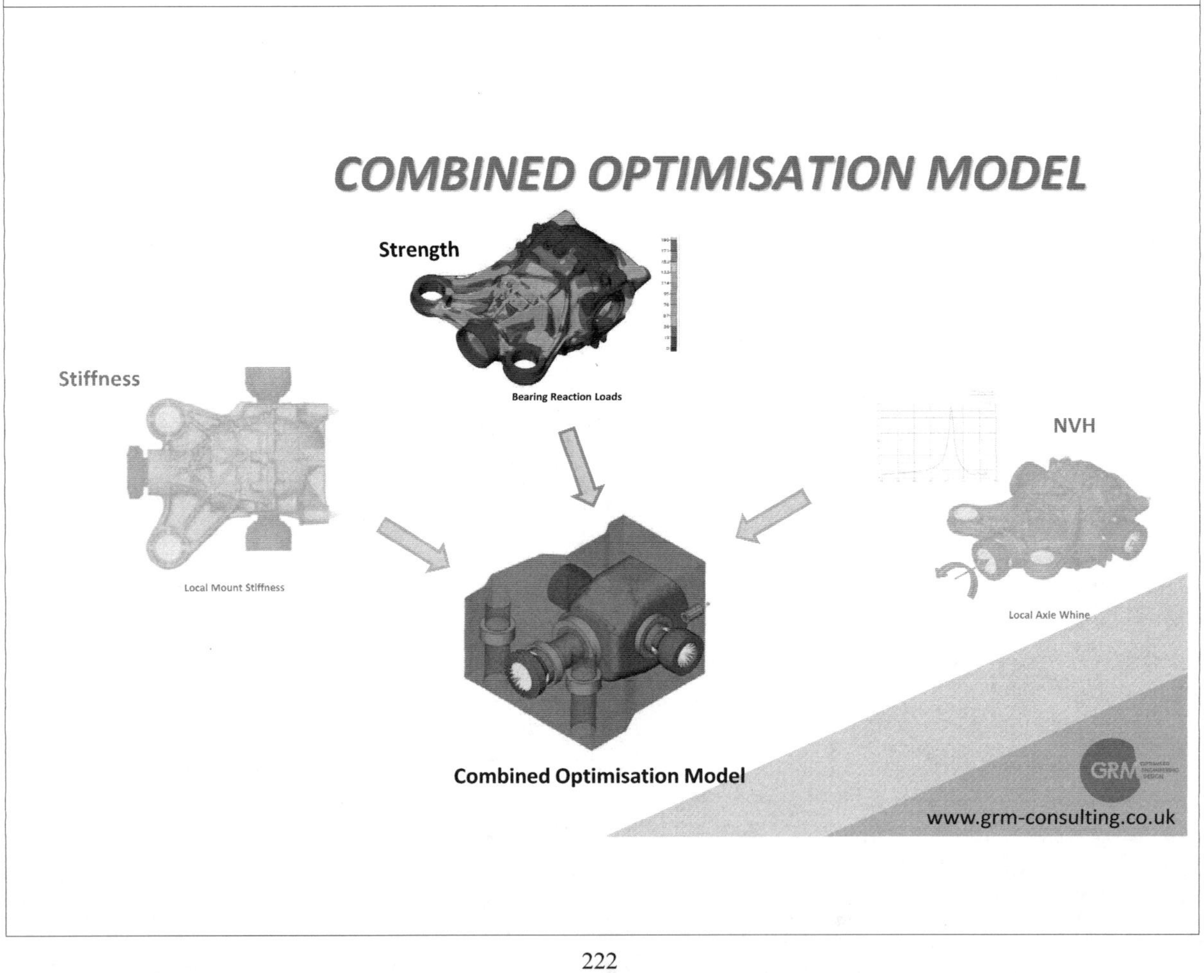
COMBINED OPTIMISATION MODEL
Strength
Bearing Reaction Loads
Stiffness
Local Mount Stiffness
NVH
Local Axle Whine
Combined Optimisation Model
GRM
www.grm-consulting.co.uk

BEARING REACTION LOADS

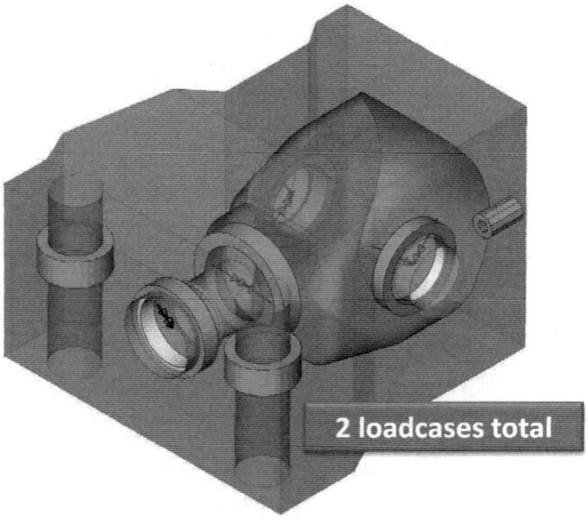

Bearing Reaction Loads

- Represents forces and moments transferred to the casing as a result of different torque loads (forwards, reverse, abuse)
- Consists of two loadcases:
  - Forwards abuse torque
  - Reverse abuse torque
- Optimisation target:
  - Stress to be below material yield, in this case aluminium A380 (250MPa).
- Baseline performance:

Loadcase	Stress (MPa)
Forwards Abuse Torque	217
Reverse Abuse Torque	207

www.grm-consulting.co.uk

COMBINED OPTIMISATION MODEL

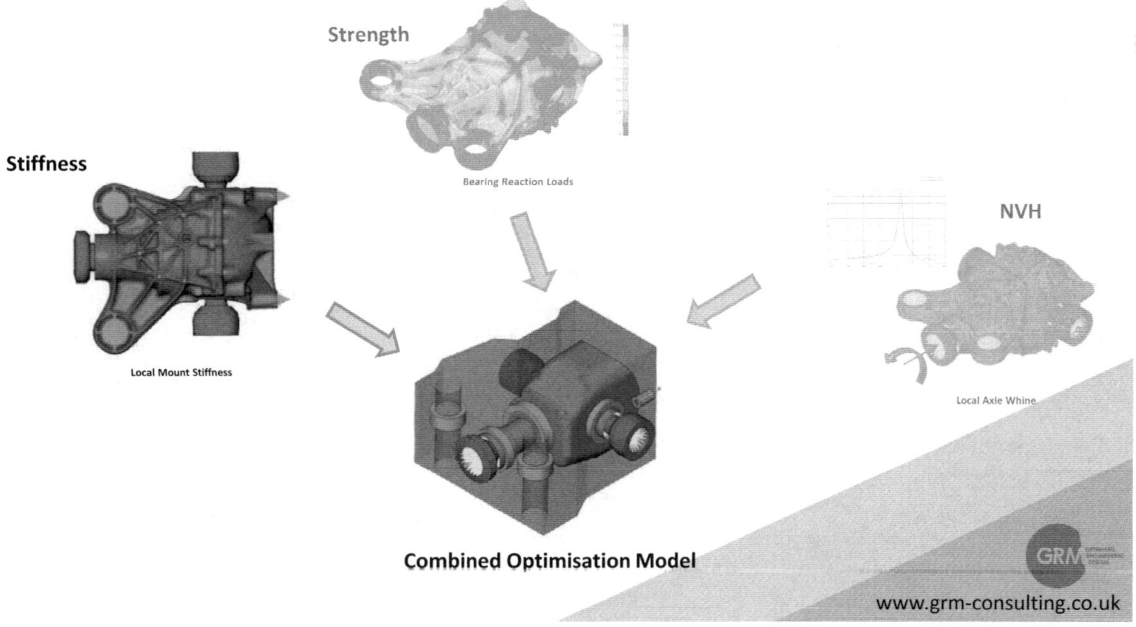

Combined Optimisation Model

www.grm-consulting.co.uk

LOCAL MOUNT STIFFNESS

- Mounts are forced statically in X, Y and Z directions while remaining mounts are constrained in all degrees of freedom.
- Mounts must be stiffer than targets listed below.

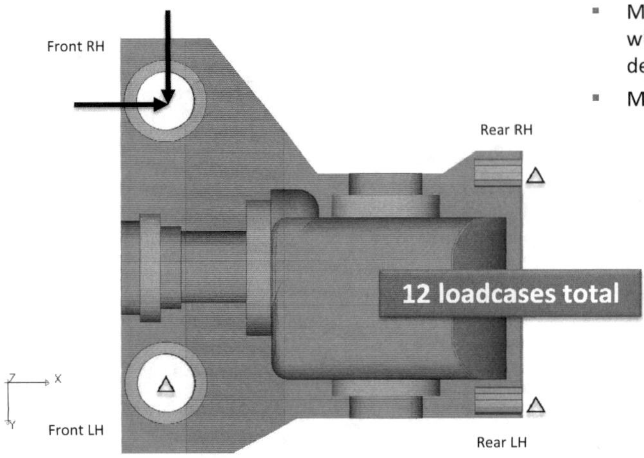

Mount Stiffness

	Static Targets (N/mm)		
	x	y	z
Front LH	6000	3300	11000
Front RH	13500	3500	17100
Rear LH	1900	15000	6200
Rear RH	1900	15000	6200

	Optimisation Targets* (mm)		
	x	y	z
Front LH	0.167	0.303	0.091
Front RH	0.074	0.286	0.058
Rear LH	0.526	0.067	0.161
Rear LH	0.526	0.067	0.161

*Note: Optimisation targets are based on a static load of 1000N

www.grm-consulting.co.uk

COMBINED OPTIMISATION MODEL

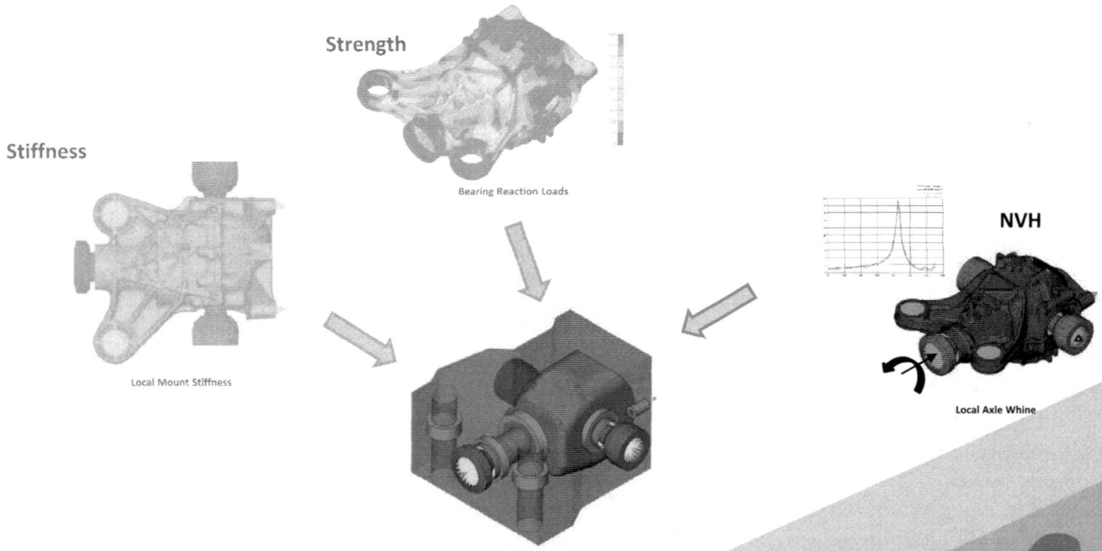

Combined Optimisation Model

www.grm-consulting.co.uk

LOCAL AXLE WHINE

- The system is excited at the centre of the drive unit flange by a constant dynamic torque of 1Nmm applied from 200Hz to 850Hz.
- There is an acceleration target at each mount.

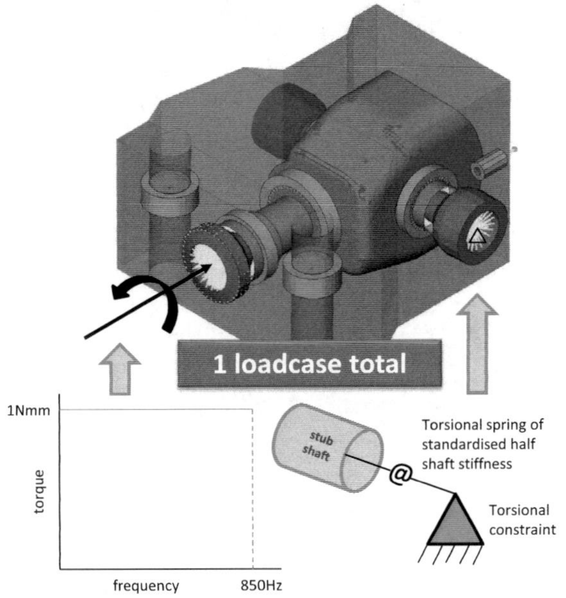

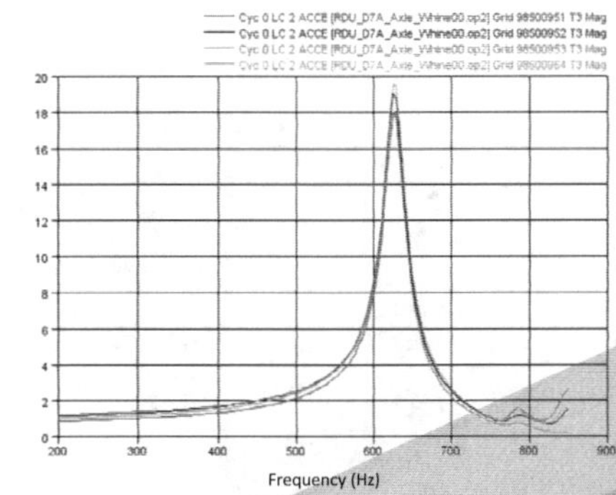

Chart showing accelerations at four mounting points

www.grm-consulting.co.uk

COMBINED OPTIMISATION MODEL

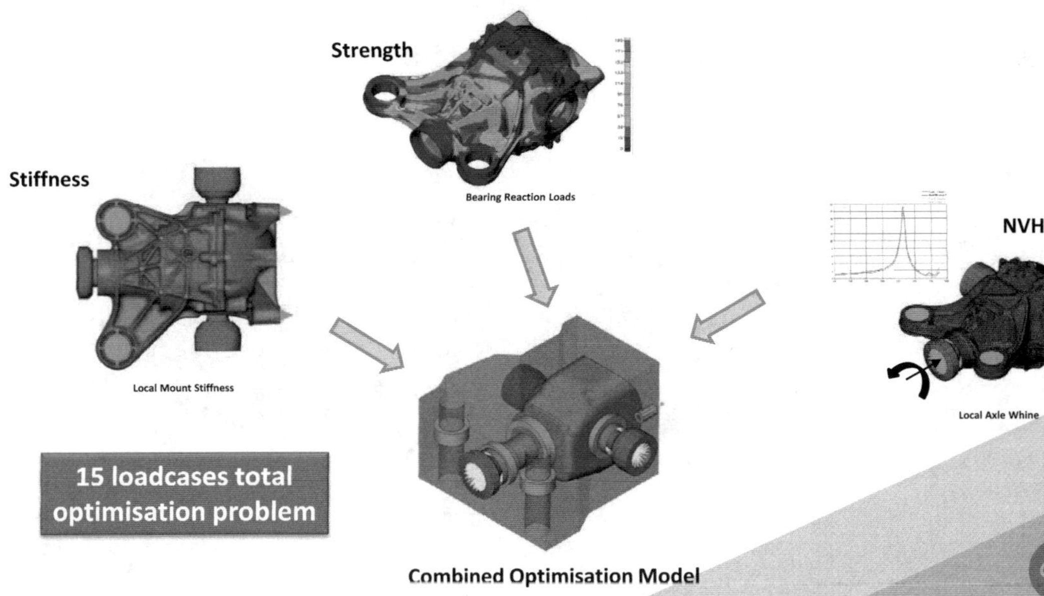

www.grm-consulting.co.uk

FIXED MASS STIFFNESS OPTIMISATION

- Indicates the optimum placement of material to increase stiffness of casing.
- Stresses decrease as casing mass is increased, however values are far in excess of targets because they are not being considered during the optimisation.

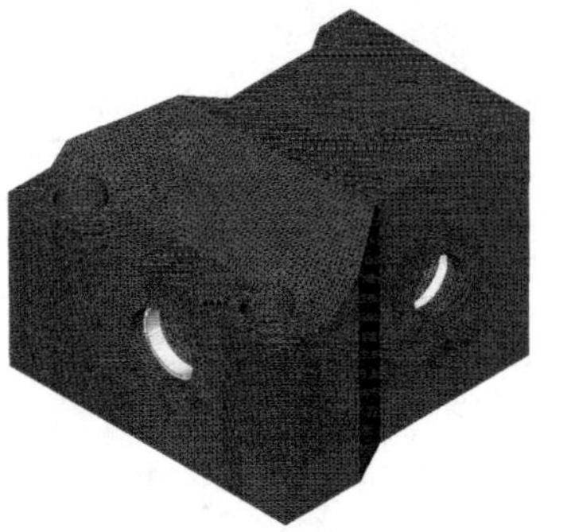

Animation of 5.3kg optimisation

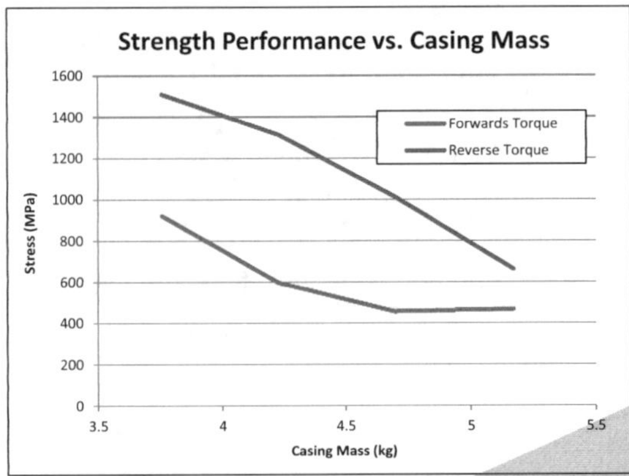

Chart showing peak stresses plotted against various casing masses

www.grm-consulting.co.uk

MINIMISE MASS STRESS OPTIMISATION

- Indicates the optimum placement of material to increase stiffness **AND** strength.
- Allows for minimum mass optimisations to be conducted, meaning mass of casing is determined by **VR&D GENESIS** as opposed to the engineer.

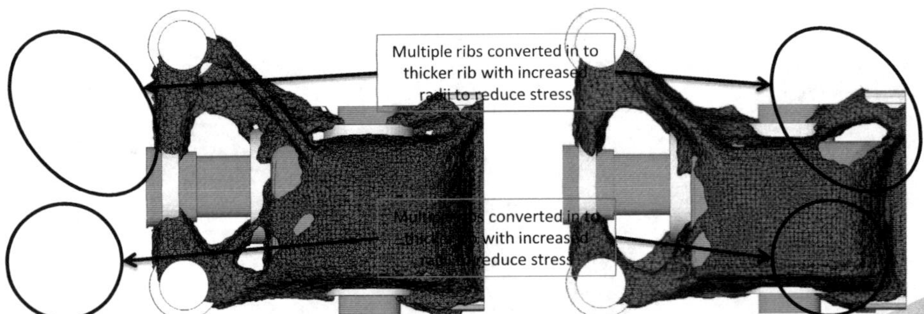

Fixed mass optimisation considering stiffness only

Minimise mass optimisation considering stress and stiffness

www.grm-consulting.co.uk

STIFFNESS & STRENGTH OPTIMISATION

- Minimise mass optimisation based on stiffness **AND** strength means that **VR&D GENESIS** can determine the mass of the casing.
- Stress values are far more reasonable and in line with baseline RDU performance.

Mass = 5.3kg
Stress = 622MPa

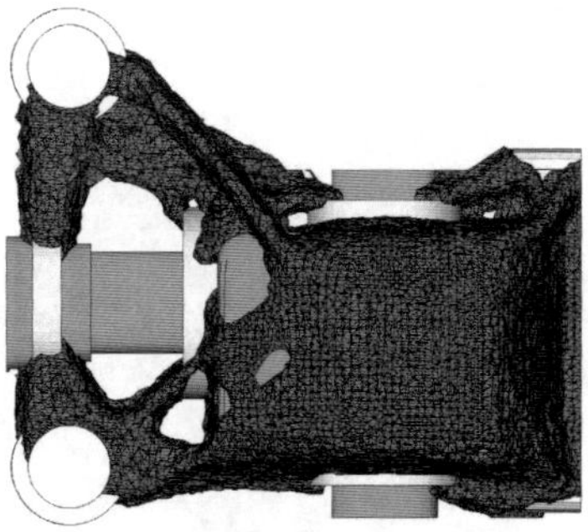

Fixed mass optimisation utilising strain energy means that mass of the RDU casing is determined by the engineer.

Mass = 5.3kg
Stress = 274MPa

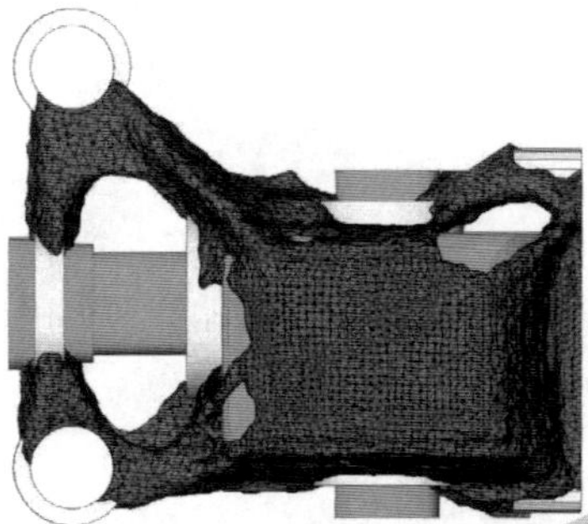

Minimise mass optimisation based on stress and stiffness mean that optimiser can determine mass of casing.

www.grm-consulting.co.uk

COMBINED OPTIMISATION RESULTS

Loadcase	Baseline (MPa)	Stiffness Opt (MPa)	Stiffness & Strength Opt (MPa)
Forwards Abuse Torque	217	467	226
Reverse Abuse Torque	207	662	274

Isosurface | **Forwards Torque** | **Reverse Torque**

Stiffness Only

Stress & Stiffness

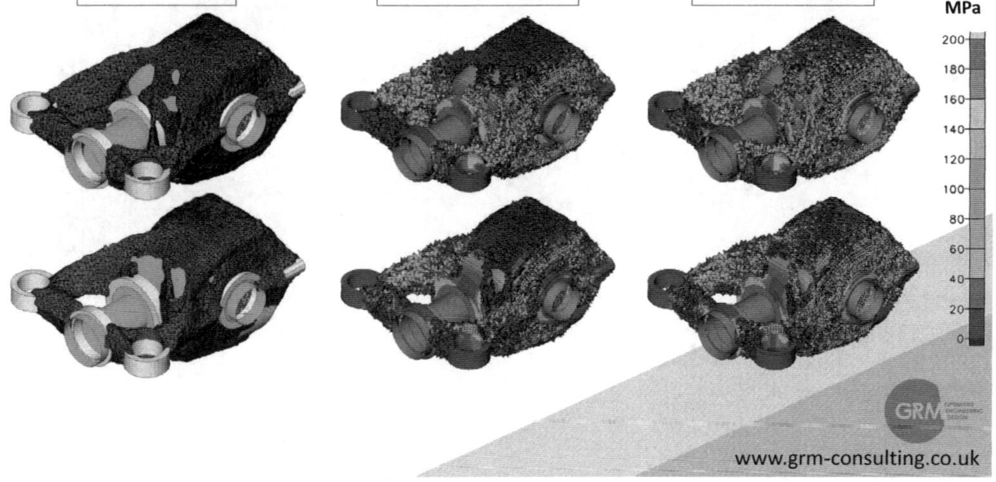

www.grm-consulting.co.uk

COMBINED OPTIMISATION RESULTS

- Stress hotspots are clearly visible in the traditional fixed mass stiffness optimisation (results are from the reverse abuse loadcase).
- When stresses are considered on the design region during the optimisation process, these stress concentrations are significantly reduced.

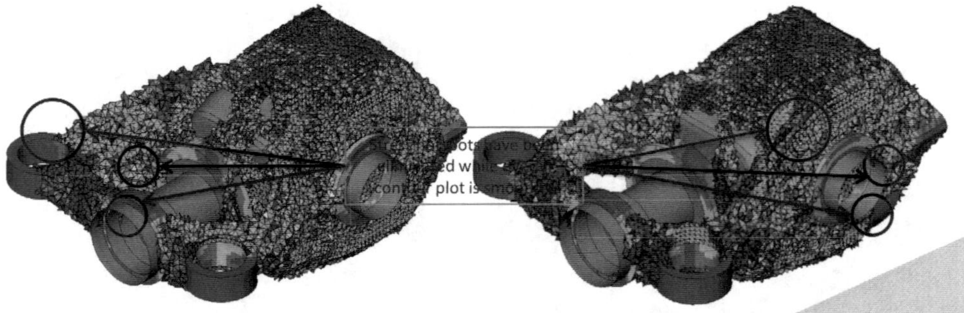

Fixed mass optimisation considering stiffness only

Minimise mass optimisation considering stress and stiffness

www.grm-consulting.co.uk

OPTIMISATION CASE STUDY RESULTS

- RDU casing developed using optimisation methodology that incorporates stress constraints.
- RDU casing stresses are reasonable and comparable to baseline performance.
- As well as casing stresses, strict stiffness targets and NVH targets have been included and met.
- All this has been accomplished with a final mass saving of 32% over the baseline.

Optimised Mass (Kg)	Change (%)
5.3	32

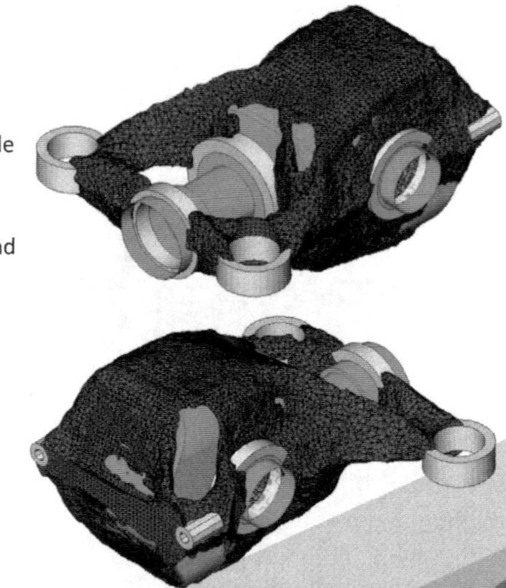

www.grm-consulting.co.uk

POWERTRAIN OPTIMISATION METHODS

- Utilising the latest developments in **VR&D GENESIS**, available to ANSYS users as GTAM, an advanced powertrain optimisation methodology was developed.
- This methodology allows us to start with a blank canvas, and utilise a CAE led approach to design and development.
- By using optimisation as a brainstorming tool a lightweight concept design has been developed for multiple attributes.
- The methodology developed using this case study has subsequently been utilised in a CAE led design process to design and deliver a prototype RDU.

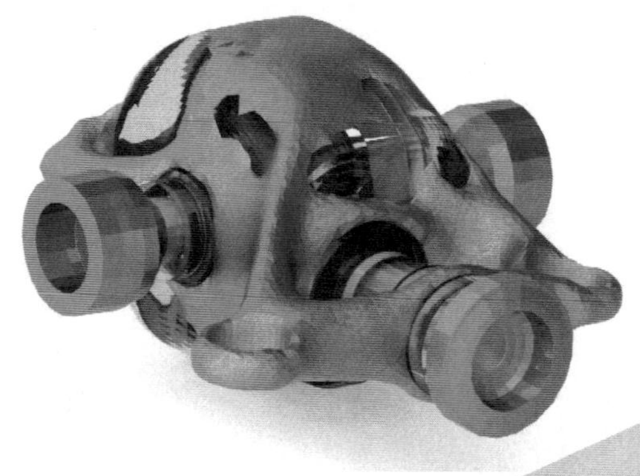

www.grm-consulting.co.uk

ULTRAN PROJECT OUTCOMES

Project Outcomes

- An RDU weight saving of 25% (8kg)
- RDU has been manufactured and is currently being tested

Optimisation Innovations

- Multi-disciplinary topology optimisation method incorporating stress constraints
- Optimisation method utilised to determine structure of skeletal casing

ULTRAN Rear Drive Unit: **25.6kg**

Baseline Rear Drive Unit: **34.1kg**

www.grm-consulting.co.uk

SUMMARY

- GTAM / GSAM
  - Integrated optimisation extensions for ANSYS Mechanical
  - Most complete suite of optimisation capabilities available for Ansys community
- ULTRAN
  - Enhanced existing & developed new optimisation methods
  - Developed revolutionary, lightweight RDU
- **GENESIS and the integrated ANSYS toolsets are meeting the needs of automotive manufacturers by helping to reduce mass and CO2 emissions.**

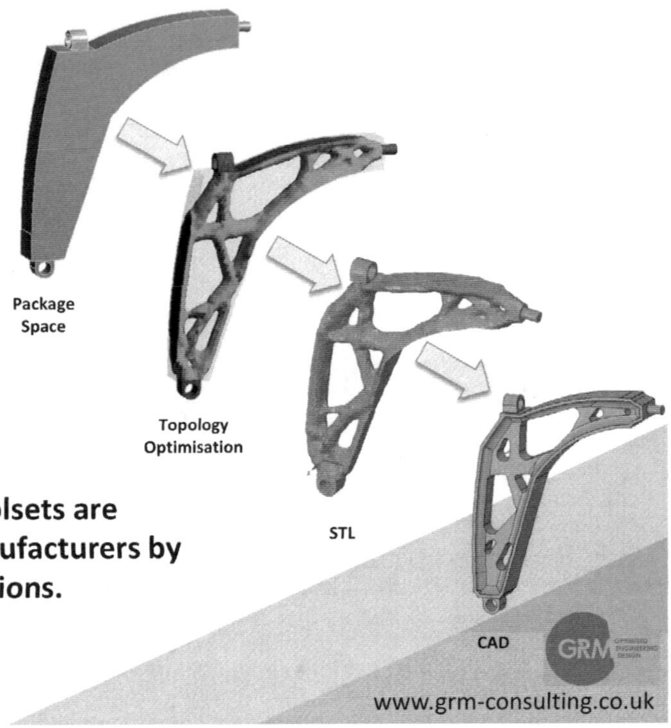

Automotive Simulation
WORLD CONGRESS

June 7-8, 2016 Munich

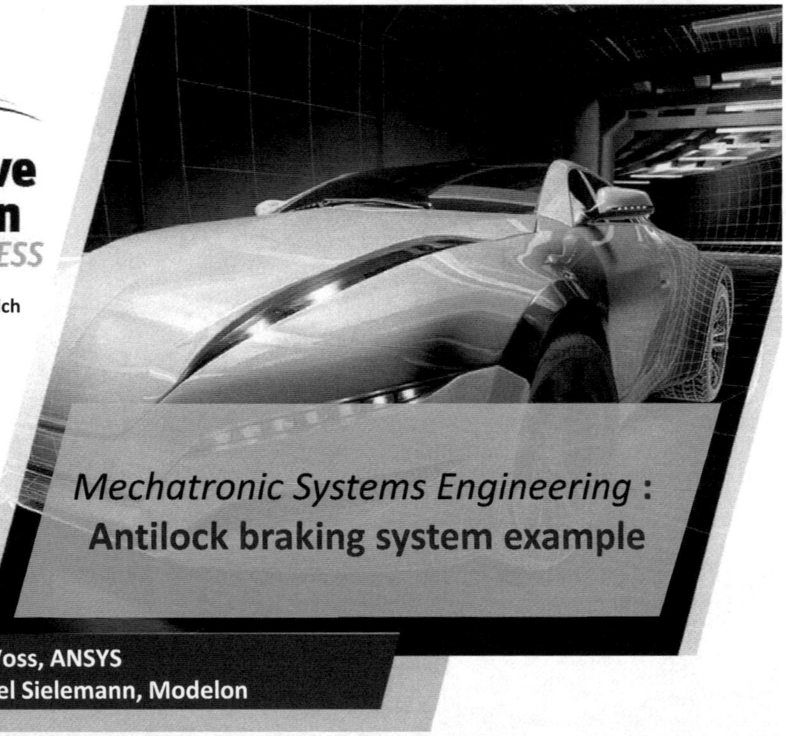

Mechatronic Systems Engineering : Antilock braking system example

Dr.-Ing. Leon Voss, ANSYS
Dr.-Ing. Michael Sielemann, Modelon

Overview

- Motivation
- ANSYS & Modelon solutions for System Simulation
- Example System Model of an ABS Sub-system
- Modelica and FMI as enablers for System Simulation
- Challenges and Outlook

Motivation
Mechatronic Systems Engineering

- Innovation in Automotive technology increases complexity of system design. Drivers are:
  - Functional Safety
  - Electrification
  - Cloud Connectivity
  - Advanced driver-assistance systems (ADAS)
- Higher number of features results in exploding effort for specification and test
- System integration requires closer collaboration between suppliers and OEMs through product lifecycles

Automotive Simulation WORLD CONGRESS

Virtual Prototypes
Mechatronic Systems Engineering

- Virtual prototypes representing mechanical, electrical and software aspects are critical to:
  - study system interactions in a dynamic environment to enable early verification of requirements
  - automate the validation effort and reduction of physical testing through virtual testing
- Example: ABS Study of vehicle dynamics + control algorithm

Automotive Simulation WORLD CONGRESS

Challenges

Mechatronic Systems Engineering

- Collaboration
  - across disciplines
  - across locations/timezones
  - across departments/companies
- IP Protection
  - model exchange technology
  - licensing/access priviledges
- Model/Data Consistency
  - version/lifecycle management
  - model interfaces
  - model reduction

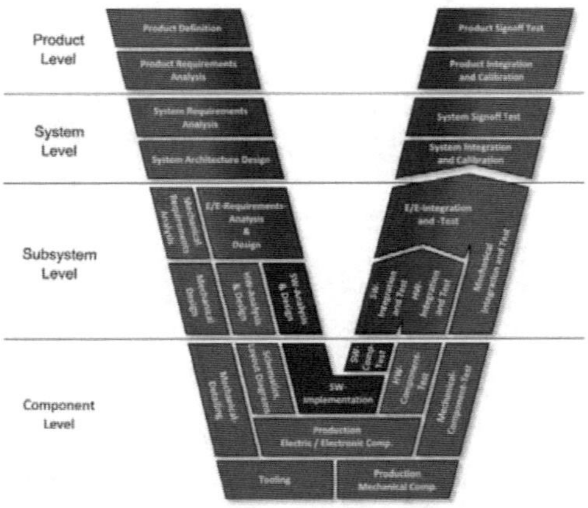

Source: ProStep iViP RECOMMENDATION, Smart Systems Engineering Behavior Model Exchange, Version 1.0

ANSYS Technology Overview

Comprehensive Component-Level Design & Simulation ...

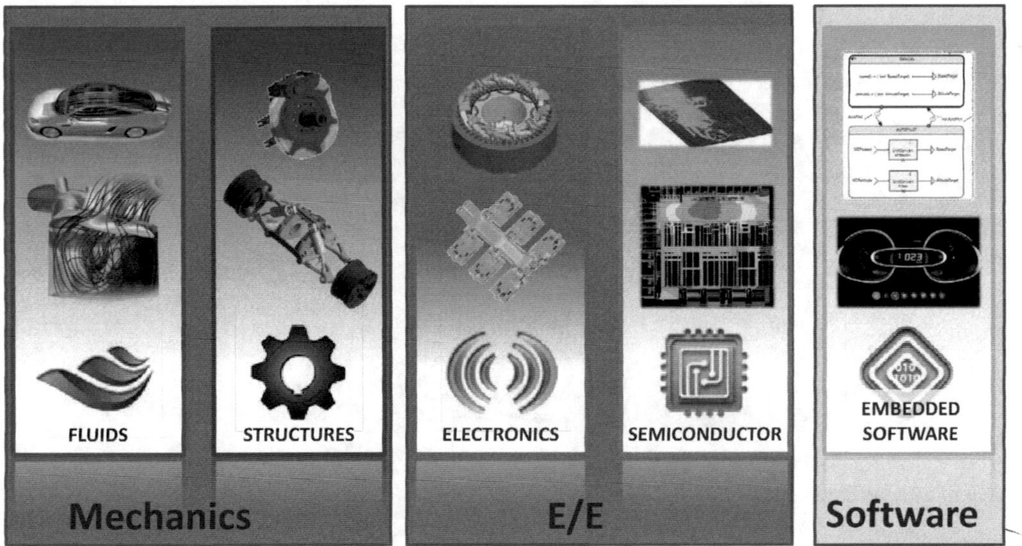

ANSYS Modelon Partnership

Enables best-in-class mechatronics system simulation

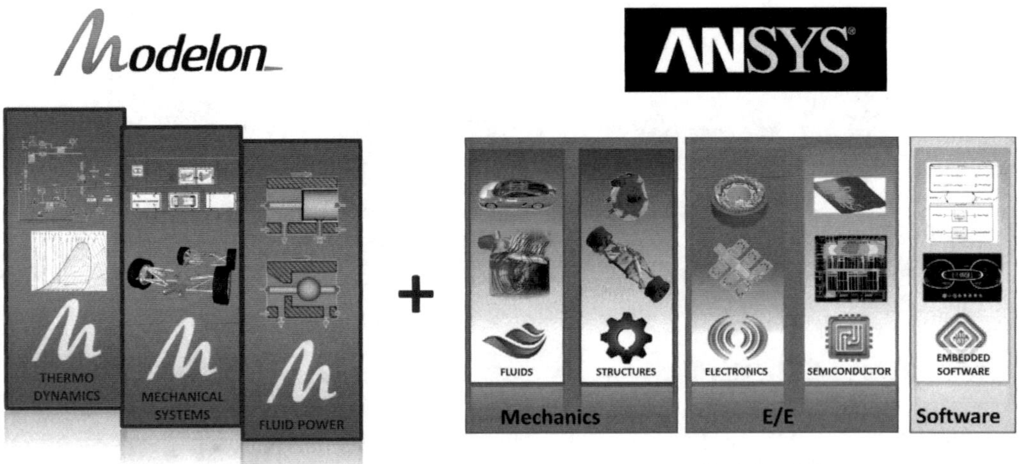

Best-in-class
Modelica Technology for
Fluids, Mechanics and Thermal

Best-in-class
FEM, CFD, EDA & Software
Design Technology

Collaborative System Development

Data Backbone for System Simulation

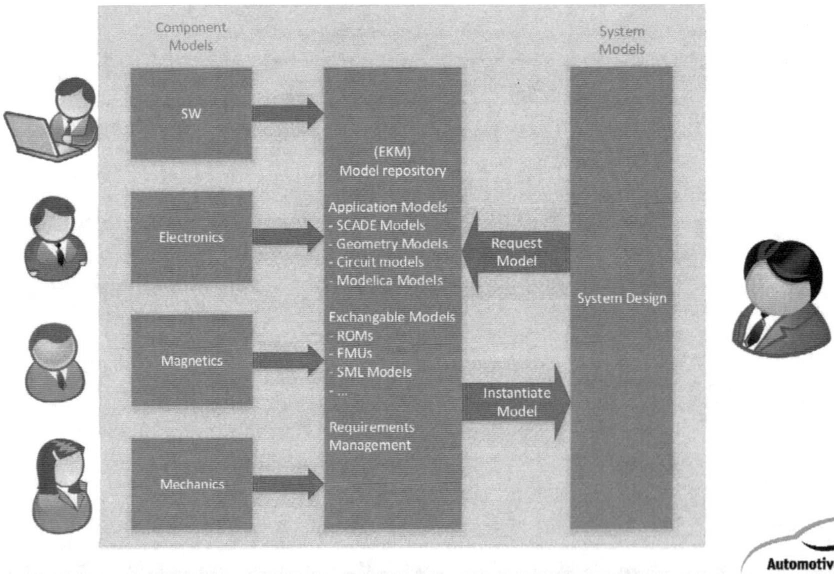

ANSYS Systems Engineering Solutions

ROM = Reduced-order model
FMI = Functional mock-up interface
CoSim = Co-Simulation

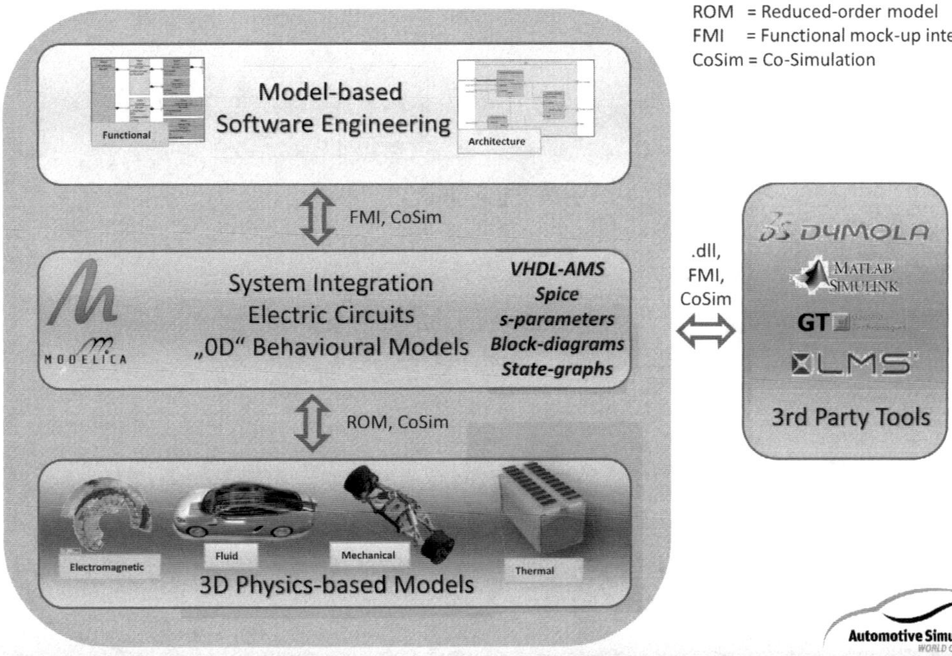

Tie Yourself to Standards, Not Tools

MODELICA

Modelica® (www.modelica.org)

- Easy-to-use multi-domain modeling
- Graphical and textual language
- First principles (physics)
- Open standard
- Large community and ecosystem

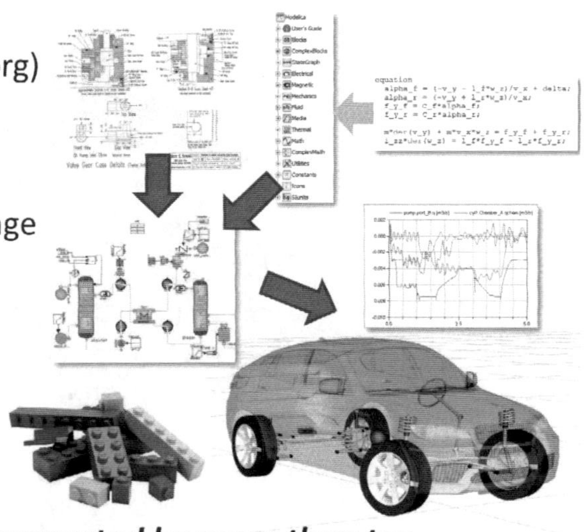

A <u>modelling language</u> supported by more than ten commercial and open source simulation packages

Tie Yourself to Standards, Not Tools (cont.) fmi

Functional Mock-up Interface® (www.fmi-standard.org) is a tool independent standard for model exchange and co-simulation based on xml-files and compiled C-code

- Model exchange:

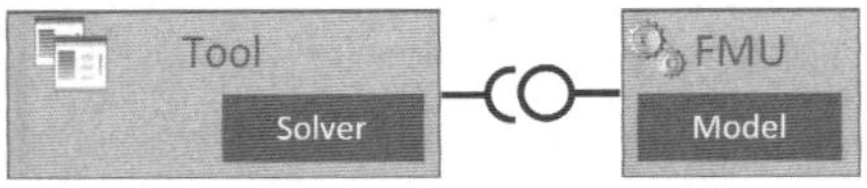

- Co-simulation:

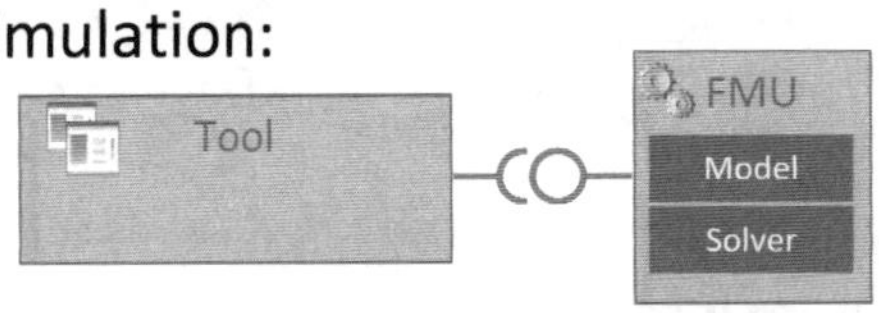

An <u>executable format</u> supported by more than 80 commercial and open source tools

Tie Yourself to Standards, Not Tools (cont.)

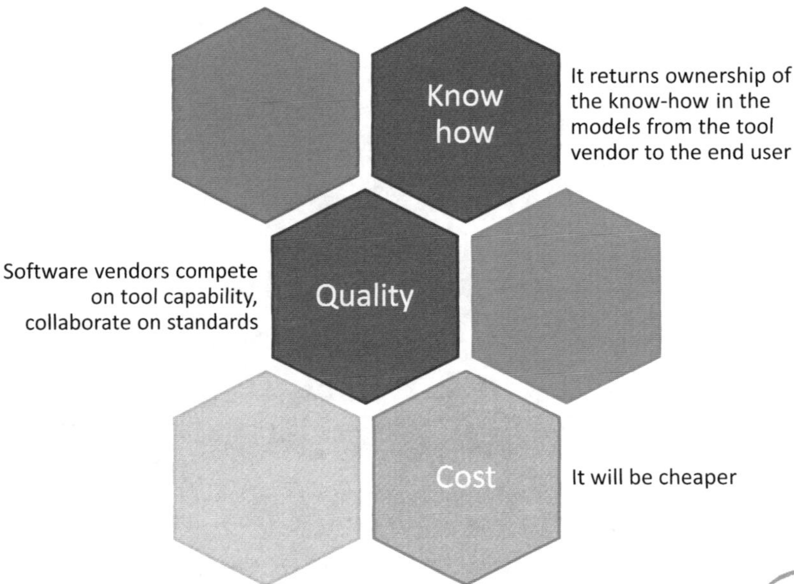

Example System – ABS
ANSYS Simplorer as an Integration Platform

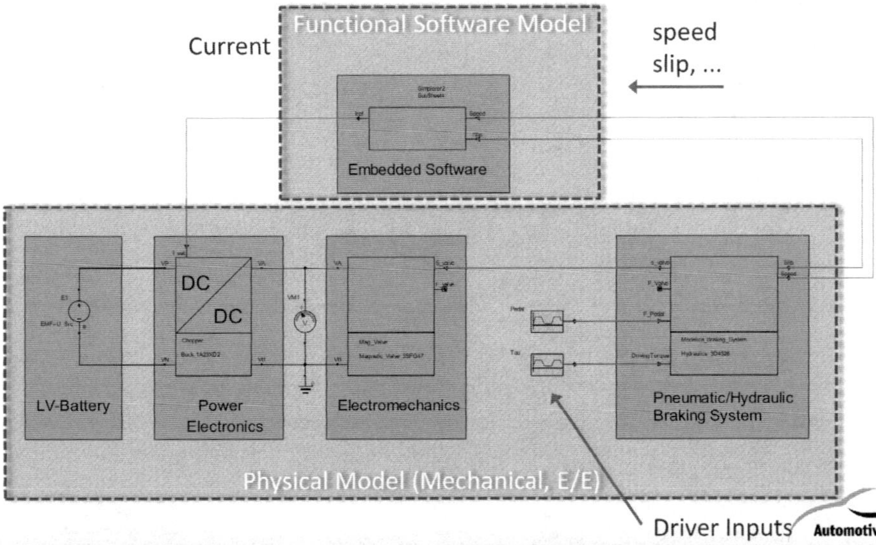

Automotive Simulation WORLD CONGRESS

Magnetic Modelling

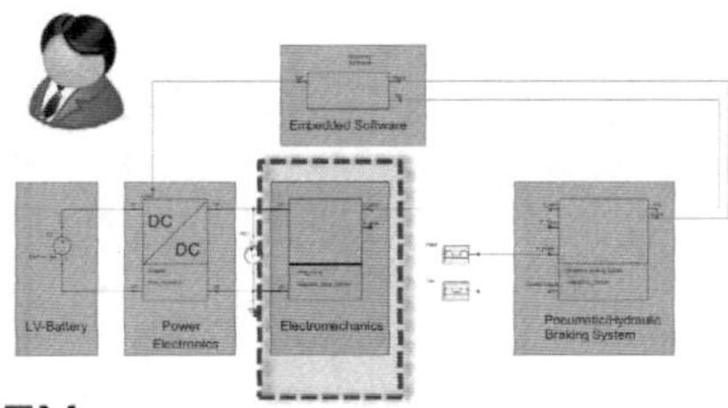

- System Modelling Approaches
  - Reduced-order Model (ROM) from FEM
  - Co-Simulation with FEM
- Modelling of spring and mass quantities as network elements

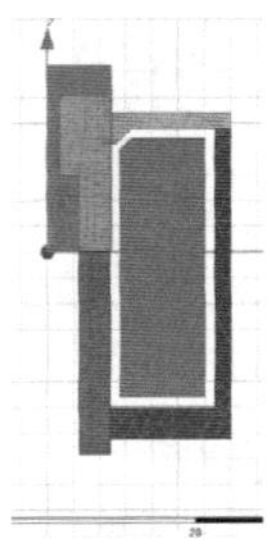

Magnetic FEM Model

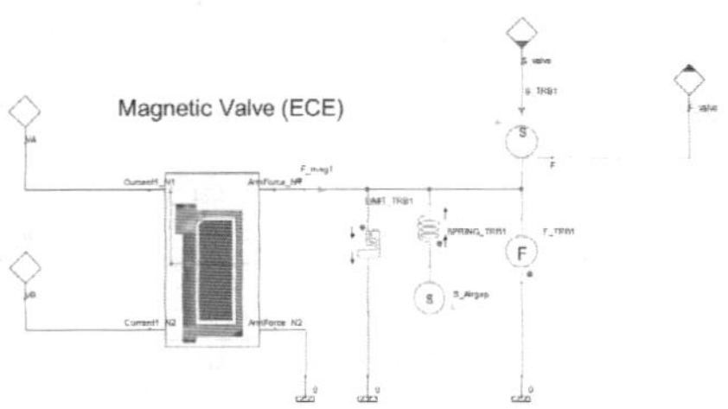

Electrical/Mechanical Network Model

Automotive Simulation WORLD CONGRESS

Power Electronics Modelling

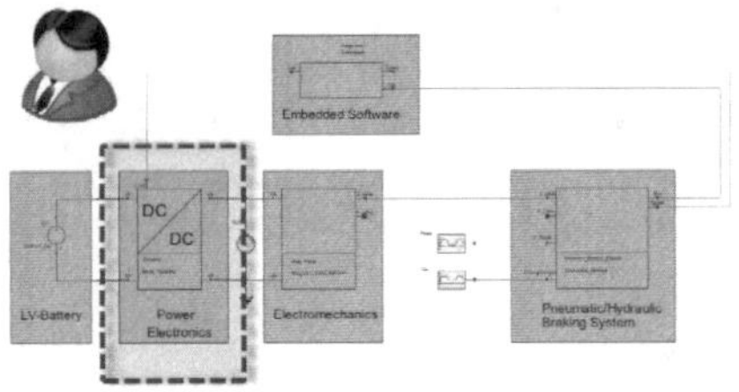

- Circuit modelling approaches
  - Simplorer Native
  - VHDL-AMS
  - Spice
- Abstraction Levels
  - Static
  - Electrothermal
  - Electrical Dynamic

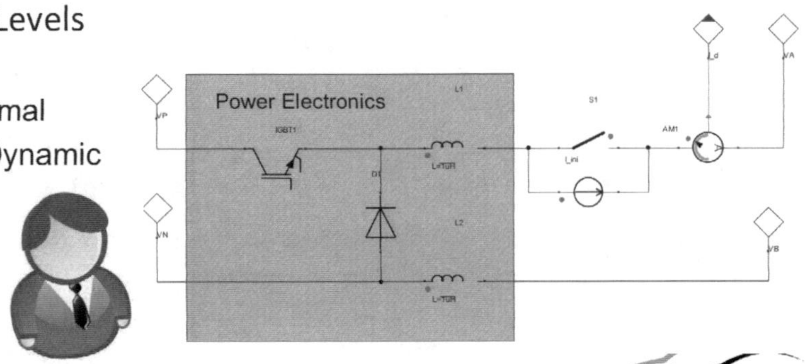

Model-based Embedded Software Design and Validation

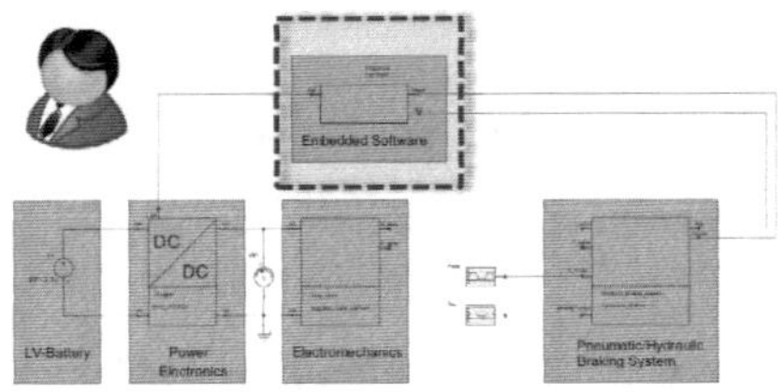

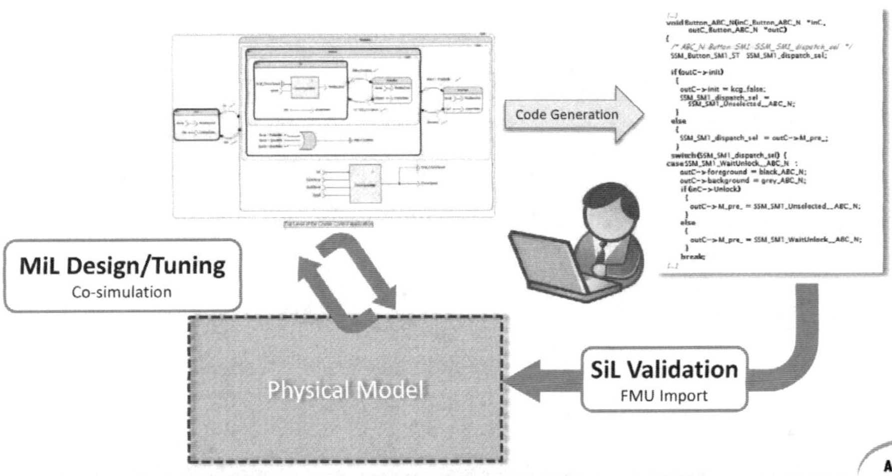

Pneumatic/Hydraulic Model

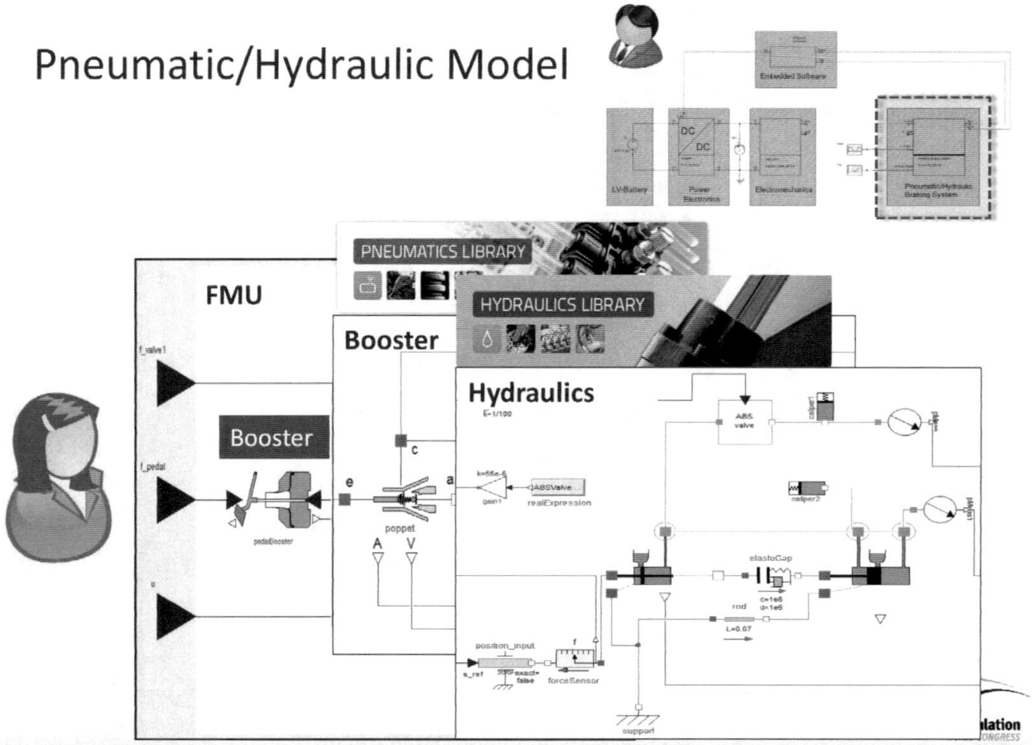

Example System – ABS
ANSYS Simplorer as an Integration Platform

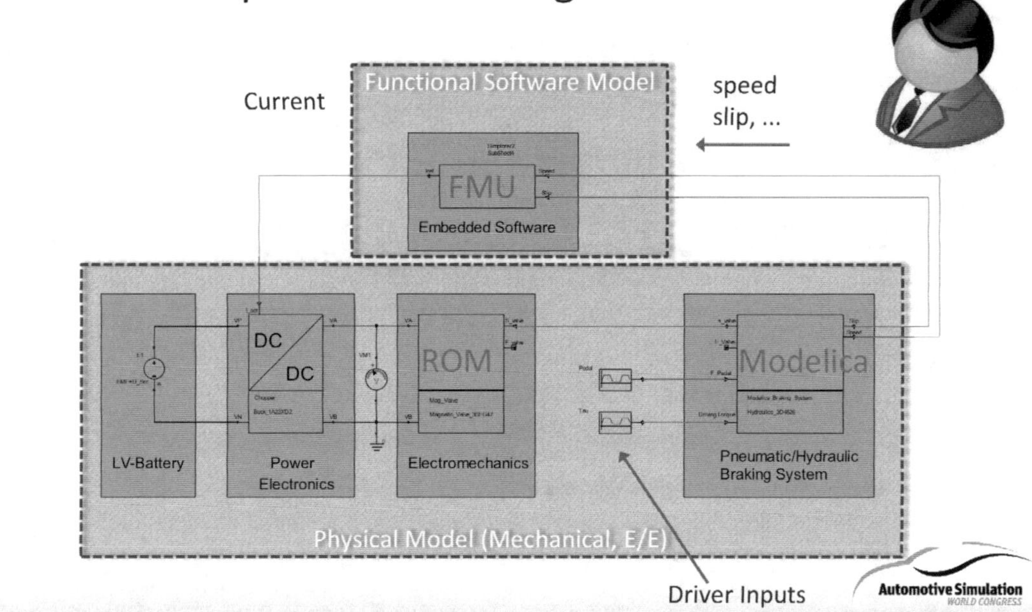

Simulation results with ABS Function
Braking transient from 15m/s

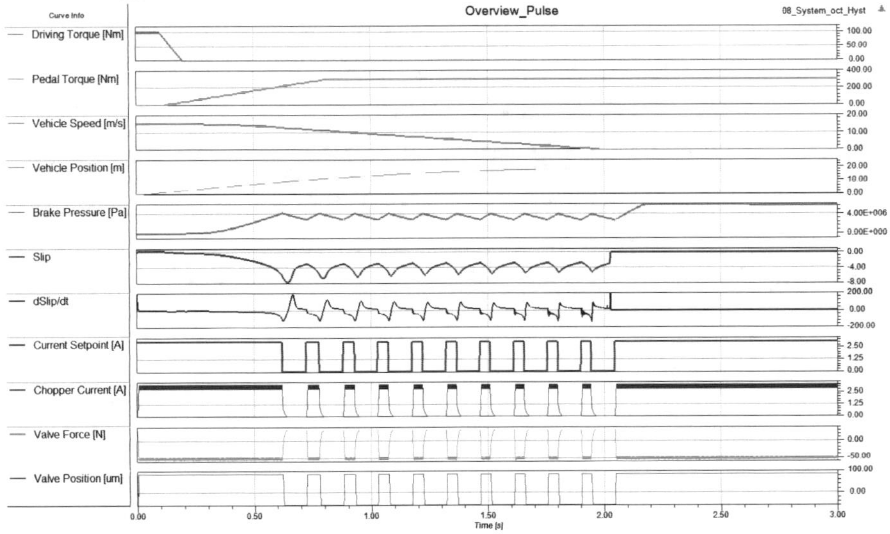

Simulation results without ABS Function
Braking transient from 15m/s

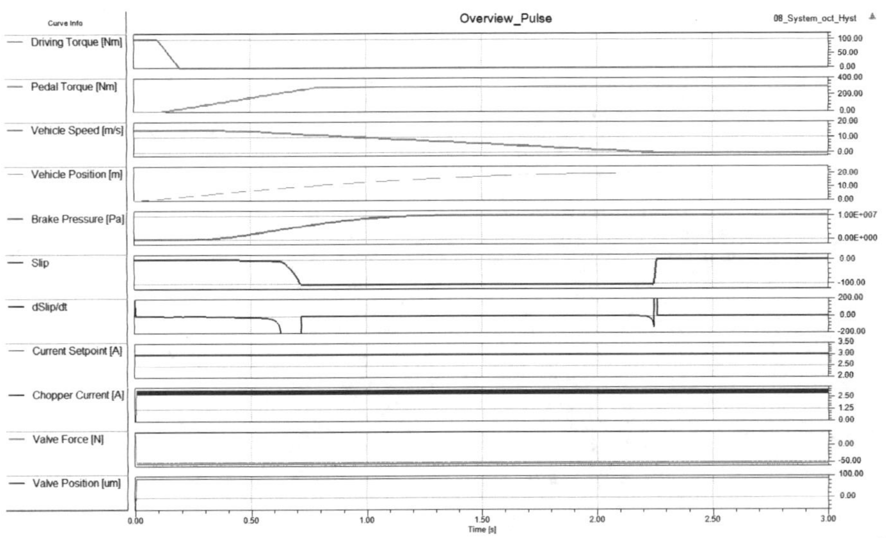

Speed Comparison with/without ABS Braking transient from 15m/s

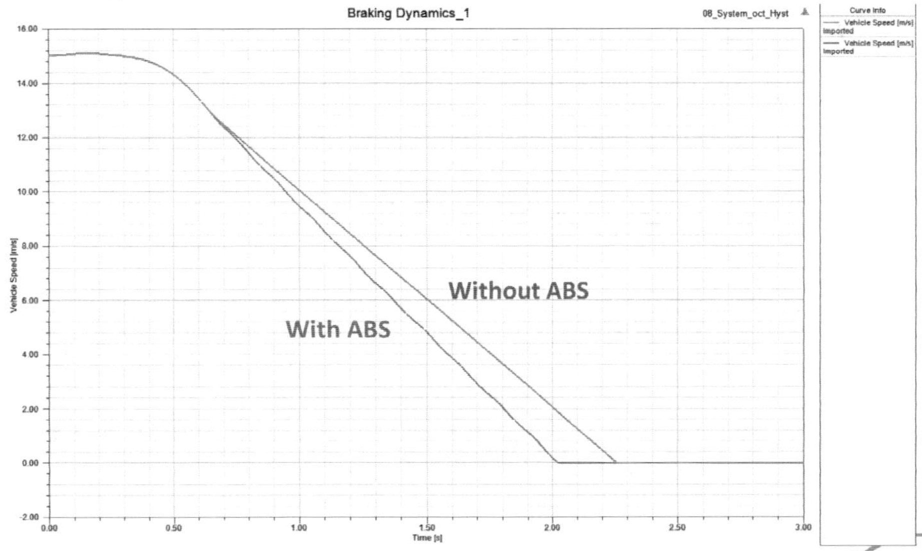

Automotive Simulation WORLD CONGRESS

Position Comparison with/without ABS Braking transient from 15m/s

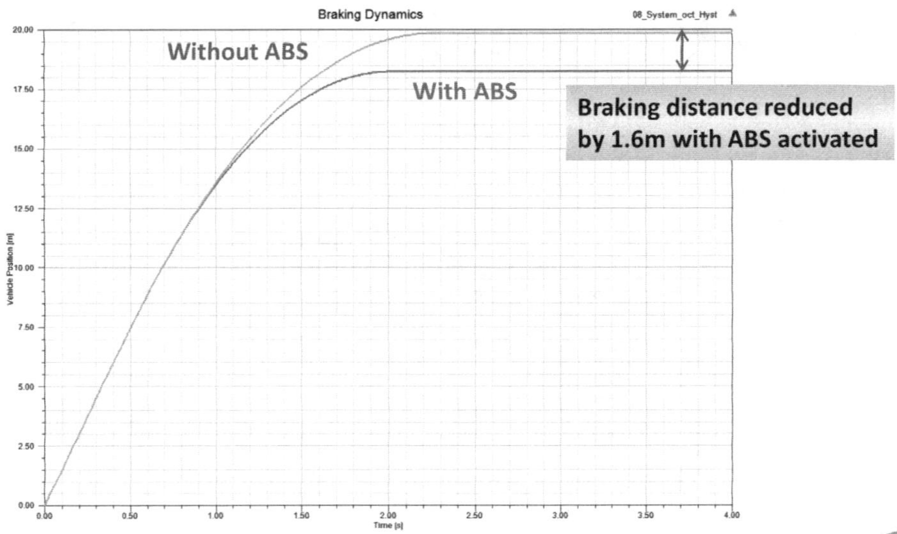

Automotive Simulation WORLD CONGRESS

Summary

- ANSYS and Modelon technology enable simulation of complex mechatronic systems such as ABS.
- Collaborative development (e.g. supplier/OEM) presents challenges of data/tool consistency and IP protection
- Standards for modelling and model-exchange such as FMI and Modelica support collaborative development
- Virtual prototyping helps manage the complexity of developing innovative automotive systems

Automotive Simulation WORLD CONGRESS

25 June 7-8, 2016 Munich

Thank you !

ANSYS

Modelon

leon.voss@ansys.com michael.sielemann@modelon.com

Automotive Simulation WORLD CONGRESS

26 June 7-8, 2016 Munich

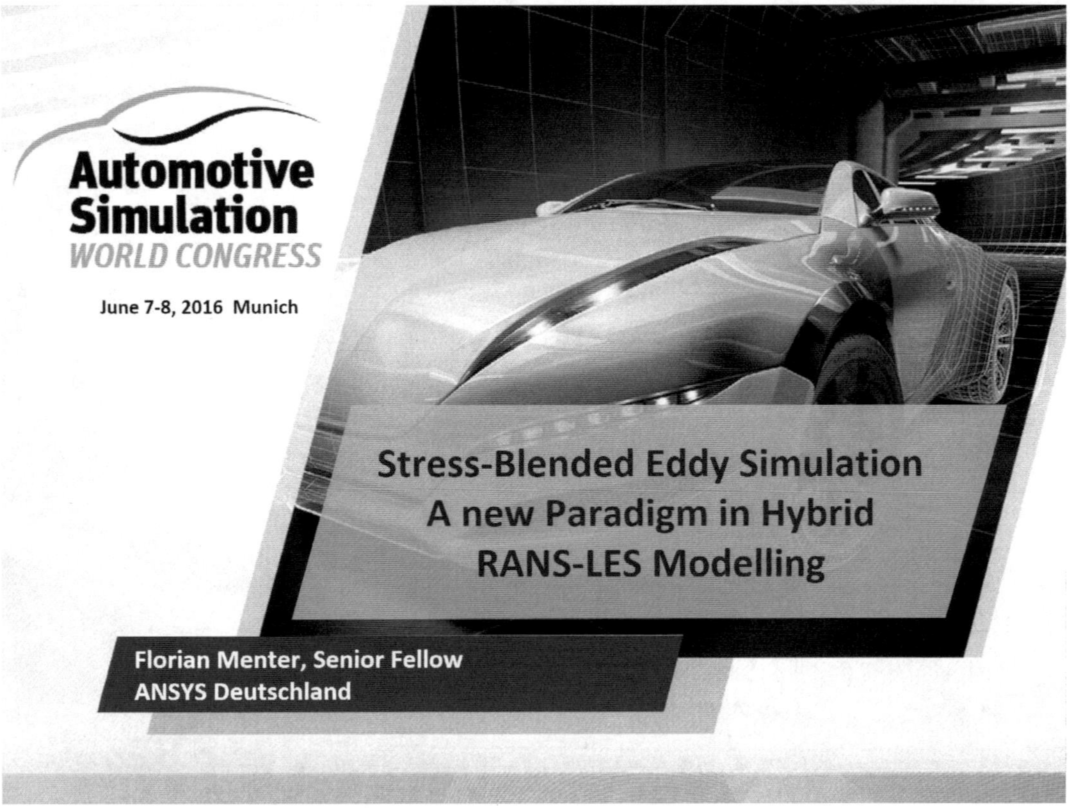

Scale –Resolving Simulation (SRS) in Automotive

- External aerodynamics
  - Formula 1 – flow mixing behind bluff wheels
  - Mixing of wake behind car/truck
  - Unsteady forces on car body
- Internal combustion engines (ICE)
  - Better prediction of shear layers emanating from valve
  - Improved mixing and tumble prediction
  - Unsteady flow details near ignition zone
- Aero-acoustics
  - Side mirror noise prediction
  - Cabin noise (from internal or external sources)
  - Muffler
- ....

ANSYS

Why Hybrid RANS-LES Models

- RANS models are very well suited for wall boundary layers
  - Highly tuned for flows up to separation
  - RANS models fail in zones of strong mixing away from walls
- Large Eddy Simulation (LES) of wall-bounded flows is extremely expensive
  - High resolution per boundary layer volume (several thousand cells)
  - Strong scaling with Reynolds number
  - Small time scales
- Solution – Hybrid Models
  - Utilize RANS for wall boundary Layers and LES away from the walls

ANSYS

Hybrid RANS-LES Methods

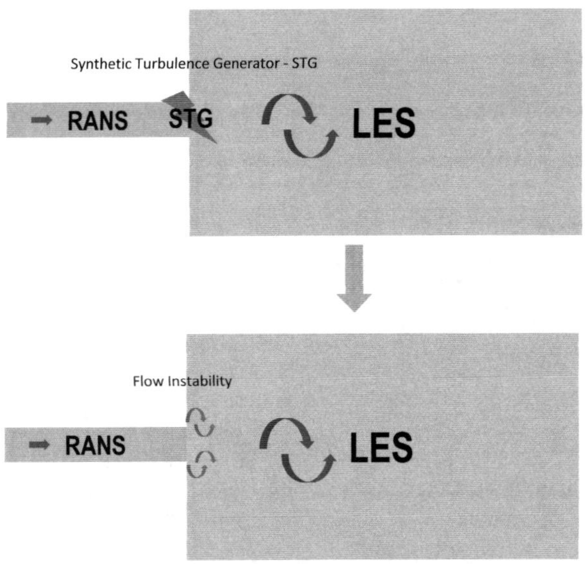

Embedded RANS-LS:

- User selects RANS model and RANS zone
- User selects LES model and LES zone
- At interface STG transfers modelled turbulence into resolved turbulence

Global Hybrid Models:

- User selects RANS model and method selects RANS zone
- User selects LES model and method selects LES zone
- Resolved turbulence is generated by flow instability

ANSYS

Hybrid Scenarios

- Flow detaches at corner
  - Boundary layer in RANS mode is essential
  - Free shear layers in LES

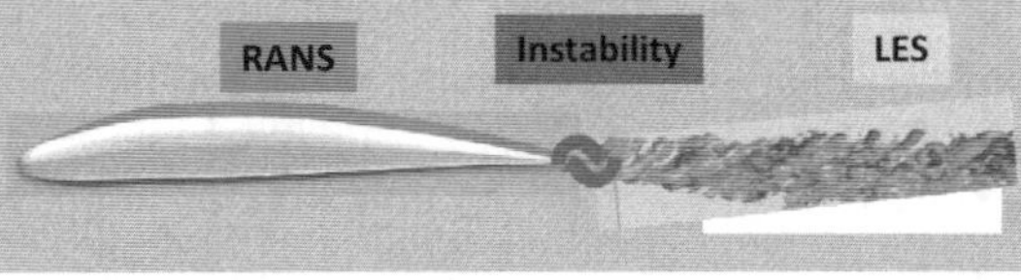

- Flow Separates and reattaches
  - Flow should remain in LES downstream of step
  - Recover to RANS on coarse grids

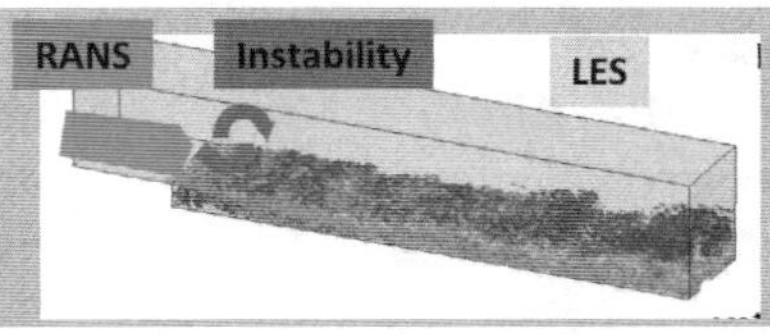

- Flow is Wall bounded
  - Should hybrid model do WMLES?

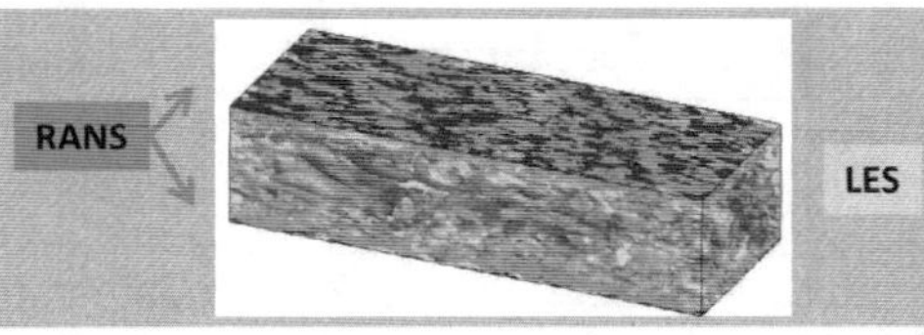

ANSYS

Detached Eddy Simulation (DES) for SST – Strelets (2000)

- **k-equation RANS**

$$\frac{\partial(\rho k)}{\partial t} + \frac{\partial(\rho \bar{U}_j k)}{\partial x_j} = P_k - \rho \frac{k^{3/2}}{L_t} + \frac{\partial}{\partial x_j}\left[(\mu + \frac{\mu_t}{\tilde{\sigma}_\kappa})\frac{\partial k}{\partial x_j}\right] \qquad L_t = \frac{\sqrt{k}}{\beta^* \omega}$$

- **k-equation LES**

$$\frac{\partial(\rho k)}{\partial t} + \frac{\partial(\rho \bar{U}_j k)}{\partial x_j} = P_k - \rho \frac{k^{3/2}}{C_{DES}\Delta} + \frac{\partial}{\partial x_j}\left[(\mu + \frac{\mu_t}{\tilde{\sigma}_\kappa})\frac{\partial k}{\partial x_j}\right] \qquad \Delta = \max(\Delta x, \Delta y, \Delta z)$$

- **k-equation DES**

$$\frac{\partial(\rho k)}{\partial t} + \frac{\partial(\rho \bar{U}_j k)}{\partial x_j} = P_k - \rho \frac{k^{3/2}}{\min\left(L_t; C_{DES}\Delta\right)} + \frac{\partial}{\partial x_j}\left[(\mu + \frac{\mu_t}{\tilde{\sigma}_\kappa})\frac{\partial k}{\partial x_j}\right]$$

ANSYS

Grid Sensitivity with DES Model

- Fine grids in boundary layer destroy RANS mode if:

$$L_t > C_{DES}\Delta_{max} \rightarrow \Delta_{max} < \delta_{BL}$$

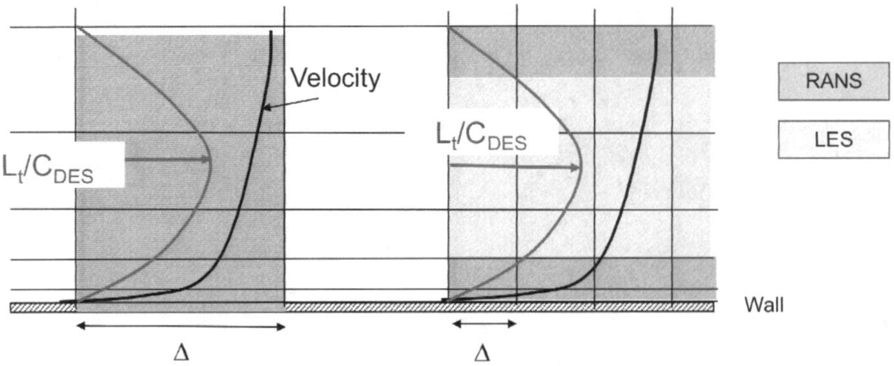

$$\varepsilon_{DES} = \frac{k^{3/2}}{\min(L_t, C_{DES}\Delta_{max})} = \varepsilon_{RANS} F_{DES}; \quad F_{DES} = \max\left(\frac{L_t}{C_{DES}\Delta_{max}}, 1\right)$$

ANSYS

Grid Induced Separation

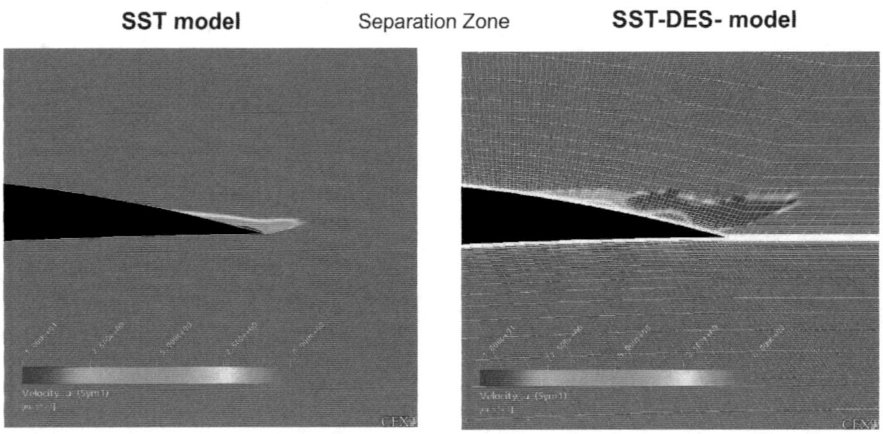

- Fine grid spacing in boundary layer leads to unphysical separation (Grid Induced Separation – GIS).
- To avoid this problem – shielding of the boundary layer required – > Delayed DES (DDES)

ANSYS

DES/DDES of Separated Flow around a realistic Car model exposed to Crosswind

Model	Exp.	DDES	DES	LES
Drag (SCx)	0.70	0.71	0.75	0.69

U=40 m/s Yaw angle 20°
$Re_H \sim 10^6$

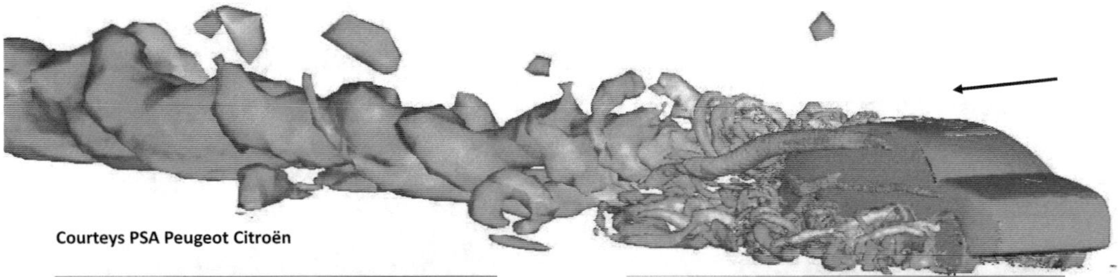

Courteys PSA Peugeot Citroën

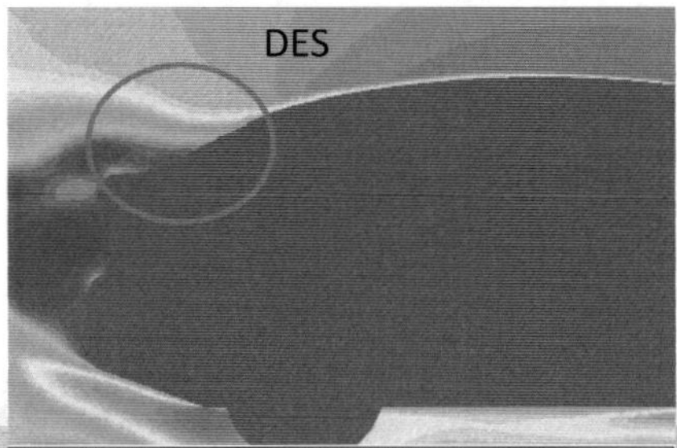

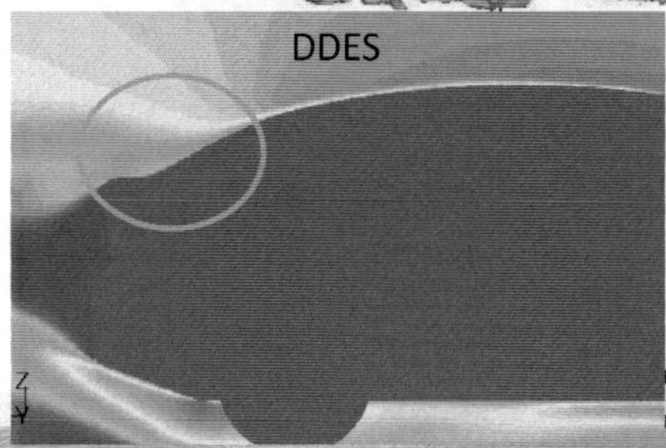

DDES for SST

- **k-equation DDES additional source term**

$$\varepsilon_{DDES} = -\beta^* \rho k \omega \cdot F_{DDES} \; with \quad F_{DDES} = \left[\max\left(\frac{L_t}{C_{DES}\Delta_{DES}} (1 - f_{DDES}), 1 \right) - 1 \right]$$

- **Shielding function**

$$f_{DDES} = \tanh\left[(C_{d1} r_d)^{C_{d2}} \right] \quad r_d = \left(\frac{L_t}{y} \right)^2 \quad L_t = \frac{k^{1/2}}{c_\mu \omega}$$

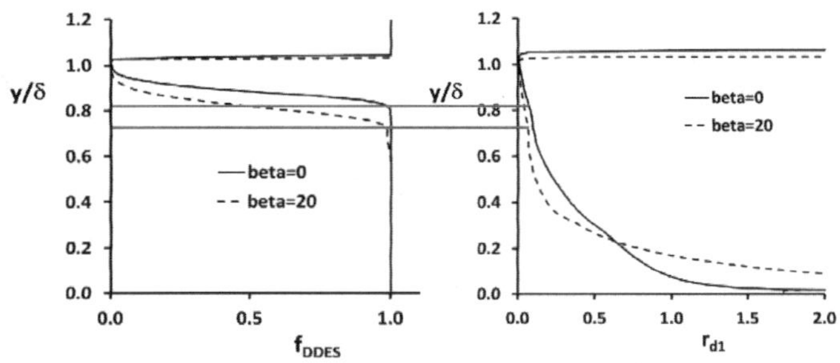

DDES Shielding

- Combination of natural shielding and shielding function
- Impact of grid refinement is 'delayed' to about $\Delta_{max} < 0.2\delta$
- However, impact of grid is much more drastic
  - Marginal change of refinement leads to strong change in eddy-viscosity
  - Non-linear feedback from DES term on shielding function
  - Exaggerated by self-similar boundary layer equations – assuming infinite running length of boundary layer

$$r = \frac{\Delta_{max}}{\delta}$$

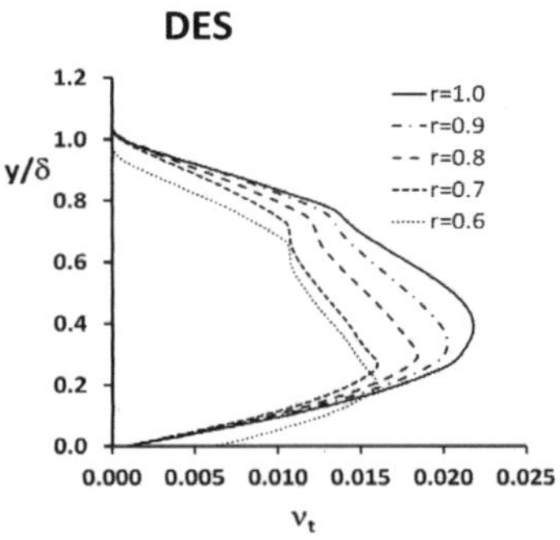

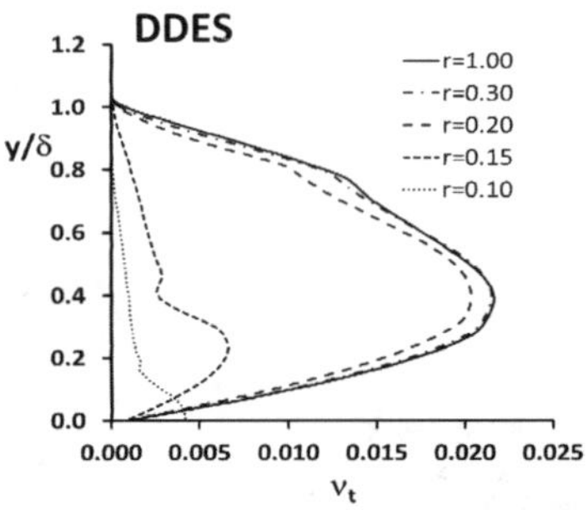

Goals

- Shielded Detached Eddy Simulation (SDES)
  - New shielding function for DES models
    - Much improved shielding under grid refinement
    - Shielding function which 'dissolves' under LES conditions
    - Improved definition of mesh length scale and calibration
- Stress Blended Eddy Simulation (SBES)
  - Based on SDES shielding function, switch from selected RANS model to selected LES model
  - Provide shielding function to clearly define/show regions of RANS and regions of LES

Shielding SDES vs DDES (1D Code)

- SDES/SBES shielding function covers entire boundary layer
- Shielding down to very fine mesh spacing

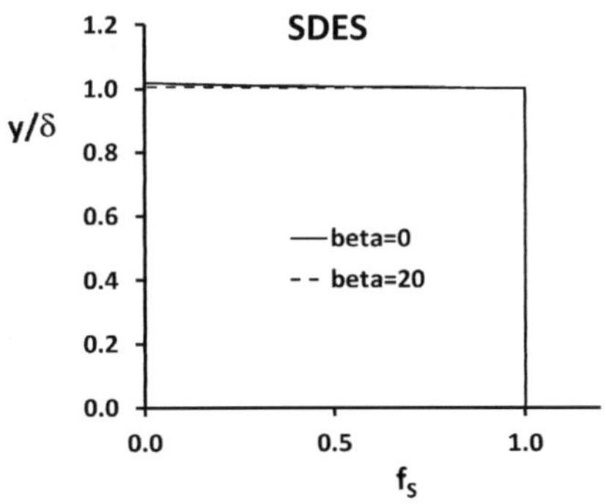

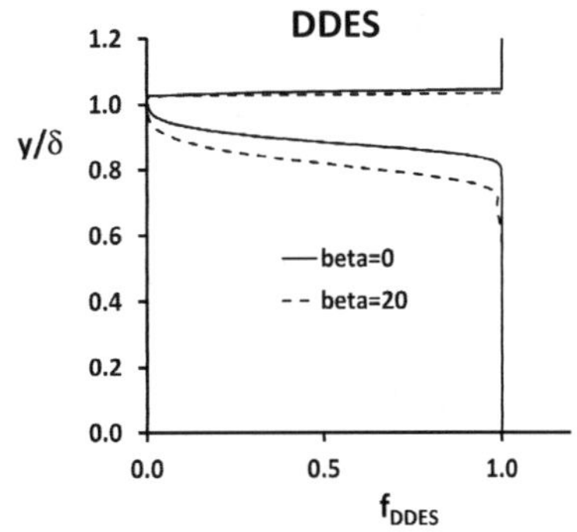

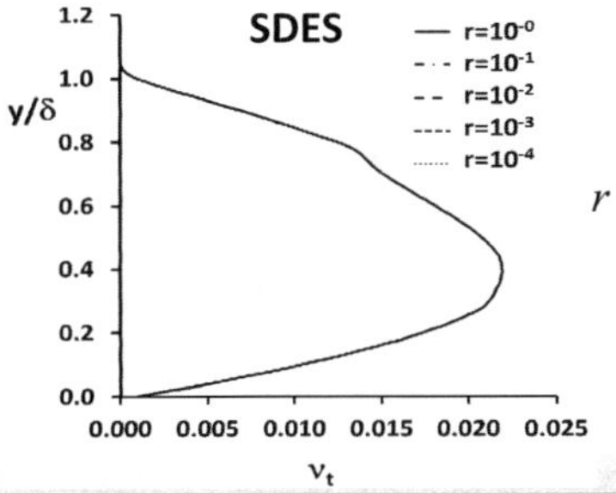

$$r = \frac{\Delta_{max}}{\delta}$$

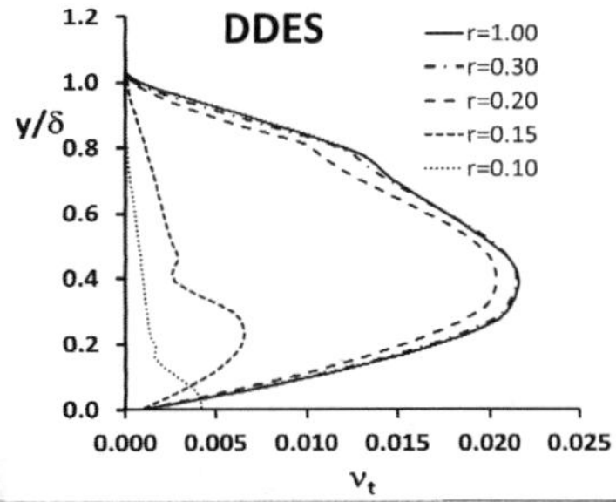

LES Portion of DES

- Source Term Equilibrium k-ω:

k-equation

$$P_k = \varepsilon \rightarrow \frac{k}{\omega} S^2 = \frac{k^{3/2}}{C_{DES}\Delta_{max}}$$

ω-equation

$$\alpha S^2 = \beta \omega^2 \rightarrow \omega = \sqrt{\frac{\alpha}{\beta}} S$$

$$\nu_t = \left(\left(\frac{\beta}{\alpha} \right)^{3/4} C_{SDES} \Delta_{max} \right)^2 S$$

$$\nu_t = \left(\tilde{C}_S \Delta_{max} \right)^2 S; \qquad \tilde{C}_S = \left(\frac{\beta}{\alpha} \right)^{3/4} C_{SDES} = 0.175$$

- (D)DES models lead to Smagorinsky LES model
- Problem is that instead of conventional LES grid scale ($Vol^{(1/3)}$) it features Δ_{max}
- On non-isotropic grids this leads to much larger eddy-viscosities than conventional LES models

Formulation of SBES

- Blending on the Stress level:

$$\tau_{ij}^{SBES} = \tau_{ij}^{RANS} \cdot f_{SBES} + \tau_{ij}^{LES}(1 - f_{SBES})$$

- For Eddy Viscosity models that translates to:

$$\nu_t^{SBES} = \nu_t^{RANS} \cdot f_{SBES} + \nu_t^{LES}(1 - f_{SBES})$$

- The underlying RANS and LES models are unaffected
- In principle, and RANS model can be combined with any alg. LES model
- All complexity is in the blending function f_{SBES}

ANSYS

Testing of Shielding Efficiency

- Steady state solutions of DDES and SBES are compared for both fully attached and separated flows
- Boundary layer shielding is estimated based on distributions of the skin friction coefficient and sectional maximum of the eddy viscosity

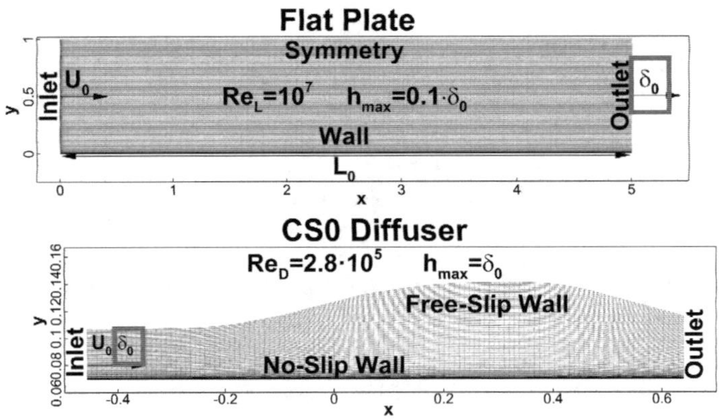

- **Grid resolution study is based on $\frac{\Delta_{max}}{\delta_0}$ see Figure for δ_0.**

ANSYS

ShieldingTest Cases: Flat Plate

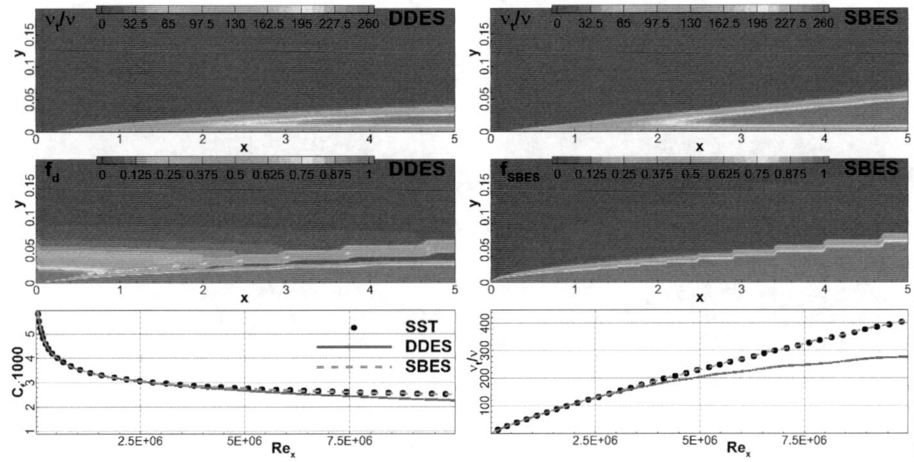

- SBES provides perfect shielding in the entire domain
- DDES protects the boundary layer only up to ReX=2.5·10^6

ANSYS

Testing of Shielding Efficiency

- Steady state solutions of DDES and SBES are compared for both fully attached and separated flows
- Boundary layer shielding is estimated based on distributions of the skin friction coefficient and sectional maximum of the eddy viscosity

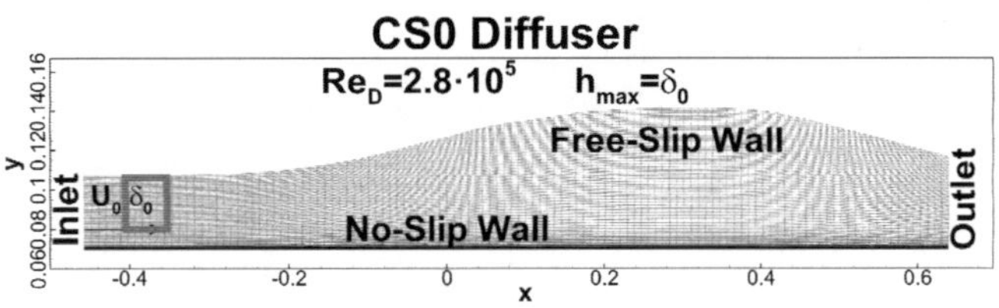

- Grid resolution study is based on $\frac{\Delta_{max}}{\delta_0}$ see Figure for δ_0.

ANSYS

Testing for RANS Test Cases: CS0 Diffuser

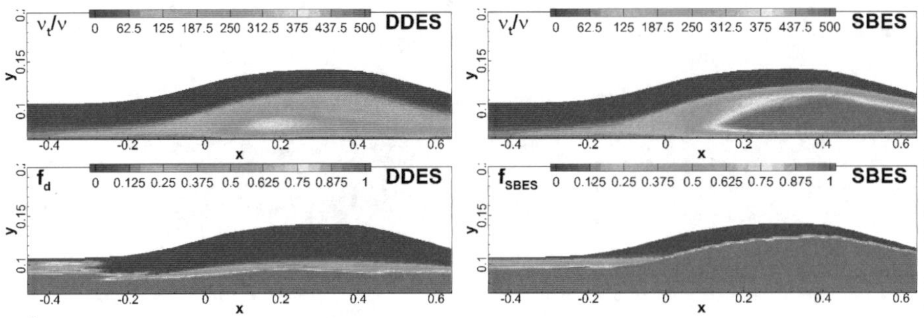

- SBES solution is identical to the SST one in the entire domain
- DDES provides insufficient shielding downstream of the reattachment point

Influence of the Grid Step Diffuser

The boundary layer protection of SBES is much stronger than for DDES

DDES

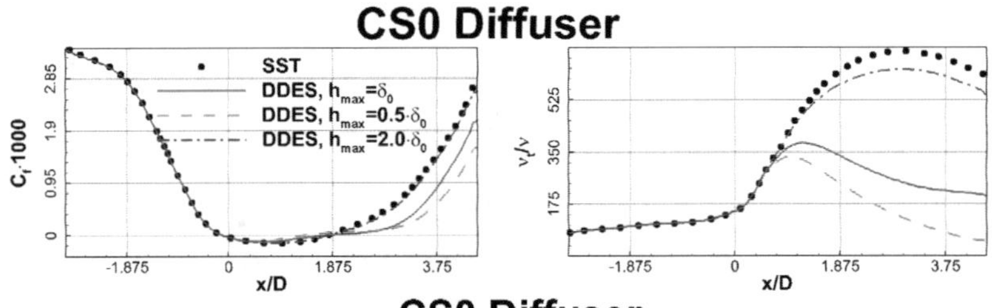

SBES

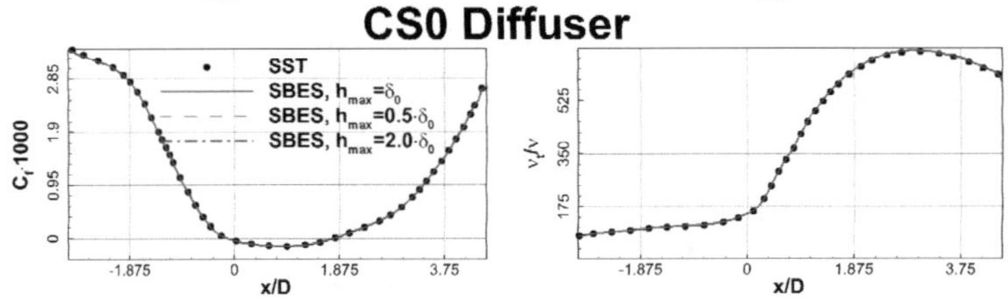

Single Stream Mixing Layer:

Mesh 1

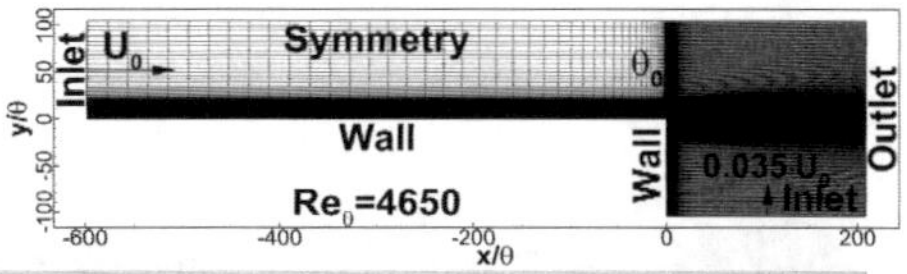

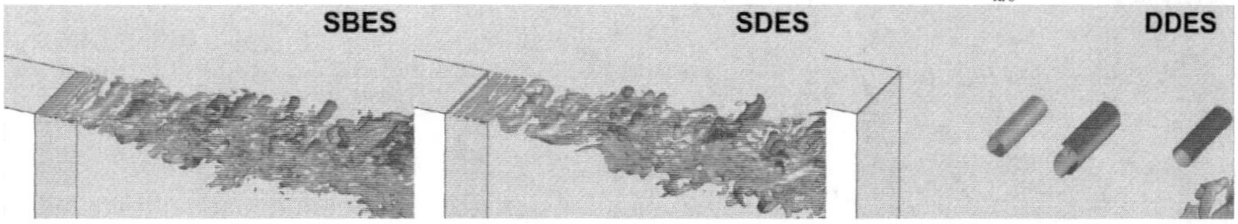

Mesh 2

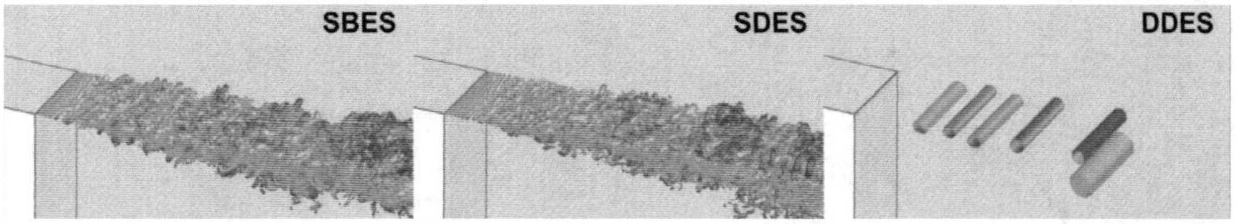

Single Stream Mixing Layer: Velocity Profiles

Better spreading rate for initial mixing layer with SBES

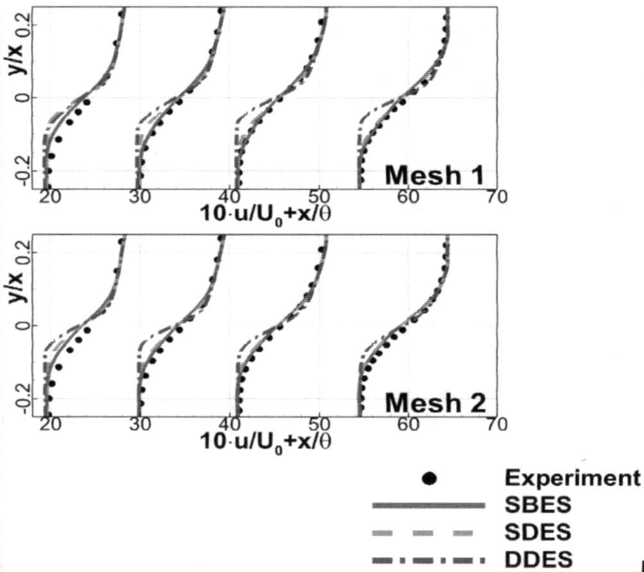

Testing for SRS Test Cases: Mixing Layer

Shielding Functions

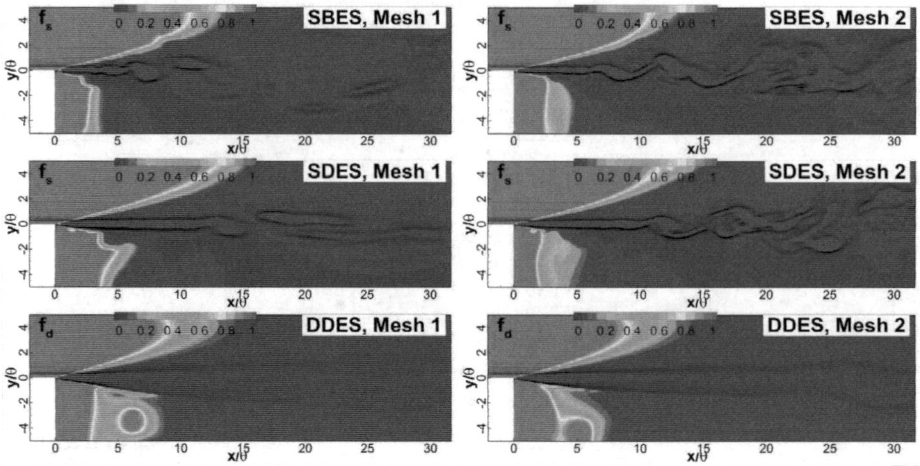

Testing for SRS Test Cases: Backward Facing Step

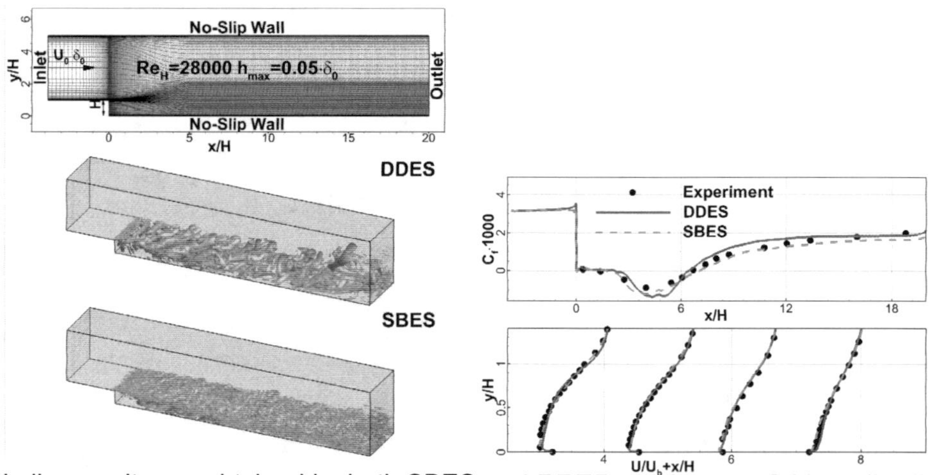

- Similar results are obtained by both SBES and DDES which agree fairly well with the experimental data
- SBES provides smaller structures in the mixing layer
- No resolved turbulence is observed near the upper wall for both models

Testing for SRS Test Cases: Backward Facing Step

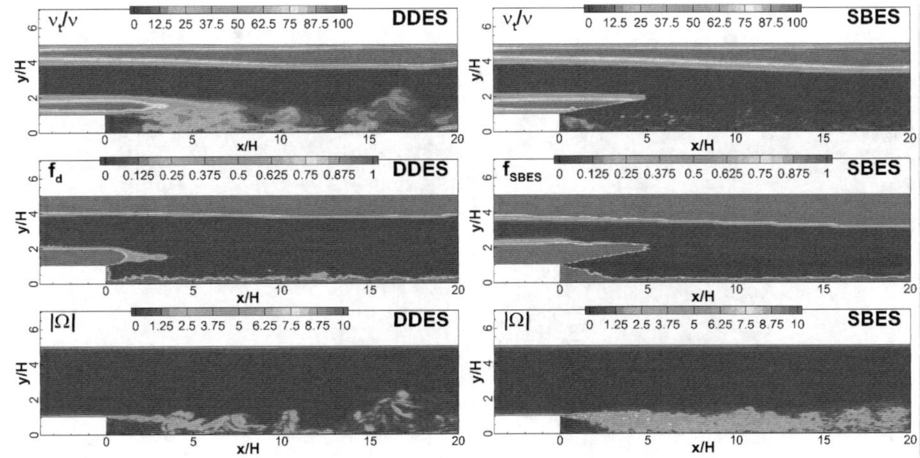

- Both f_d and f_{SBES} functions covers the boundary layer at the upper wall
- f_{SBES} function covers noticeably larger area near the step than f_{DDES}
  - However, the development of 3D turbulence in the mixing layer is less rapid for DDES

Testing for SRS Test Cases: Periodic Hills

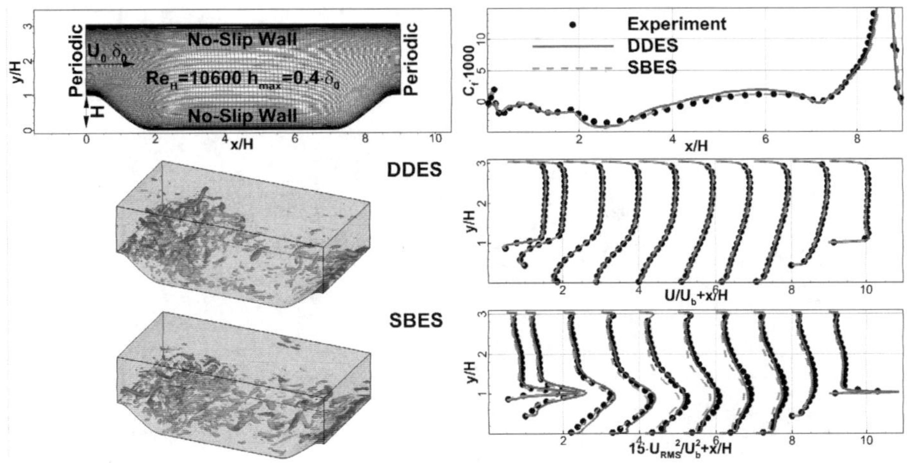

- Both SBES and DDES solutions are in a good agreement with the experimental data
  - Similar turbulent structures are obtained by both models as the grid is of LES quality

Testing for SRS Test Cases: Periodic Hills

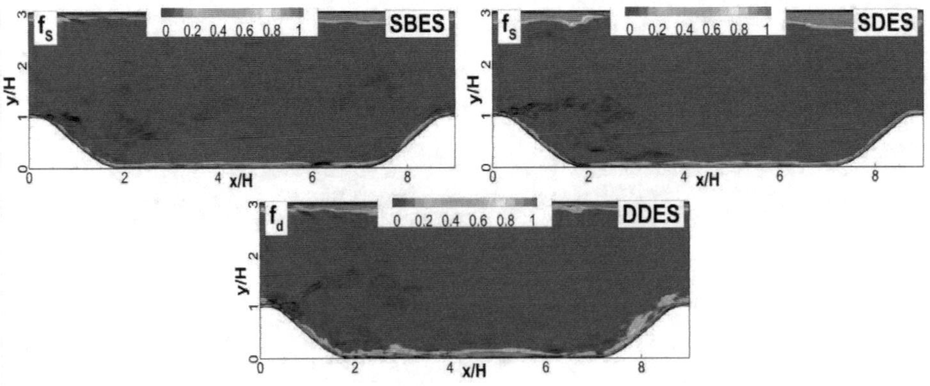

Shielding functions for DDES and SDES/SBES are very similar for this case

Comparison of SBES and DDES Thermal Mixing in a T-Junction

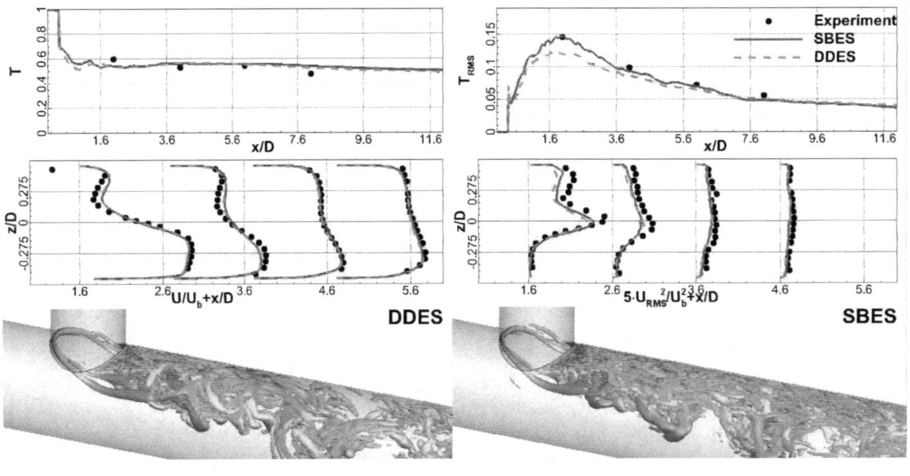

- The mean and RMS velocity profiles of SBES are close to those of DDES
- SBES provides better agreement with the experiment for the RMS wall temperature

Testing for SRS Test Cases: Thermal Mixing in the T-Junction Flow

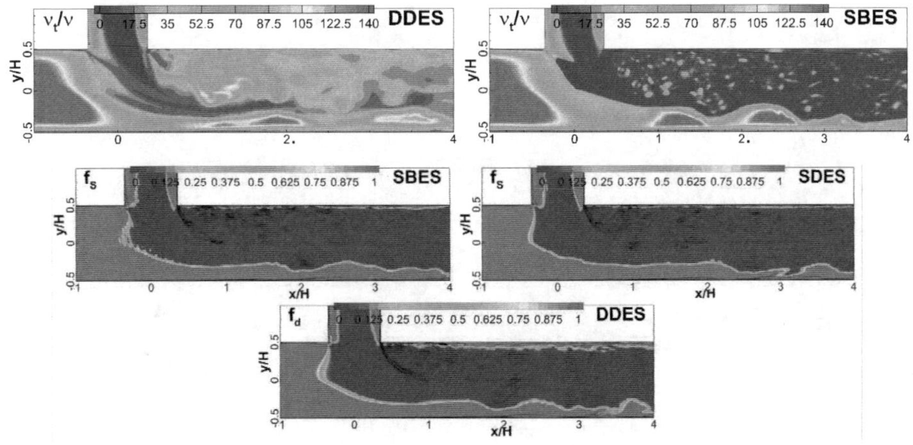

- Lower Eddy-Viscosity for SBES model in LES zone due to switch to WALE model
- Shielding functions all remain in LES mode in and downstream of mixing zone

Testing for SRS Test Cases: Subsonic Round Jet

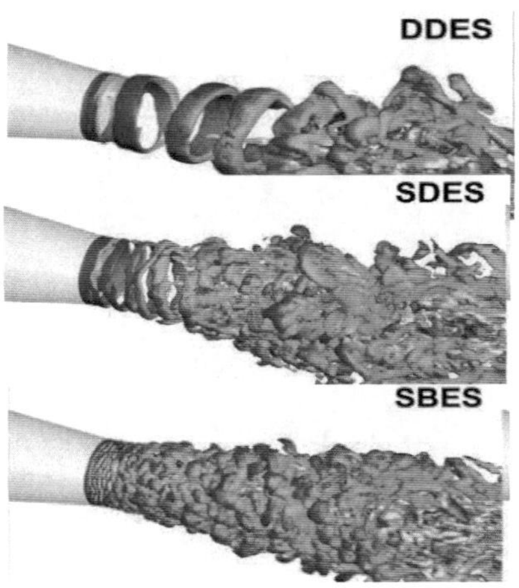

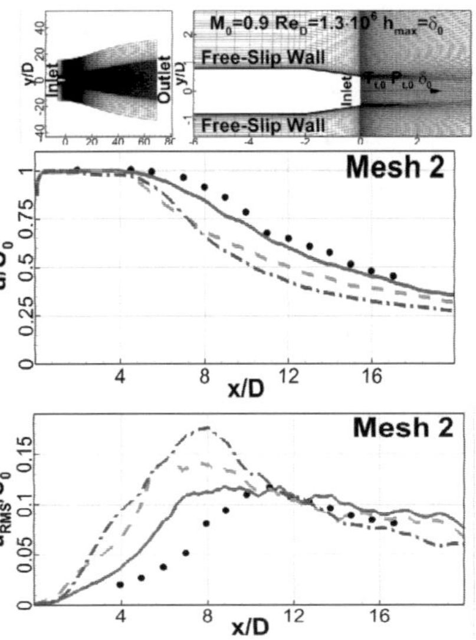

- **Much faster 'Transition' to LES with SDES and SBES**
- **SBES gives lower RMS and more realistic spreading**

Testing for SRS Test Cases: Subsonic Round Jet

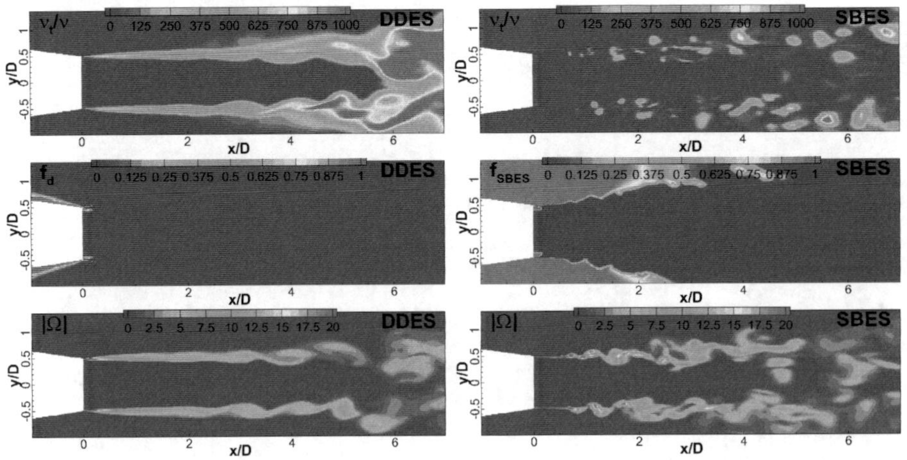

- DDES has larger eddy viscosity in the mixing layer which suppresses the development of 3D turbulence

---

Case Description

- **Incompressible flow**
  - $Re_c = \frac{c \cdot U_\infty}{\nu} = 9.36 \cdot 10^5$
- **Computational domain**
  - Domain size: 6.14c × 0.9c × 0.4c
- **Boundary conditions**
  - Developed velocity profile and turbulence properties from SST RANS computations
  - Pressure-outlet boundary with constant pressure
  - Slip wall boundary condition
  - No-slip wall boundary condition
  - Periodic boundary conditions in z-direction
- **Computational aspects**
  - SBES without any synthetic turbulence generator
  - Non-iterative time advancement scheme (time step size $\Delta t = 0.001 \cdot c/U_\infty$)
  - BCD advection interpolation scheme

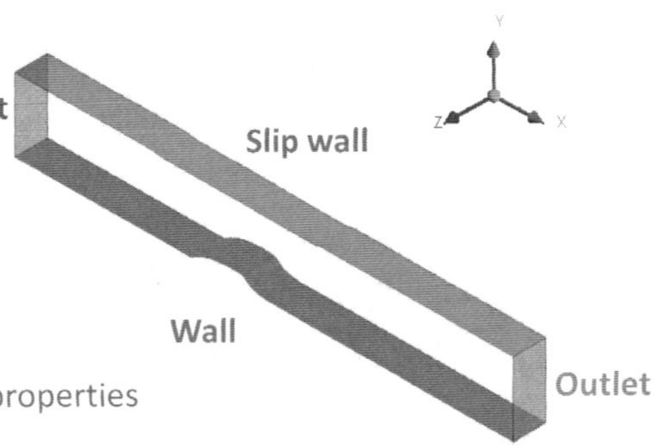

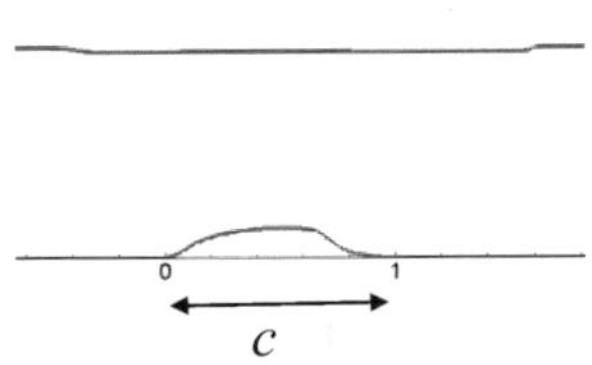

Computational mesh

– **NTS-mesh from the Go4Hybrid project**
  - mesh size ≈ $5 \cdot 10^6$ cells
  - $L_z = 0.4c$, $N_z = 81$, $\Delta z = 5 \cdot 10^6$

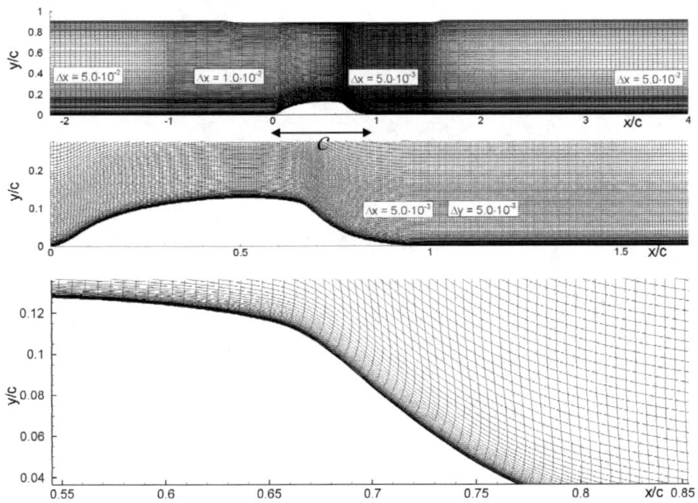

Instantaneous flow pattern

Computational time = $17 \cdot c/U_\infty$

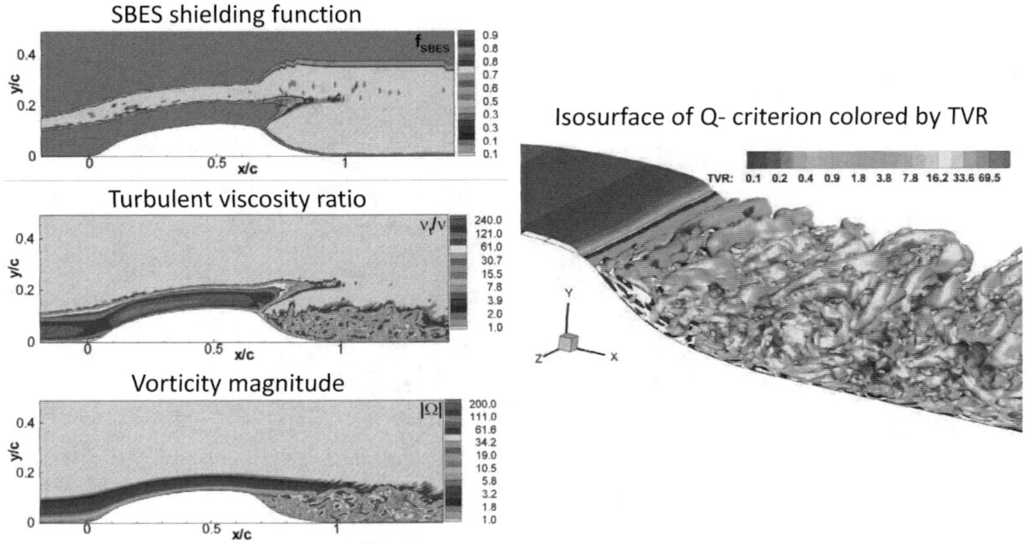

- Boundary layer is fully protected by shielding function until separation
- After separation, shielding function is getting equal to zero, switching SBES to LES mode
  – It promotes to the development of turbulent structures;

Skin friction and pressure distribution

Computational time = $17 \cdot c/U_\infty$, time of averaging = $7 \cdot c/U_\infty$

Skin friction coefficient

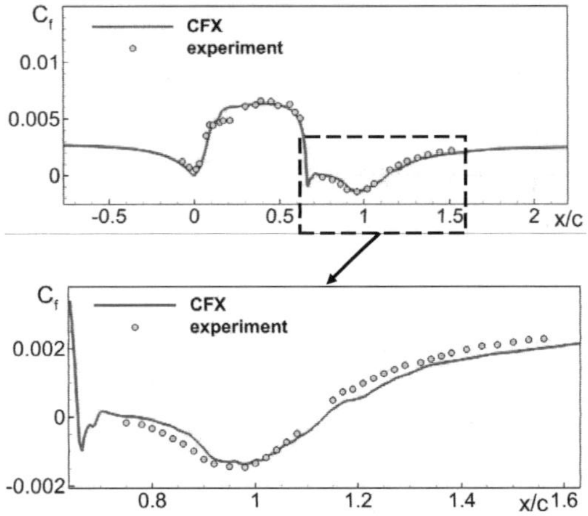

Pressure coefficient

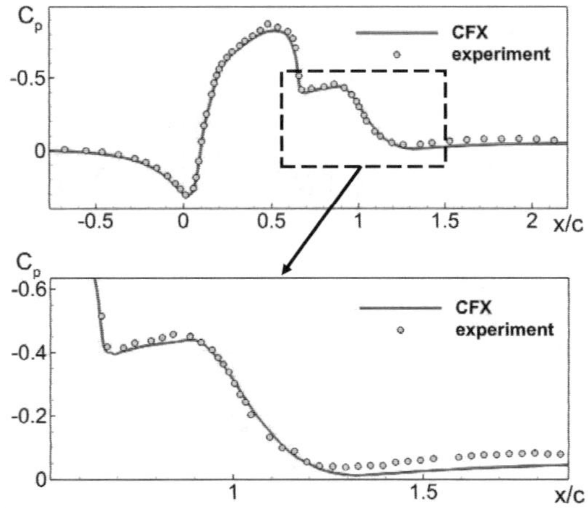

- CFX skin friction and pressure distributions fit well with the experimental data
- The length of separation zone is virtually identical with experimental one

ANSYS

Periodic Channel - WMLES

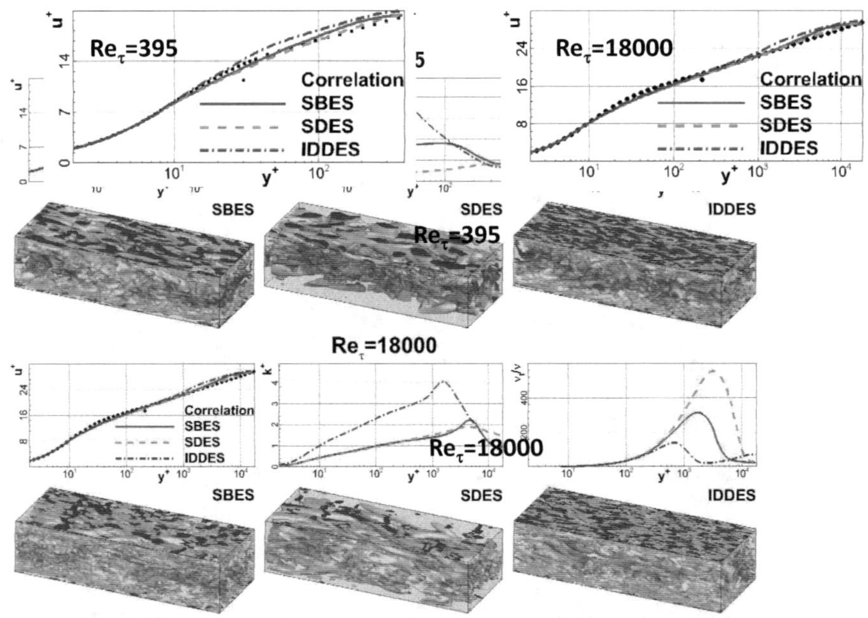

ANSYS

Summary

- SDES/SBES blending function clearly shows the zones of RANS and LES usage
- In RANS mode SDES/SBES effectively protects attached boundary layers regardless of the used grid
- SBES allows the selection of user specified RANS and LES models
  - Fluent since R16.1
  - CFX since R17
- For free-shear flows SBES provides much more rapid development of 3D turbulence in mixing layers and the results are in a good agreement with the experimental data
- SBES is already in use for complex automotive applications

Automotive Simulation
WORLD CONGRESS

June 7-8, 2016 Munich

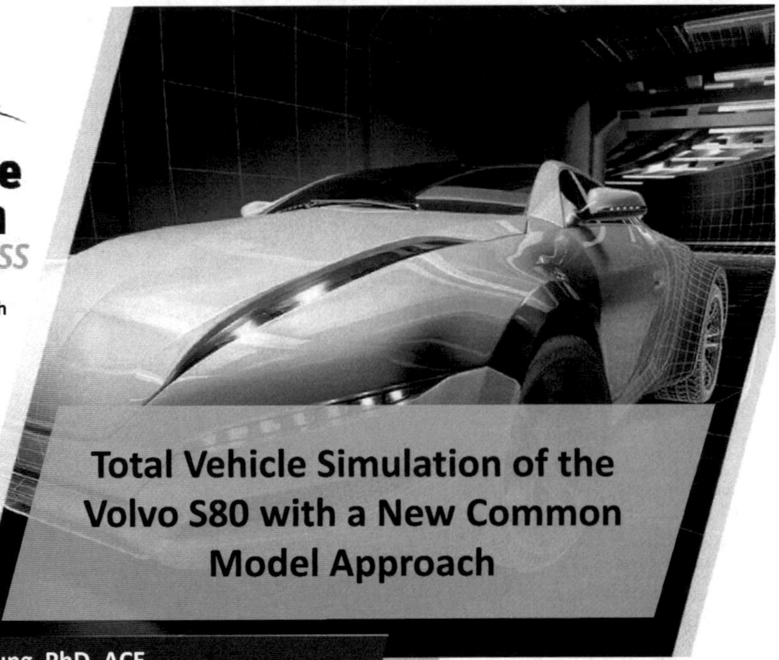

Total Vehicle Simulation of the Volvo S80 with a New Common Model Approach

Torbjörn Virdung, PhD, ACE
ANSYS Sweden

Agenda

- Motivation
- Workflow
- Current Study
- Results
- Potential Future Work

- Xingshi Wang, PhD, ANSYS North America
- Aditya Kshatriya, ANSYS India
- Torbjörn Virdung, PhD, ANSYS Sweden
- Peyman Davoudabadi, PhD, ANSYS North America

Motivation

- **Within the automotive OEMs, it is common to have separate teams for different areas of simulation focus**
  - Aerodynamics
  - Underhood Thermal Management
  - Aero Acoustics
  - Engine CHT
  - Brakes
  - Exhaust After-treatment
  - Climate Comfort
  - Contamination (Snow, Dust, Rain)
  - ...

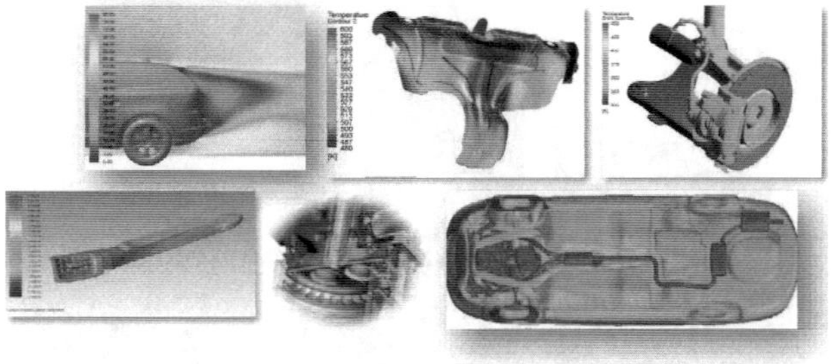

- **This allows each team to focus on individual areas.**
- **However, potential inefficiencies with this setup are**
  - Duplication of CAD work (Cleanup/assembly)
  - Duplication of Meshing
  - Inconsistency and excessive need of boundary conditions and material data
  - Potential compromise on accuracy due to interdependencies
  - Non-optimal simulation times

Motivation

- **Using a common multi-objective model allows for**
  - Minimizing time spent on CAD assembly and CAD cleanup within the company
  - Including all relevant physics within one simulation
  - Minimizing the need for gathering boundary conditions and material data
    - E.g. for the engine CHT, only one or two boundary conditions are needed
  - Using a common material database for all simulations
  - Consistent meshing strategy
  - No compromise/assumptions on boundary conditions at the interface of differenet attributes
  - Respecting the coupled nature of the various subsystems as in the real world
  - Separate mesh strategies
  - Common model MDO

- Specific simulations, where only selected physics or attributes are to be included, can be performed by using the common CAD model with
  - Seperate surface meshing strategies, e.g. refining/coarseing the mesh
  - Separate volume meshing strategies, e.g. redistributing volume mesh refinement and/or excluding prism layers on exterior if Aerodynamics is excluded

Model per Attribute

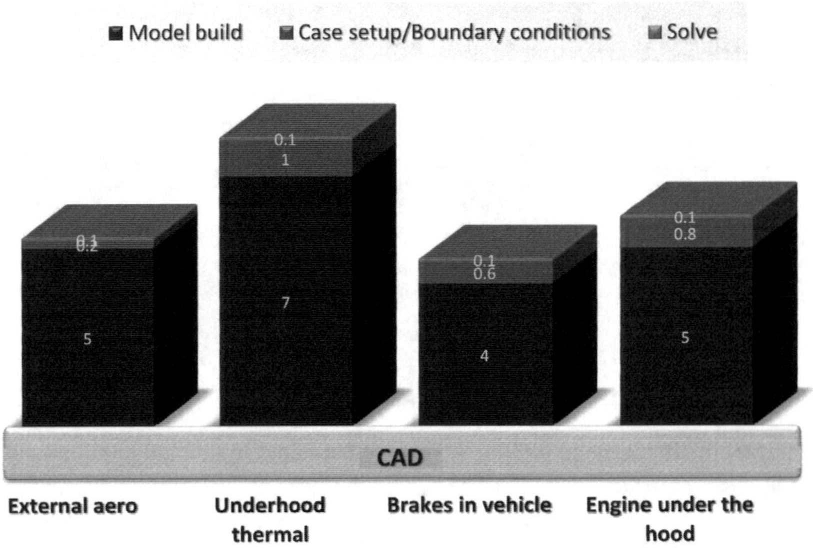

Common Multi-Objective Model

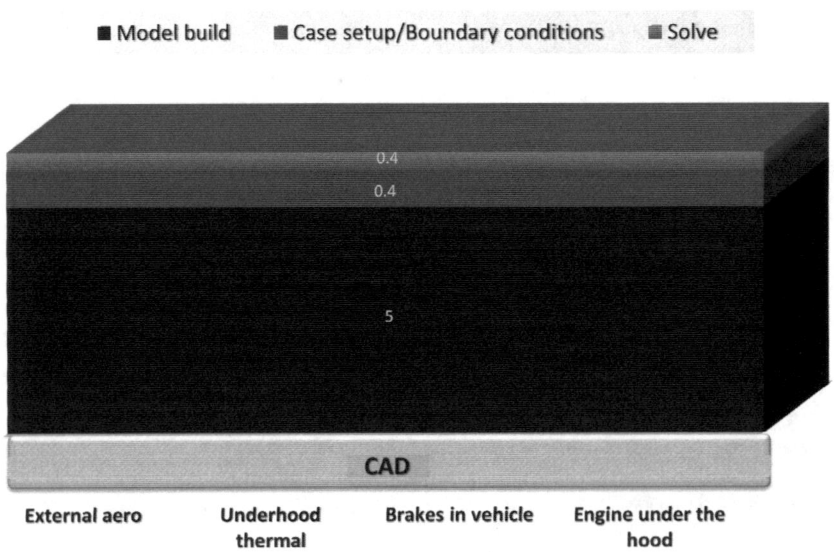

Ideal Workflow (Current Study)

CAD → **Conformal HQ surface mesh** → **Volume Mesh** → **Case Setup**

- **CAD**
  - CAD/Tesselated representation of the complete vehicle
- **Conformal HQ surface mesh**
  - Surface mesh suitable for all attributes
- **Volume Mesh**
  - Volume mesh suitable for all attributes
- **Case Setup**
  - Automated set up for all attributes

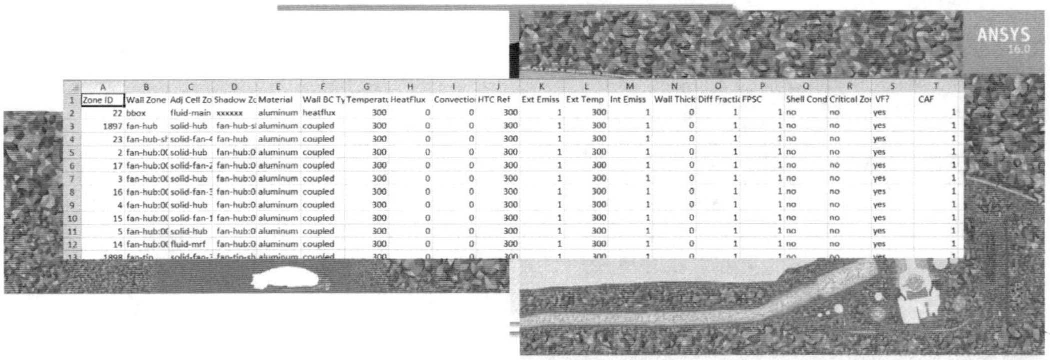

	A	B	C	D	E	F	G	H	I	J	K	L	M	N	O	P	Q	R	S	T
1	Zone ID	Wall Zone	Adj Cell Zo	Shadow Zc	Material	Wall BC Ty	Temperatu	HeatFlux	Convectio	HTC Ref	Ext Emiss	Ext Temp	Int Emiss	Wall Thick	Diff Fractio	FPSC	Shell Cond	Critical Zo	VF?	CAF
2	22	bbox	fluid-main	xxxxxx	aluminum	heatflux	300	0	0	300	1	300	1	0	1	1	no	no	yes	1
3	1897	fan-hub	solid-hub	fan-hub-s	aluminum	coupled	300	0	0	300	1	300	1	0	1	1	no	no	yes	1
4	23	fan-hub-sh	solid-fan-4	fan-hub	aluminum	coupled	300	0	0	300	1	300	1	0	1	1	no	no	yes	1
5	2	fan-hub:0(	solid-hub	fan-hub:0	aluminum	coupled	300	0	0	300	1	300	1	0	1	1	no	no	yes	1
6	17	fan-hub:0(	solid-fan-2	fan-hub:0	aluminum	coupled	300	0	0	300	1	300	1	0	1	1	no	no	yes	1
7	3	fan-hub:0(	solid-hub	fan-hub:0	aluminum	coupled	300	0	0	300	1	300	1	0	1	1	no	no	yes	1
8	16	fan-hub:0(	solid-fan-3	fan-hub:0	aluminum	coupled	300	0	0	300	1	300	1	0	1	1	no	no	yes	1
9	4	fan-hub:0(	solid-hub	fan-hub:0	aluminum	coupled	300	0	0	300	1	300	1	0	1	1	no	no	yes	1
10	15	fan-hub:0(	solid-fan-1	fan-hub:0	aluminum	coupled	300	0	0	300	1	300	1	0	1	1	no	no	yes	1
11	5	fan-hub:0(	solid-hub	fan-hub:0	aluminum	coupled	300	0	0	300	1	300	1	0	1	1	no	no	yes	1
12	14	fan-hub:0(	fluid-mrf	fan-hub:0	aluminum	coupled	300	0	0	300	1	300	1	0	1	1	no	no	yes	1
13	1898	fan-tip	solid-fan-[illegible]	fan-tip-sh	aluminum	coupled	300	0	0	300	1	300	1	0	1	1	no	no	yes	1

Workflow Alternatives

CAD → **Conformal surface mesh** → **Volume Mesh** → **Case Setup**

- **CAD**
  - CAD/Tesselated representation of the complete vehicle
  - ↓
    - Fine-tuned surface mesh for selected attributes
    - Allows for centralized CAD assembly and clean-up (LCC, Junior Engineers)
- **Conformal surface mesh**
  - Surface mesh suitable for all attributes
  - ↓
    - Fine-tuned volume mesh per attribute
    - Prism layers
    - Refinement zones
    - Solids
    - Part Swapping
- **Volume Mesh**
  - Volume mesh suitable for all attributes
  - ↓
    - Modified volume mesh per attribute
    - Solids
- **Case Setup**
  - Automated set up for all attributes
  - ↓
    - Modifications per load case
    - Morphing
    - MDO
    - 0D/1D couplings
    - Multiphysics

Common Model

- **Volvo S80 D5 MY2012**
- **2.4-litre, 5-cylinder Twin-turbo Diesel Engine, 215 hp and 440 Nm**

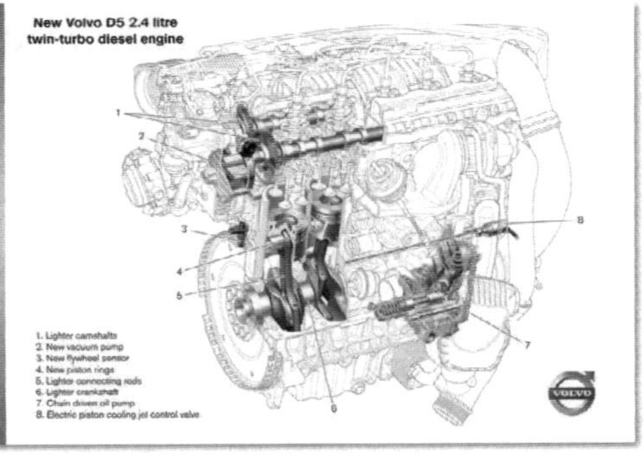

Common Model

- **Attributes included**
  - Aerodynamics
  - Underhood Thermal Management
  - Engine CHT/Under the hood
  - Cooling system
  - Brake system
  - Air intake system
  - Exhaust system
- **~400 Solids**
- **~50 Shells (readily extensible to 1000's)**

- **Use case**
  - Vehicle moving at 110 kph
  - Engine rpm: 2050
  - Thermal sources:
    - Combustion heat in cylinders
    - Hot exhaust gas into exhaust system and environment
    - Frictional heat in brake rotor
  - Brakes engaged
  - Turbo and pump
  - CRFM (Fans and HXs)

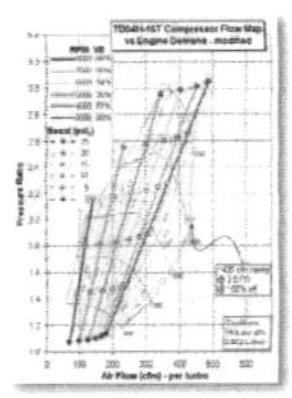

Model Build Process

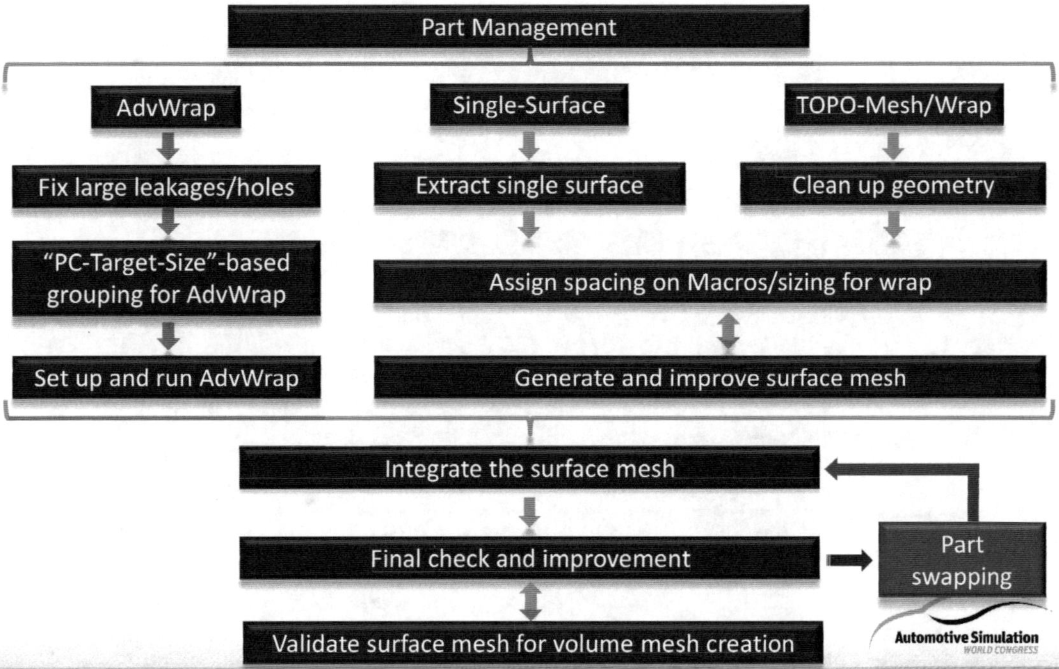

Turnaround

Sub System/Model	Model Build Time (days)
Body, grille	8
Powertrain: Entire engine solids Air intake Cooling circuit	 2.5 1 .1
UTM: Exhaust, turbo, EGR Mounts CRFM Shields	 2 .5 .25
Brakes	1.5
Misc. (CRFM, chassis, suspension, transmission, etc.)	.25
MRFs and Checks	1
Integration	4
Volume Mesh	.5
Detailed Case setup	2
Postprocess	2
Total	**26 (~5 weeks)**

- Full vehicle with all details
- One common model built
- CAD-to-Solution time reduced significantly with no compromise on accuracy
- Maximizing efficiency across multiple teams
- Foundation for common model MDO

~400 Solid zones
~50 Shell zones
~265 M conformal cells

Sub Systems

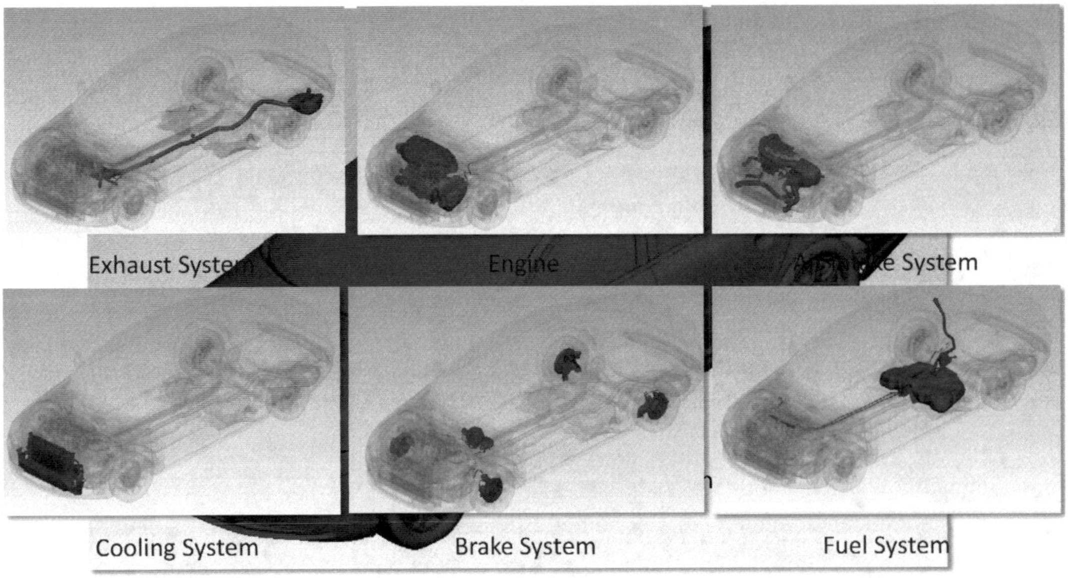

Simulation Details

- One single simulation
- Realizable k-ε Turbulence Model with Enhanced Wall Function
- Second Order Upwind discretization, Coupled Solver, GGNB
- Surface to Surface (S2S) Radiadion Model
- Secondary streams
  - Exhaust
  - Air intake system
  - Cooling

  (Exhaust and Air intake system: Mix at EGR)
- Ideal gas
- Turbocharger modeled using MRF, 140k rpm
- 110 kph, 35°C inlet
- Fans, RPM from supplied data
- Water pump modeled with MRF, 2050 rpm
- 3000 iterations

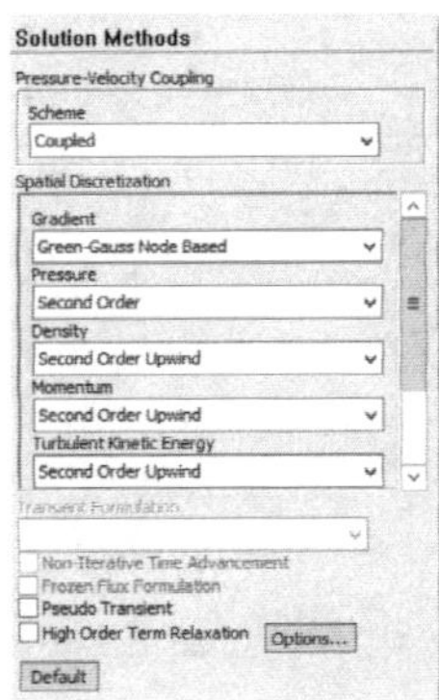

CAD, Mesh, Results – Aerodynamics

- Detailed capture of all styled surfaces with no deviation from the exterior geometry
- 12 prism layers on styling surfaces with first height 0.02mm
- Rotating rims (MRF)
- Moving ground
- Refinement regions at wake and flow separation regions
- Cd = 0.27

Automotive Simulation
WORLD CONGRESS

CAD, Mesh, Results – Exterior

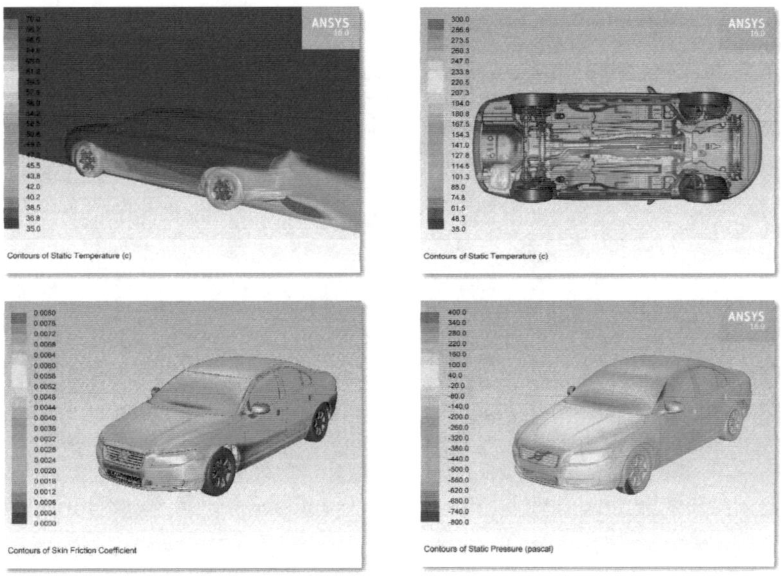

Automotive Simulation
WORLD CONGRESS

CAD, Mesh, Results – Exhaust

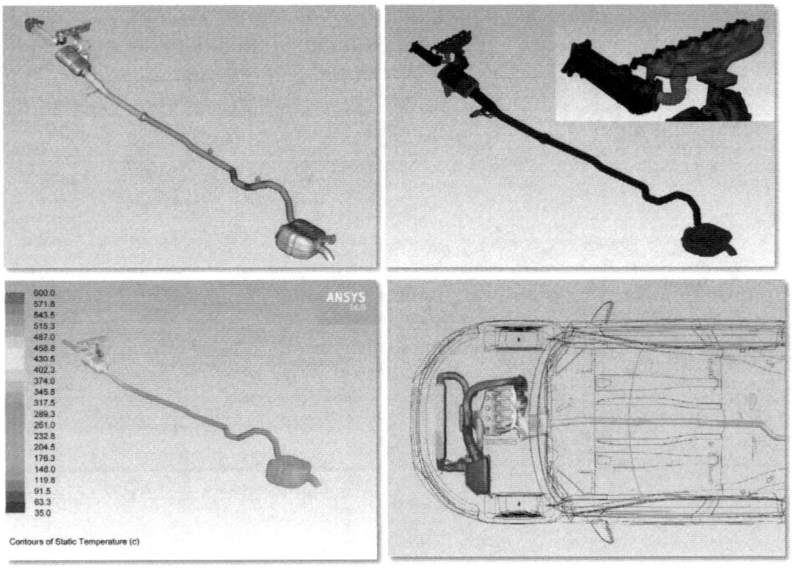

- Complete exhaust system from ports through tailpipe including manifold, turbo and EGR
- Gas flow mixes with AIS stream at EGR
- Includes hangers, brackets, bars as solids

CAD, Mesh, Results – Engine

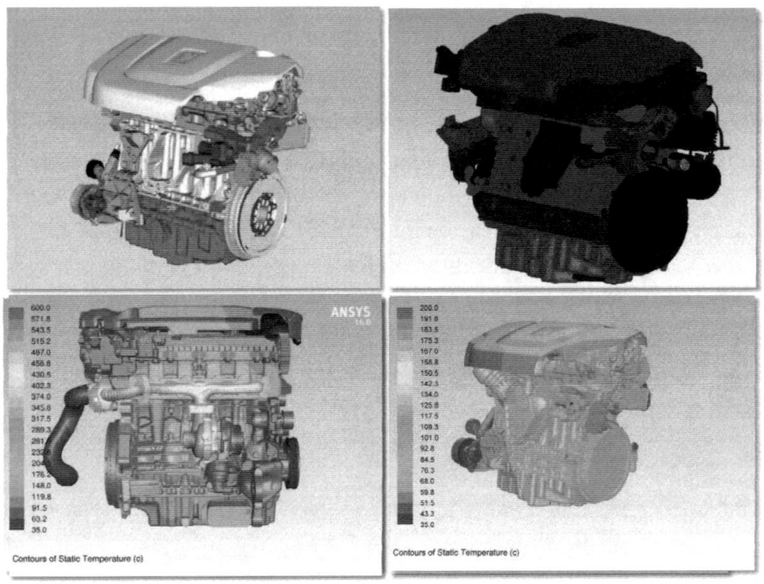

- Internal components included: head, block, crank and case, camshaft, valves, pistons, pins, rods, pulleys, gear sets, oil sump, fuel pump, oil separator, mounts
- Manifold heat shield is laminated steel/ceramic

CAD, Mesh, Results – Air Intake System

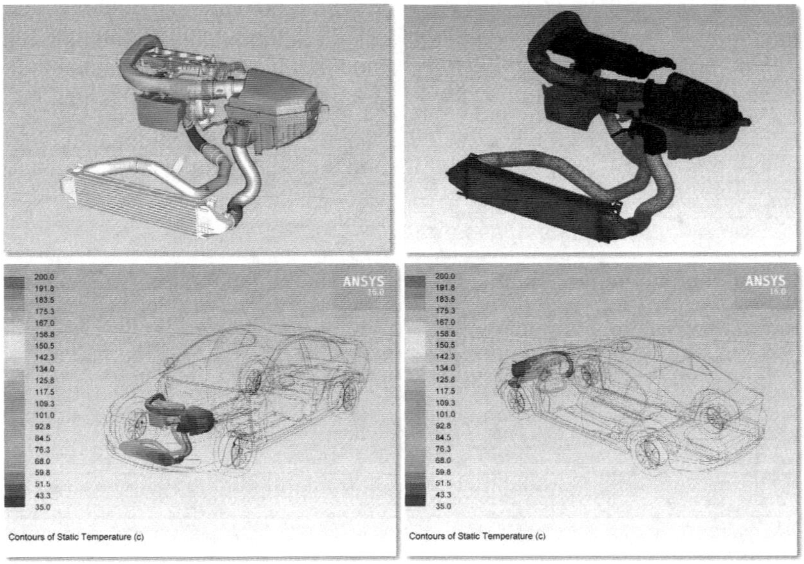

- From ambient to ports including turbo, CAC, EGR manifold
- Mix with Exhaust stream at EGR

CAD, Mesh, Results – CRFM

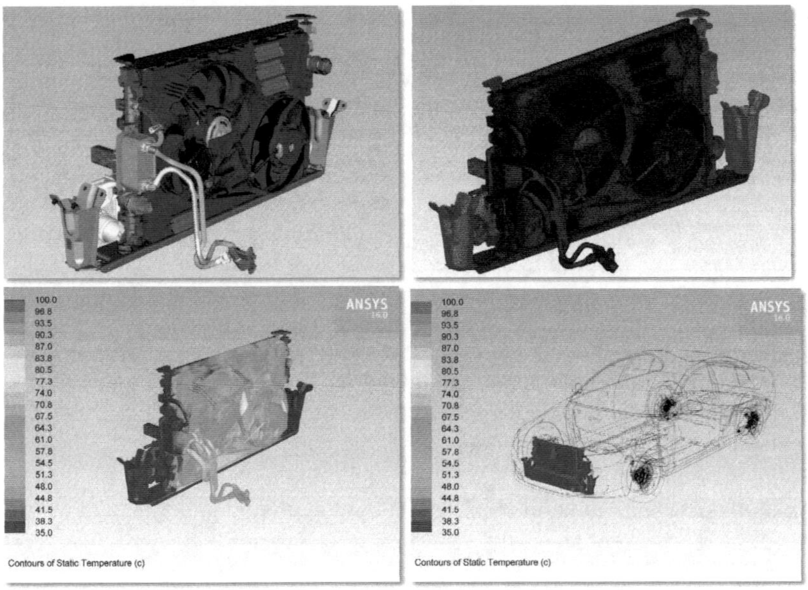

- Dual-Fan setup
- Fans modeled with MRF

CAD, Mesh, Results – Brakes

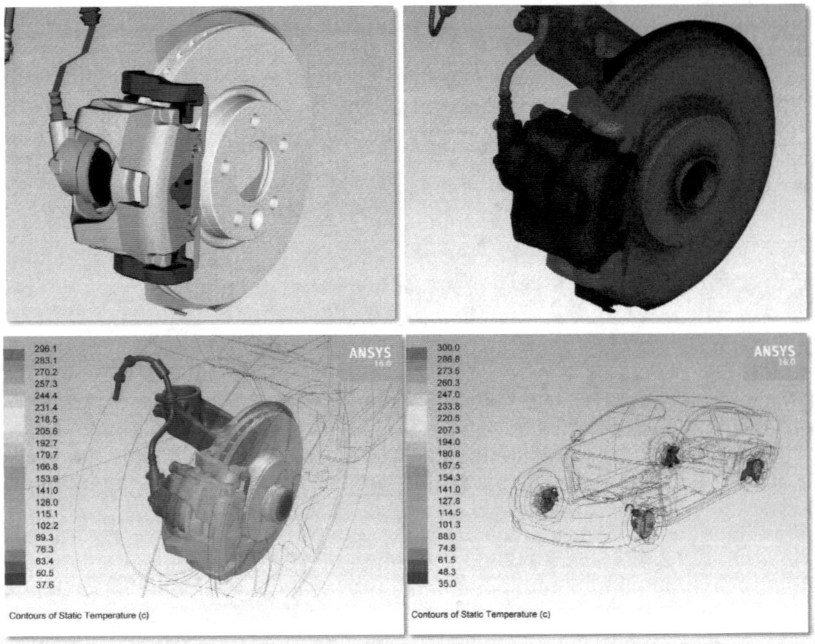

- ~200 solid zones
- All components accounted for
- MRF zones for the cooling slots (2x40)
- Frictional heat applied in the brake rotor

Automotive Simulation WORLD CONGRESS

CAD, Mesh, Results – Fuel Tank

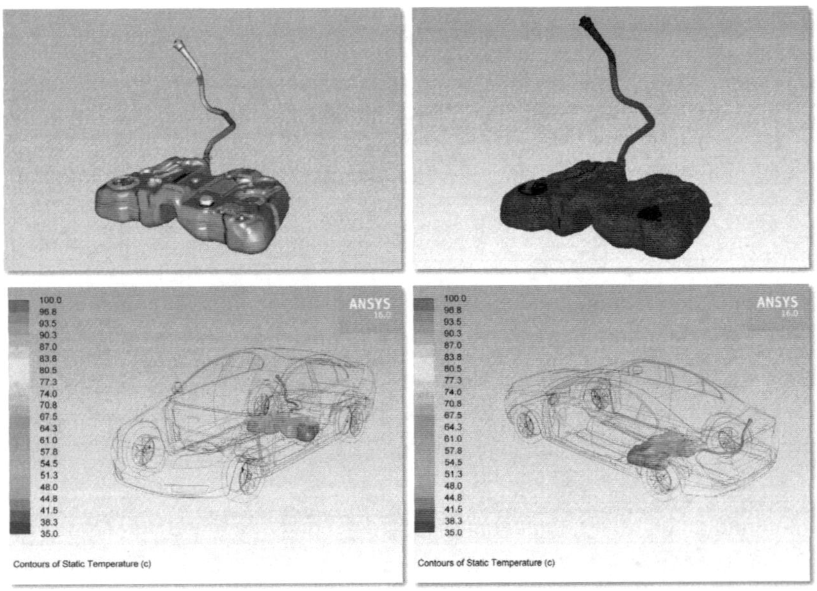

- Fluid-like material in the fuel tank
- Simplified internal components
- Accounts for accumulation of heat within the fuel tank

Automotive Simulation WORLD CONGRESS

Results – Misc

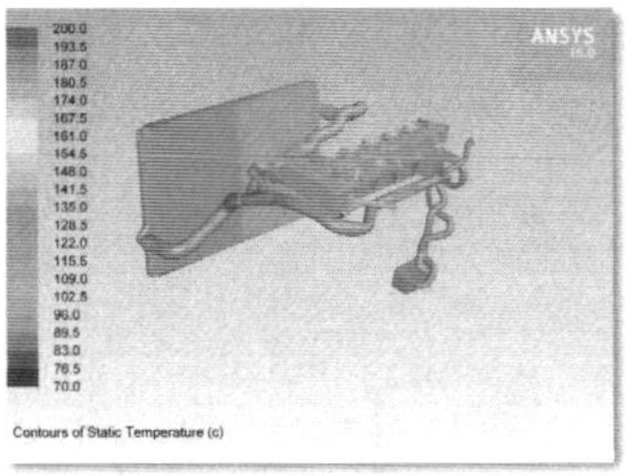

Coolant Stream

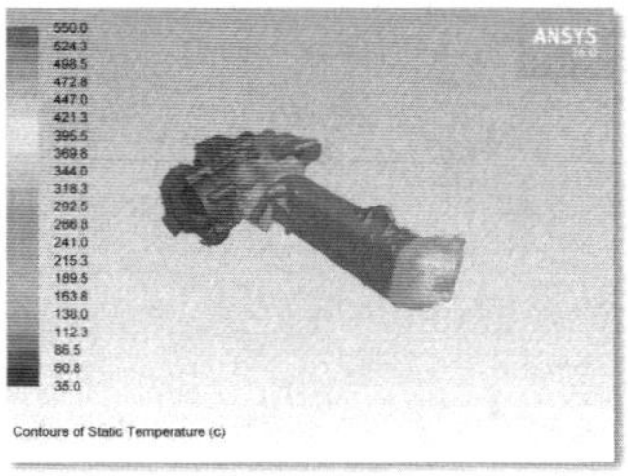

EGR

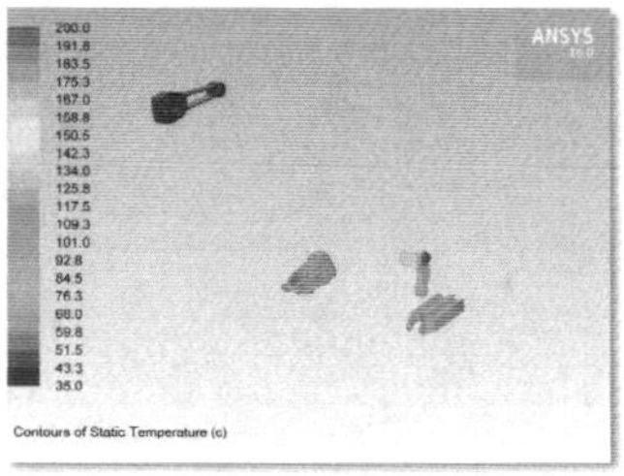

Mounts

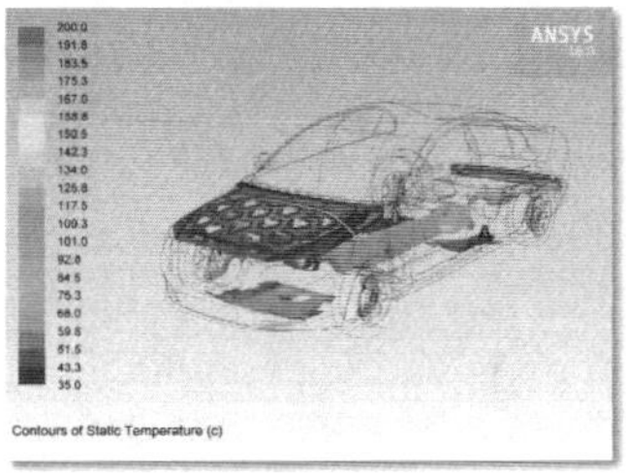

Shields

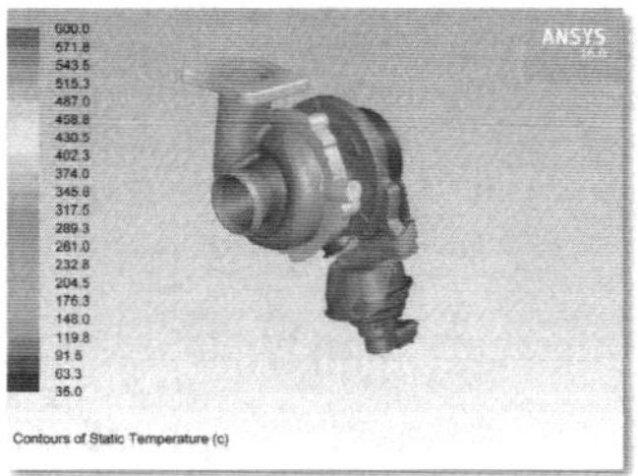

Turbo

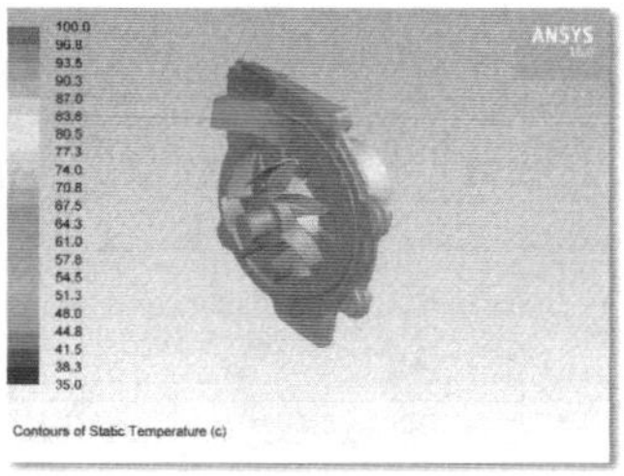

Water Pump

Potential Future Work

- **Integrated, common model MDOs**
- **Approach is natively compatible with full FSI**
  - **Aeroelasticity**
  - **Thermo-mechanical+life simulations of components and subassemblies under realistic loads from the above 3d system level simulations**
- **In-cylinder combustion integration**
- **Transient studies**
  - **Aero Acoustics**
  - **Thermal Soak**
  - **Driving Cycles**
  - **Lubrication**
  - **Splashing/Wading with thermal effects**
- **Exhaust gas entrainment into the cabin**

Questions?

- If you need more information/have further questions, please find me during the congress

Automotive Simulation
WORLD CONGRESS

June 7-8, 2016 Munich

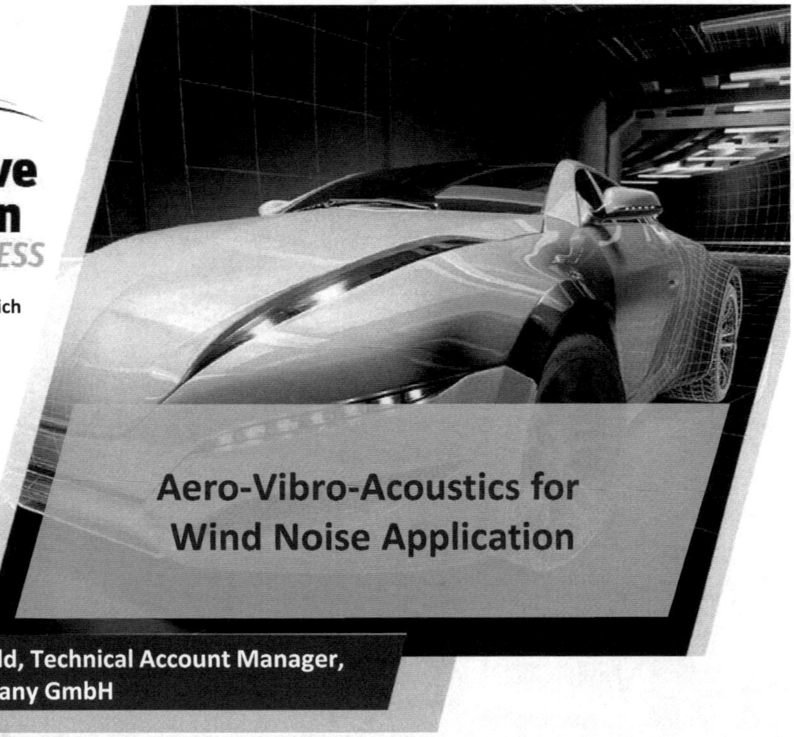

Aero-Vibro-Acoustics for Wind Noise Application

Marco Oswald, Technical Account Manager,
ANSYS Germany GmbH

Outline

- Motivation
- Problem Description
- Simulation Methodology
- Setup
- Results
- Solution Performance
- Summary
- References

Motivation (1)

Wind noise is aerodynamically generated noise perceived in the vehicle interior. Important quality concern for car makers.

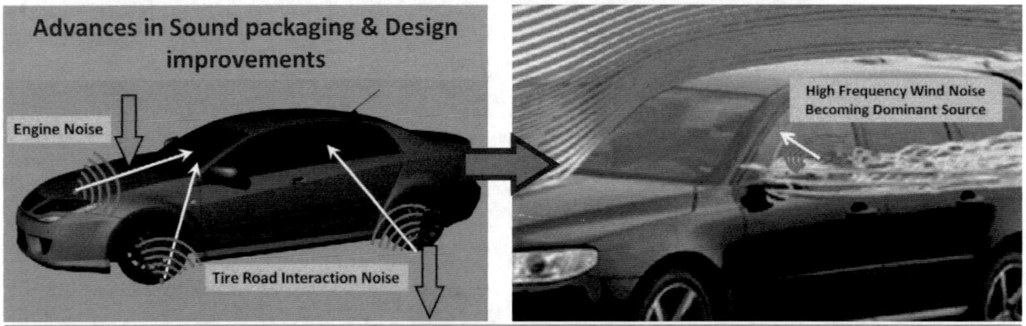

J.D. Power 2014 U.S. Vehicle Dependability Study report lists Excessive Wind Noise amongst the top 5 problems most commonly experienced by vehicle owners

Noisy Breaks, Engine Noises, Excessive Fuel Consumption, Engine Loses Power

Motivation (2)

- Wind noise is aerodynamically generated noise perceived in the vehicle interior. Important quality concern for car makers.

Cavity Resonance

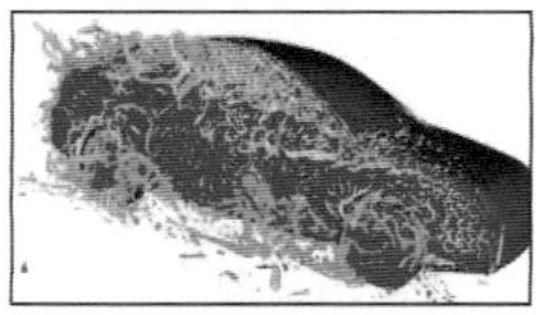

A-Pillar Vortex

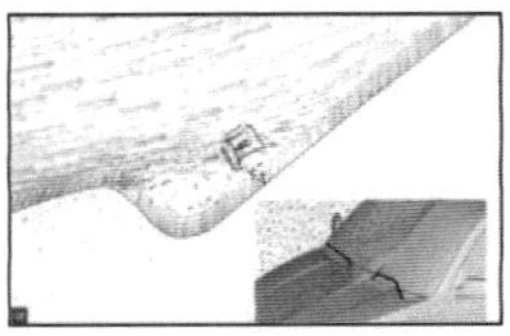

Flow Separation/ Vortex Shedding

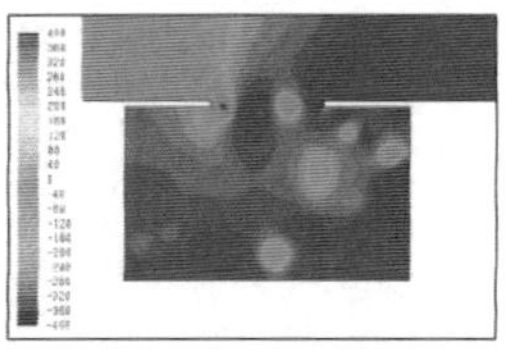

Cavity Resonance

Flow Separation/ Vortex Shedding

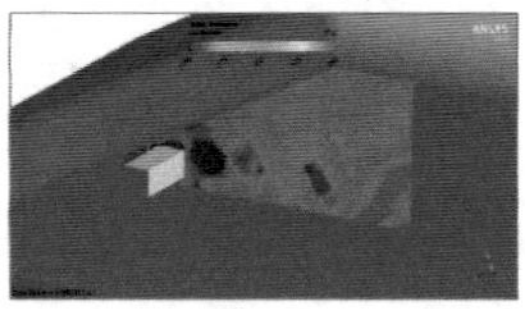

Side-View Mirror Vortex

Problem Description

- Demonstrate aero-vibro-acoustics coupling to predict noise at SAE-Body with mirror by means of a deterministic method
- Validate with extensive experimental data [1] from Friedrich-Alexander University in Erlangen/Germany

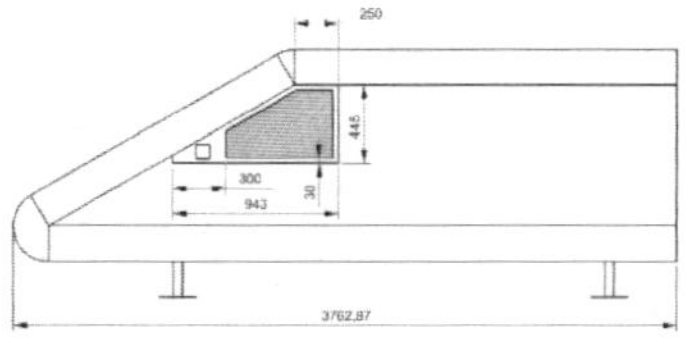

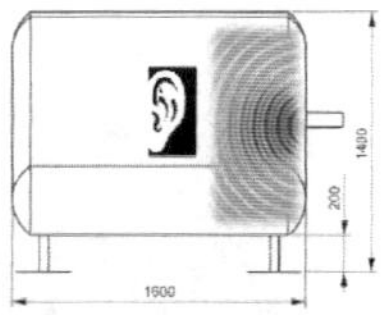

Courtesy of FAU Erlangen-Nuremberg

Sound Source Transient separated flow at side mirror → **Transfer Path** Side glass → **Receiver** Ear of a driver

Simulation Methodology (1)

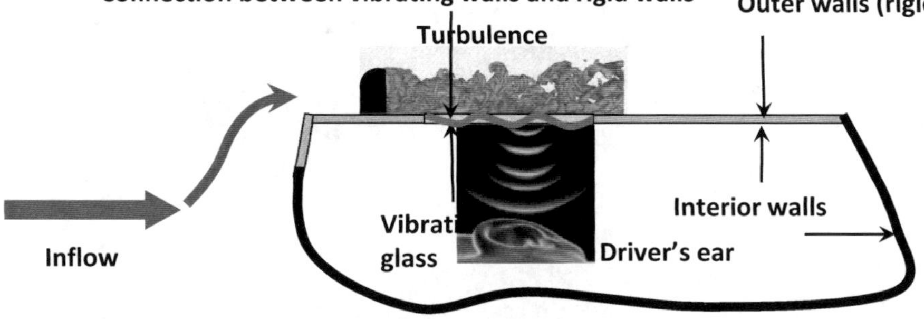

External CFD Model Transient Flow → Vibrating Surfaces (Side Glass) → Acoustics Model (Car Interior)

Compressible CFD modeling | Vibroacoustics Modeling

Simulation Methodology (2)

- From source to ears: SAE-Body with side view mirror

Solve Transient CFD → Time → Freq. Domain Transform → Mapping Freq. Domain Pressure Loading → Solve Vibro-Acoustics Model

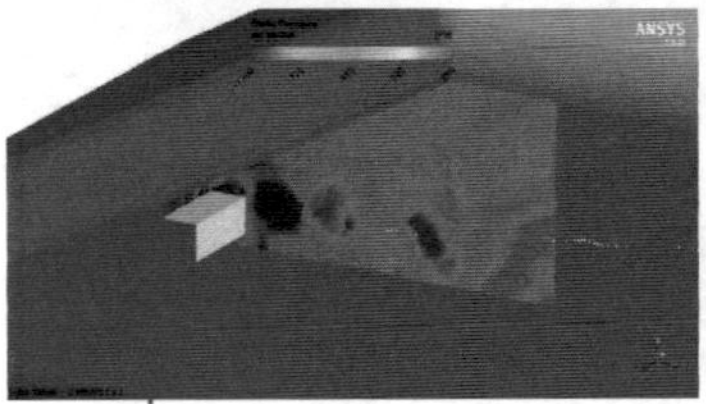
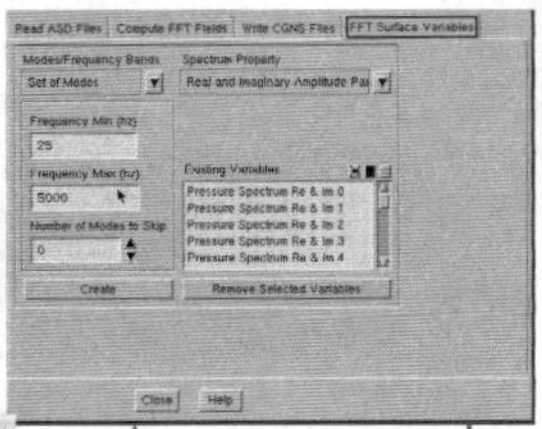
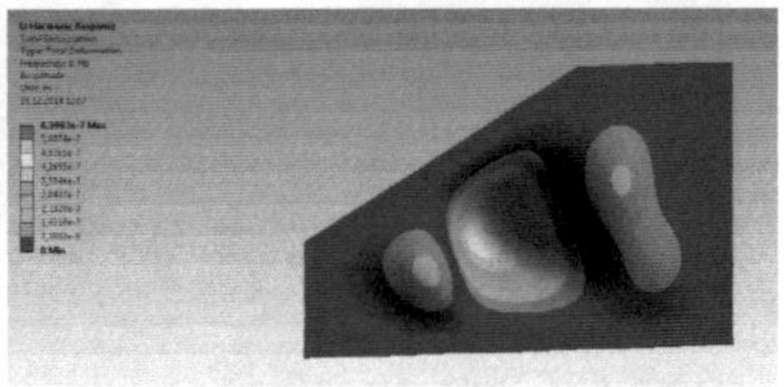

SAE_fine.index	13 KB
SAE_fine1000.asd	30.365 KB
SAE_fine1025.asd	80.971 KB
SAE_fine1089.asd	80.971 KB
SAE_fine1153.asd	80.971 KB
SAE_fine1217.asd	80.971 KB
SAE_fine1281.asd	80.971 KB
SAE_fine1345.asd	80.971 KB
SAE_fine1409.asd	80.971 KB
SAE_fine1473.asd	80.971 KB
SAE_fine1537.asd	80.971 KB
SAE_fine1601.asd	80.971 KB
...	

final.cgns	20.656 KB
final_1.cgns	253.116 KB
final_2.cgns	253.116 KB
final_3.cgns	253.116 KB
final_4.cgns	253.116 KB
final_5.cgns	253.116 KB
final_6.cgns	253.116 KB
...	

Setup – Mesh (1)

- CFD
  - ANSYS FluentMeshing
  - 87 prism/tet cells
  - 15 prism-layers
  - First prism-layer height: 5e-02 mm
  - Surface mesh size:
  - Mirror+window: 1 mm
  - A-pillar: 3 mm
  - SIM1

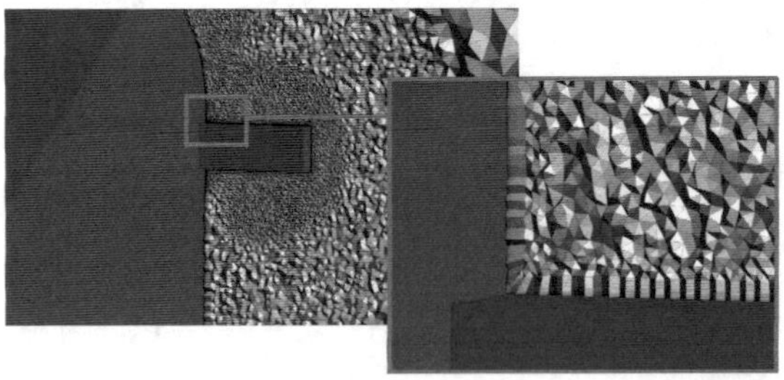

Setup – Mesh (2)

- CFD
  - ANSYS FluentMeshing
  - 37 prism/hexcore cells
  - 15 prism-layers
  - First prism-layer height: 5e-02 mm
  - Surface mesh size:
  - Mirror+window: 1 mm
  - A-pillar: 3 mm
  - SIM2

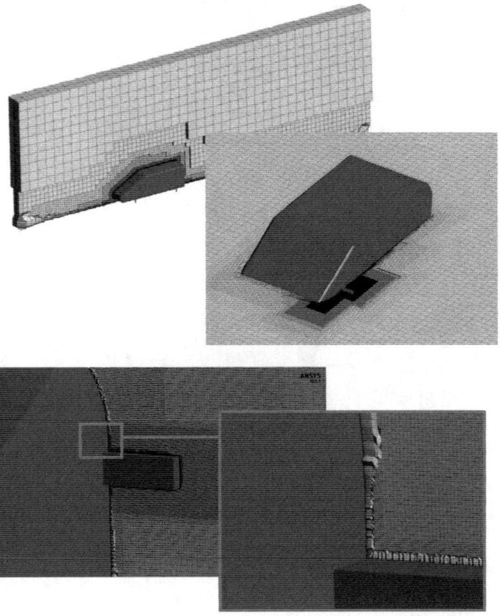

Setup – Mesh (3)

- Vibro-Acoustics
  - ANSYS Mechanical
  - 660,000 hex+tet+pyramid cells
  - Glass+sealing frame: 3 mm (hex cells)
  - Hex+tet+pyramid cells for cabin
  - 12 linear elements per λ_{1kHz} inside cabin

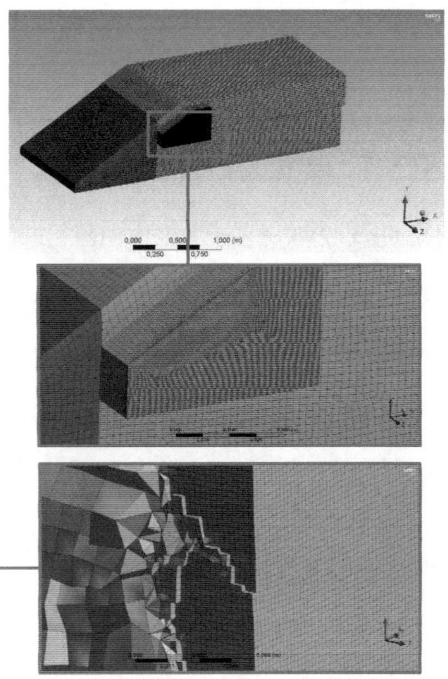

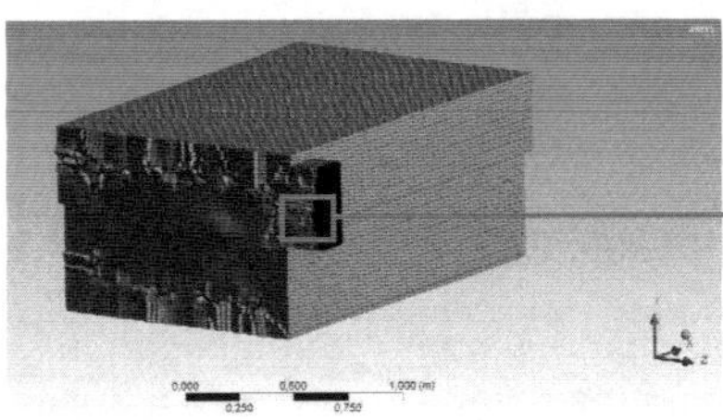

Setup - Boundary Conditions

- CFD domain consists of external SAE-Body surfaces, side window, mirror and wind tunnel boundaries
- Velocity-Inlet: v = 150 km/h, Tu = 0.19%
  - Use measured profile (see below)
- Pressure-Outlet: p = 1 atm
- Tunnel top, floor, sides: no-slip walls

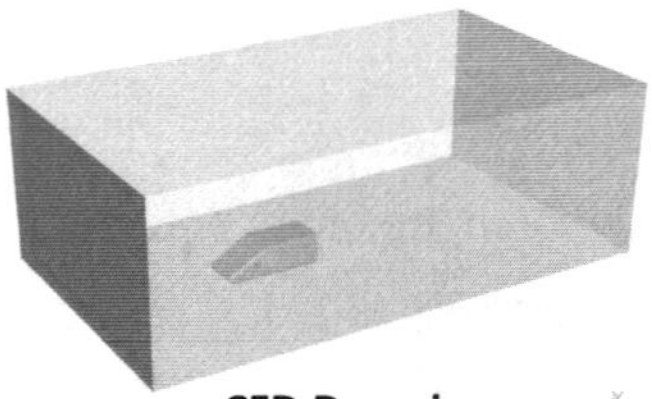

CFD Domain

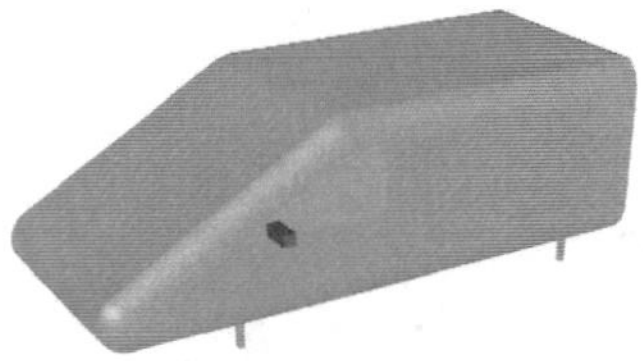

SAE-Body

Inlet Velocity Profile

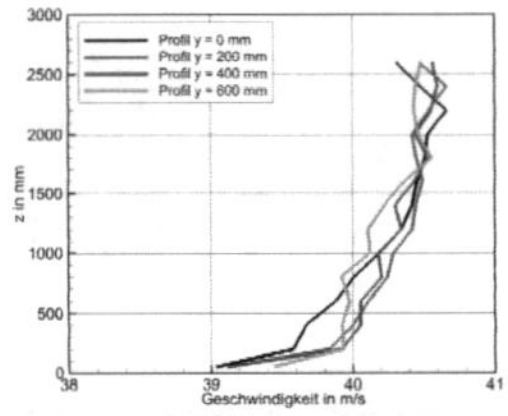

Setup – CFD-Simulation

- ANSYS Fluent
- Material: Air as Ideal Gas
- Turbulence Model Steady State: SST K-Omega
- Turbulence Model Transient: DDES – SST K-Omega
- Spatial Discretization
  - Momentum : Bounded Central Differencing (for DDES)
  - All others: Second Order Upwind
- Time Discretization
  - 2nd Order Implicit, Δt= 3e-5 s
- Record data on side-window every time-step

Setup – Vibro-Acoustics-Simulation (1)

- ANSYS Mechanical
- Frequency resolution: 3.2 Hz
- Strong coupling:
  - Full Vibro-acoustics harmonic analysis from 3 to 1000 Hz
- Import complex pressure in frequency domain
- Setup with ACT extension

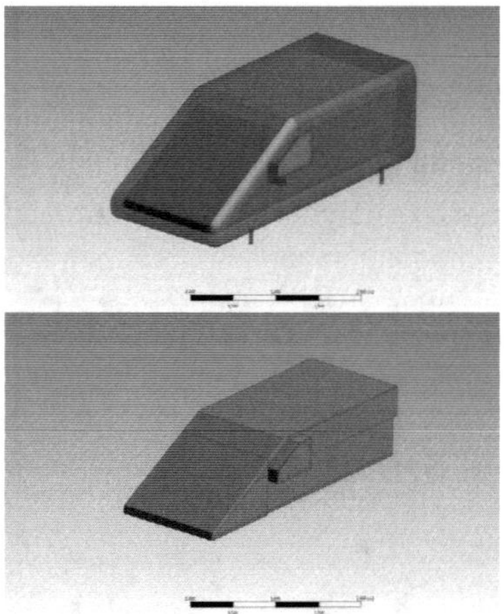

Vibro-Acoustics Domain

Setup – Vibro-Acoustics-Simulation (2)

- Material Properties
  - Complex mounting system including sealant
  - Frame with high stiffness attached to synthetic rubber

PROPERTIES	Floatglas [5], [7]	Sealant [4], [5], [6]
Density (kg/m³)	2,500	1,800
Young's Modulus (MPa)	70,000	50
Poisson's Ratio	0.2	0.38
Thickness	2.85	2.85

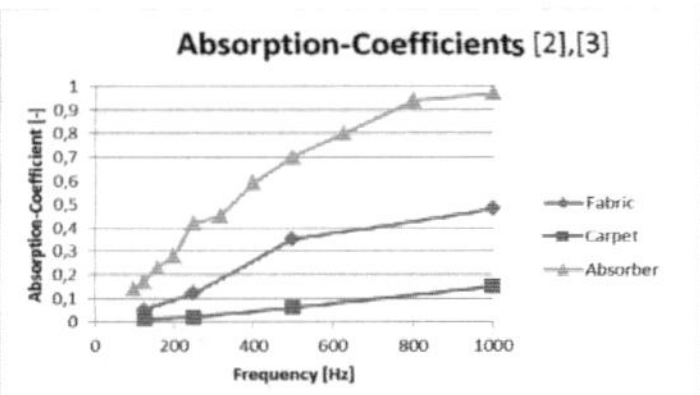

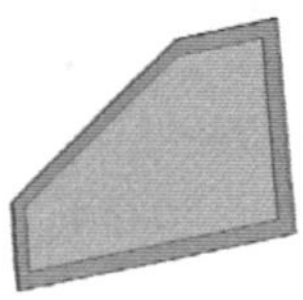

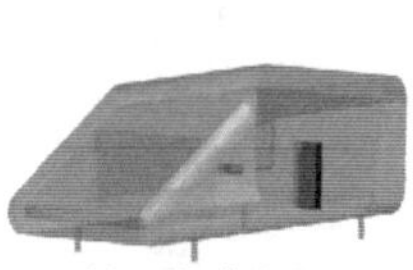

Vertical Cut

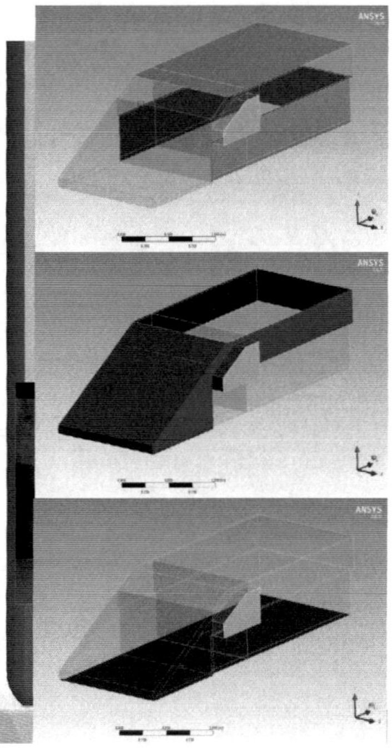

Experimental Data (1)

- External flow pressure spectra at 39 probe positions are available from measurement
- Sampling frequency: 44.1 kHz
- $\Delta t \approx 2.268\text{e-}05$ s
- Sampling time T = 180 s

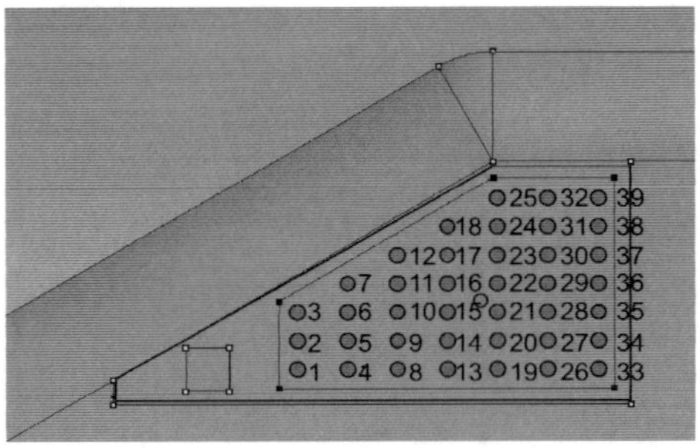

Experimental Data (2)

- Laser-scanning vibrometer measurements to visualize displacements
- 1426 measuring points
- Spectral amplitudes of displacement
- Frequency resolution: 1.5 Hz
- Resolved up to 10 kHz

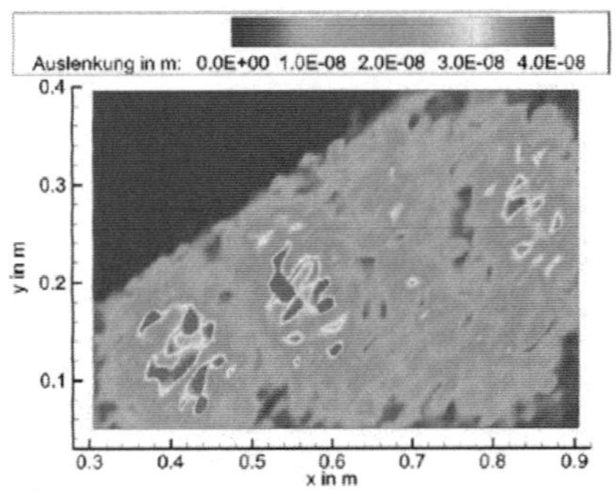

Displacements at 412 Hz

Experimental Data (3)

- Sound pressure data available at driver's ear
- Sampling frequency: 96 kHz
  - $\Delta t \approx 1.042e\text{-}05$ s
- Sampling time:
  - T = 60 s

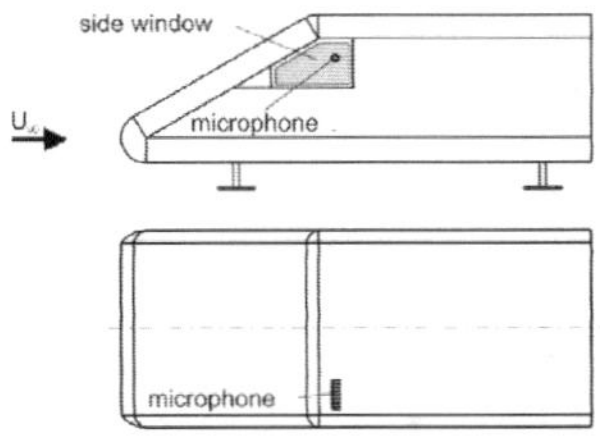

Micro @ driver's ear

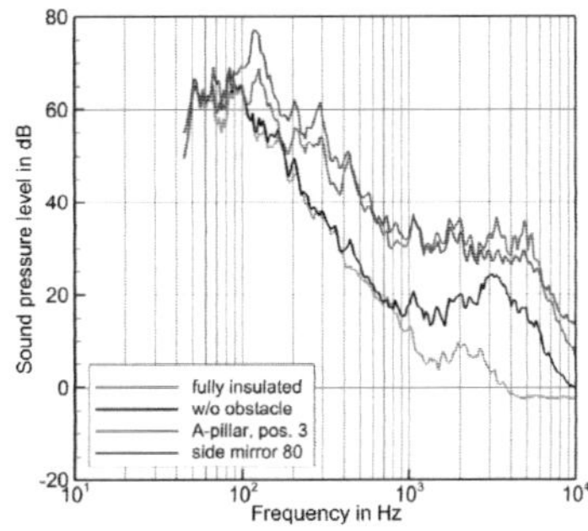

SPL at driver's ear

Results - Flow Field

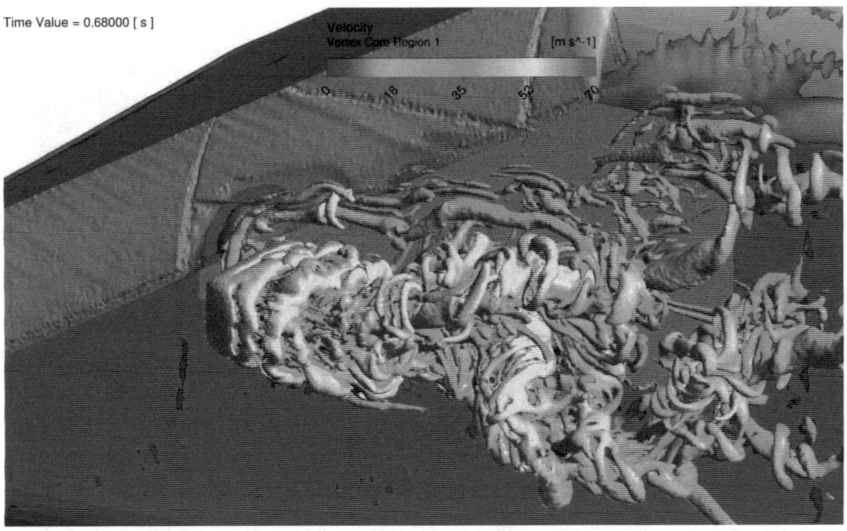

Iso-surface of Q-Criterion colored by velocity magnitude

Results – External Flow Wall Pressure – 1/3-Octave

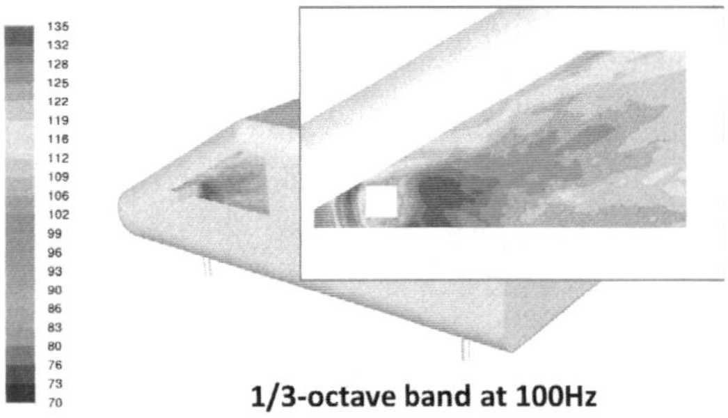

1/3-octave band at 100Hz

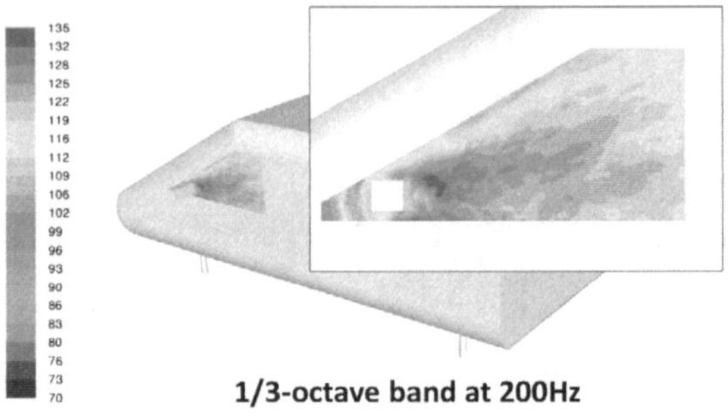

1/3-octave band at 200Hz

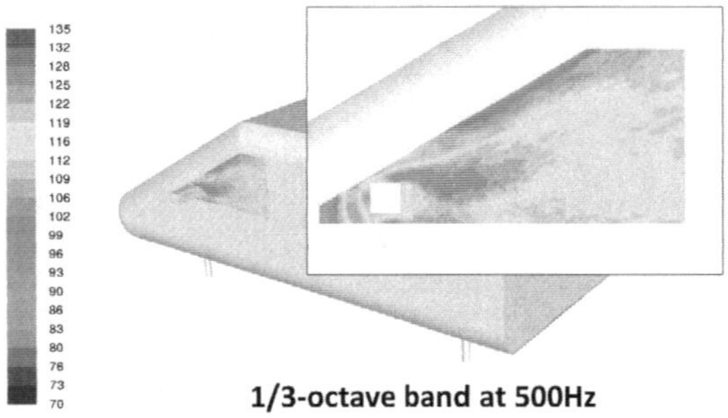

1/3-octave band at 500Hz

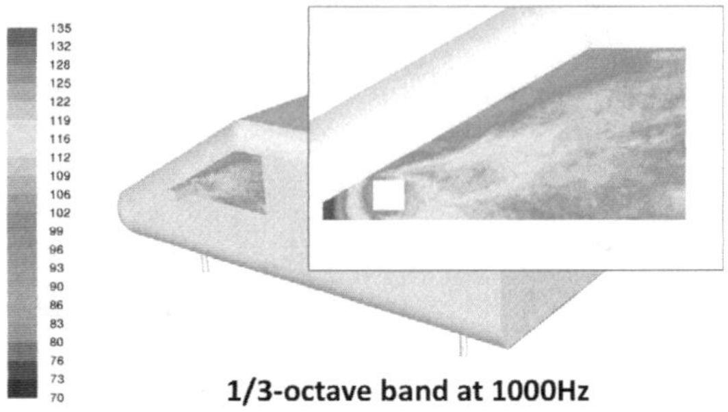

1/3-octave band at 1000Hz

Results – External Flow Wall Pressure – Complex Pressure

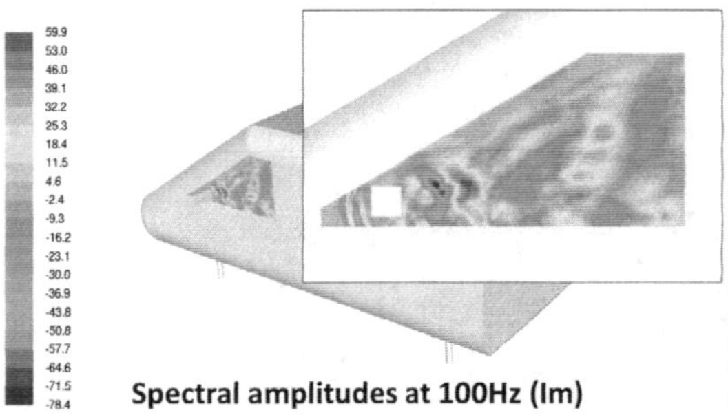

Spectral amplitudes at 100Hz (Im)

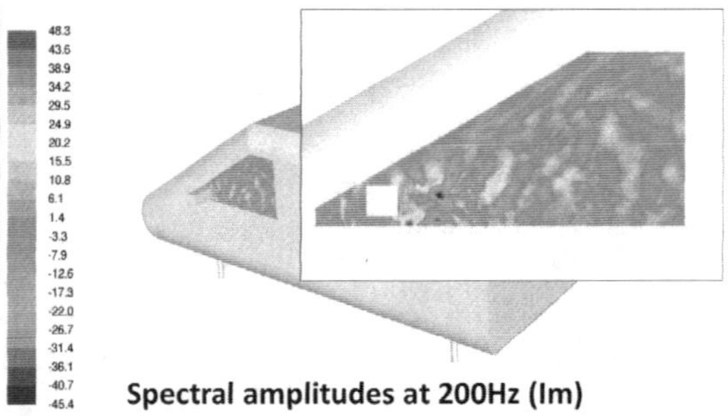

Spectral amplitudes at 200Hz (Im)

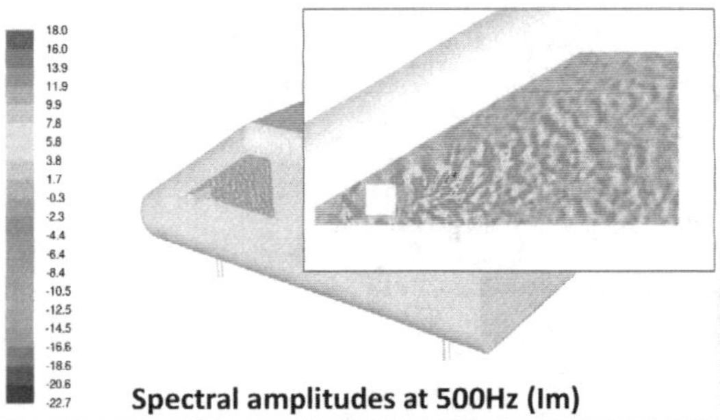

Spectral amplitudes at 500Hz (Im)

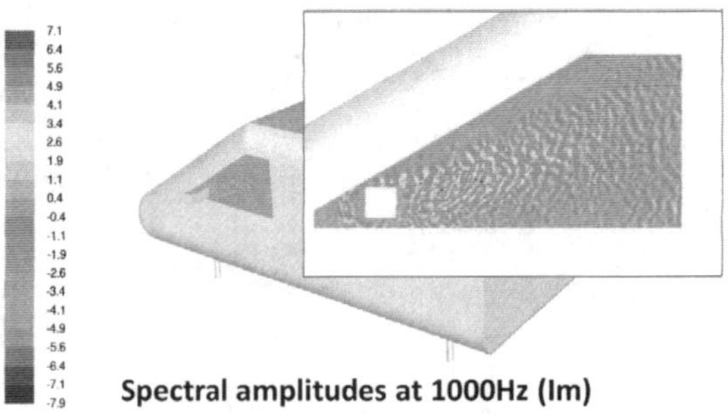

Spectral amplitudes at 1000Hz (Im)

Results – Pressure Mapping

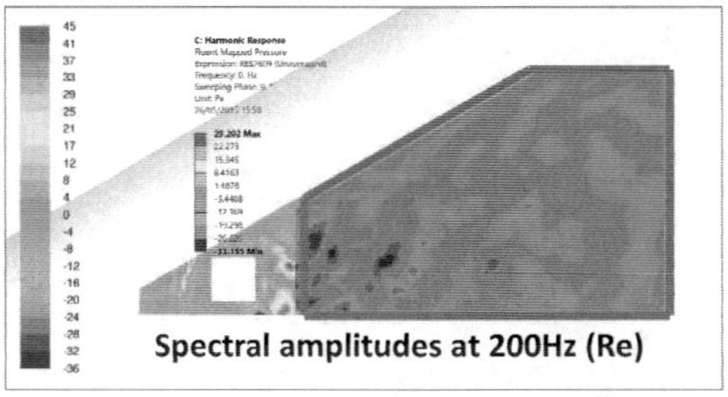

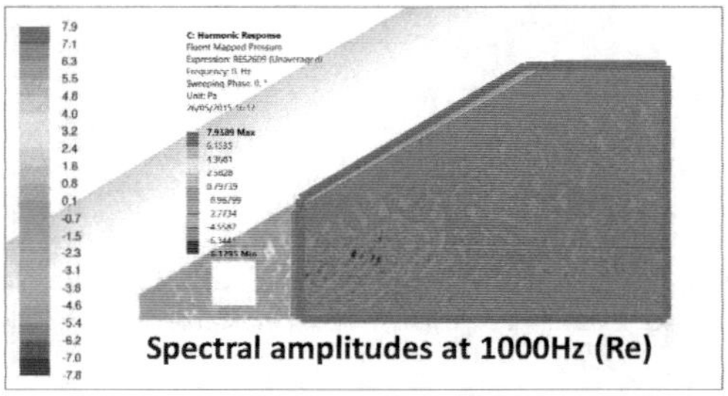

Automotive Simulation WORLD CONGRESS

Flow Pressure Spectrum @ Probes

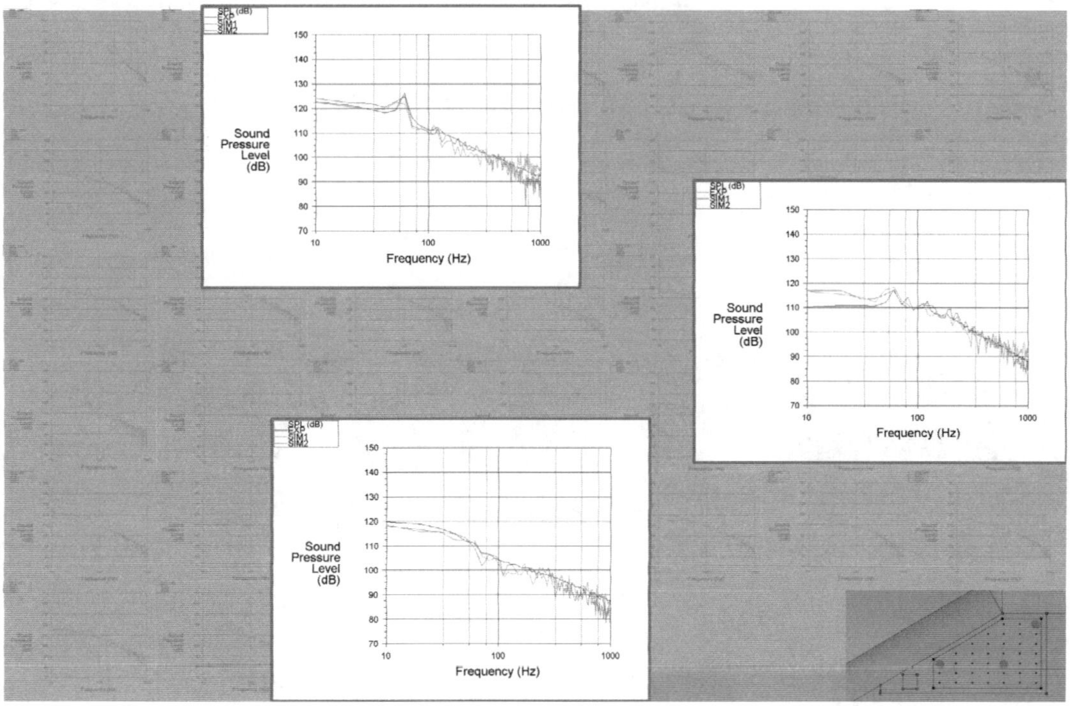

Results – Displacements SIM1

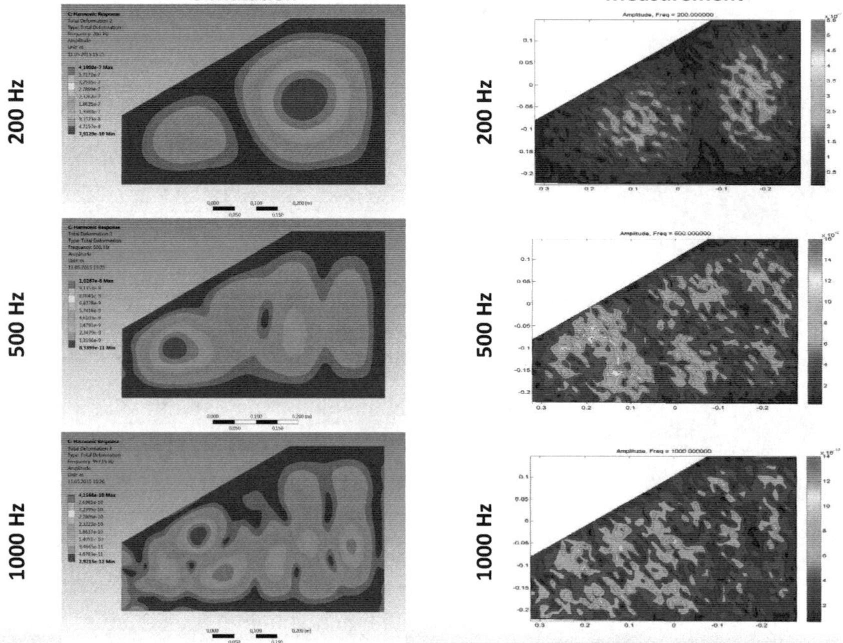

Results – Acoustic Pressure SIM1

- Deformation colored by acoustic pressure at 500 Hz

Acoustic pressure at 500 Hz

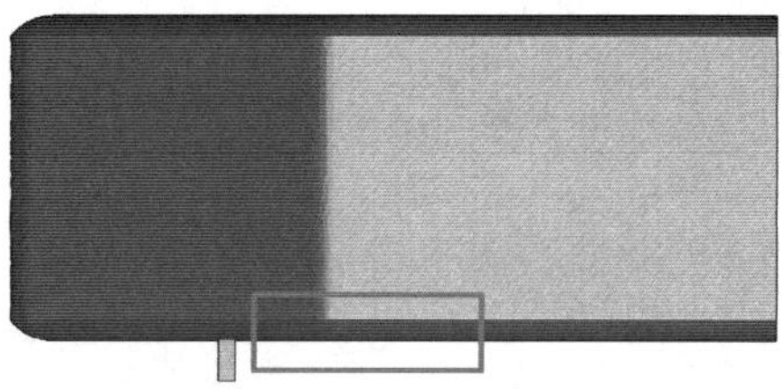

Deformations at 500 Hz

Results - SPL(dB) vs 1/3-Octave in Cabin

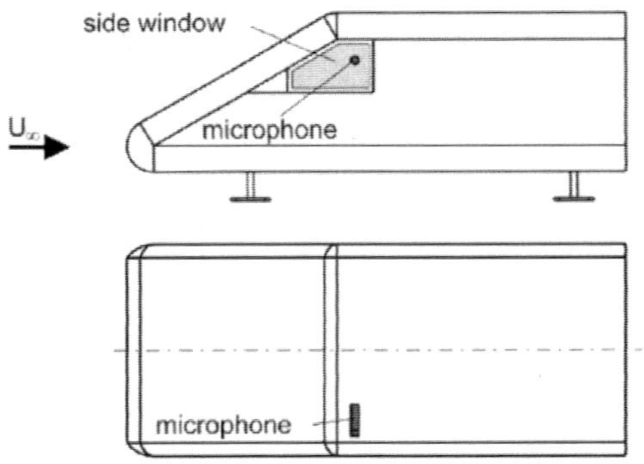

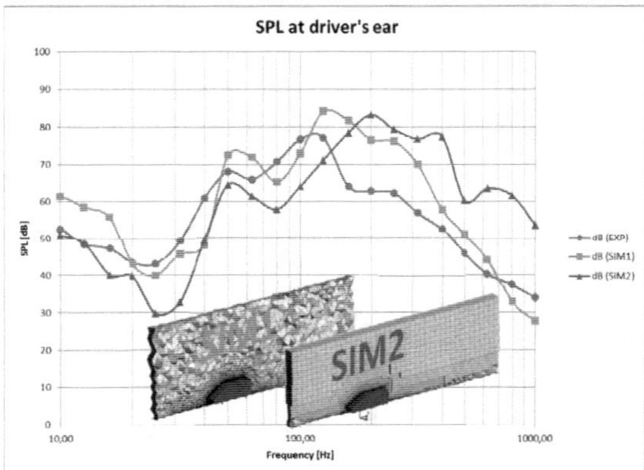

Automotive Simulation WORLD CONGRESS

Solution Performance SIM1

- CFD: AMD Opteron, 2.3 GHz with Infiniband
- Vibro-Acoustics: Intel Xeon, 3.4 GHz with shared memory

	Simulation procedure	Wall clock time 8 cores	Wall clock time 200 cores	Wall clock time 312 cores
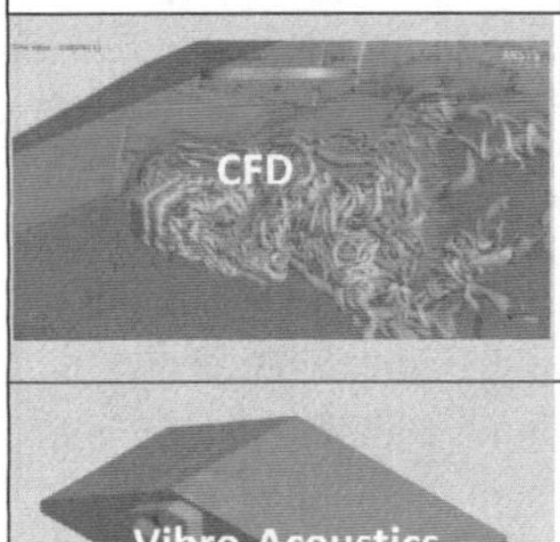 CFD	Steady state (coupled solver)	-	9.3h	-
	Initial transient (ramp down Δt from 1e-03 to 3e-05 for 1,000*Δt)	-	53.9h	-
	Final transient (acoustic sampling for 11,000*Δt)	-	-	284.0h
 Vibro-Acoustics	Fully Coupled (3-1000 Hz)	11.1h	-	-

Summary

- Complete Aero-Vibro-Acoustics workflow
  - From turbulent compressible flow through vibrating side window into the cabin
  - One virtual environment including CFD-, Structural- and Acoustics-Simulations
- CFD results are showing excellent agreement with measured pressure spectra on the window
- Glass deformation well predicted up to 500 Hz
- Computed sound spectra inside cabin shows good trends as compared with test data up to 1000 Hz
- Structural damping and sealant can be fine-tuned

dynardo

Sensitivity analysis and optimization of an electric motor

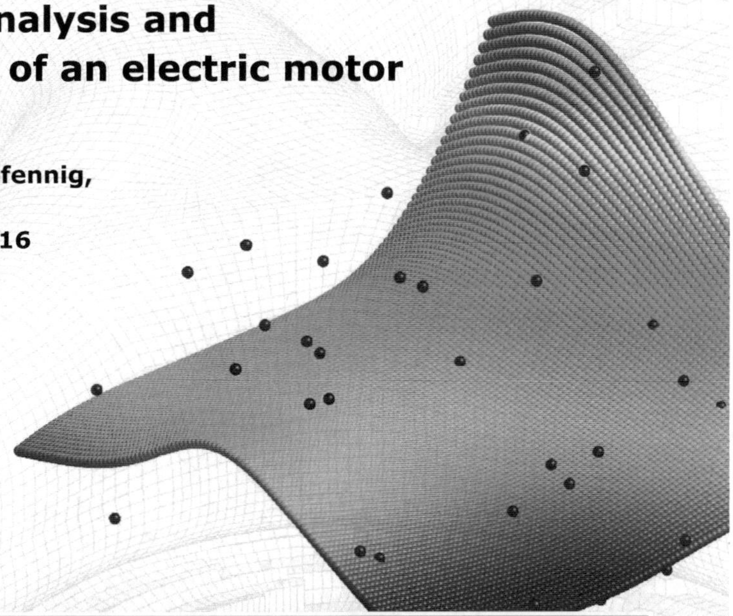

<u>Markus Stokmaier</u>,
Michael Schimmelpfennig,
Thomas Most
Dynardo GmbH, 2016

dynardo

Outline

Intro: Dynardo & optiSLang

The motor simulation

- ANSYS Maxwell 2D model and parametrization

Sensitivity analysis with optiSLang

- Problem understanding
- Identification of influential parameters
- Identification of tradeoffs

Optimization with optiSLang

- Single-objective optimization maximizing efficiency and minimizing torque ripples

Summary

Introduction:

Dynardo & optiSLang

Dynardo

- Founded: 2001 (Will, Bucher, CADFEM International)
- More than 50 employees, offices at Weimar and Vienna
- Leading technology companies Daimler, Bosch, E.ON, Nokia, Siemens, BMW are supported

Software Development

multiPlas

Dynardo is engineering specialist for CAE-based sensitivity analysis, optimization, robustness evaluation and robust design optimization

CAE-Consulting

- Mechanical engineering
- Civil engineering & Geomechanics
- Automotive industry
- Consumer goods industry
- Power generation

Software Development

dynardo optiSLang

SoS • Statistics on Structure

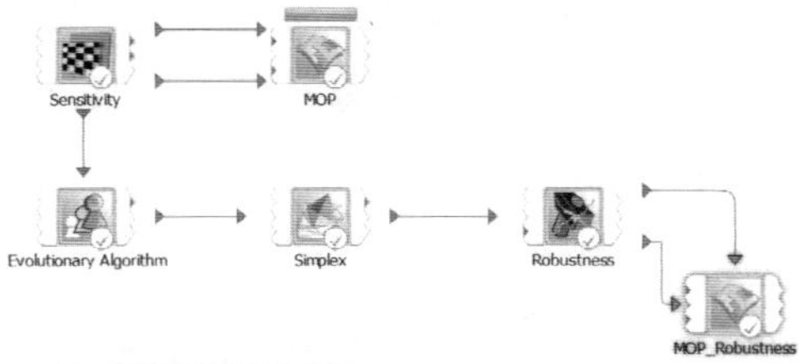

multiPlas • material library

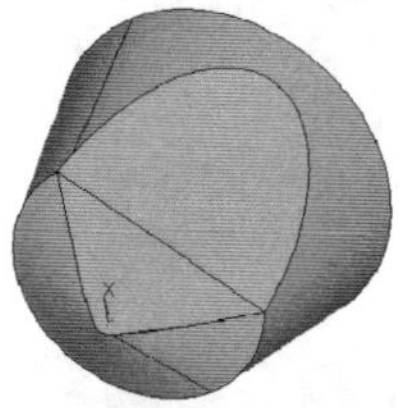

INSIDE

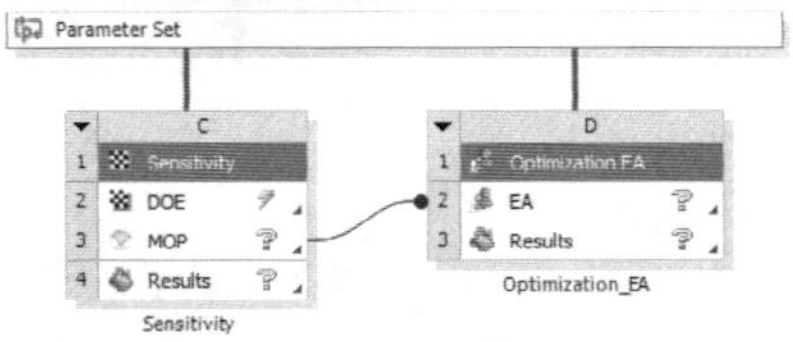

Full integration of optiSLang in Ansys Workbench

- optiSLang modules *Sensitivity, Optimization* and *Robustness* are directly available in ANSYS Workbench

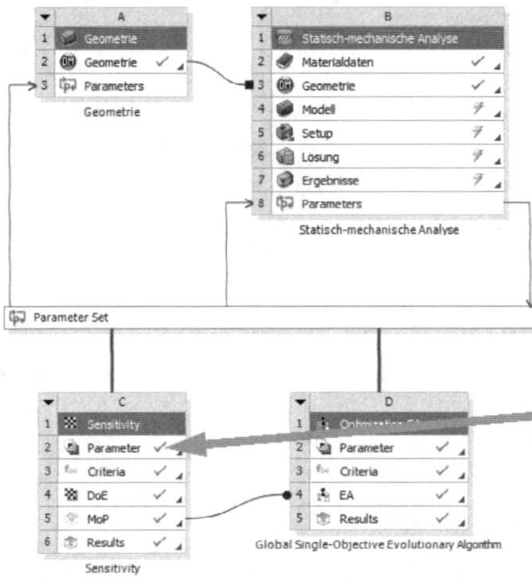

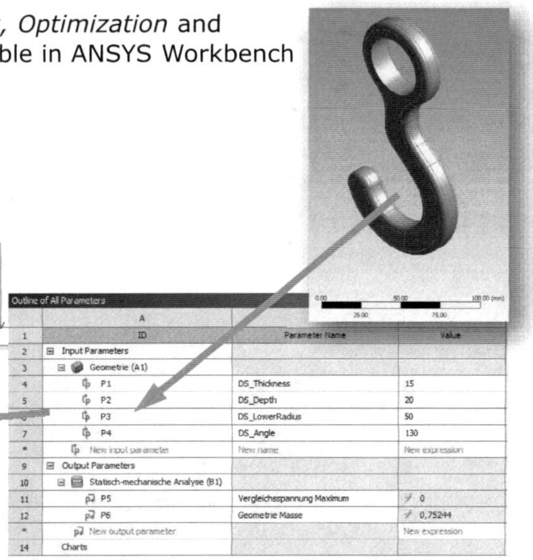

Robust Design Optimization with optiSLang

1st Sensitivity Analysis
Quantification of parameter importance

2nd Multidisciplinary Optimization
Adaptive Response Surface, Evolutionary Algorithm, Pareto Optimization

3rd Robustness Evaluation
Variance based robustness analysis

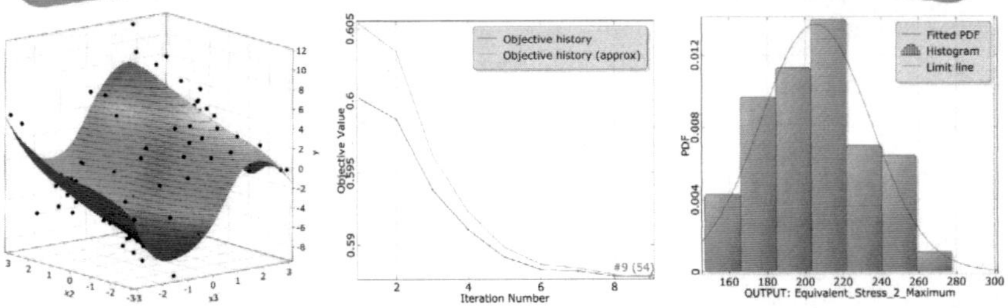

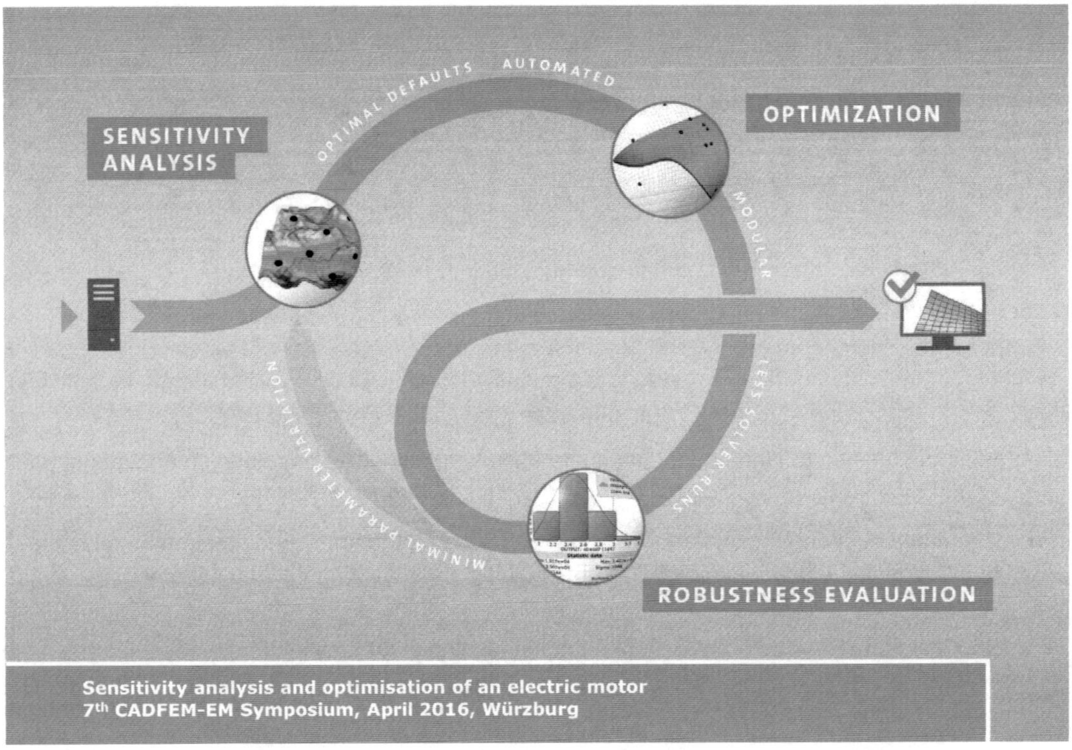

dynardo

The Motor Simulation

dynardo

Commutator motor: working principle

What creates the driving torque?

B-field from magnets

B-field from coils

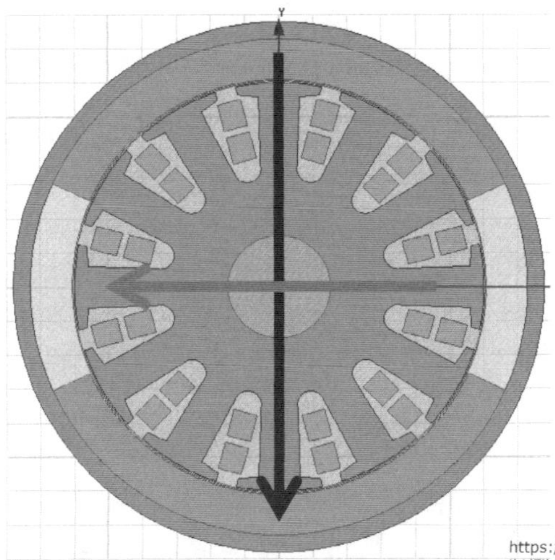

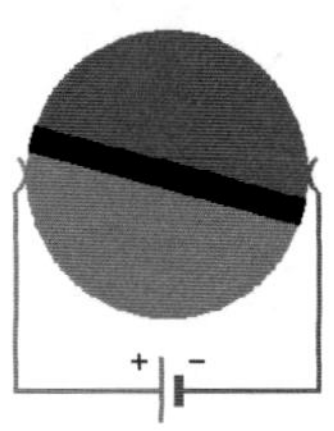

https://commons.wikimedia.org/wiki/File:Kommutator_animiert.gif

Maxwell 2D model setup by Lester Pena-Gomez, CADFEM Germany

© Dynardo GmbH

The model: 2D commutator motor FE-simulation

Motor characteristics

- Commutator principle
- 12 lamellae & coils
- One current branch
  U_0 = 12 V
- Fixed outer diameter
  OD = 78 mm

Desired output quantities

- Torque
- Losses
- Efficiency
- Measures of torque ripples

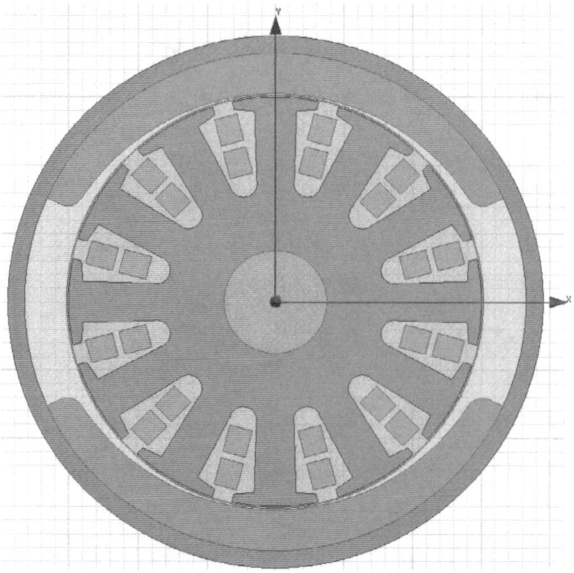

Maxwell 2D model setup by Lester Pena-Gomez, CADFEM Germany

© Dynardo GmbH

The model: 2D commutator motor FE-simulation

Desired output quantities

- Torque
- Losses
- Efficiency
- Measures of torque ripples

Data extraction:

- Key properties extracted from last cycle
- Access to scalar output variables in ANSYS Workbench
- Access to signals via text files

$$CV = \frac{\text{standard deviation}}{\text{mean value}}$$

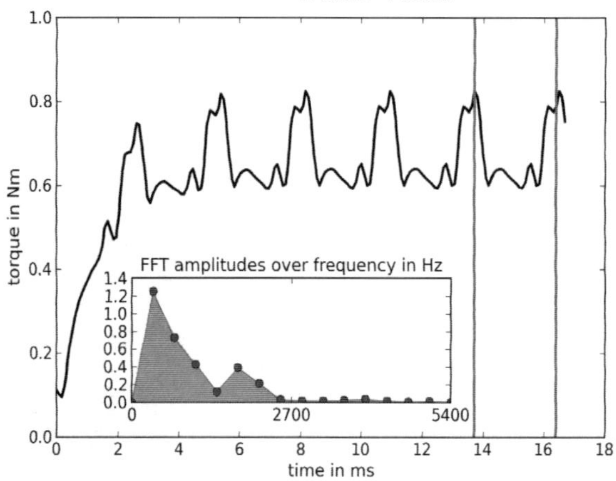

Parametrization

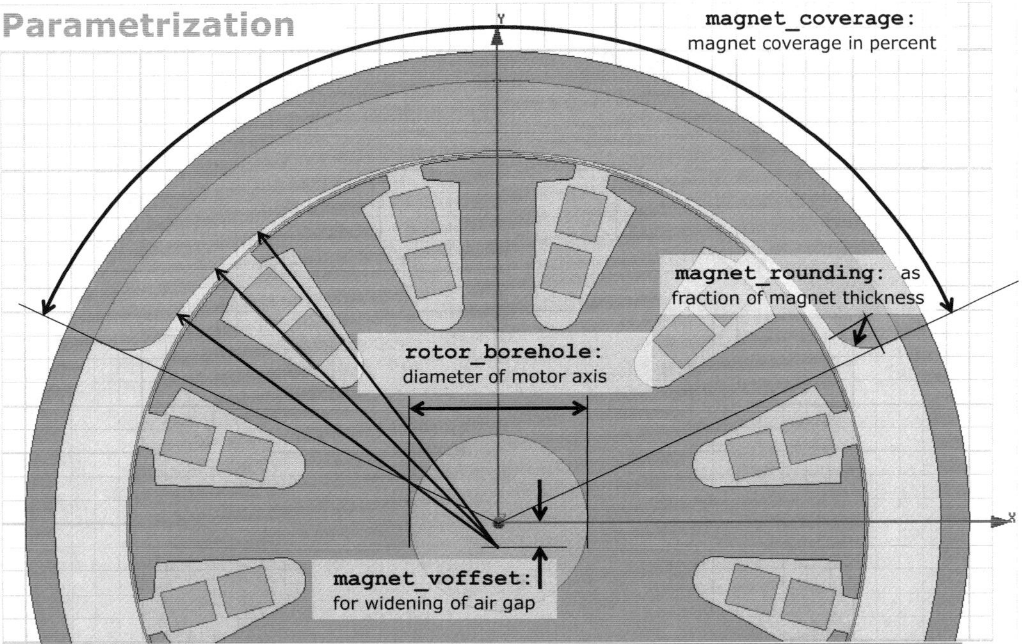

Parametrization

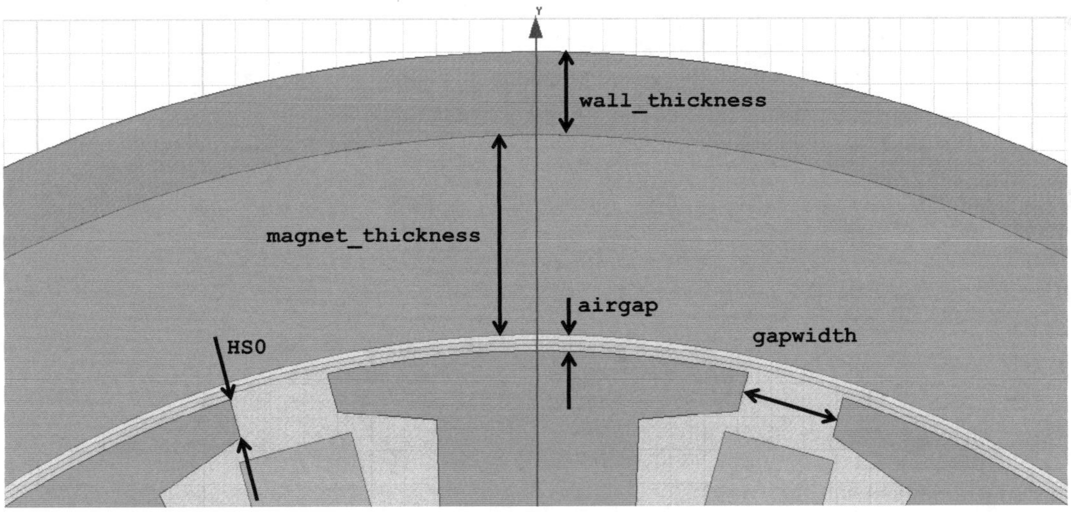

Calling Maxwell from optiSLang

Process Integration

Parametric model as base for

- User-defined optimization (design) space
- Naturally given robustness (random) space

Design variables
Entities defining the design space

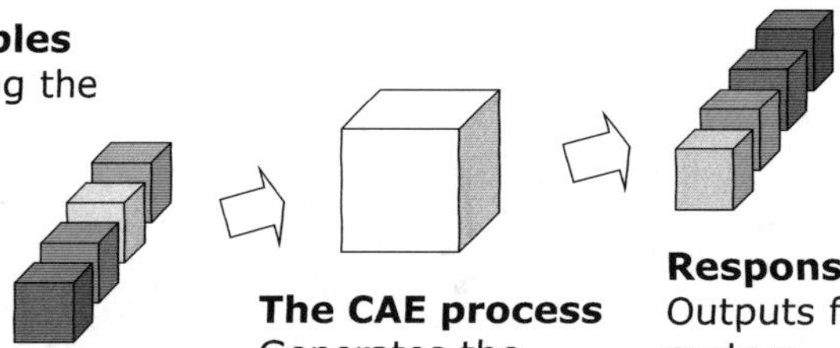

Scattering variables
Entities defining the robustness space

The CAE process
Generates the results according to the inputs

Response variables
Outputs from the system

optiSLang Integrations & Interfaces

Direct integrations

- ANSYS Workbench
- MATLAB
- Excel
- Python
- AMESim
- SimulationX

Supported connections

- ANSYS APDL
- Maxwell
- Abaqus
- Adams
- AMESim
- ...

Arbitary connection of ASCII file based solvers

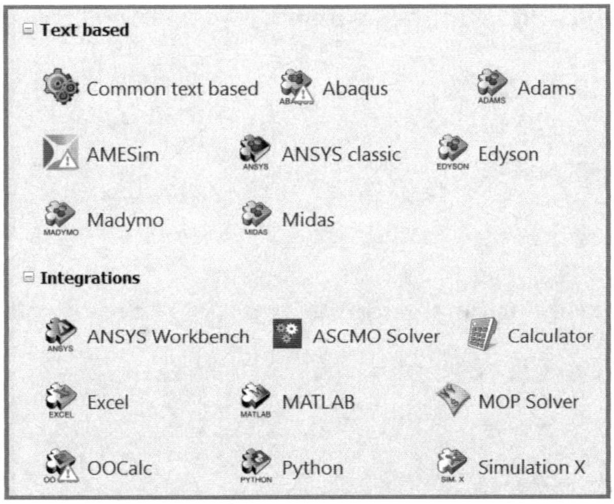

Process Integration: Maxwell & optiSLang

A) optiSLang inside Workbench

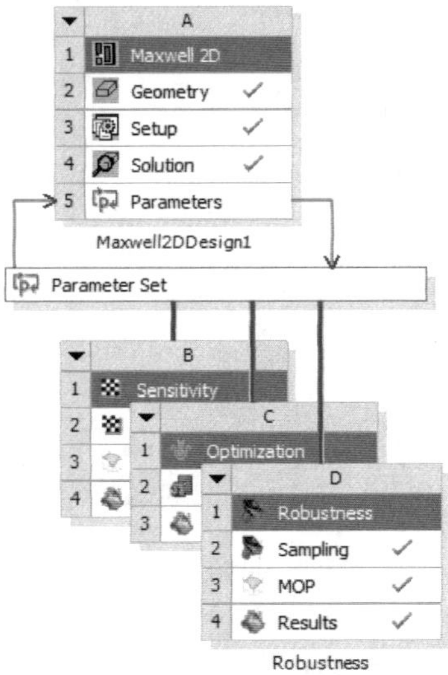

B) Direct integration via text files

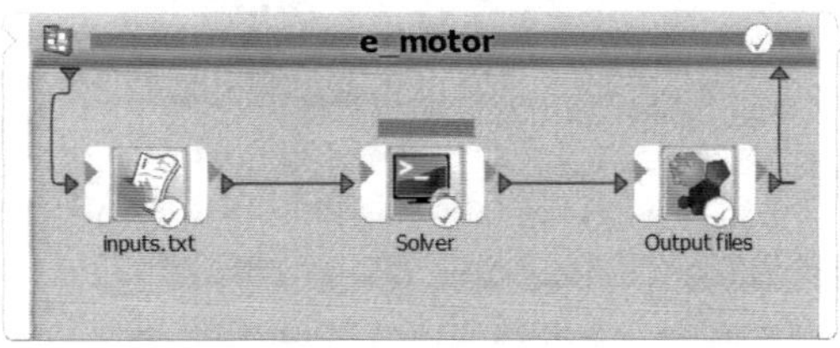

C) Large-Scale DSO

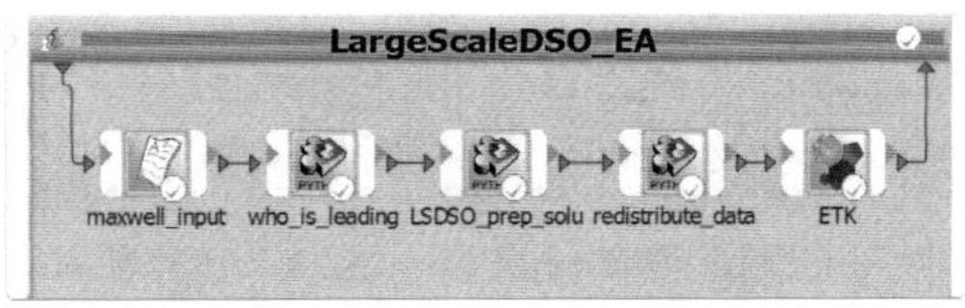

© Dynardo GmbH

dynardo

Process Integration: Maxwell & optiSLang

A) optiSLang inside Workbench

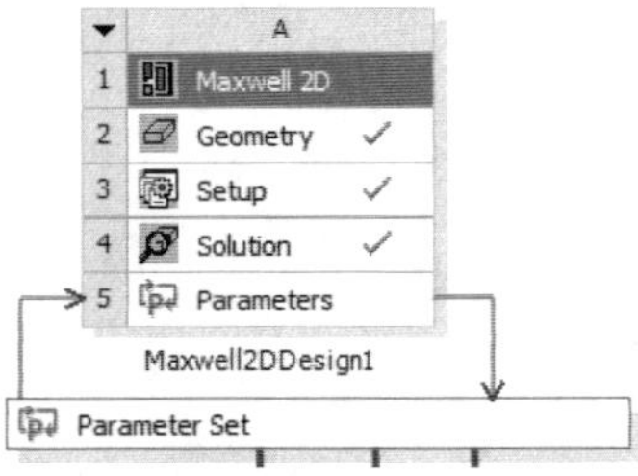

B) Direct integration via text files

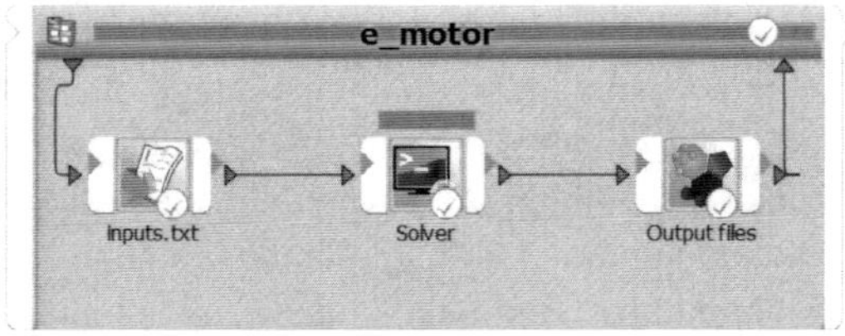

A2) Workbench inside optiSLang

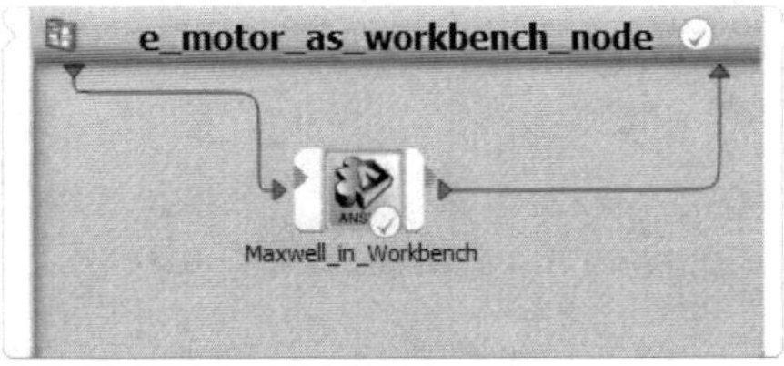

C) Large-Scale DSO

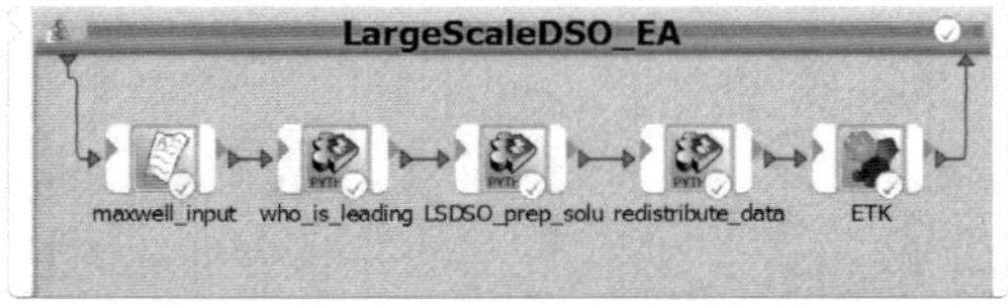

© Dynardo GmbH

dynardo

Method A: using ANSYS Workbench

In Maxwell

- Export **scalar output variables** to the Workbench **Parameter Set**
- Parallel design computation with Optimetrics → Parametric

Parallel/distributed computation:

- Take advantage of the whole ANSYS parallelization functionality

Alternative:

- optiSLang full version and a "Workbench node"

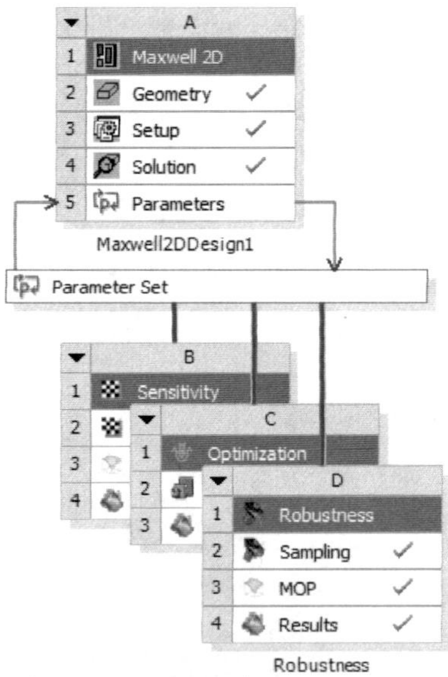

Method A: using ANSYS Workbench

In Maxwell

- Export **scalar output variables** to the Workbench **Parameter Set**
- Parallel design computation with Optimetrics → Parametric

Parallel/distributed computation:

- Take advantage of the whole ANSYS parallelization functionality

Alternative:

- optiSLang full version and a "Workbench node"

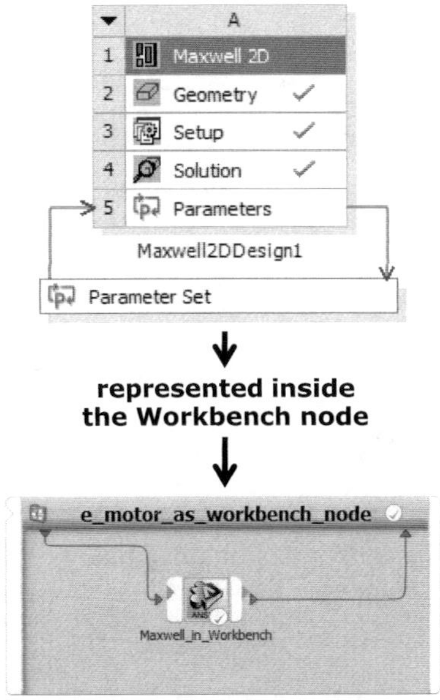

Method B: scripting and text files (direct coupling)

Maxwell

- Batch job including Python script
- Write **transient reports** into files
  → signal data accessible

optiSLang

- Text-based batch job node

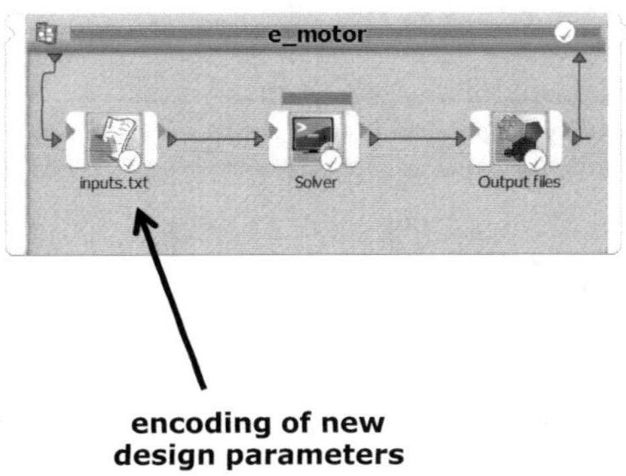

Method B: scripting and text files (direct coupling)

Maxwell
- Batch job including Python script
- Write **transient reports** into files
  → signal data accessible

optiSLang
- Text-based batch job node

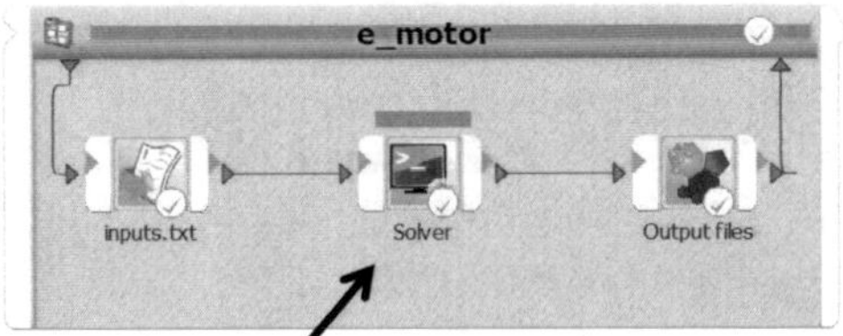

batch/shell script

```
$MAXWELL_CMD  -runscriptandexit $EXE_SCRIPT -log $PRJ_FILE
```


Method B: scripting and text files (direct coupling)

Maxwell
- Batch job including Python script
- Write **transient reports** into files
  → signal data accessible

optiSLang
- Text-based batch job node

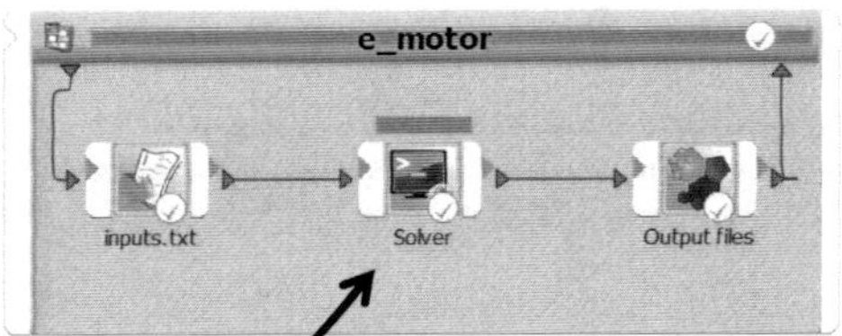

batch/shell script

```
$MAXWELL_CMD  -runscriptandexit $EXE_SCRIPT -log $PRJ_FILE
python $PLOT_SCRIPT
```

© Dynardo GmbH

dynardo

Method B: scripting and text files (direct coupling)

Maxwell

- Batch job including Python script
- Write **transient reports** into files
  → signal data accessible

optiSLang

- Text-based batch job node
- Extract signal data with ETK
- Signal data → free mathematical computations inside any optiSLang system

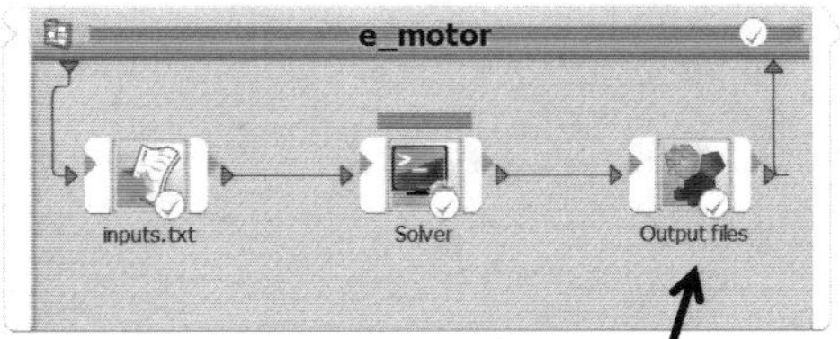

ETK node
(ETK: Extraction Tool Kit)

Parallel/distributed computation:

- optiSLang spawns Maxwell batch jobs

Sensitivity analysis and optimisation of an electric motor
Automotive Simulation World Congress, June 7-8 2016, Munich 25

© Dynardo GmbH

dynardo

Method C: Scripted Large-Scale DSO

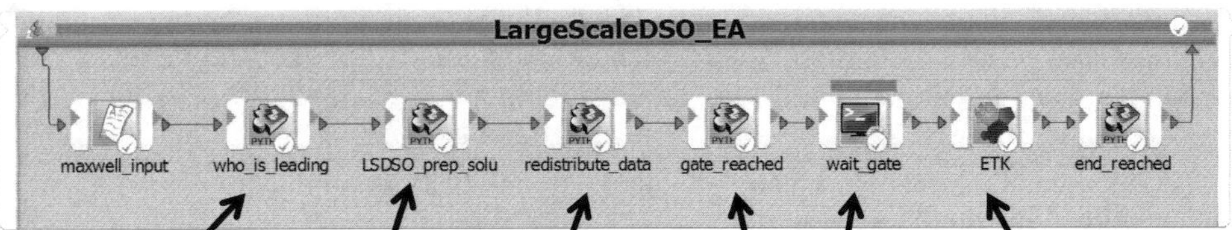

The lead design will execute the LSDSO job while the other designs of the same iteration stay passive.

The lead design takes the data files it finds in the LSDSO result folder and distributes them among the other design folders.

As usual, each design analyses data it finds in its own folder for evaluating the responses.

Each design appends a central design table in a *.csv file. The lead design waits for completion, creates an Optimetrics setup based on the CSV table, and then executes the LSDSO job.

Before response extraction via ETK all designs have to wait for the lead design to finish with simulation and data redistribution. A gate should open only after each design is waiting in front of it.

Sensitivity analysis and optimisation of an electric motor
Automotive Simulation World Congress, June 7 8 2016, Munich 26

Method C: Scripted Large-Scale DSO

Benefits:

- The whole job distribution functionality and flexibility of the Large-Scale DSO approach can be leveraged for simultaneous and parallelized computation
- Signal traces saved as text files in dedicated folders
- Job execution in the background
- The setup works for Sensitivity, Optimization, and Robustness evaluation, i.e. it does not matter whether the algorithm goes through a single iteration (sensitivity & robustness) or through repetitive iterations (optimizer)
- Consecutive iterations can be of different size (amount of design variations)

Requirements:

- Unique names for all signal traces
- A few script adjustments

Process Integration: Maxwell & optiSLang

A) optiSLang inside Workbench

- Straightforward wizards extend the WB-environment
- Simultaneous design solution possible via DSO if only scalar outputs are investigated
- Signal analysis possible in case of serial execution

A2) Workbench node in OSL

- Power of WB: multi-physics (and convenience)
- Power of OSL: freedom to define complex workflows

B) Direct integration via text files

- The OSL full version offers the freedom to build manifold custom workflows (nested systems, performance maps, ...)
- Straightforward parallelization using multiple Maxwell instances
- Signal analysis

C) Large-Scale DSO

- Enables signal analysis in combination with simultaneous design evaluation via Optimetrics-DSO
- Designs of one iteration of any OSL algorithm are bundled into one single LS-DSO job
- Simulation in background

Sensitivity analysis

What is Sensitivity Analysis?

regression
→ simplest form of model

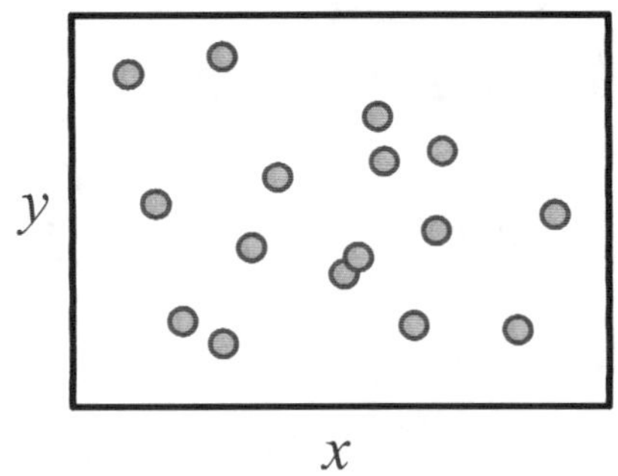

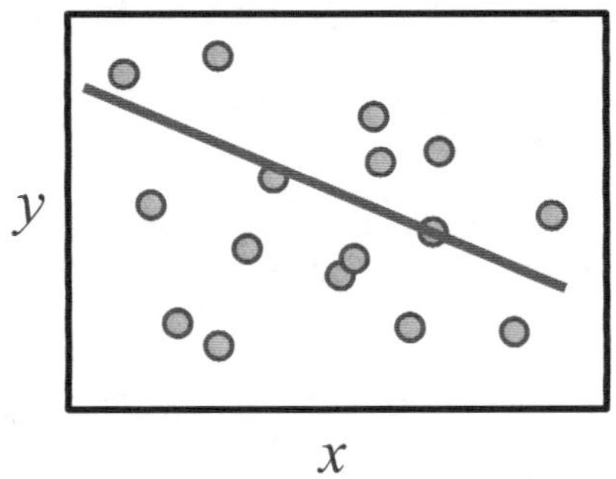

random scattering
no influence
no correlation

detectable correlation

Sensitivity Analysis Flowchart

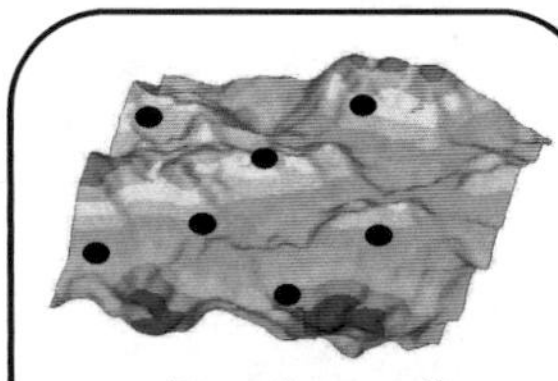

Design of Experiments
- Deterministic
- (Quasi)Random

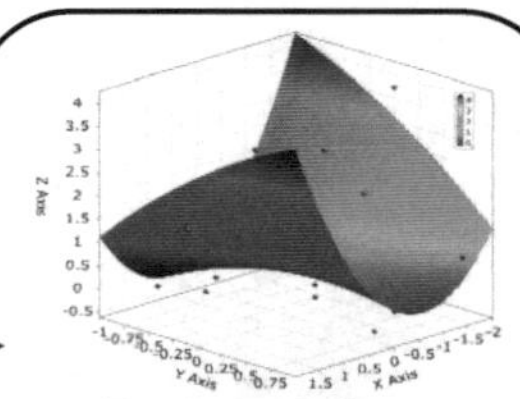

Regression Methods
- 1D regression
- nD polynomials
- Sophisticated metamodels

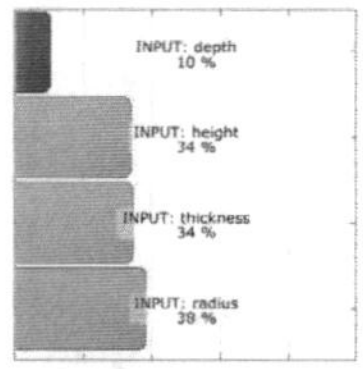

Sensitivity Evaluation
- Correlations
- Reduced regression
- Variance-based

Solver

1. Design of Experiments generates a specific number of designs, which are all evaluated by the solver
2. Regression methods approximate the solver responses to understand and to assess its behavior
3. The variable influence is quantified using the approximation functions

Response Surface Method

- Approximation of response variables as explicit function of all input variables
- Approximation function can be used for sensitivity analysis and/or optimization
- Global methods (**Polynomial regression**, Neural Networks, ...)
- Local methods (Spline interpolation, **Moving Least Squares**, Radial Basis Functions, **Kriging**, ...)
- Approximation quality decreases with increasing input dimension
- Successful application requires objective measures of the prognosis quality

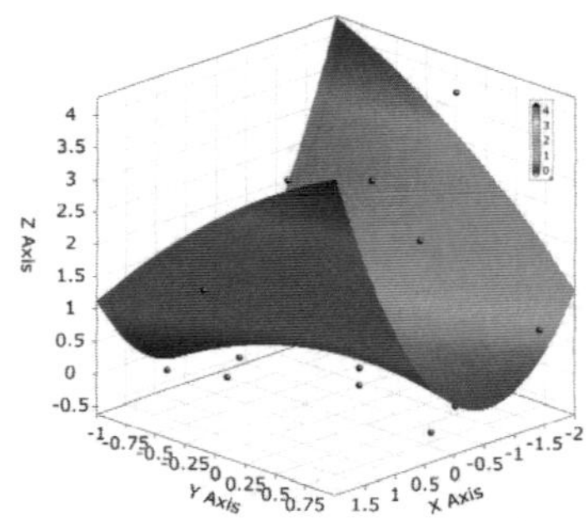

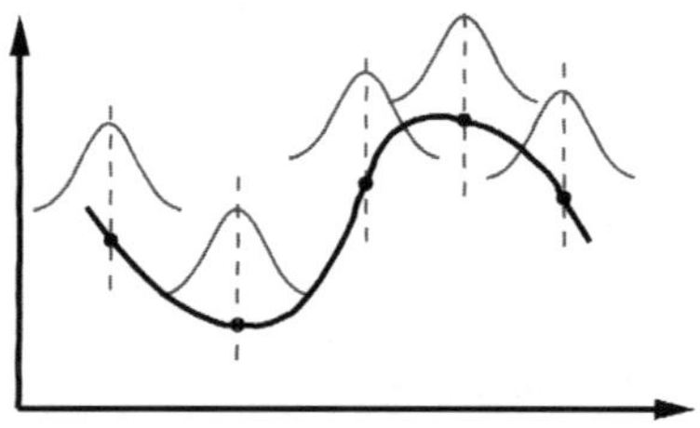

Metamodel of Optimal Prognosis (MOP)

- Approximation of solver output by fast surrogate model
- Reduction of input space to get best compromise between available information (samples) and model representation (number of inputs)
- Determination of optimal approximation model
- Assessment of approximation quality
- Evaluation of variable sensitivities

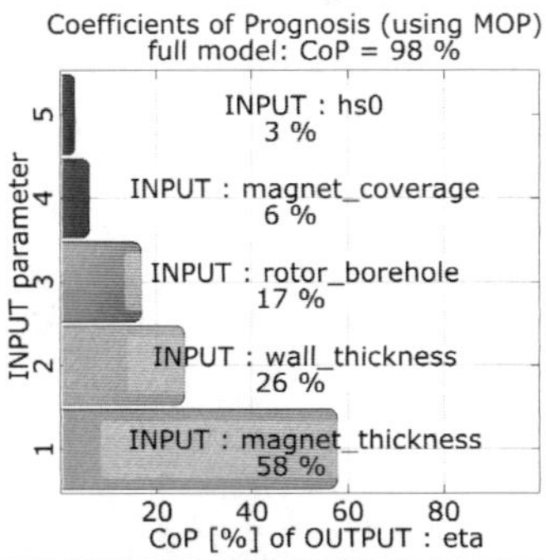

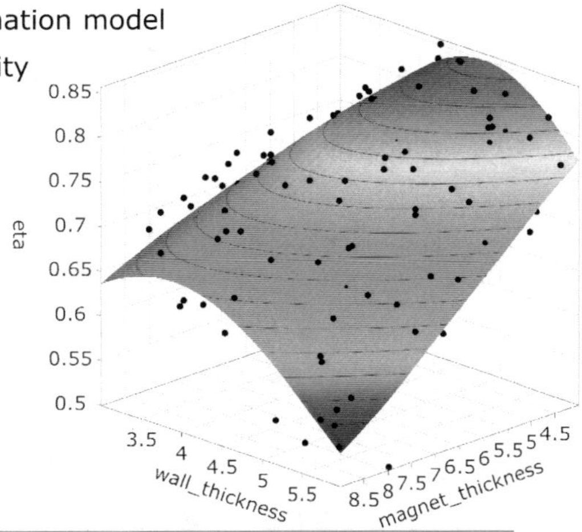

Metamodelling: CoP matrix

- Very well explainable: eta, P_loss, P_mech, P_el, torque ⟶ CoP high
- Less well explainable: size of torque ripples ⟶ CoP lower

Models \ Parameter	magnet_coverage	wall_thickness	hs0	magnet_thickness	magnet_voffset	rotor_borehole	Total
eta	5.8 %	26.0 %	2.9 %	59.0 %		17.9 %	98.7 %
torque_amps_sum_normed	25.4 %	25.1 %			19.0 %		60.6 %
torque_cv	34.7 %	25.1 %			15.1 %		61.9 %
P_loss	14.0 %	25.4 %	3.4 %	51.0 %		17.1 %	98.5 %
P_mech	6.4 %	28.1 %	1.7 %	53.5 %		22.7 %	97.5 %
P_el	6.3 %	26.5 %	3.1 %	54.6 %		20.7 %	97.9 %
torque	6.6 %	27.3 %	1.9 %	52.7 %		22.0 %	97.5 %

- 1st insight: 6 out of 9 parameters identified as highly influential
- 2nd insight: there are two subgroups with high and lower total CoP

Metamodelling: understanding the CoP values ...

... by investigating the matrix of parameter correlations

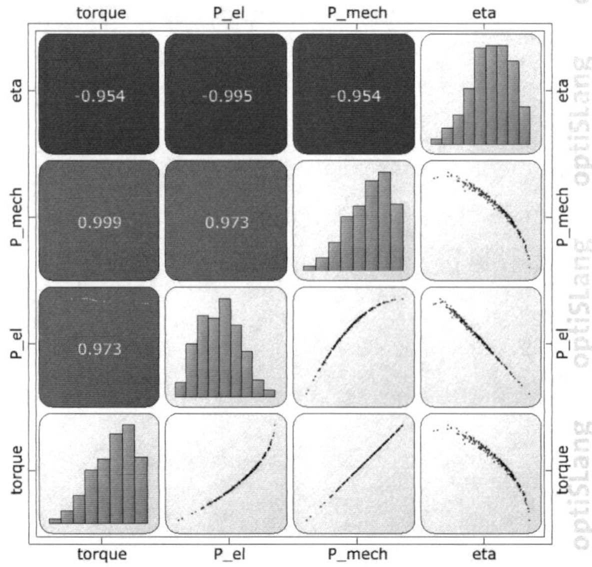

Metamodelling: understanding the CoP values

- P_mech & torque contain the same information

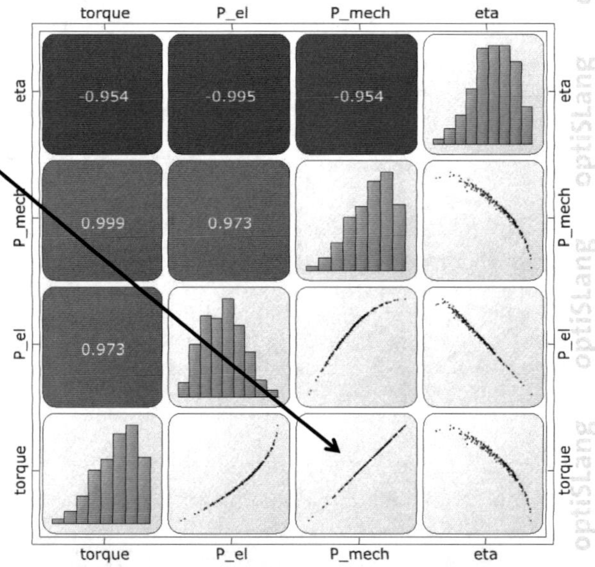

Metamodelling: understanding the CoP values

- P_mech & torque contain the same information
- P_mech follows P_el until losses increase

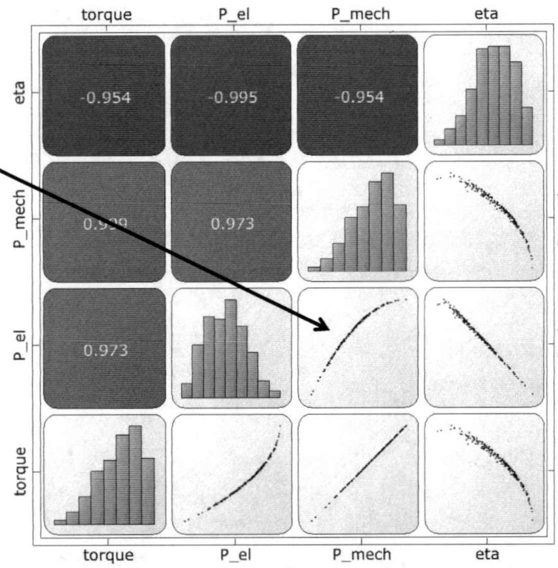

Metamodelling: understanding the CoP values

- P_mech & torque contain the same information
- P_mech follows P_el until losses increase
- eta decreases as losses increase

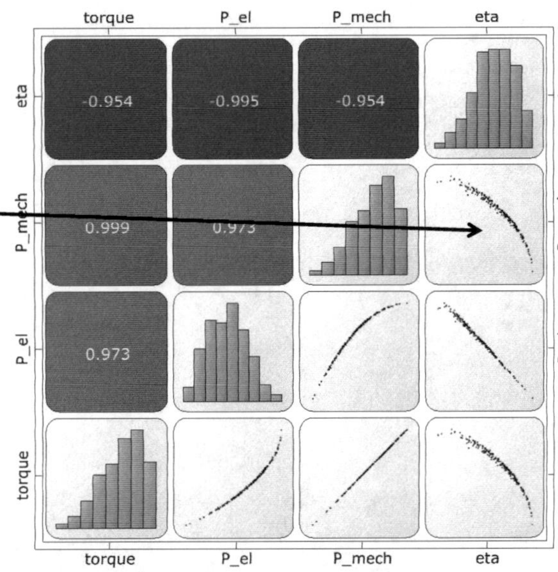

© Dynardo GmbH

Metamodelling: understanding the CoP values

- P_mech & torque contain the same information
- P_mech follows P_el until losses increase
- eta decreases as losses increase

CoP values high because these relationships get fully captured by the metamodel

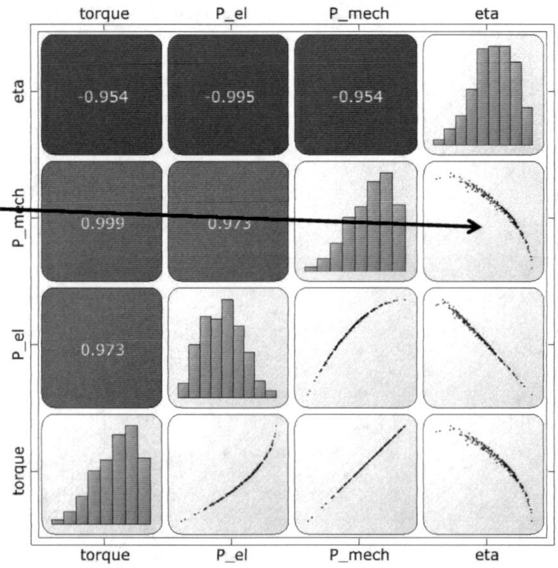

Sensitivity analysis and optimisation of an electric motor
Automotive Simulation World Congress, June 7-8 2016, Munich

© Dynardo GmbH

Metamodelling: torque and eta

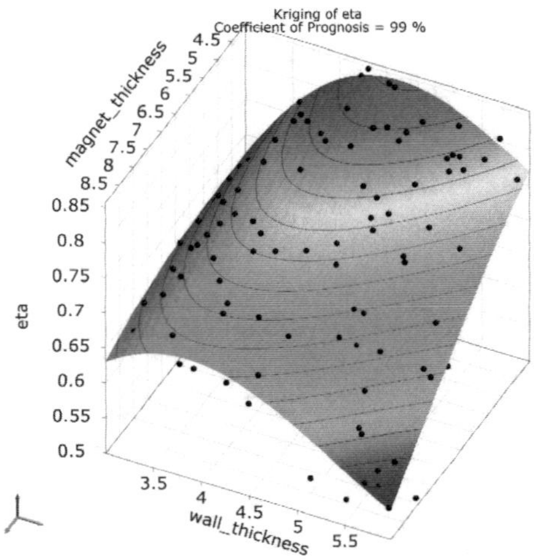

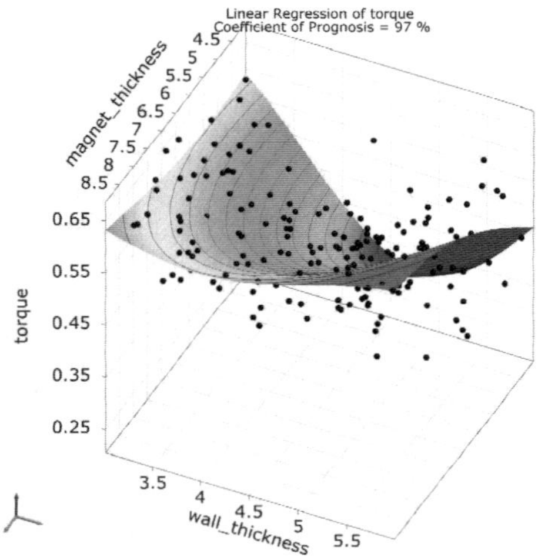

- Above: the torque-eta goal conflict as represented in the response surfaces
- High CoP: MOP information can be relied on for analyzing Pareto-optimal designs

Sensitivity analysis and optimisation of an electric motor
Automotive Simulation World Congress, June 7-8 2016, Munich

© Dynardo GmbH

dynardo

Metamodelling: understanding the CoP values

There is no simple relationship between the torque ripples and the other output variables.

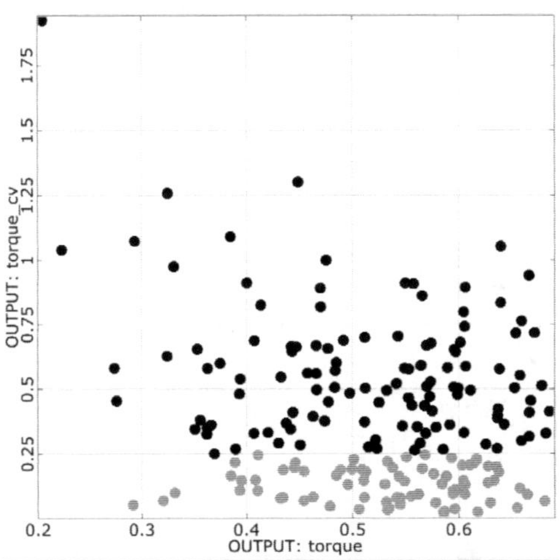

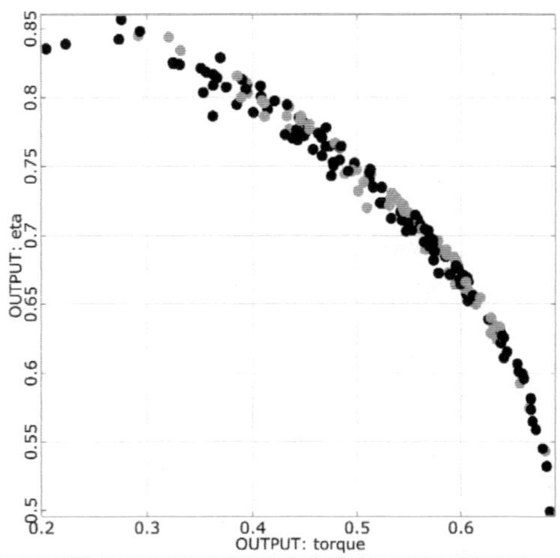

© Dynardo GmbH

dynardo

Metamodelling: Response Surfaces

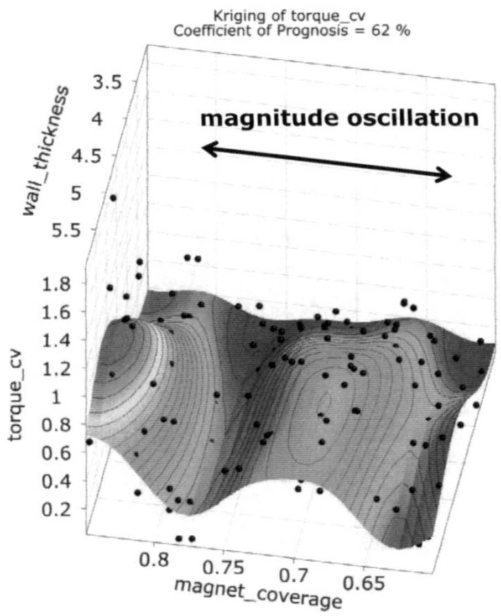

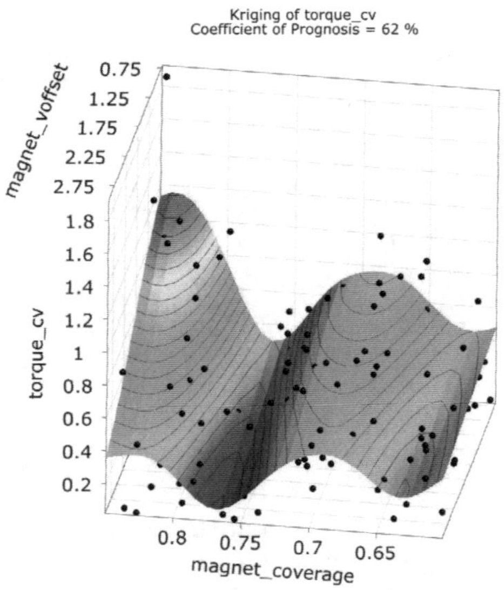

Metamodelling: Response Surfaces

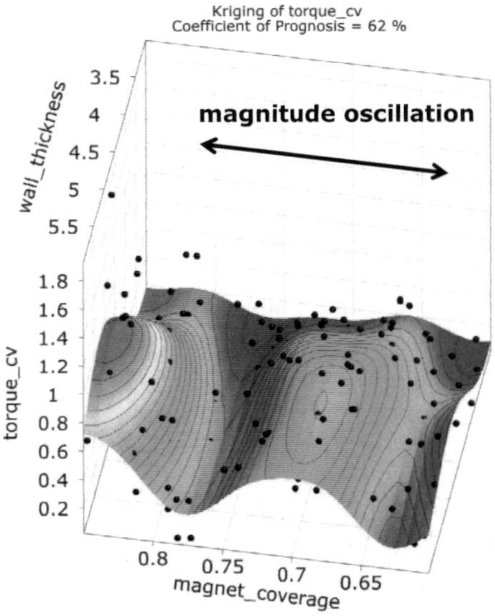

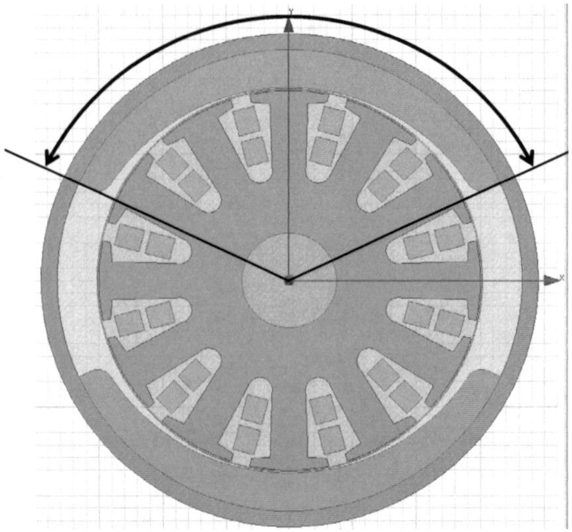

Metamodelling: CoP matrix

- Very well explainable: eta, P_loss, P_mech, P_el, torque → CoP high
- Less well explainable: magnitude of torque ripples → CoP lower

Models	magnet_coverage	wall_thickness	hs0	magnet_thickness	magnet_voffset	rotor_borehole	Total
eta	5.8 %	26.0 %	2.9 %	59.0 %		17.9 %	98.7 %
torque_amps_sum_normed	25.4 %	25.1 %			19.0 %		60.6 %
torque_cv	34.7 %	25.1 %			15.1 %		61.9 %
P_loss	14.0 %	25.4 %	3.4 %	51.0 %		17.1 %	98.5 %
P_mech	6.4 %	28.1 %	1.7 %	53.5 %		22.7 %	97.5 %
P_el	6.3 %	26.5 %	3.1 %	54.6 %		20.7 %	97.9 %
torque	6.6 %	27.3 %	1.9 %	52.7 %		22.0 %	97.5 %

Parameter

High CoP values → guideline for parameter reduction

Here: → torque ripples are highly nonlinear
→ influences only partially captured
→ continuing optimization with all 9 parameters

dynardo

Optimization

dynardo

Optimization problem definition

For

eta ⟷ torque

or

eta ⟷ P_mech

- the tradeoff is already well captured in the random sampling,
- little information gain can be expected from further Pareto-optimization (CoP near 100%).

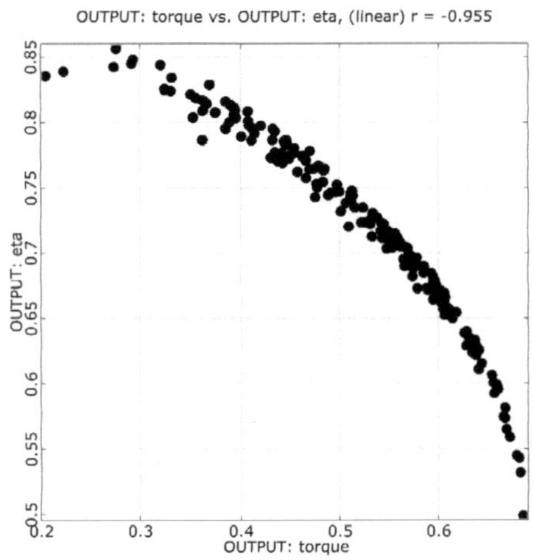

Optimization problem definition

For

eta ↔ torque

or

eta ↔ P_mech

- the tradeoff is already well captured in the random sampling,
- little information gain can be expected from further Pareto-optimization (CoP near 100%).

However, for

torque_cv ↔ other goals

there remains optimization potential.

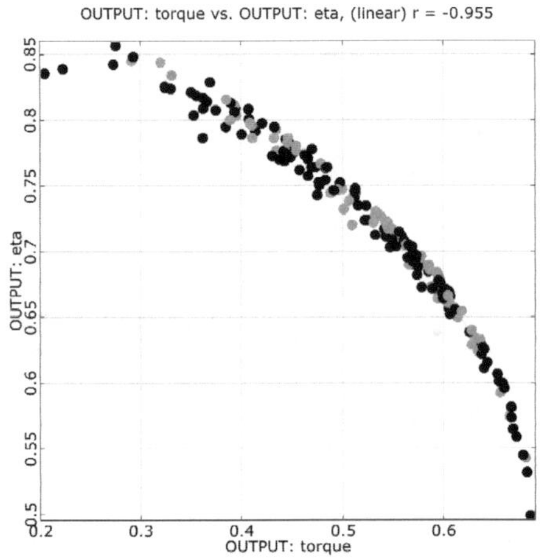

Single-Objective Local Optimization

Choosing the starting point

- We want to maximize eta and minimize torque_cv while torque > 0.5

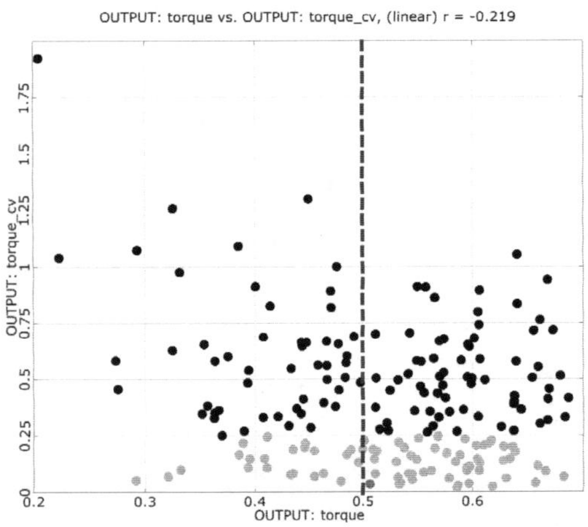

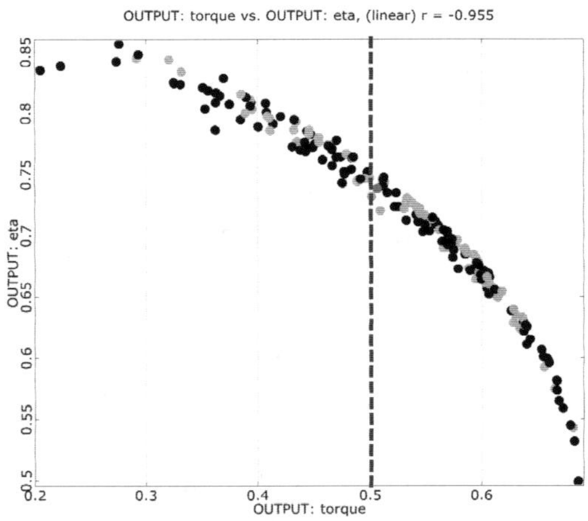

Single-Objective Local Optimization

Choosing the starting point

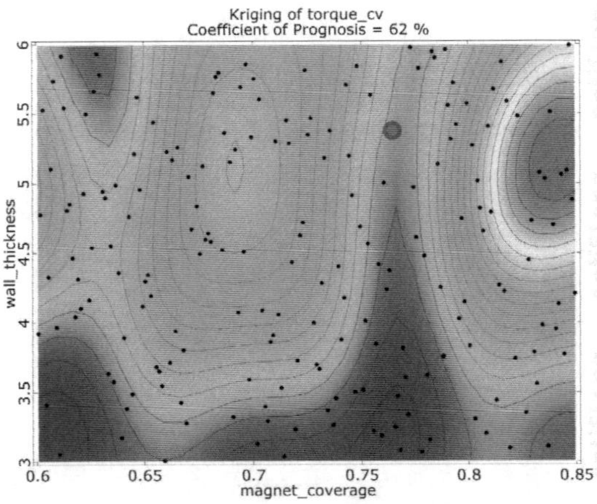

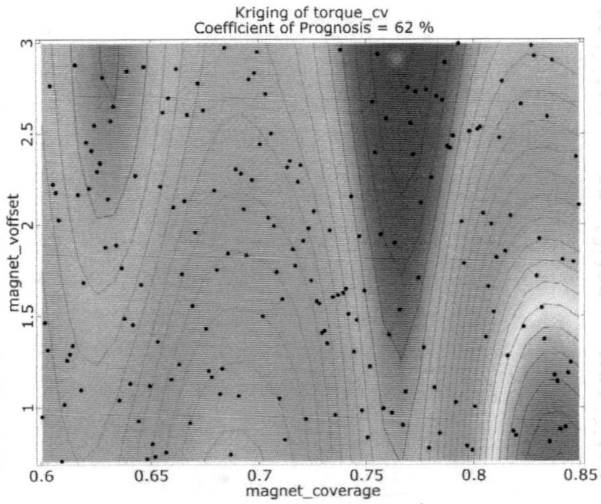

optiSLang Optimization Algorithms

Gradient-based Methods

- Most efficient method if gradients are accurate enough
- Consider its restrictions like local optima, only continuous variables, and noise

Adaptive Response Surface Method

- Attractive method for a small set of continuous variables (<20)
- Adaptive RSM with default settings is the method of choice

Nature-inspired Optimization

- GA/EA/PSO imitate mechanisms of nature to improve individuals
- Method of choice if gradient or ARSM fails
- Very robust against numerical noise, non-linearity, number of variables,...

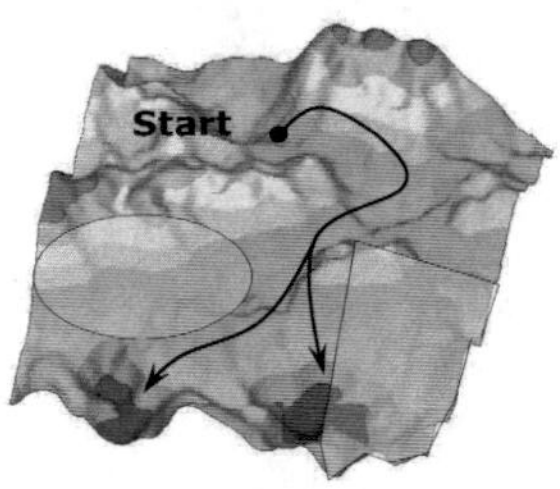

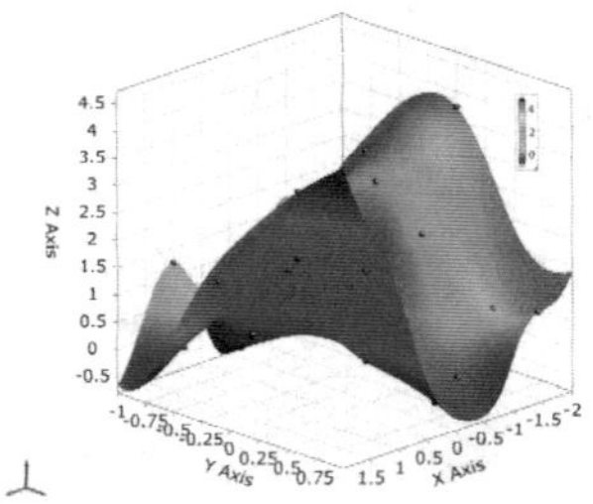

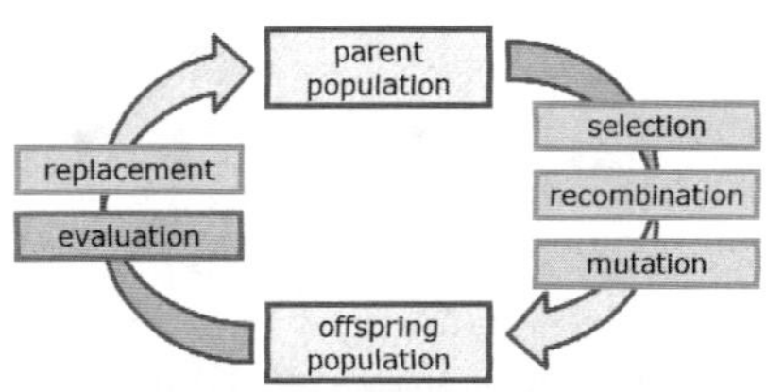

dynardo

Single-Objective Local Optimization

Minimize

`(1-eta) + 0.4*torque_cv`

Constraint:

`torque ≥ 0.5`

Algorithm:
Adaptive Response Surface Method (ARSM)

eta: + 2.6%
torque_cv: - 14%

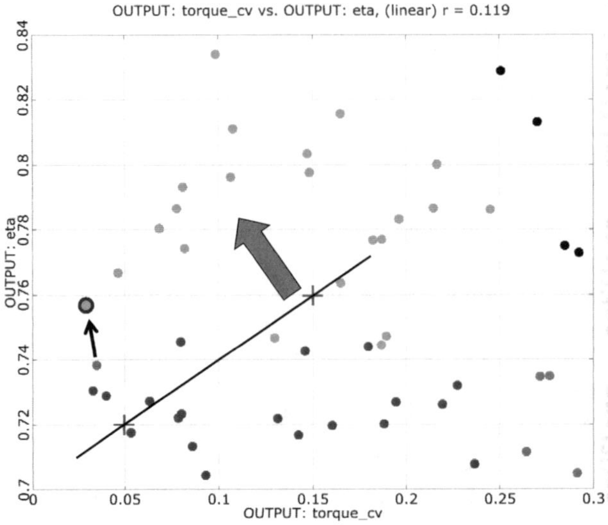

dynardo

Optimization result

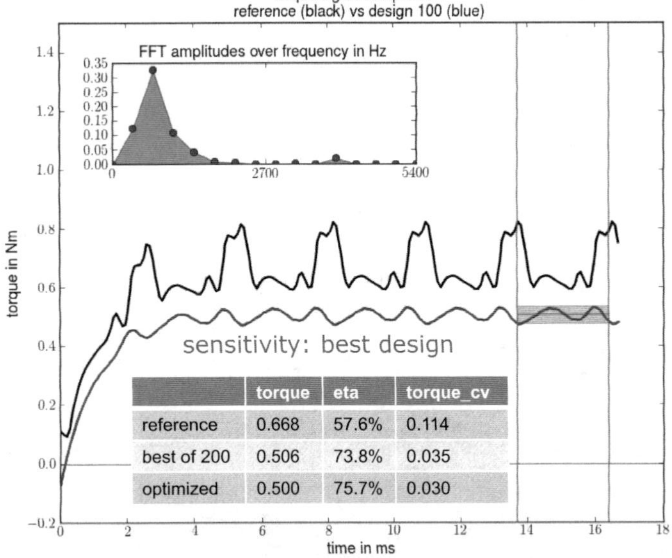

	torque	eta	torque_cv
reference	0.668	57.6%	0.114
best of 200	0.506	73.8%	0.035
optimized	0.500	75.7%	0.030

quantities averaged over last cycle
(ratios with respect to reference)

eta: + 28%
torque_cv: - 69%

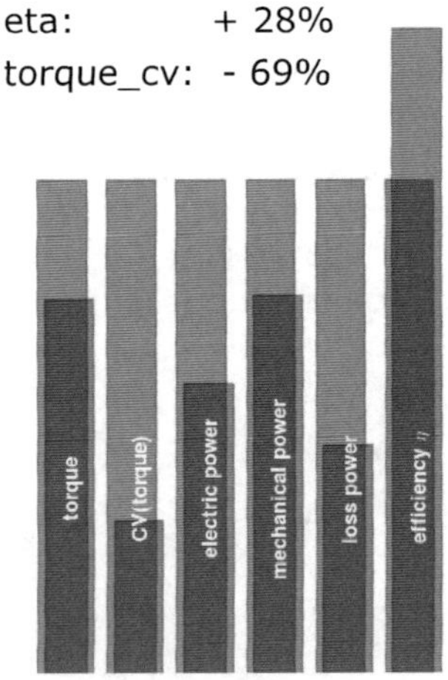

Optimization result

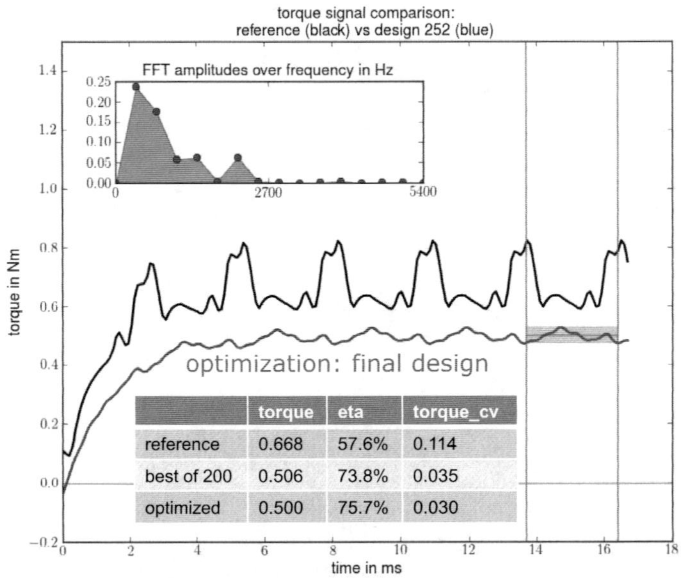

	torque	eta	torque_cv
reference	0.668	57.6%	0.114
best of 200	0.506	73.8%	0.035
optimized	0.500	75.7%	0.030

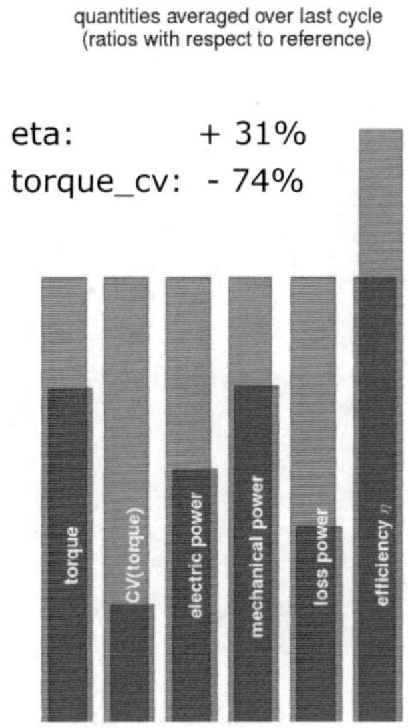

© Dynardo GmbH

Optimization result

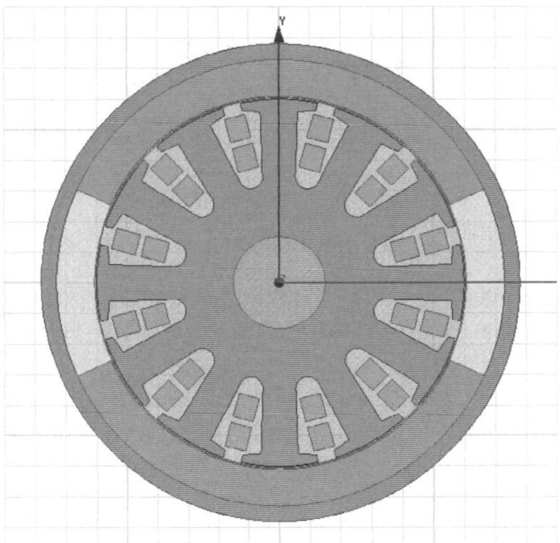

reference
design

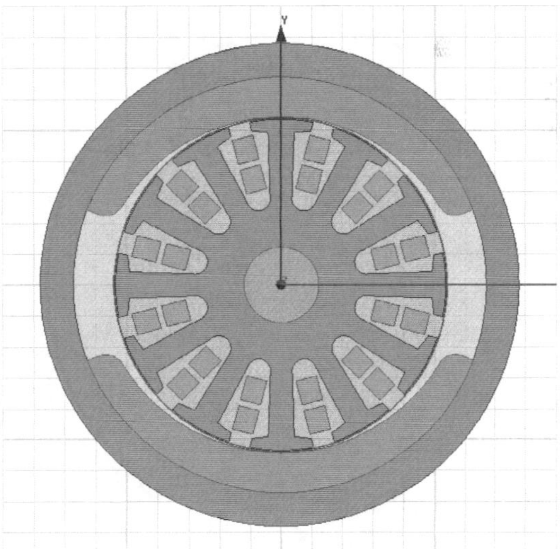

sensitivity:
best design

© Dynardo GmbH

Optimization result

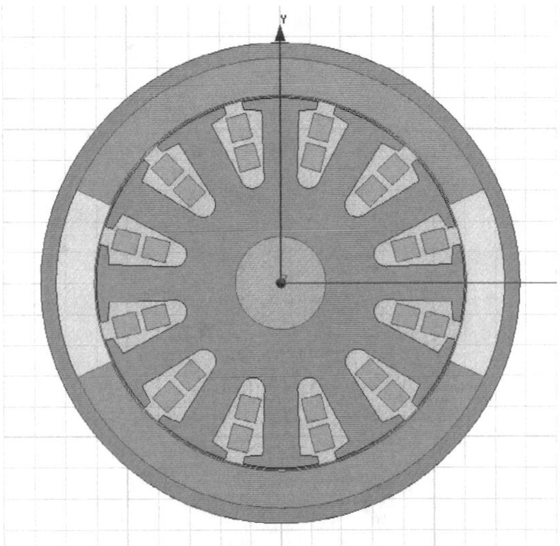

reference design

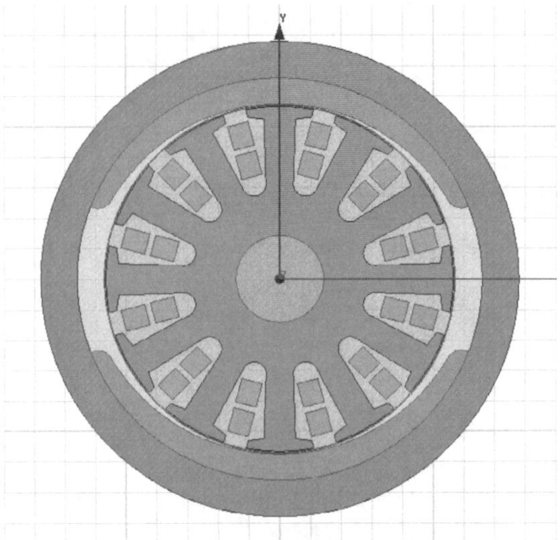

optimization: final design

Sensitivity analysis and optimisation of an electric motor
Automotive Simulation World Congress, June 7-8 2016, Munich 55

© Dynardo GmbH

Summary

Sensitivity analysis and optimisation of an electric motor
Automotive Simulation World Congress, June 7-8 2016, Munich 56

dynardo

Summary

Coupling Maxwell with optiSLang

- Via Workbench (node) ⟶ multi-physics
- Via text files ⟶ powerful signal processing
  ⟶ Large-Scale DSO

Sensitivity analysis

- Identification of important parameters and correlations
- Exploring tradeoffs and optimization potentials
- Metamodel of Optimal Prognosis (MOP):
  ⟶ can be used for optimization
  ⟶ visualization ⟶ gain knowledge about nonlinear interactions

Optimization

- ARSM: efficient & robust algorithm for optimization directly on simulation
- Torque ripples reduced by 74%, efficiency increased by 31%
- Play with parametrization and goals
  ⟶ fast gain of engineering intuition

dynardo

Appendix: backup slides

The setup in detail

The setup: Maxwell design variables

selection considered for sensitivity and optimization

	Name	Value	Unit	Evaluated Value
Δt ⟷ band edge el. size	ClickingFaktor	1		1
RPM	Drehzahl	1800		1800
magnet_coverage	Embr	0.726		0.726
band edge element size	fractions	1		1
wall_thickness	GDicke	2.5	mm	2.5mm
hs0	HS0	1.2	mm	1.2mm
hs1 → impact on coil cross sec.	HS1	1	mm	1mm
magnet_thickness	MagDicke	6	mm	6mm
magnet_rounding	MagRParameter	0		0
airgap	Luftspalt	0.5	mm	0.5mm
fixed: motor size	ODia	78	mm	78mm
gapwidth	OeffnungNut	3	mm	3mm
magnet_voffset	Offs	0	mm	0mm
rotor_borehole	RIDia	15	mm	15mm
dependent: inner diameter	IDia	ODia-2*GDicke-2*MagDicke		61mm
dependent: rounding radius	MagRKante	MagDicke*MagRParameter		0mm
dependent: rotor size	RODia	IDia-2*Luftspalt		60mm

accessible in Workbench Parameter Set or ...

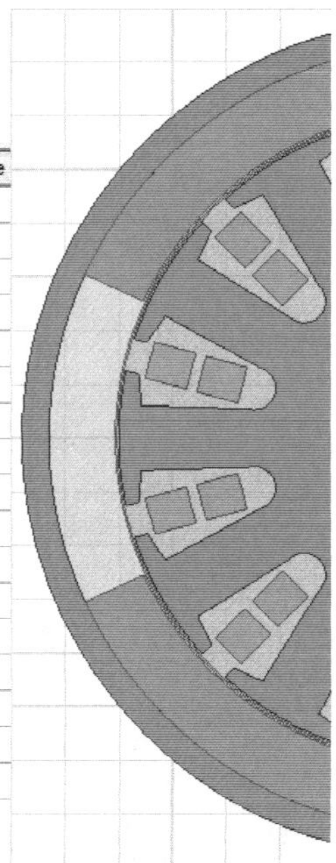

The setup: Maxwell design variables

generate a reference input file

```
names = oDesign.GetVariables()
f=open(os.path.join(reference_dir,"inputs.txt"),"w")
for n in names :
    f.write("%s = %s\n" %(n,oDesign.GetVariableValue(n)))
f.close()
```

The setup: Maxwell design variables

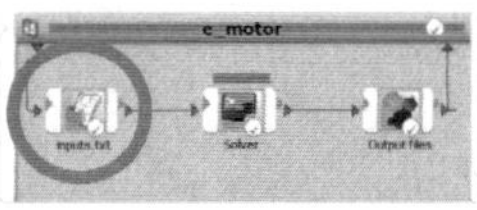

optiSLang text input node

↓

creates modified input files

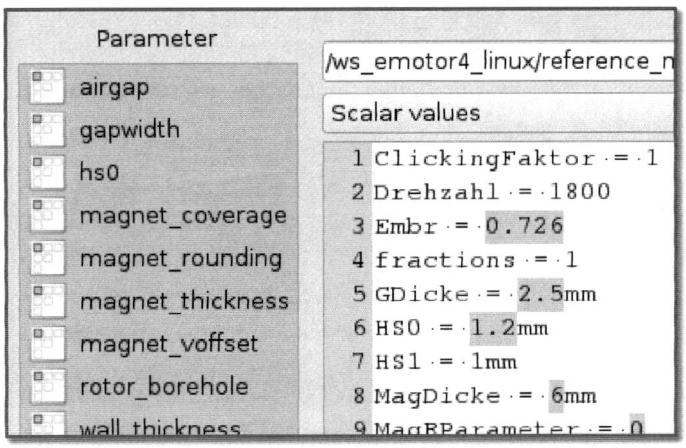

The setup: Maxwell design variables

How Maxwell incorporates the new design variable settings:

```
f= open(os.path.join(ppath,'inputs.txt'),'r')

for line in f.readlines():
    words = line.split('=')
    #AddInfoMessage(str(words))
    oDesign.ChangeProperty(
        [
            "NAME:AllTabs",
            [
            "NAME:LocalVariableTab",
            [
                "NAME:PropServers",
                "LocalVariables"
            ],
            [
                "NAME:ChangedProps",
            [
                "NAME:"+words[0],"Value:=",words[1]
            ]
            ]
            ]
        ])
f.close()
```


The setup: Maxwell output variables

scalar output variables

↓

Workbench Parameter Set

DesignXplorer Setup

General | Table | Calculations | Goals | Options

Name	Solution	Calculation	Calculation Range
torque	Setup1 : Transient	mean(Moving1.Torque)	Time(8.2407ms:10.926ms)
P_loss	Setup1 : Transient	mean(P_loss)	Time(8.2407ms:10.926ms)
P_mech	Setup1 : Transient	mean(P_mech)	Time(8.2407ms:10.926ms)
P_el	Setup1 : Transient	mean(P_el)	Time(8.2407ms:10.926ms)
eta	Setup1 : Transient	mean(P_mech)/mean(P_el)	Time(8.2407ms:10.926ms)
torque_cv	Setup1 : Transient	stddev(Moving1.Torque)/mean(Moving1.Torque)	Time(8.2407ms:10.926ms)

Setup Calculations... | Delete

OK | Abbrechen

The setup: Maxwell output variables

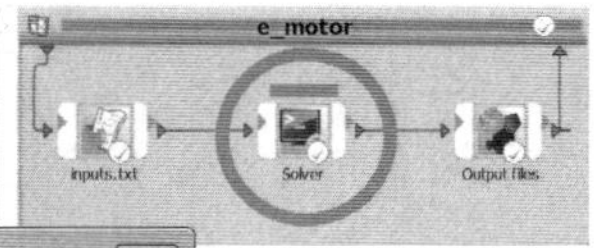

Transient output variables

↓

Basis for transient report generation & analysis in Maxwell

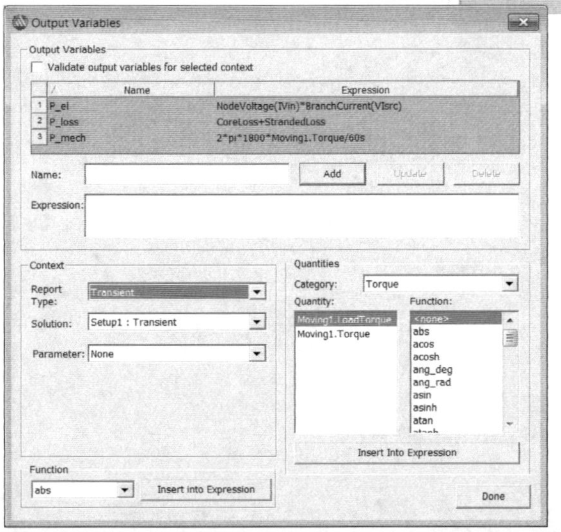

The setup: Maxwell output variables

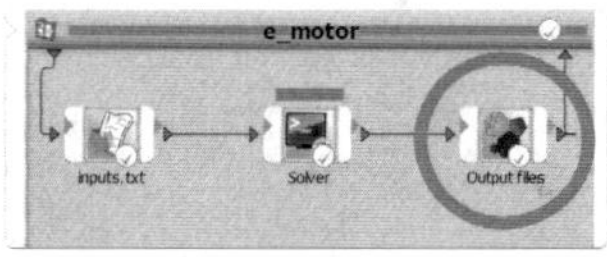

transient reports

↓

scripted export command

↓

ASCII files

↓

ETK

```
oModule = oDesign.GetModule("ReportSetup")
names = oModule.GetAllReportNames()
for nam in names:
    oModule.ExportToFile(nam, os.path.join(reference_dir,nam)+".tsv")
```

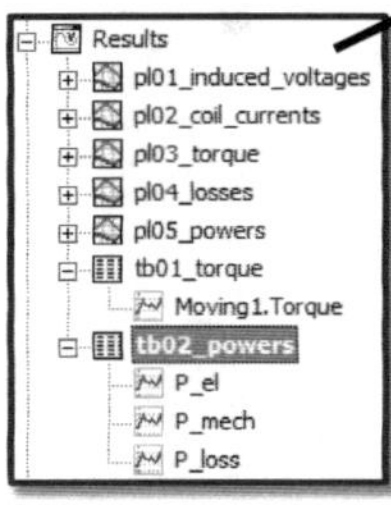

pl01_induced_voltages.tsv	29.08.2015 21:24	TSV-Datei	66 KB
pl02_coil_currents.tsv	29.08.2015 21:24	TSV-Datei	51 KB
pl03_torque.tsv	29.08.2015 21:24	TSV-Datei	11 KB
pl04_losses.tsv	29.08.2015 21:24	TSV-Datei	31 KB
pl05_powers.tsv	29.08.2015 21:24	TSV-Datei	27 KB
tb01_torque.tsv	29.08.2015 21:24	TSV-Datei	3 KB
tb02_powers.tsv	29.08.2015 21:24	TSV-Datei	4 KB

The setup: Maxwell output variables

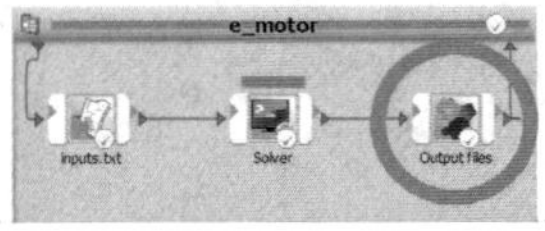

signal processing and output variable definition in ETK

ETK | Variables

	ID	Type	Value	Base path mode	File	Expression
1	powers_signal	REAL SIGNAL	[3:30]	Relative to working dir	tb02_powers.tsv	constant
2	torque_signal	REAL XYDATA	[1:30]	Relative to working dir	tb01_torque.tsv	constant
3	torque	REAL	0.667806			mean(extract(torque_signal,1))
4	P_el	REAL	218.371			mean(extract(powers_signal,1))
5	P_mech	REAL	125.878			mean(extract(powers_signal,2))
6	P_loss	REAL	10.4643			mean(extract(powers_signal,3))
7	torque_cv	REAL	0.113561			$SIG_STDDEV(torque_signal)/torque
8	sig_len	REAL	30			$SIG_SIZE(torque_signal)
9	time_equidistant	REAL VECTOR	[30]			linspace(extract(torque_signal,0)[0]/1e9,extract(torque_signal,0)[29]/1e9,sig_len)
10	torque_fft_magnitudes	REAL XYDATA	[1:30]			abs(fft(xydata(time_equidistant,extract(torque_signal,1))))
11	torque_amps_sum_normed	REAL	4.95787			sum(extract(torque_fft_magnitudes,1)[1:sig_len/2+1])/torque
12	eta	REAL	0.576444			P_mech/P_el

Automotive Simulation
WORLD CONGRESS

June 7-8, 2016 Munich

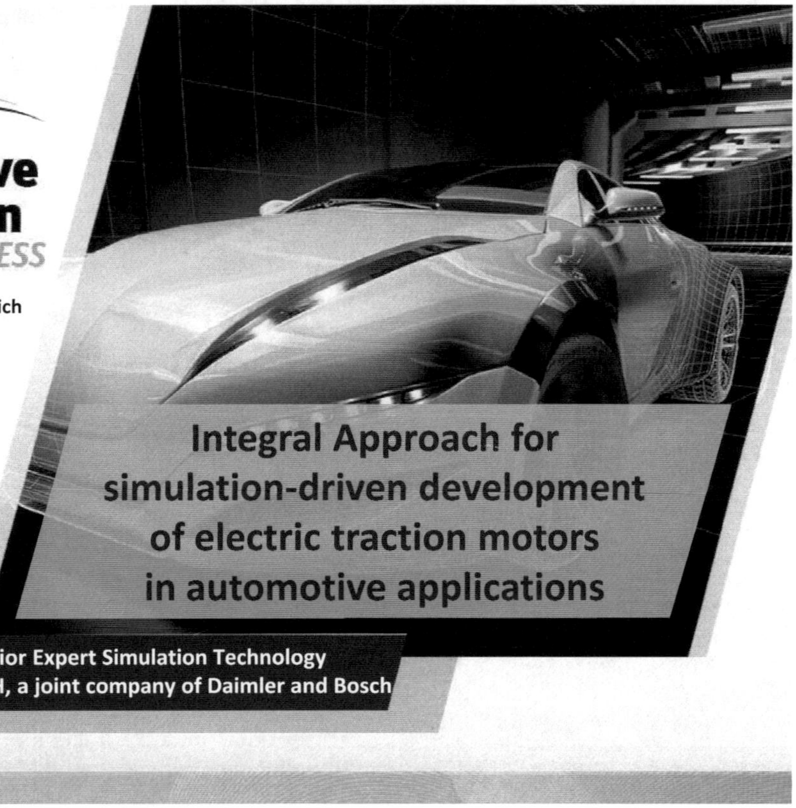

Integral Approach for simulation-driven development of electric traction motors in automotive applications

Marc Brück, Senior Expert Simulation Technology
EM-motive GmbH, a joint company of Daimler and Bosch

Table of contents

- EM-motive GmbH
- Customer requirements / engineering constraints
- Material properties
- Components vs. systems engineering
- Last-minute requirement-modification
- Demands on virtual product development
- Resulting workflow
- Summary and outlook

EM-motive GmbH – A joint company of Daimler & Bosch

- **More than 270 000 e-motors manufactured since 2012**
- **Full-range portfolio for EVs, plugin-hybrids and hybrids**
- **Customer vehicles with EM-motive e-motor**

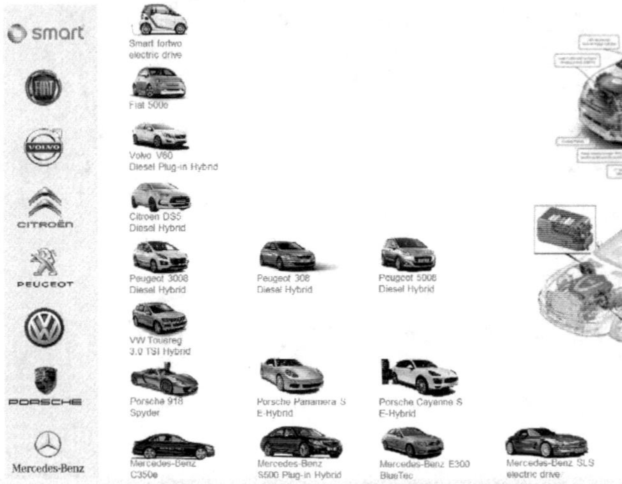

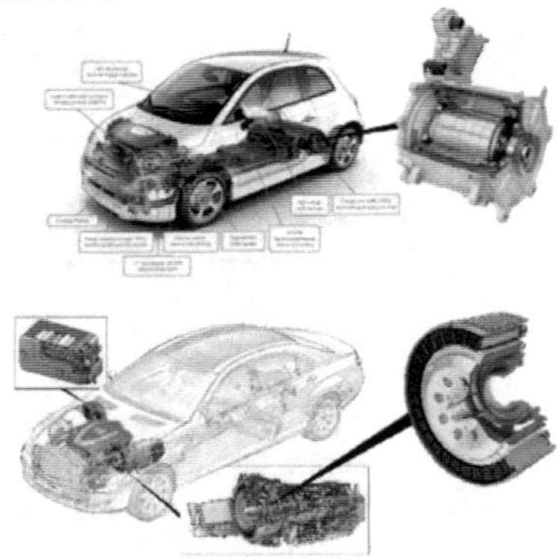

Customer requirements / engineering constraints

- **Customer requirements for different physical domains**

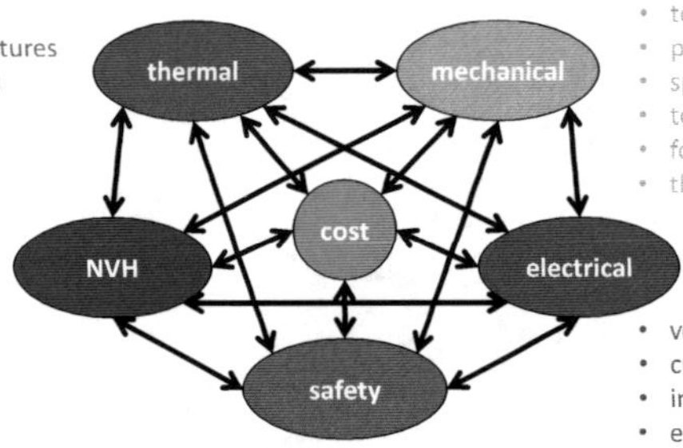

- coolant flow-rate
- coolant temperature
- environment temperatures
- winding temperatures
- magnet temperatures
- tbc

- mounting space
- torque
- power
- speed
- tolerances to other parts
- forces on bearings
- tbc

- airborne noise
- structure-born noise
- tbc

- voltage
- current
- inverter-specific properties
- efficiency
- tbc

- back-emf
- short-circuit current
- inverter-specific properties
- tbc

- **Main engineering constraints**
  - Cost
  - Mounting space
  - Cooling
  - Inverter

Material properties

- **... relating to different physical domains and state variables**

  - dynamic viscosity μ
  - specific heat capacity c_p
  - thermal conductivity λ
  - emissivity ε
  - density ρ

  - Young's modulus E
  - shear modulus G
  - Poisson's ratio ν
  - yield strength R_e
  - tensile strength R_m
  - compressive strength R_d
  - bending strength R_b
  - hardness HV
  - thermal expansion coefficient α
  - endurance limit, s-n-curves (Wöhler)
  - friction coefficient μ_r
  - tbc

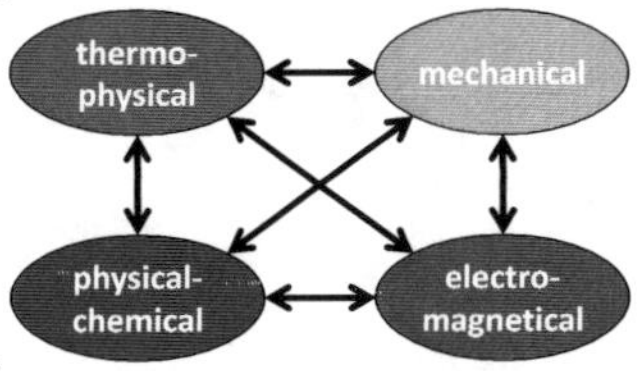

  - cleanness
  - contact angle Θ
  - surface tension σ
  - surface roughness R_a
  - tbc

  - specific electric resistivity ρ
  - coercivity H_{cJ}
  - remanence B_r
  - B-H-curves
  - specific loss-curves
  - stacking factor

  - **often f(pressure, temperature, orientation, stress/strain, shear rate, ...)**
  - **statistical scattering (batch, production process,...)**
  - **different commodity prices (lot size scenarios)**

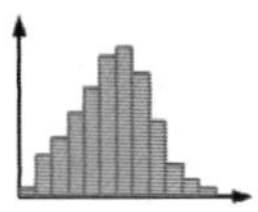

- **<u>All</u> information essential for design optimization including material choice**

Automotive Simulation WORLD CONGRESS

Component vs. (or better and?) systems engineering

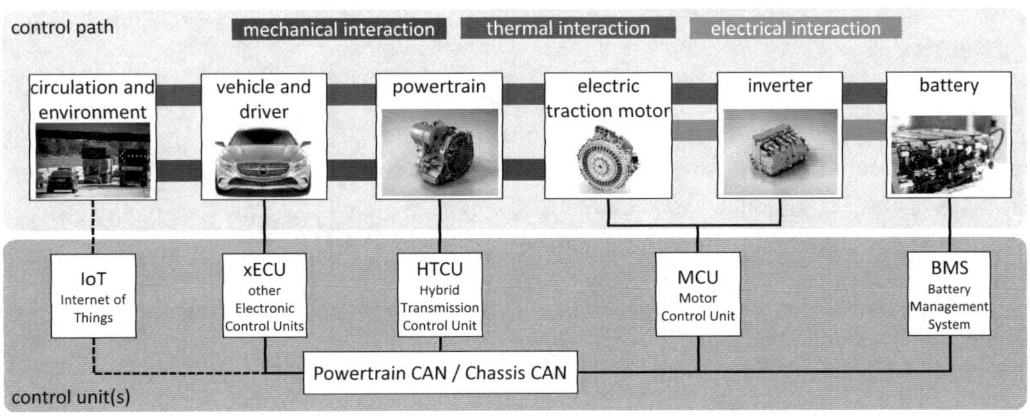

- **Components strongly interact**
- **Static boundary conditions not suitable for dynamic interactions**
- **Systems engineering necessary to handle the complexity**

Automotive Simulation WORLD CONGRESS

Example for last-minute requirement-modification

- Customer: „Max. speed to be increased by 1000 rpm"
- Engineer: „Centrifugal forces ~ speed^2, actual rotor design will fail"
- Customer: „Design the rotor a little more robust"

- Consequences for engineering
  - bridge thickness to be increased
    → increased magnetic flux leakage
    → reduced torque and power
  - Customer: „Not acceptable!!"
  - current to be increased
    → increased Ohmic losses
    → reduced efficiency
  - Customer: „Not acceptable!!"

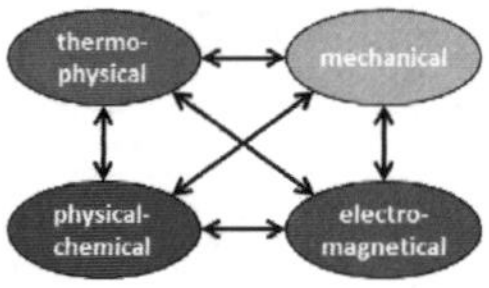

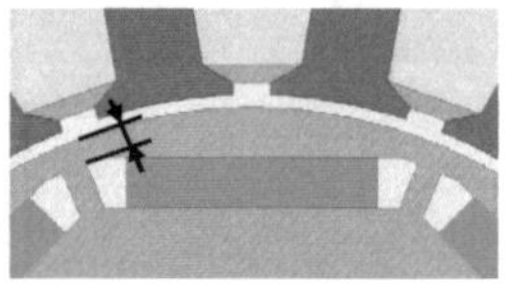

→ **Back to start, completely new design of motor in all physical domains**

- Customer: „Expect updated quotation till end of the week, any delay in global project schedule is not acceptable"

Demands on virtual product development

- Cost-effective
- Fast ...
  - ... standard analysis (procurement phase)
  - ... price indication
  - ... detailed analysis (component development)
  - ... reduced-order-model for systems engineering
- Reliable, applicable and reproducible
  - Product must comply with all requirements
  - Virtual product development = integral part of the engineering process
  - Evaluation of performance „at a glance"
- How to achieve?
  - Highly automated workflow
  - Unique database for material properties
  - Coupling of physical domains, systems engineering
  - Minimum number of software packages
  - Commercial software with high-quality support

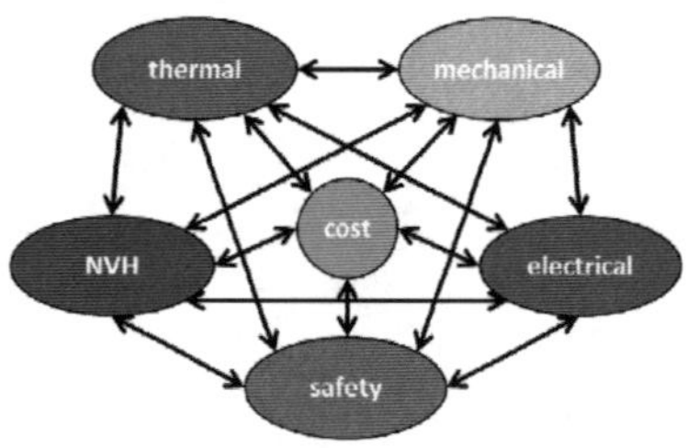

Resulting workflow for virtual product develop.

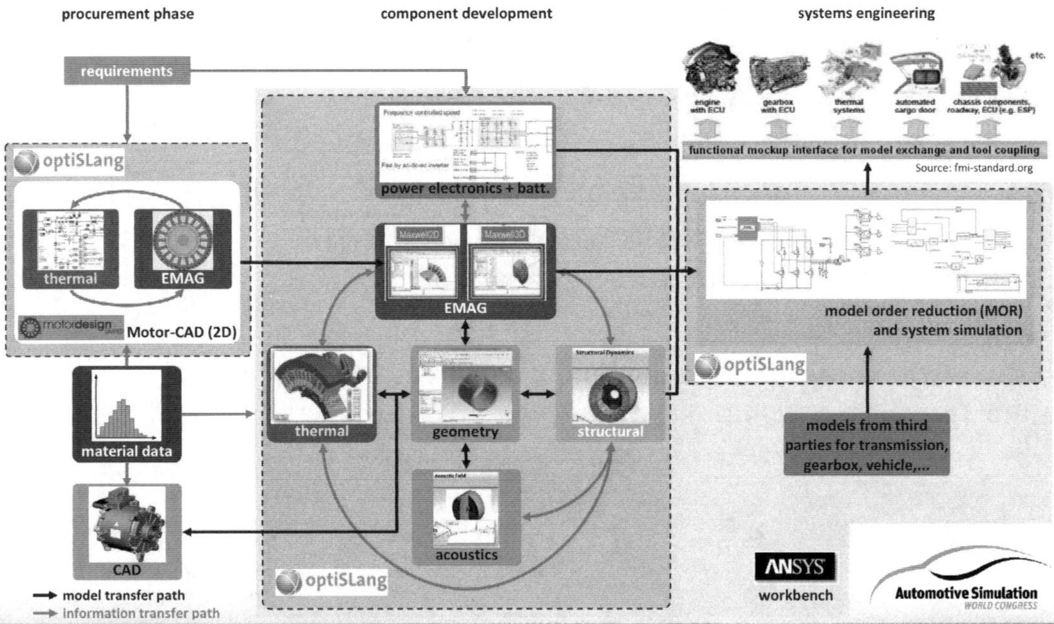

Automotive Simulation WORLD CONGRESS

Resulting workflow for virtual product develop.

- Postprocessing with evaluation of performance „at a glance“
- Requirements and performance indicators in different domains
  - Cost
  - Thermal
  - Mechanical
  - EMAG
  - Control
  - NVH
  - Environment
  - tbc

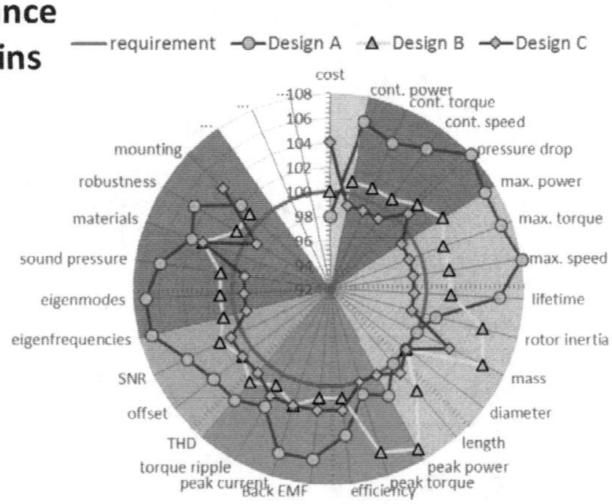

- Radar chart
  - Dimensionless representation
  - Requirements as 100 % reference (outside the red circle = requirement fulfilled)
  - All information in one single chart
  - Easy comparison of several designs

Automotive Simulation WORLD CONGRESS

Summary and outlook

- Requirements cover different physical domains and interact
- Entire material properties essential for design optimization
- Systems engineering necessary to handle the complexity
- Last-minute-modifications lead to a restart in engineering
- Challenging workflow for virtual product development needed
- Workflow using Motor-CAD, ANSYS and optiSLang software and postprocessing via Radar chart presented
- Workflow will be refined and tuned
- Intensive discussions with customer about requirements, cost and technical feasability will improve quality and time to market

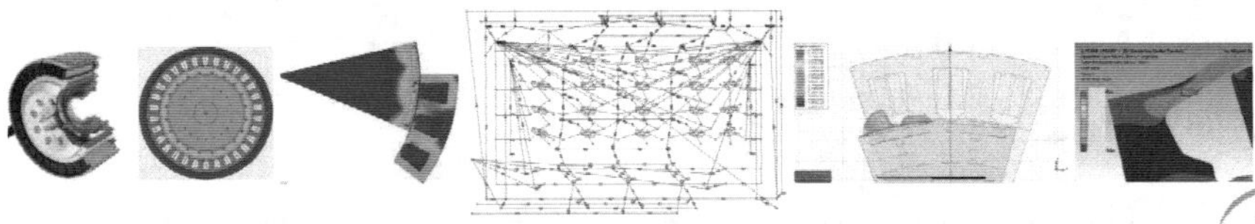

Automotive Simulation WORLD CONGRESS

11 June 7-8, 2016 Munich

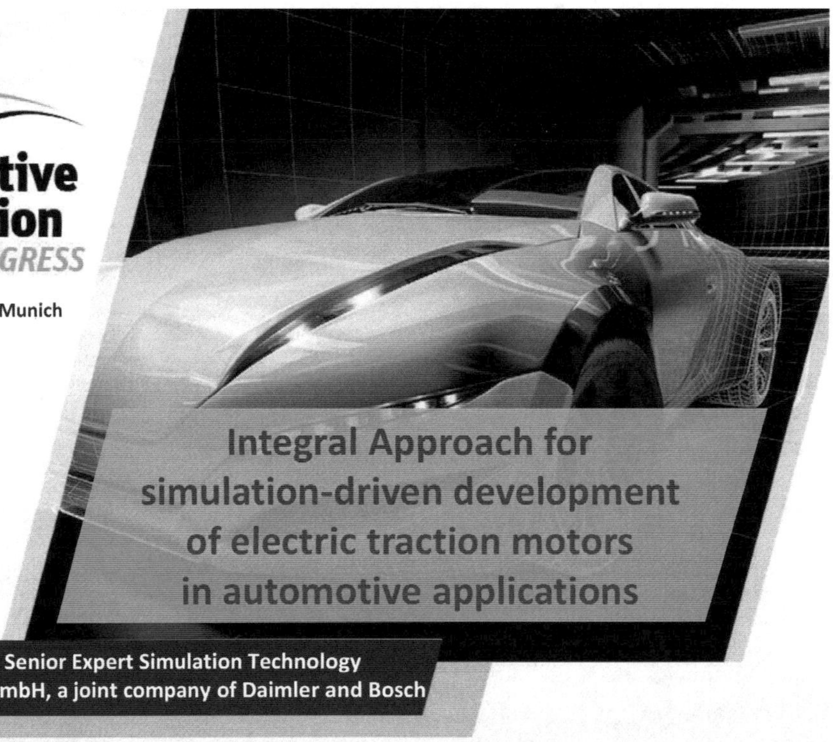

Automotive Simulation
WORLD CONGRESS

June 7-8, 2016 Munich

Integral Approach for simulation-driven development of electric traction motors in automotive applications

Marc Brück, Senior Expert Simulation Technology
EM-motive GmbH, a joint company of Daimler and Bosch

Efficiency map and drive cycle simulation for PMSMs

7th June 2016

Dr. James Goss
Motor Design Ltd (MDL)
www.motor-design.com

Overview

1. Introduction to Motor Design Ltd
2. Motor Design Ltd software and links to the Ansys toolchain
3. Modelling efficiency maps and performance over driving cycles
4. Validation examples

1. Motor Design Ltd (MDL)

- MDL founded in 1998
- **Strategic Partner of ANSYS**
- Develop Motor-CAD & Motor-LAB software
- High level of support to users
- Motor design software developed by motor designers
- Involved in a number of funded research projects:
  - **Evoque_e** – electric vehicle development with JLR
  - **HVEMs** –high volume manufacturing techniques with Make Like Production prototyping facility with JLR
  - **ELETAD** – helicopter electric tail rotor
  - **HERRB** – helicopter main rotor brake/generator

Continuous Software Development

- Software has been under continual development over the past 18 years with feedback from its hundreds of users worldwide
  - New features driven by user requirements
  - Work closely with our customers

- Many users in various industrial sectors
  - Automotive (Bosch, BMW, Daimler, JLR, Nissan, Porsche, Renault, Williams ......)
  - Aerospace (Ametek, BAE, Eaton, Thales, Safran, Goodrich, UTC ....)
  - Industrial (ABB, Emerson, Regal Beloit, SEW, Siemens ....)
  - Traction (ABB, Bombardier, Caterpillar, GE, Komatsu.......)
  - Renewable (Alstom, Gamesa, GE, Siemens, Vestas ......)
  - Universities (Bristol, Manchester, Newcastle, Nottingham, Sheffield ......)

Evoque_e project

$25M UK Funded Project Started October 2013 to design and build:

1. Battery Electric Vehicle (BEV): 0g/km >100 mile range
2. Plug in Hybrid Electric Vehicle (PHEV): 40g/km > 30 mile range
3. Mild Hybrid Electric Vehicle (MHEV): 95g/km

Project completed March 2016.

Project partners: **Jaguar Land Rover**, GKN Driveline (GKN), Zytek Automotive, AVL Powertrain UK Ltd (AVL), TATA Steel, Delta Motorsport, Drive System Design (DSD). **Motor Design Ltd** (MDL), Cranfield University, Bristol University and Newcastle University.

2. MDL Software for Motor Design

- Integrated software for fast motor performance analysis
  - **Motor-CAD EMag:** 2D FEA electromagnetic analysis and loss calculation
  - **Motor-CAD Therm:** Network/FEA Thermal Analysis
  - **Motor-LAB**: fast & accurate prediction of efficiency maps and drive cycles

➢ Motor design software developed by motor designers

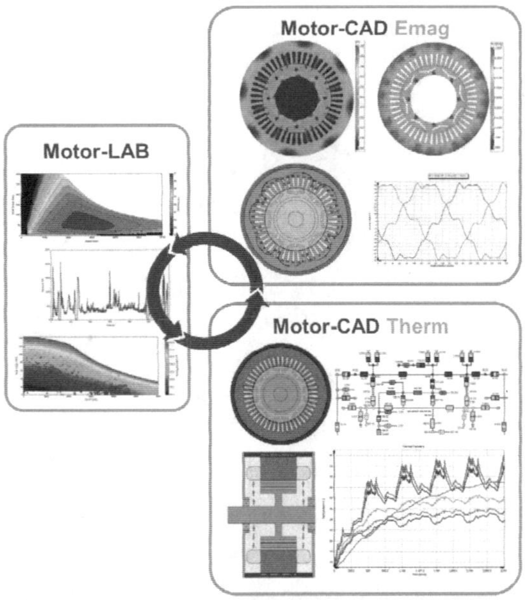

MDL and Ansys Software for Motor Design

- Integrated software for fast motor performance analysis
  - **Ansys Maxwell:** 2D and 3D electromagnetic FEA solver
  - **Motor-CAD Therm:** Network/FEA Thermal Analysis
  - **Motor-LAB**: fast & accurate prediction of efficiency maps and drive cycles

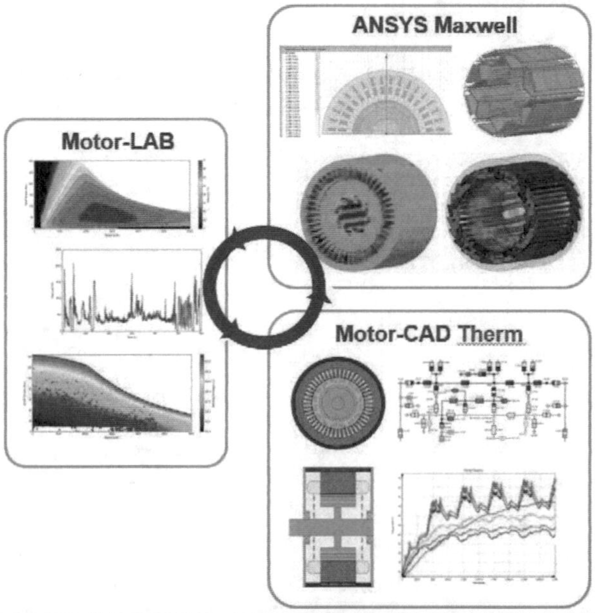

➢ Motor design software developed by motor designers

Motor-CAD – Maxwell Links

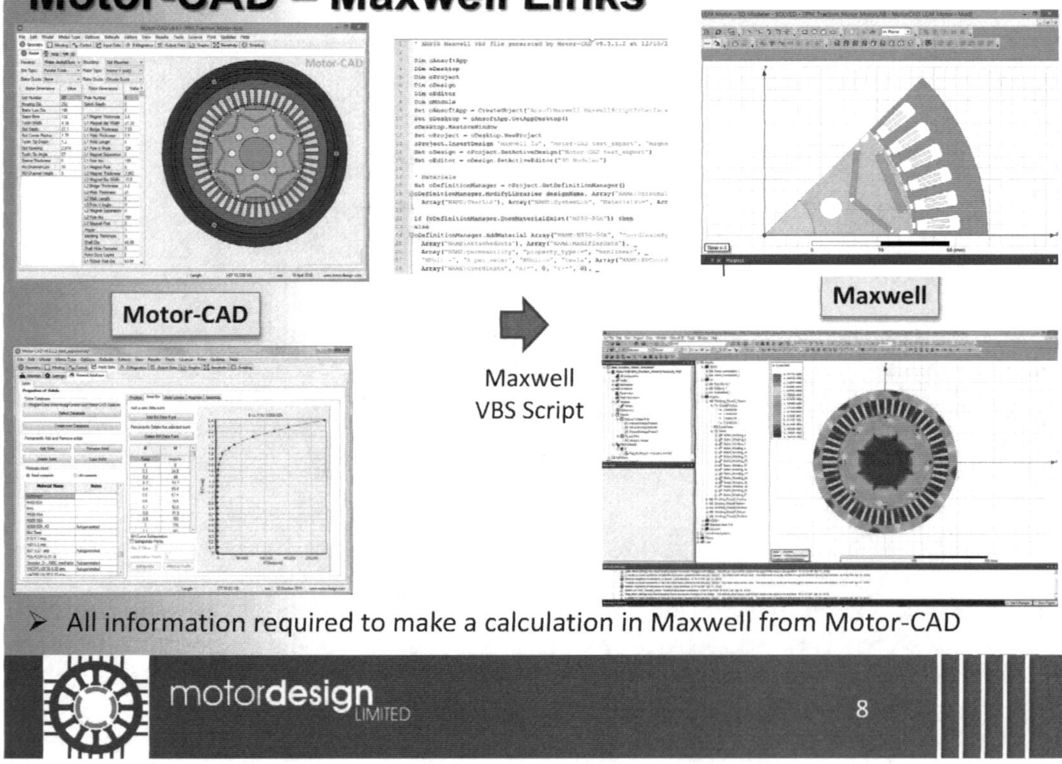

➢ All information required to make a calculation in Maxwell from Motor-CAD

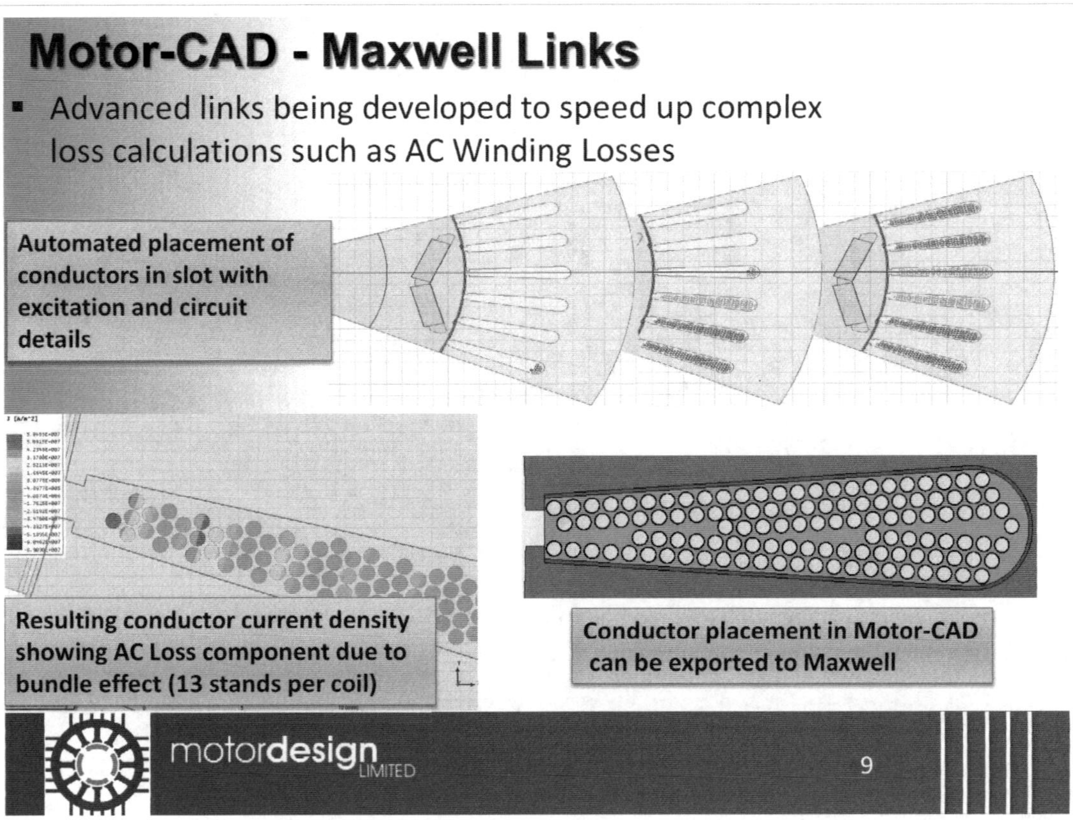
Motor-CAD - Maxwell Links
Advanced links being developed to speed up complex loss calculations such as AC Winding Losses
Automated placement of conductors in slot with excitation and circuit details
Resulting conductor current density showing AC Loss component due to bundle effect (13 stands per coil)
Conductor placement in Motor-CAD can be exported to Maxwell
motordesign LIMITED
9

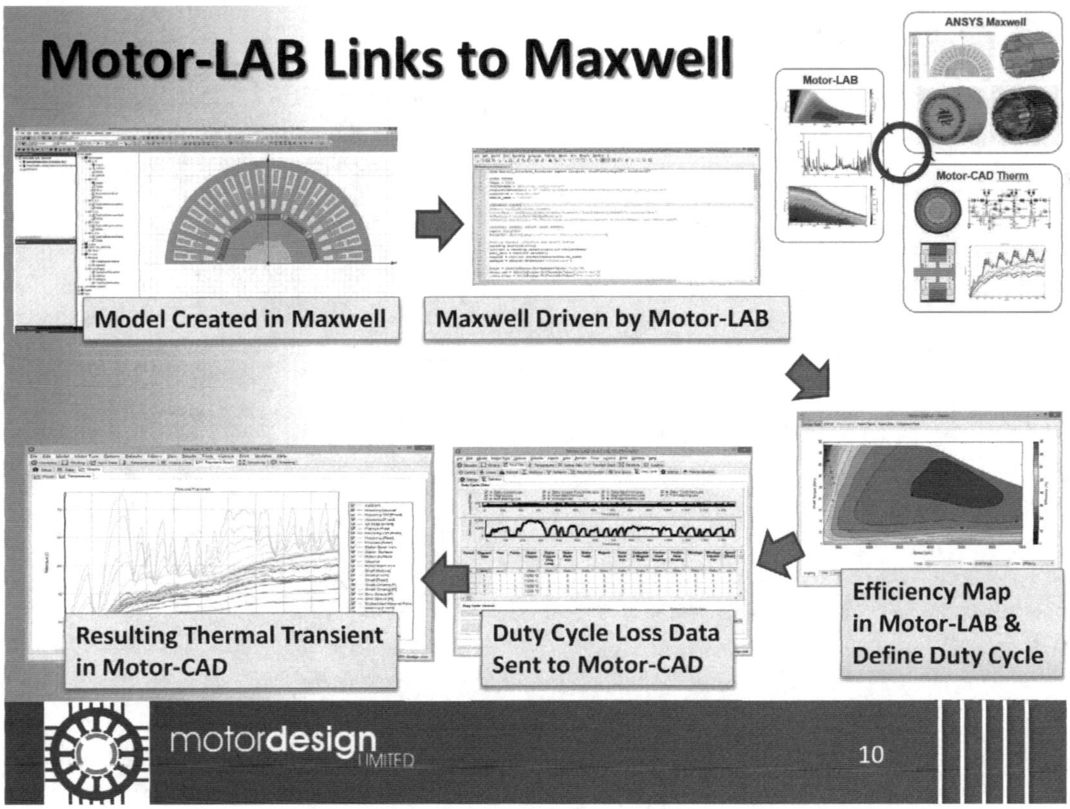
Motor-LAB Links to Maxwell
ANSYS Maxwell
Motor-LAB
Motor-CAD Therm
Model Created in Maxwell
Maxwell Driven by Motor-LAB
Efficiency Map in Motor-LAB & Define Duty Cycle
Duty Cycle Loss Data Sent to Motor-CAD
Resulting Thermal Transient in Motor-CAD
motordesign LIMITED
10

Motor-CAD 3D Links to ANSYS Tools

- Motor-CAD 3D Geometry exported to ANSYS Design Modeler

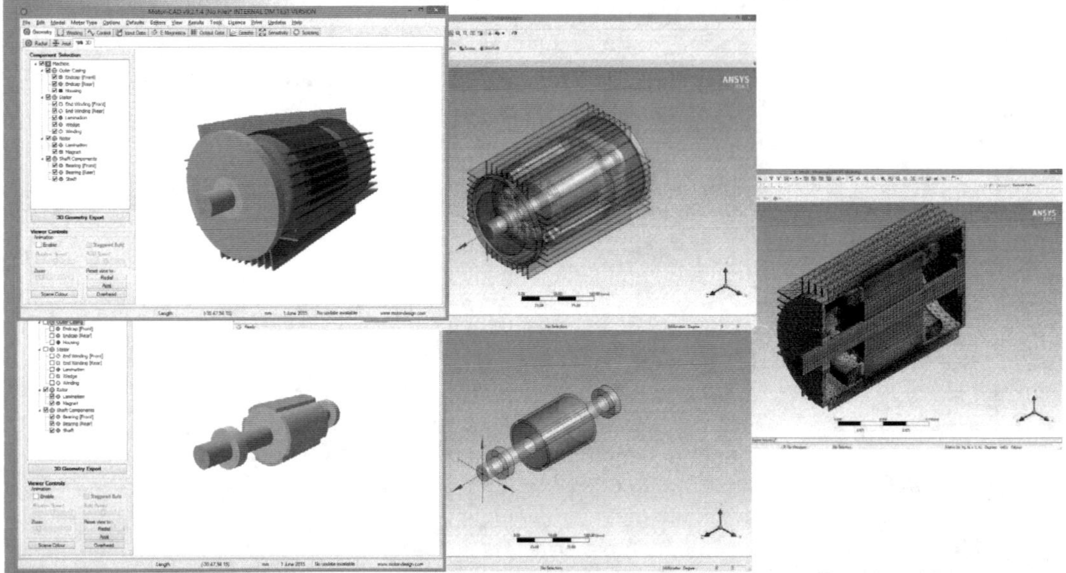

motordesign LIMITED 11

Motor-CAD and Fluent

- Motor-CAD and Fluent CFD complement each other
- Motor-CAD fast & quick to put together
- CFD more time consuming to construct models and make calculations
- Useful to filter designs using Motor-CAD and do CFD on final candidate
- Then use results to calibrate Motor-CAD convection & flow formulations
- Assistance in CFD model setup using links to Motor-CAD being developed
  - Speed up the creation of CFD models

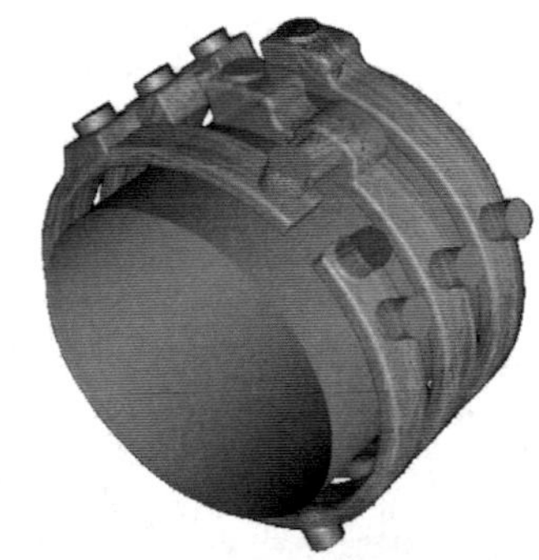

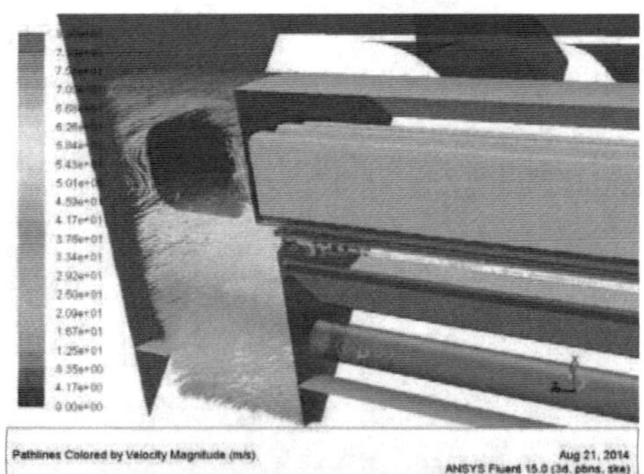

motordesign LIMITED 12

Motor-CAD – Icepak Hybrid Model

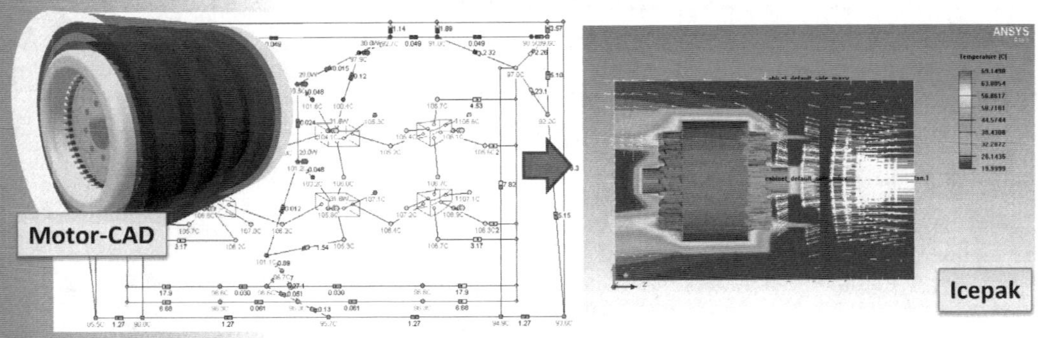

- Automatically pass the thermal resistance network from Motor-CAD to Icepak for the main body of the machine
- Allows extremely fast combined lumped circuit/CFD simulations times
- Currently under development

3. Modelling performance of BPM machines over complex driving cycles

PMSMs for EV applications

- PM topology makes up the large majority of traction motors in BEV, PHEV and MHEV vehicles
- These operate over a large number of operating conditions and duty cycles
- Electromagnetic and thermal modelling across the full operating envelope is required.

Bosch integrated motor/generator

Chevrolet Bolt

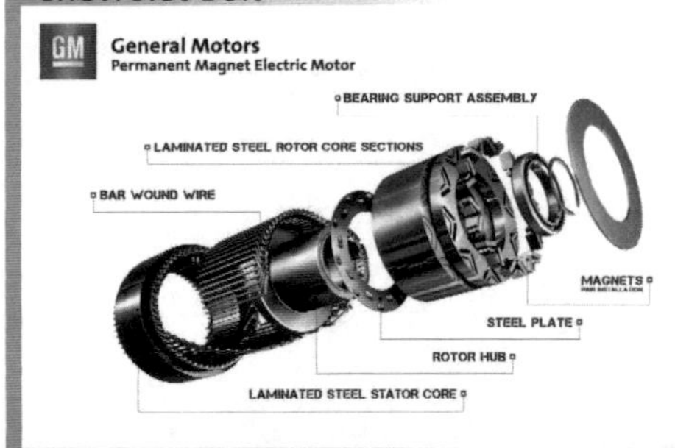

Nissan Leaf

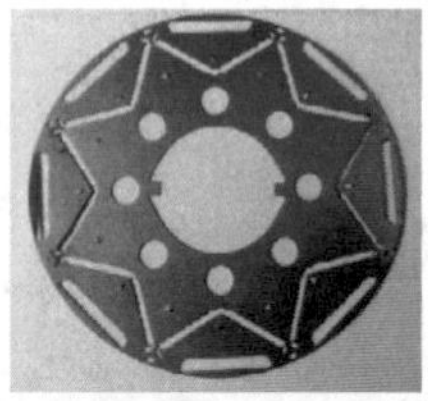

Salient PMSM phasor diagram

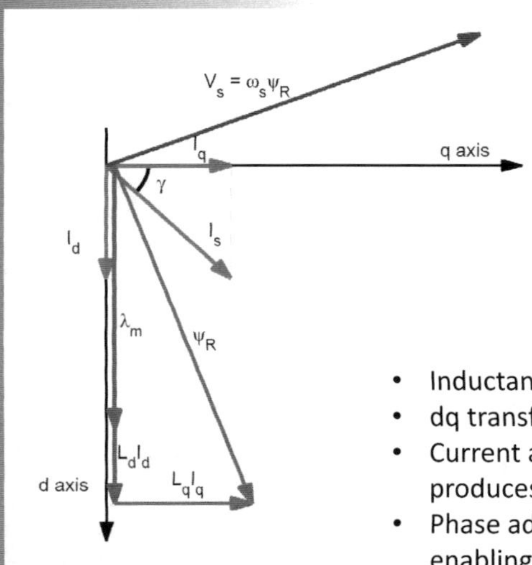

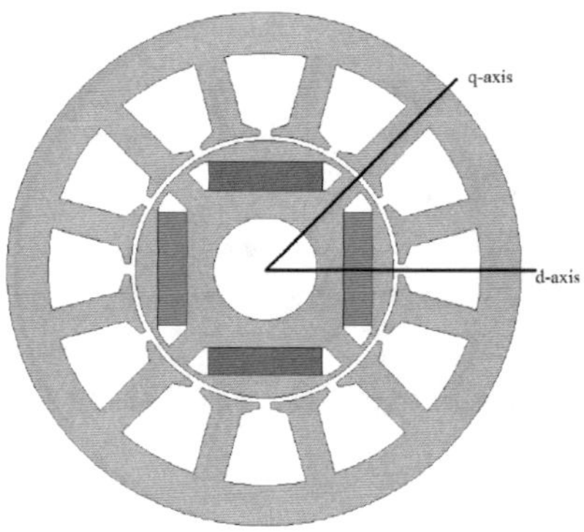

- Inductance varies with rotor position
- dq transformation very useful!
- Current angle advanced in phase relative to the EMF produces reluctance torque
- Phase advance also weakens d-axis flux and voltage, enabling higher speed operation
- Significant saturation and cross saturation requires FEA analysis to model

Definition of Motor Operating Point

$$T = \frac{m}{2} p(\lambda_m I_q + I_q I_d (L_d - L_q))$$

$$V = \omega_s \sqrt{(\lambda_m + L_d I_d)^2 + (L_q I_q)^2}$$

Where T is the torque, m the number of phases, p the number of pole pairs, $I_{d,q}$ the direct and quadrature components of the peak phase current I_s, V is the peak phase voltage, ω_s is the electrical rotational speed, $L_{d,q}$ are the direct and quadrature axis inductances λ_m the permanent magnet flux linkage and γ *is* the current advance angle.

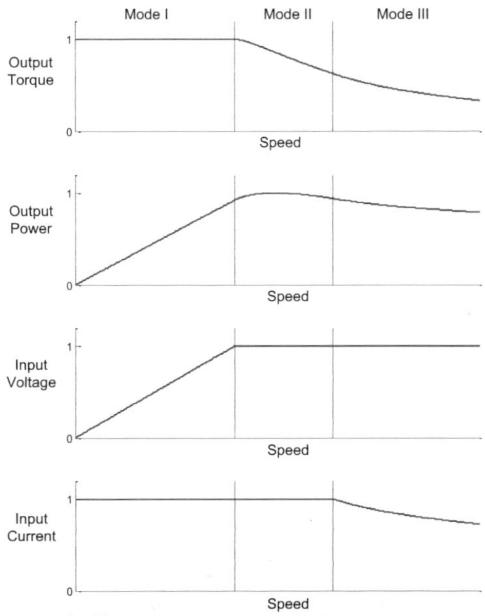

motordesign LIMITED 17

Accounting for Saturation and Cross Coupling Effects

- The flux linkages ψ_d and ψ_q do not simply vary linearly with I_d and I_q.
- q-axis is highly saturable, d-axis has cross-saturation effects

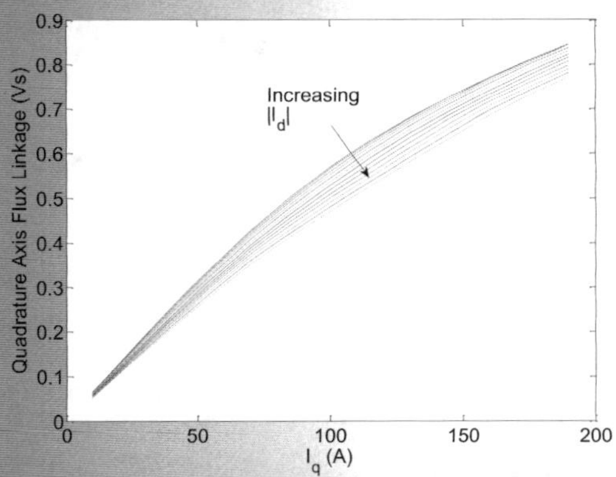

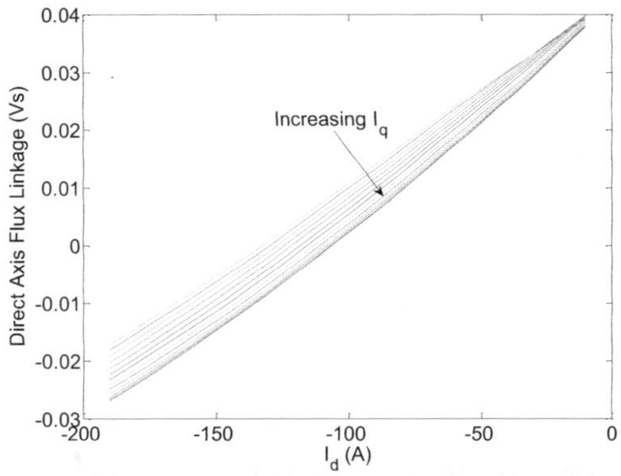

motordesign LIMITED 18

FEA derived functional maps

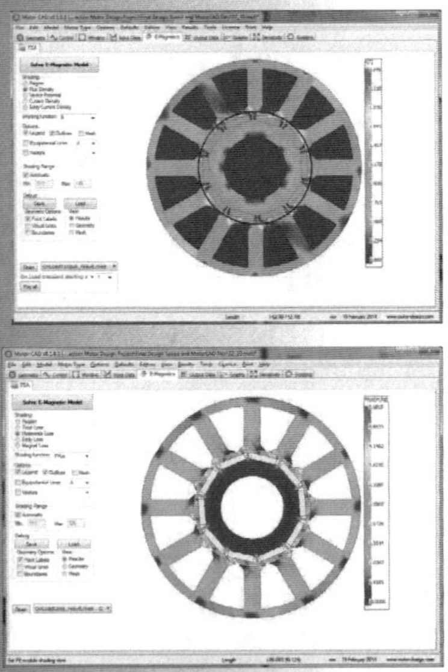

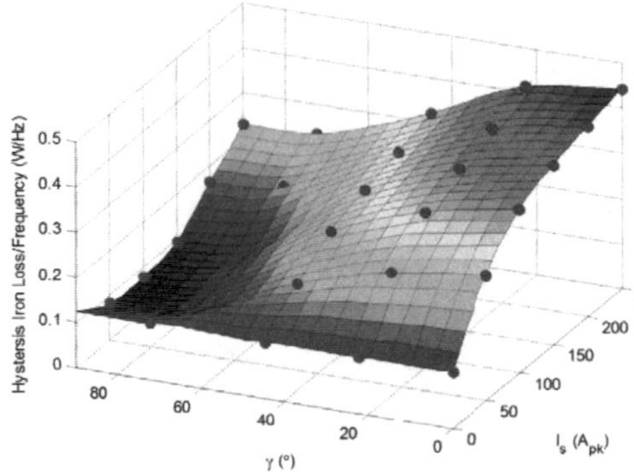

Flux linkages -> (Torque and Voltage)
Hysteresis and eddy current iron losses
Magnet eddy current losses
AC winding losses

Modelling performance over complex driving cycles

- Using these pre-built models the performance at each operating point calculated using maximum torque per amp or maximum efficiency control strategies. This is achieved with the application of non-linear optimisation algorithms.

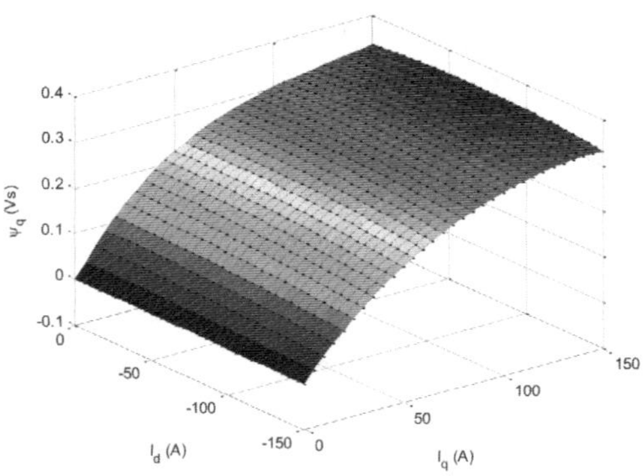

Operating Point Computation

Maximum torque per amp		**Maximum efficiency**	
minimise	$I_s = \sqrt{I_d^2 + I_q^2}$,	minimise	$W_{total} = W_{cu} + W_{fe} + W_{magnet}$,
subject to	$T_{shaft} - T_{demand} = 0$	subject to	$T_{shaft} - T_{demand} = 0$
and	$V_{lim} \geq 2\pi f \sqrt{\psi_d^2 + \psi_q^2}.$	and	$V_{lim} \geq 2\pi f \sqrt{\psi_d^2 + \psi_q^2}.$

- In most cases these gives very similar performance
- Functional maps for ψ_d, ψ_q, W_{cu}, W_{fe} and W_{magnet} are used to solve the equations.

motor**design** LIMITED 21

Modelling performance over complex driving cycles

- Thousands for operating points can be solved rapidly, allowing simulation of efficiency maps and losses over drive cycles

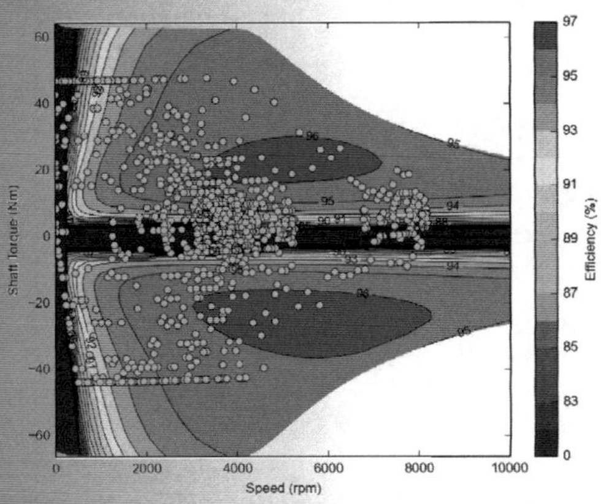

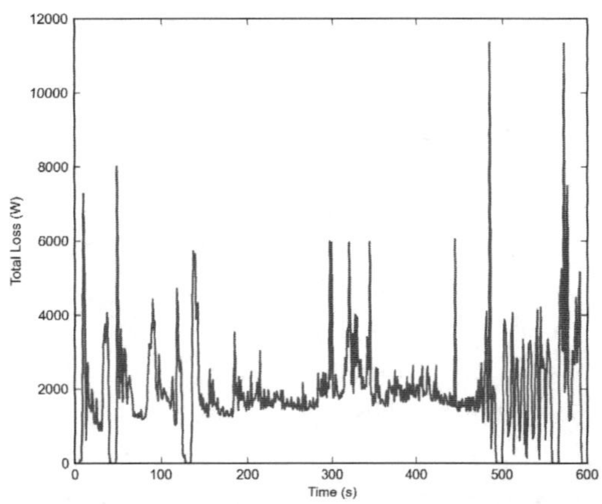

motor**design** LIMITED 22

Modelling performance over complex driving cycles

- The losses for each component can then be coupled with the lumped parameter thermal model to calculate temperature rise over the duty cycle.

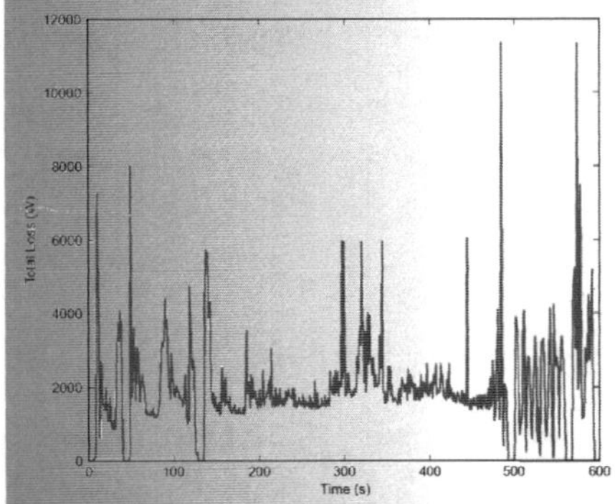

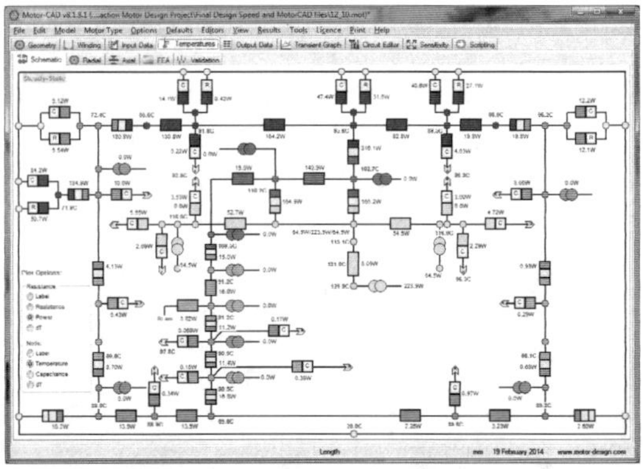

Modelling performance over complex driving cycles

- This process is fast and hence can be applied iteratively
- It allows machines to be sized based on their transient performance. Rather the continuous operation or using rules of thumb.
- This results in minimum size and cost for the design.

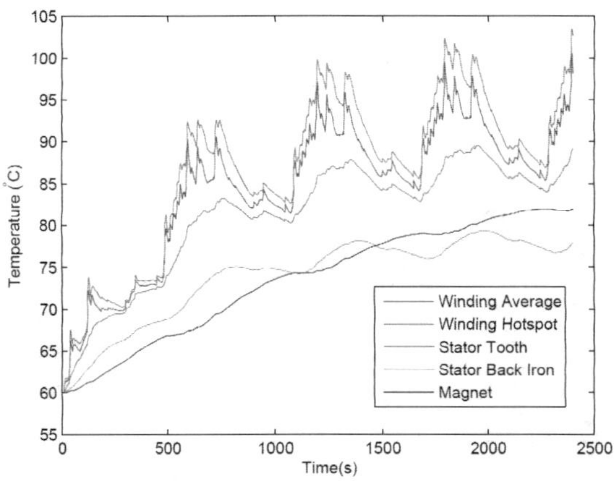

4. IPM Example

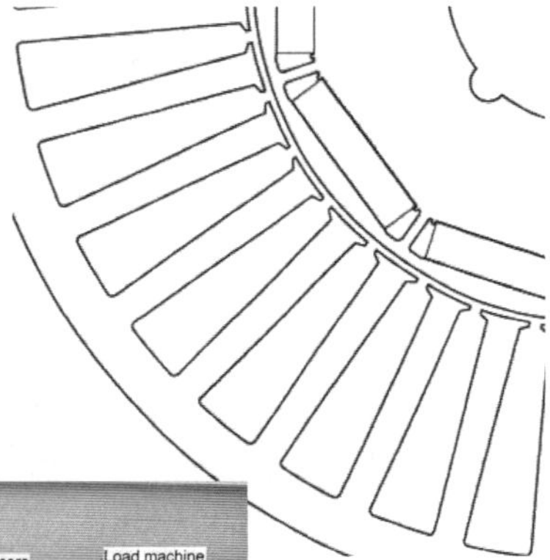

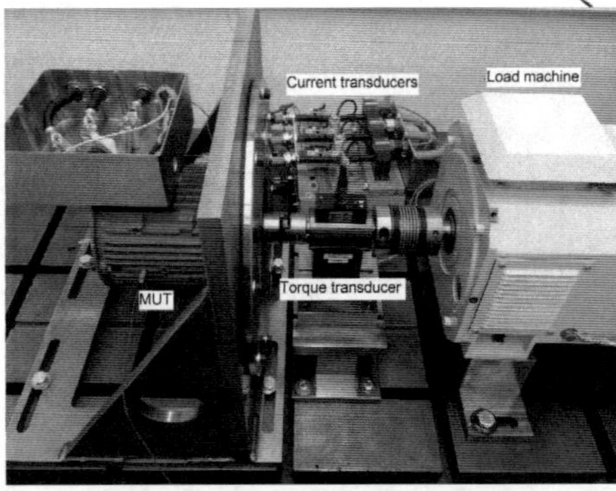

36 slot 10 pole IPM

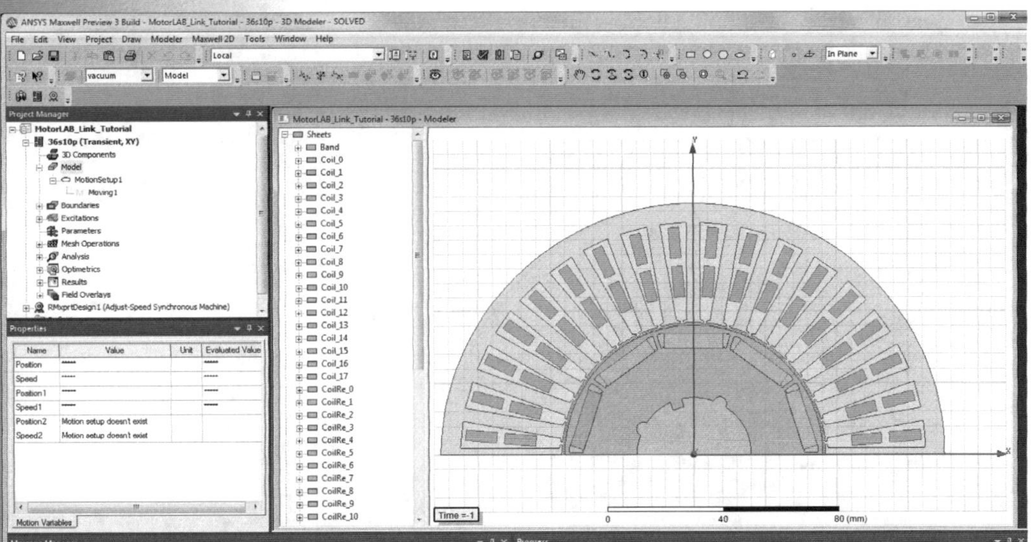

Efficiency Map Comparison

Experimental

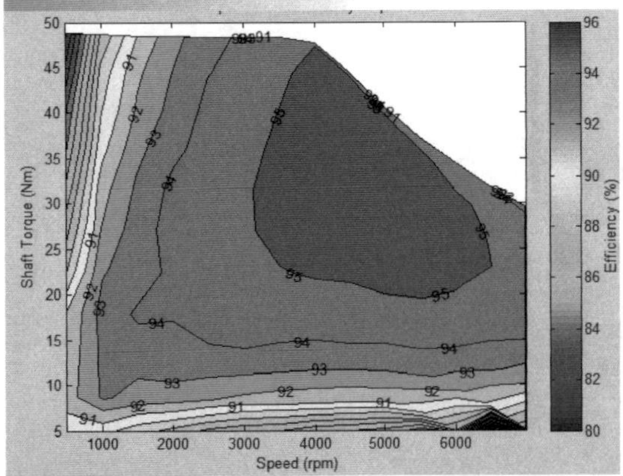

Maxwell->Motor-LAB model

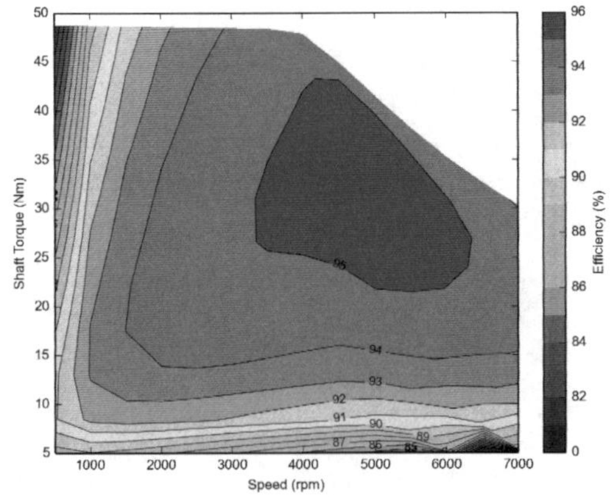

motordesign LIMITED 27

Modelling of Nissan LEAF Motor

- Much data available on web relating to the Nissan LEAF motor
- Modelled LEAF to demonstrate and validate multi-physics modelling methods for traction applications

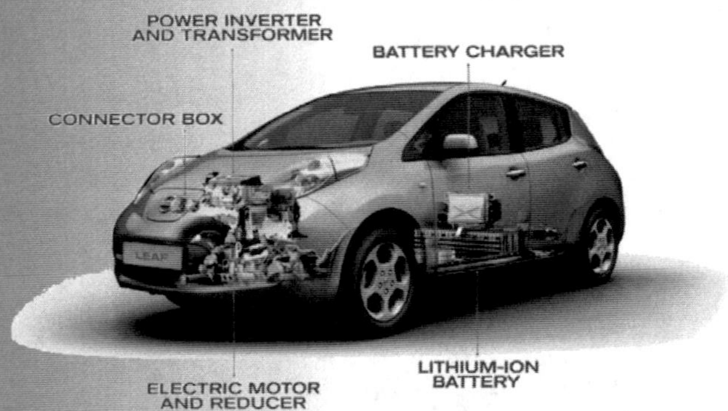

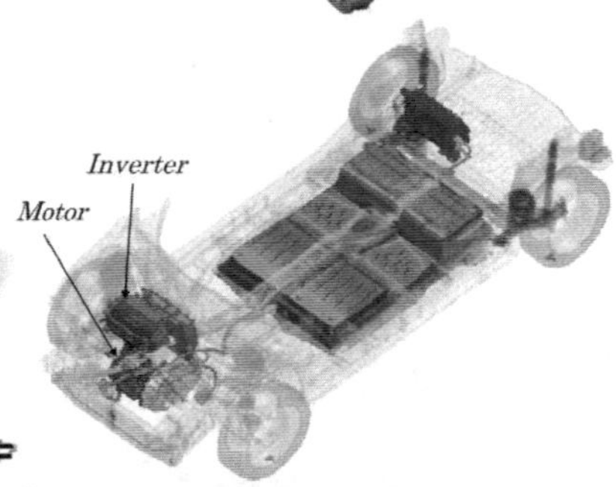

LEAF

motordesign LIMITED 28

Electromagnetic Models for LEAF Motor

- Motor-CAD EMag used to calculate electromagnetic performance
- Can also use ANSYS Maxwell

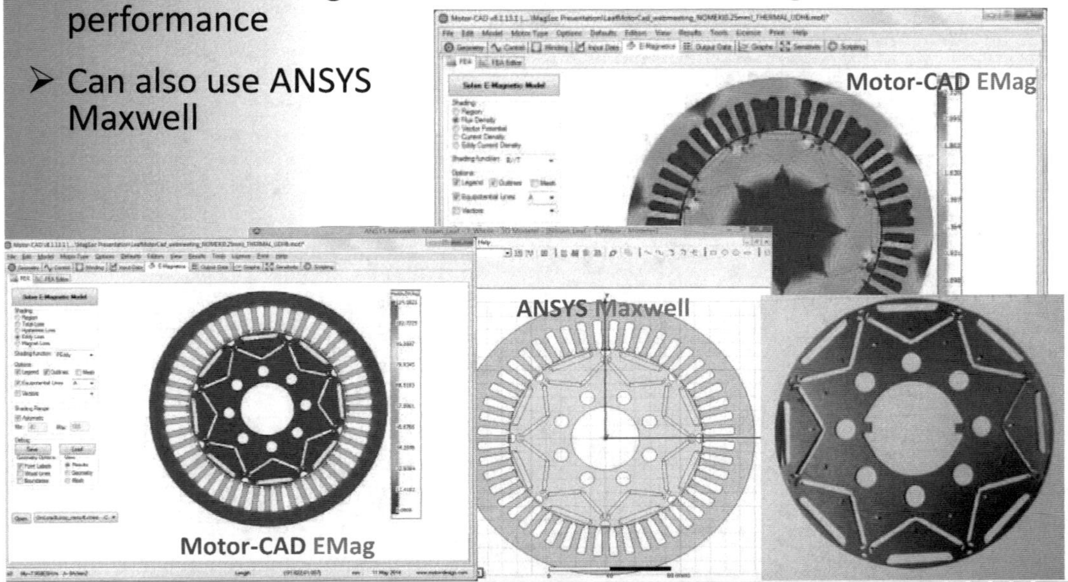

motordesign LIMITED 29

Electromagnetic Model Validation

- Motor-LAB automatically calculates the optimum phase advance for either maximum torque/amp or maximum efficiency control
  - In this case also apply a 80kW max. power limit as in the test procedure
- Optimum use of a minimum number of Motor-CAD EMag FEA calculations to model saturation and losses gives a very fast calculation
  - Map calculated within a few minutes
- Very good match with measured torque/speed characteristic and efficiency map showing accurate calculation of electromagnetic performance and losses in Motor-CAD EMag

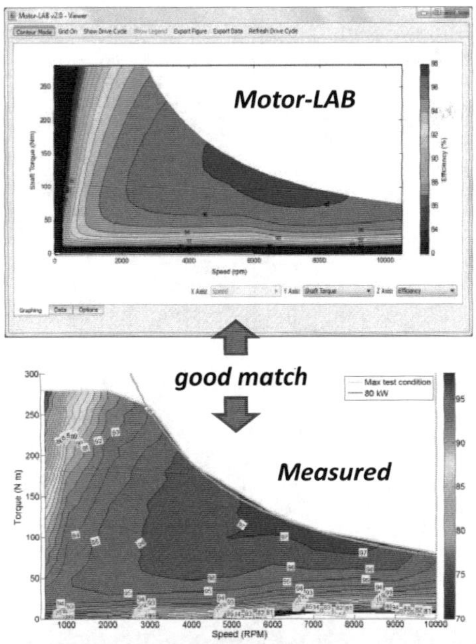

motordesign LIMITED 30

50kW, 60kW, 70kW and 80kW Duty Cycle

- 50, 60, 70 & 80kW thermal transient set up in Motor-LAB
- Loss data then transferred to Motor-CAD Therm model

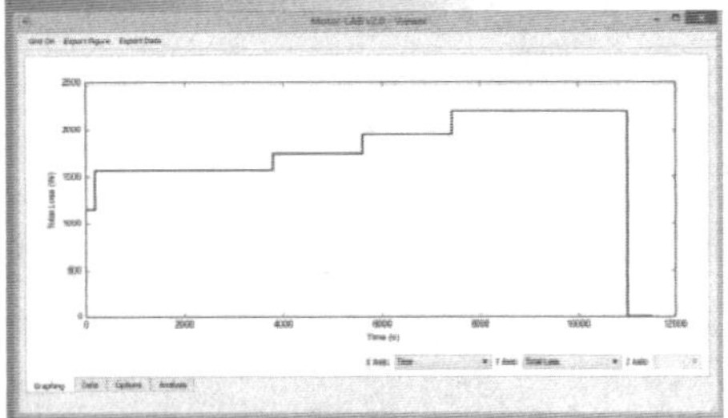

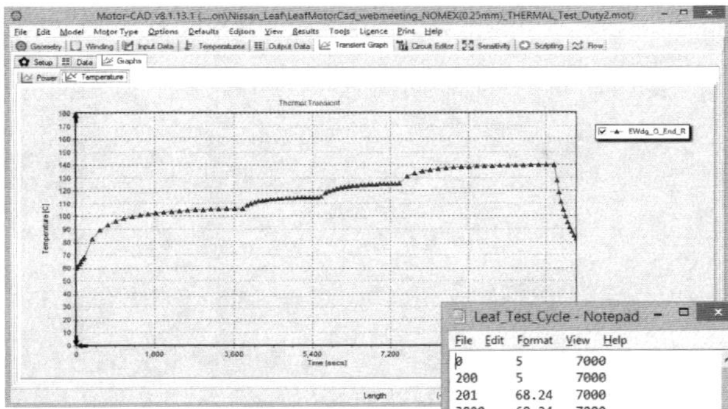

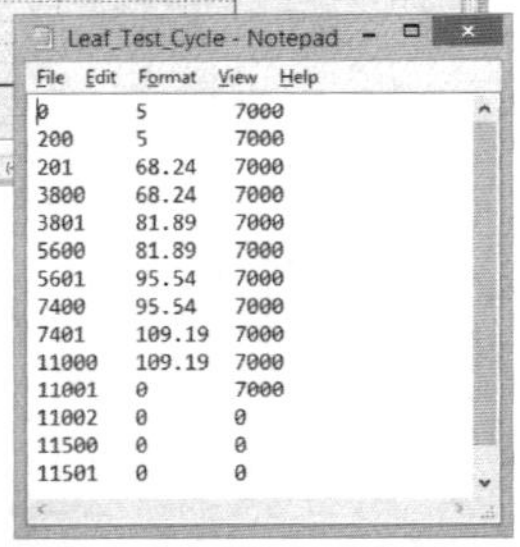
Leaf_Test_Cycle - Notepad

File Edit Format View Help

```
0      5       7000
200    5       7000
201    68.24   7000
3800   68.24   7000
3801   81.89   7000
5600   81.89   7000
5601   95.54   7000
7400   95.54   7000
7401   109.19  7000
11000  109.19  7000
11001  0       7000
11002  0       0
11500  0       0
11501  0       0
```

Period	Elapsed Time	Time	Points	Stator Copper	Stator Back Iron	Stator Tooth	Magnet	Rotor Back Iron	Embedded Magnet Pole	Friction Front Bearing	Friction Rear Bearing	Windage	Windage Exterior Fan	Speed [Start]	Speed [End]	Fault
Units	secs	secs		Watts	Watts	Watts	Watts	Watts	Watts	Watts	Watts	Watts	Watts	rpm	rpm	
1	200	200	5	41.52	583.5	366	0.4861	46.52	0	0	0	0	0	7000	7000	☐
2	201	1	1	41.52	583.5	366	0.4861	46.52	0	0	0	0	0	7000	7000	☐
3	3800	3599	20	404.5	625.4	385.7	1.924	40.16	0	0	0	0	0	7000	7000	☐
4	3801	1	1	404.5	625.4	385.7	1.924	40.16	0	0	0	0	0	7000	7000	☐
5	5600	1799	20	554.1	644.7	395.9	2.476	38.5	0	0	0	0	0	7000	7000	☐

motordesign LIMITED

31

Thermal Model for LEAF Motor

- Thermal model validated by comparing with thermal transient tests at 50, 60, 70 and 80kW

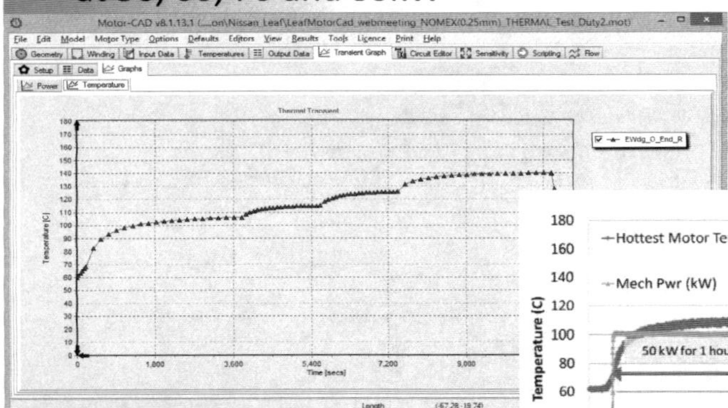

- Losses from Motor-LAB efficiency/loss maps

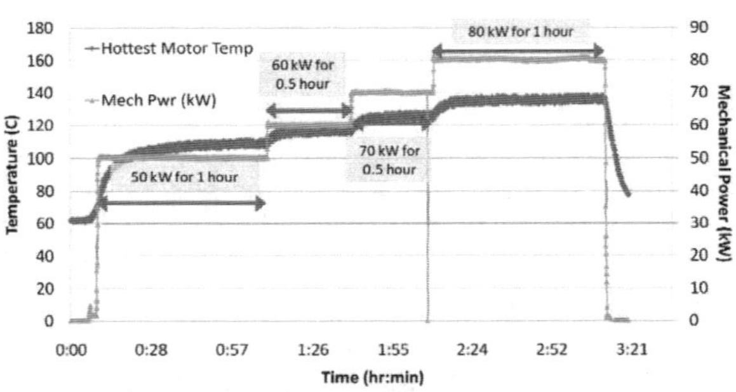

Fig. 15. Nissan LEAF continuous test at 7,000 rpm.

- Excellent match between Motor-CAD and measured thermal transient

motordesign LIMITED

32

Motor Design Limited

Thank you for your attention!

Questions are welcomed

motordesign LIMITED

33

Automotive Simulation
WORLD CONGRESS

June 7-8, 2016 Munich

Auslegung einer permanenterregten Synchronmaschine für ein Motocross-eBike

Prof. Dr. Steinhart, Swen Bosch, Matthias Hauber
STZ: El. Antriebstechnik und Leistungselektronik

STZ: Leistungselektronik und el. Antriebstechnik

- **Dienstleistungsangebot**
  - Entwicklung, Beratung, Simualtion, Aufbau von Prototypen
  - Simulation von PV-Wechselrichtern
    - Simulation der Regelung
    - Zertifizierung
  - Implementation von digitalen Regelungen auf DSPs, Mikrocontroller
    - PV-Wechselrichter
    - El. Antriebe
    - Blindleistungskompensation
  - Auslegung von elektrischen Maschinen
- **Schwerpunkt**
  - Aktorik
  - Elektrische Antiebstechnik
  - Leistungselektronik
  - Regelungstechnik

STZ: Leistungselektronik und el. Antriebstechnik

Energieübertragung in ein rotierendes System

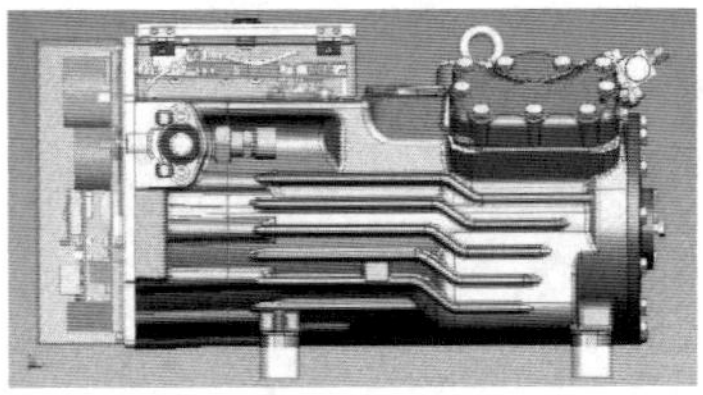

Integrierte Elektronik

Kompletter Antrieb in einem HSK63-Kopf

Antrieb für Eiskernbohrer: Rotor

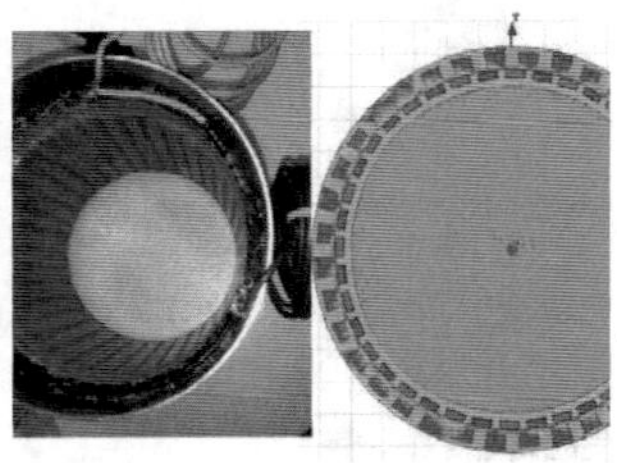

Antrieb für Eiskernbohrer: Stator

Drehmoment-Drehzahlcharakteristik von elektrischen Maschinen

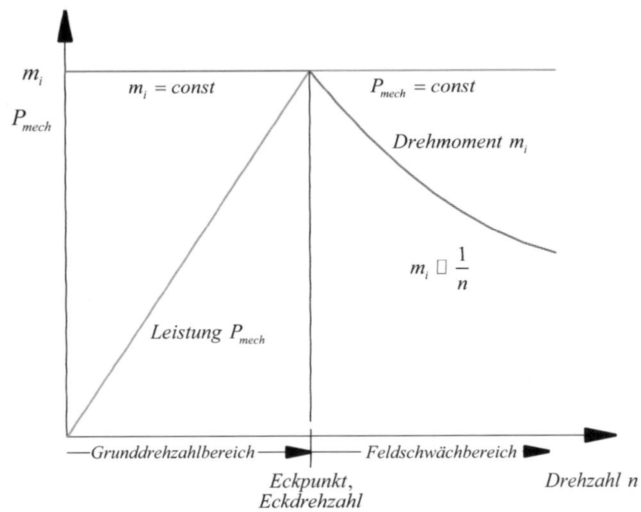

Kraft und Drehmomententwicklung

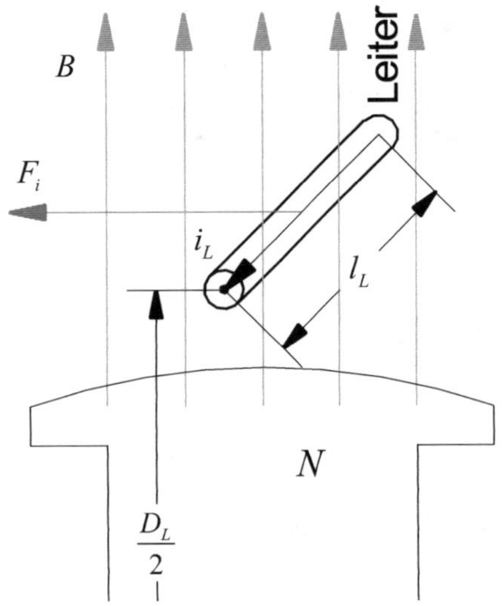

$$F_{L,i} = B \cdot l_L \cdot i_L$$

$$F_T = N \cdot B \cdot l_L \cdot i_L$$

$$M_W = F_T \cdot \frac{D_L}{2} = N \cdot B \cdot l_L \cdot i_L \cdot \frac{D_L}{2}$$

Aufteilung der einzelnen Verluste

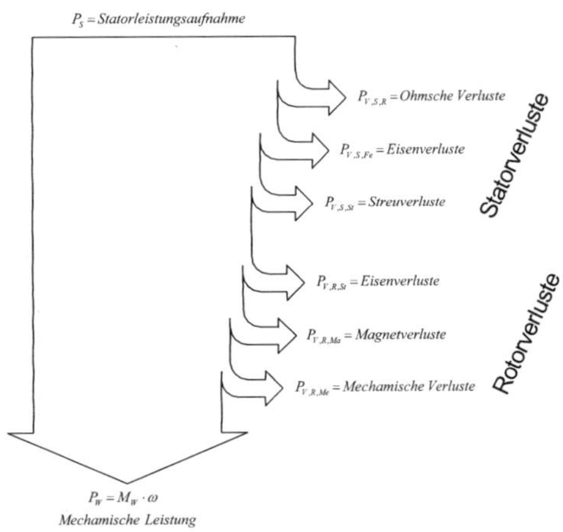

$$\boldsymbol{P_{Hy} = f_S \cdot B^2}$$

$$\boldsymbol{P_{Wb} = {f_S}^2 \cdot B^2}$$

$$\boldsymbol{P_{V,S,Fe} = P_{Fe} + P_{Wb}}$$

Hysterese- und Wirbelstromverluste

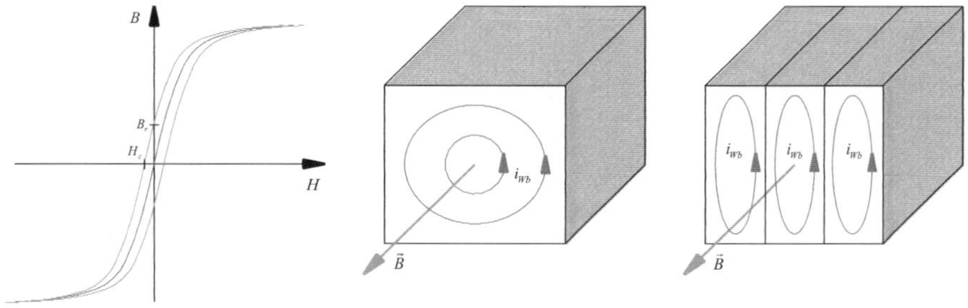

$$P_{Wb} = f_S{}^2 \cdot B^2$$

Automotive Simulation WORLD CONGRESS

Streuverluste

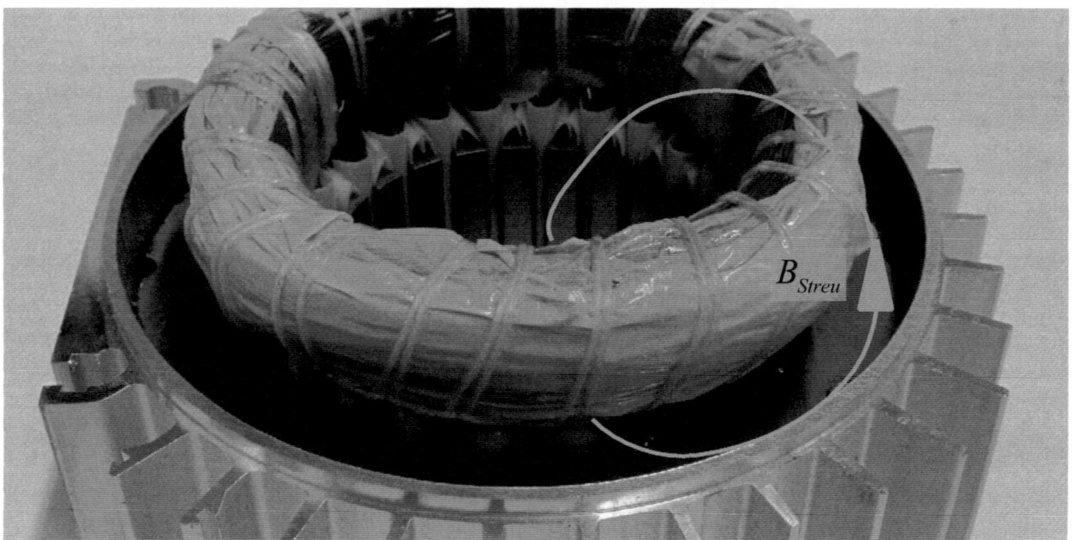

$$P_{V,St} = f_S{}^2 \cdot B^2$$

Automotive Simulation WORLD CONGRESS

Dimensionierung des Antriebs

- **Maximale Drehzahl:** **4500 U/min**
  - Mechanischer Aufbau
  - Eisenverluste
  - Statorfrequenz (375 Hz bei p = 5)
- **Drehmoment:** **35 Nm, kurzfristig 70 Nm**
- **Eckpunkt für die Feldschwächung:** **4500 U/min**
- **Kühlung des Motors:** **Wasser**
- **Kühlung des Umrichters:** **Wasser**
- **Getriebeübersetzung:** **1:5,75**
- **Max. Geschwindigkeit des eBikes:** **100 km/h**
- **Hohe Beschleunigung bzw. Steigfähigkeit: 87%**

Aufbau des Stators

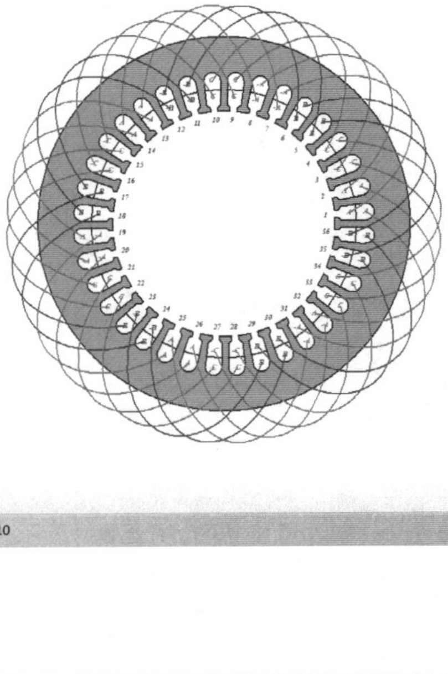

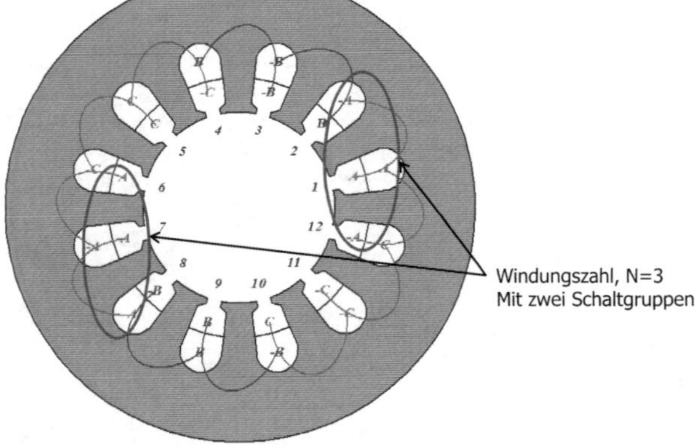

Aufbau der Maschine mit eingebetteten Magneten

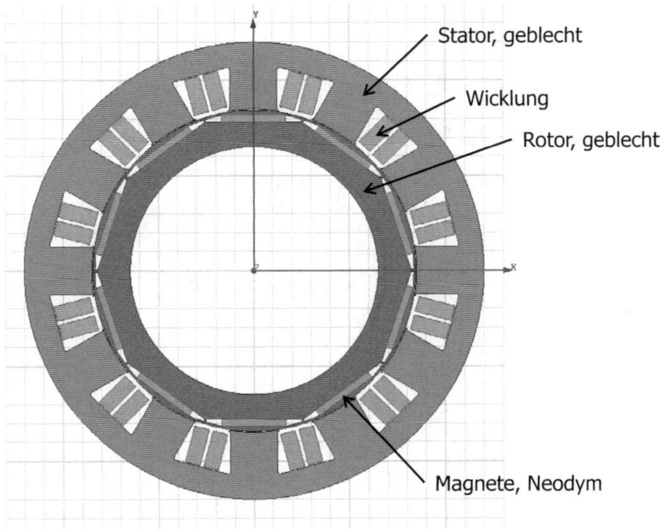

Automotive Simulation WORLD CONGRESS

Drehmoment als Funktion des Statorstroms

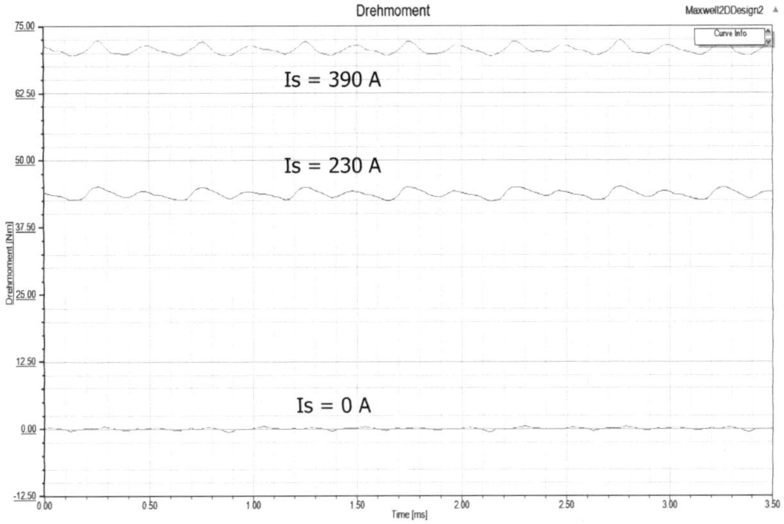

Automotive Simulation WORLD CONGRESS

Magnetische Flussdichte bei 230 A Statorstrom

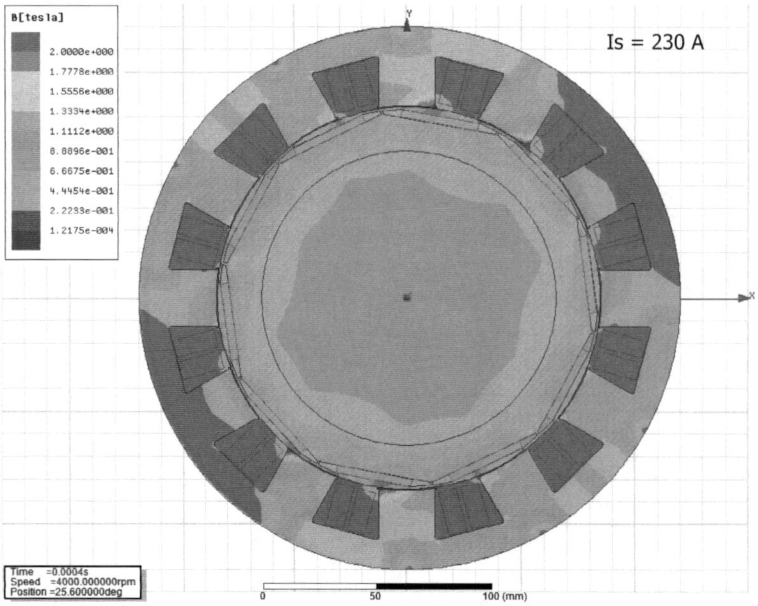

Leistungsabgabe über das Hinterrad als Funktion der Geschwindigkeit

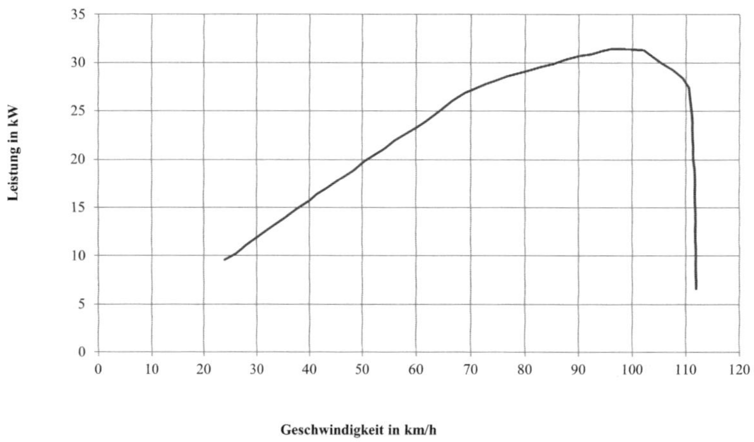

Technische Daten

Versorgungsspannung	60 V (DC)
Nennstrom	230 A, kurzfristig 390 A
Polpaarzahl	5
Kupferdrahtfläche	20 mm^2
Windungszahl	3
Statorinduktivität	L_{sd} = 27 µH, L_{sq} = 39 µH
Statorwiderstand	2,8 mΩ
Nenndrehzahl	4000 U/min
Außendurchmesser	240 mm
Bohrdurchmesser	170 mm
Aktive Länge	50 mm
Luftspalt	0,5 mm
Magnetmaße (Länge, Breite, Höhe)	50 mm, 34 mm, 3,5 mm
Remanenzflussdichte B_r der Magneten	1,23 T
Transformatorblech	M 250-35A
Nennmoment	35 Nm, kurzfristig 70 Nm
Wirkungsgrad im Nennpunkt	92%

Automotive Simulation WORLD CONGRESS

Test der ersten Prototypen im Feld

Automotive Simulation WORLD CONGRESS

Zusammenfassung: Auslegung PM-Maschine

- **Grundsätzliche Überlegungen zum Maschinendesign**
- **Entwurf der PM-Maschine**
  - **Verluste**
- **Aufbau der Prototyp-Maschine**
  EMT Ingenieurbüro GmbH
  Oberer Wasen 23
  D-73630 Remshalden
- **Aufbau des Motocross-eBikes**
  Fa. Christ in Memmingen
- **Test der E-Maschine auf dem Prüfstand**
- **Feldtests mit Motocross-eBike**

Automotive Simulation WORLD CONGRESS

17 June 7-8, 2016 Munich

CADFEM®

Simulation is more than Software®

Numerical Vibration Analysis of Electric Drives

Martin Hanke, Marold Moosrainer, CADFEM GmbH

CADFEM®

ANSYS EM Vibration Workflow

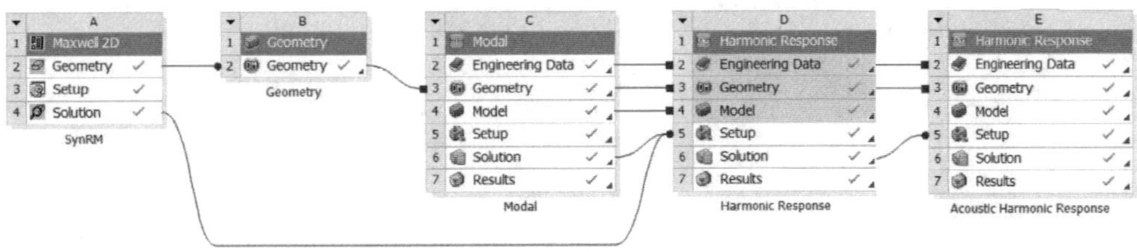

CADFEM

ANSYS EM Vibration Workflow

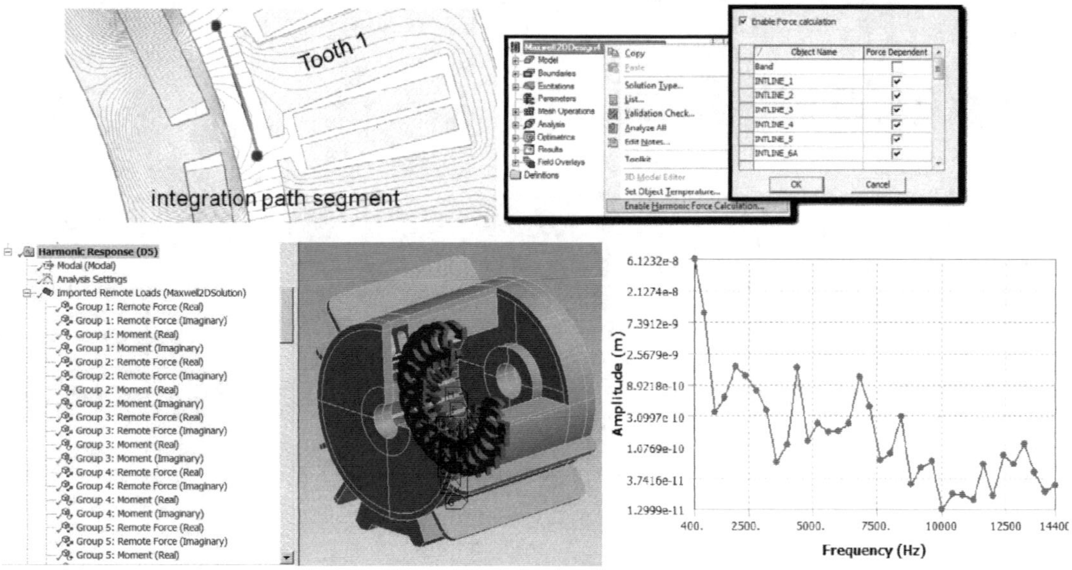

CADFEM

CADFEM Workflow: Rotational Speed Interpolation

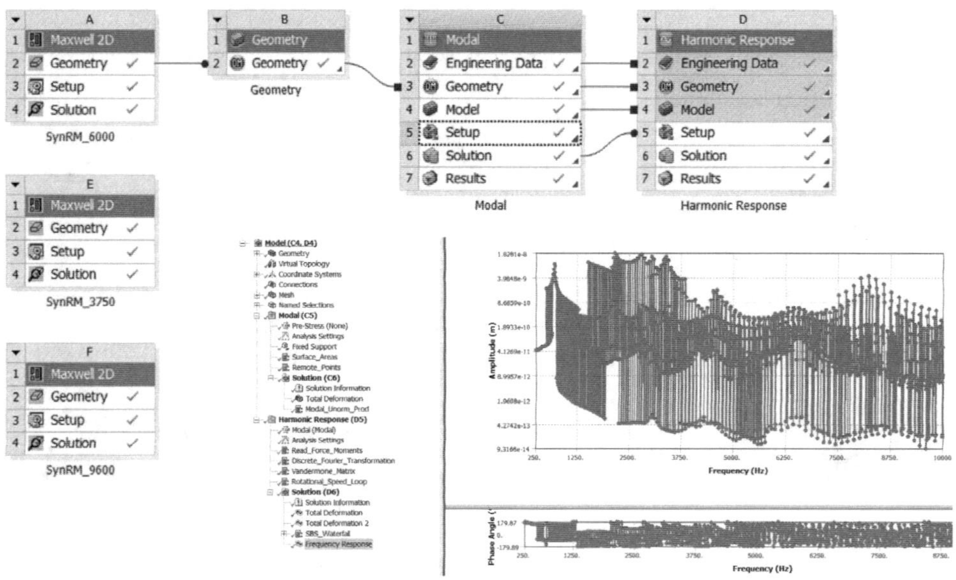

CADFEM

ERP from MCF

Equivalent Radiated Power from Modal Coefficient File

$$\text{Acoustic spectral power}: \quad P = c \cdot \rho \cdot \iint v_n^2 \cdot dA / 2$$

$$v_n = \omega \cdot \sum_m c_m \cdot u_{m,n}$$

$$v_n^2 = \omega^2 \cdot \sum_m \sum_{m'} c_{m'} \cdot c_m \cdot u_{m,n} \cdot u_{m',n}$$

$$P = c \cdot \rho \cdot \omega^2 / 2 \cdot \sum_m \sum_{m'} c_{m'} \cdot c_m \cdot \iint u_{m,n} \cdot u_{m',n} \cdot dA$$

$$P = c \cdot \rho \cdot \omega^2 / 2 \cdot \sum_m \sum_{m'} c_{m'} \cdot c_m \cdot U_{mm',n}$$

- To calculate ERP modal coefficients c_m from mcf file and the matrix $U_{mm'}$ are needed
- No expansion pass in harmonic required
- dB = 10*$\log_{10}$(P/1e-12)

CADFEM

Magnetic Force Calculation from Maxwell's Stress Tensor

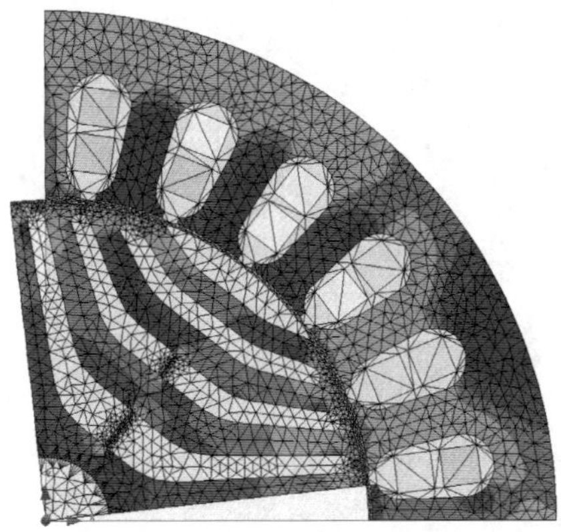

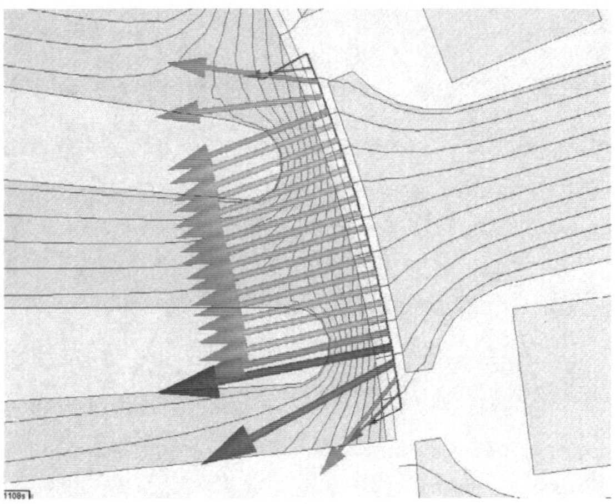

CADFEM

Forces in Time and Frequency Domain

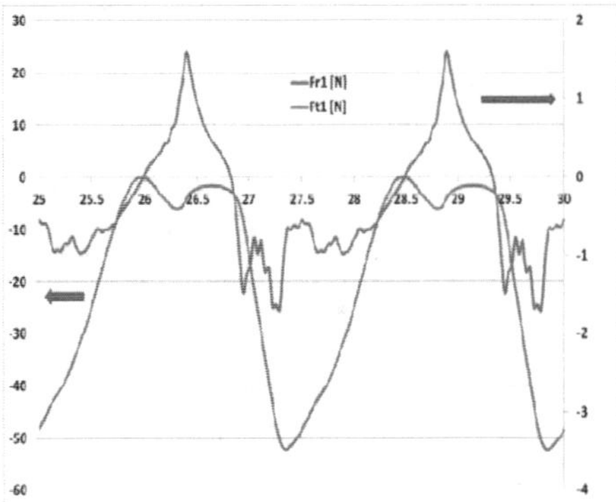

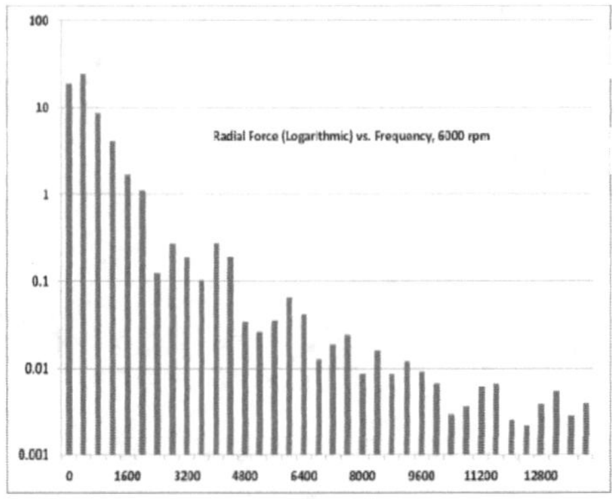

CADFEM

Remote Force Application

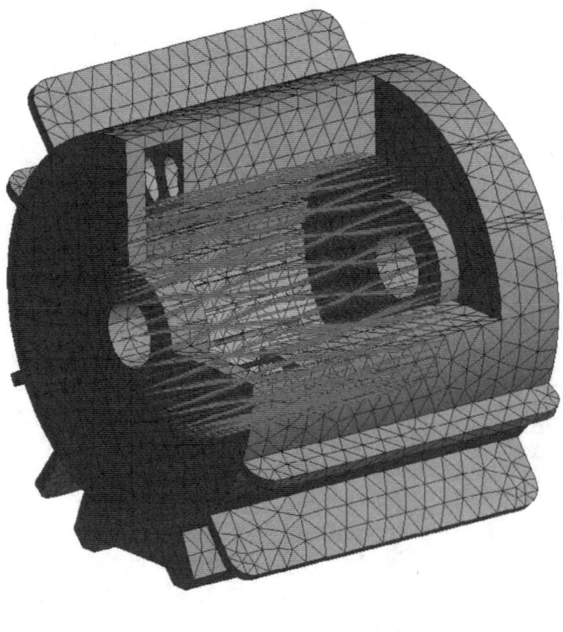

Material Homogenization

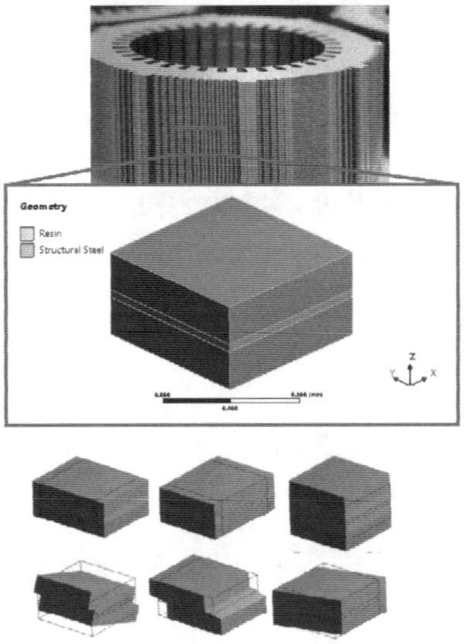

$$K \cdot u = F$$

$$\begin{bmatrix} K_{11} & K_{12} & K_{13} & K_{14} & K_{15} & K_{16} \\ K_{21} & K_{22} & K_{23} & K_{24} & K_{25} & K_{26} \\ K_{31} & K_{32} & K_{33} & K_{34} & K_{35} & K_{36} \\ K_{41} & K_{42} & K_{43} & K_{44} & K_{45} & K_{46} \\ K_{51} & K_{52} & K_{53} & K_{54} & K_{55} & K_{56} \\ K_{61} & K_{62} & K_{63} & K_{64} & K_{65} & K_{66} \end{bmatrix} \cdot \begin{bmatrix} 0 \\ 1 \\ 0 \\ 0 \\ 0 \\ 0 \end{bmatrix} = \begin{bmatrix} F_1^{(2)} \\ F_2^{(2)} \\ F_3^{(2)} \\ F_4^{(2)} \\ F_5^{(2)} \\ F_6^{(2)} \end{bmatrix}$$

$$W_{\text{int}} = V \cdot \sigma \cdot \varepsilon = V \cdot c \cdot \varepsilon^2 = W_{ext} = F \cdot u$$

With $u = \varepsilon = 1$:

$$V \cdot c = F \Rightarrow c = F / V$$

Table of Properties Row 2: Anisotropic Elasticity

	A	B	C	D	E	F
1	(MPa)	(MPa)	(MPa)	(MPa)	(MPa)	(MPa)
2	2.0264E+05	62744	6105.2	0	0	0
3	62744	2.0264E+05	6105.2	0	0	0
4	6105.2	6105.2	13405	0	0	0
5	0	0	0	69946	0	0
6	0	0	0	0	1893.3	0
7	0	0	0	0	0	1893.3

CADFEM ANSYS

Simulation is more than Software®

Force Pattern Identification from Tooth Forces

CADFEM

Force Shifting

- **Phase shift of Fourier coefficients:**

Periodic function $f(t)$ in range $t \in [0, p]$,

Fourier row with complex coefficients $(i^2 = -1)$

$$f(t) = \sum_j c_j \cdot e^{i \cdot j \cdot (2\pi / p) \cdot t}$$

Insert a phase shift s :

$$f(t - s) = \sum_j c_j \cdot e^{i \cdot j \cdot (2\pi / p) \cdot (t - s)}$$

$$f(t - s) = \sum_j \boxed{c_j \cdot e^{-i \cdot j \cdot (2\pi / p) \cdot s}} \cdot e^{i \cdot j \cdot (2\pi / p) \cdot t} := \sum_j \boxed{c_j^{'}} \cdot e^{i \cdot j \cdot (2\pi / p) \cdot t}$$

CADFEM

Breathing Mode Excitation

- Fourier coefficients of shifted forces are equal if j*s/p is integer
- First nontrivial order with equal coefficients is j=p/s for
- Picture shows radial force waves as funtion of frequency order

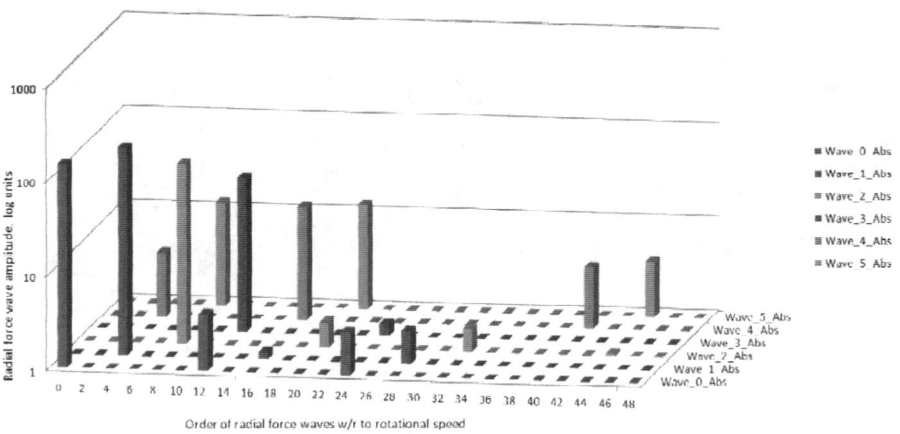

CADFEM

Breathing Mode Identification

- From the scalar product of modes with breathing shape the contributions can be found and identified
- 8100Hz as 24th order would result in 20500 RPM for resonance
- Bigger or more flexible machines may have that resonance in working region

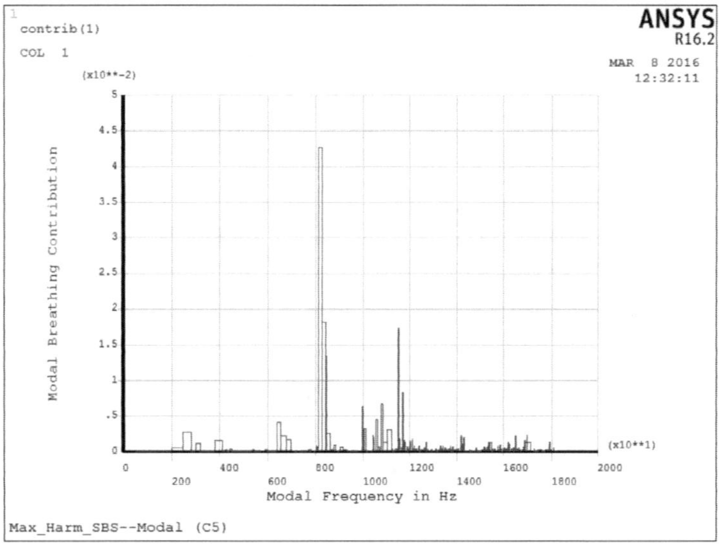

CADFEM

Maximum Mode, Acoustic Radiation

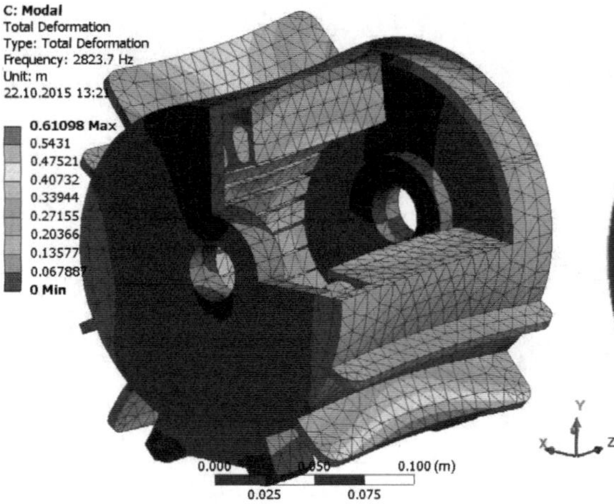

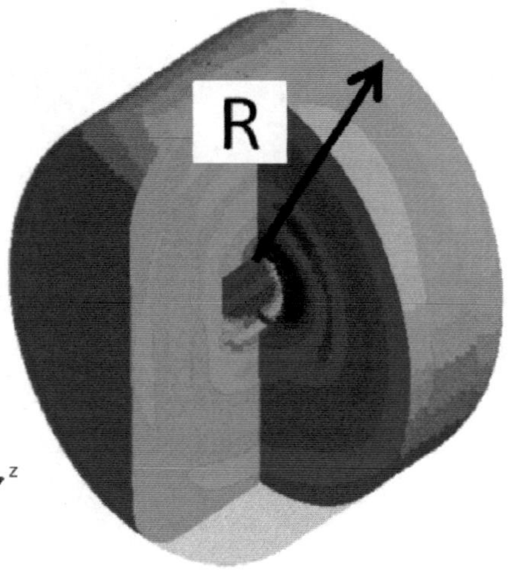

CADFEM

ERP Spectrum

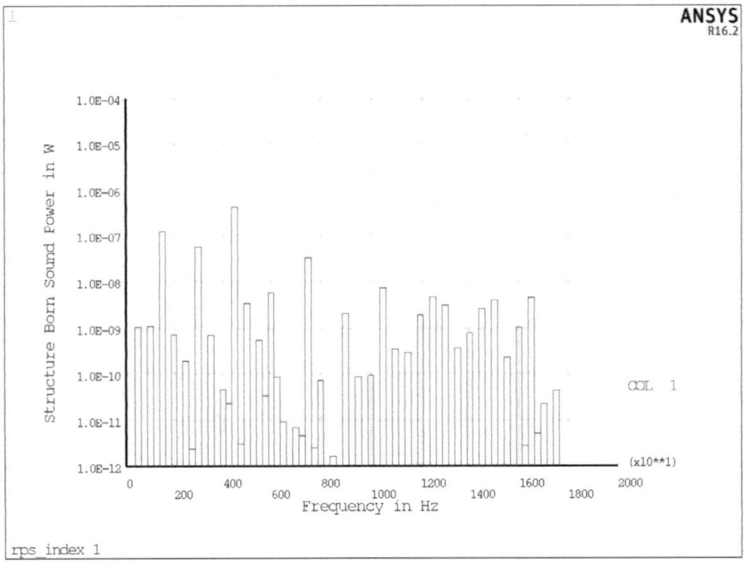

CADFEM

Waterfall Diagrram of ERP

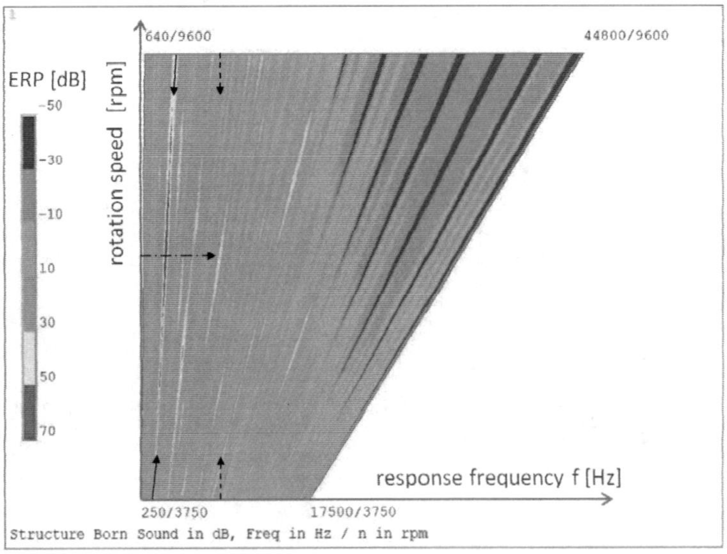

CADFEM

Option: Periodic Excitation from Gears

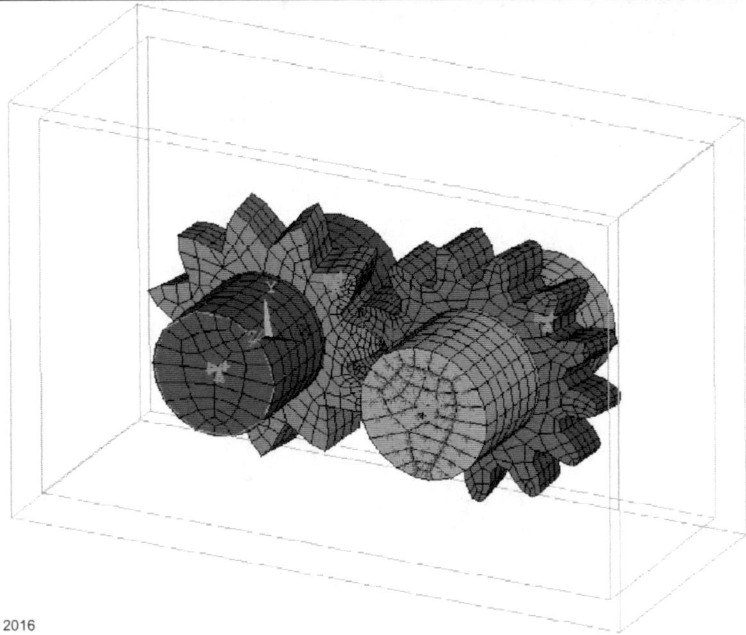

CADFEM

Summary

- Ansys workflow for single rotational speed
- Variation of rotational speeds requires data reuse
  - Force Interpolation
  - Harmonics with MSUP w/o expansion pass
  - Precalculation of ERP
- Allows waterfall diagram

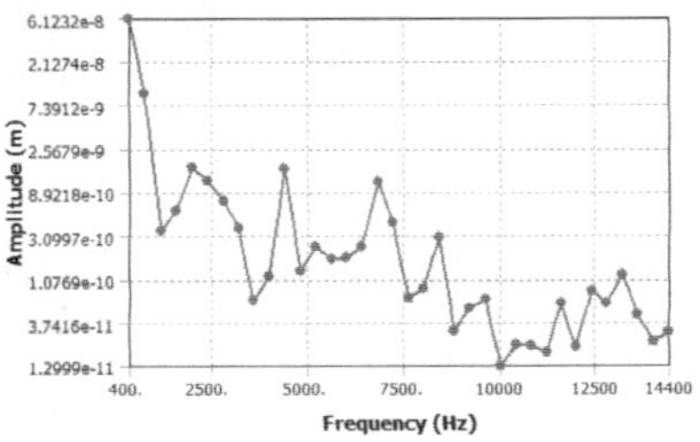

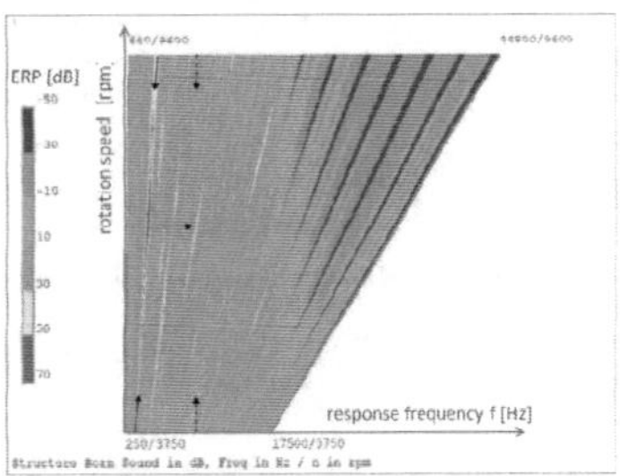

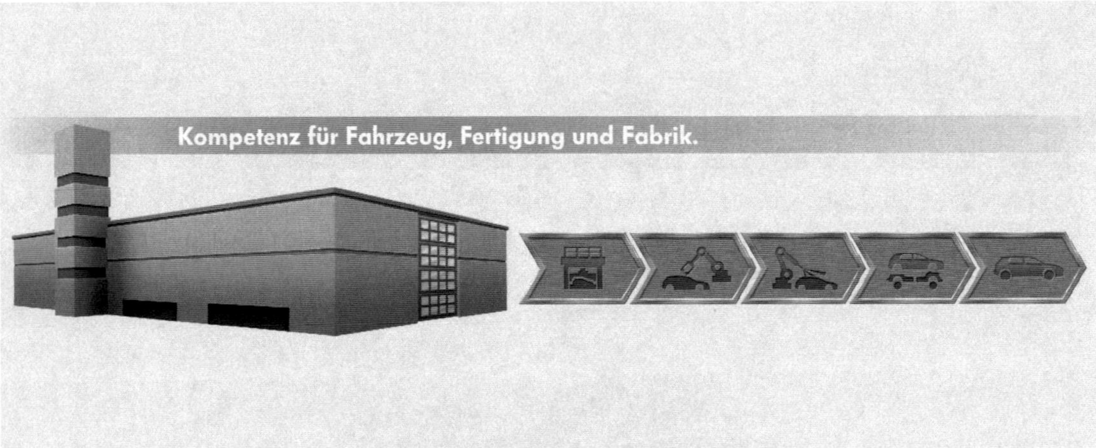

Inductive heating of shaped blanks for hot stamping of car body parts

Automotive Simulation World Congress 2016

Tobias Vibrans (Volkswagen AG)
Dr. Jörg Neumeyer, Dr. Jürgen Wibbeler (CADFEM GmbH)

Conventional heating in hot stamping
Motivation

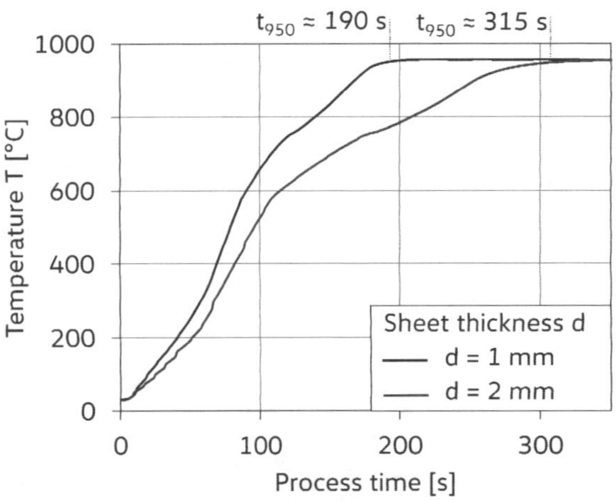

- Slow heating rates lead to high requirements in furnace lengths
- Alternative heating methods allow faster heating [1-6].

Prototypal production line including an inductive heating device – „near series scaled"

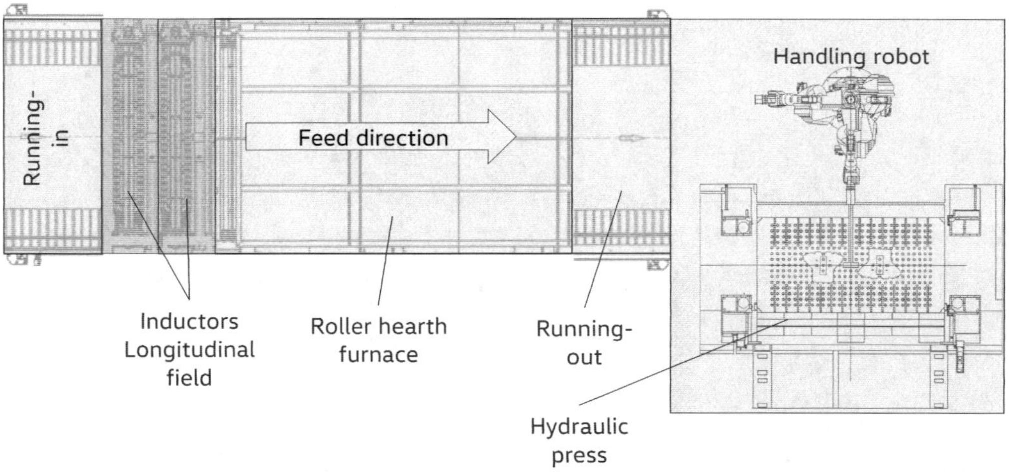

The prototypal production line allows investigation of inductive heating characteristic and influences on pressed series parts.

Volkswagen AG | Technology Development – Tool- and Pressshop | T. Vibrans p. 3

Prototypal production line including an inductive heating device – „near series scaled"

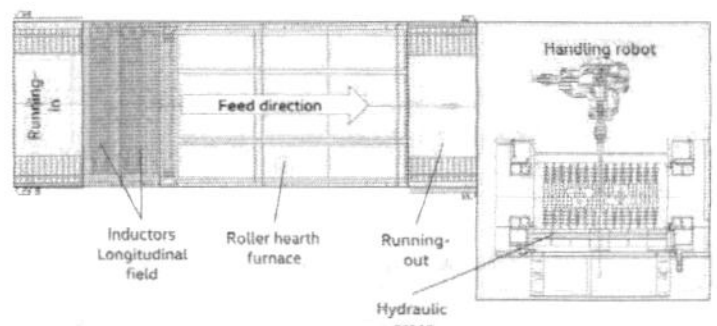

The prototypal production line allows investigation of inductive heating characteristic and influences on pressed series parts.

Volkswagen AG | Technology Development – Tool- and Pressshop | T. Vibrans p. 4

Prototypal production line

Potential for fast heating

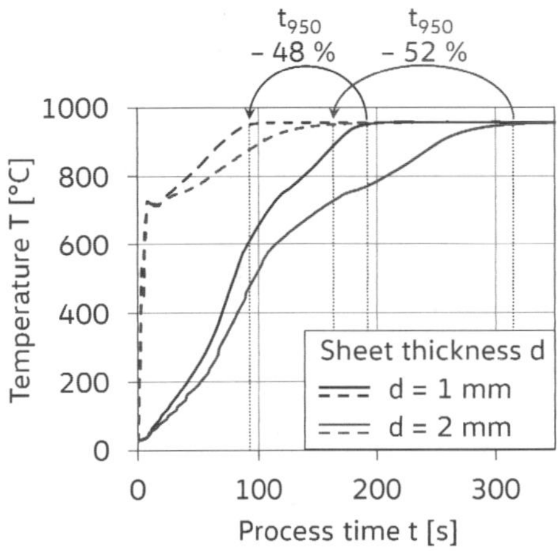

Longitudinal field induction heating enables time savings of ~50 % for heating blanks (22MnB5+AS150) to 950 °C.

Experimental evaluation inductive heating

Blanks of a series production vehicle

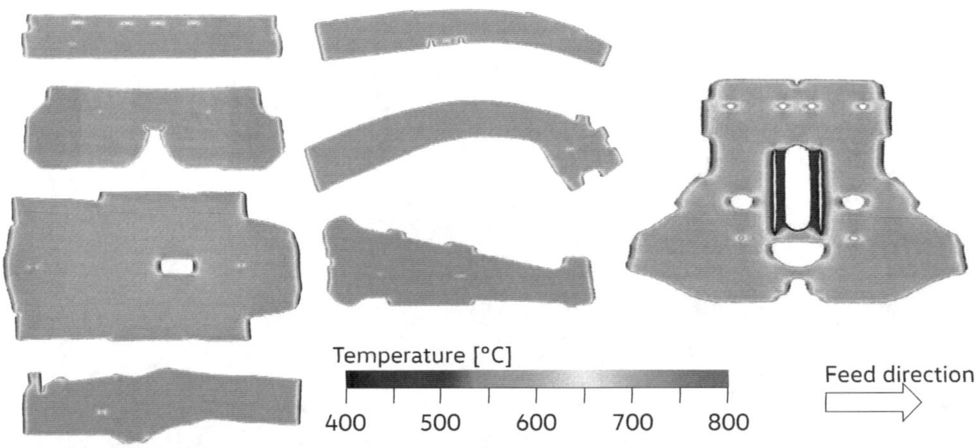

- Homogeneous heating to $T_{Curie} \approx 730$ °C nearly independent of blank geometry
- Lower temperatures at edges perpendicular to feed direction and long inner bridges

Simulation procedure
Layout

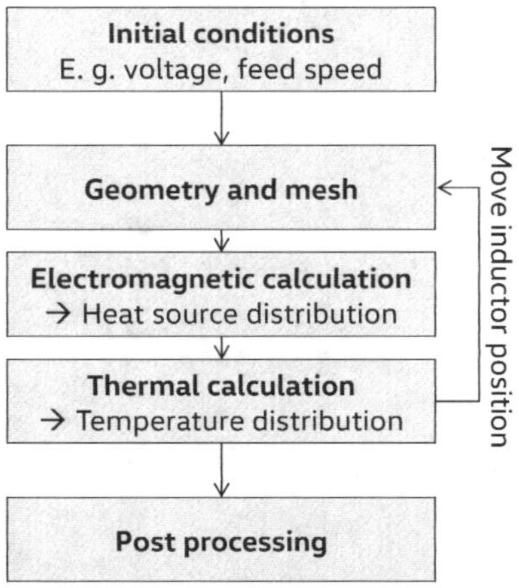

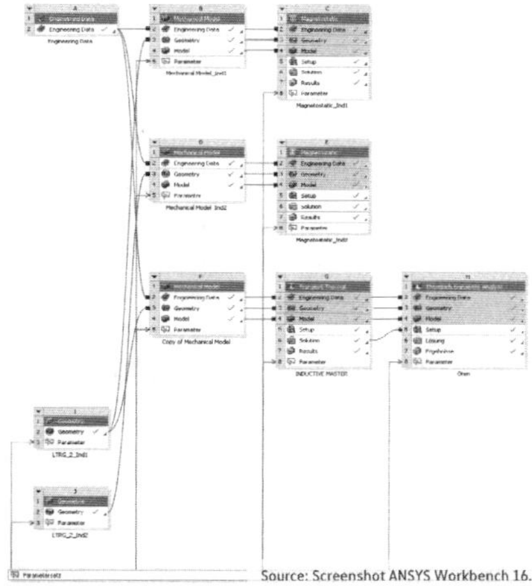

Source: Screenshot ANSYS Workbench 16.2

- Used Software: Ansys Workbench 16.2 with APDL snippets
- Indirect coupling of electromagnetic and thermal calculation

Simulation procedure
Geometry and mesh

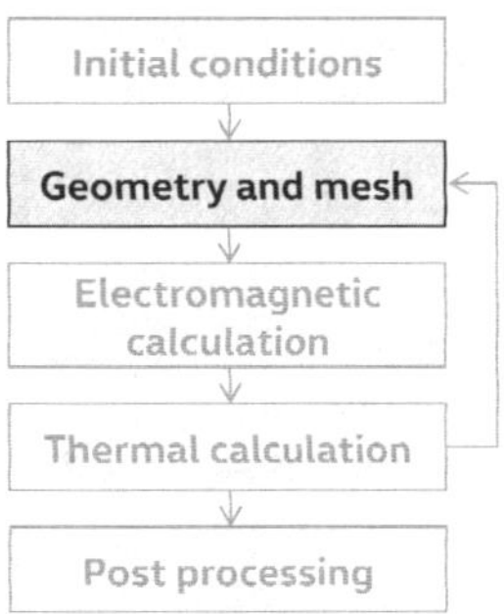

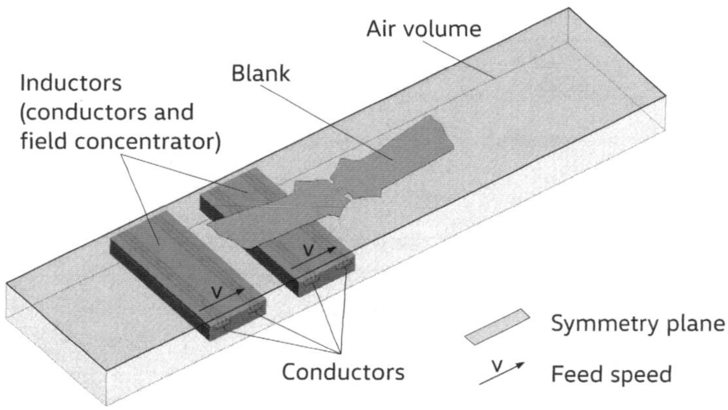

- Modelling of inductors and blanks based on CAD data of experimental device
- Decrease of calculation time by using symmetry

Simulation procedure
Geometry and mesh

- Initial conditions
- **Geometry and mesh**
- Electromagnetic calculation
- Thermal calculation
- Post processing

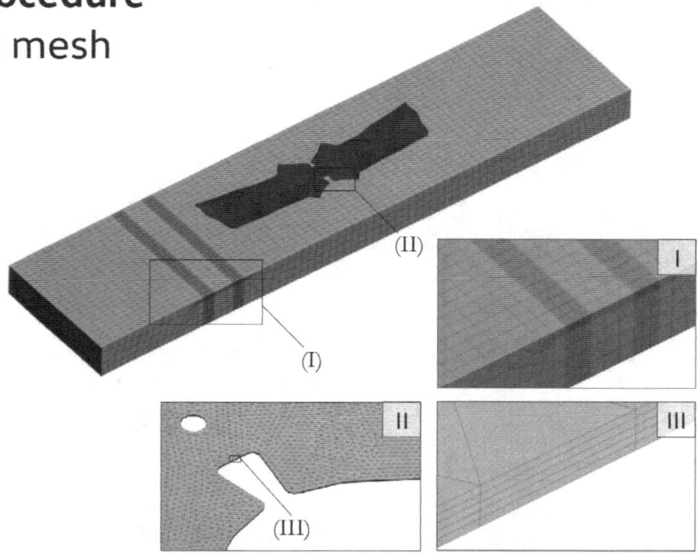

- Electromagnetic calculation: Solid236 – Thermal calculation: Solid90
- Mesh refinement perpendicular to sheet plane for depicting skin effect

Simulation procedure
Electromagnetic calculation

- Initial conditions
- Geometry and mesh
- **Electromagnetic calculation**
- Thermal calculation
- Post processing

Magnetization curve

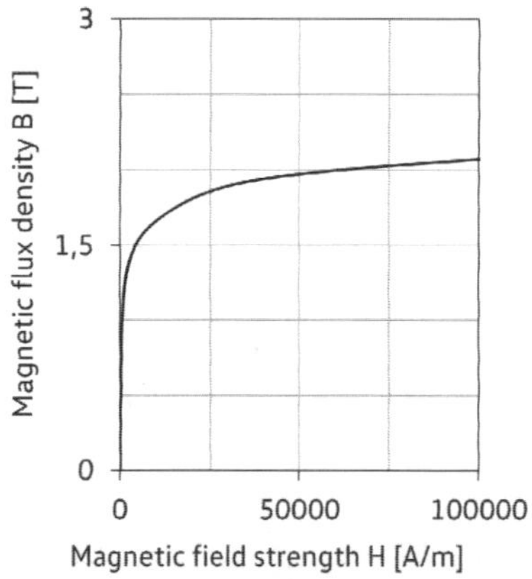

Electric conductivity

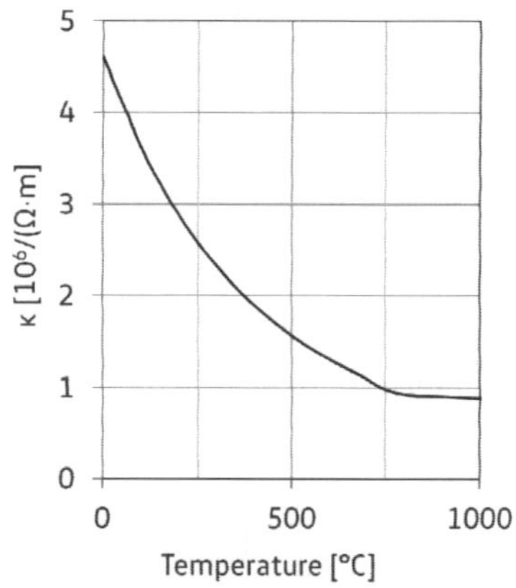

- Blank-specific material properties have been measured and implemented

Simulation procedure
Electromagnetic calculation

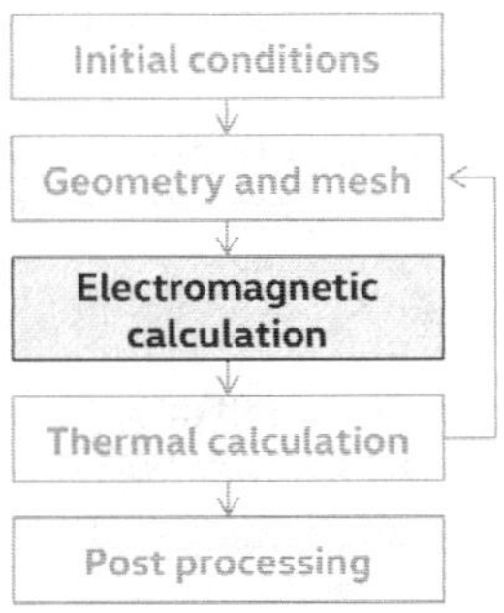

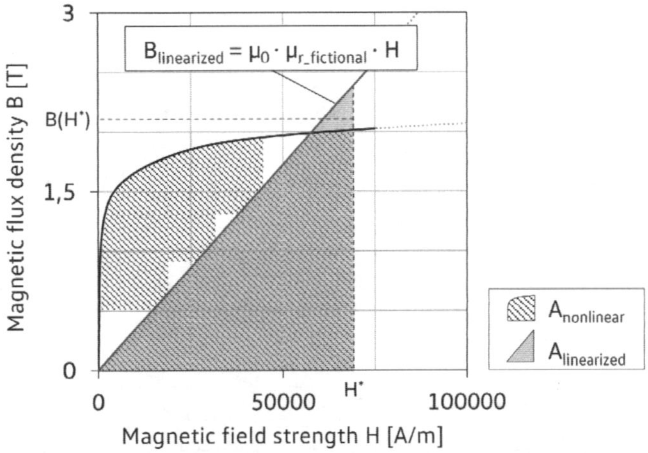

- Linearizing the magnetization curve for harmonic calculation (method by [7])

Simulation procedure
Thermal calculation

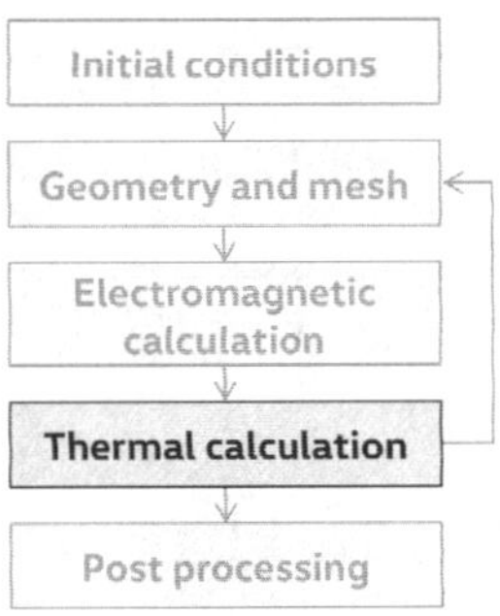

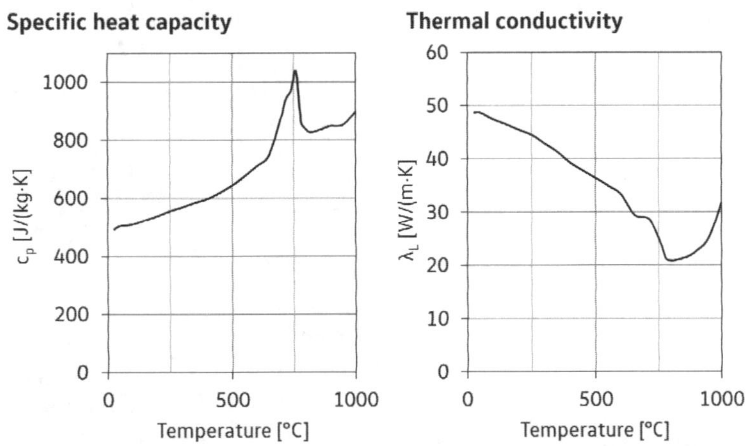

- Blank-specific material properties have been measured and implemented

Simulation procedure
Thermal calculation

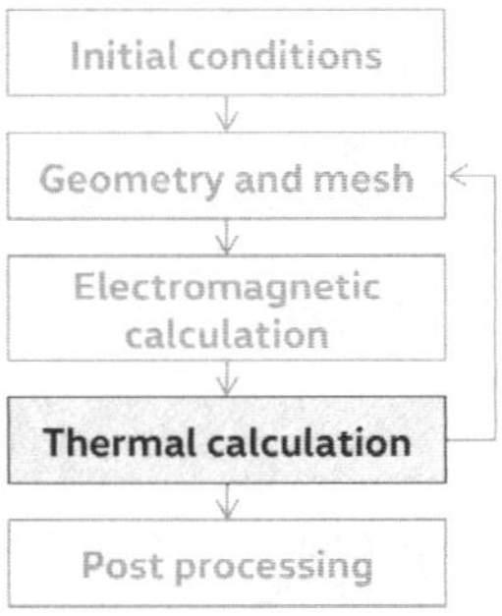

Temperature dependent heat flow

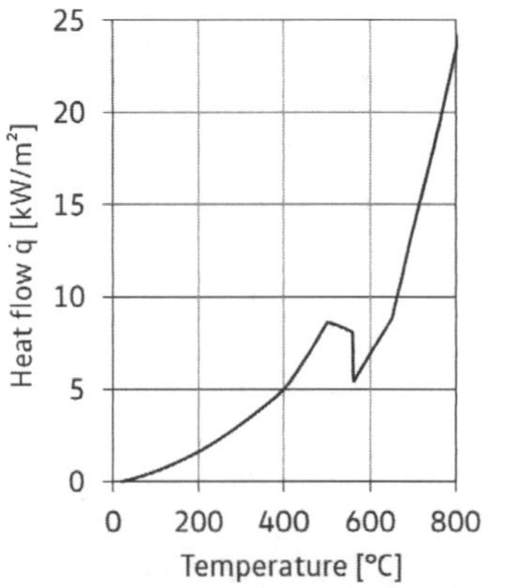

Validation by cooling experiments

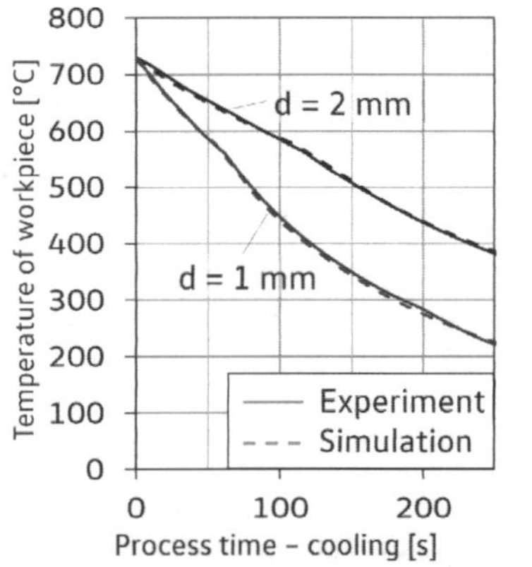

- Determination of heat transfer parameters by experimental obtained cooling curves
- Validity for different sheet thicknesses has been proved

Simulation procedure
Post processing

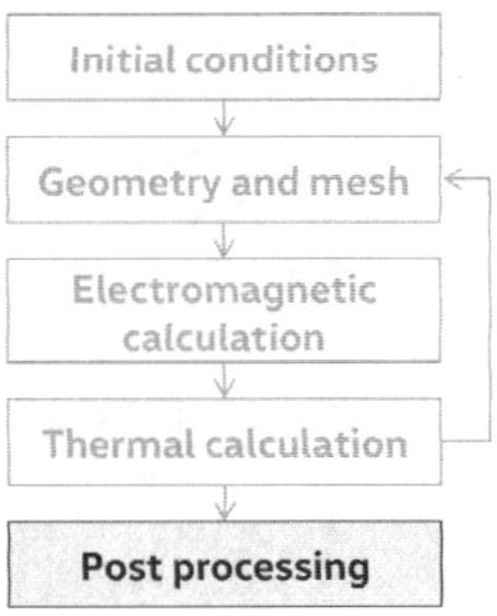

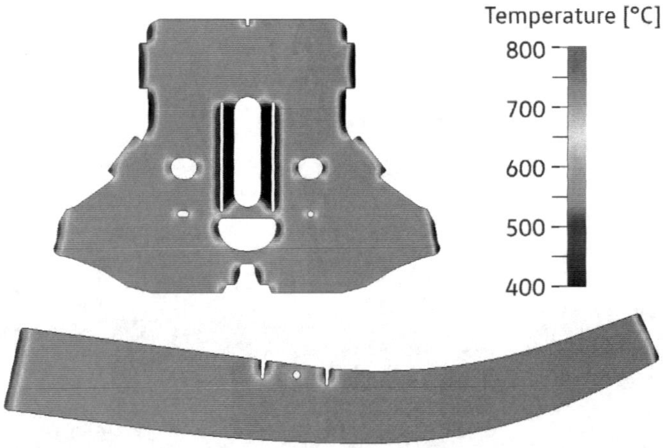

- Analysis and interpretation of heating progression and resulting temperature distributions

Simulation validation
Experiment vs. simulation

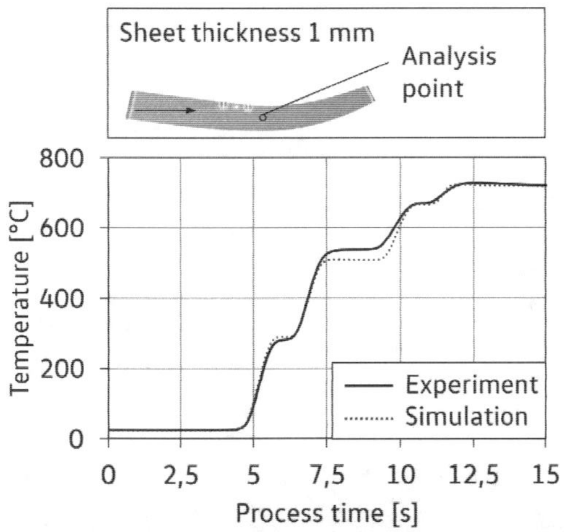

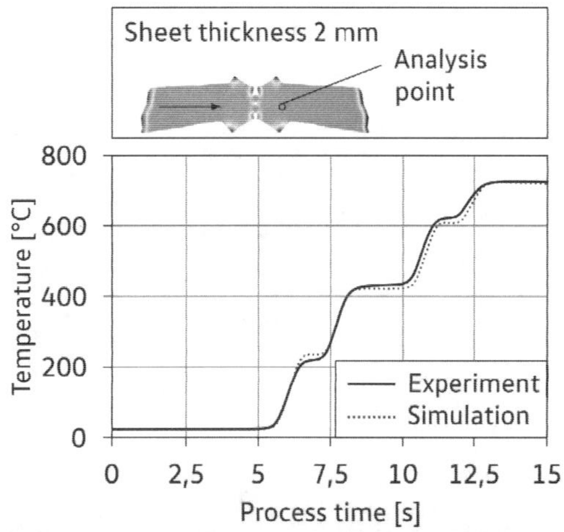

- Average temperature differences (experiment vs. simulation) < 10 K
- Valid simulation of heating progress during inductive heating

Volkswagen AG | Technology Development - Tool- and Pressshop | T. Vibrans p. 15

Simulation validation
Experiment vs. simulation

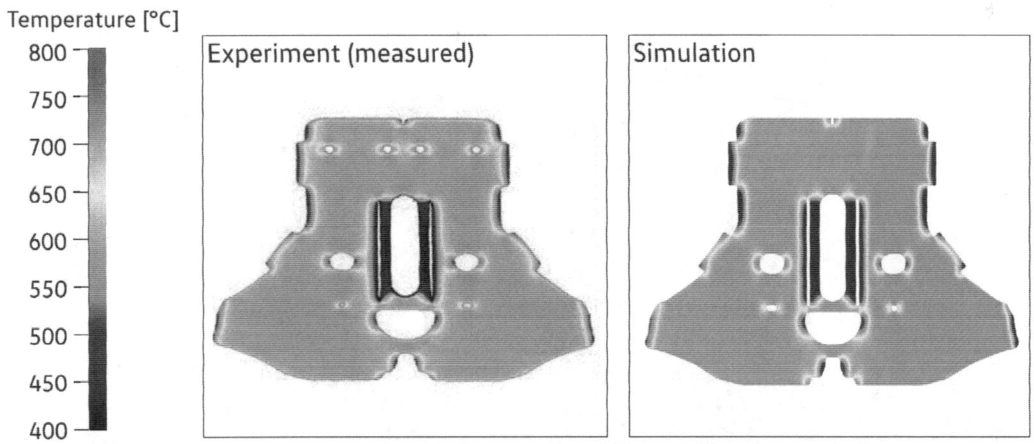

- Valid simulation of heat distribution after inductive heating
- Prediction of inductive heatability already in stage of part development possible

Volkswagen AG | Technology Development - Tool- and Pressshop | T. Vibrans p. 16

Simulation applications
Example 1: Blank position optimization

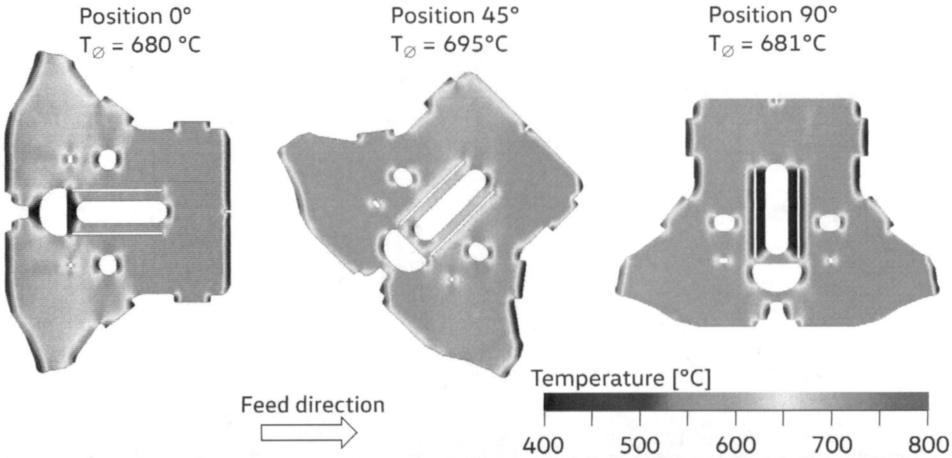

- Optimization of inductive heating process with simulation
- Temperature distribution depends on the relation of blank position to feed direction

Simulation applications
Example 2: Heatability of Tailor Rolled Blanks

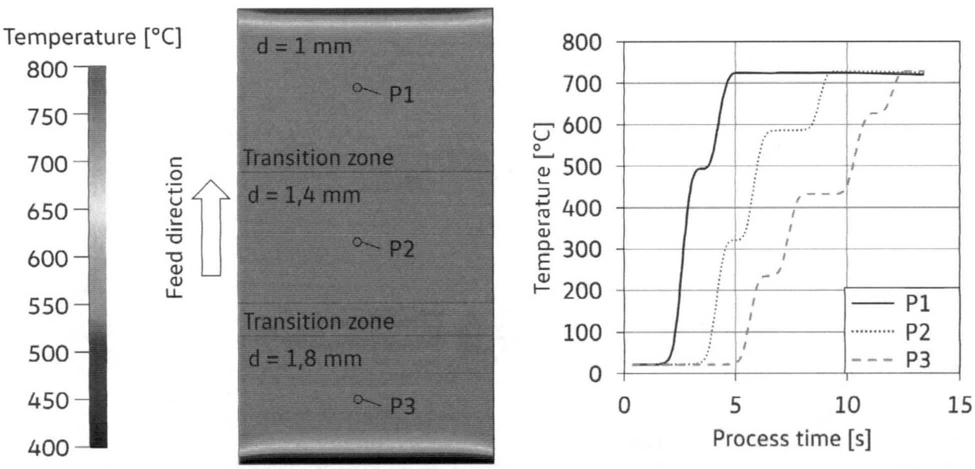

- T_{Curie} is reached homogenously for sheet thicknesses from 1,0 mm to 1,8 mm
- Inductive heating of TRBs is feasable

Longitudinal field inductive heating
Conclusion and outlook

Advances by simulation
- ✓ Comprehension of inductive heating process
- ✓ Identification of influencing parameters
- ✓ Optimization of heating process
- ✓ Heatability analysis of complex shaped blanks

Advances by longitudinal inductive heating in hot stamping (22MnB5 + AS150)
- ✓ Decrease of heating time (to 950 °C): ~50 %
- ✓ Decrease of furnace length: ~35 %

Outlook
- ✓ Research on inductive blank heating with transversal field and transverse movable inductors
- ✓ Connected research in conflict area: substrate – coating – heating

Evolution steps of heating in hot stamping

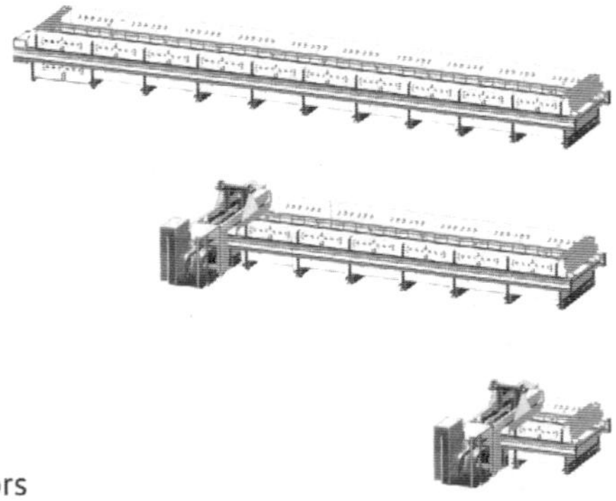

I

II

III

Inductive heating of shaped blanks.
Thank you for your attention.

References

[1] Tröster, T.; Marten, T.; Adelbert, S.; Kadim, A.: „Wirbelbetterwärmung von Platinen für das Presshärten" FOSTA Forschungsvereinigung Stahlanwendung e. V., 2012

[2] Tröster, T.; Niewel, J.: „Einsatz des induktiven Erwärmens von Platinen und Ermittlung der entsprechenden Prozessfenster für das Presshärten"; FOSTA Forschungsvereinigung Stahlanwendung e.V., 2014

[3] Schülbe, H.; Jestremski, M.; Nacke B.: „Numerische Untersuchung induktiver Erwärmungsprozesse für das Presshärten" in elektrowärme international. 2011/3: S. 283–88.

[4] Veit, R.; Kolleck, R.; Gasser, C.; Wilfinger, G.: "Development of a Close-to-production Prototype of an Induction Heating Device for Hot Stamping of Boron Alloyed Steels" in SCT 2011.

[5] Behrens, B.-A.; Hübner, S.: „Konduktive Erwärmung von Formplatinen für das Presshärten" EFB-Forschungsbericht Nr. 400, 2014.

[6] Vasily Ploshikhin et al.: „Flexible Wärmebehandlung zur gezielten Gestaltung von Bauteileigenschaften und zur Erhöhung der Energieeffizienz der Prozesskette Warmumformen (FlexWB)"; BMBF-Forschungsbericht 2012

[7] Jürgen Wibbeler: „Simulation Methods for Inductive Annealing of Steel" in ANSYS Conference & 33rd CADFEM Users' Meeting 2014 (Bremen, Germany)

CADFEM®

ANSYS

Simulation ist mehr als Software®

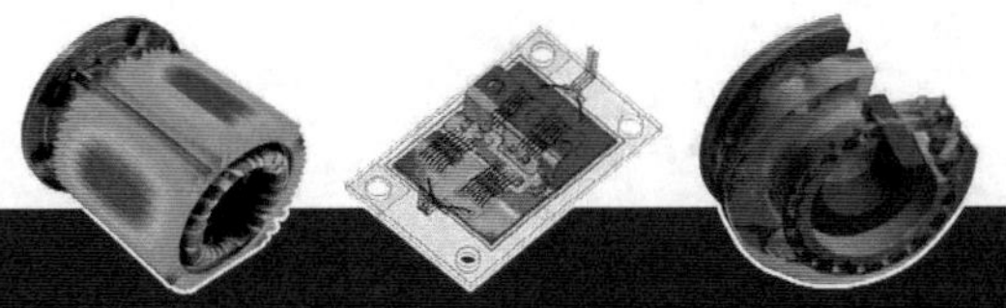

Topology optimization driven additive manufacturing at the example of a wheel rim

Dipl.-Ing. Markus Kellermeyer, CADFEM GmbH

Automotive Simulation World Congress 2016 , June 7-8 in Munich

CADFEM®

ANSYS

Table of contents

- Introduction
- ANSYS Topology Optimization
- Workflow – From Design Region to Additive Manufacturing
- Wheel Rim – Topology Optimization
- Lattice Structure for Wheel Rim

CADFEM

ANSYS

CADFEM International - 30+ companies, 400+ engineers, 15 countries

CAE Services:
Software Distribution, Support, Training, IT-Services, Consulting, Customization, Educational Programs

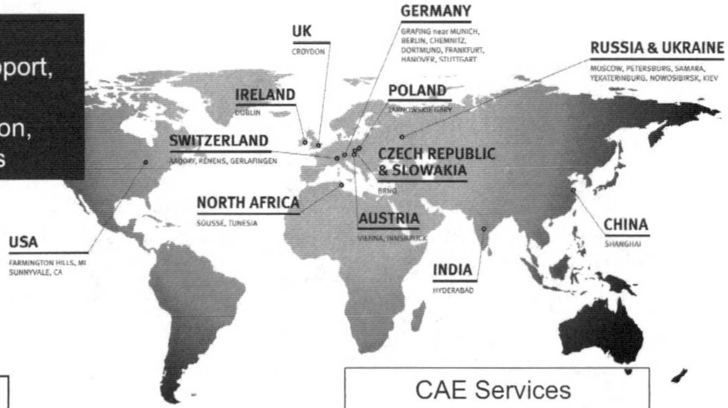

MicroConsult

ihf

initTech

AdCo ENGINEERING GW

dynardo

virtualcitySYSTEMS

CAE Services
ANSYS Channel Partner

Complimentary Services

Start-ups

Automotive Simulation WORLD CONGRESS

CADFEM

ANSYS

Introduction

Automotive Simulation WORLD CONGRESS

CADFEM

Introduction

- A different way to a good design

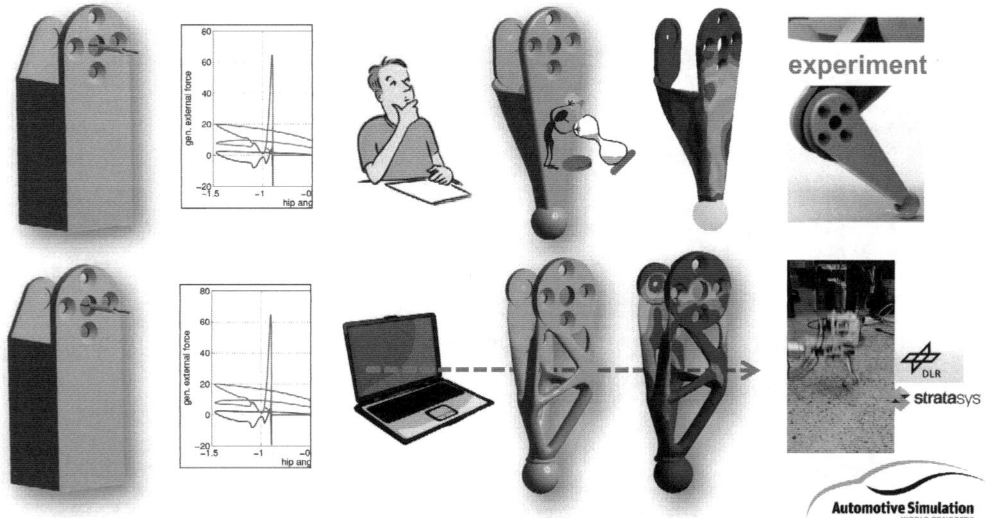

Automotive Simulation WORLD CONGRESS

CADFEM

ANSYS Topology Optimization

Automotive Simulation WORLD CONGRESS

CADFEM

ANSYS

ANSYS Topology Optimization

- The simulation process in ANSYS Mechanical

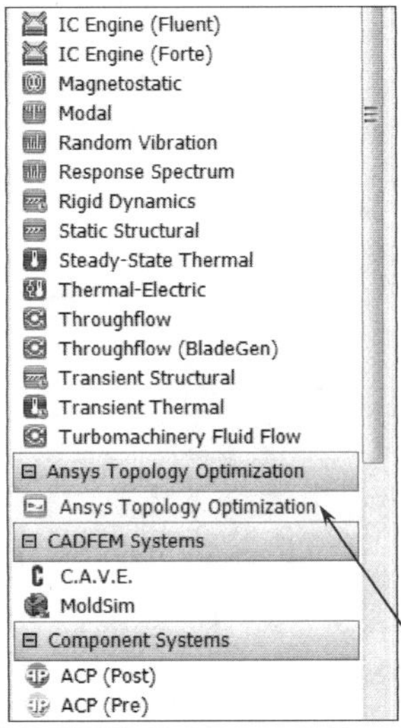

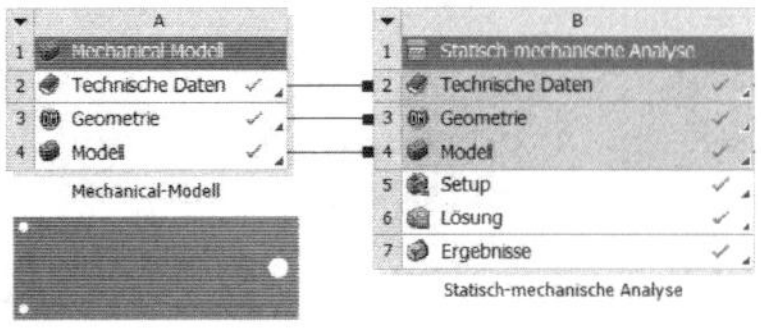

System:
ANSYS Topology Optimization

Automotive Simulation WORLD CONGRESS

CADFEM

ANSYS

ANSYS Topology Optimization

- Definitions in Topology Optimization (so far):

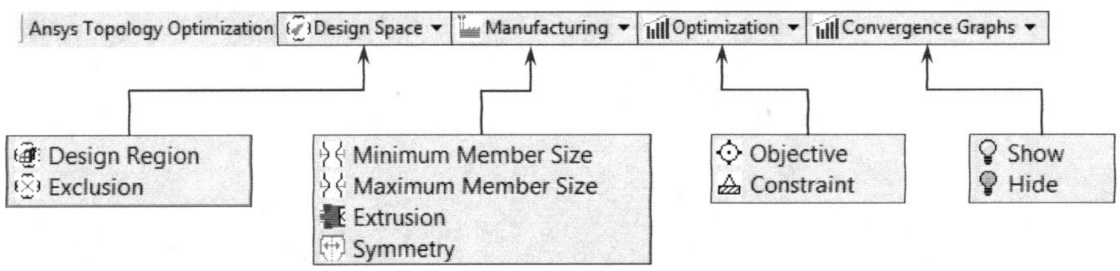

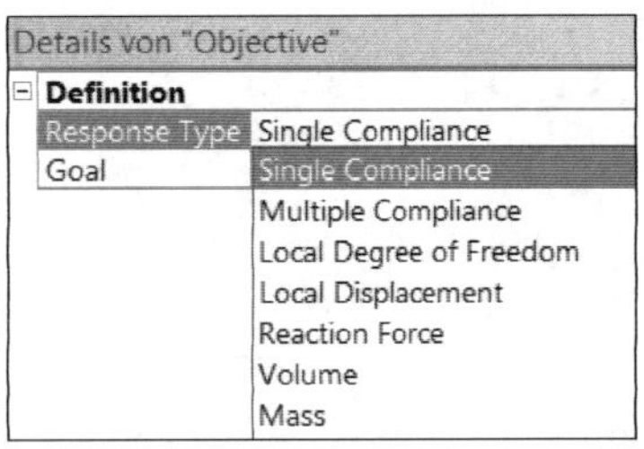

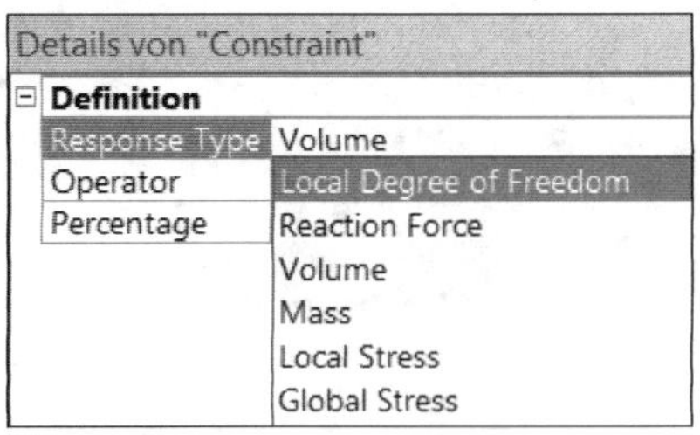

Automotive Simulation WORLD CONGRESS

CADFEM

ANSYS Topology Optimization

- How „Minimum and Maximum Member Size“ affect the resulting structure:

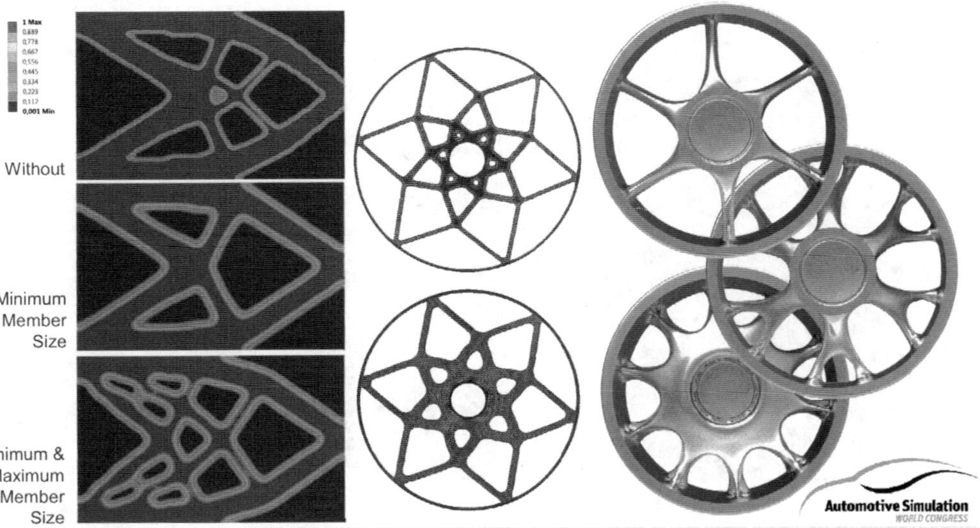

CADFEM

ANSYS Topology Optimization

- Postprocessing
  - Unimportant regions can be capped
  - Results can be displayed as averaged smooth values or as element values

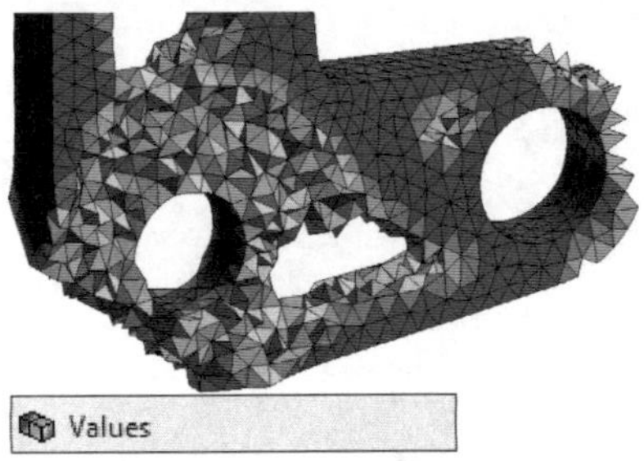

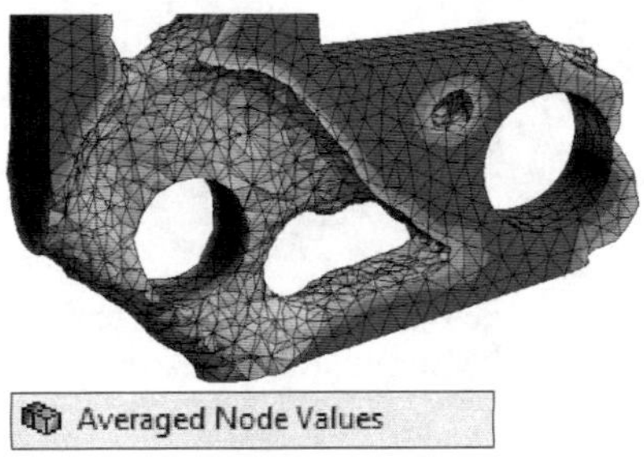

- The results can directly be exported as they are seen in the main window → STL

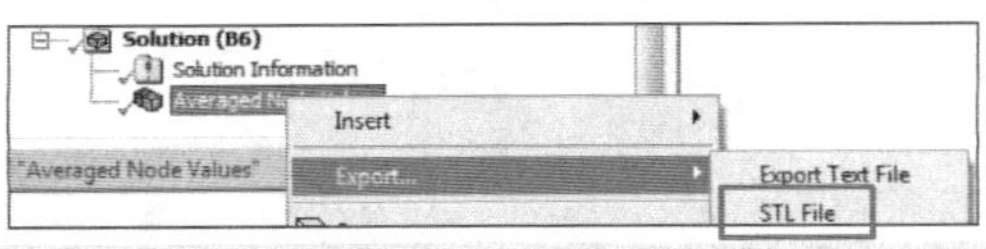

CADFEM

ANSYS Topology Optimization - STL and CAD preparation

- Combination of STL&CAD; Smoothing steps

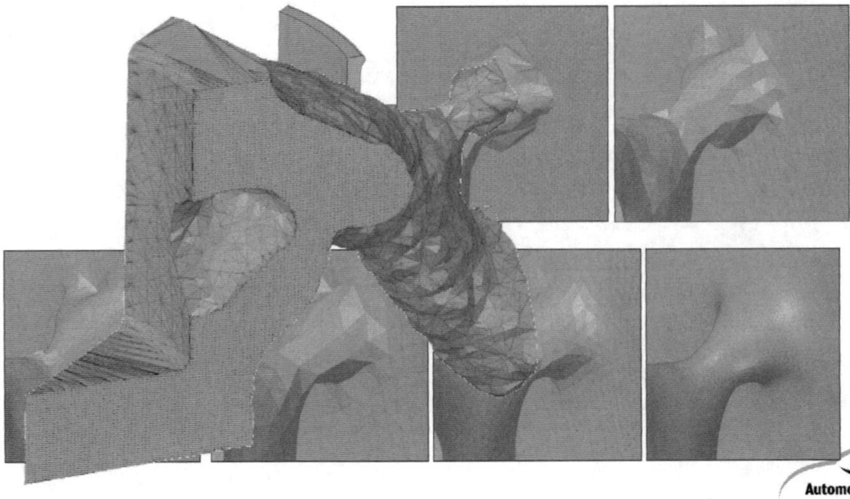

Automotive Simulation WORLD CONGRESS

CADFEM

ANSYS Topology Optimization - STL and CAD preparation

- 2 ways to CAD
  - Directly convert the STL to CAD (just one click but large CAD size), good for quick evaluation in the simulation

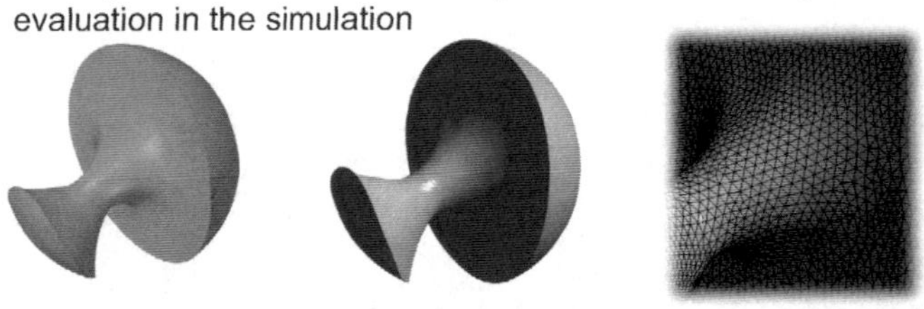

  - Define your own rules how the faces should be created (quick but manually, small CAD size), good for integration in existing assemblies

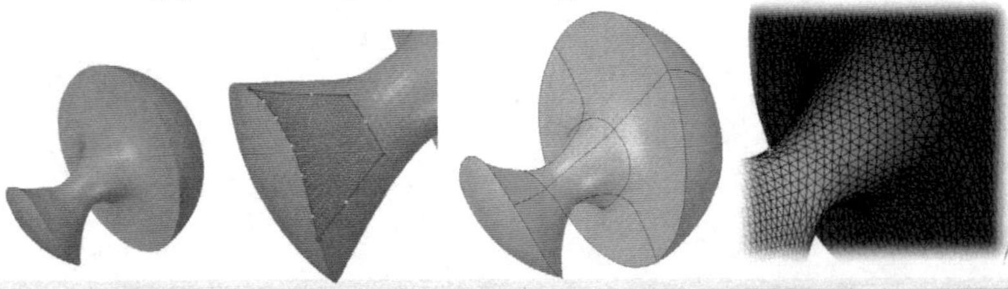

Automotive Simulation WORLD CONGRESS

CADFEM

ANSYS

Workflow – From Design Region to Additive Manufacturing

Automotive Simulation WORLD CONGRESS

CADFEM

ANSYS

Workflow – From Design Region to Additive Manufacturing

- Scheme for simulation and 3D Printing

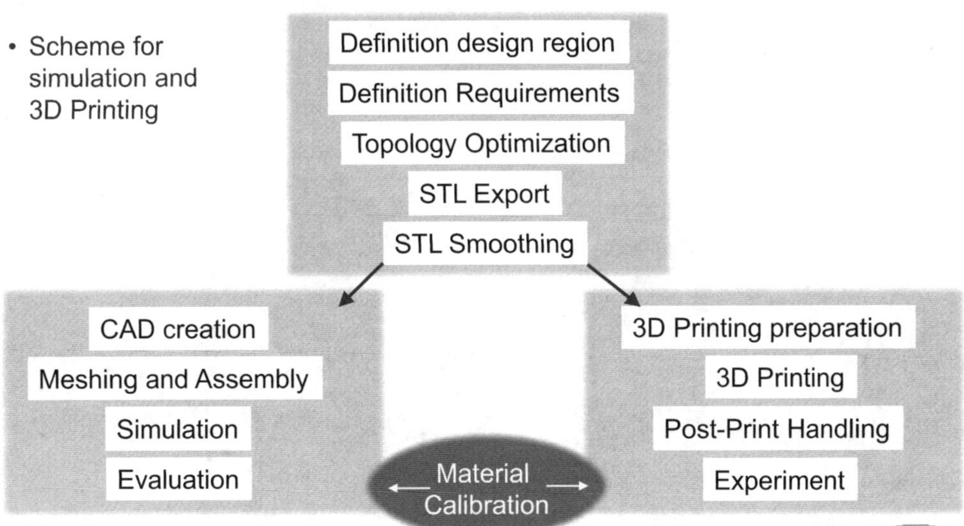

Automotive Simulation WORLD CONGRESS

June 7-8, 2016 Munich

Wheel Rim

Demonstrator: wheel rim

- Model setup:

load case 1

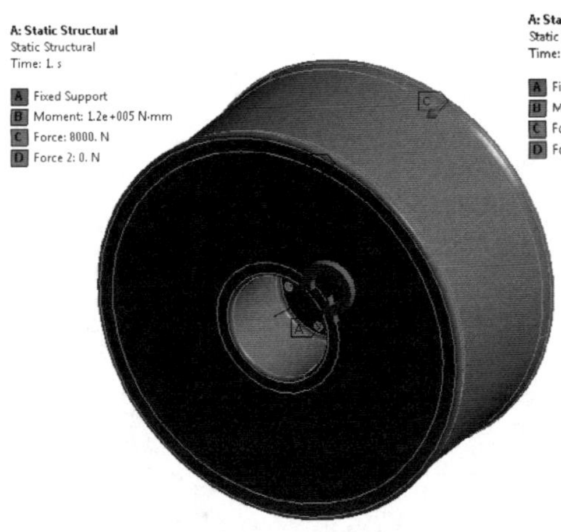

load case 2

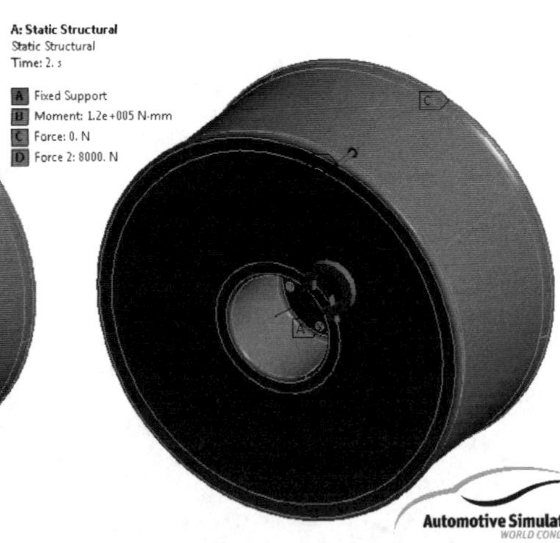

CADFEM

Wheel rim – design decision

- Different variants created through different manufacturing constraints

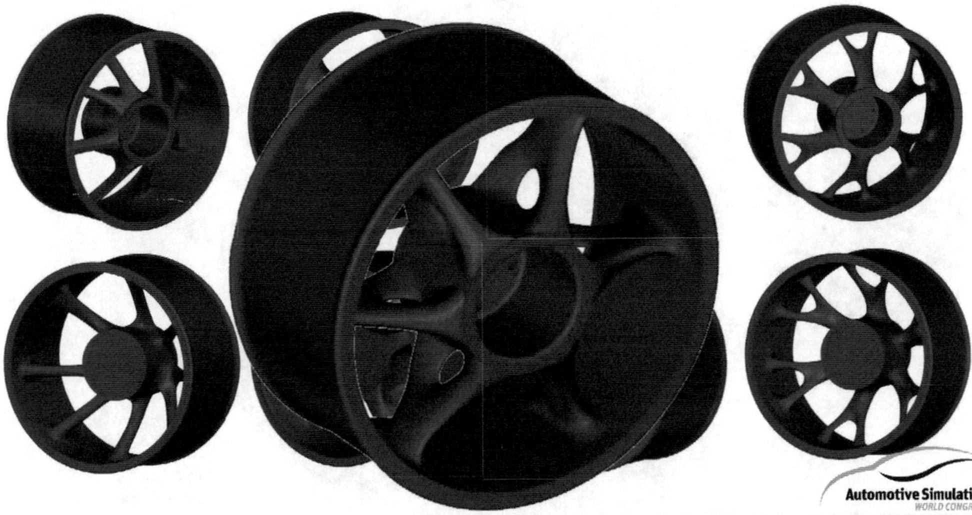

Automotive Simulation WORLD CONGRESS

17 June 7-8, 2016 Munich

CADFEM

Wheel rim – validation cycle

- Quick test for „Variant 3“

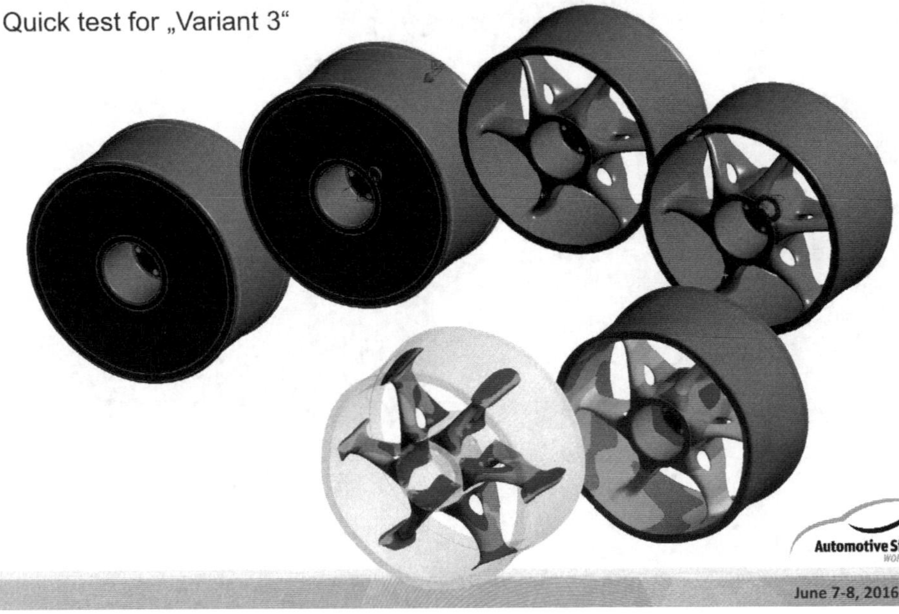

Automotive Simulation WORLD CONGRESS

18 June 7-8, 2016 Munich

CADFEM

Wheel rim – validation cycle

- Quick test for „Variant 4“

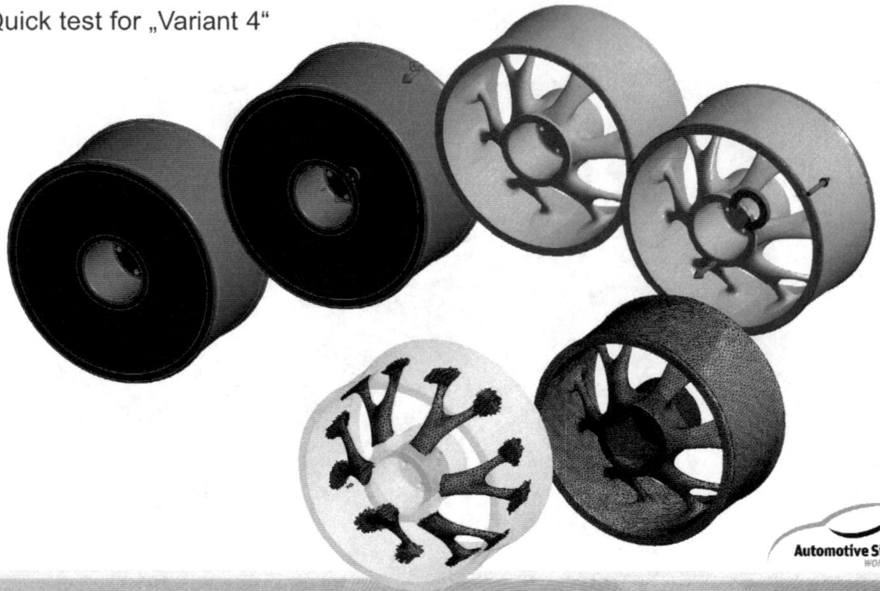

Automotive Simulation WORLD CONGRESS

June 7-8, 2016 Munich

CADFEM

Wheel rim – Chosen design

- Final variant – design region & result from topology optimization (and exclusions)

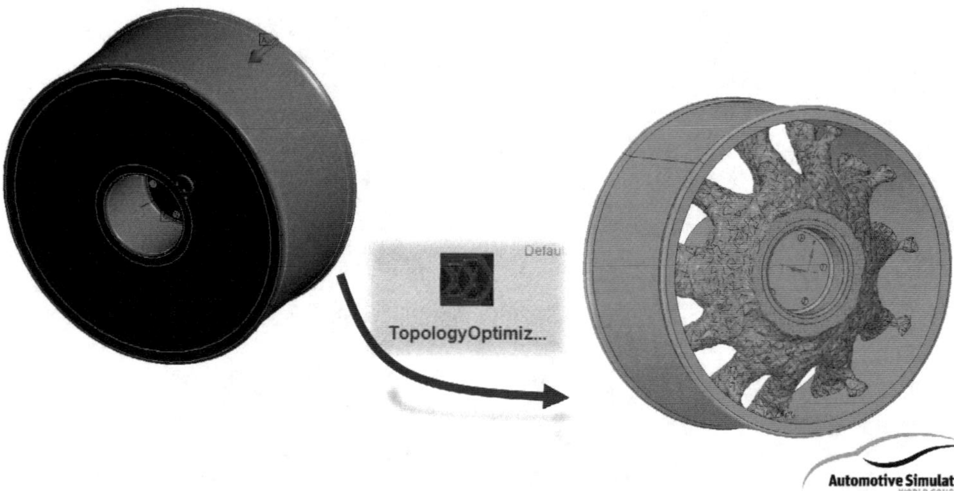

Automotive Simulation WORLD CONGRESS

June 7-8, 2016 Munich

CADFEM

Wheel rim - How to define regions for lattice

- Make it lighter; how to identify the regions that are less important

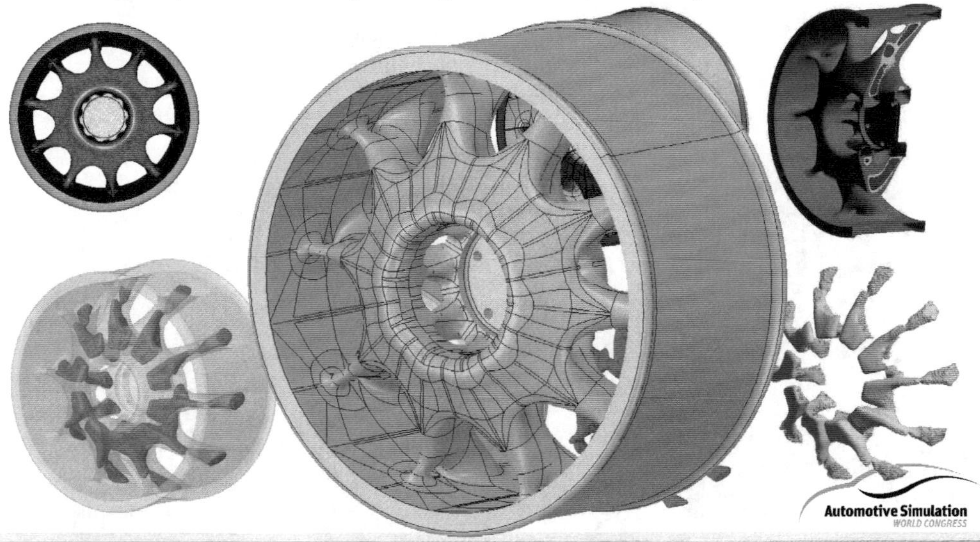

Automotive Simulation WORLD CONGRESS

21

June 7-8, 2016 Munich

CADFEM

Lattice Structure for Wheel Rim

Automotive Simulation WORLD CONGRESS

22

June 7-8, 2016 Munich

CADFEM

Wheel rim - Lattice structure

- How a lattice structure can be created

- Mesh nodes are connected by beam elements
- Meshing options can be used to control the density

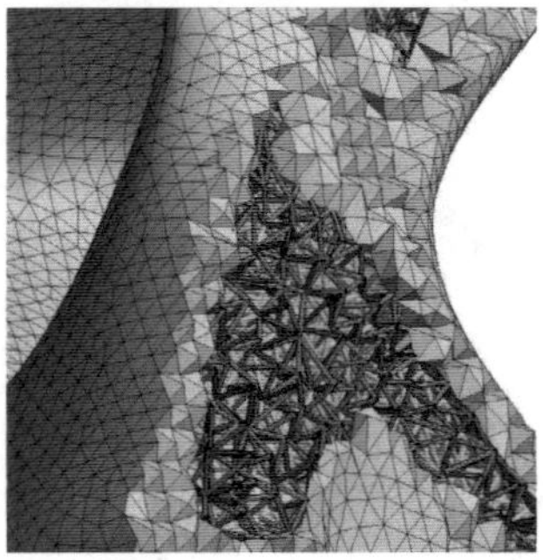

Automotive Simulation WORLD CONGRESS

CADFEM

Wheel rim - Lattice structure

- Stress Evaluation

Total Deformation

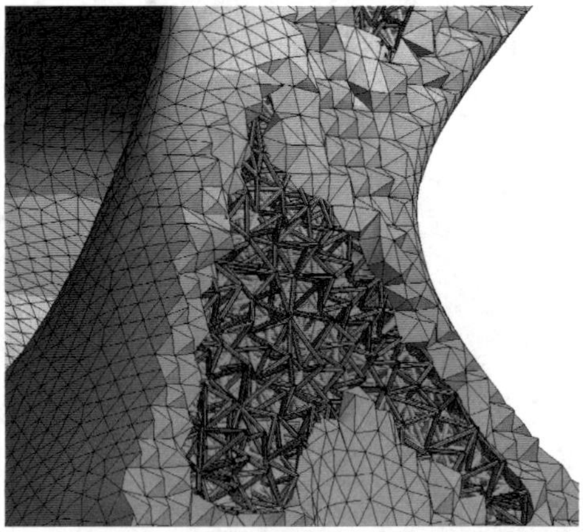

Automotive Simulation WORLD CONGRESS

CADFEM

Wheel rim - Lattice structure

- Lattice structure, touching the outer surface

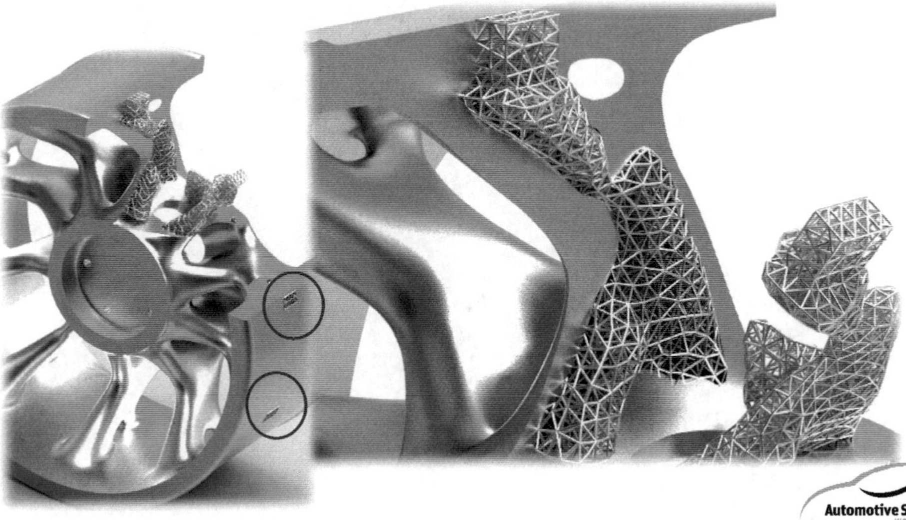

CADFEM

Wheel rim - Lattice structure

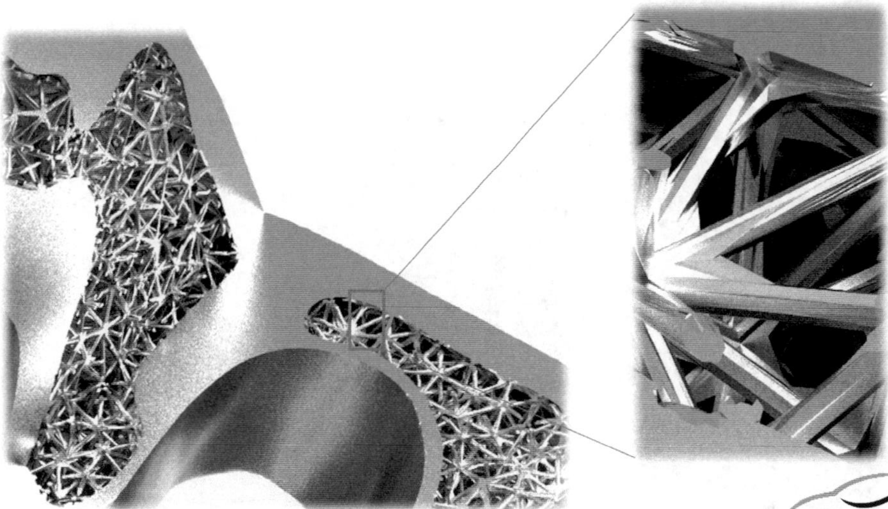

CADFEM ANSYS

Wheel rim - Evaluation of critical details

- Beam-to-Solid Submodeling

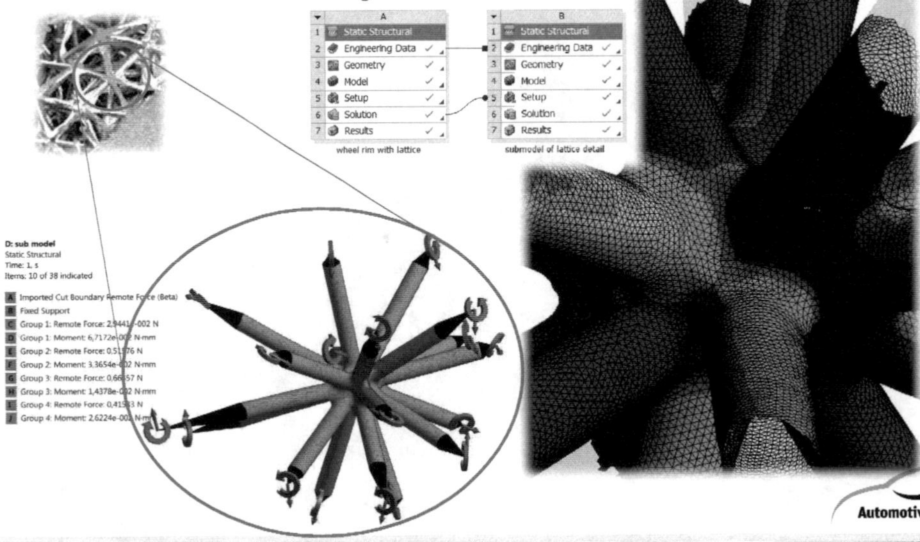

Automotive Simulation WORLD CONGRESS

CADFEM ANSYS

Wheel rim – Rendered wheel and rendered lattice structure

Automotive Simulation WORLD CONGRESS

CADFEM

Wheel rim – 3D Print

- Rendered vs. printed

printed by stratasys

Automotive Simulation WORLD CONGRESS

June 7-8, 2016 Munich

CADFEM

Wheel rim – 3D Print

- PolyJet-based 3D Print
  - Layers consisting of photopolymer are printed onto a building platform
  - Cures immideately upon exposure to UV light

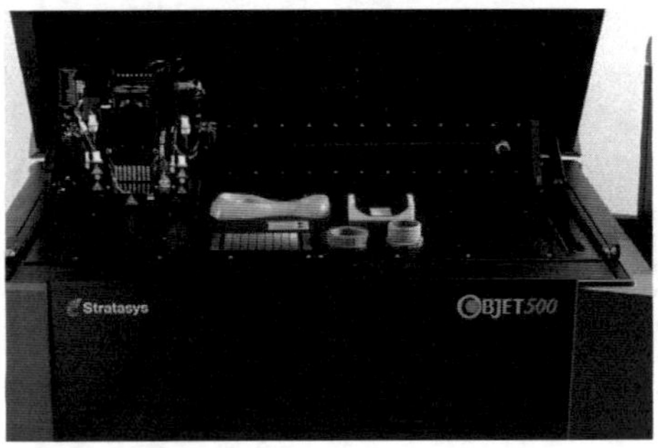

stratasys

Automotive Simulation WORLD CONGRESS

June 7-8, 2016 Munich

CADFEM®

ANSYS

Simulation ist mehr als Software®

Thank you for your attention

Markus Kellermeyer
mkellermeyer@cadfem.de

Automotive Simulation WORLD CONGRESS

31

June 7-8, 2016 Munich

Automotive Simulation WORLD CONGRESS

June 7-8, 2016 Munich

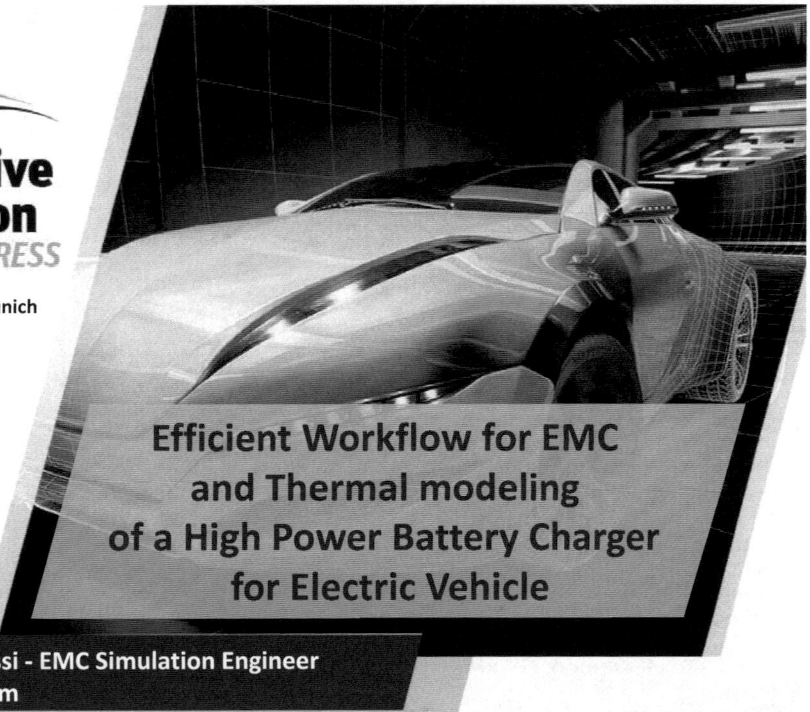

Efficient Workflow for EMC and Thermal modeling of a High Power Battery Charger for Electric Vehicle

Daniel Grossi - EMC Simulation Engineer
Meta System

EXAMPLE OF A HIGH POWER SWITCHING CONVERTER: BATTERY CHARGER FOR FULL ELECTRIC VEHICLE

DUT description:

up to 7kW Battery Charger

From 250 to 400 VDC output

Input 110/230 VAC

Efficiency up to 95%

PFC for AC input

LLC Type Converter for DC Output

Water cooling

WORKFLOW FOR EMC AND THERMAL MODELING

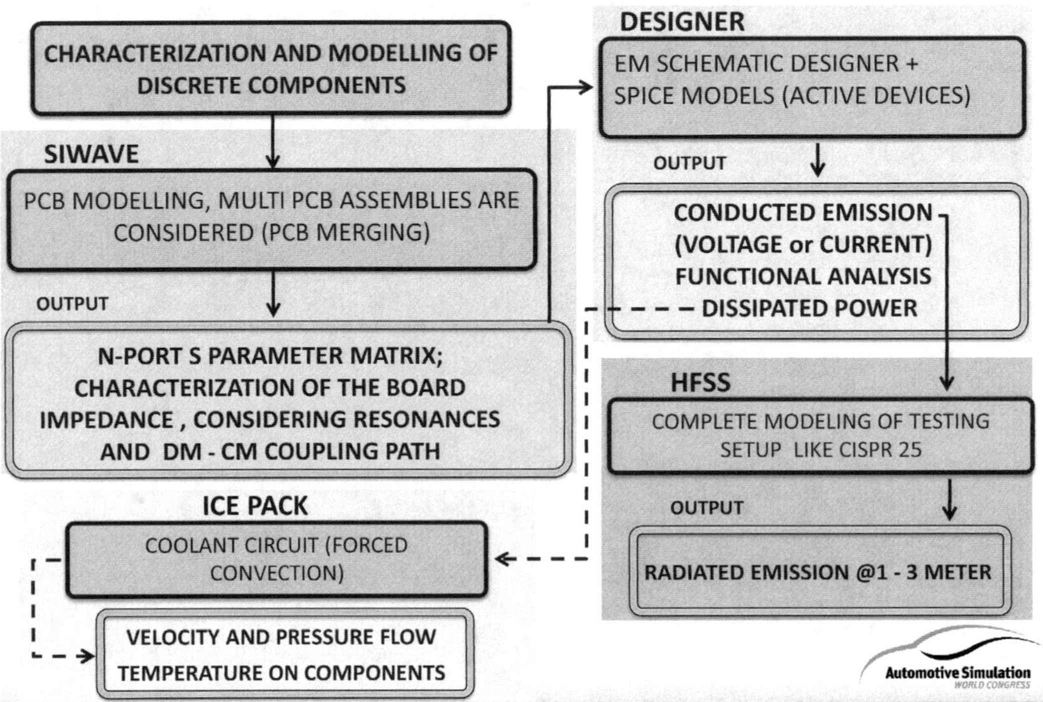

CUSTOM COMPONENT MEASUREMENT

Discrete and custom components (X and Y capacitors, common mode chokes and snubber resistors) are modeled with **Scattering parameters** while the Measurments are carried out with VNA (typical frequency range 10kHz-100MHz) **DC point and passivity check are needed.**

2 port component
Series connection

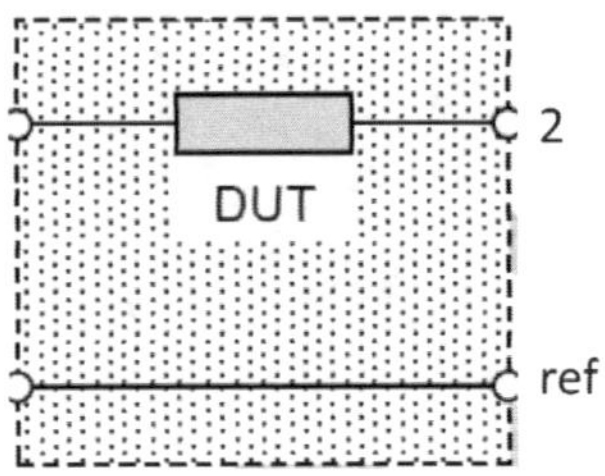

4 port component

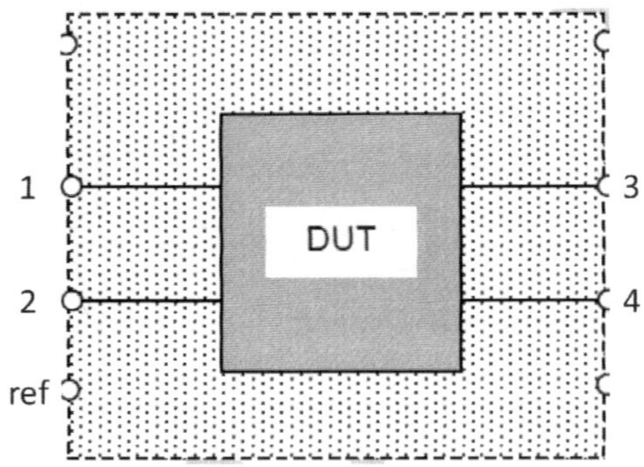

Reference: EN 55017 Methods of measurements of the suppression characteristics of passive EMC filtering devices

SIWAVE – PCB SOLVER

PCB in Siwave	
Important Tools	**Analysis to be carried out**
Import of real components (S parameter)	Filter attenuation, resonances (LNA)
Adjacent PCB board and enclosure effect **(PCB Merging)**	Impedance matrix at ports
	Near field from PCB

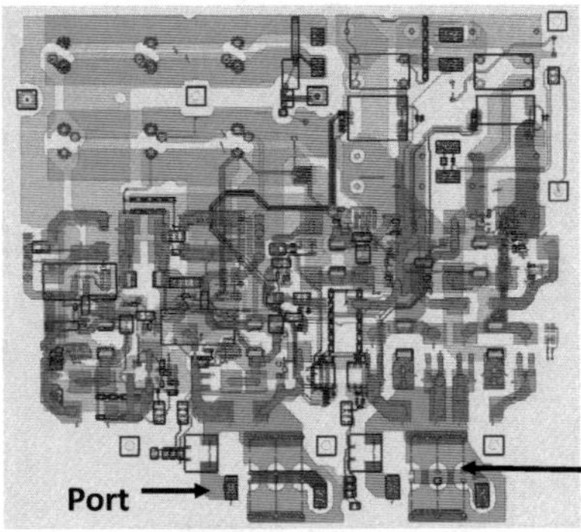

Ports are the connection to EM schematic software.

BATTERY CHARGER – PCB MERGING

The connection points, between boards or to with the enclosure, are modelled with **Via**

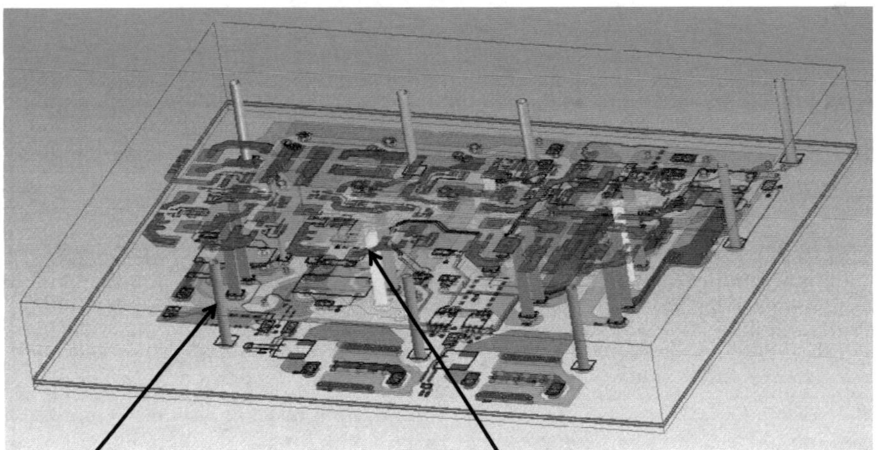

DESIGNER - EM SCHEMATIC

EM Schematic in Designer	
Important Tools	**Analysis to be carried out**
Import of active components (SPICE models)	Functional analysis
Effect of LISN and custom components (S parameters)	Conducted emission level
Linear and Transient solver	Sources for near field analysis

EM schematic considers the parasitic effect of PCB and frequency behavior of components.

Common mode return path is considered inside the Siwave PCB model

Currents at the ports are assigned as near field sources

Automotive Simulation WORLD CONGRESS

BATTERY CHARGER - EM SCHEMATIC

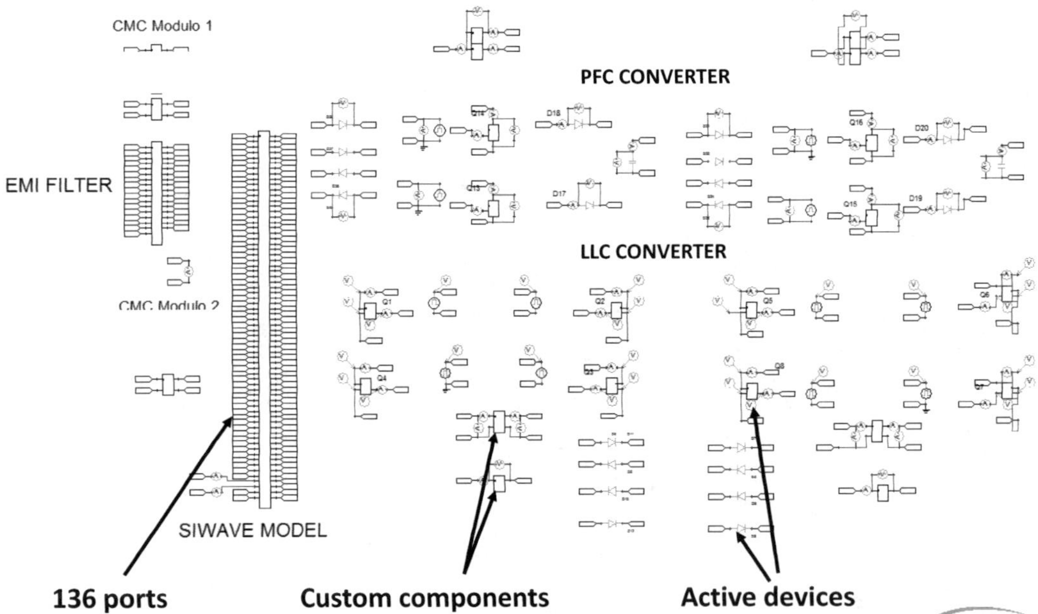

Automotive Simulation WORLD CONGRESS

HFSS – 3D COUPLING and RADIATED EMISSION

HFSS 3D layout - HFSS MCAD	
Important Tools	**Analysis to be carried out**
Import of Near field from SIWave	Internal field distribution
Internal coupling between field, cables and shield	Radiated emission
	Effect of internal radiated field on conducted emission

The precise modelling of the internally radiated field coupling can be carried out using HFSS.
Input for this type of simulation is the distribution of near field from Designer and Siwave.
Starting from the modelling of the testing setup and RF current on cables, the field level @ 1 meter or 3 meters can be estimated.

Automotive Simulation WORLD CONGRESS

BATTERY CHARGER – NEAR FIELD SOURCE AND 3D COUPLING

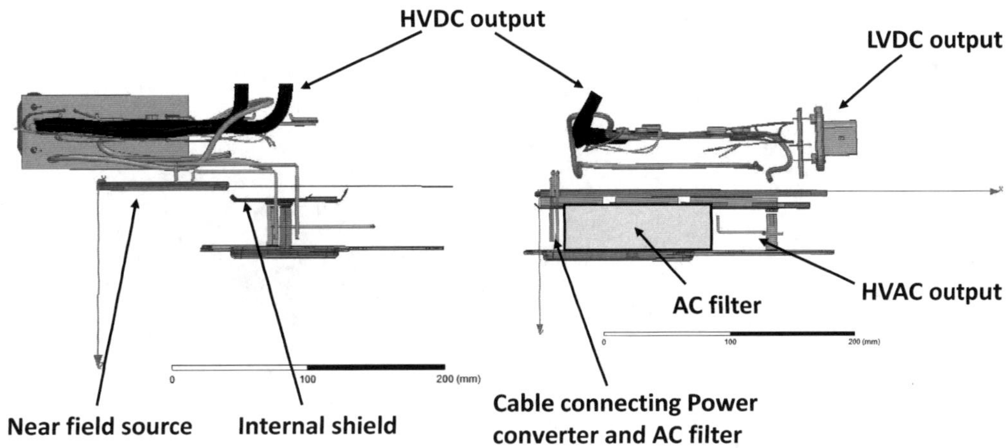

Automotive Simulation WORLD CONGRESS

COMPARISON BTW SIMULATION - MEASUREMENT

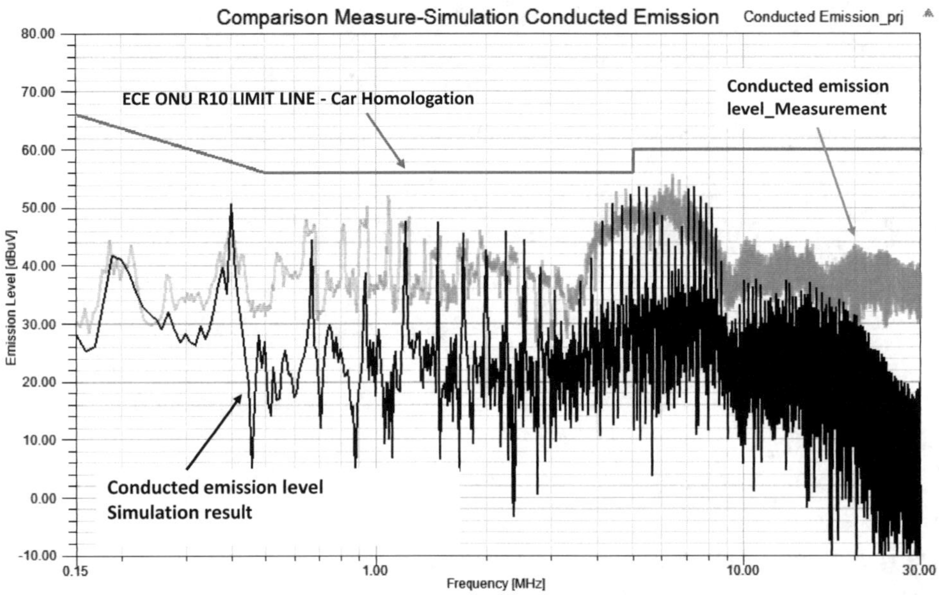

RESULTS: FUNCTIONAL ANALYSIS

LLC - Resonant current and gate voltage

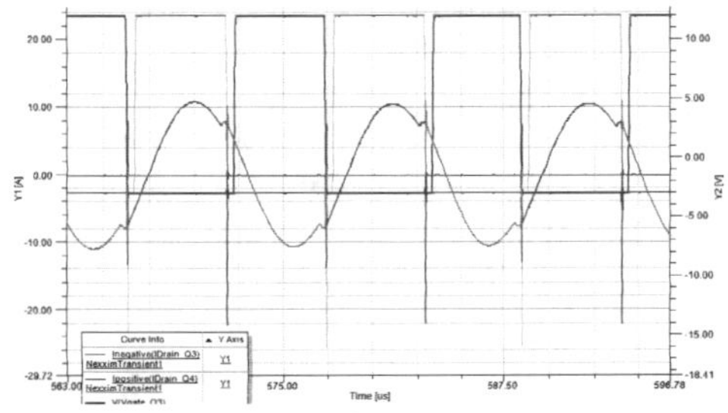

PFC – V bus

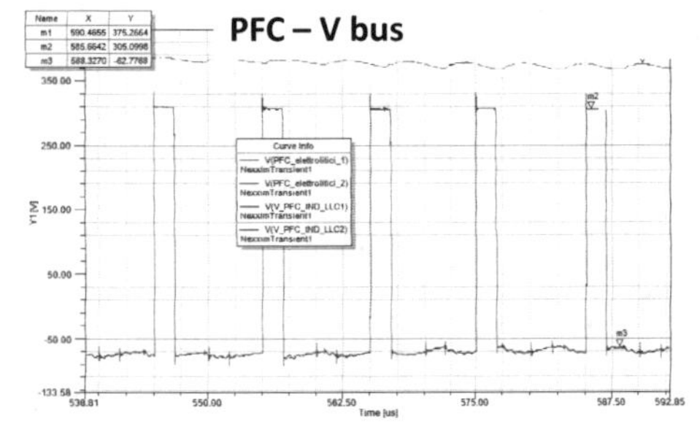

Signal level at input AC filter

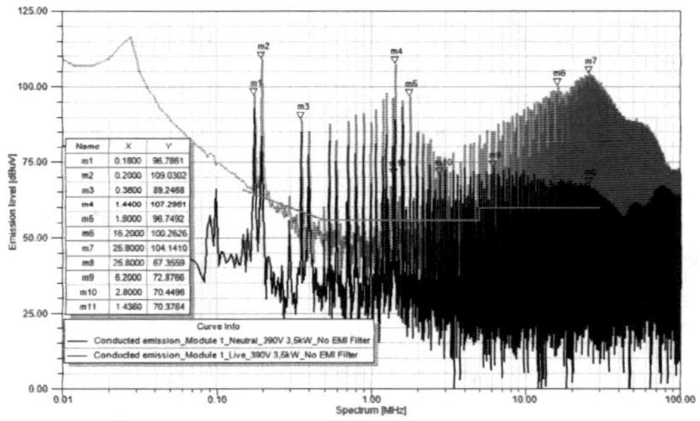

EMC PROBLEM SOLVING - CONDUCTED EMISSION

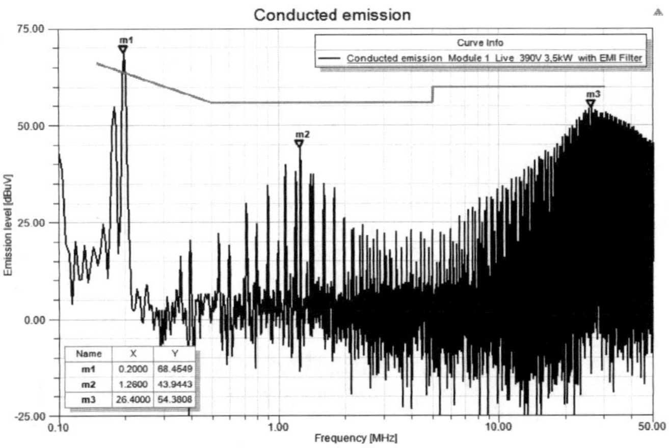

Graph shows the result of conducted emission simulation on HVAC output (230V mains connection).
@200kHz the harmonics of the PFC converter has a level that exceeds the limit line.
@1,25 and 25MHz,the conducted emission of the Converter show resonances quite near the limit line.

EMC PROBLEM SOLVING - INTERNALY RADIATED COUPLING

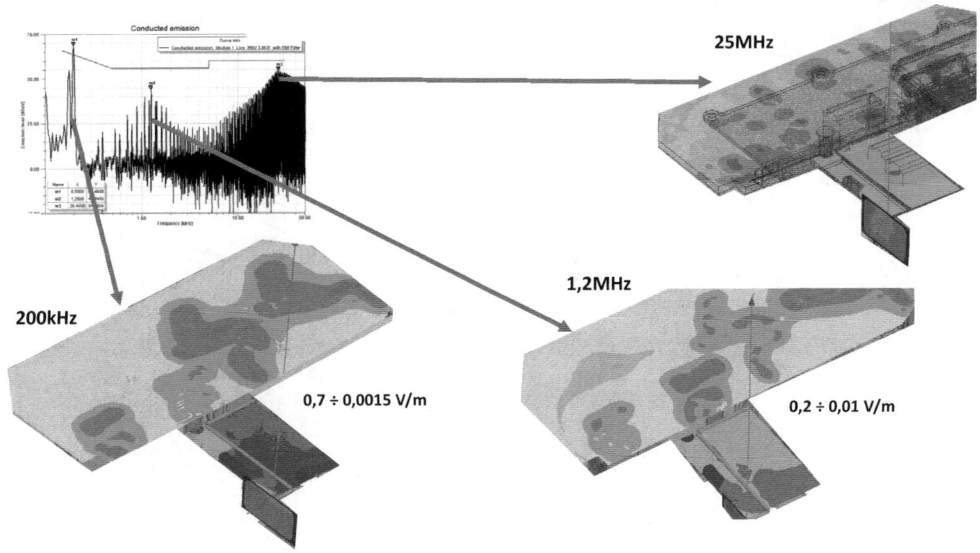

The coupling between the near field sources and output cables, have been calculated at 200kHz, 1,25MHz and 25MHz in order to evaluate the radiated field effect on the conducted emission simulation.

EMC PROBLEM SOLVING – AC FILTER NEW DESIGN

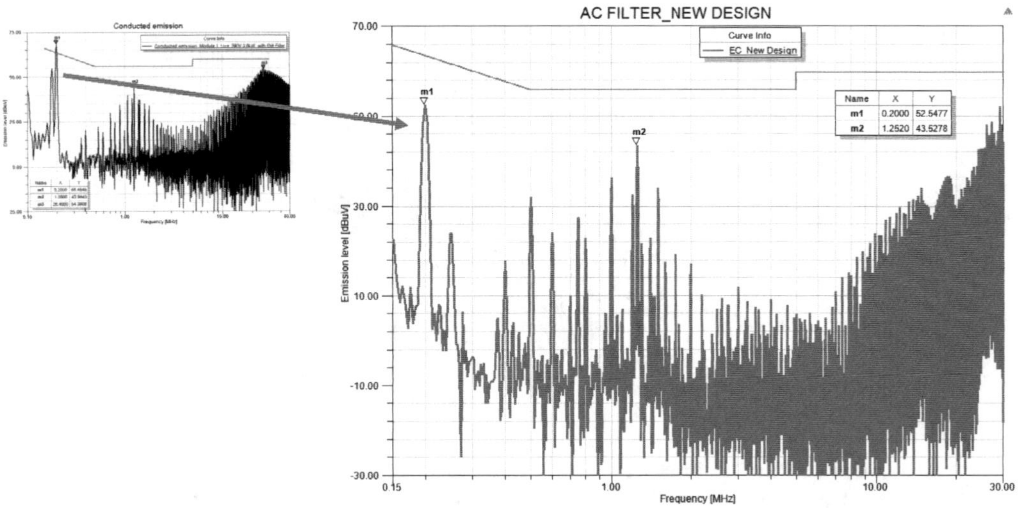

The Differential mode noise has been reduced with a DM filter stage
The number of Y Capacitors has been modified in order to lessen the ground path between noise source and the output.

EMC PROBLEM SOLVING – REDUCTION OF THE RESONANCE

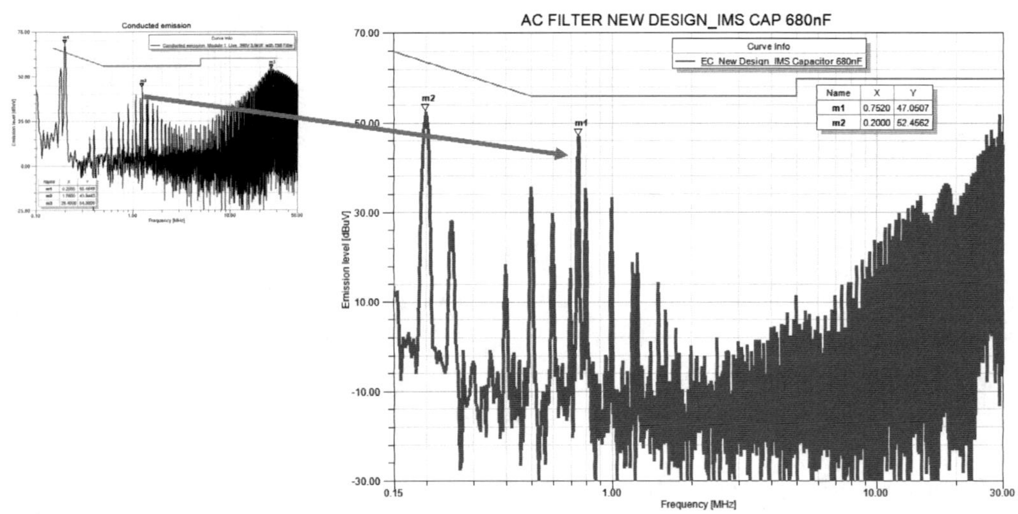

The interaction between ceramic capacitors and the parasitic elements of the boards create the resonance (@ 0,75MHz with 680nF capacitor).

EMC PROBLEM SOLVING – REDUCTION OF THE RESONANCE

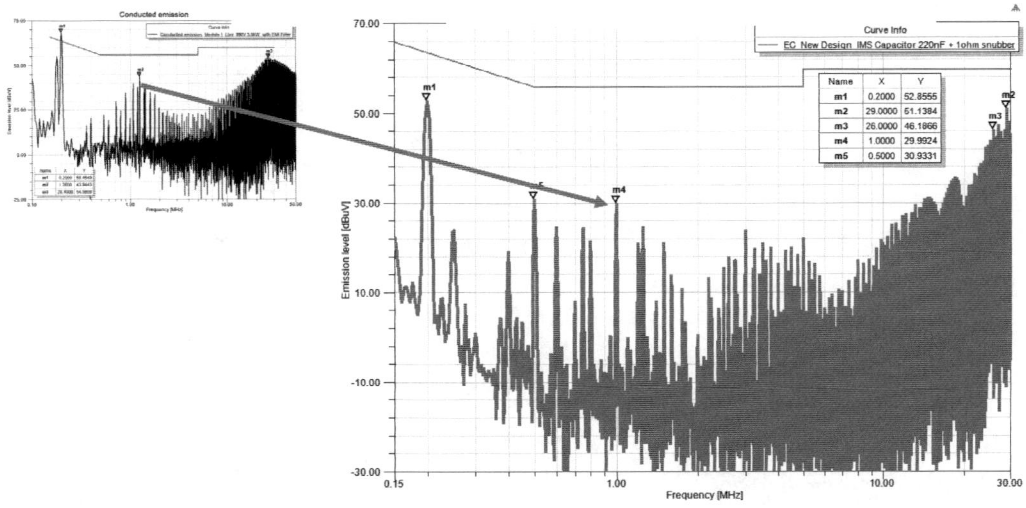

The proposed solution results in diminution of the resonance by more than 10dB. Not to mention the confirmation of this improvement through testing.

EMC PROBLEM SOLVING – CURRENT ON CAPACITORS

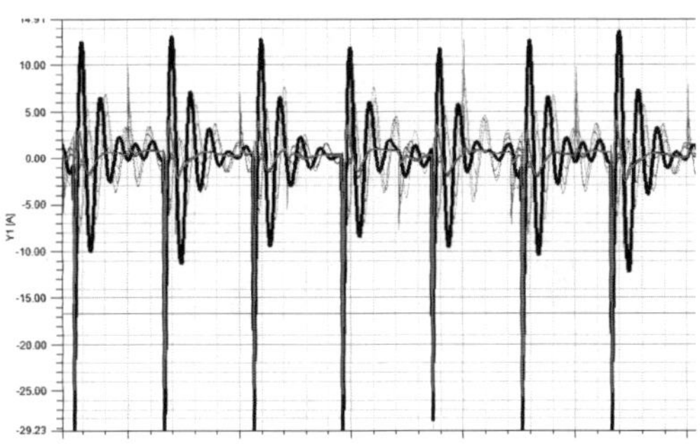

220nF

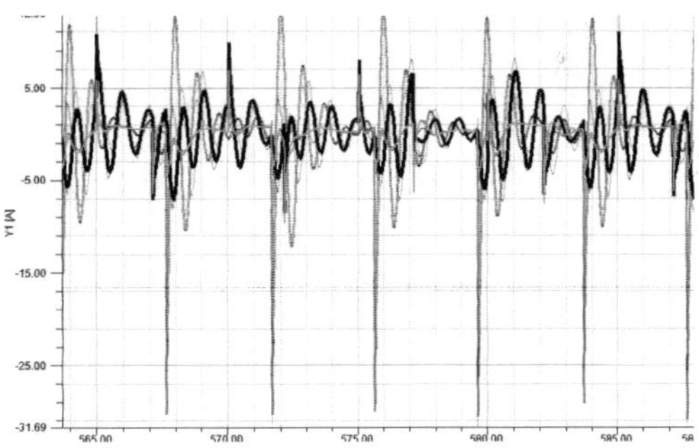

220nF + snubber resistor

RF current on capacitors and the dissipated power on the snubber resistor must be evaluated, in order to avoid any functional problems.

THERMAL ANALYSIS, TEMEPRATURE ON COMPONENTS: COMPARISON BTW MEASUREMENT AND SIMULATION

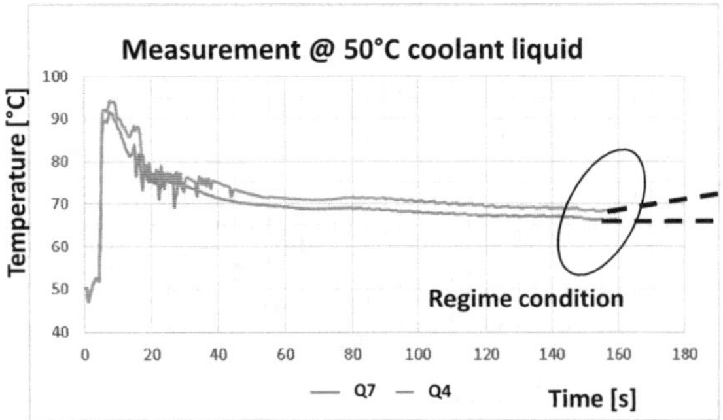

Simulation – Temperature on PCB and case

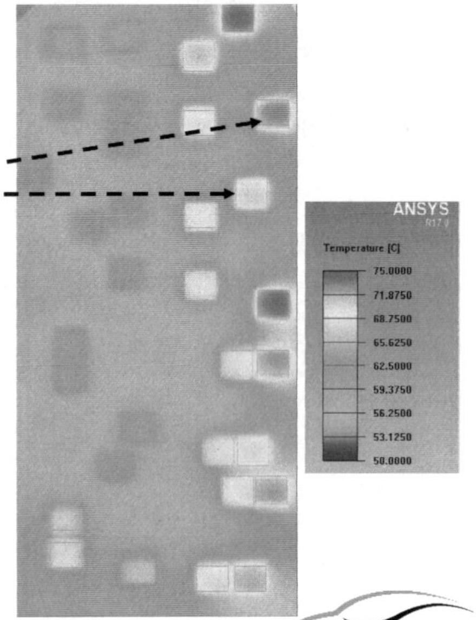

Component	Measured Temperature [°C]	Simulated temperature [°C]	ΔT [°C]
Q4	69	72	3
Q7	67	69	2

INTEGRATION BTW EMC SIMULATION @ COMPONENT LEVEL AND VEHICLE SIMULATION

The integration between EMC simulation at component level and that at Vehicle level is interesting for the HEV Manufacturers.

A complete overview of the EMC behavior, at Vehicle level, can be estimated considering the information from the simulation of OBC and the electrical environment of the Vehicle, such as the type of the cables (shielded or not), the routing of the cable, and grounding.

Needed information from component EMC simulation:

Voltage or Currents on the cables → **PWL Source, time domain**

EM field, closely to the OBC case → **ASCII format, Frequency domain**

Output impedance (Differential mode and Common mode impedance)
→ **SnP file, Frequency domain**

CONCLUSION

- The proposed workflow, for the functional and **EMC and Thermal analysis** of a **high power Battery Charger**, implicate the measurement of components and the simulation utilizing Ansys Electronic desktop software; it allows the **functional analysis** (voltages and currents inside the device), the **simulation of conducted emission along with the calculation of near field from PCB and the radiated emission along with the evaluation of temperatures on components.**

- The workflow is based on the following points**: the real behavior of passive components as a function of frequency,** the modelling of the noise path (**CM and DM coupling path**), the detection of the **resonances** and the **noise** generated by the active devices. The comparison with measurements demonstrates a good match; the **envelope of the noise** contains all the relevant harmonics and temperature on critical components are within a small deviation.

- The model has been used for **EMC problem solving, with a significant cutback in the testing time.**

- The planned next step incorporates the i**ntegration of the workflow with Simplorer, for a rather upgraded and elaborated modelling of active devices (Characterization tool). Coupling of Simplorer and Ice pack, for Thermo-Functional simulation and Worst Case Analysis for EMC.**

Automotive Simulation WORLD CONGRESS

21 June 7-8, 2016 Munich

Automotive Simulation WORLD CONGRESS

June 7-8, 2016 Munich

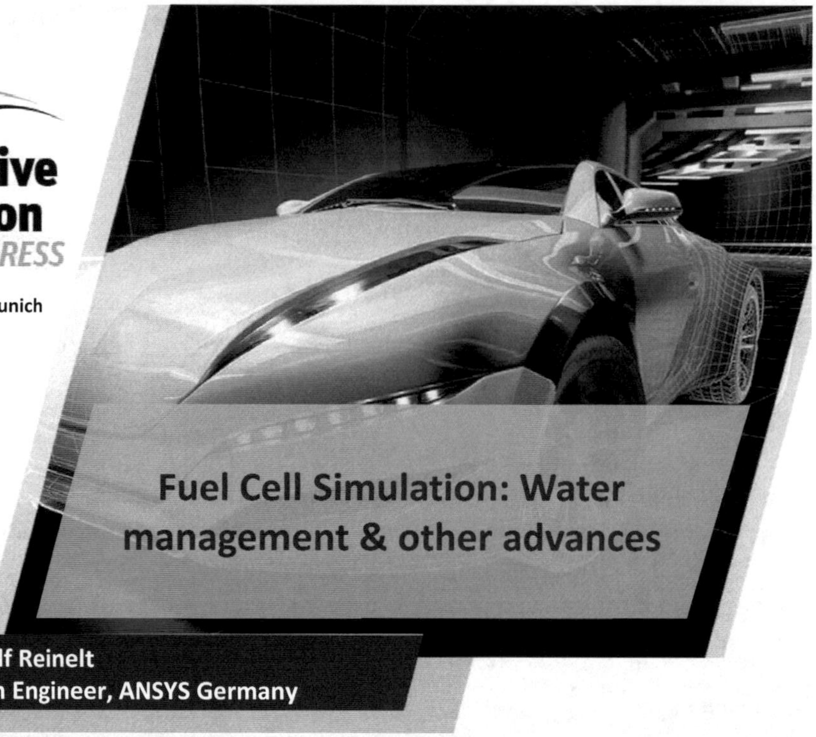

Fuel Cell Simulation: Water management & other advances

Dr.-Ing. Rolf Reinelt
Application Engineer, ANSYS Germany

Motivation

	ICE	Battery	Fuel Cell
Infrastructure	+	o	-
Price	+	o	-
Range	+	o	+
Loading time	+	o	+
CO2	-	+	+

How a fuel cell works

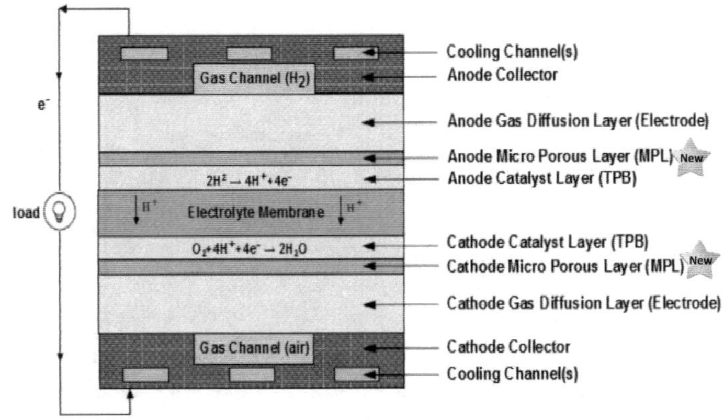

- **Membrane can transport protons H+, but blocks gases H2 and O2**
- **Oxydation of hydrogen is split in 2 reactions**
- **External transport of electrons can be used to drive the car**

Challenges

- **3D**
- **Highly nonlinear**
- **Species, Flow, Temperature, Liquid Water, Dual Potentials**
- **Water management**
  - **Membrane needs to be wet to transport protons**
  - **Condensated water blocks precious metals and blocks transport of gases**
- **Transport of gases and electrons in the same porous media**
- **Reactions and condensation sensitive to temperature**
- **Fuel cell needs to be compact and lightweight**

Highlights of 17.0

- **R&D since 2013**
- **Cooperation with external sponsor**
- **Improved water modeling**
  - Micro Porous Layer (MPL)
  - Convergence rates
  - Accuracy
- **New (and old) model is included now in Fluent License**

Fluent Addon Modules
0. None
1. MHD Model
2. Fiber Model
3. Fuel Cell and Electrolysis Model
4. SOFC Model with Unresolved Electrolyte
5. Population Balance Model
6. Adjoint Solver
7. Single-Potential Battery Model
8. Dual-Potential MSMD Battery Model
9. PEM Fuel Cell Model
10. Macroscopic Particle Model

Automotive Simulation WORLD CONGRESS

New Water Management Concept

➢ Water can exist in 3 forms
- Dissolved in the ionomer
- Water vapor in the porous media and gas channels
- Liquid water in the porous media and gas channels

➢ Electrochemical reaction in cathode catalyst layer generates water in dissolved phase

➢ Deviation from equilibrium as driving force for phase changes

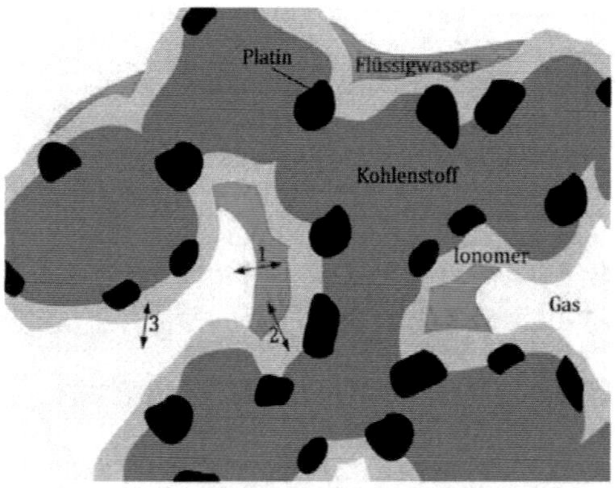

- 1: Evaporation/Condensation (source term: S_{gl})
- 2: Absorption from/desorption to liquid phase (S_{ld})
- 3: Absorption from/desorption to gas phase (S_{gd})

H. Scholz *Modellierung und Untersuchung von Flutungsphänomenen in Niedertemperatur-PEM-Brennstoffzellen*

Automotive Simulation WORLD CONGRESS

Water model

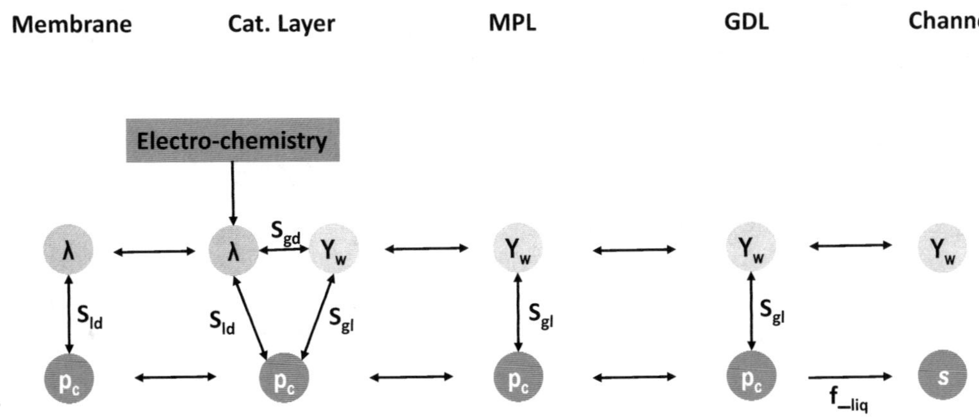

λ = water content p_c = capillary pressure Y_w = mass fraction of water specie

Liquid Water Transport in Porous Media

- **Implemented new liquid water equation in Porous Media**
  - **--- solve capillary pressure equation in MEM, CAT, MPL, GDL**
  - **--- determine saturation from p_c-s relations.**

$$\frac{\partial}{\partial t}(\varepsilon_i \rho_l s) = \nabla \cdot \left(\frac{K K_{rl}}{\nu_l} \nabla (p_c + p_g) \right) + S_{gl} - S_{ld}$$ New

Gas-to-Liquid New

$$S_{gl} = \begin{cases} \gamma_{gl} \varepsilon s D_{gl} \frac{M_{w,H_2O}}{RT} p \ln\left(\frac{p-p_{sat}}{p-p_{wv}}\right), & p_{wv} <= p_{sat} \\ \gamma_{gl} \varepsilon (1-s) D_{gl} \frac{M_{w,H_2O}}{RT} p \ln\left(\frac{p-p_{sat}}{p-p_{wv}}\right), & p_{wv} > p_{sat} \end{cases}$$

$$f_{liq} = \max\left[\Theta \varepsilon s p_c, 0\right]$$

from GDL to channel

$$p_c = \begin{cases} \sigma \cos\theta_c \sqrt{\frac{\varepsilon}{K}} J(1-s) & if\ \theta_c < 90^o \\ -\sigma \cos\theta_c \sqrt{\frac{\varepsilon}{K}} J(s) & if\ \theta_c > 90^o \end{cases}$$

$$J(x) = 1.417x - 2.12x^2 + 1.263x^3$$

Liquid Water Transport in Gas Channels (optional)

- **Solve liquid saturation** in gas channels **to obtain meaningful values to model** pressure drop **through viscous momentum resistance, using a user-modifiable function in** *pemfc_user.c*

$$\frac{\partial}{\partial t}(\rho_l s) + \nabla \cdot (\rho_l \vec{v}_l s) = \nabla \cdot (D_{liq} \nabla s) \qquad \vec{v}_l = \chi \vec{v}_g$$

default resistance: $R_{momentum} = f(s) = \alpha s^{\beta}$

liquid saturation in cathode channel

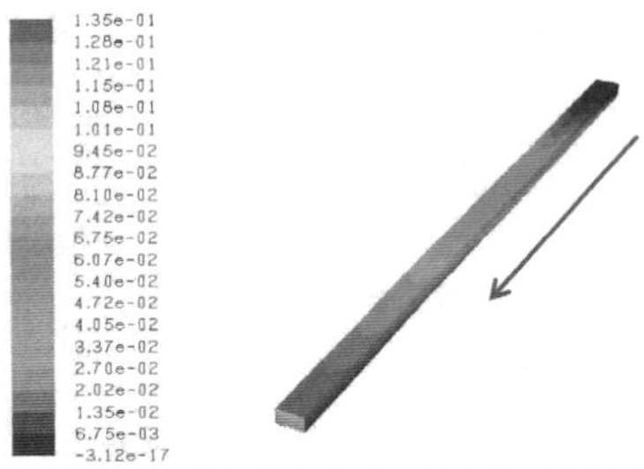

static pressure

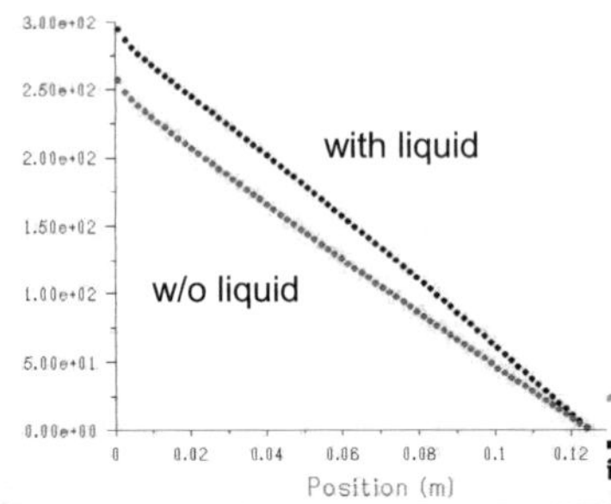

Dissolved Phase Water Transport in MEA

- **Transport equation for** dissolved (3rd phase) **water in MEA (catalyst-membrane-catalyst assembly) (Wu 2009)**

$$\frac{\partial}{\partial t}\left(\varepsilon_i M_{w,H2o} \frac{\rho_i}{EW} \lambda\right) + \nabla \cdot \left(M_w \frac{n_d}{F} \vec{i}_m\right) = \nabla \cdot \left(M_w \frac{\rho_l}{EW} D_w \nabla \lambda\right) + S_\lambda + S_{gd} + S_{ld}$$

Reaction $S_\lambda = \frac{M_{W,H2O}}{2F} R_{cat}$

Old: $\nabla \cdot \left(M_w \frac{\rho_l}{EW} D_w \nabla \lambda\right) = 0$ solved only in membrane

--- **new** osmotic drag **treatment through the entire MEA**
--- **water is produced in** dissolved **phase (*rather than vapor phase*)**
--- mass transfer **with other two phases:** vapor **and** liquid **(absorption/desorption)**
--- **mass conservation is always observed**

Addition of Micro Porous Layer (MPL)

➢ **Include MPL as separate layers with different properties from GDLs**

New

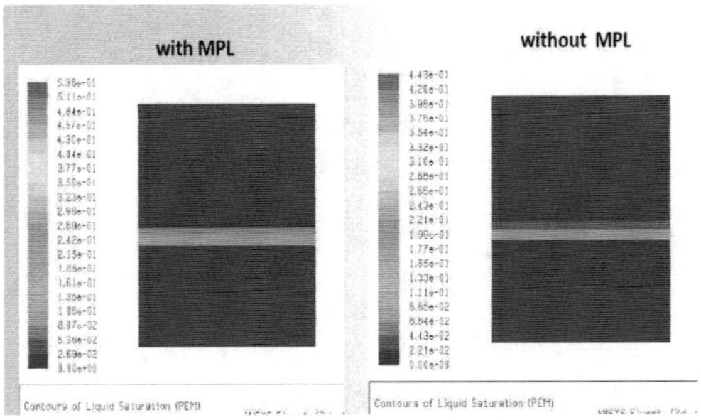

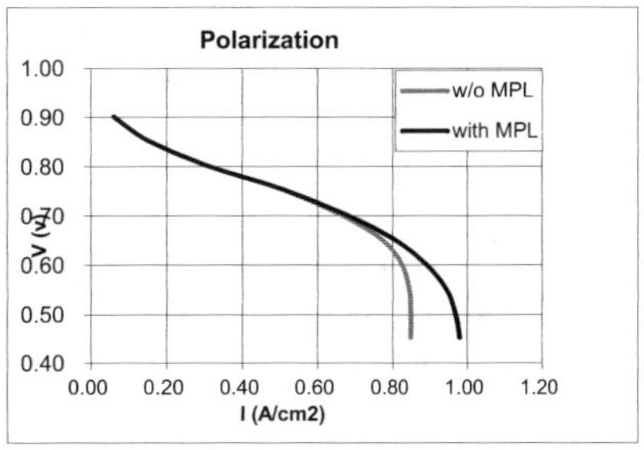

Automotive Simulation WORLD CONGRESS

Other Improvements

1. **Calculation of thermal conductivity for triple-phase fluid**
2. **Computing Half-Cell voltages using Nernst Equation**
3. **Variable coolant properties**
4. **Use Entropy to compute reaction heat**
5. **Variable phase change rates**
6. **Thermal contact resistance**
7. **Cathode Particle resistance model**
8. **Temperature-dependent transfer current**

Automotive Simulation WORLD CONGRESS

Model Validation

- **global & local polarizations**
- **current density distribution**

50 cm² MEA of Mench et al (2004)

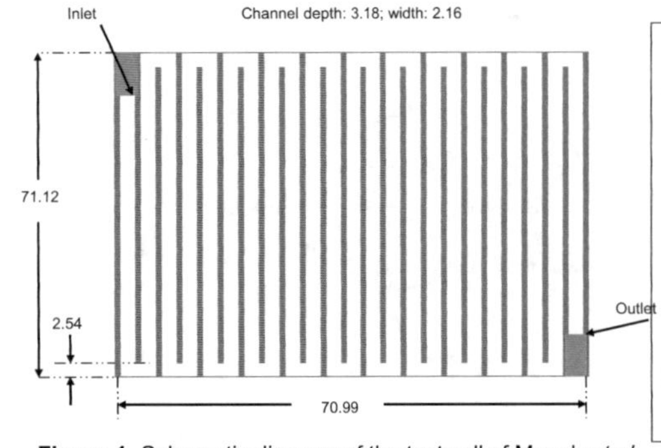

Figure 1. Schematic diagram of the test cell of Mench *et al* [1]: all numbers in *mm*

- 48 gold-plated steel current collecting ribs embedded into an insulating slab
- $T_{cell} = 80\ ^{o}C$; $P = 1.5 atm$; $T_a = 90^{o} C$; $T_c = 80^{o} C$
- $\xi_{an} = 1.875$; $\xi_{ca} = 1.5,\ 2.25$
- mass flow rate was kept constant in each cathode stoichiometry run

Workbench

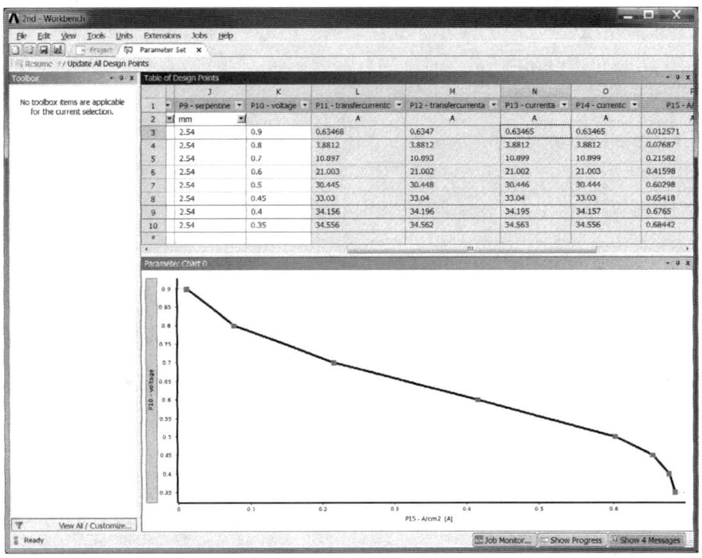

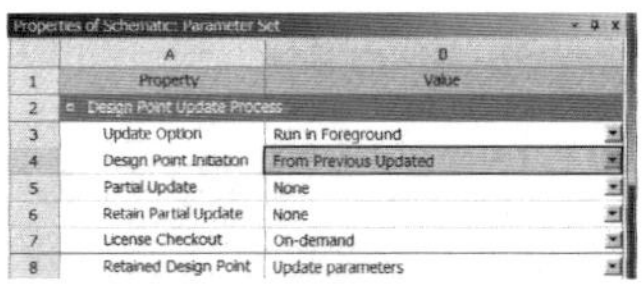

Species Mass Fractions in Gas Channels

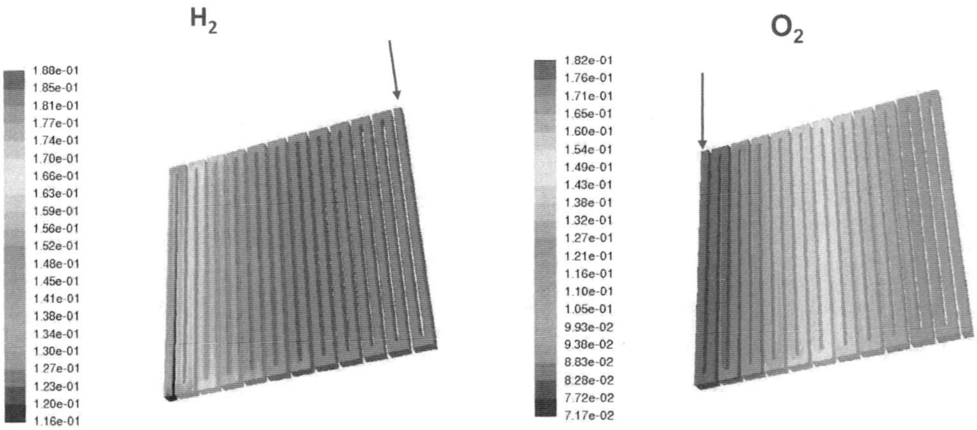

$\xi_{ca} = 1.5 \qquad V_{cell} = 0.45$

Automotive Simulation WORLD CONGRESS

Liquid Saturation at Different Cell Voltage

location: mid-plane of cathode gas diffusion layer

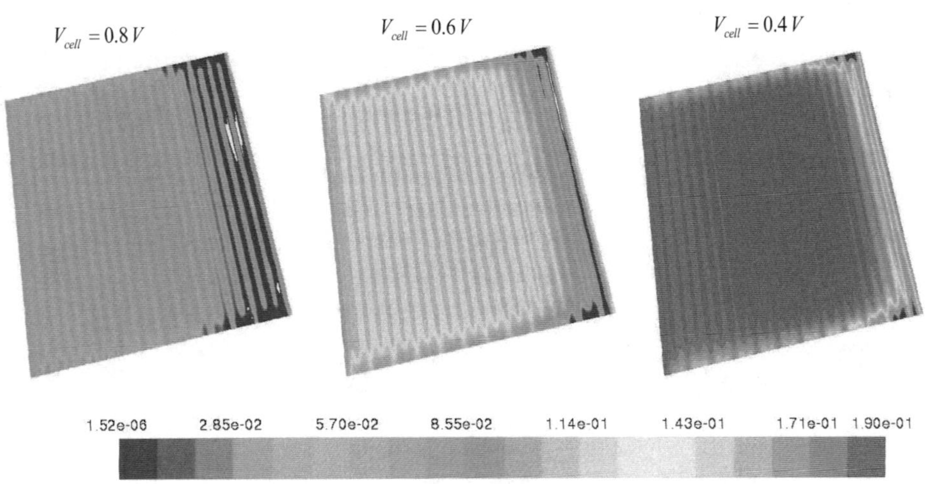

$\xi_{ca} = 1.5$

Automotive Simulation WORLD CONGRESS

Global Polarization

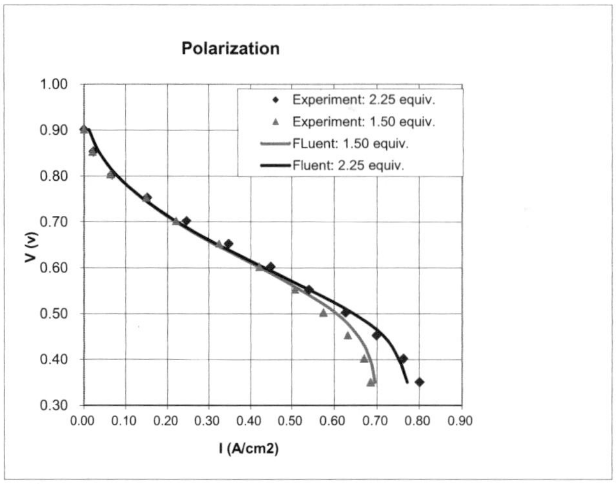

Computed (lines) and measured (symbols) global polarization curves for cathode stoichiometry of 1.5 and 2.25 equiv.

Local Polarization

1.5

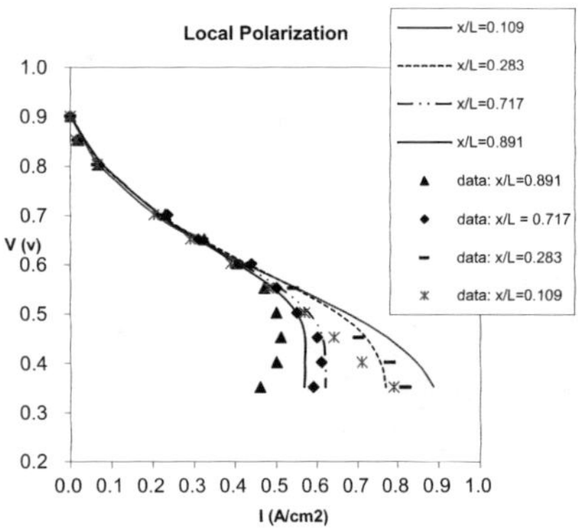

Computed (lines) and measured (symbols) local polarization curves for cathode stoichiometry of **1.5 equiv**.

2.25

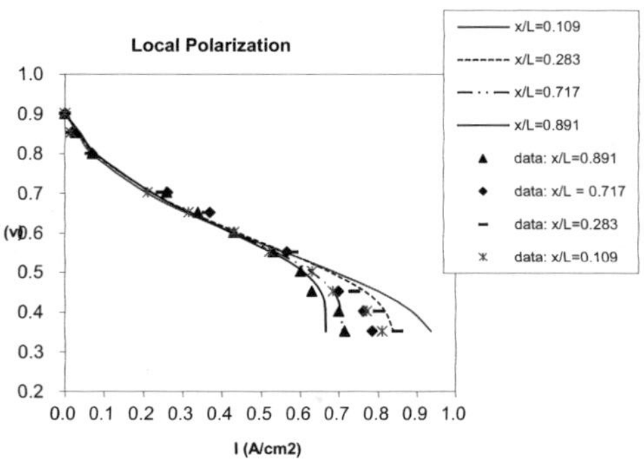

Computed (lines) and measured (symbols) local polarization curves for cathode stoichiometry of **2.25 equiv**.

Local Current Density

1.5

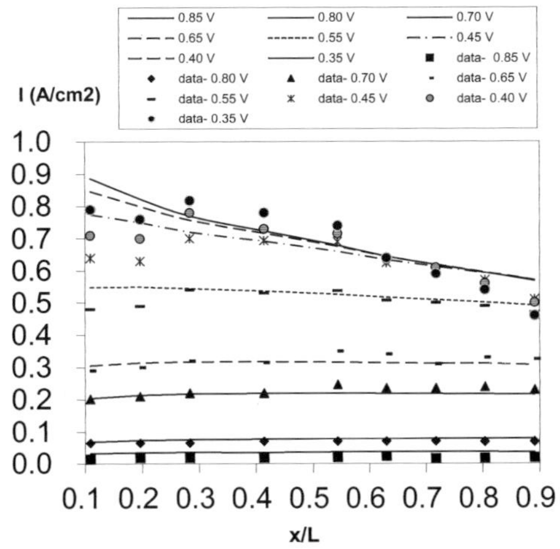

Computed (lines) , measured (symbols): cathode stoichiometry of 1.5 equiv.

2.25

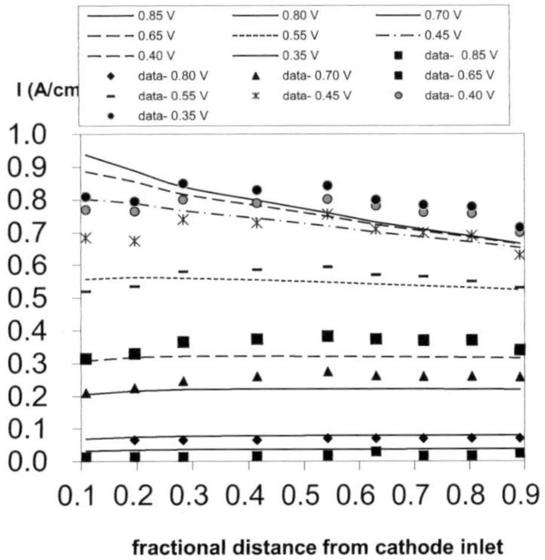

Computed (lines), measured (symbols): cathode stoichiometry of 2.25 equiv.

Automotive Simulation WORLD CONGRESS

19 June 7-8, 2016 Munich

Stack Simulation – Example

- 5 PEMFC stack with end plates (omitted in this picture).
- Close to 9.3 Mio mesh cells.

Automotive Simulation WORLD CONGRESS

20 June 7-8, 2016 Munich

Stack Simulation – Example

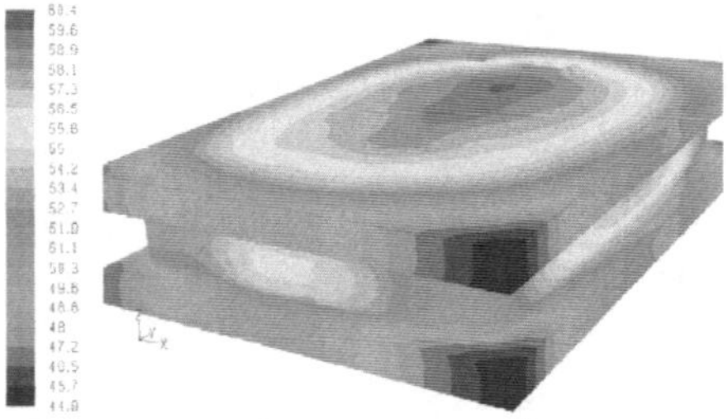

Temperature °C on outer walls

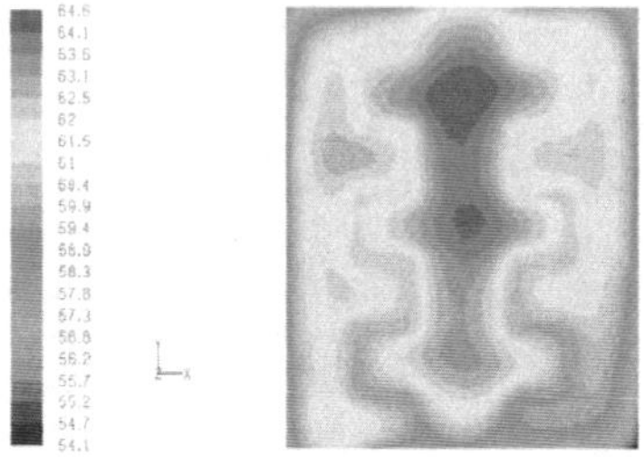

Temperature °C in an anode

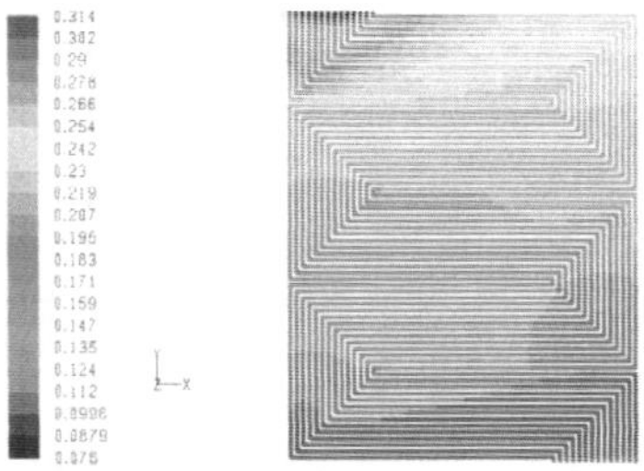

Y_{H2}

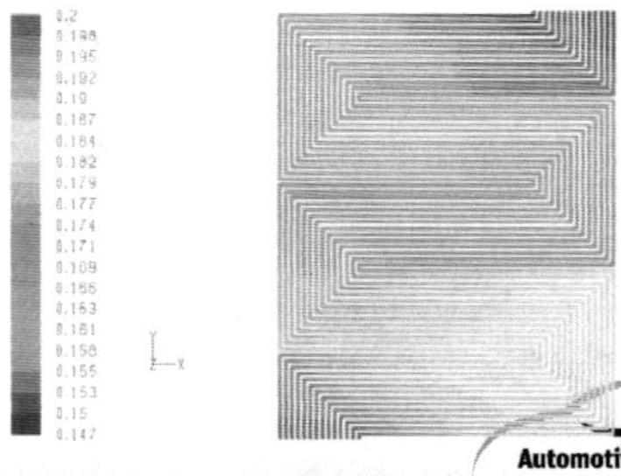

Y_{O2}

Summary

- **Better (MPL)**
- **Faster**
- **Cheaper**

Vergleich von Simulation und Messung bei Trafos und Drosseln

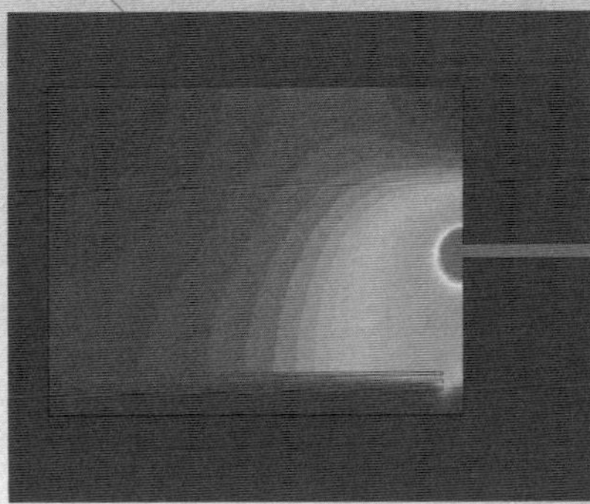

Horst Edel

Ingenieurbüro Edel

1

Übersicht

- Groß- und Kleinsignalmessungen.
- Verluste.
- Drossel mit Luftspalt.
- Sepic Trafo.

2

Kleinsignalmessungen

3

Testmessung

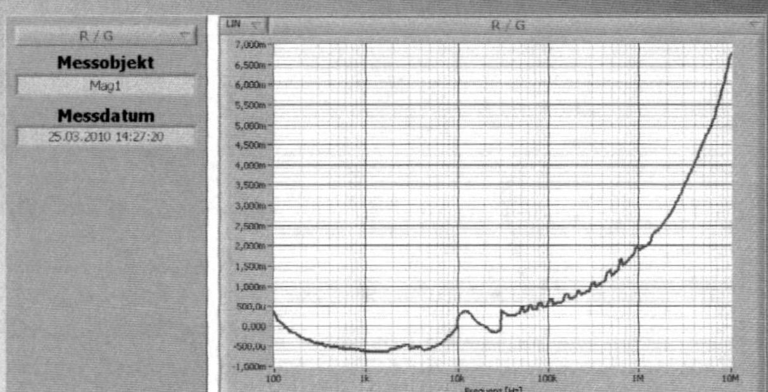

Kalibriermessung. Direkter Kurzschluß der beiden Meßkontakte.

Testmessung. Die beiden Meßkontakte wurden über einen 1cm langen Runddraht mit 1mm Durchmesser verbunden.

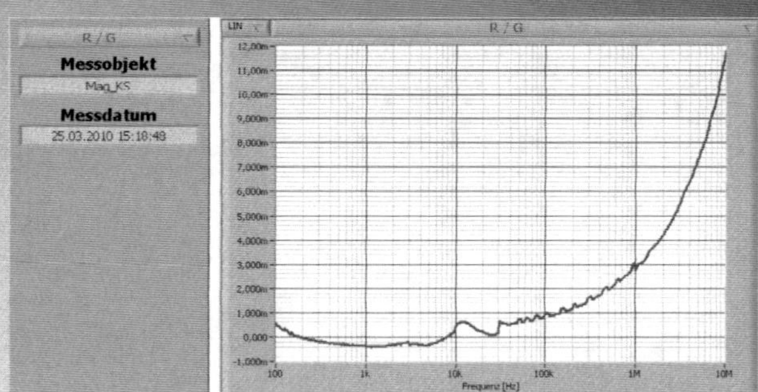

4

Testmessung

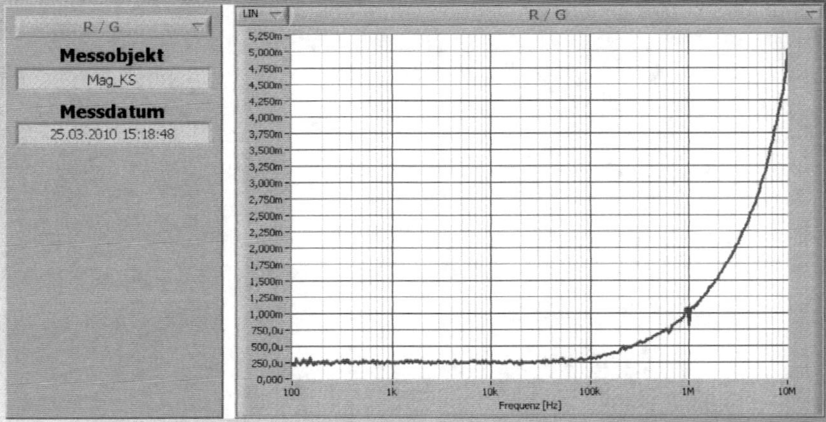

Es wurden die Kalibrierdaten des direkten Kurzschlußes von denen des 1cm langen Drahtes abgezogen. Es ergibt sich die Impedanz des Runddrahtes.

Berechnung des Drahtwiderstandes bei Gleichstom:

$$R' = \frac{1}{\sigma \cdot A_{Cu}} = \frac{1}{5.68 \cdot 10^{7}\, A/(Vm) \cdot (0.5 \cdot 10^{-3}\, m)^2 \pi} = 22.4m\frac{\Omega}{m}$$

$$R = R' \cdot l = 22.4m\frac{\Omega}{m} \cdot 10 \cdot 10^{-3}\, m = 224\mu\Omega$$

5

Großsignalmessungen

6

Großsignalmessungen in Resonanzwandler

Direkte Messung des Stroms und der Spannung über der Drossel.
Zur Strom- und Spannungsmessung wurden bewußt nur passive Tastköpfe verwendet.
Signaldarstellung und -aufbereitung mit 12 Bit Oszillograf HRO 66Zi von Fa. LeCroy.

7

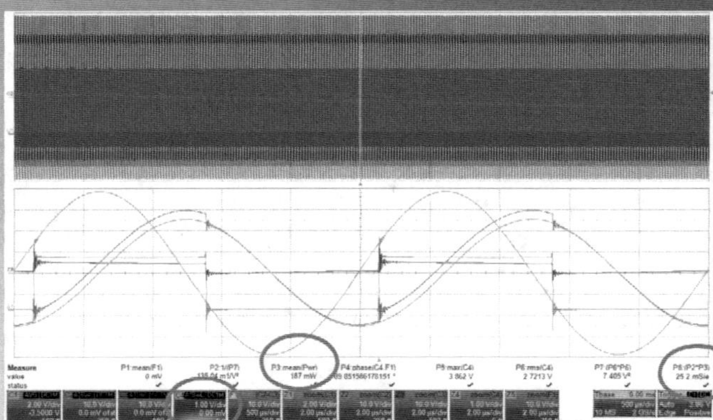

Darstellung der Spannungen und Ströme.

Resonanzstrom C4 hat eine Skalierung von 1A/Tlg.

Pv = 187mW

Rv = 25.2mΩ

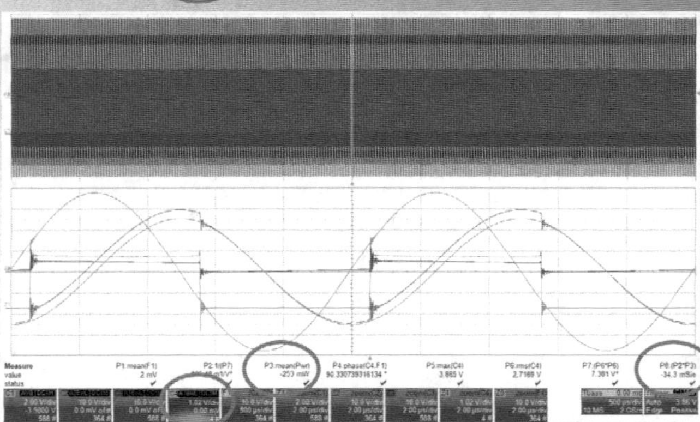

Darstellung der Spannungen und Ströme.

Resonanzstrom C4 hat eine Skalierung von 1.02A/Tlg.

Pv = -253mW

Rv = -34.3mΩ

8

Direkte Messung des Stroms und der Spannung über der Drossel

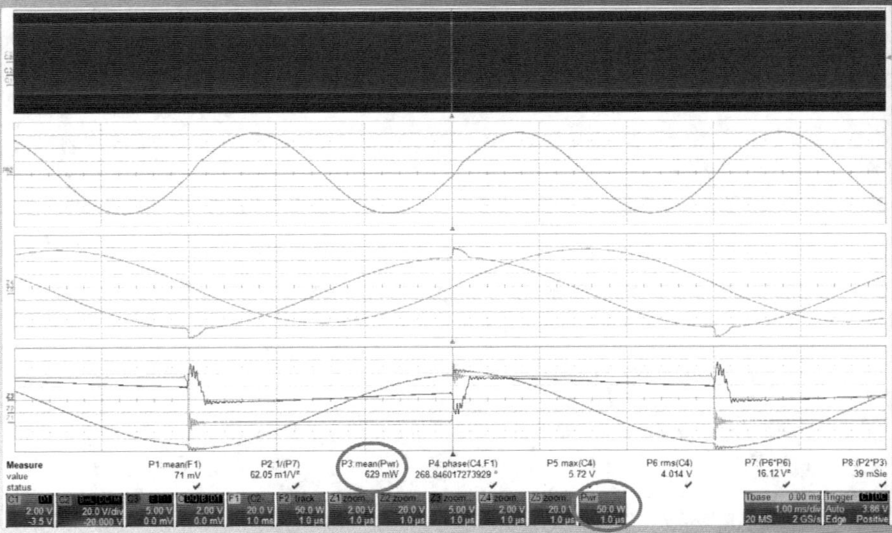

Hier ist das Problem deutlich zu sehen. Die Augenblicksleitung Pwr hat Spitzenwerte von +/- 150W. Der Mittelwert mean(Pwr) ist 0.6W, das entspricht 0.4%. Außerdem geht die Phasenbeziehung zwischen Spg und Strom sehr stark in das Ergebnis ein. Wenn man annimmt, daß die Eingangsverstärker und A/D-Wandler des LeCroy sicherlich keine größere Genauigkeit als 1% haben, ist es klar, daß es hier Schwierigkeiten gibt.

9

Messung des Stroms und der Spannung des Resonanzkreises

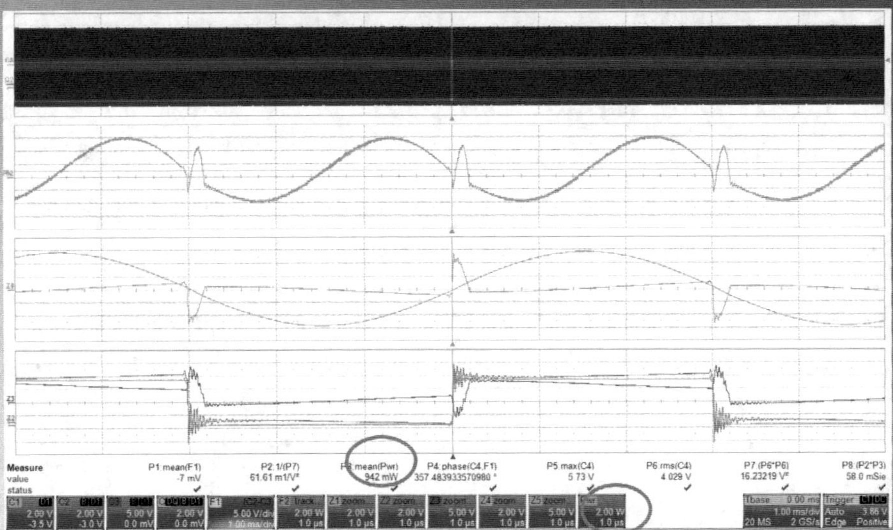

Jetzt hat die Augenblicksleitung Spitzenwerte von +/- 6W. Wenn man die Signale so darstellt, daß sie die maximale Auflösung haben, werden die Verlustleistungen und Widerstände reproduzierbar !!

Mit der Annahme, daß die verwendeten COG Resonanzkondensatoren geringe und konstante Verluste haben, sind hiermit vergleichende Messungen machbar.

10

Messungen mit Leistungsmeßgerät von Fa. ZES Zimmer

11

Messungen mit Leistungsmeßgerät von Fa. ZES Zimmer

Bezeichnung:	**Lr11**
Kern:	EE22
Material:	N87
Windungszahl:	8
Drahtstärke:	90 x 0.1mm
Drahtlänge:	460mm
Meßwicklung:	
Windungszahl:	8
Drahtstärke:	1 x 0.18mm

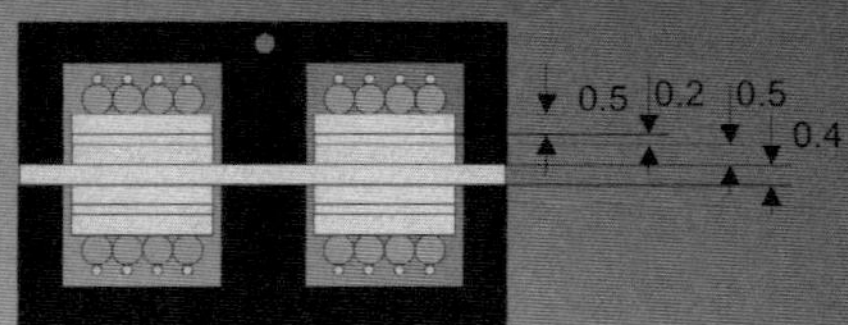

Messung	Ue [V]	6	7	8	9	10	11
1	Pv1 [W]	0.37	0.54	0.73	1.00	1.27	1.49
	φ1 [°]	89.84	89.83	89.83	89.81	89.80	89.81
	Rv1 [Ω]	27	29	30	33	34	32
2	Pv2 [W]	0.59	0.83	1.11	1.48	1.85	2.18
	φ2 [°]	89.76	89.74	89.74	89.72	89.72	89.73
	Rv2 [Ω]	43	44	46	48	49	48
	Rv2-Rv1 [Ω]	16	15	16	15	15	16
3	Pv3 [W]	0.17	0.23	0.30	0.32	0.39	0.56
	φ3 [°]	-89.93	-89.93	-89.93	-89.94	-89.94	-89.93
	Rv3 [Ω]	12	12	13	11	11	13

Messung 1: Spannungsmessung an der Meßwicklung
Messung 2: Spannungsmessung an der Leistungswicklung
Messung 3: Spannungsmessung am 100nF COG Kondensatorblock

Vorsicht bei Meßbereichsumschaltungen und bei anderen Meßbereichen !!! 12

Verluste

13

Daten für MnZn Ferritmaterial [Ferroxcube Datenbuch]

Soft Ferrites — Ferrite materials survey

Properties specified in this section are related to room temperature (25 °C) unless otherwise stated. They have been measured on sintered, non ground ring cores of dimensions ∅25 × ∅15 × 10 mm which are not subjected to external stresses. Products generally comply with the material specification. However, deviations may occur due to shape, size and grinding operations etc. Specified product properties are given in the data sheets or product drawings.

PERMITTIVITY

The basic permittivity of all ferrites is of the order of 10. This is valid for MnZn and NiZn materials. The isolating material on the grain boundaries also has a permittivity of approximately 10. However, if the bulk permittivity of a ferrite is measured, very different values of apparent permittivity result. This is caused by the conductivity inside the crystallites. The complicated network of more or less leaky capacitors also shows a strong frequency dependence.

Table 5 Permittivity as a function of frequency for MnZn ferrites

FREQUENCY (MHz)	PERMITTIVITY (ε_r)
0.1	$\approx 2.10^5$
1	$\approx 10^5$
10	$\approx 5.10^4$
100	$\approx 10^4$

14

Daten für MnZn Ferritmaterial [Ferroxcube Datenbuch]

RESISTIVITY

Ferrite is a semiconductor with a DC resistivity in the crystallites of the order of 10^{-3} Ωm for a MnZn type ferrite, and about 30 Ωm for a NiZn ferrite.

Since there is an isolating layer between the crystals, the bulk resistivity is much higher: 0.1 to 10 Ωm for MnZn ferrites and 10^4 to 10^6 Ωm for NiZn and MgZn ferrites.

This resistivity depends on temperature and measuring frequency, which is clearly demonstrated in Tables 1 and 2 which show resistivity as a function of temperature for different materials.

Table 3 Resistivity as function of frequency for MnZn ferrites

FREQUENCY (MHz)	RESISTIVITY (Ωm)
0.1	≈2
1	≈0.5
10	≈0.1
100	≈0.01

Table 1 Resistivity as a function of temperature of a MnZn-ferrite (3C94)

TEMPERATURE (°C)	RESISTIVITY (Ωm)
−20	≈10
0	≈7
20	≈4
50	≈2
100	≈1

15

Daten für 3F3 Ferritmaterial [Ferroxcube Datenbuch]

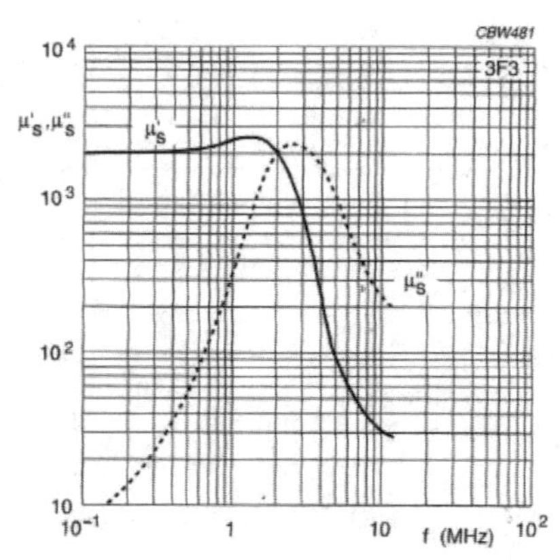

Fig.1 Complex permeability as a function of frequency.

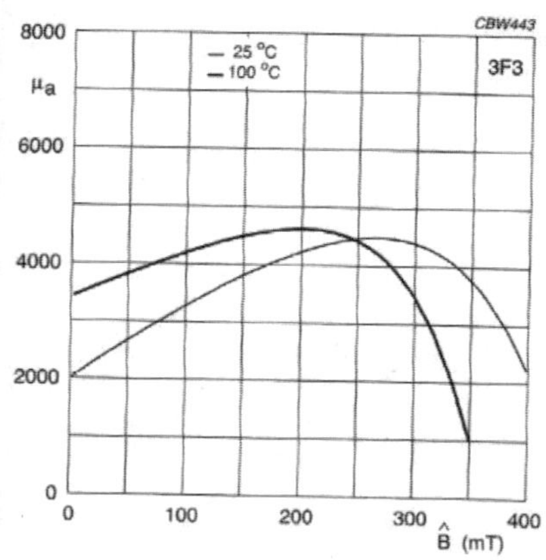

Fig.4 Amplitude permeability as function of peak flux density.

Komplexe Permeabilität: $\mu = \mu_s' - j\mu_s''$

16

Daten für 3F3 Ferritmaterial [Ferroxcube Datenbuch]

3F3 SPECIFICATIONS

A medium frequency power material for use in power and general purpose transformers at frequencies of 0.2 - 0.5 MHz.

SYMBOL	CONDITIONS	VALUE	UNIT
μ_i	25 °C; ≤10 kHz; 0.25 mT	2000 ±20%	
μ_a	100 °C; 25 kHz; 200 mT	≈ 4000	
B	25 °C; 10 kHz; 1200 A/m	≈ 440	mT
	100 °C; 10 kHz; 1200 A/m	≈ 370	
P_V	100 °C; 100 kHz; 100 mT	≤80	kW/m³
	100 °C; 400 kHz; 50 mT	≤150	
ρ	DC; 25 °C	≈ 2	Ωm
T_C		≥200	°C
density		≈ 4750	kg/m³

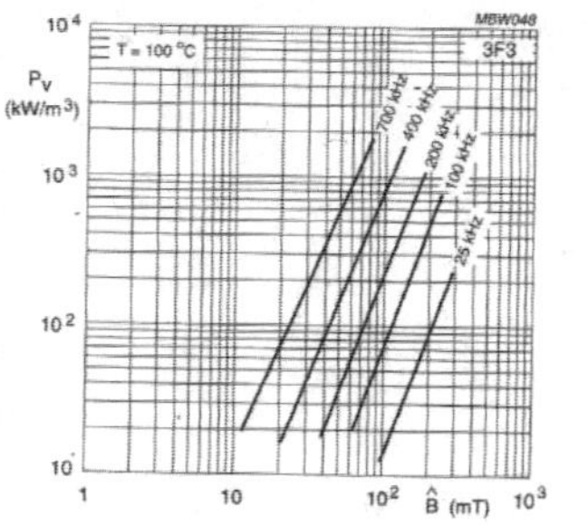

Fig.6 Specific power loss as a function of peak flux density with frequency as a parameter.

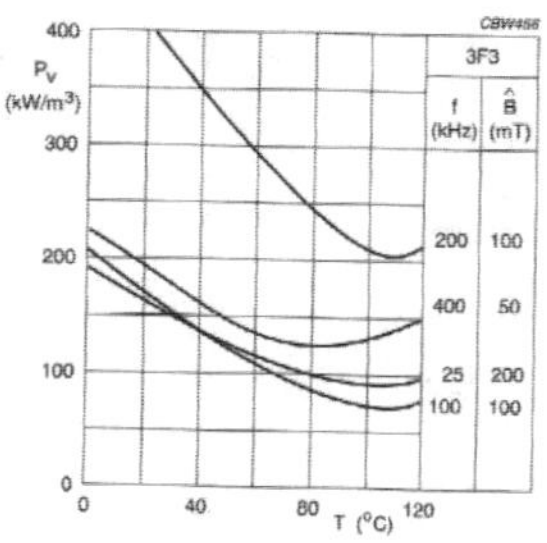

Fig.7 Specific power loss for several frequency/flux density combinations as a function of temperature.

ρ ist der spezifische Widerstand. Bulk conductivity σ ist der Leitwert mit $\sigma = 1/\rho$.

Diagramm zur Ermittlung der Steinmetz Parameter C, X und Y.

Abhängigkeit der Verluste von der Temperatur.

17

Daten für 3F36 Ferritmaterial [Ferroxcube Datenblatt 2016]

3F36

3F36 is the superior material for power conversion applications dedicated to the 300–700kHz frequency range. Its outstanding performance can be observed comparing the temperature performance charts with 3F35 and 3F3 materials.

Choose 3F36 for efficient power conversion design over a wide temperature range.

SYMBOL	CONDITIONS	UNIT	3F36	3F35	3F3
µi	25°C; 10kHz; 0.25mT		**1600 ± 20%**	1400 ± 20%	2000 ± 20%
µa	100°C; 25kHz; 200mT		**≈ 2400**	≈ 2400	≈ 4000
B	25°C; 10kHz; 1200A/m	mT	**≈ 520**	≈ 500	≈ 440
			≈ 420	≈ 420	≈ 370
Pv	100°C; 400kHz; 50mT	kW/m³	**≈ 50**	≈ 60	≈ 140
	100°C;500kHz;50mT		**≈ 90**	≈ 90	≈ 220
	100°C;500kHz;100mT		**≈ 700**	≈ 700	≈ 1200
ρDC	25°C	Ωm	**≈ 12**	≈ 10	≈ 2
Tc		°C	**≥ 230**	≥ 240	≥ 200
density		kg/m³	**≈ 4750**	≈ 4750	≈ 4700

Data refers to typical values of ring T25/15/10

18

Daten für 3F36 Ferritmaterial [Ferroxcube Datenblatt 2016]

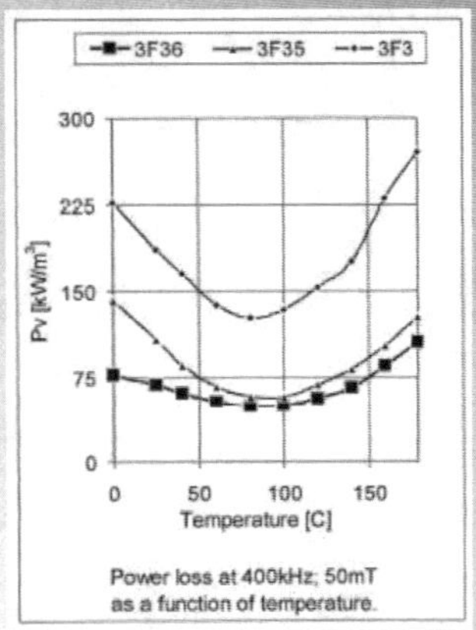

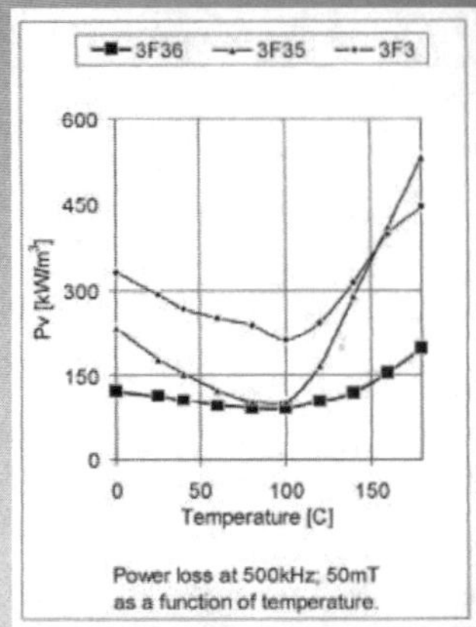

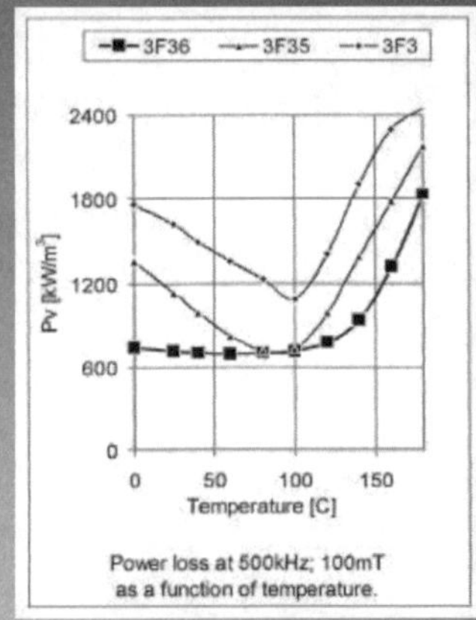

<table>
<tr><td>Material</td><td>Einführung</td><td></td><td></td><td></td></tr>
<tr><td>3F3</td><td>< 1988</td><td rowspan="2">12 Jahre</td><td></td><td rowspan="3">28 Jahre</td></tr>
<tr><td>3F35</td><td>< 2000</td><td rowspan="2">16 Jahre</td></tr>
<tr><td>3F36</td><td>2016</td><td></td></tr>
</table>

19

Aufteilung der Drosselverluste

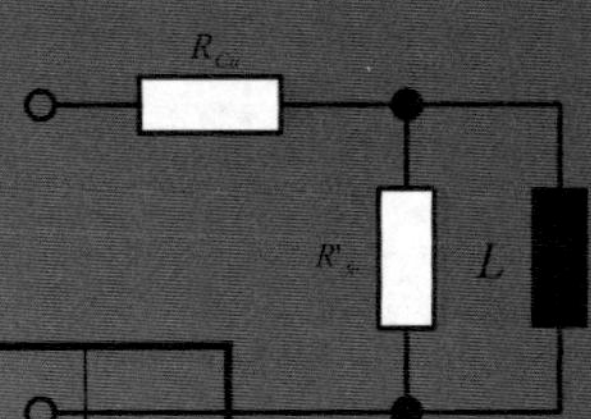

<table>
<tr><td rowspan="4">R_{Cu}</td><td rowspan="3">besteht aus</td><td>Gleichstromkupferwiderstand</td><td></td><td></td></tr>
<tr><td rowspan="2">Induktionseffekten in der Wicklung</td><td>Stromverdrängung durch benachbarte Leiter.
Stromverdrängung durch Kern ($\mu_r > 1$)</td><td></td></tr>
<tr><td>Stromverdrängung durch Luftspalt.</td><td></td></tr>
<tr><td colspan="3">weitgehend unabhängig von Temperatur und Aussteuerung. Kleinsignalmessung</td><td></td></tr>
<tr><td rowspan="5">R_{fe}</td><td rowspan="4">besteht aus</td><td>spezifischen Verlusten nach Steinmetz (Verluste/Volumen)</td><td>$P_{sm}''' = C_m \cdot f^x \cdot \hat{B}^y$ Material spezifisch, gemessen an Ringkernen.</td><td>$\sigma = 0$</td></tr>
<tr><td>zusätzlichen Verlusten durch andere Kernform</td><td>Keine Gleichverteilung der Flußdichte über dem Volumen.</td><td>$\sigma = 0$</td></tr>
<tr><td rowspan="2">Wirbelstromverlusten (abhängig von Kernform und -größe.)</td><td>Verluste durch Kernleitfähigkeit.</td><td>$\sigma \neq 0$</td></tr>
<tr><td>Verluste durch Volumenresonanz.</td><td>$\sigma \neq 0$
μ_r, ε_r</td></tr>
<tr><td colspan="3">sehr abhängig von Temperatur und Aussteuerung. Großsignalmessung</td><td></td></tr>
</table>

20

Simulation einer EE38 Drossel

Wicklung: 4 Windungen mit 100μm Kupferstärke

Kerndaten:

Effektive Weglänge:	l_e = 52.4mm
Effektive Fläche:	A_e = 194µm²
Effektives Volumen	V_e = 10200nm²

Mit einem Luftspalt von 0.4mm ergibt sich ein A_L = 630nH und ein μ_e = 136.
Steinmetz Koeffizienten für das Material 3F3: C_m = 0.75; x = 1.5; y = 2.5;

21

Berechnung der Induktivität, der Aufmagnetisierung und der Steinmetzverluste für f = 158KHz

$$L = \frac{\mu_0 \mu_e}{l_e} A_e N^2 = \frac{4 \cdot 10^{-7} \cdot \pi \cdot 136}{52.4m} 194\mu \cdot 4^2 = 10.1\mu H$$

$$B_s = \frac{\mu_0 \mu_e}{l_e} N I_s; \qquad P_{sm}''' = C_m \cdot f^x \cdot \hat{B}^y$$

I_s [A]	B_s [mT]	P_{sm}''' [W/m³]	P_{sm} [W]
10m	0.13	9.1m	92.6n
1	13	907.6	9.3m
6	78	80.0K	816m

Bei einer Kleinsignalmessung mit I_s = 10mA ergeben sich keine Steinmetzverluste !!??

22

Berechnung der Verluste

Steinmetz - Verlustleistungsdichte: $P_{sm}''' = C_m \cdot f^x \cdot B_s^y$

Steinmetz - Verlustleistung: $P_{sm}' = \iiint\limits_{V_{fe}} C_m \cdot f^x \cdot B_s^y \, dV_{fe}$

	Name	Value	Unit	Evaluated Value	Description
	$Is	1	A	1A	Erregerstrom
	$Cm	0.75		0.75	Steinmetzfaktor 3F3
	$X	1.5		1.5	Steinmetzfaktor-f 3F3
	$Y	2.5		2.5	Steinmetzfaktor-B 3F3
	$sigmaFe	0.5	S_per_m	0.5S_per_m	Leitfähigkeit 3F3
	$sigmaCu	58000000	S_per_m	58000000S_per_m	Leitfähigkeit Cu

Strom - Verlustleistungsdichte: $P_{ohm}''' = \frac{\underline{J} \cdot \underline{J}^*}{2\sigma}$

Wirbelstrom - Verlustleistung: $P_{ohmFe} = \iiint\limits_{V_{Fe}} \frac{\underline{J} \cdot \underline{J}^*}{2\sigma_{Fe}} dV_{Fe}$

Kupfer - Verlustleistung: $P_{ohmCu} = \iiint\limits_{V_{Cu}} \frac{\underline{J} \cdot \underline{J}^*}{2\sigma_{Cu}} dV_{Cu}$

23

Berechnung des Widerstandes

$$R = \frac{P_{ges}}{I_{eff}^2} = 2 \frac{P_{ges}}{\hat{I}^2}$$

Berechnung der Induktivität

Magnetische Energie: $w = \iiint\limits_{V_{Fe}} \frac{1}{4} \underline{\hat{B}} \cdot \underline{\hat{H}}^* dV = \frac{1}{4} L \hat{I}^2$

Induktivität: $L = \frac{4}{\hat{I}^2} w$

Berechnung der mittleren Aufmagnetisierung

$$\overline{B}_{mag} = \frac{1}{A_{Fe}} \cdot \iint\limits_{A_{Fe}} B_{mag} \, dA_{Fe}$$

24

Simulationsmodell EE38 Drossel mit mittigem Luftspalt

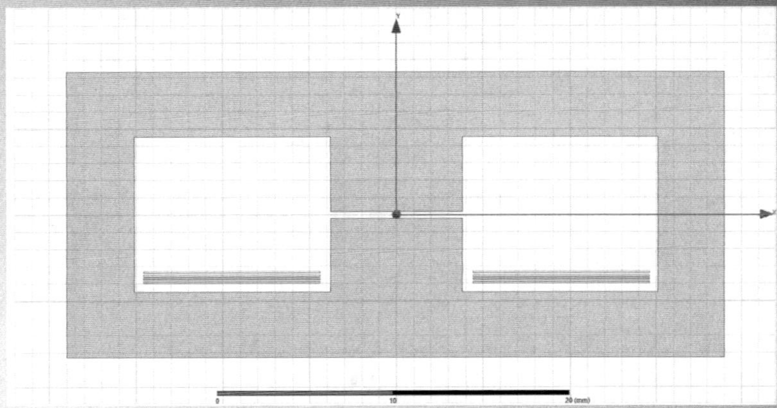

Modellierung im kartesischen Koordinatensystem.

25

Darstellung der simuliertern Werte

Werte — Maxwell2DDesign1

	Freq [kHz]	Iswirk	Psm	Pohmfe	Pcu	Pges	Rges	L	Bmag
1	158.000	1.6129	8.822E-003	1.630E-003	1.055E-002	2.101E-002	4.201E-002	1.029E-005	0.011896

Für f = 158KHz und I_s = 1A ergibt sich eine mittlere Aufmagnetisierung von 12mT, Steinmetzverluste von 8.8mW und eine Induktivität von 10.3μH.

I_s [A]	B_s [mT]	P_{sm} [W]	L [H]
1	13	9.3m	10.1μ

Zum Vergleich die analytisch berechneten Werte.

26

Darstellung der Verluste mit Is = 10mA

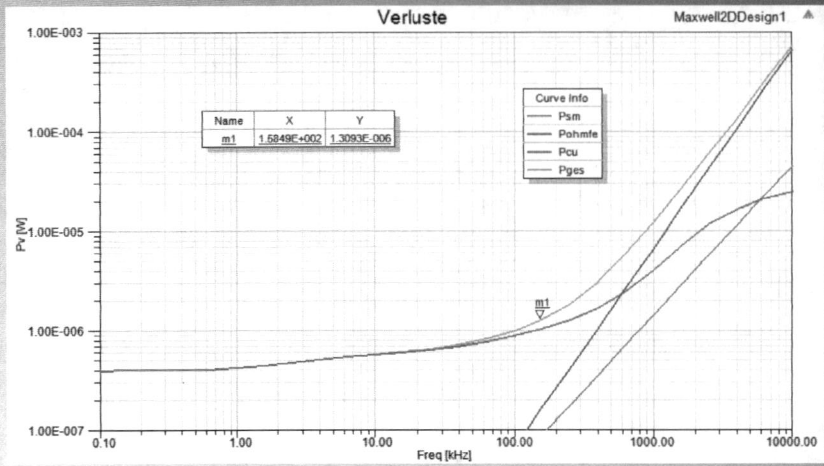

Bei kleinen Frequenzen ergeben sich nur Kupferverluste. Ab ca. 100KHz übernehmen die Wirbelstromverluste. **Bei Kleinsignalmessungen überwiegen im Ferrit die Wirbelstromverluste.**

27

Darstellung des Verlustwiderstandes mit Is = 10mA

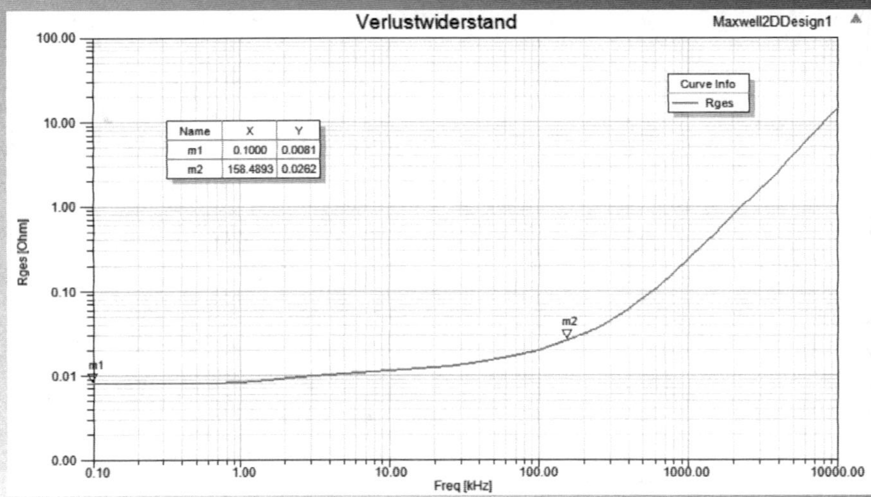

28

Vergleich Simulation - Messung

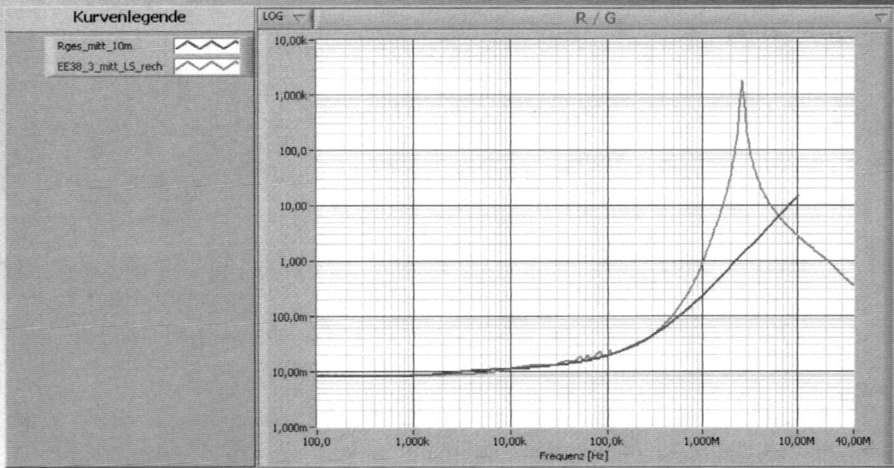

Rote Kurve: Messung mit HP4194A
Blaue Kurve: Simulation mit I_s = 10mA

29

Darstellung der Verluste mit I_s = 6A

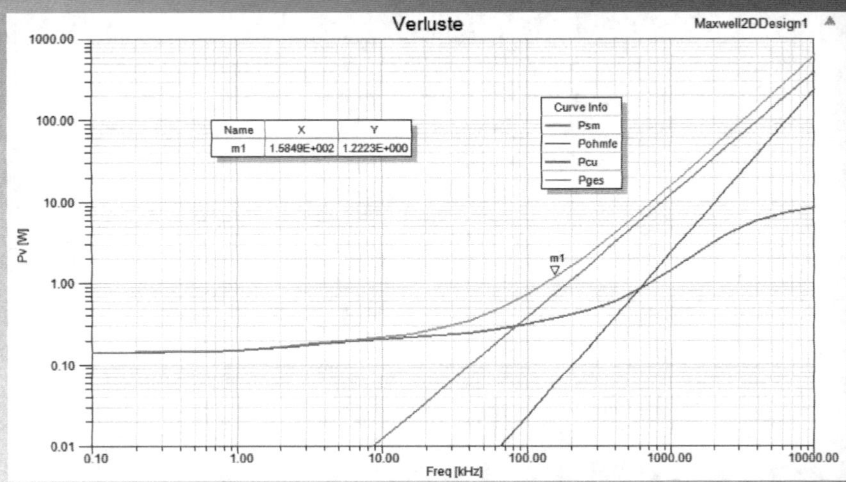

Bei kleinen Frequenzen ergeben sich nur Kupferverluste. Ab ca. 10KHz übernehmen jetzt die **Steinmetz**verluste. Bei Großsignalmessungen überwiegen im Ferrit die Steinmetzverluste. Bei f = 158KHz ergeben sich Gesamtverluste von Pges = 1.2W.

30

Darstellung des Verlustwiderstandes mit $I_s = 6A$

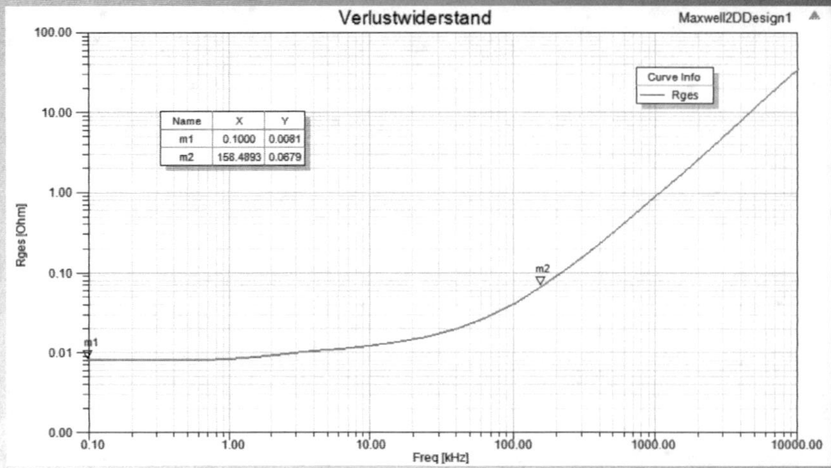

Rges ist proportional zu den Gesamtverlusten Pges. Er ist deutlich größer als der bei der Kleinsignalmessung.

31

Darstellung des Realteilbetrags H_r

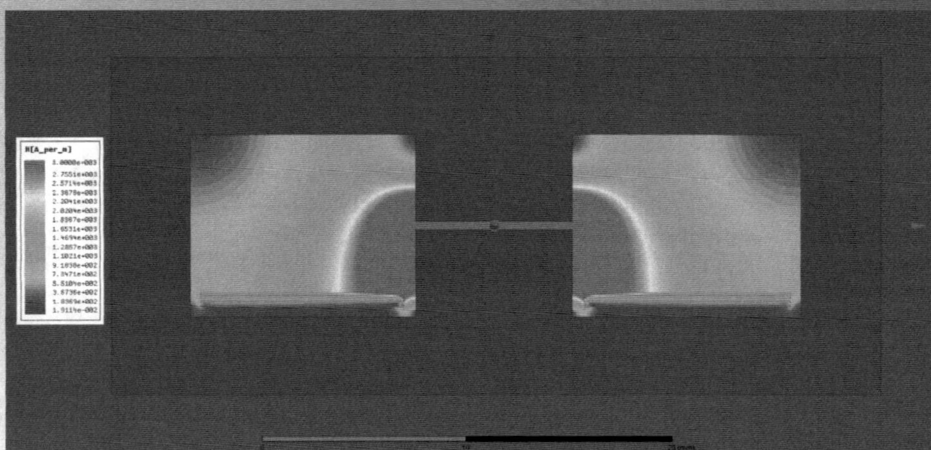

32

Darstellung des Realteilbetrags B_r

Obwohl das B-Feld sehr ungleich über der Fläche verteilt ist, paßt der Mittelwert gut mit dem analytisch berechneten Wert überein.

33

Darstellung des Realteilvektors J_{sr}

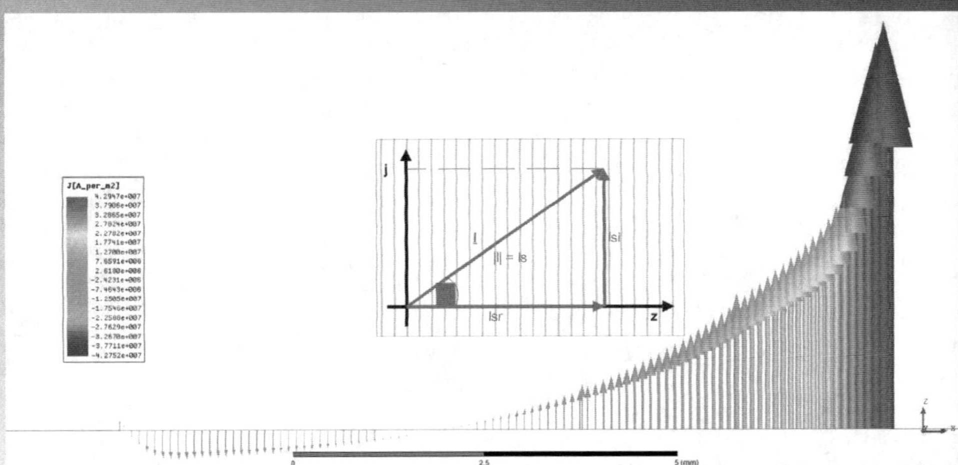

Darstellung des linken Wickelfensters bei f = 158KHz. Skalierung ist nicht begrenzt. Die Zeichenebene wurde um 90° gedreht. Die Stromdichte ist wegen dem Luftspalt rechts sehr groß und wird links negativ.

34

Vergleich der Verlustwiderstände

	Kleinsignal	Großsignal
Messung	24.7mΩ	98mΩ
Simulation	26.2mΩ	68mΩ

Messung und Simulation passen für die Kleinsignalbetrachtungen sehr gut überein. Im Großsignalbereich sind sie unterschiedlich.

35

Simulation eines EE38 Sepic Trafos

Primärwicklung: 16 Windungen

Primärinduktivität: 160μH

Primäre Streuinduktivität: 6μH

Sekundärwicklung: 4 Windungen

36

Simulation eines EE38 Sepic Trafos

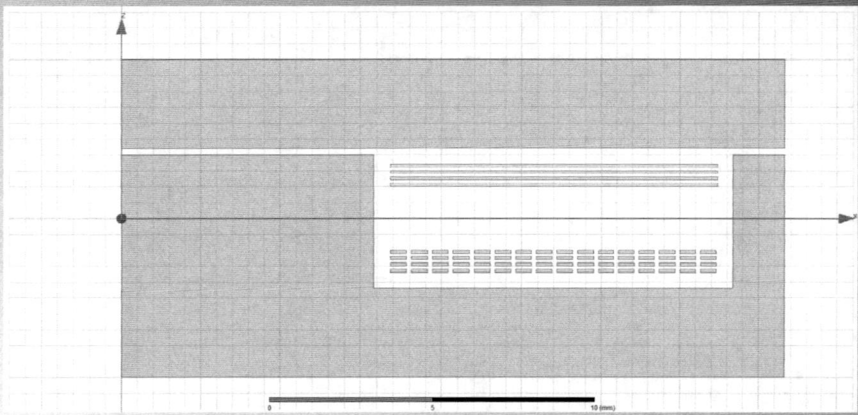

Modellierung im zylindrischen Koordinatensystem.

Primärwicklung: 16 Windungen pro Lage, 4 Lagen parallel

Sekundärwicklung: 4 Windungen in Reihe

37

Charakterisierung eines Trafos

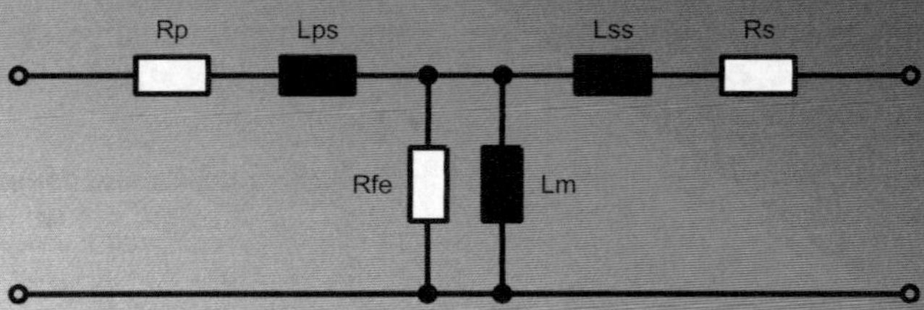

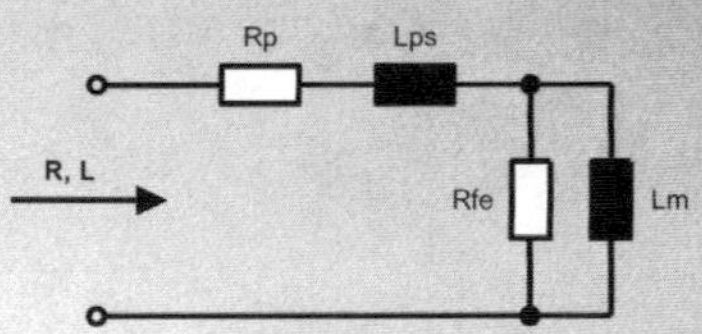

Leerlauf

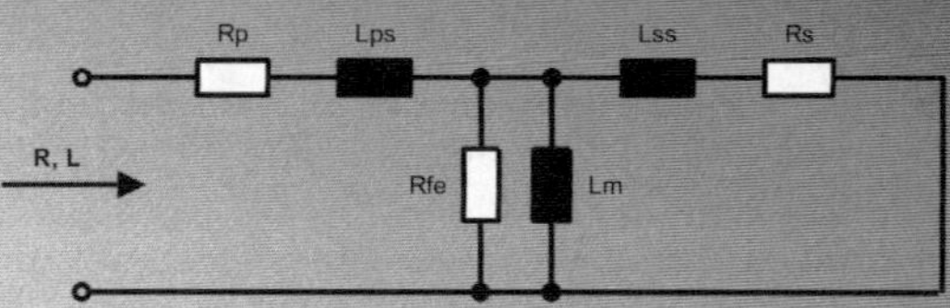

Kurzschluß

38

Leerlauf

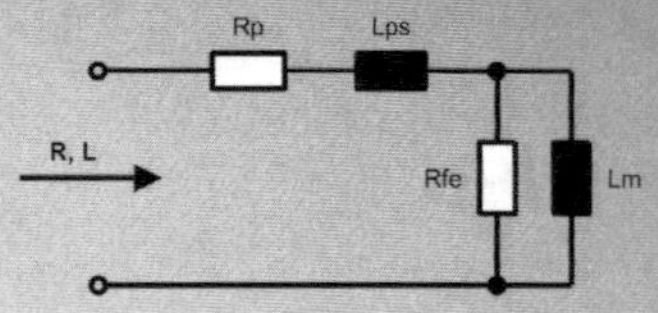

R = Rp ... Rp + Rfe

L = Lps + Lm

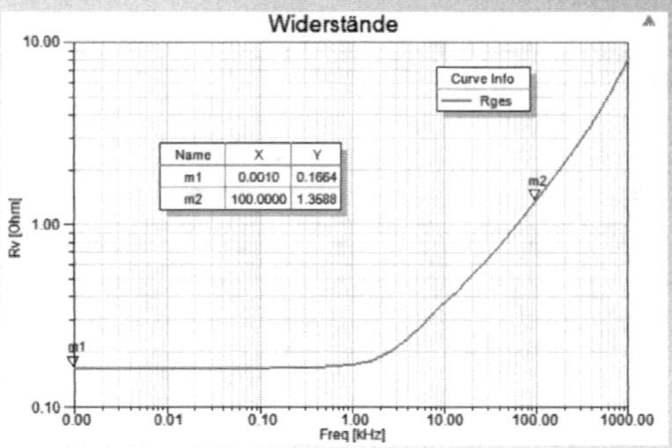

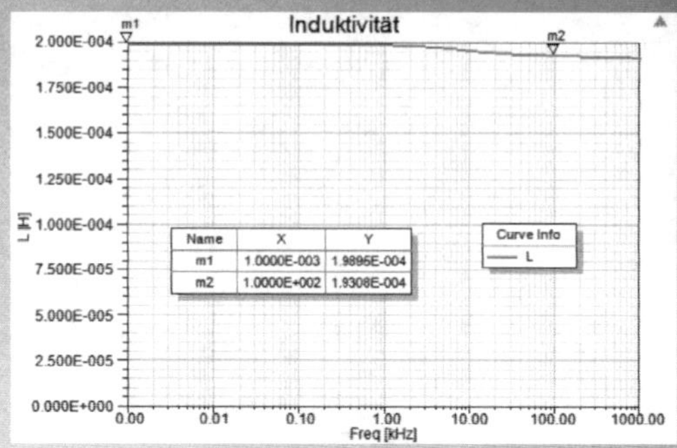

39

Vergleich Simulation Messung im Leerlauf

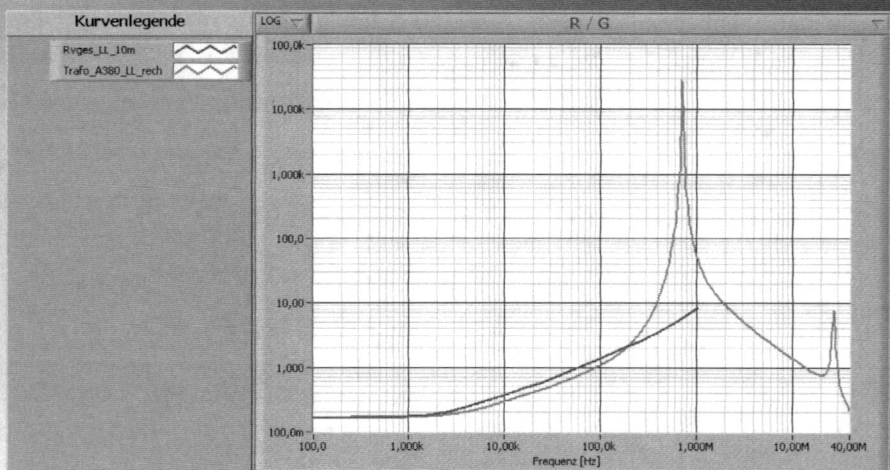

40

Darstellung des Realteilbetrags H_r im Leerlauf

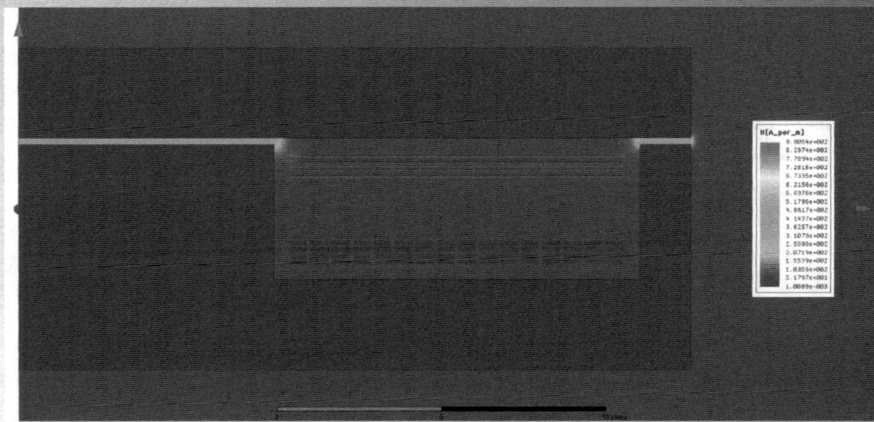

41

Kurzschluß

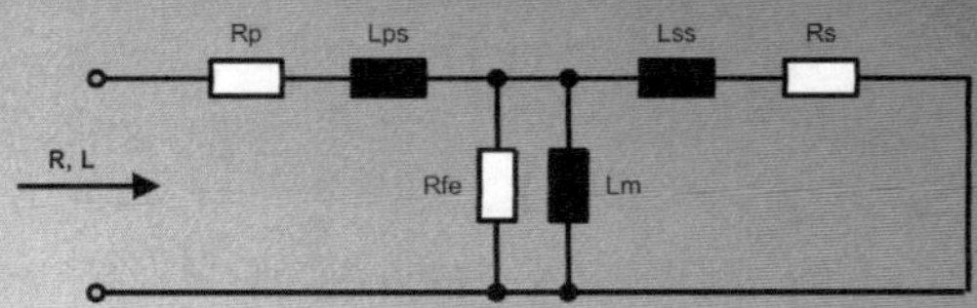

R = Rp ... Rp + Rs

L = Lps + Lm ... Lps + Lss

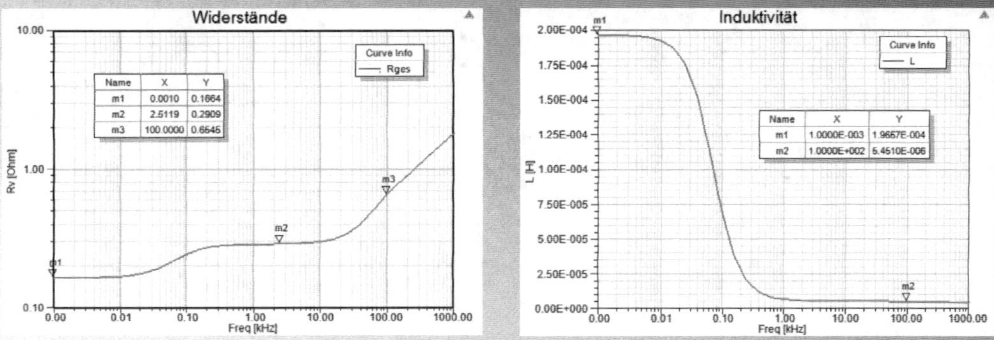

42

Vergleich Simulation Messung im Kurzschluß

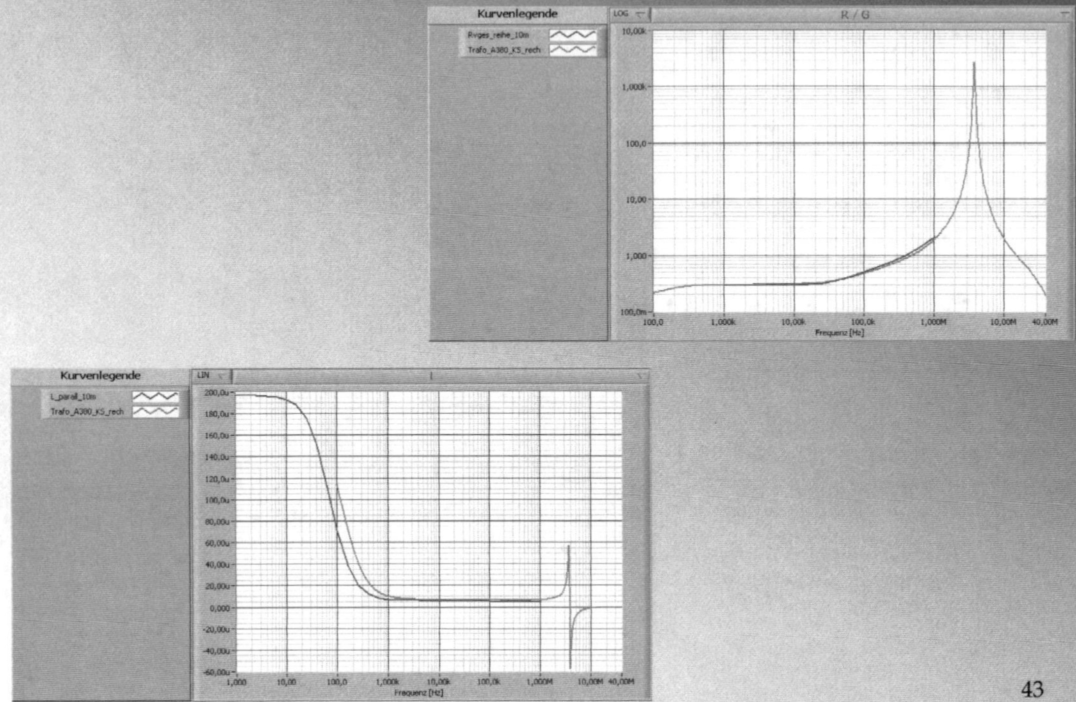

43

Darstellung des Realteilbetrags H_r im Kurzschluß

44

Wenn Gott gewollt hätte, daß die Menschen den Magnetismus verstehen, hätte er die Feldlinien sichtbar gemacht.

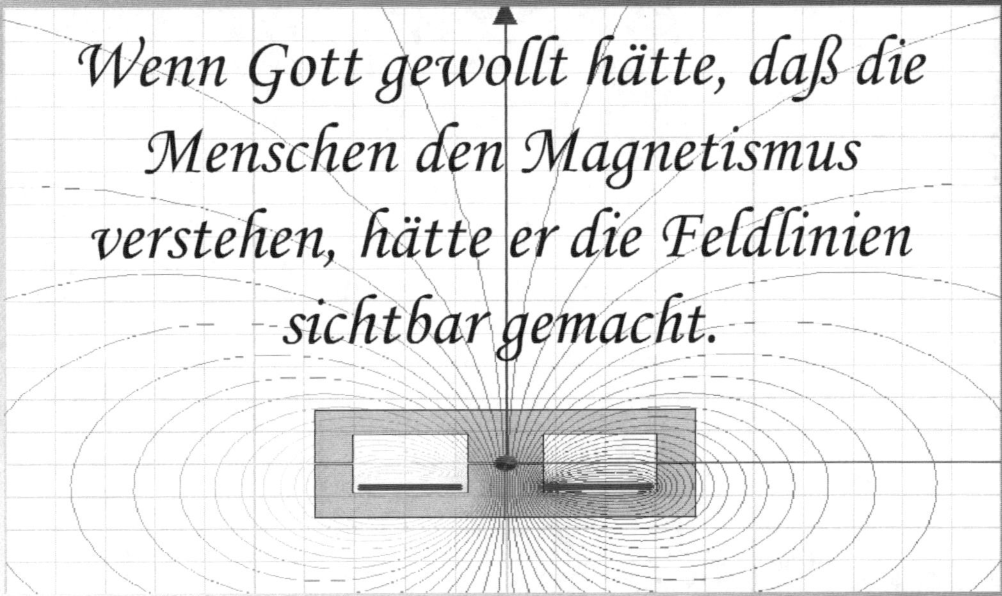

Quelle: unbekannter, leicht frustrierter Netzteilentwickler.

45

Vergleich von Simulation und Messung bei Trafos und Drosseln

Horst Edel

Ingenieurbüro Edel

46

Automotive Simulation
WORLD CONGRESS

June 7-8, 2016 Munich

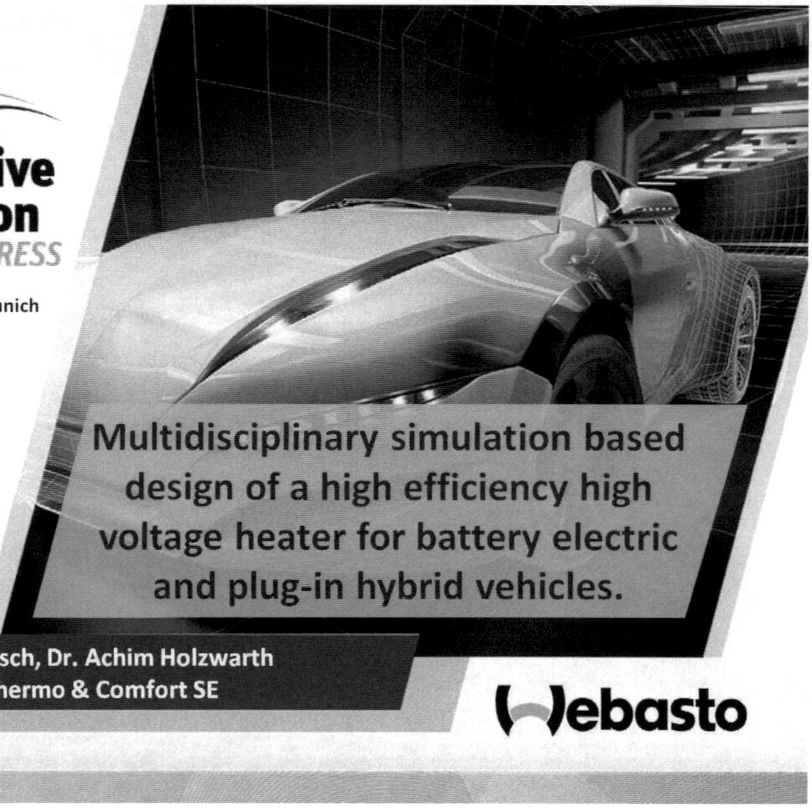

Multidisciplinary simulation based design of a high efficiency high voltage heater for battery electric and plug-in hybrid vehicles.

Florian Larisch, Dr. Achim Holzwarth
Webasto Thermo & Comfort SE

Webasto

Overview

Webasto
Feel the Drive

- The Webasto HVH
- Heat exchanger design
- Heat layer layout
- IGBT dimensioning
- Thermal deformation and stresses

Automotive Simulation
WORLD CONGRESS

2 June 7-8, 2016 Munich

Our focus – electric vehicles and plug-in hybrids

Webasto
Feel the Drive

Automotive Simulation WORLD CONGRESS

Electric range and vehicle heating

Webasto
Feel the Drive

Tendencies related to vehicle concept

	ICE	Hybrid	Plug-in hybrid	Range extender	BEV
Electric range		•	●	⬤	⬤
Engine waste heat	⬤	⬤	●	•	
Required add. heater power	•	●	●	⬤	⬤

Mobility concepts that require additional heat sources

• = very low

⬤ = very high

ICE = internal combustion engine
BEV = battery electric vehicle

Automotive Simulation WORLD CONGRESS

Vehicle heating is passenger comfort and safety

Webasto
Feel the Drive

HVH = Series equipment for BEV and PHEV

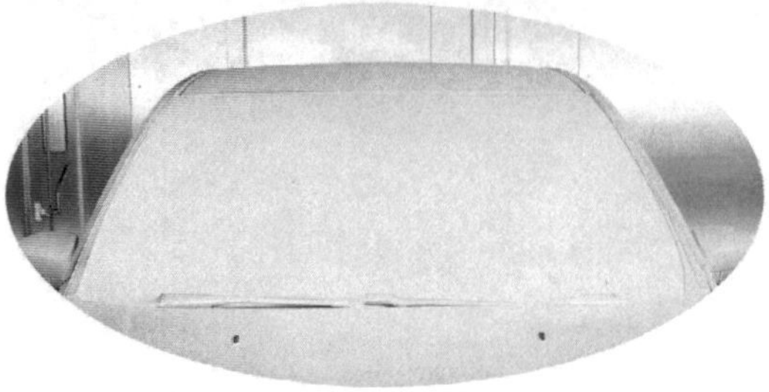

Defrosting
Fast de-icing before and during driving

Comfort
Comfortable cabin temperatures even at high external air rates

Defogging
Clear windows while driving under any ambient conditions

Demist
Internal moisture dissipation (high external air rates)

Automotive Simulation WORLD CONGRESS

5 June 7-8, 2016 Munich

The Webasto High Voltage Heater - HVH

Webasto
Feel the Drive

AWARD WINNER

INDUSTRIE PREIS 2014 BEST OF

Facts and figures:

- Heating output: 5 kW / 7 kW
  200 W min, 50 W steps
- Dimensions: 284 x 200 x 54 mm
  1.700 cm³
- Weight: 2,0 kg
- Efficiency: > 95 %

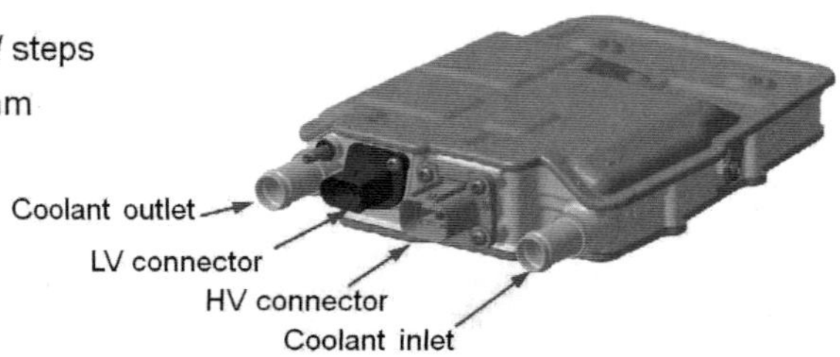

▶ **Compact design**

▶ **All connectors on one side**

▶ **Developed for highest efficiency**

Automotive Simulation WORLD CONGRESS

6 June 7-8, 2016 Munich

Webasto heat layer technology

Webasto Feel the Drive

Main concept of Webasto heat layer technology:

- Heating **layer sprayed on heat exchanger**
  - Directly bonded onto heat exchanger
  - No thermal loss towards neighboring components
- **Even heat up** of coolant over **entire surface**
- **Minimized heat capacity** with **extreme small thermal resistance** between heating element and coolant leading to ΔT of less than 60K
- **Radiated heat** to environment **minimized by thermal insulation** of covers from heat exchanger

Outline of Webasto heat layer structure

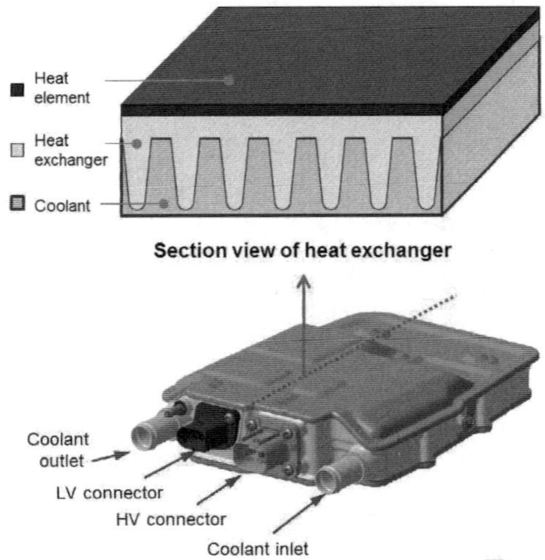

Simulations performed during development

Webasto Feel the Drive

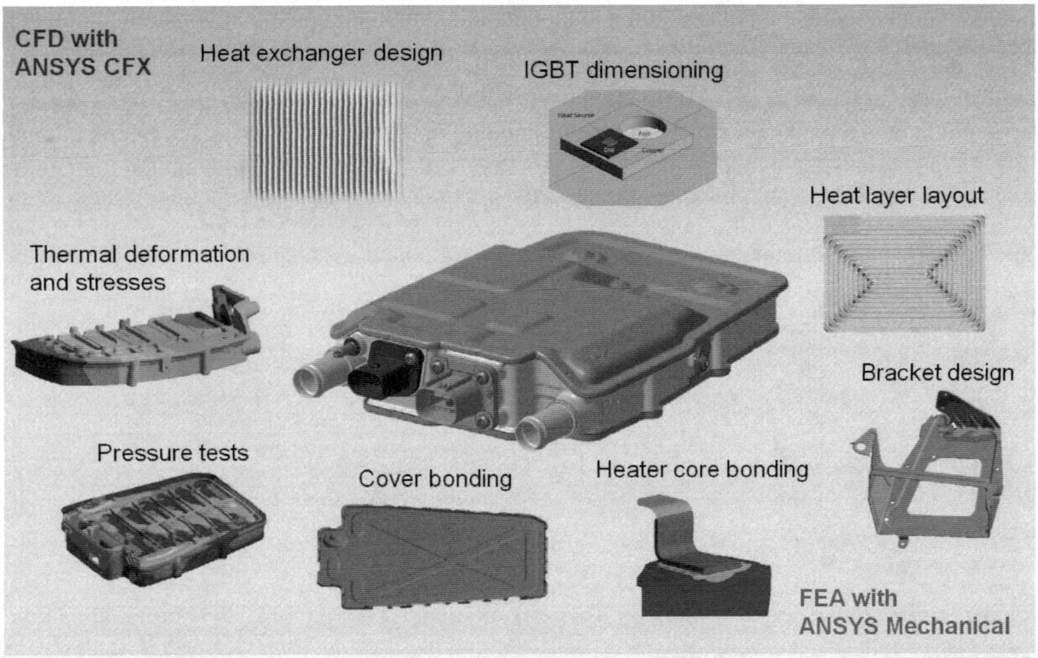

Heat exchanger design

Webasto
Feel the Drive

Objectives:

Low pressure drop
No boiling
IGBT cooling
Fast heat up
Compact design
Manufacturability / Costs

Software:

Design Modeler
ANSYS Meshing / ICEM
ANSYS CFX

Automotive Simulation
WORLD CONGRESS

Front loading
Concept studies

Webasto
Feel the Drive

Simulation was an essential part from the very beginning:

Eight different concepts involving various heating technologies (tubular heater, PTC, heating layer).

Dozens of variants within each concept.

Hundreds of individual simulations.

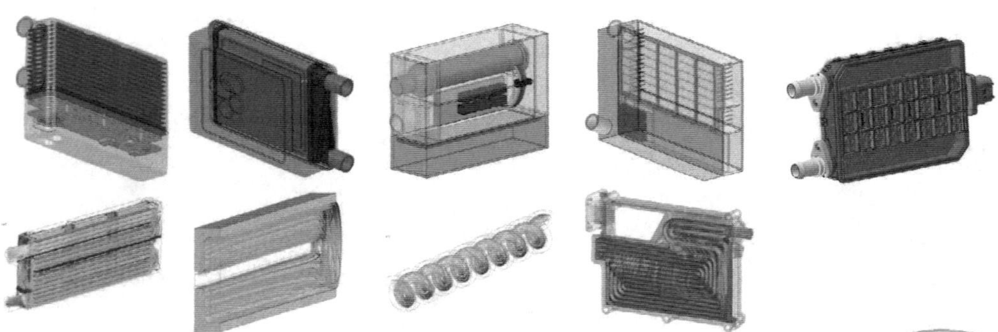

Automotive Simulation
WORLD CONGRESS

Concept variants

Webasto
Feel the Drive

Example of concept variant simulations (discarded concept):
Influence of channel design on pressure drop.
Evaluation locations have to be chosen with care - no 'blind' design optimization.

Pressure drop

Automotive Simulation
WORLD CONGRESS

Final concept
Design optimization

Webasto
Feel the Drive

Optimization of channel geometry:

Parameter study on single segment.

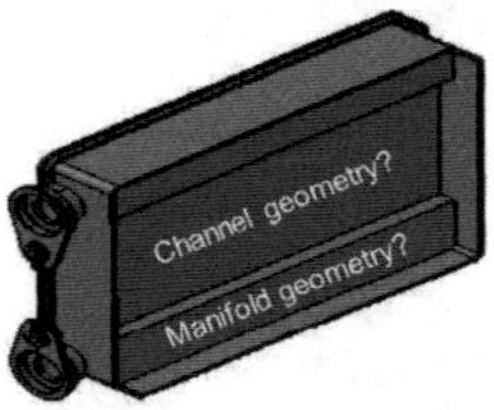

Optimization of even flow distribution on all segments:

Parameter study of inlet manifold and channel inlet geometry.

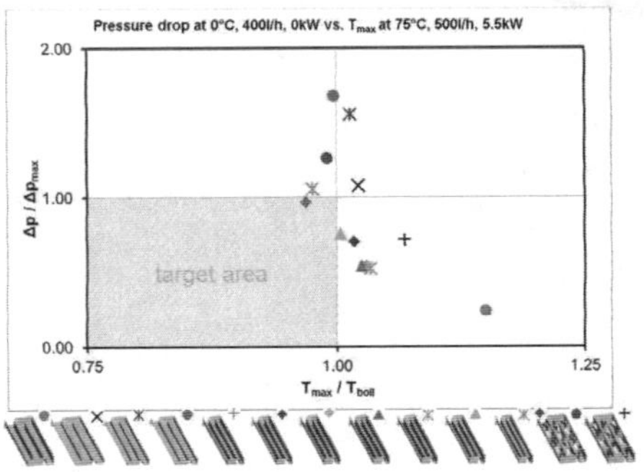

	deviation from mean [%]
Measurement Position 1	-9.1
Measurement Position 2	-4.4
Measurement Position 3	-0.4
Measurement Position 4	0.7
Measurement Position 5	2.0
Measurement Position 6	2.8
...	...
Measurement Position n	3.7

Successful simulation based design of the heat exchanger geometry that meets/exceeds the requirements

Automotive Simulation
WORLD CONGRESS

Final concept Validation

Webasto
Feel the Drive

Validation of the final concept

Pressure drop between inlet and outlet.

Simulation vs. measurement data

Maximum coolant inlet / outlet temperature allowed to avoid boiling.

Simulation vs. measurement data

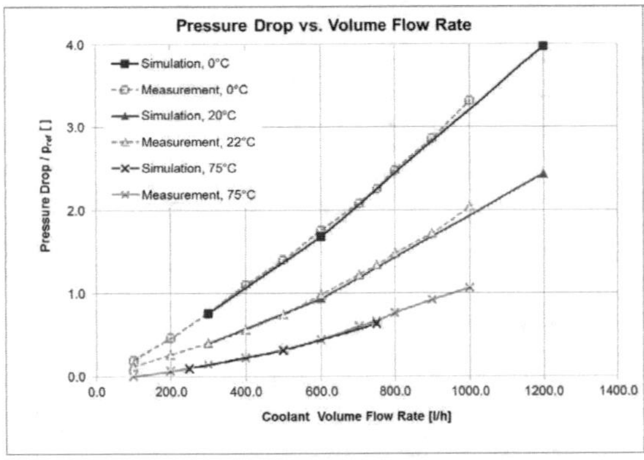

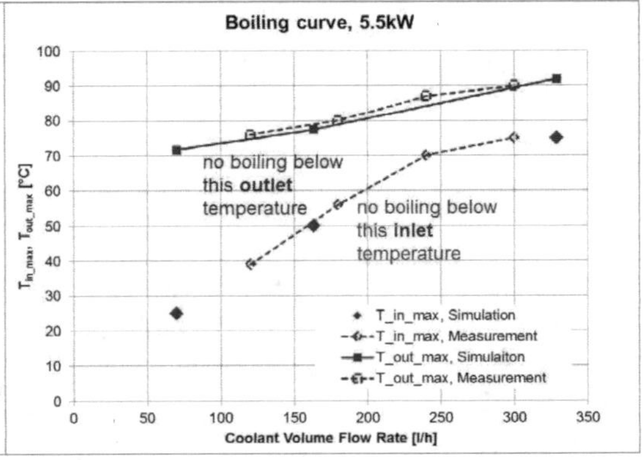

Excellent agreement of predicted and measured data.

Automotive Simulation WORLD CONGRESS

Heat layer layout

Webasto
Feel the Drive

Objectives:

Even temperature distribution
Reduce local temperature peaks
Even current distribution
Efficient manufacturing

Software:

Design Modeler
ANSYS Meshing
ANSYS CFX
ANSYS Mechanical

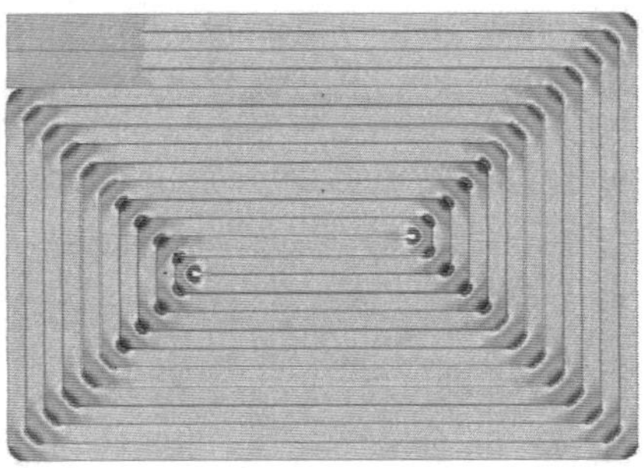

Automotive Simulation WORLD CONGRESS

Heat layer concept

Webasto Feel the Drive

Concept:

- Surface: ~200cm²
  - Cohesive connection to heat exchanger
  - No thermal losses at boundaries
  - Even heat up of coolant over entire surface
- Total thickness: 0.6mm
- Metal heating element with almost constant ohmic resistance.

Outline of Webasto heat layer structure

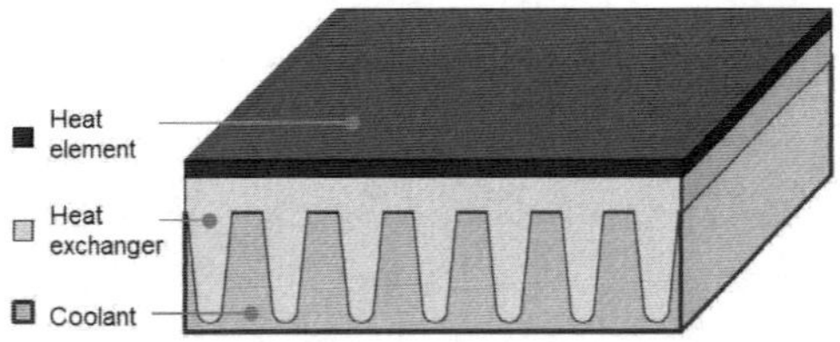

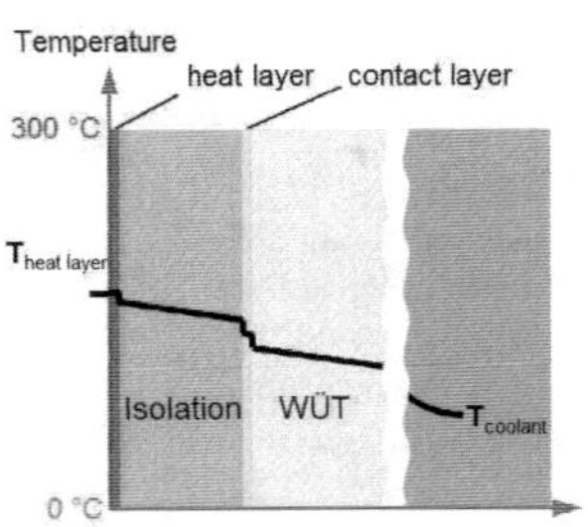

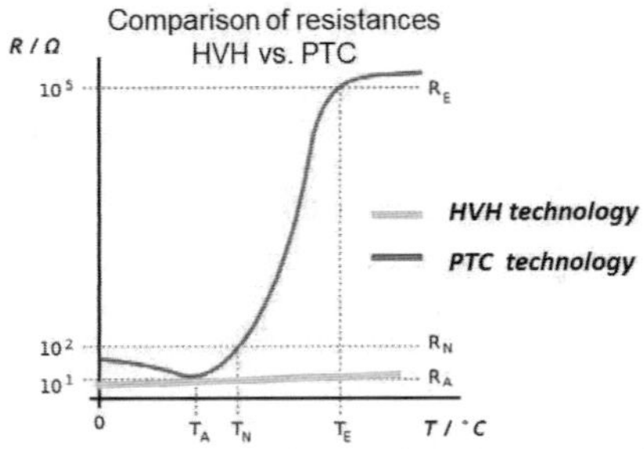

Heat layer layout
Pre tests

Webasto Feel the Drive

Pilot tests on test setup to assess the applicability and accuracy of simulation based heat layer layout.

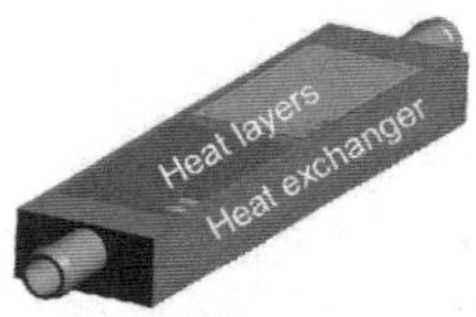

Thermography

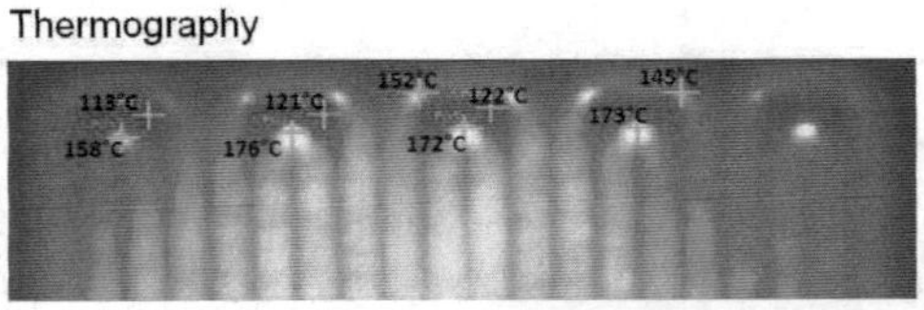

Simulation

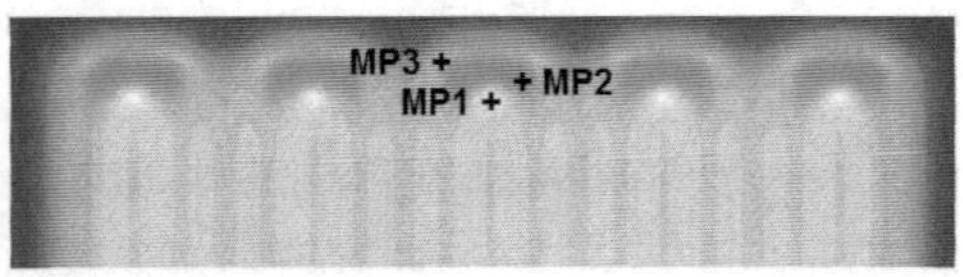

	Measurement [°C]	Simulation [°C]
M P1	172	176
MP 2	122	128
MP 3	152	157

Simulation proved to be a suitable tool for the design of the heat layer. Further heat layer optimization was done by simulation.

Heat layer layout
Optimization example

Webasto Feel the Drive

Problem: Temperature peaks at turning points lead to high thermal stresses in the heat layer structure.

Simplified simulation model was used to find design with reduced temperature peak values.

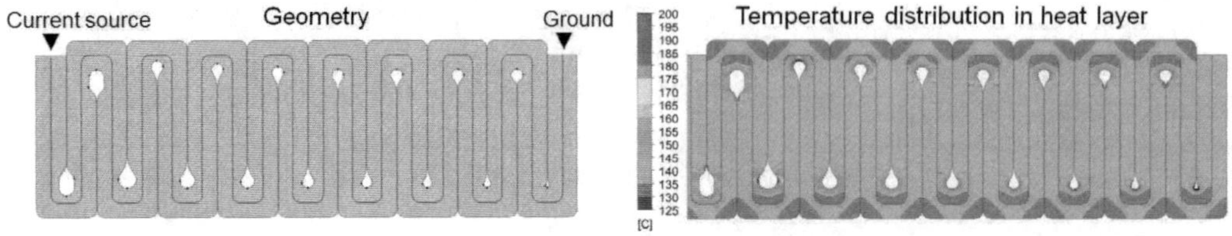

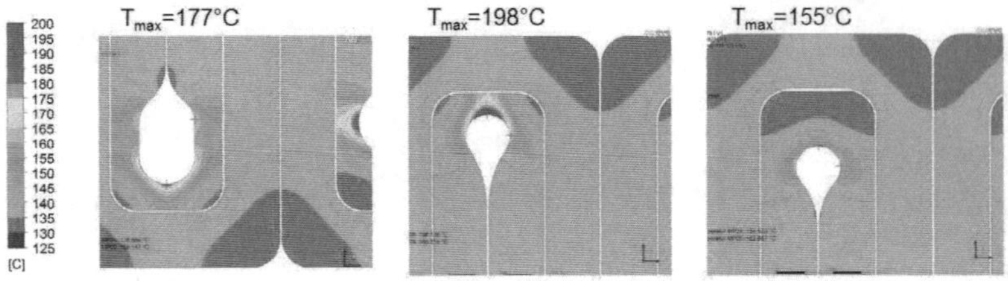

Detail view of temperature distribution in heat layer showing the effect of different designs

Automotive Simulation World Congress

IGBT dimensioning

Webasto Feel the Drive

Objectives:

Predict IGBT load spectrum to select appropriate IGBT
Predict active power cycling
Predict passive thermal cycling

Software:

Design Modeler
ANSYS Meshing
ANSYS CFX

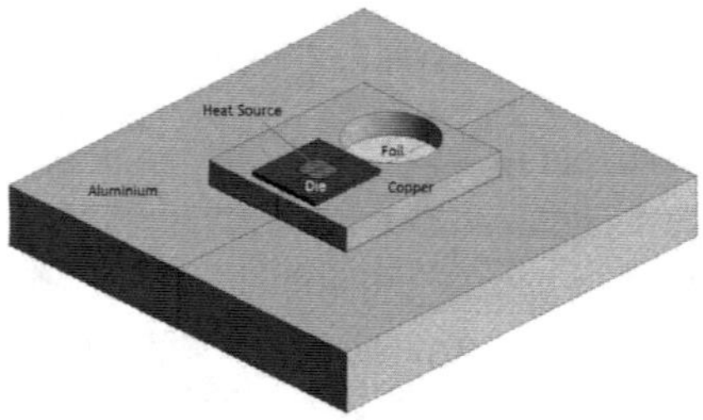

Automotive Simulation World Congress

IGBT dimensioning
Relevant load cycles

Webasto Feel the Drive

IGBT lifetime is determined by its...

... passive thermal cycling – external thermal cycles induced by the application the IGBT is used in.

Stresses die-substrate and bond wire connections.

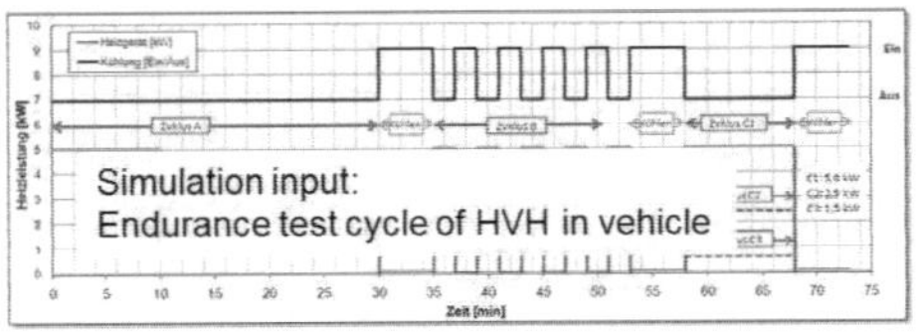

→

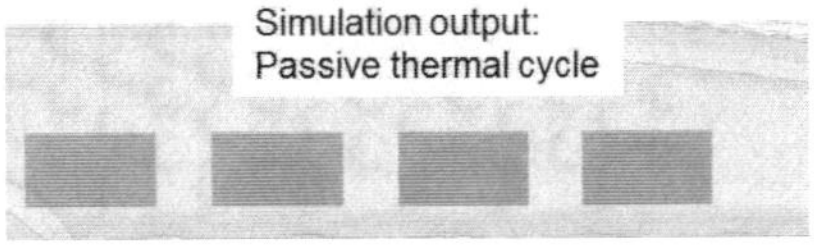

... active power cycling – internal heating of dies due to switching and steady operation.

Stresses solder between substrate and baseplate, leads to substrate solder delamination.

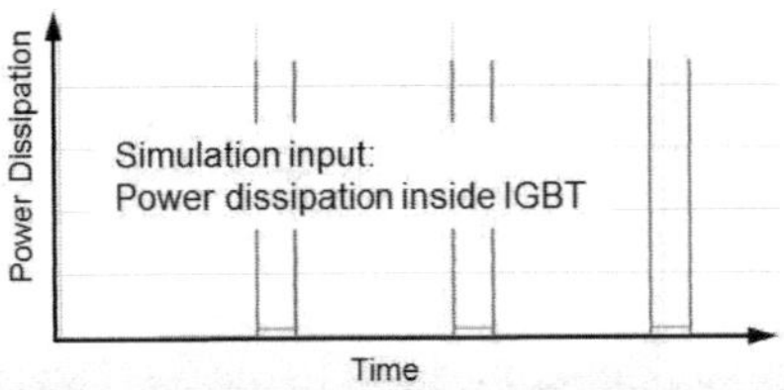

→

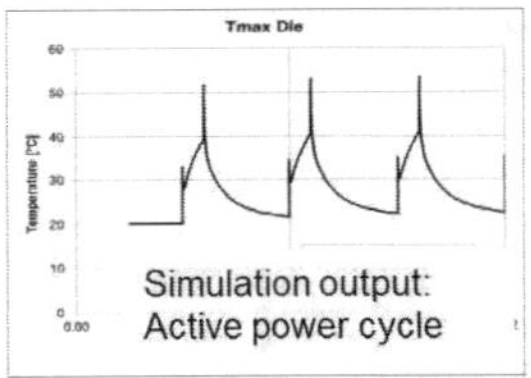

Automotive Simulation World Congress

IGBT dimensioning
Active power cycle

Webasto Feel the Drive

Simulation of IGBT internal heating during active power cycle.

PMW control induces extremely short but high power switching losses in addition to usual power dissipation during power-on.

Transient temperature distribution in cross sectional view through IGBT and simplified heat exchanger

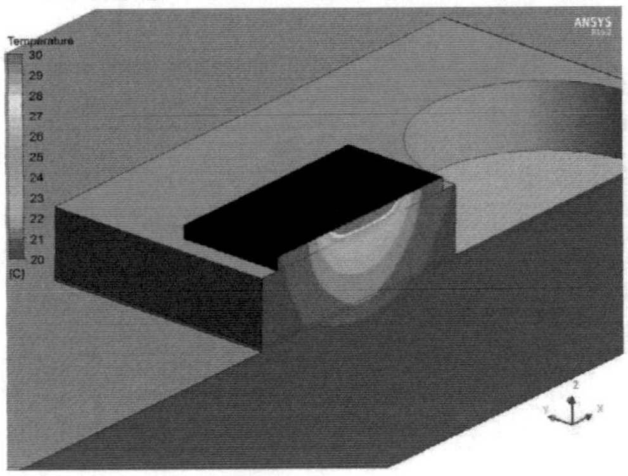

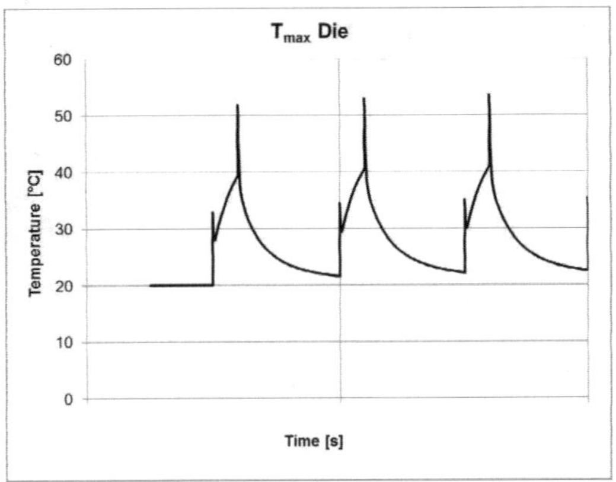

Automotive Simulation World Congress

IGBT dimensioning
Combining the results

Webasto Feel the Drive

Endurance test cycle of HVH in vehicle

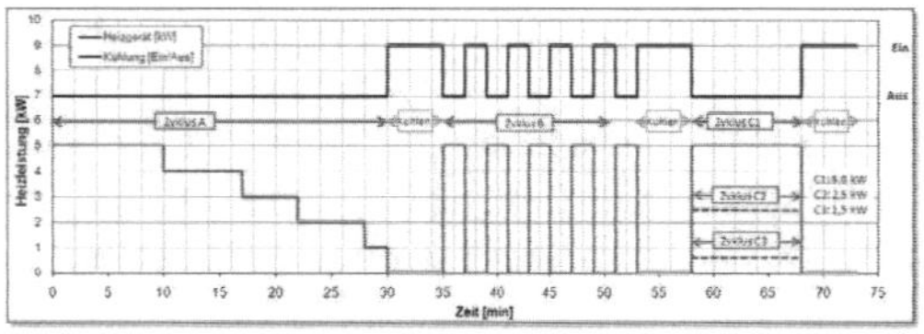

\+

Passive thermal cycles

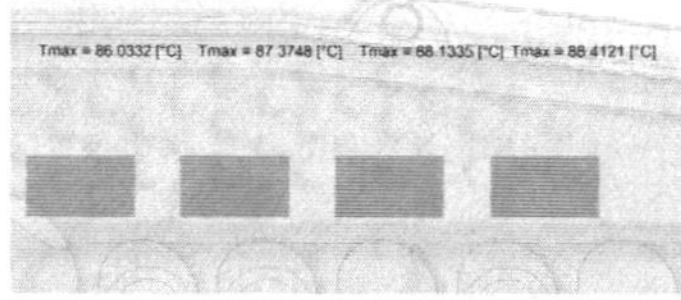

\+

Active power cycles

\+

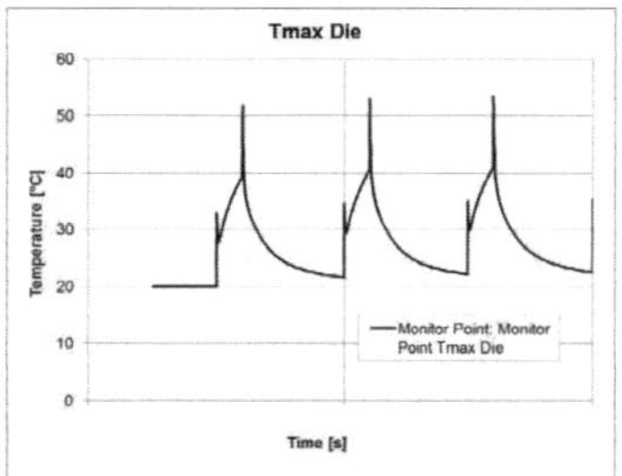

→ Load spectrum of IGBTs over lifetime

Simulation provided not measurable information that was essential for IGBT selection.

Thermal deformation and stresses

Webasto Feel the Drive

Objectives:
Predict HVH deformation
Predict heat layer deformation and curvature
Predict heat layer stresses

Software:
Design Modeler
ANSYS Meshing
ANSYS Mechanical
ANSYS CFX

Thermal deformation and stresses

Webasto
Feel the Drive

Different thermal expansion coefficients of heat layer materials induce stresses during heat up / cool down.

Deformation of the HVH and the connected heat layer due to heat up / cool down add additional stresses.

Thermal electric + static structural simulation combined with CFD Results (to include the coolants heat removal) to predict stresses.

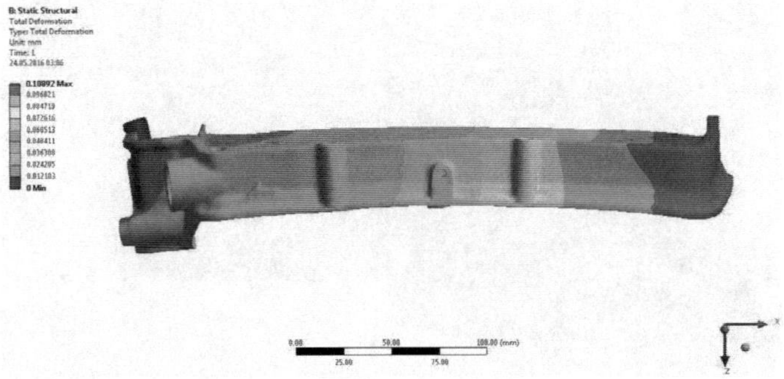

Total Deformation of HVH during heat up from 22°C to ~150°C

Automotive Simulation WORLD CONGRESS

Webasto
Feel the Drive

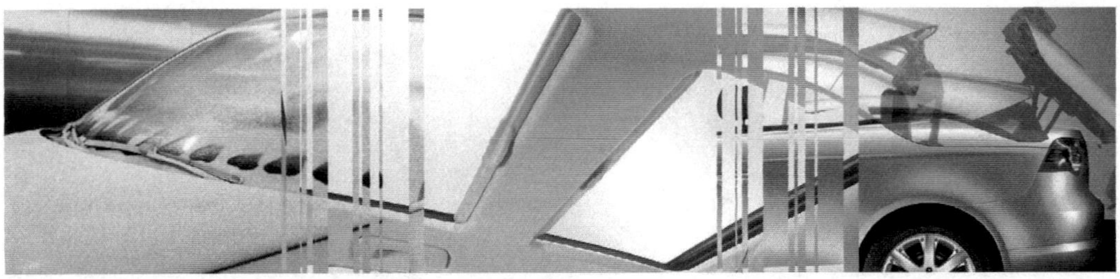

Thank you for your attention…
… Questions welcome!

Contact
Florian Larisch
Simulation R&D

Telephone +49 89 - 85794 1830
E-Mail florian.larisch@webasto.com

Automotive Simulation WORLD CONGRESS

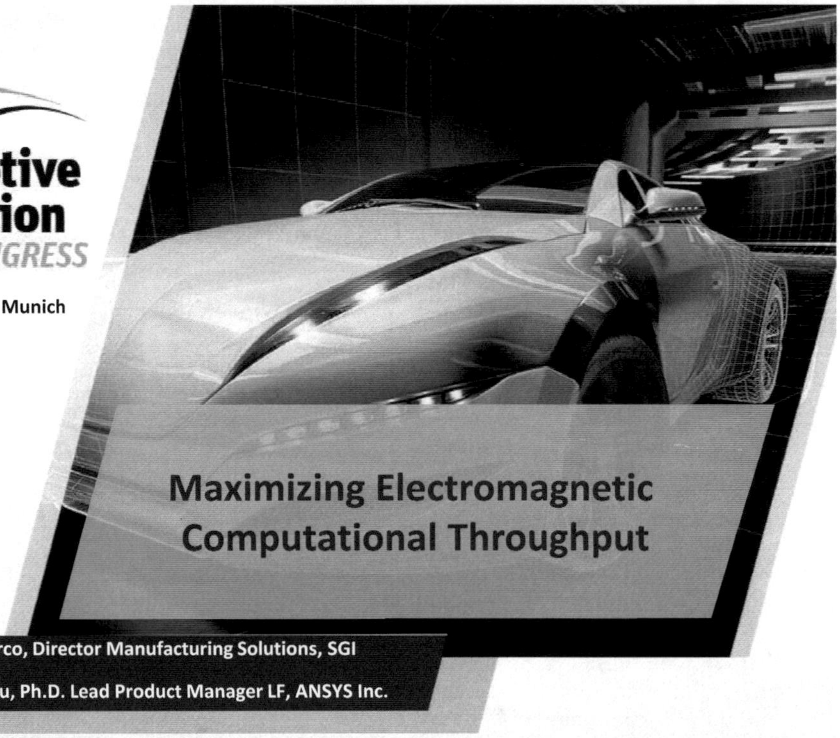

Automotive Simulation
WORLD CONGRESS

June 7-8, 2016 Munich

Maximizing Electromagnetic Computational Throughput

Tony DeVarco, Director Manufacturing Solutions, SGI

Marius Rosu, Ph.D. Lead Product Manager LF, ANSYS Inc.

Agenda

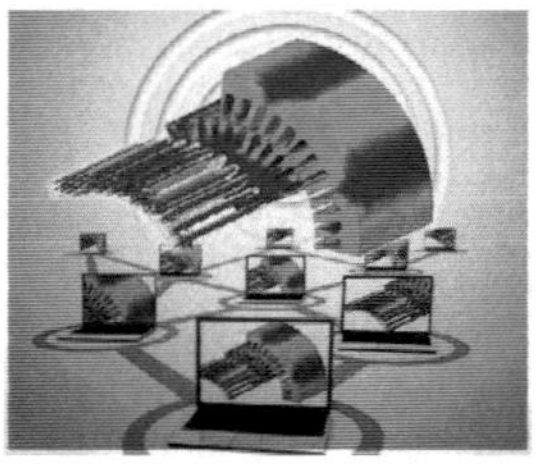

- **Engineering Insights Provided**
- **High Performance Computing**
  - Challenges
  - ANSYS LF Solution: "Time-Decomposition-Method"
  - Case-Studies
  - SGI® UV™ Overview
  - Resources

Engineering Insights Provided

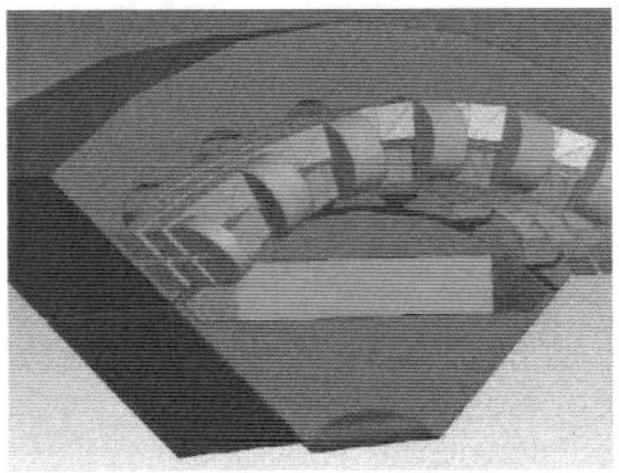

Electric Motors

Power Transformers

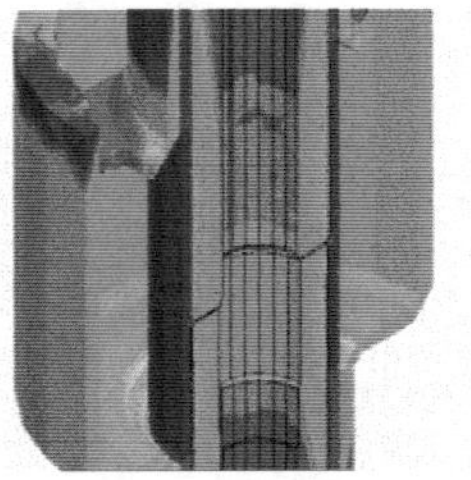

Actuators

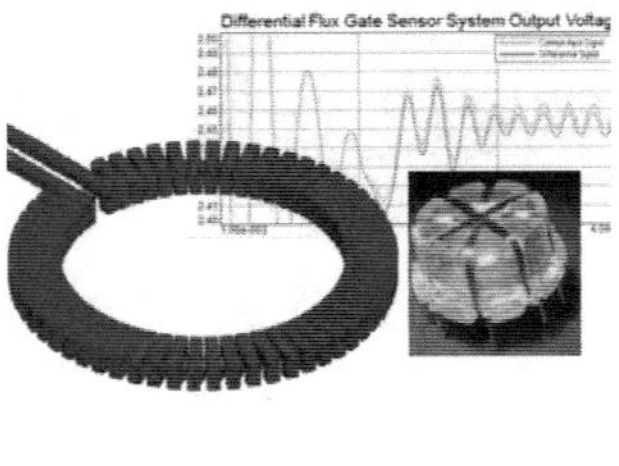

Sensors

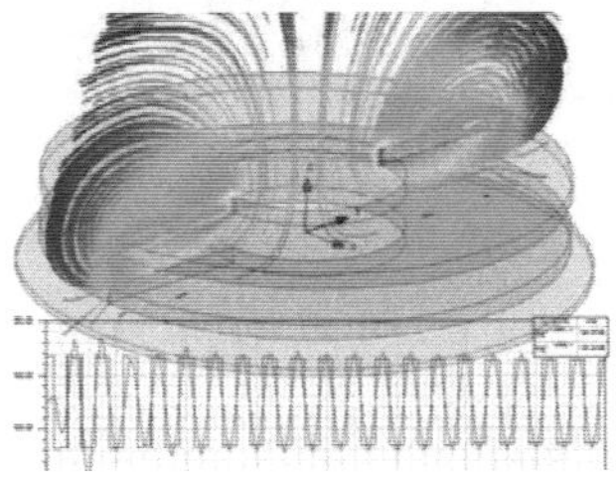

Wireless Power Transfer

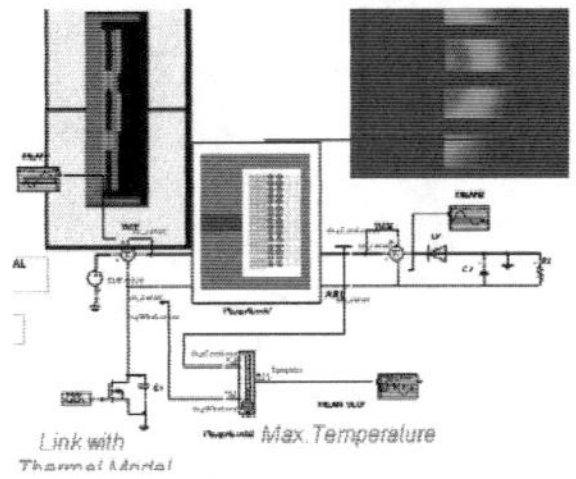

Power Electronics

High Performance Computing

Hardware

One Machine

2-8 Cores

Multiple Machines

Up to 32 Cores

Cluster or Shared Memory

32 to 1000's of Cores

Cloud

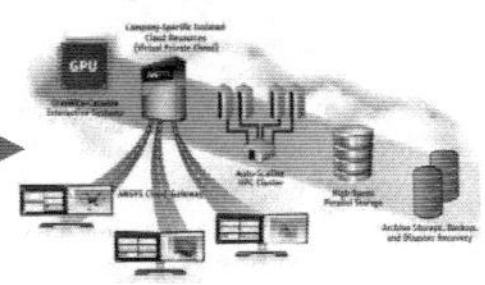

Elastic Number of Cores

Software

MPI-1 → MPI-2 → MPI-3 - - - →

Challenges for Commercial Software Development

Scalability	Stability	Robustness	Efficiency

Patent Pending

Time Decomposition Method

- Current approach – solve all time steps sequentially:

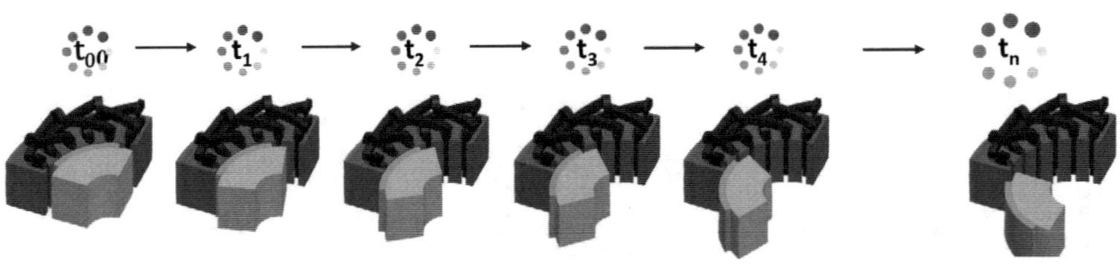

New approach (TDM)– solve all time steps simultaneously:

Patent Pending

Challenges for LF Electromagnetics

Nonlinear B-H curve

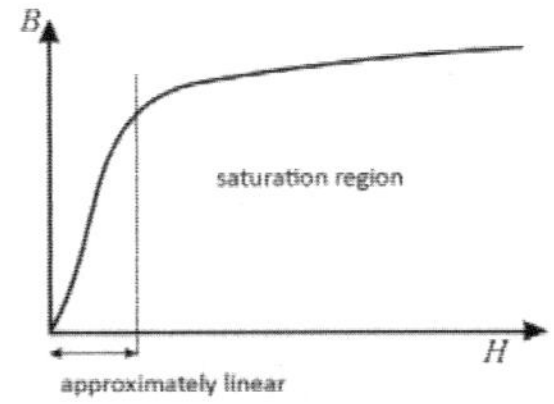

Motions

External circuits

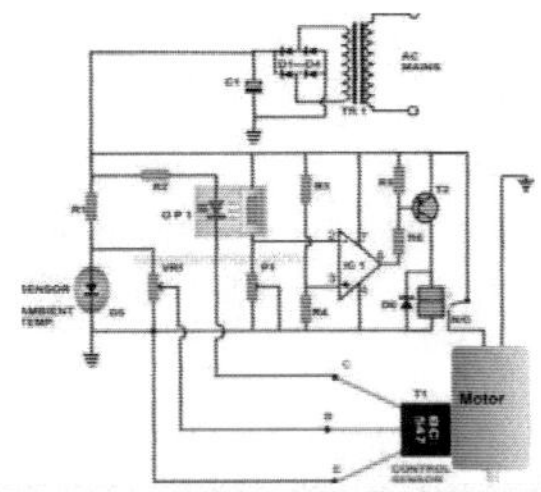

Automotive Simulation WORLD CONGRESS

Patent Pending

Multi-level HPC for Speed and Scale

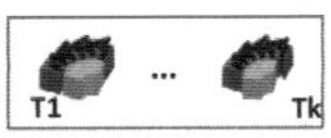

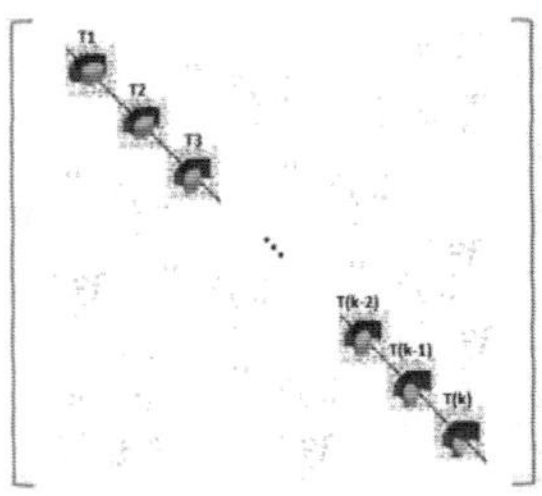

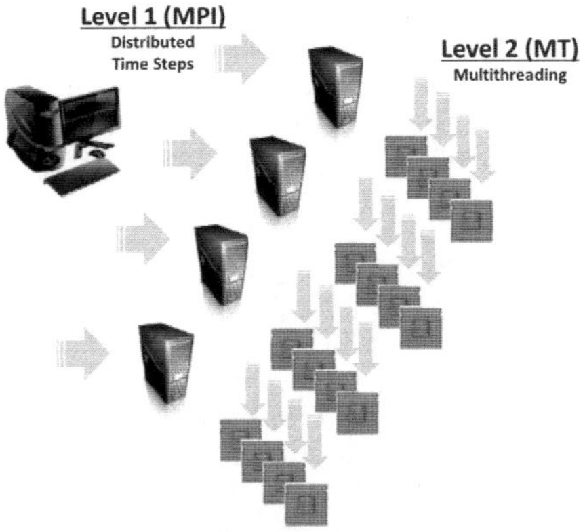

Game Changing
HPC Technology for ANSYS Maxwell - makes LF electromagnetic analysis the fastest commercial solver of its kind

Automotive Simulation WORLD CONGRESS

TDM Architecture

Patent Pending

Hybrid Parallel:
Low level: Shared memory parallel (MT)
High level: Distributed memory (MPI)

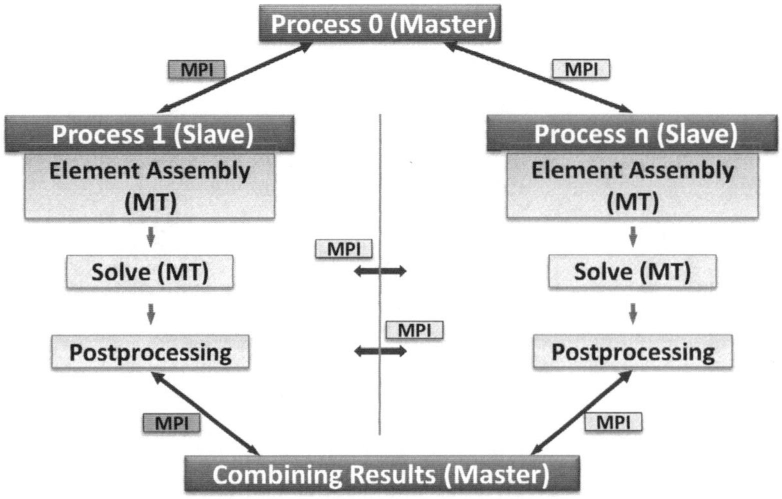

Distributed parallel: element assembly, solve and postprocessing

Patent Pending

Time Decomposition Method - Description

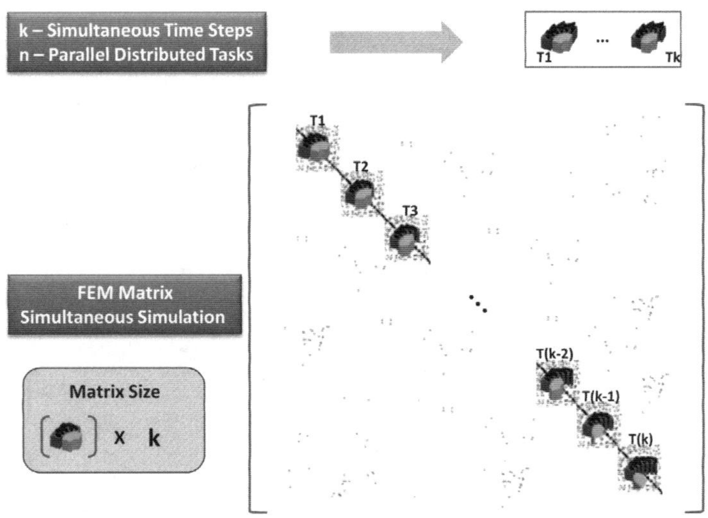

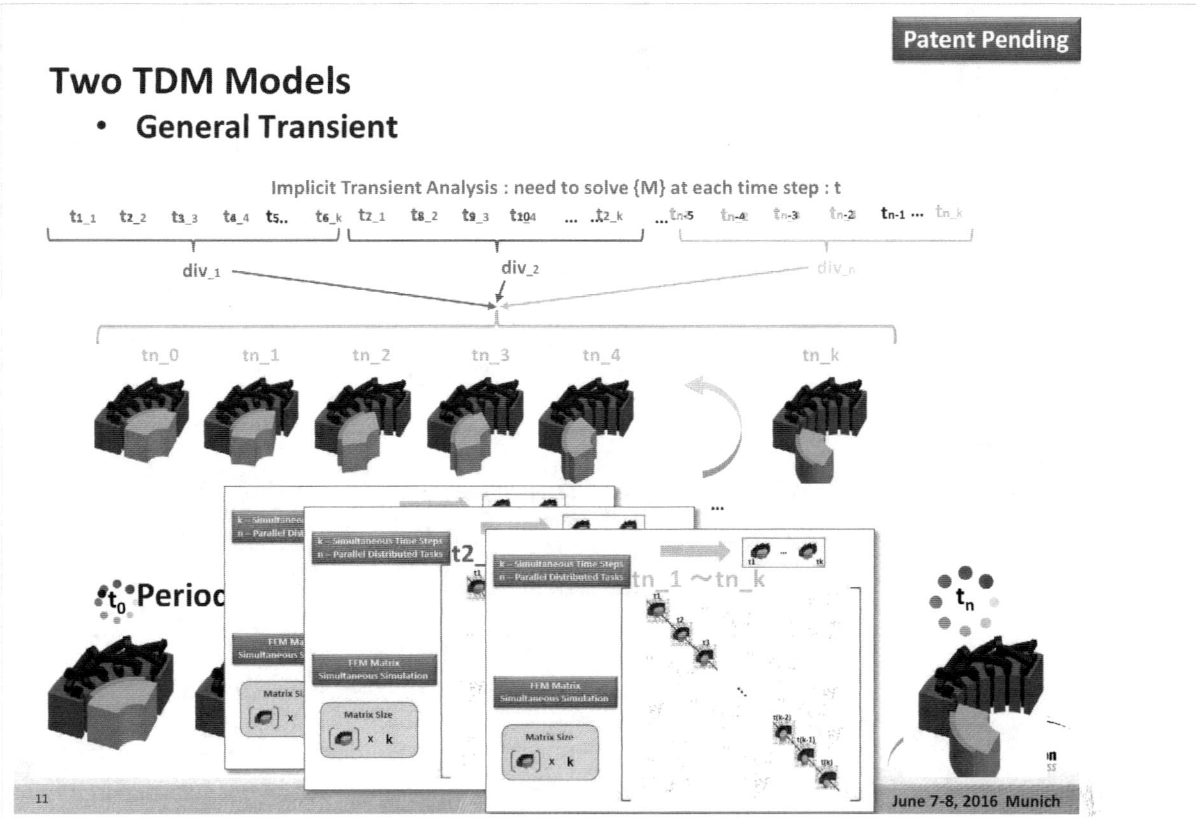
Patent Pending
Two TDM Models
• General Transient
Implicit Transient Analysis : need to solve {M} at each time step : t
div_1
div_2
tn_0
tn_1
tn_2
tn_3
tn_4
tn_k
Periodic
FEM Matrix Simultaneous Simulation
Matrix Size
k - Simultaneous Time Steps
n - Parallel Distributed Tasks
tn_1 ~ tn_k
11
June 7-8, 2016 Munich

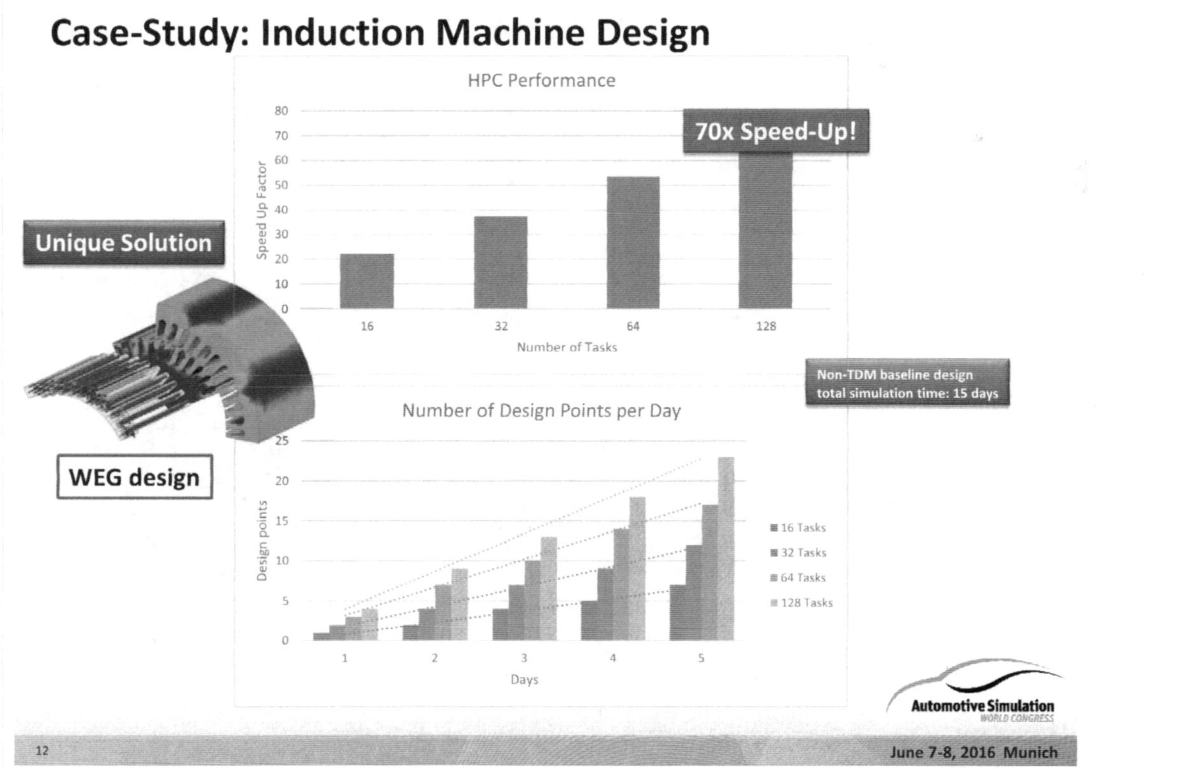
Case-Study: Induction Machine Design
HPC Performance
70x Speed-Up!
Speed Up Factor
Number of Tasks
16
32
64
128
Unique Solution
WEG design
Non-TDM baseline design total simulation time: 15 days
Number of Design Points per Day
Design points
Days
16 Tasks
32 Tasks
64 Tasks
128 Tasks
Automotive Simulation World Congress
12
June 7-8, 2016 Munich

Fast Solution for LF Electromagnetics Design

WEG Quote:

"This is amazing technology," said **Briam Cavalca Bork, Product Engineer**. "ANSYS Maxwell with Time Domain Method allows us to fully utilize our HPC hardware

Simulation time has improved by a factor of 20x but more importantly we have gained more in-depth and timely design insight that will allow us to deliver industry leading, innovative machine designs.

complex analyses on a greater variety of scenarios."

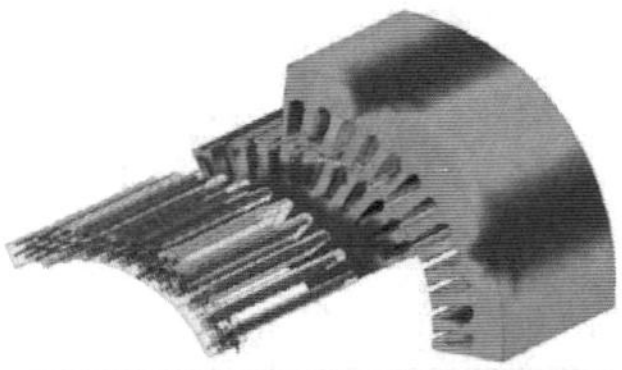

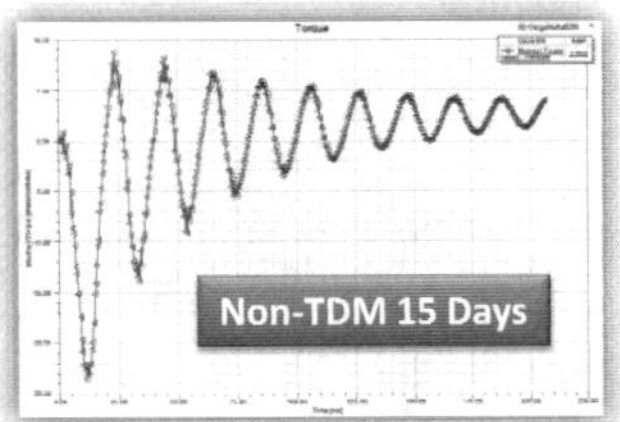

# Tasks	# Total Cores	TDM Matrix Size	TDM Simulation Time
16	128	16.5 Million	16h:10min
32	256	34.1 Million	9h:39min
64	512	69.3 Million	6h:44min
128	1024*	139.7 Million	5h:14min

Benchmark system used: SGI UV 3000; 4 TB memory @ 2133MHz; E5-4627v3 @ 2.6 GHz; 64 sockets; 640 cores (10 per socket)

* Benchmark system used: SGI UV 3000; 8 TB memory @ 2133 MHz; E5-4667v3 @ 2.0 GHz; 64 sockets; 1024 cores (16 per socket)

Case-Study: 8kW BLDC Motor 2000 rpm Design

880,000 Tets	1,700,000 DoF
17 GB RAM	1000 Time Steps

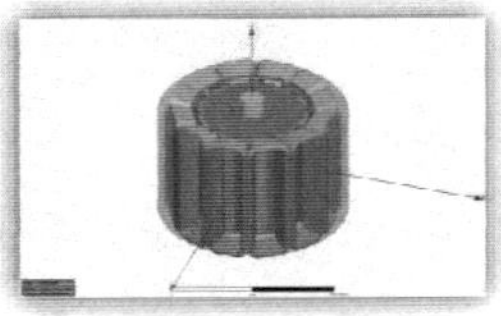

NodesXCores	Total Cores	Tasks	Subdiv.	DOFs/Subdiv.	Solve Time	Speed Up
1X8	8	-	1	1.7Mil	270hours18min	-

SGI UV 3000

# Tasks	# Total Cores	TDM Matrix Size	TDM Simulation Time	Speed up
99	99 (1 core/task)	166.6 Million	9h:00min	30X
99	396 (4 cores/task)	166.6 Million	4h:50min	60X

30x

60x

2.6 GHz; 64 sockets; 640 cores (10 per socket)

SGI HPC Products

Large scale-out, distributed memory models:
Best in Class speed, scale and efficiency

SGI ICE Platform

Fat-node cluster, large models.
Ease of use for small depts & businesses

SGI UV 30/3000

Customers that value factory integrated solutions for Technical Computing and Big Data.

SGI Rackable

For user files, fast scratch, active archiving

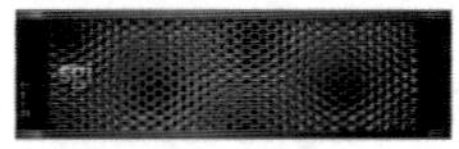

SGI InfiniteStorage

Seven Generations of Shared Memory Systems: 1996-2015

Origin 2000, 1996
MIPS R10K

Origin 3000, 1999
MIPS R12K

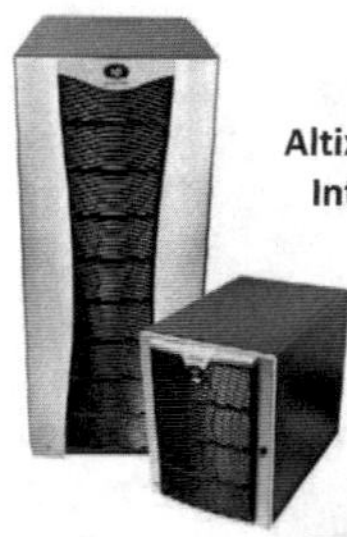

Altix 3000, 2003
Intel Itanium

Altix 4700, 2006
Intel Itanium

UV 1000, 2010
Intel E7

UV 2000/3000, 2012/2015
Intel E5

UV 300, 2014
Intel E7

ANSYS

sgi

SGI® UV™ 3000

Suggested for CAE Workloads

- Up to 64 sockets (640 cores), 16 TB shared memory/rack
- Intel® Xeon® Processor E5-4627 v3 10-core (2.60 GHz)
- NVIDIA® Tesla® and Intel® Phi™ support
- NVIDIA® Quadro K5200 or M6000 graphics support
- SGI NUMALink™ 6 Interconnect
- SUSE or RedHat
- Altair PBS Professional Batch Scheduler
- SGI® VizServer™ available for remote visualization

Individual Rack Unit (IRU) and Blades

Notes
- IRU: 10U high by 19" wide by 27" deep
- 8 blades – 16 sockets – per IRU
- 3 UV1 12V Power Supplies
- Nine 12V cooling fans N+1

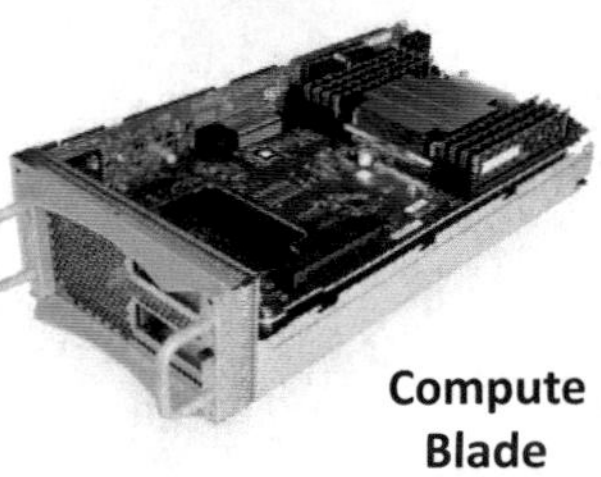

Compute Blade
Front

Accelerator Blades

ANSYS

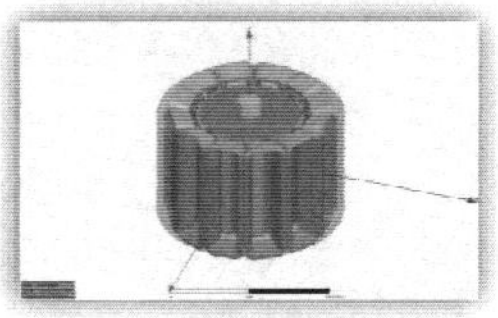

SGI UV 3000 Benchmark System used:
4 TB of Shared Memory @ 2133MHz
64 sockets; 640 cores Intel® Xeon ® E5-4627v3 @ 2.6 GHz;

ANSYS Maxwell and SGI ® UV™ 3000 Maximizes Electromagnetic Computation Throughput

http://www.ansys.com/Products/Release-Highlights/electronics-10x-faster-transient-electromag-field-simulation

sgi

Contact Information

- **Tony DeVarco, SGI** tdevarco@sgi.com
- **Marius Rosu, Ph.D., ANSYS** marius.rosu@ansys.com

Automotive Simulation
WORLD CONGRESS

June 7-8, 2016 Munich

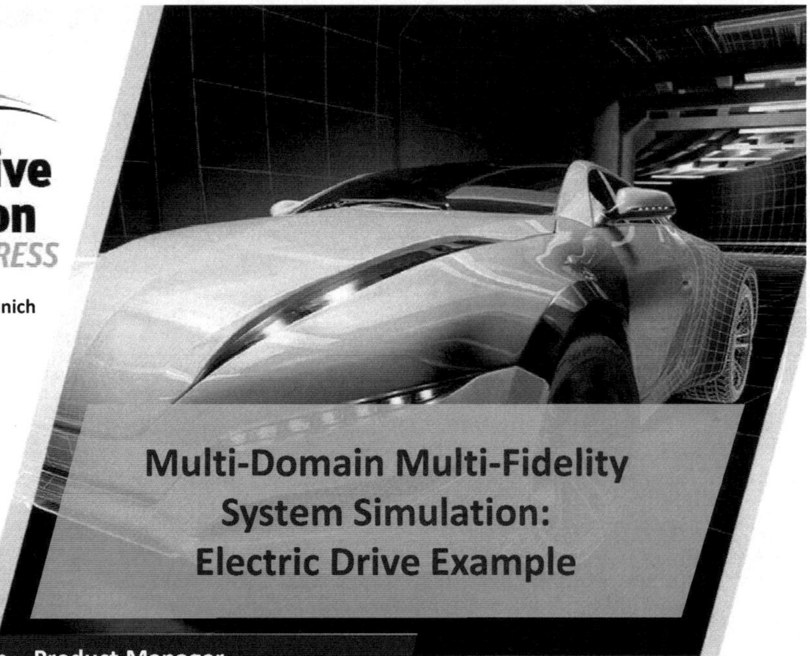

Multi-Domain Multi-Fidelity System Simulation: Electric Drive Example

Lee Johnson – Product Manager
ANSYS – California, USA

Agenda

- A System Example
- Keys to Model Assembly
- Detail & Fidelity
- Model Selection: Driven by Analysis

Electric Drive Systems

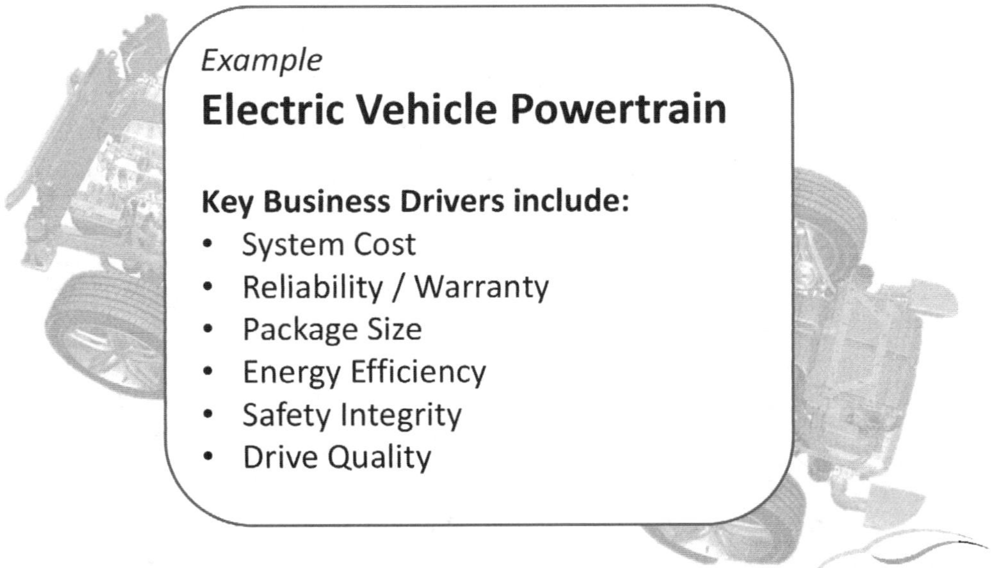

Example

Electric Vehicle Powertrain

Key Business Drivers include:

- System Cost
- Reliability / Warranty
- Package Size
- Energy Efficiency
- Safety Integrity
- Drive Quality

Electric Drive System

Key Components

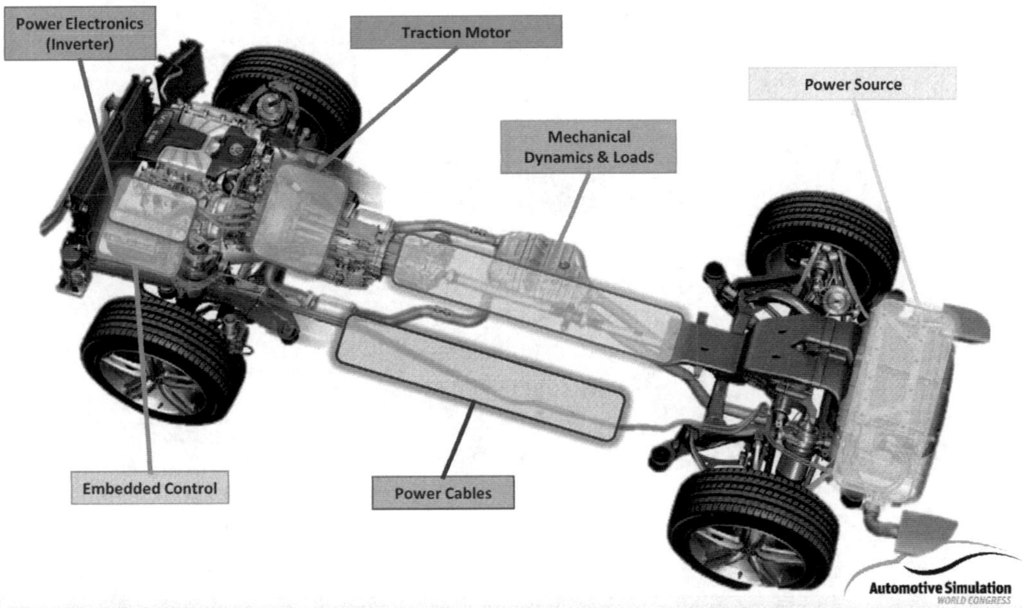

Electric Drive System

As a System Model

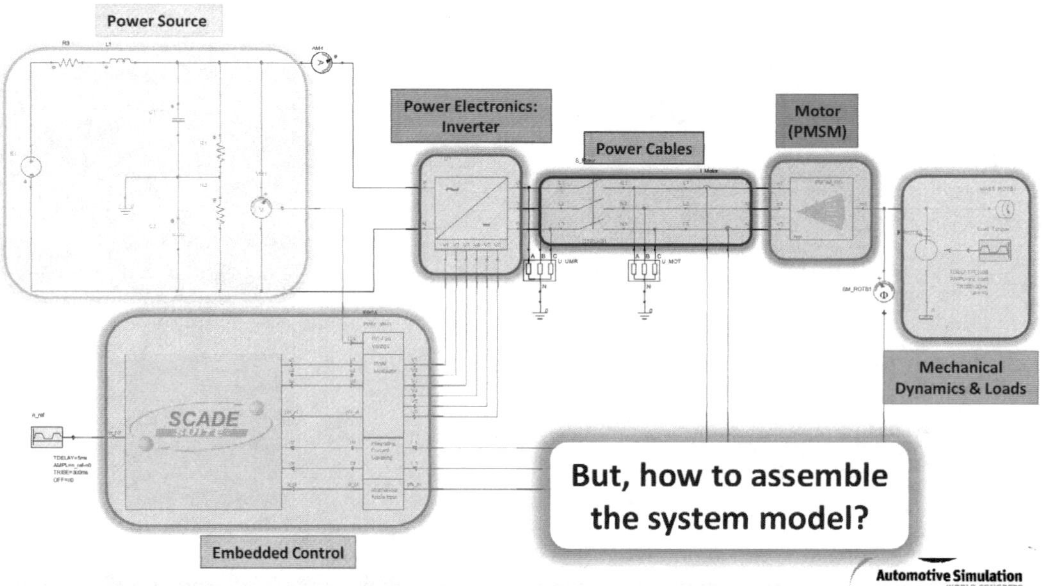

Flexibility, Reusability & Interoperability are Keys

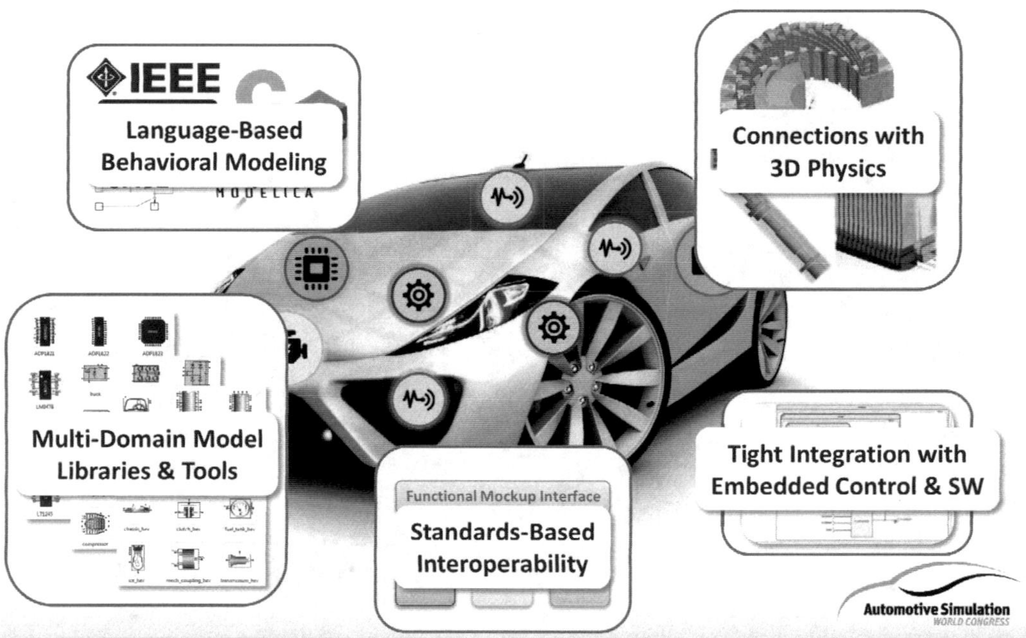

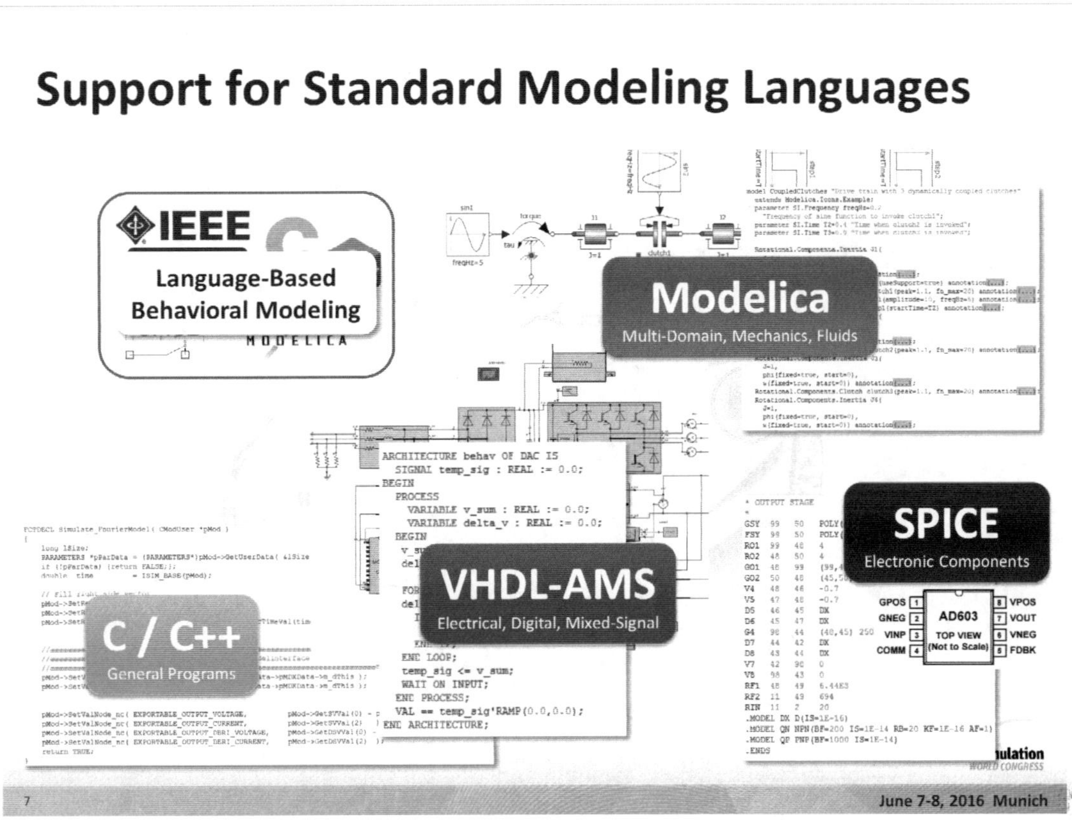
Support for Standard Modeling Languages
IEEE
Language-Based
Behavioral Modeling
Modelica
Multi-Domain, Mechanics, Fluids
SPICE
Electronic Components
VHDL-AMS
Electrical, Digital, Mixed-Signal
C / C++
General Programs
AD603
TOP VIEW
(Not to Scale)
GPOS
GNEG
VINP
COMM
VPOS
VOUT
VNEG
FDBK
ARCHITECTURE behav OF DAC IS
SIGNAL temp_sig : REAL := 0.0;
BEGIN
PROCESS
VARIABLE v_sum : REAL := 0.0;
VARIABLE delta_v : REAL := 0.0;
BEGIN
END LOOP;
temp_sig <= v_sum;
WAIT ON INPUT;
END PROCESS;
VAL == temp_sig'RAMP(0.0,0.0);
END ARCHITECTURE;
7
June 7-8, 2016 Munich

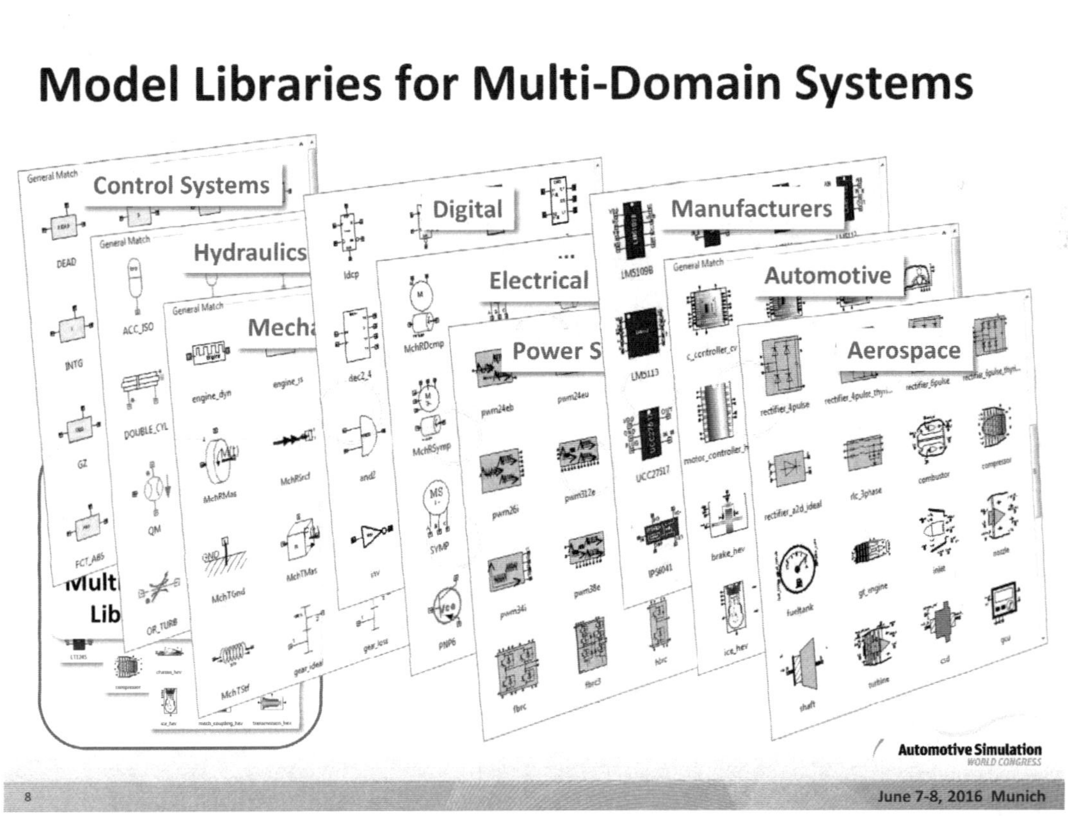
Model Libraries for Multi-Domain Systems
Control Systems
Hydraulics
Mecha
Digital
Electrical
Power S
Manufacturers
Automotive
Aerospace
Automotive Simulation
WORLD CONGRESS
8
June 7-8, 2016 Munich

Reduced-Order Modeling

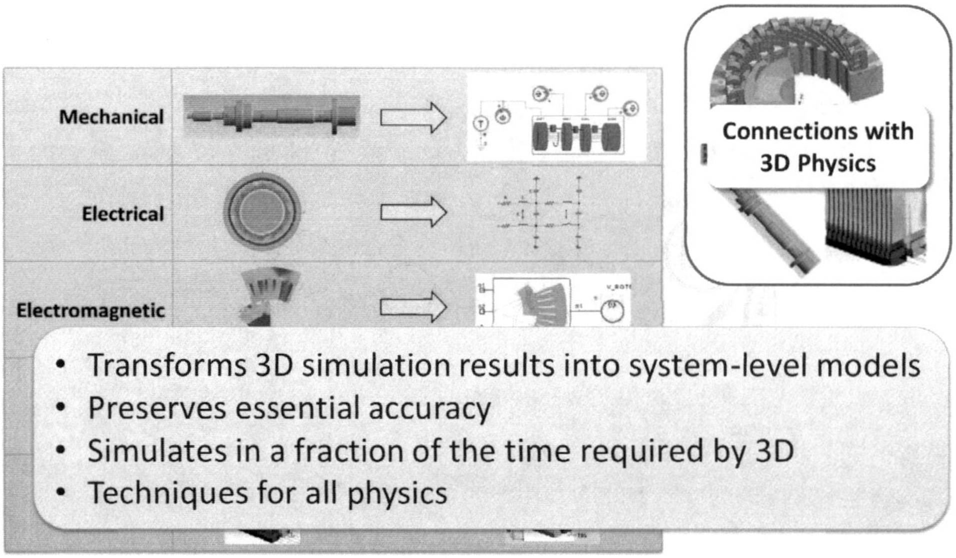

- Transforms 3D simulation results into system-level models
- Preserves essential accuracy
- Simulates in a fraction of the time required by 3D
- Techniques for all physics

Physical Modeling for Embedded Control

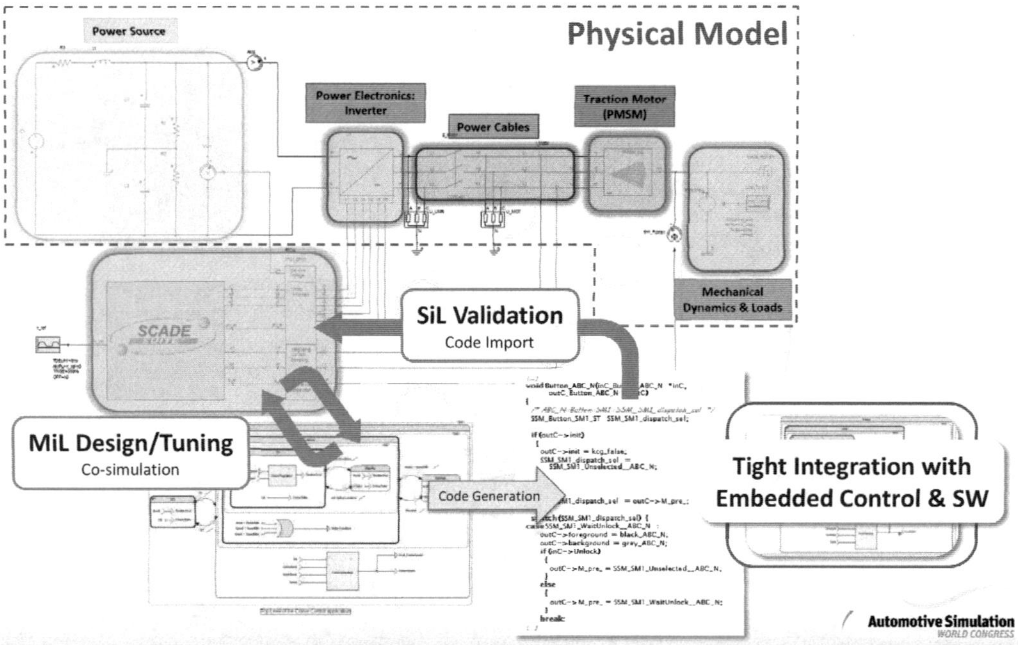

Standards-Based Interoperability

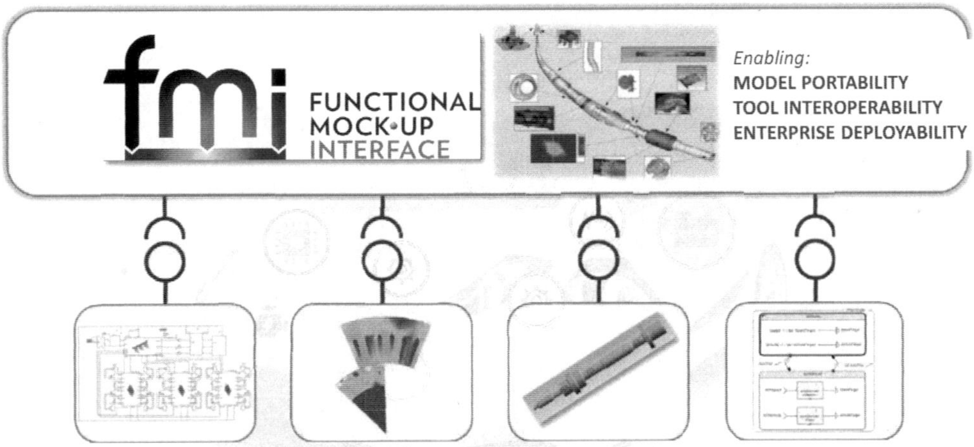

Aside: Behavioral Models

Mathematical Description of Behavior

$$L_f \frac{di_f}{dt} + R_f * i_f = v_f$$

$$I_f = \frac{V_f}{R_f}$$

$$L_a \frac{di_a}{dt} + R_a * i_a + e_a =$$

$$R_a * I_a + E_a = V_a .$$

$$E_a = K_m \phi \, w,$$

Expressed in a Formal

Characterized with Empirical Data

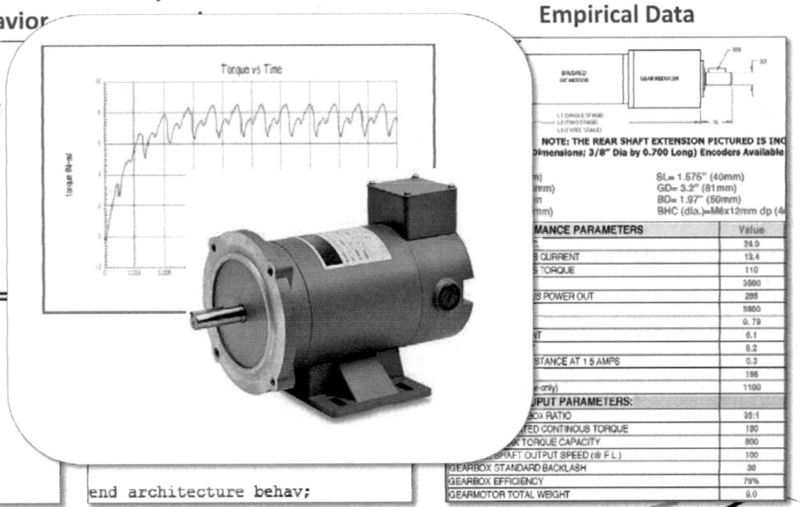

Behavioral Models in Product Development

In Many Forms...

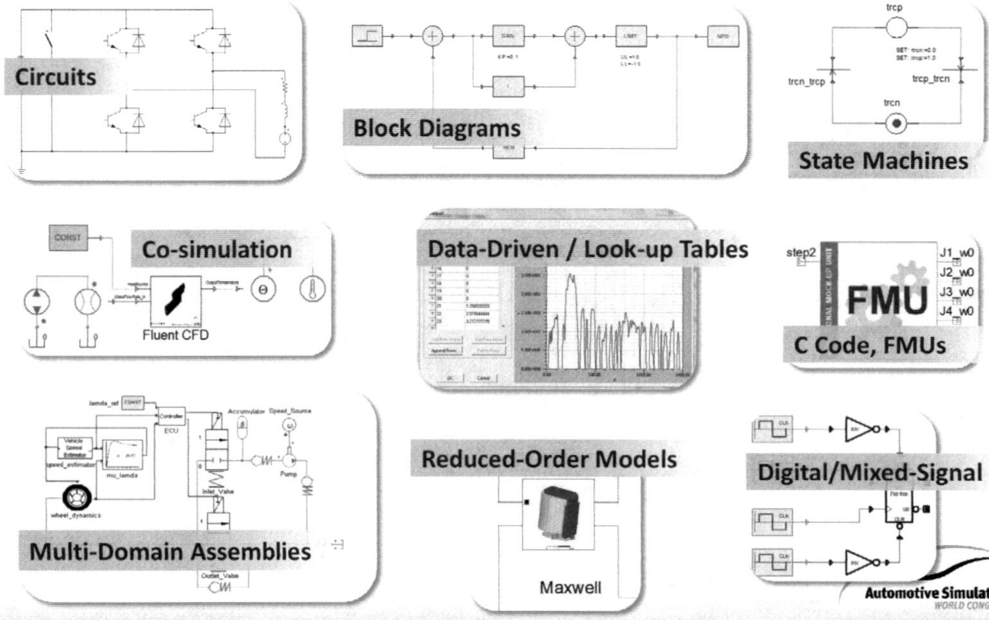

One More Concept: Model Detail & Fidelity

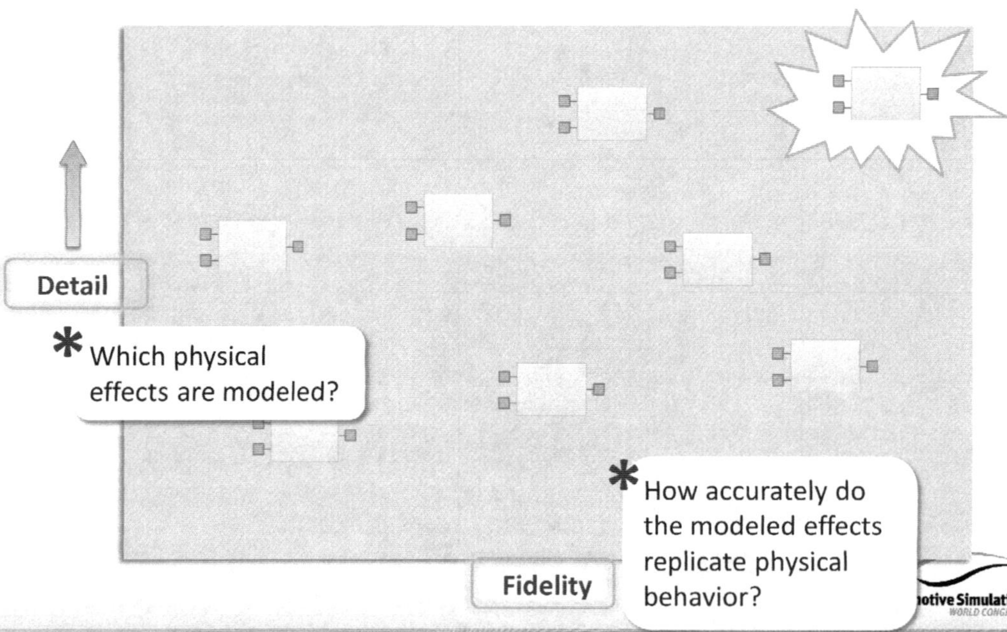

Back to the Example...

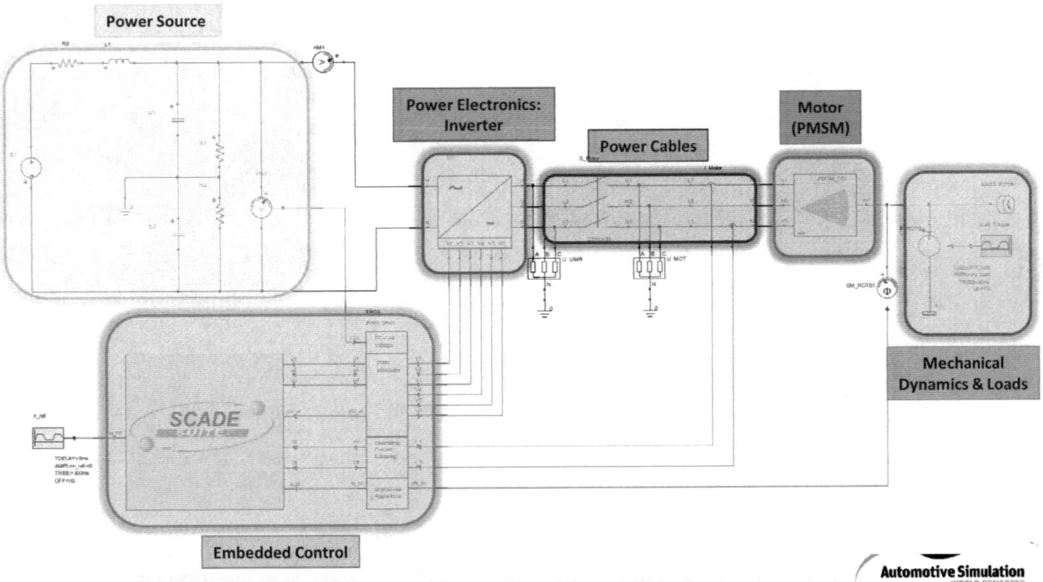

Automotive Simulation WORLD CONGRESS

Modeling the Drive System

Power Source

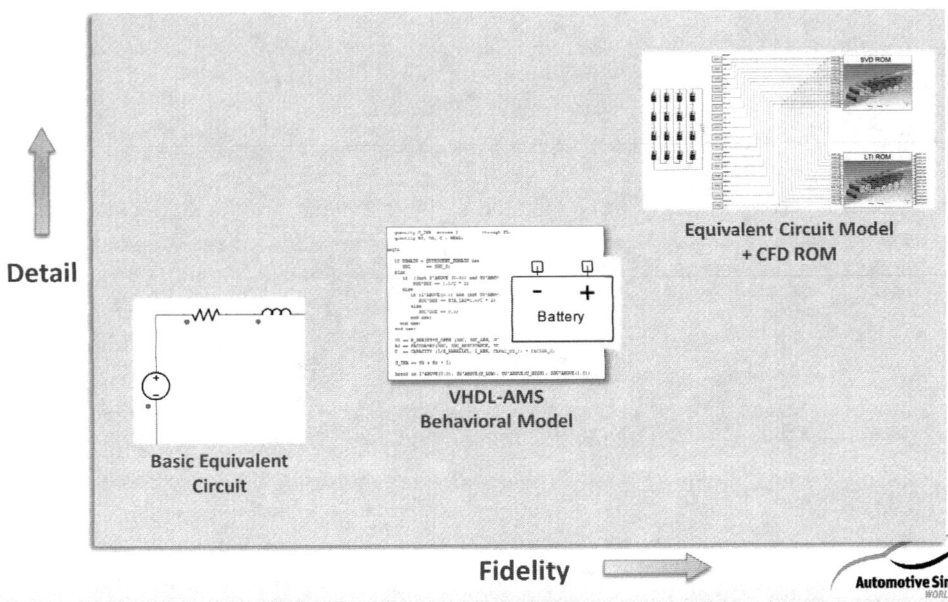

Automotive Simulation WORLD CONGRESS

Modeling the Drive System

Motor

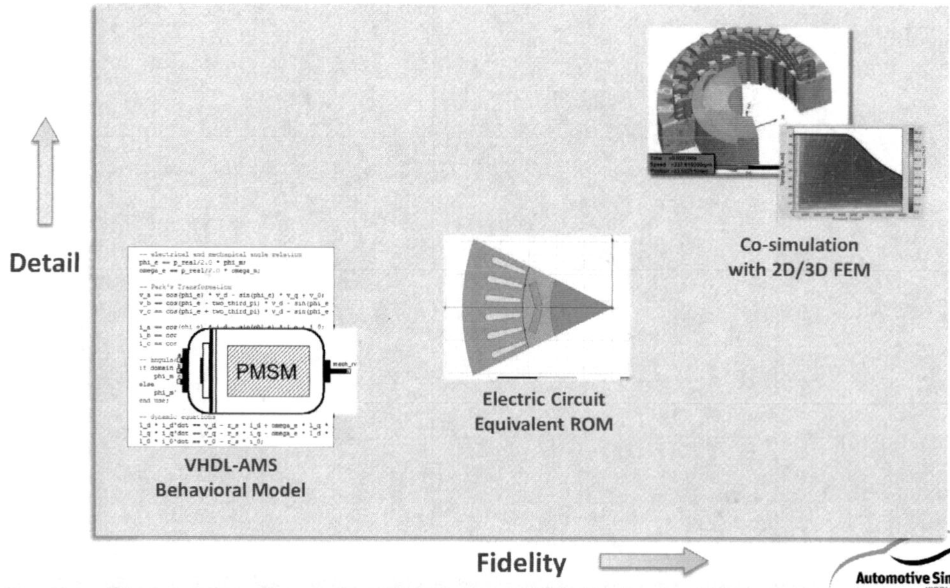

Modeling the Drive System

Power Cables

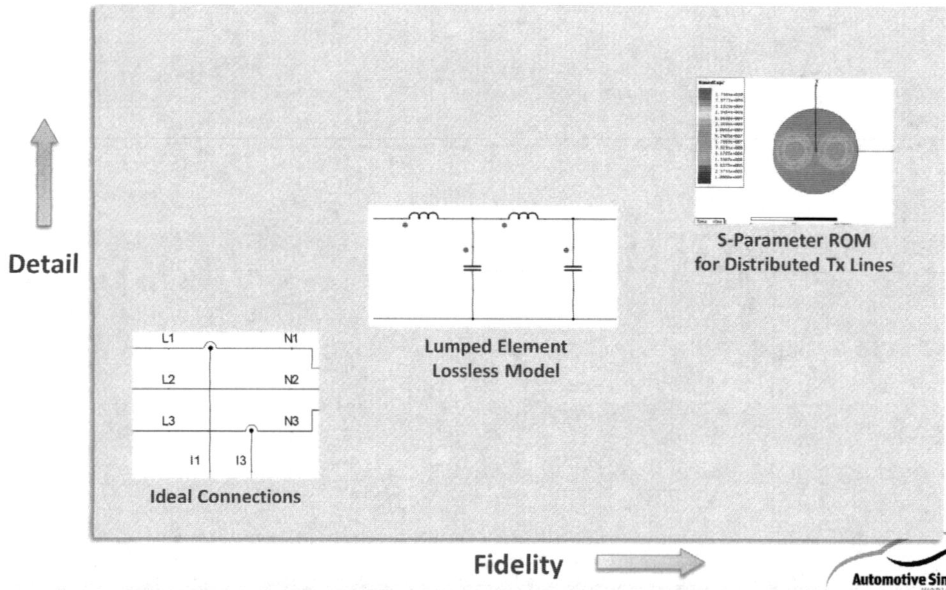

Modeling the Drive System

Mechanical Dynamics

Mechanical ROM of Flexible Shaft

Detail

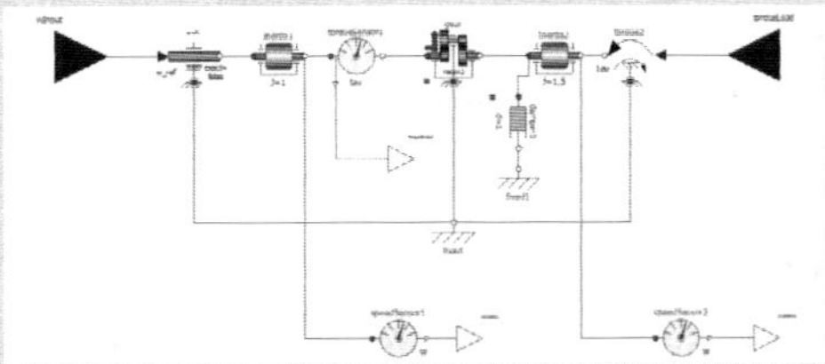

Modelica Behavioral Model

Fidelity

Modeling the Drive System

Embedded Control

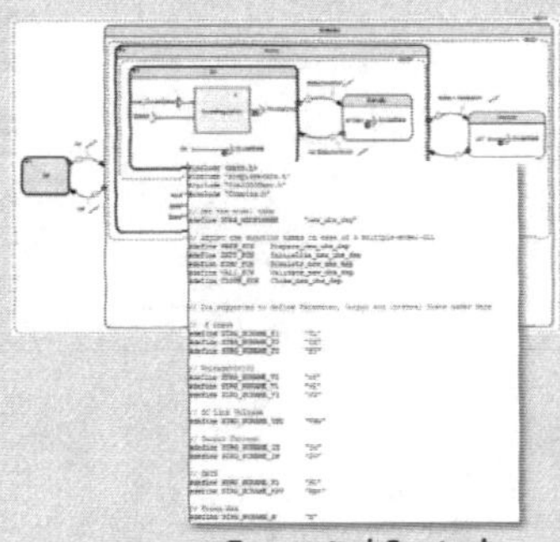

Detail

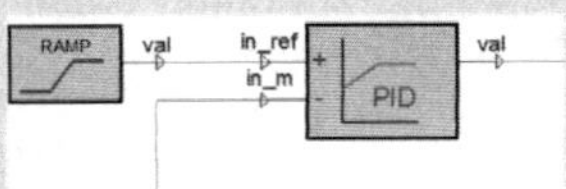

Ideal Control Blocks

Generated Control Application Code

Fidelity

Assembling & Analyzing the System

Goal: Evaluate System Architecture, Size Key Components

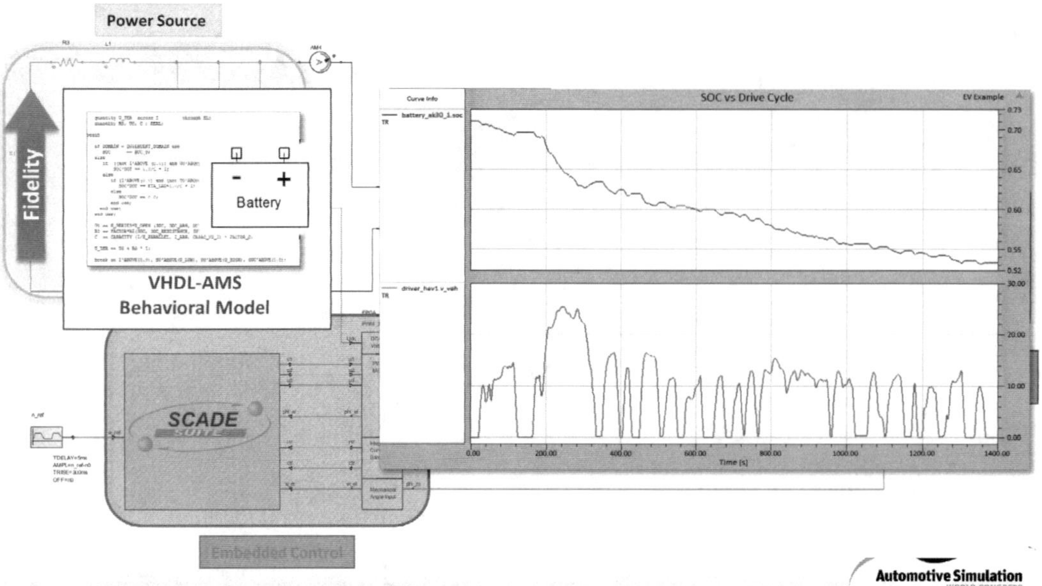

Assembling & Analyzing the System

Goal: Characterize Motor Losses

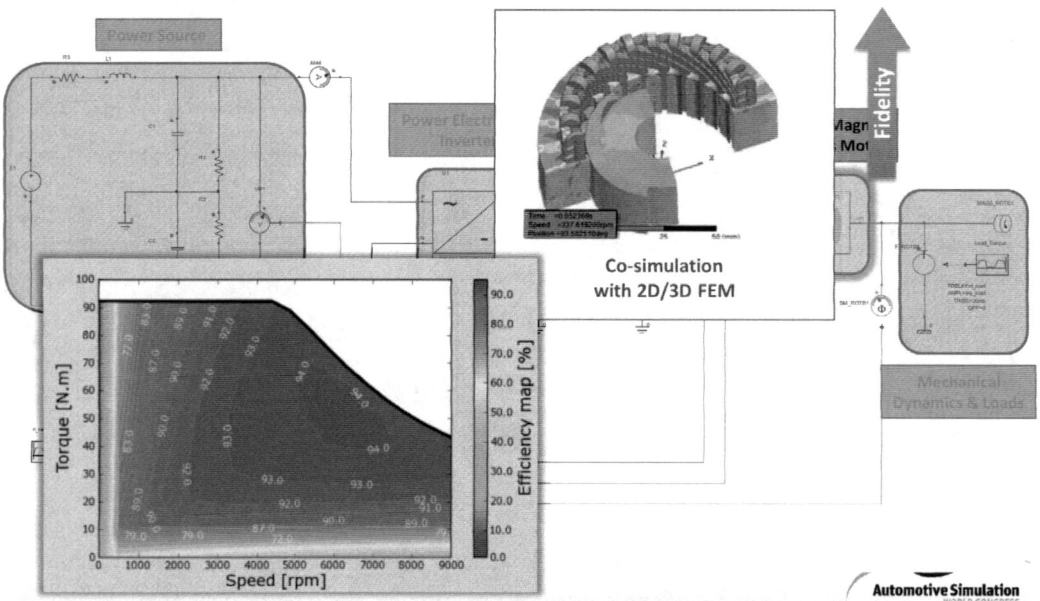

Assembling & Analyzing the System

Goal: EMC Prediction

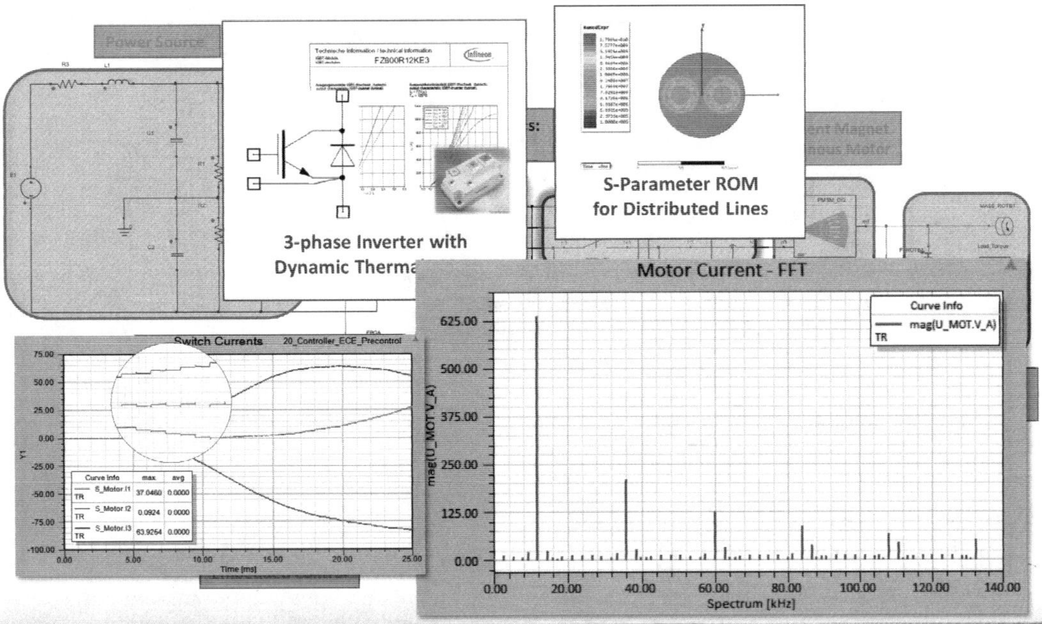

Assembling & Analyzing the System

Goal: MiL, SiL Testing / Calibration & Tuning

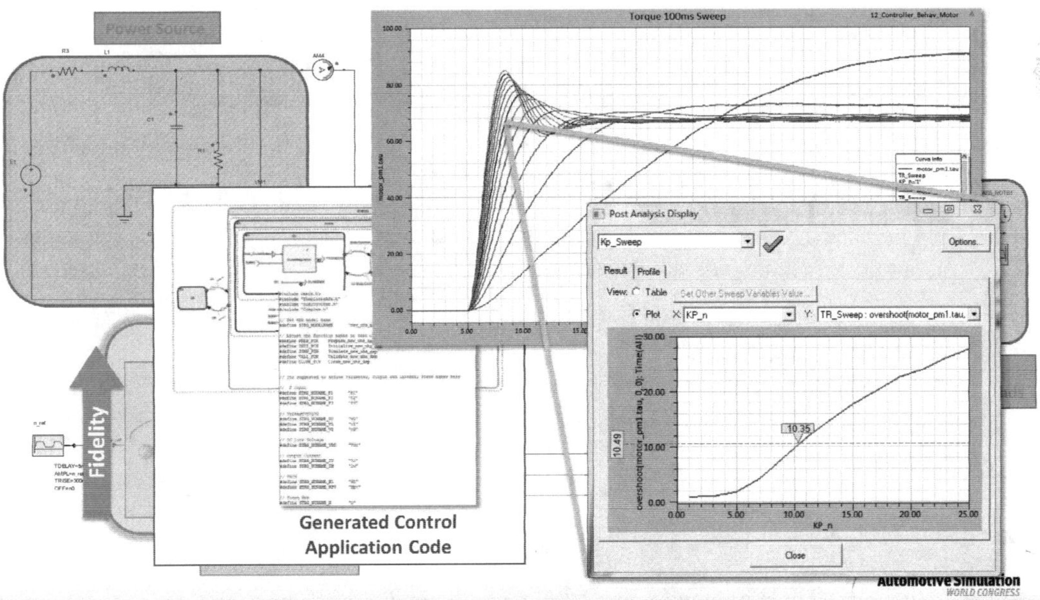

Simplorer: Platform for Virtual System Prototyping

Model, integrate, and analyze electrified systems
Power electronics, physics, and software control

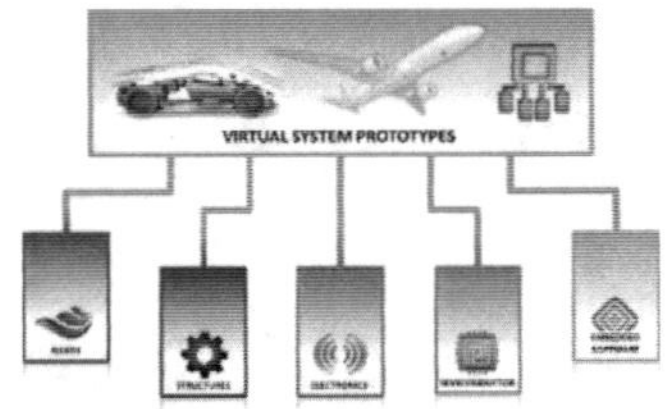

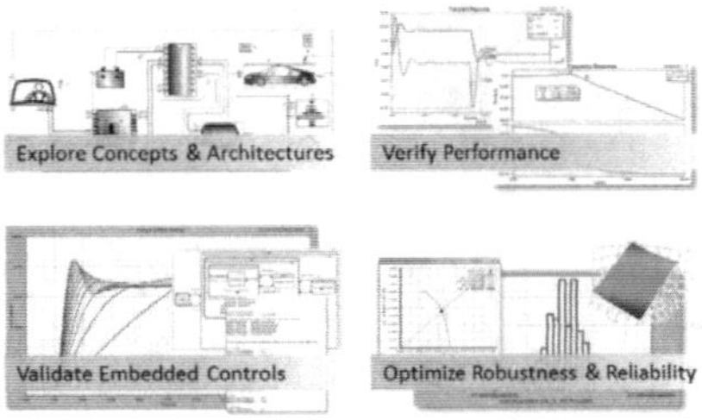

Expand system testing coverage
High-performance, multi-fidelity simulation throughout development

Optimize complex system performance
Understand interactions and tradeoffs earlier in the design process

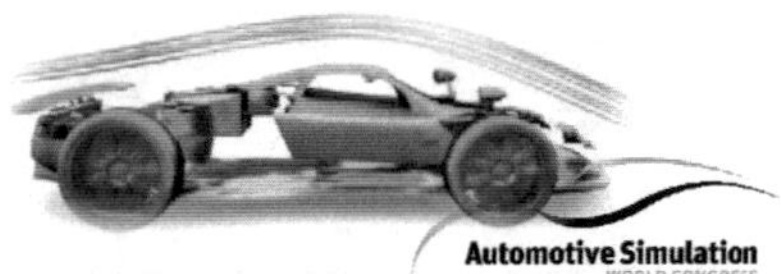

Connecting ANSYS Solutions

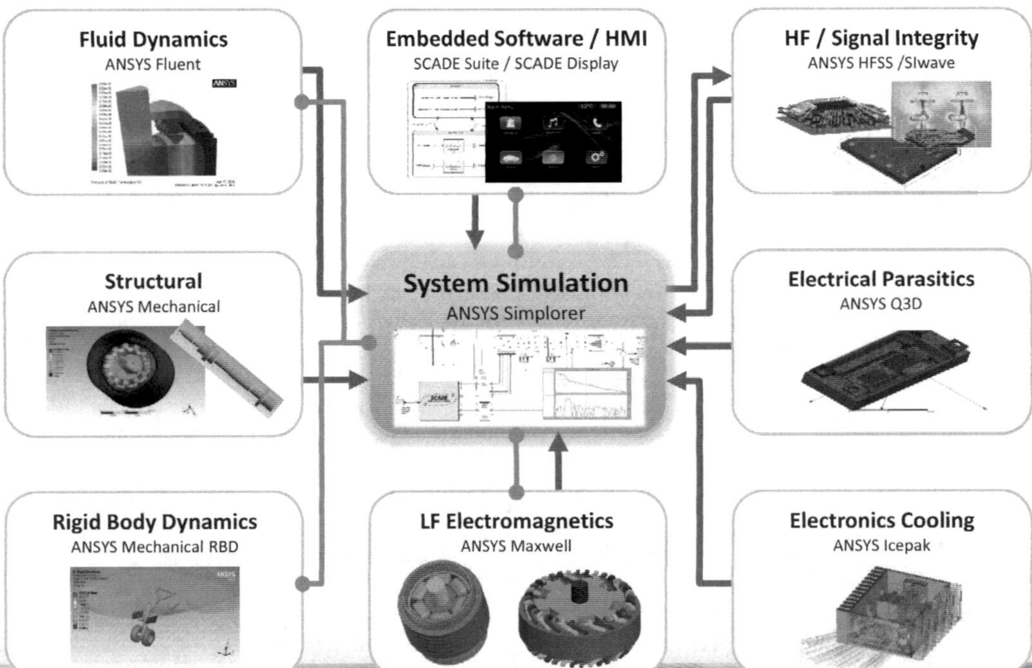

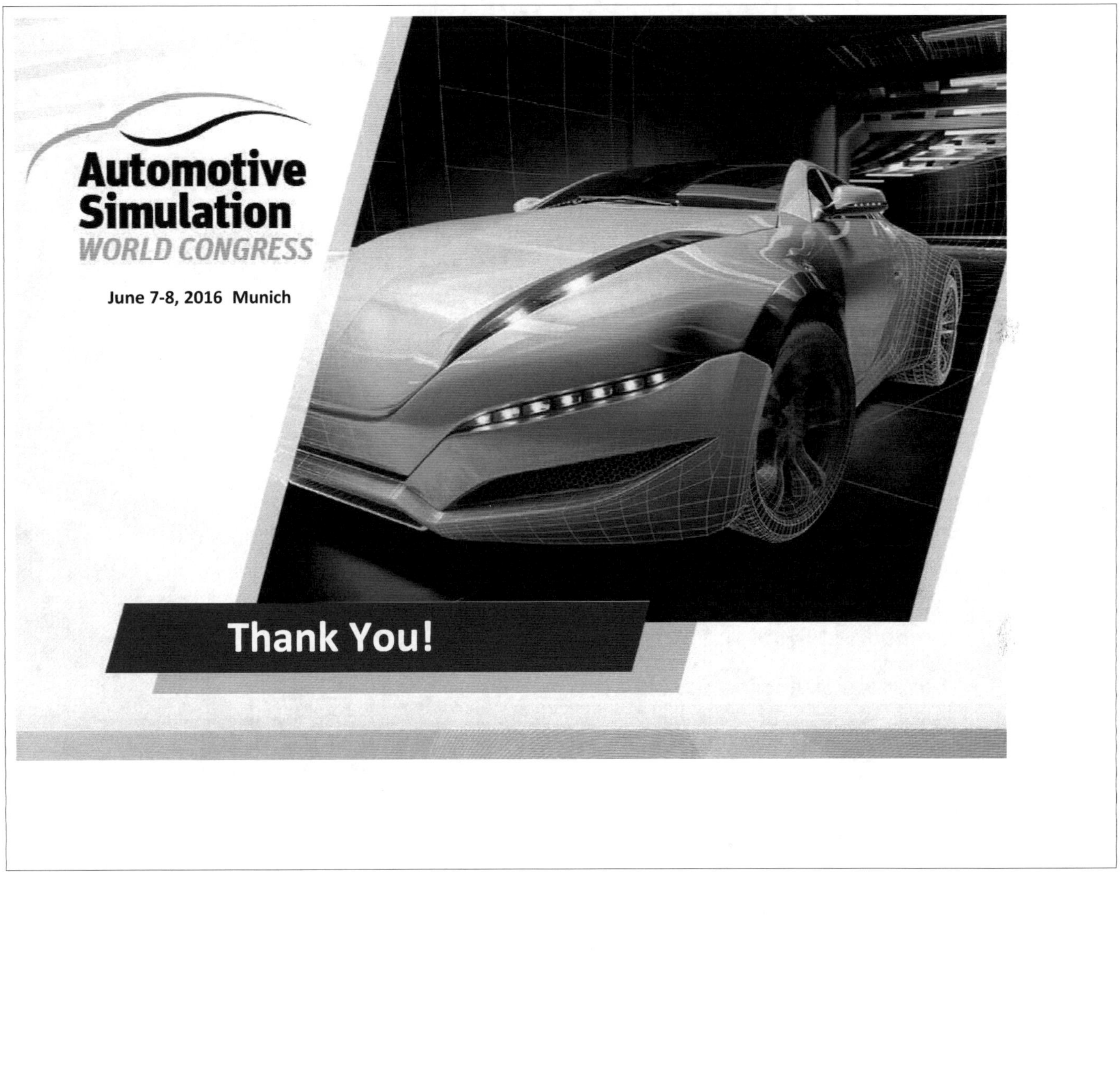
Automotive
Simulation
WORLD CONGRESS
June 7-8, 2016 Munich
Thank You!

Realize Your Product Promise®

ANSYS

Ensuring safety with ISO26262 and AUTOSAR compliant embedded software and auto-code generation

X. Fornari

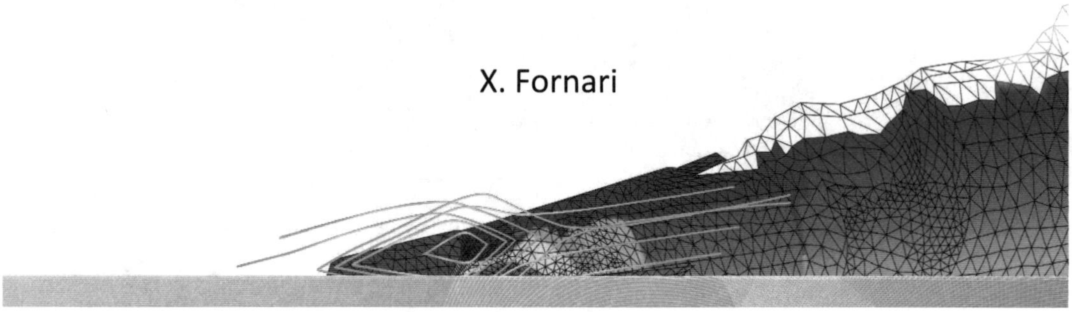

Embedded Software System Weight

- **Software is a large part of embedded System: ~50%**
- **Software size is increasing at a fast pace**
  - **10Mi in the Ford GT, 20Mi in S-Class Mercedes, ...**
  - **more than in the Boeing Dreamliner or the Airbus A380**

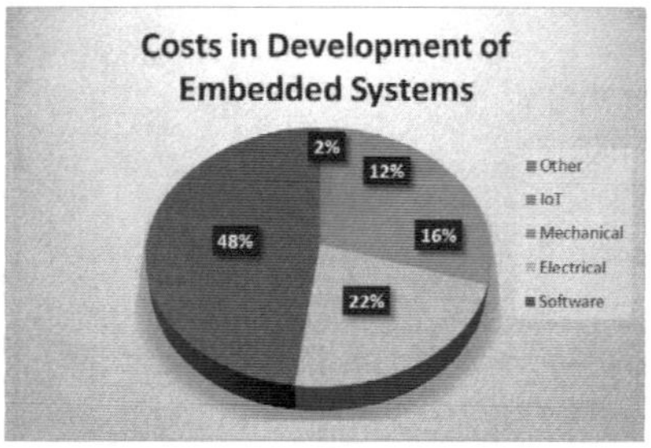

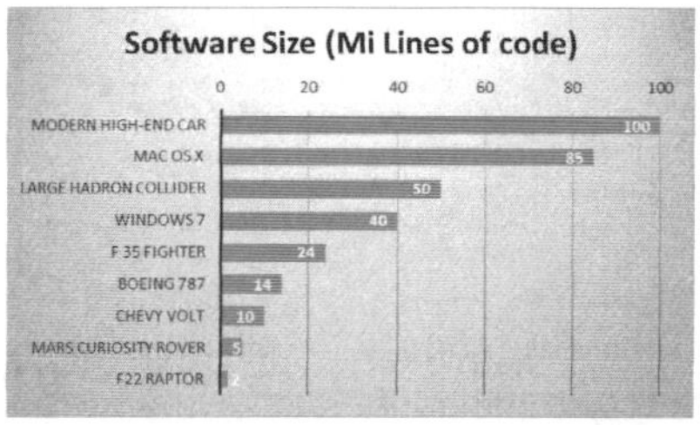

https://www.tasktop.com/blog/how-code-your-car-paving-future-software-delivery/
https://www.linkedin.com/pulse/20140626152045-3625632-car-software-100m-lines-of-code-and-counting

ANSYS

Software Applications in Automotive

Airbag Deployment	ABS	Instrument Cluster	Side Airbag	Tire Pressure	Rear Camera
Automatic Braking	Driver Monitor	V2V Com.	Remote Keys	Transmission	
Adaptive Cruise Control	Elec. Valve Timing	HUD	Alarm System	Lane Correction	
Adaptive Lighting	Throttle Control	Wiper Control	Seat Position	Blind Spot Detection	
	Battery Management	Climate Control	Cabin Lighting	Stability Control	
	Accident Recorder	GPS	Entertainment		
	Parking System				
	Drivetrain				

Safety Related | Security Related | Others

ANSYS

Advanced Driver Assist Systems (ADAS)

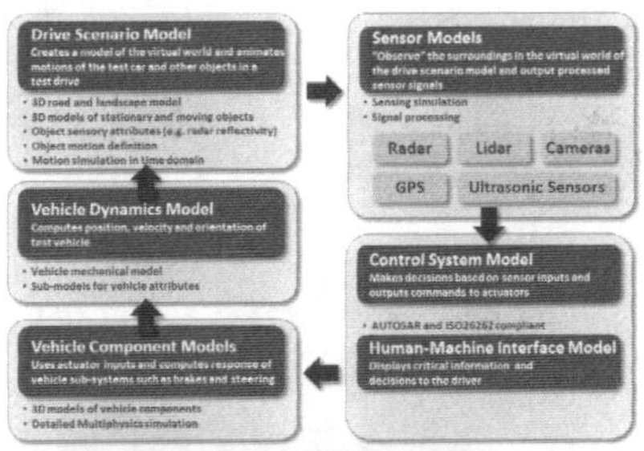

Managing System Design Complexity

Optimizing Overall System Performance

Reducing Physical Hardware, Electronics and Embedded Software Component Costs

ANSYS

Safety Challenge in Automotive

High level of safety: determinism is not an option

2014 RECALLS

Vehicle Safety & Compliance
803 Recalls
63,9 Million Vehicles
Including:
Ignition switch default
29 Million Vehicles
Airbags Takata
33.8M Airbags (Toyota, BMW, Honda,...)

COSTS

Unattended acceleration:
89 deaths 57 Injuries
$1.6B Economic Loss Class Action
$3M Compensation
Ignition switch default
51 deaths
$2.5B cost
Collaterals
Car Rental companies suffers from return
Used cars prices fall down

NHTSA is asking US Government the ability to increase up to $300M the fines, instead of $35M max.

- **1.2Bi vehicles today (3Bi 2035)**
- **3.5Bi passengers for 20 000 airplanes ➔ 641 fatalities**

ANSYS

The development now and into the future: Costs and Risks

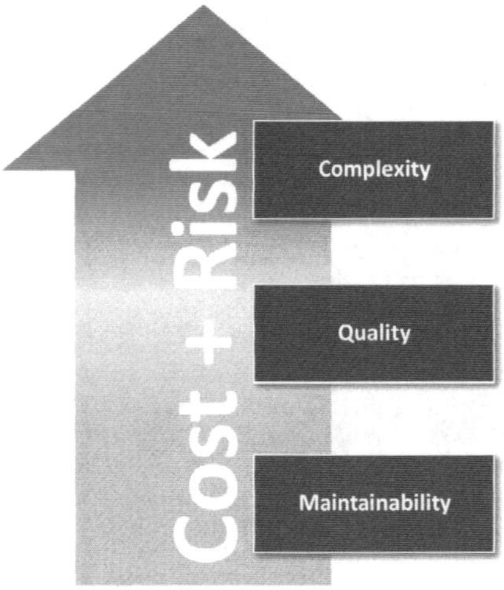

Complexity
- Roughly 10M lines of code in a modern car today
- Autonomous driving on the horizon
- Electrification of automotive started in earnest

Quality
- Need for strict rules for development, verification and validation of embedded software
- Short product cycles
- Growing responsibility for highly safety critical functions in software

Maintainability
- Software needs to be maintained not only developed
- Up to 90% of software rewritten from one car generation to the next[1]
- Distributed development between OEMs and Subcontractor

[1]: „Challenges in Automotive Software Engineering" (http://dl4a.org/uploads/pdf/p33.pdf)

ANSYS

Standardization is part of the answer…

- **Standardization is the key for:**
  - **Reliability**

Standardization is part of the answer…

- **Standardization is the key for:**
  - **Reliability**
  - **Maintainability**

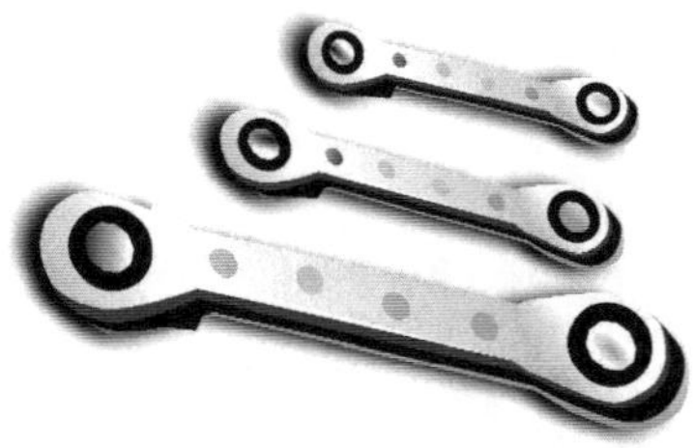

Standardization is part of the answer…

- **Standardization is the key for:**
  - **Reliability**
  - **Maintainability**
  - **Ecosystems development**

ANSYS

Standardization is part of the answer…

- **Standardization is the key for:**
  - **Reliability**
  - **Maintainability**
  - **Ecosystems development**
- **Automotive has already different standards, such as:**
  - **ISO 26262**

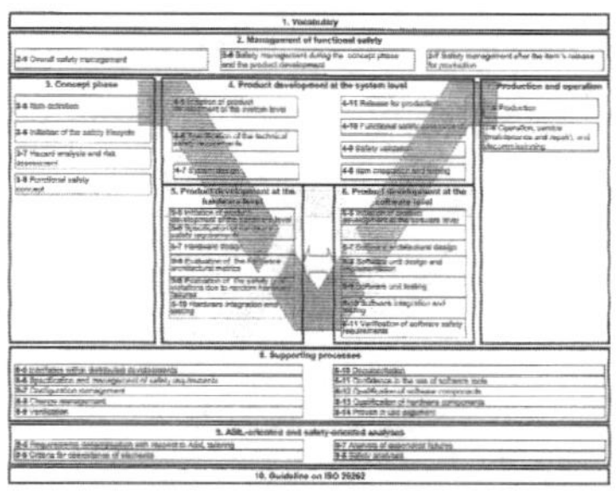

ANSYS

Standardization is part of the answer…

- **Standardization is the key for:**
  - Reliability
  - Maintainability
  - Ecosystems development
- **Automotive has already different standards, such as:**
  - ISO 26262
  - ASAM – MCD-2 DC

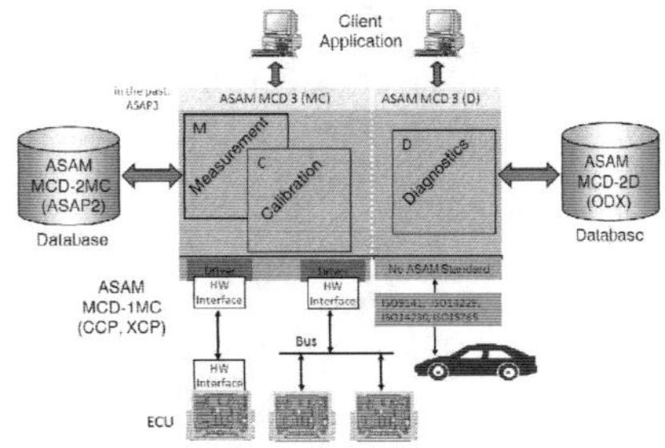

ANSYS

Standardization is part of the answer…

- **Standardization is the key for:**
  - Reliability
  - Maintainability
  - Ecosystems development
- **Automotive has already different standards, such as:**
  - ISO 26262
  - ASAM – MCD-2 DC
  - AUTOSAR

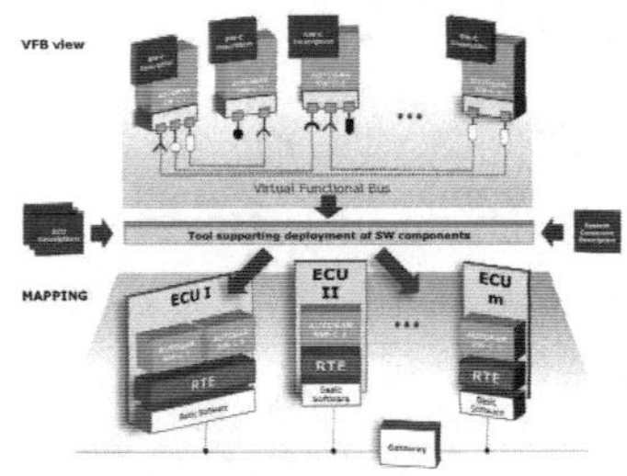

ANSYS

Another part is Tools and Methodology

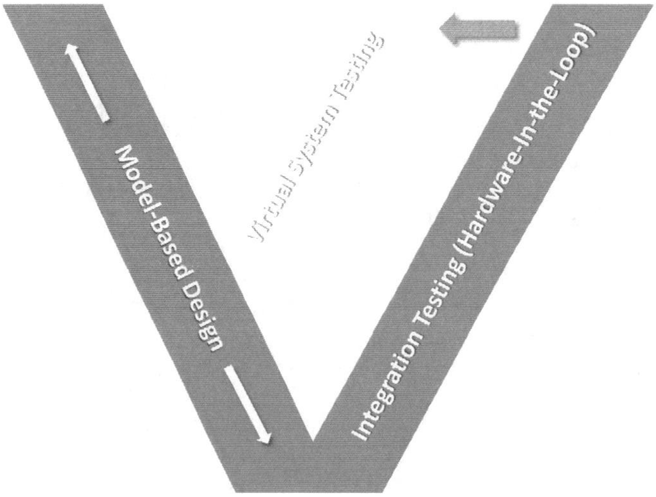

ANSYS Model-Based Engineering Solutions

Model-Based Systems Engineering **Multi-Physics & System Simulation**

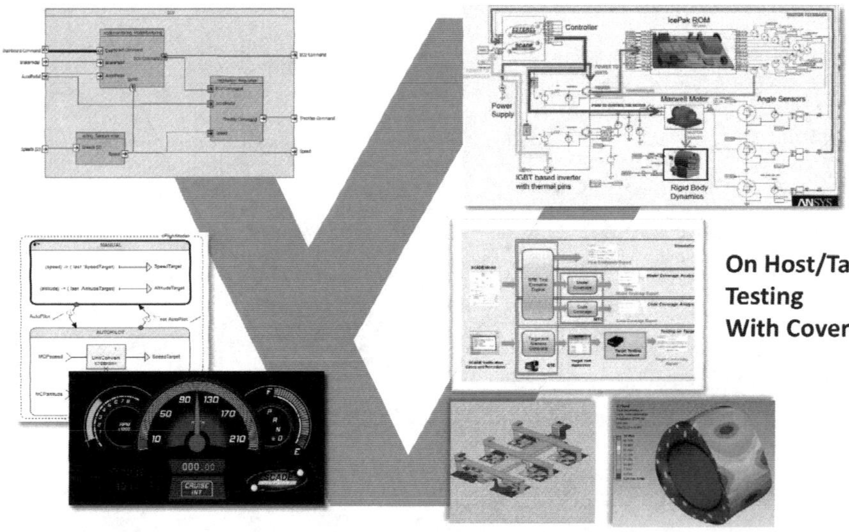

On Host/Target Testing With Coverage

Model-Based Software Engineering **3D Physical Simulation**

ISO 26262 provides…

- The definition of an automotive safety lifecycle
- A risk-based approach based on Automotive Safety Integrity Levels (ASIL), from A (lowest) to D (highest)
- The use of ASILs for specifying the applicable requirements of ISO 26262 to be met, so as to avoid unreasonable risk

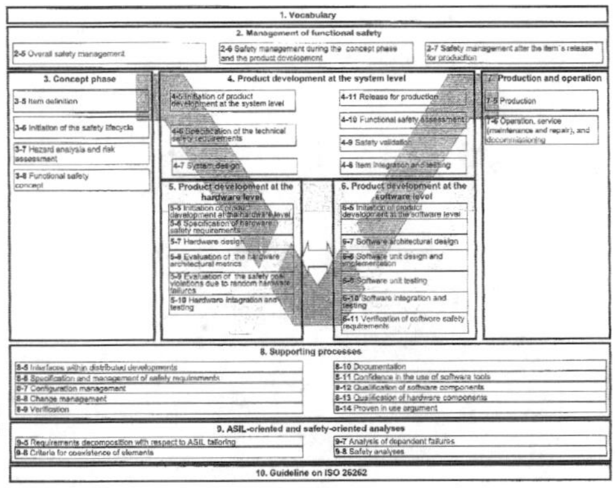

ANSYS

Following the ISO26262 Design process

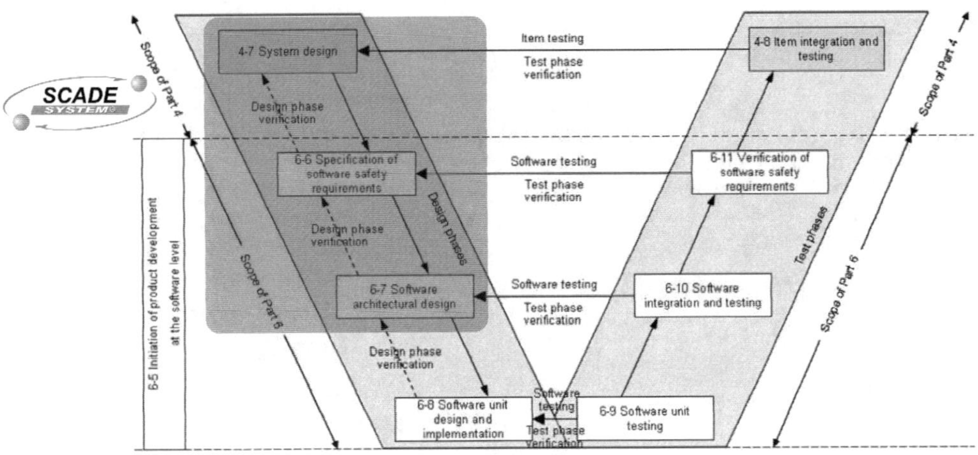

ANSYS

System Level

System Architecture and System-Level functional Design

- Allows to quickly prototype a System architecture
- In the beginning not all details are known

- **At the end a coherent architecture is defined with all information flows**

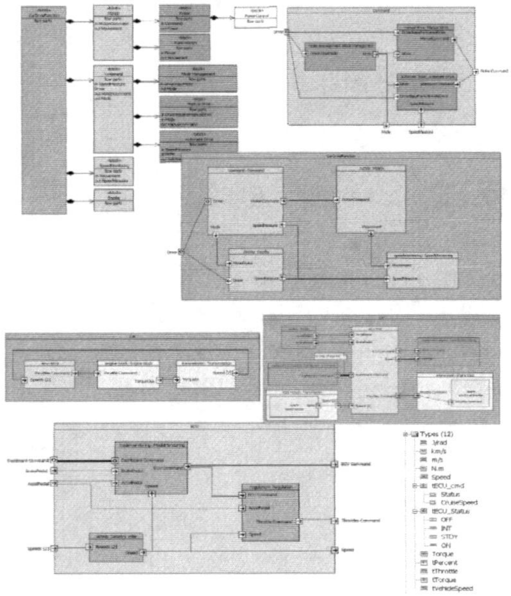

Software Unit Design and Implementation (6.8)

On this level, strict rules for modelling are needed:

- All the information for Implementation should be present
- Ambiguities are – unlike in the previous steps – not permitted

This specification is to be translated into source code!

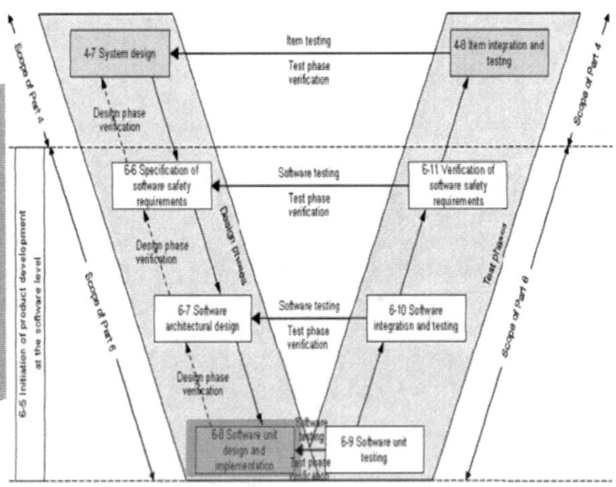

SCADE SUITE SCADE DISPLAY

Software Unit Design using MBD can be used

- **For Logic:**

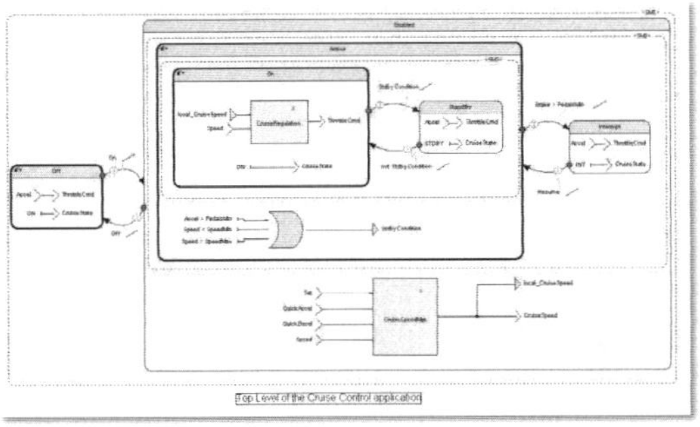

- **And Display:**

Code Generators Qualification

CERTIFICATE
No. Z10 14 09 55460 007
Holder of Certificate: Esterel Technologies SAS
Certification Mark:
Product: Software Tool for Safety Related Development
Model(s): Code Generator SCADE Suite KCG 6.4
Parameters:
Tested according to:

- **TÜV SÜD has qualified SCADE Suite and SCADE Display KCG for ISO 26262-8:2011 (ASIL D)**
- **KCG is the only COTS code generator developed/verified according to safety standards (DO-178 level A/TQL-1), IEC 61508), ensuring it does not generate unintended code**
  - Formal verification of its design and code
  - Structural coverage (MC/DC) of its code
- **KCG has also been qualified by aviation authorities (FAA, EASA) and rail authorities**

Software Unit Testing, Software Integration Testing (6.9; 6.10)

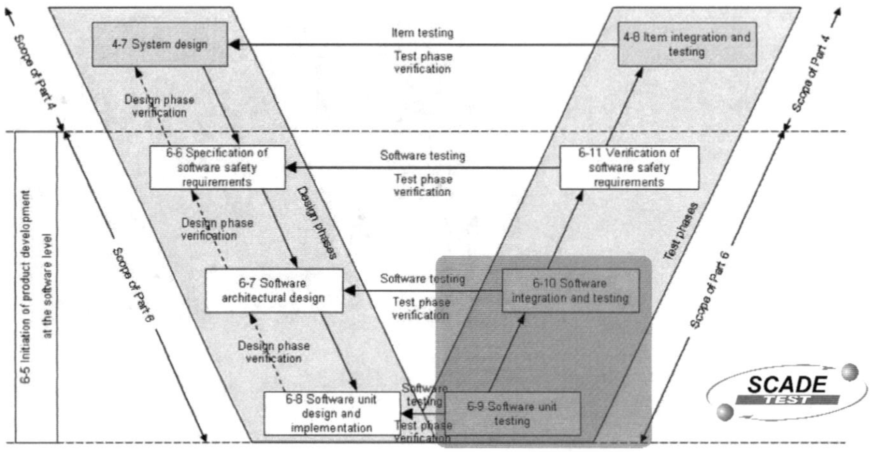

Automotive Open System Architecture AUTOSAR

- **A Methodology and an Architecture**
  - Software Components (SWC and BSW)
  - Virtual Functional Bus
  - Deployment on ECUs
- **Objectives**
  - Modularity & Configurability
  - Standardized Interfaces
  - Runtime Environment (RTE)

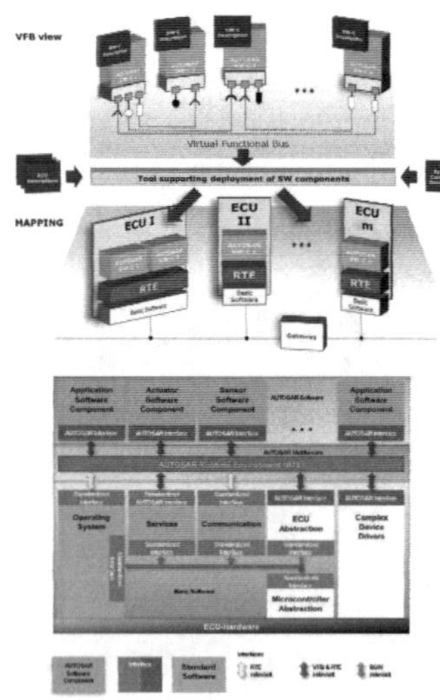

AUTOSAR SWC Design

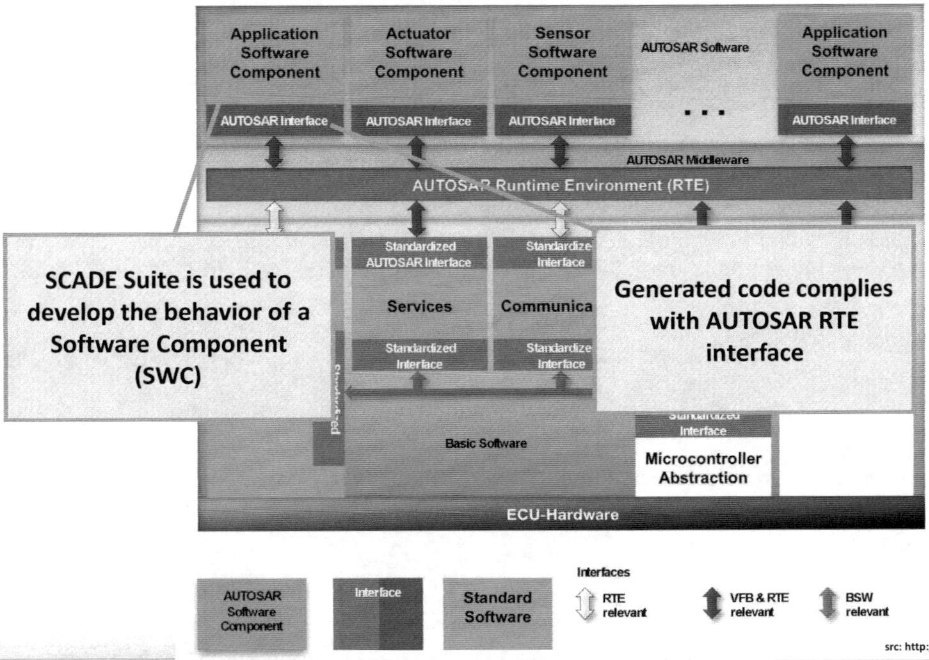

src: http://ww

Support of AUTOSAR

- **SCADE Automotive Package is supporting AUTOSAR Version 4.2.2**
- **A SWC description can be:**
  - read **for a** later **design**
  - generated **from an** existing **design**
- **With SCADE Architect and SCADE Suite, this can be done in a seamless way**

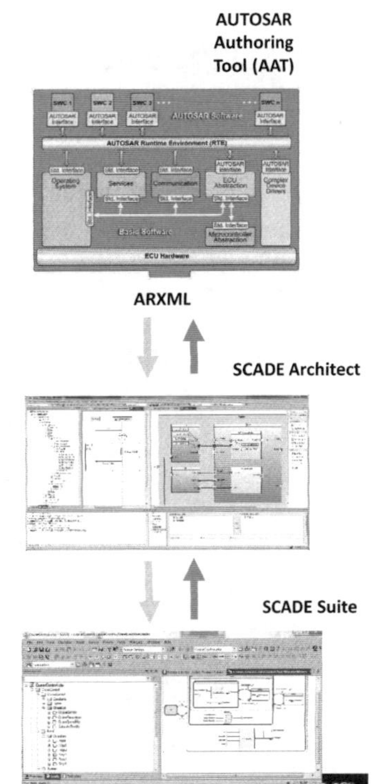

Integrated Workflow for SW-intensive Systems

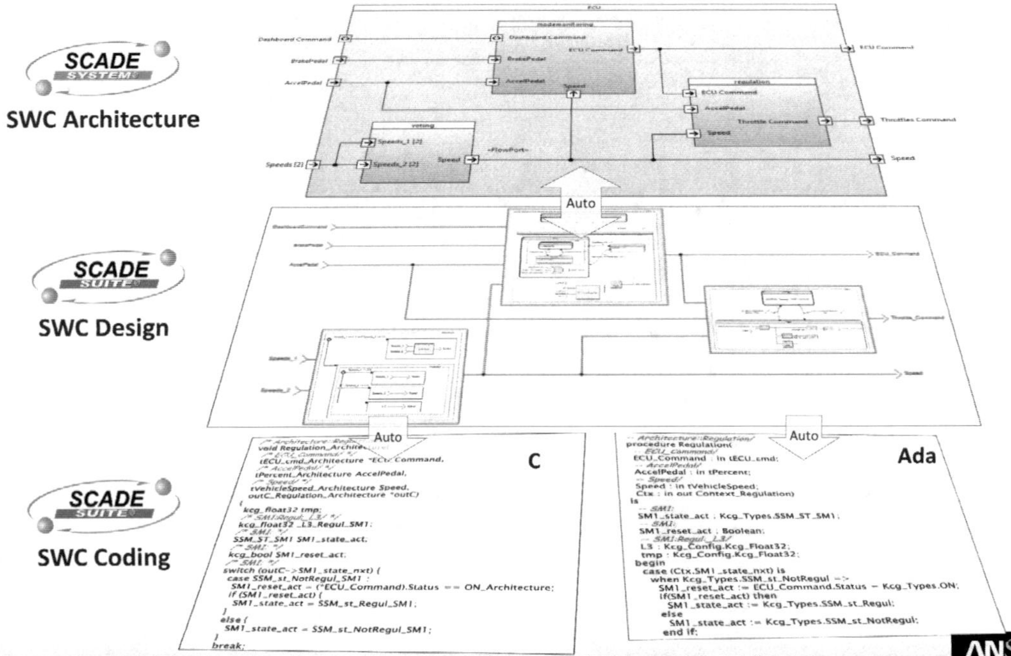

ANSYS

SCADE AUTOSAR Capabilities

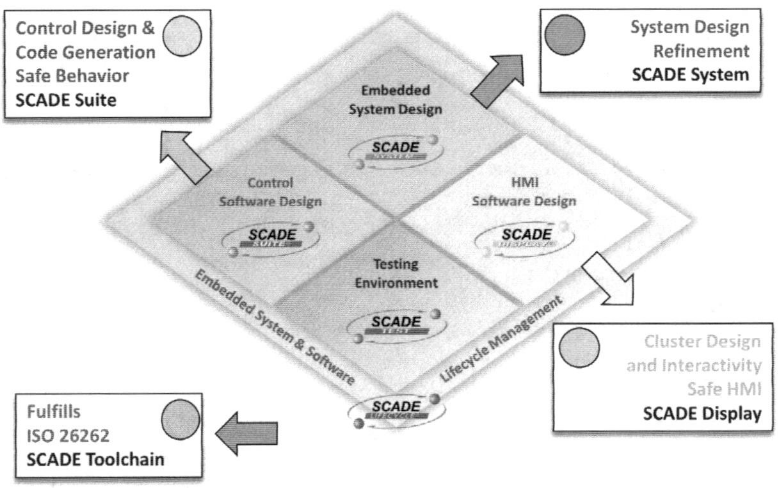

ANSYS

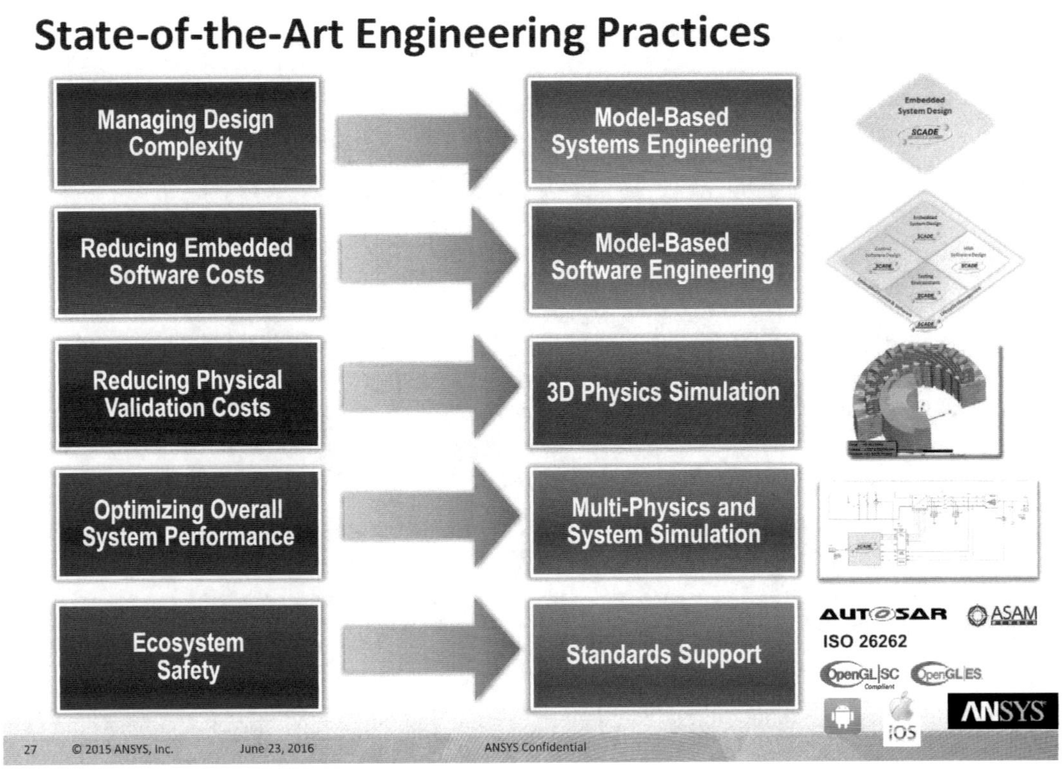
State-of-the-Art Engineering Practices
Managing Design Complexity
Model-Based Systems Engineering
Reducing Embedded Software Costs
Model-Based Software Engineering
Reducing Physical Validation Costs
3D Physics Simulation
Optimizing Overall System Performance
Multi-Physics and System Simulation
Ecosystem Safety
Standards Support
AUTOSAR
ASAM
ISO 26262
OpenGL SC
OpenGL ES
iOS
ANSYS
27 © 2015 ANSYS, Inc. June 23, 2016 ANSYS Confidential

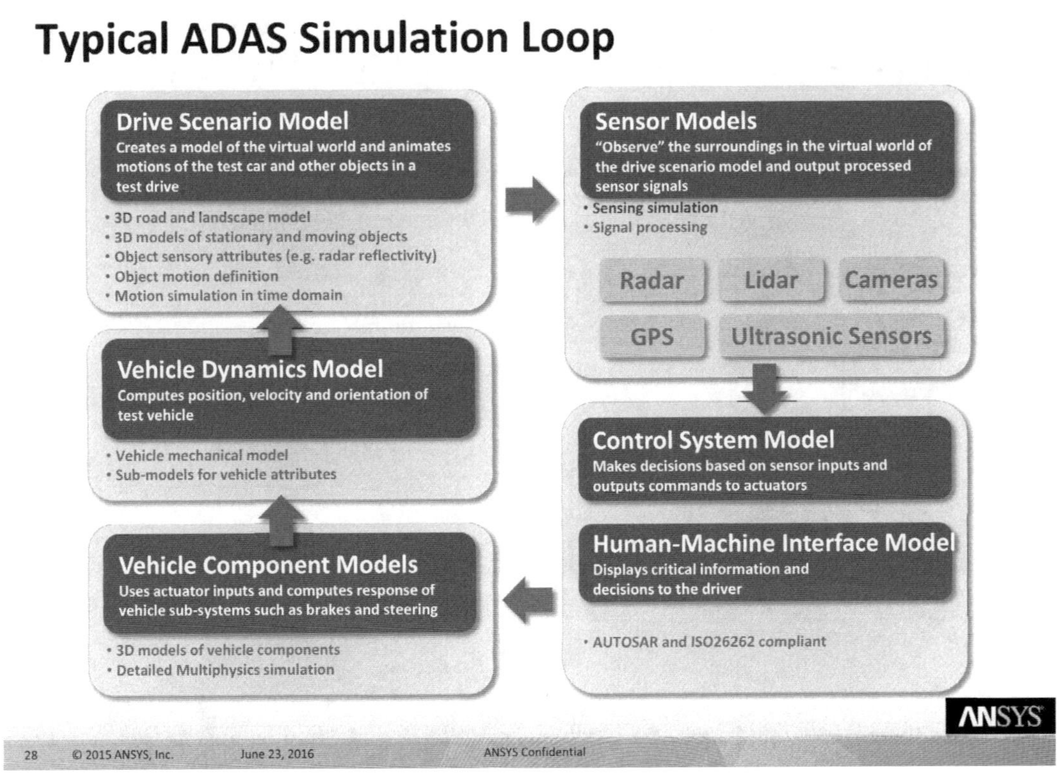
Typical ADAS Simulation Loop
Drive Scenario Model
Creates a model of the virtual world and animates motions of the test car and other objects in a test drive
• 3D road and landscape model
• 3D models of stationary and moving objects
• Object sensory attributes (e.g. radar reflectivity)
• Object motion definition
• Motion simulation in time domain
Sensor Models
"Observe" the surroundings in the virtual world of the drive scenario model and output processed sensor signals
• Sensing simulation
• Signal processing
Radar
Lidar
Cameras
GPS
Ultrasonic Sensors
Vehicle Dynamics Model
Computes position, velocity and orientation of test vehicle
• Vehicle mechanical model
• Sub-models for vehicle attributes
Control System Model
Makes decisions based on sensor inputs and outputs commands to actuators
Human-Machine Interface Model
Displays critical information and decisions to the driver
• AUTOSAR and ISO26262 compliant
Vehicle Component Models
Uses actuator inputs and computes response of vehicle sub-systems such as brakes and steering
• 3D models of vehicle components
• Detailed Multiphysics simulation
ANSYS
28 © 2015 ANSYS, Inc. June 23, 2016 ANSYS Confidential

ANSYS ADAS Simulation Tool Chain

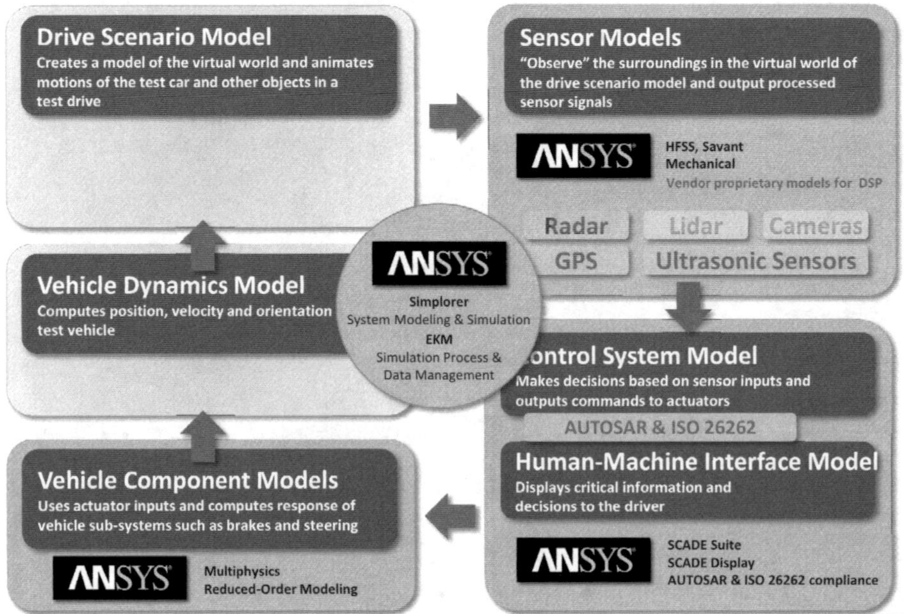

ANSYS

The development now and into the future: Reducing Costs and Risks

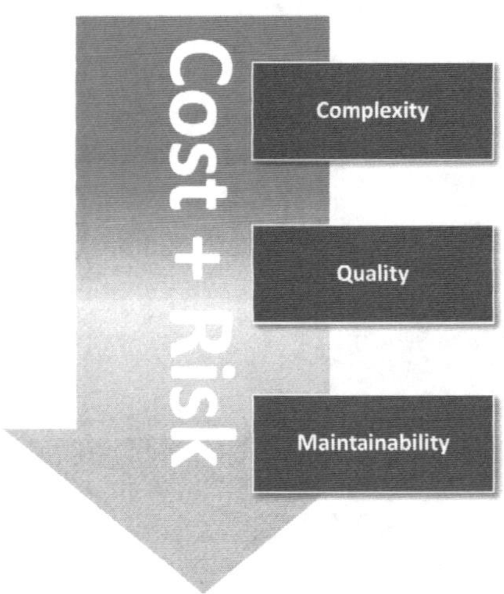

Complexity

- Validate Early
- Do Not Repeat Yourself (DRY)
- Virtual Prototypes and Higher level of abstraction (MBD)

Quality

- Fulfill standards requirement
- Software Models are easier to understand
- Efficiency of V&V at model level
- No functional V&V on Code level

Maintainability

- Model-based (languages are easier to read
- Simulation of model specifications enhances understanding of behaviors
- Short prototyping cycles enhance communication between OEM and subcontractor

ANSYS

Summary of Benefits

- SCADE 's Model-Based approach of allows meeting efficiently all requirements of ISO 26262-6
- Qualified Code Generation of SCADE Suite and SCADE Display KCG is based on using the safety standard approach of ISO-26262-8 and thus provides unique certification benefits (no code review, no low-level testing are needed, etc.)
- The SCADE toolchain, configured for AUTOSAR, provides a way to efficiently develop high-dependability software (potentially certified up to ASIL D)
- Complete integration with ANSYS 3D tools for full virtual system simulation
- Bottom line is reduction of cost and time to certification at any integrity level compared to manual or non-certified model-based approaches

ANSYS

Automotive Simulation WORLD CONGRESS

June 7-8, 2016 Munich

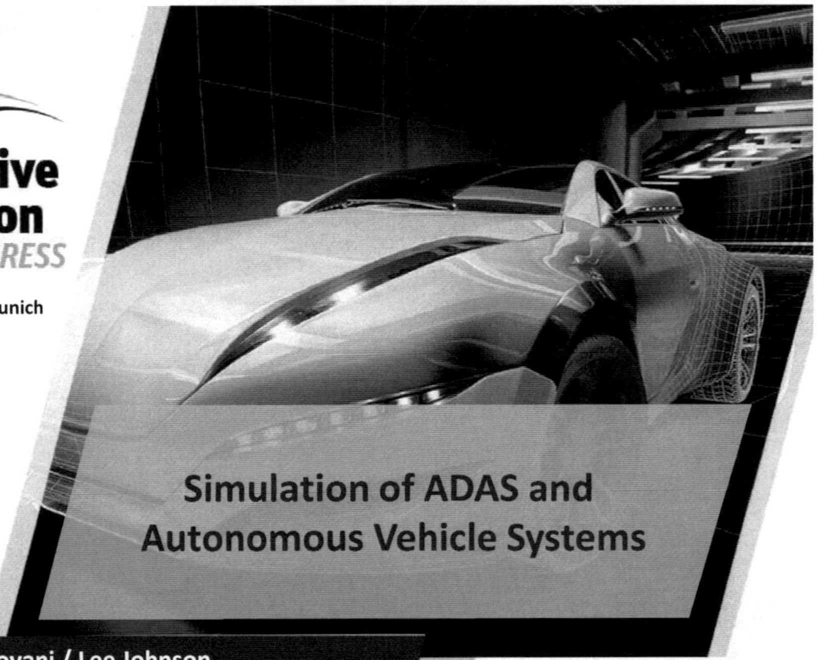

Simulation of ADAS and Autonomous Vehicle Systems

Sandeep Sovani / Lee Johnson
ANSYS

Agenda

- Introduction – Autonomous Vehicles
- Simulation Tool Chain for ADAS
- ANSYS Capabilities
- Conclusion & Outlook

Five Levels of Autonomy

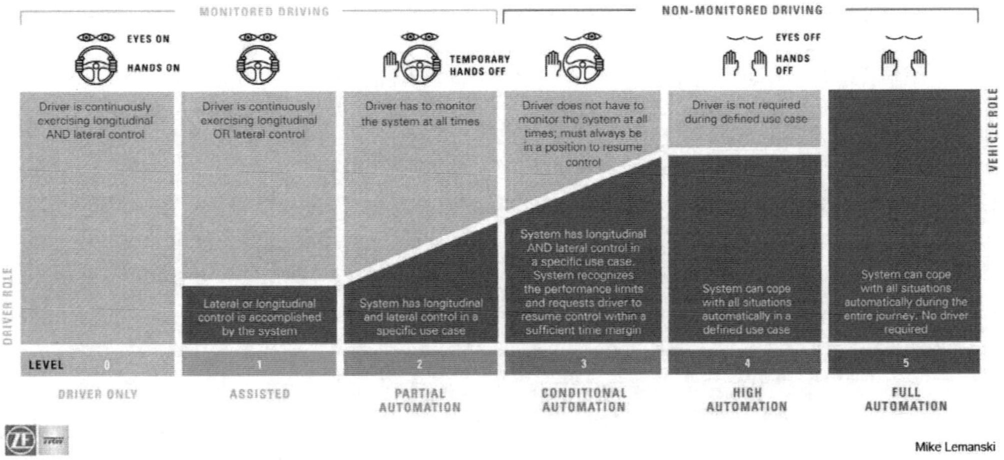

Reference: http://safety.trw.com/autonomous-cars-must-progress-through-these-6-levels-of-automation/0104/

Advanced Driver Assist Systems (ADAS)

Source: Continental

Current Focus on Levels 1-3

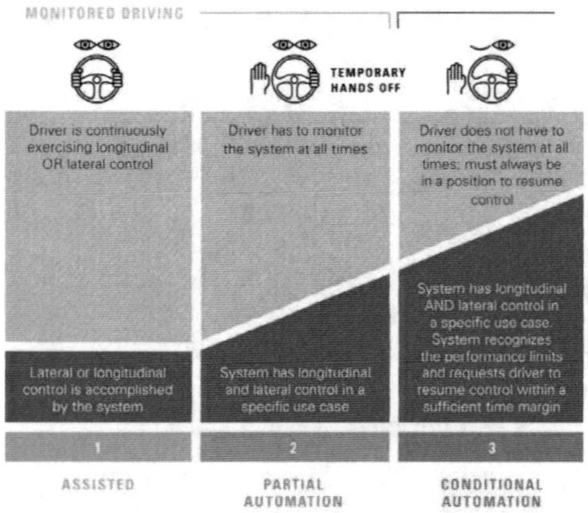

Advanced Driver Assistance Systems (ADAS) Examples –

Source: Continental

- Cooperative Driving Systems (CDS)

Reference: http://safety.trw.com/autonomous-cars-must-progress-through-these-6-levels-of-automation/0104/

Basic Control Loop

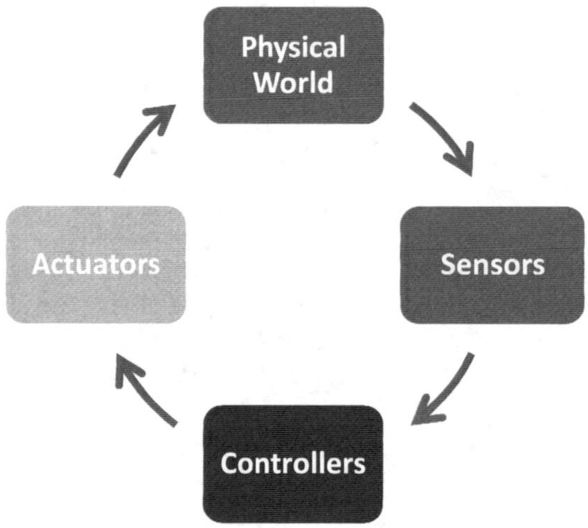

ADAS & Autonomous

Control Loop

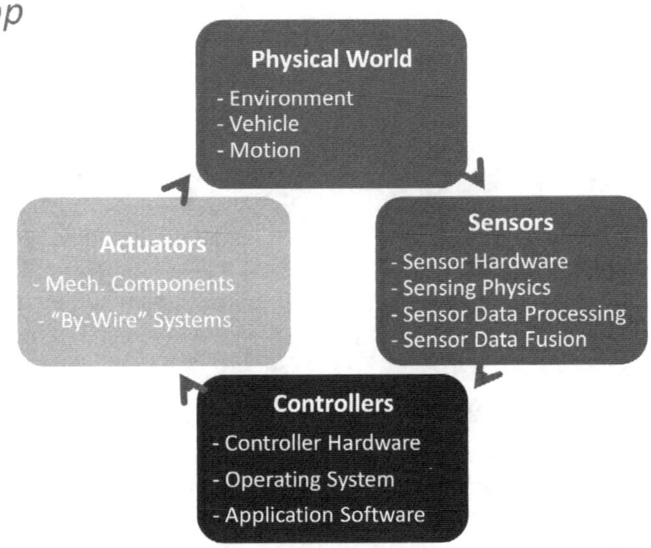

ANSYS: Simulation tool flow of complete ADAS Loop for achieving system performance

ADAS Simulation Loop

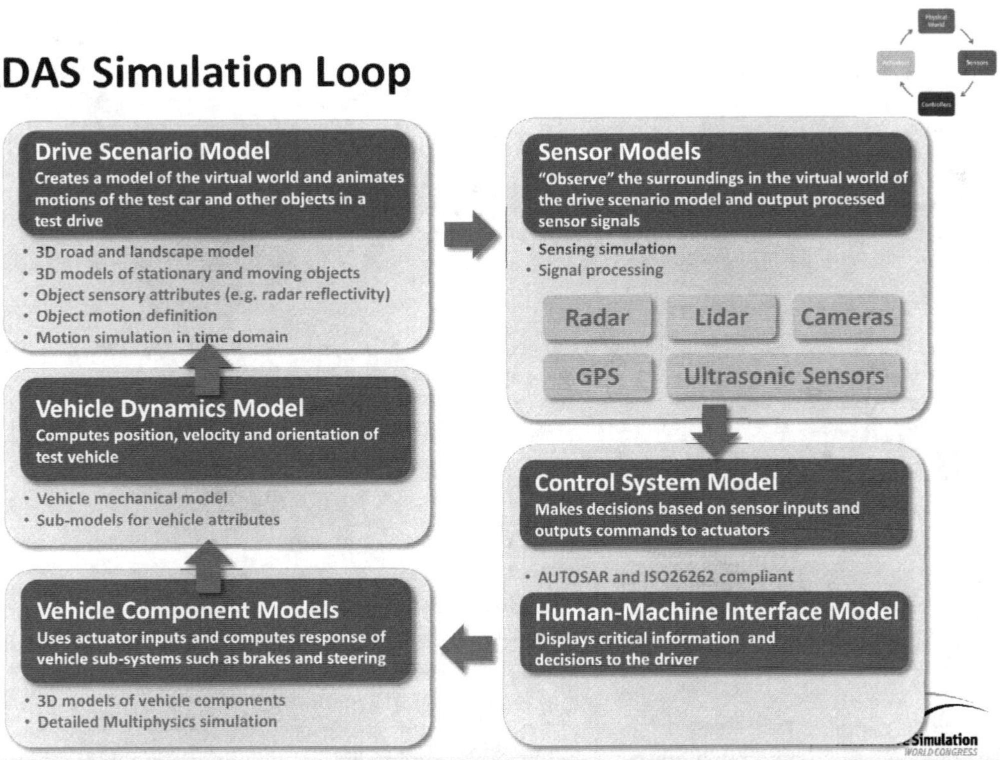

Model-Based System Development

ANSYS Capabilities

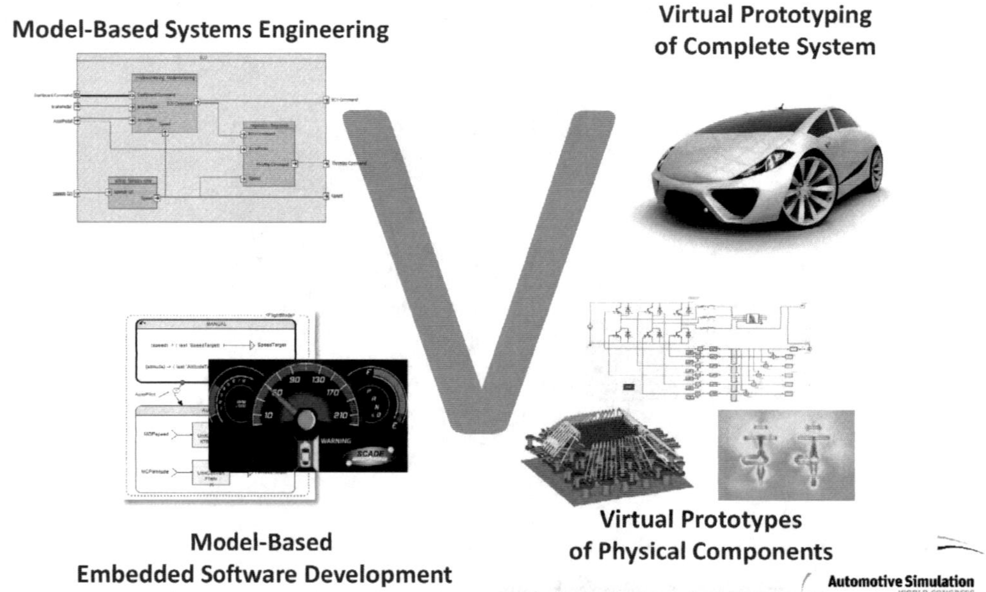

ADAS Simulation Tool Chain

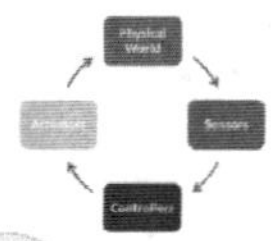

Drive Scenario Model
Creates a model of the virtual world and animates motions of the test car and other objects in a test drive

Partner software

Sensor Models
"Observe" the surroundings in the virtual world of the drive scenario model and output processed sensor signals

ANSYS HFSS, Savant Mechanical

Sensing simulation Signal processing

Radar | Lidar | Cameras | GPS | Ultrasonic Sensors

Vehicle Dynamics Model
Computes position, velocity and orientation of test vehicle

Partner software MODELICA

ANSYS
Simplorer
System Modeling & Simulation
WorkBench
3D Modeling & Simulation
EKM
Simulation Process & Data Management

Control System Model
Makes decisions based on sensor inputs and outputs commands to actuators

Human-Machine Interface Model
Displays critical information and decisions to the driver

ANSYS SCADE Suite SCADE Display

Vehicle Component Models
Uses actuator inputs and computes response of vehicle sub-systems such as brakes and steering

ANSYS Multiphysics Reduced-Order Modeling

ANSYS

Control Systems and HMI Modeling with SCADE

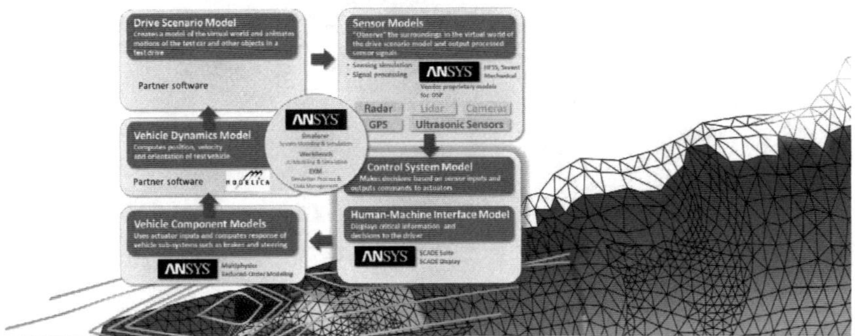

SCADE: An Embedded Software Development Toolset Proven in Aerospace ...

S afety

C ritical

A pplication

D evelopment

E nvironment

"Airbus never experienced any bug in flight in our Flight Control System software produced automatically."

... now ready for automotive systems

Automotive Simulation WORLD CONGRESS

SCADE Product Family

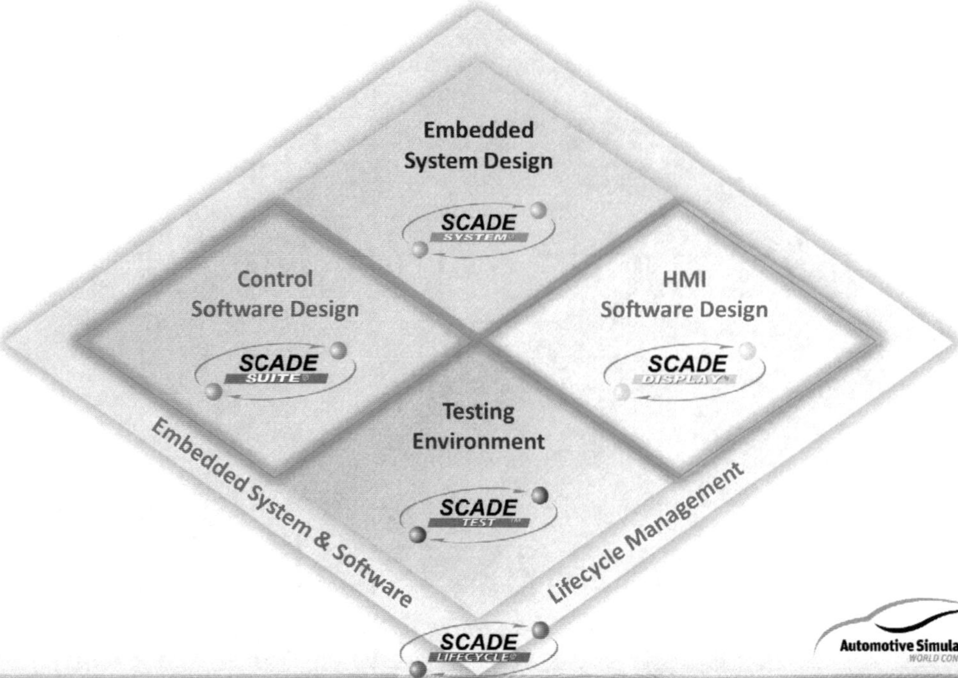

ANSYS SCADE Suite

Embedded Control Software Design

Efficient modeling of controls, logic and algorithm designs within a single environment

Integrated Suite for Prototyping, Modeling, Simulation, Verification, and Optimization

Efficient debugging and optimization of software models and code size, speed and performance

Certified Code Generation

Automatic C certified code generator
*(**ISO 26262,** IEC 61508, DO-178B/C, EN 50128)*
provides 50% embedded code production and testing cost reduction

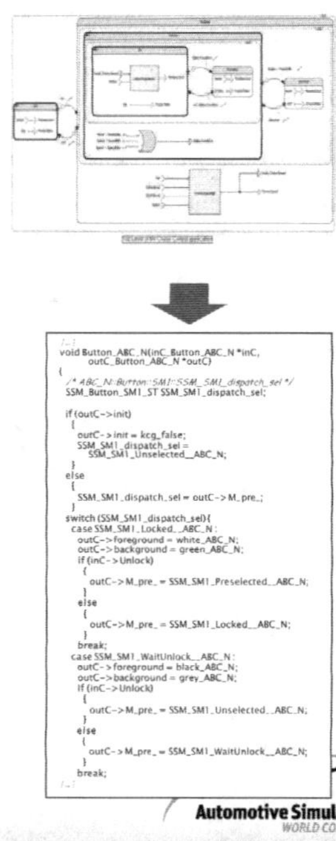

ANSYS SCADE Display

HMI Software Design
Efficient Modeling of HMI designs within an integrated environment with Logic Design

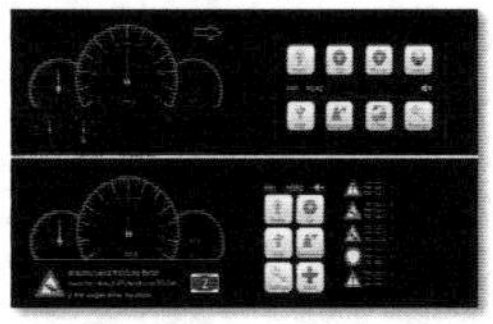

Complete GUI Prototyping, Modeling, Simulation, Verification, and Optimization
Covering Rapid Prototyping, Model Checking and Debugging, Simulation, Integration with Graphics Platforms and Human Factors Optimization

Certified Code Generation
Automatic certified code generator
*(**ISO 26262,** IEC 61508, DO-178B/C, EN 50128)*
provides 50% Embedded Code Production
and Testing Cost Reduction

PC, Android, Apple iOS and embedded graphics platforms

Lifecycle V-model of ISO 26262 (Part 6) and the SCADE products

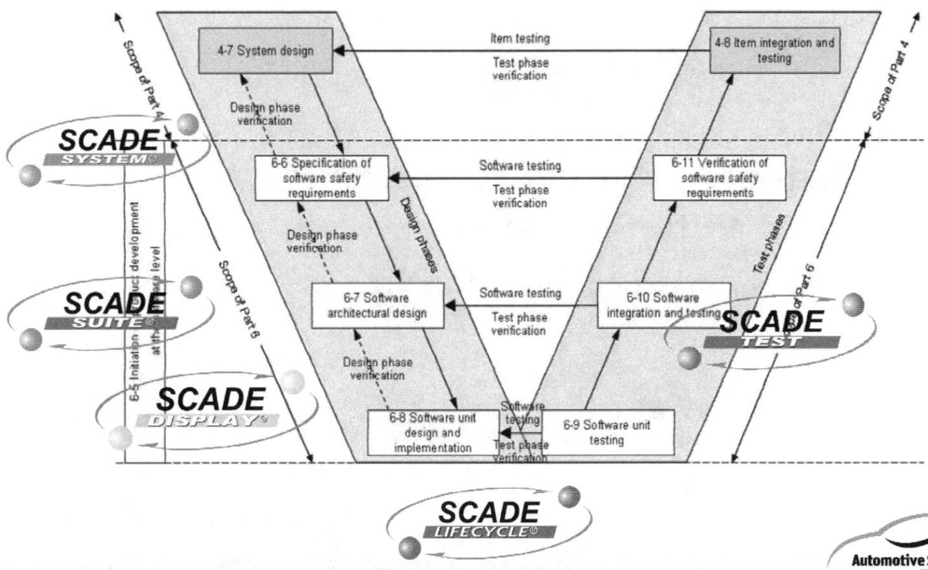

TÜV SÜD ISO 26262 Qualification

CERTIFICATE

No. Z10 14 09 55460 007

Holder of Certificate: Esterel Technologies SAS

Factory(ies): 55460

Certification Mark:

Product: Software Tool for Safety Related Development

Model(s): Code Generator SCADE Suite KCG 6.4

Parameters: The code generator, classified as T3 offline support tool according to IEC 61508-4 and EN 50128 is qualified for the use in safety-related software development according to IEC 61508, EN 50128 and ISO 26262.

The report EE86102C is a mandatory part of this certificate.

Tested according to: IEC 61508-1:2010 (SIL 3) IEC 61508-3:2010 (SIL 3) EN 50128:2011 (SIL 3/4) ISO 26262-8:2011 (ASIL D)

Test report no.: EE86102C

Valid until: 2019-09-22

Date, 2014-09-23

Page 1 of 1

TÜV SÜD has qualified the SCADE Suite and SCADE Display Code Generators for ISO 26262 for use up to the highest safety level (ASIL D)

Realize Your Product Promise®

ANSYS

Virtual System Prototyping with Simplorer

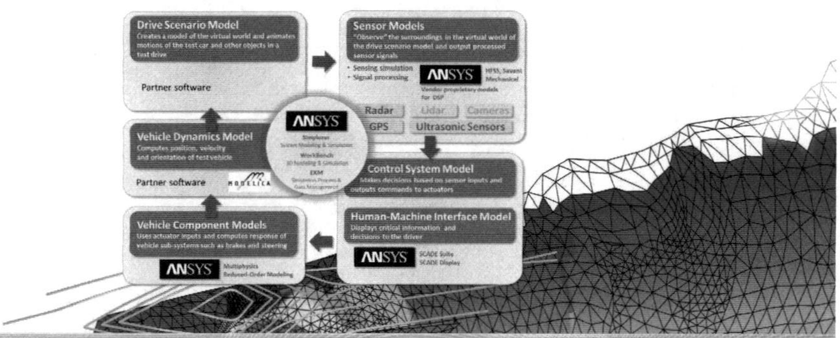

ANSYS Simplorer

Platform for Virtual System Prototyping

Model, integrate, and analyze electrified systems
Power electronics, physics, and software control

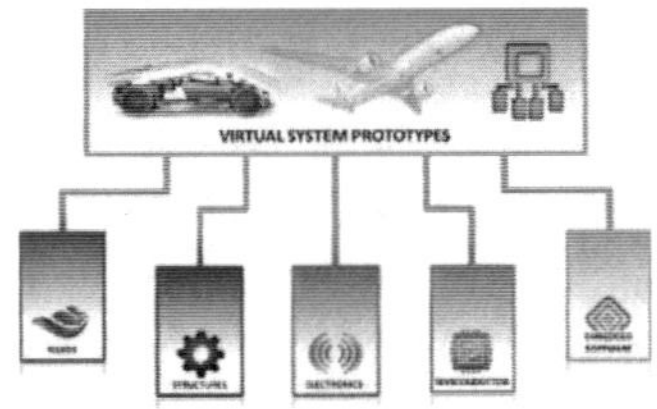

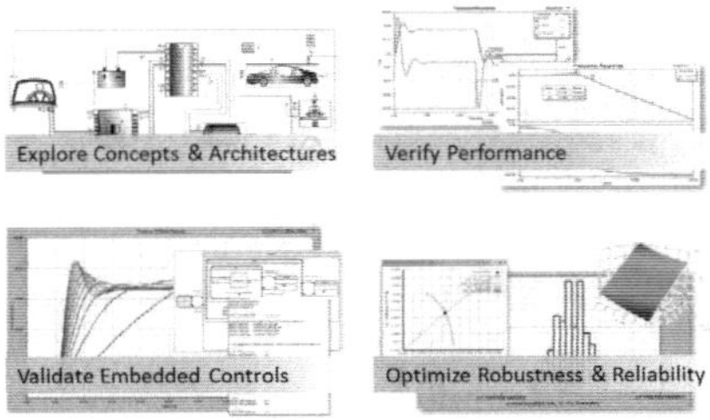

Expand system testing coverage
High-performance, multi-fidelity simulation throughout development

Optimize complex system performance
Understand interactions and tradeoffs earlier in the design process

Automotive Simulation WORLD CONGRESS

ANSYS Simplorer

Model Creation, Assembly, and Reuse

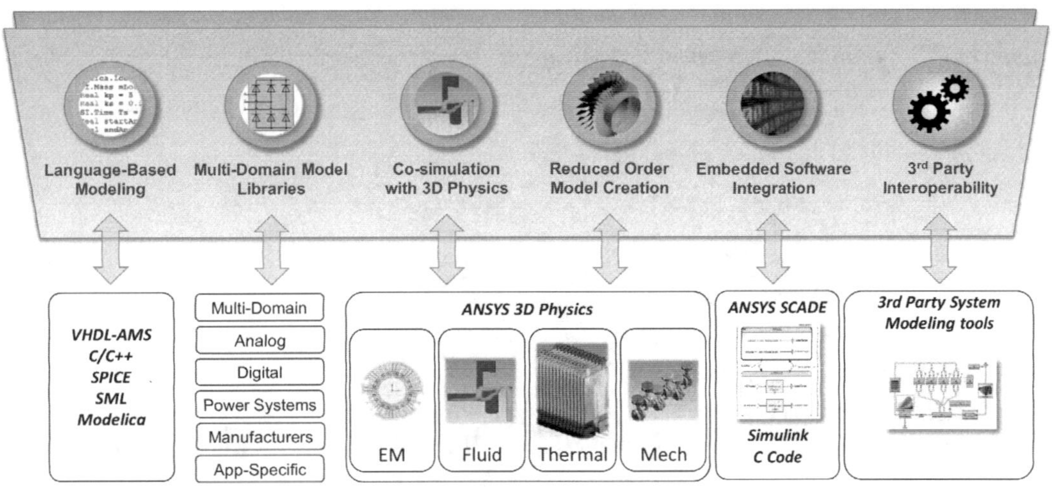

Automotive Simulation WORLD CONGRESS

ANSYS Simplorer

Platform for Testing the Virtual System

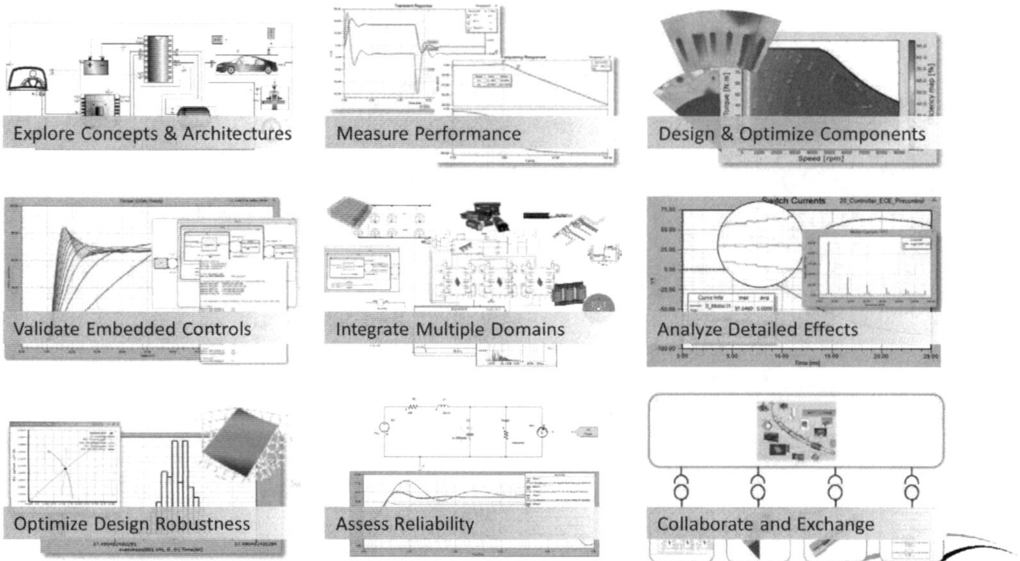

ANSYS

Vehicle Component Models with ANSYS Multiphysics and ROM

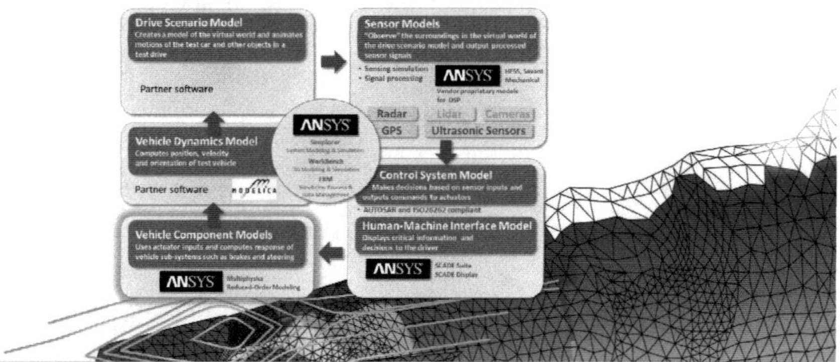

ROMs for Physical Simulation

A Reduced Order Model (ROM) is a simplification of a high-fidelity dynamical model that preserves essential behavior and dominant effects, for the purpose of reducing solution time or storage capacity required for the more complex model.

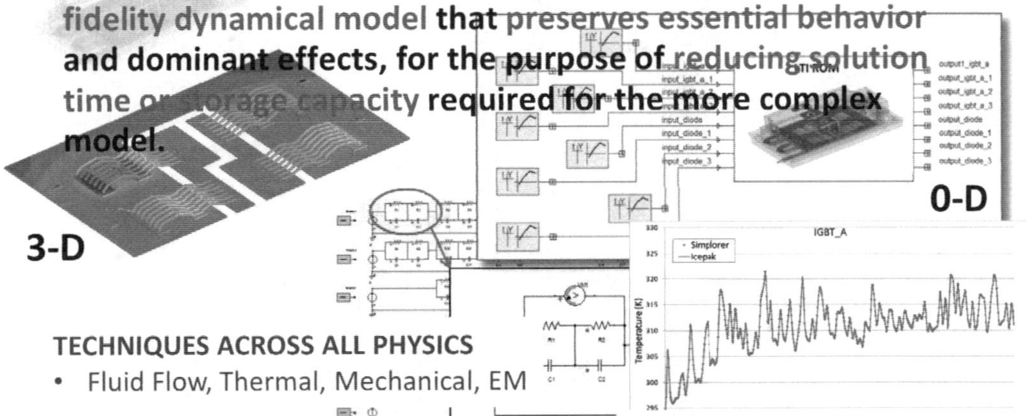

TECHNIQUES ACROSS ALL PHYSICS

- Fluid Flow, Thermal, Mechanical, EM

SOLVES IN SECONDS (VS. HOURS)

Automotive Simulation WORLD CONGRESS

ROMs & Systems: Electric Powertrain

System-Level Objectives

- Evaluate architectural selections and component choices to optimize fuel economy and cost
- Verify control strategies and calibrate control parameters
- Assess system reliability (worst-case analysis, fault injection)

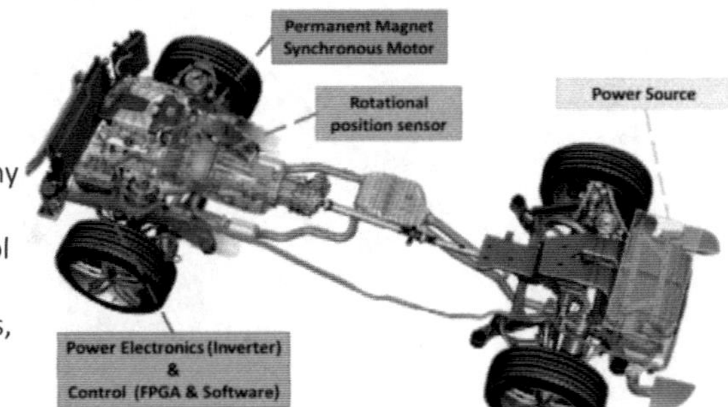

Key System-Level Models

- *ANSYS Maxwell*: Permanent magnet synchronous machine extracted as ROM
- *ANSYS SCADE*: Motor control software
- *ANSYS Simplorer*: High-power electronics (inverter), behavioral multi-domain sensors

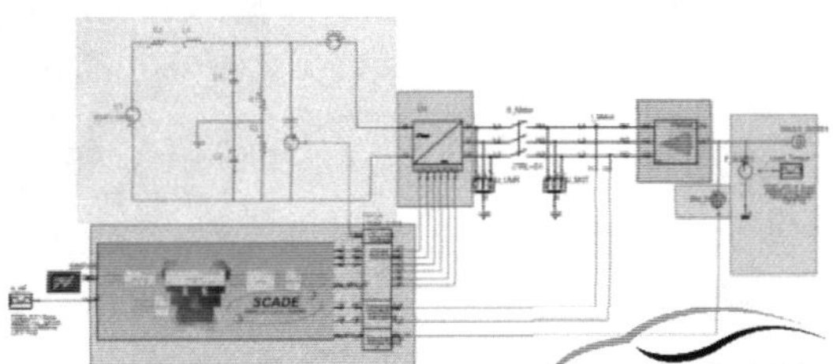

Automotive Simulation WORLD CONGRESS

Sensor Modeling with HFSS and Savant

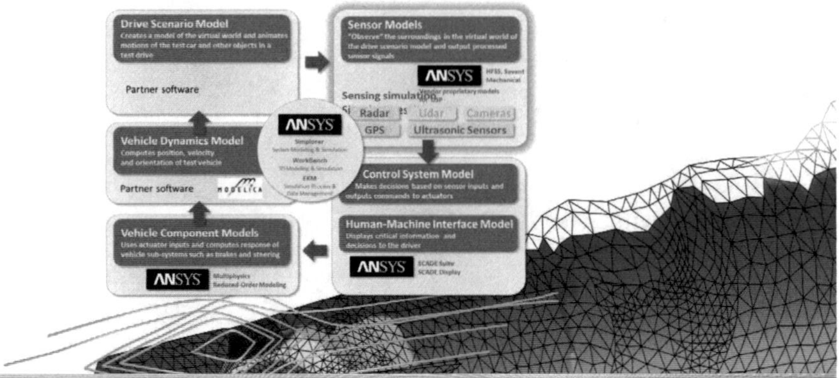

Complete Radar Simulation

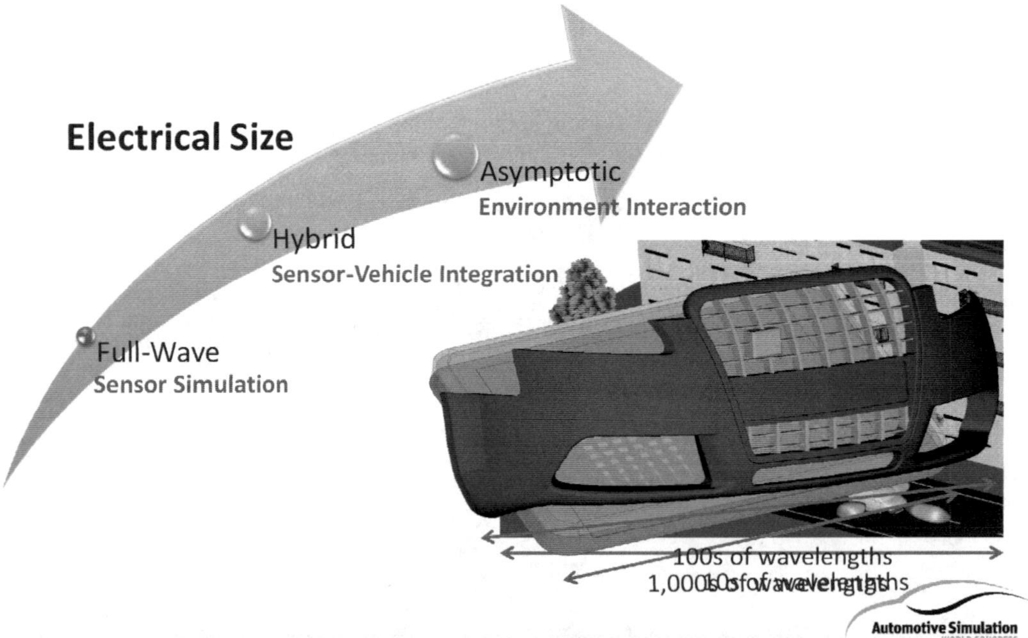

Radar in Environment

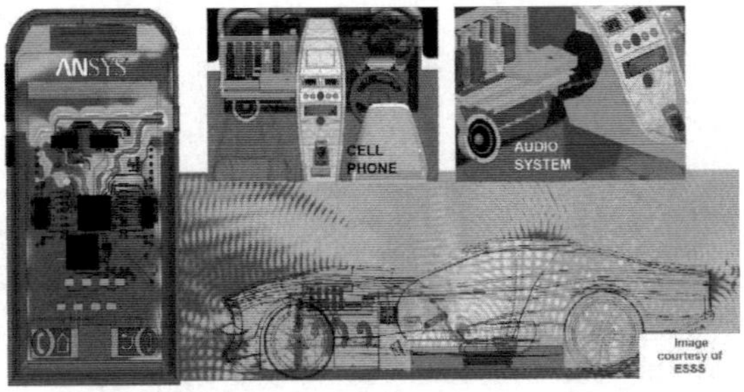

Automotive Simulation WORLD CONGRESS

27 June 7-8, 2016 Munich

Communications Electromagnetics Simulation

Automotive Simulation WORLD CONGRESS

28 June 7-8, 2016 Munich

GPS ECU and Signal Simulation

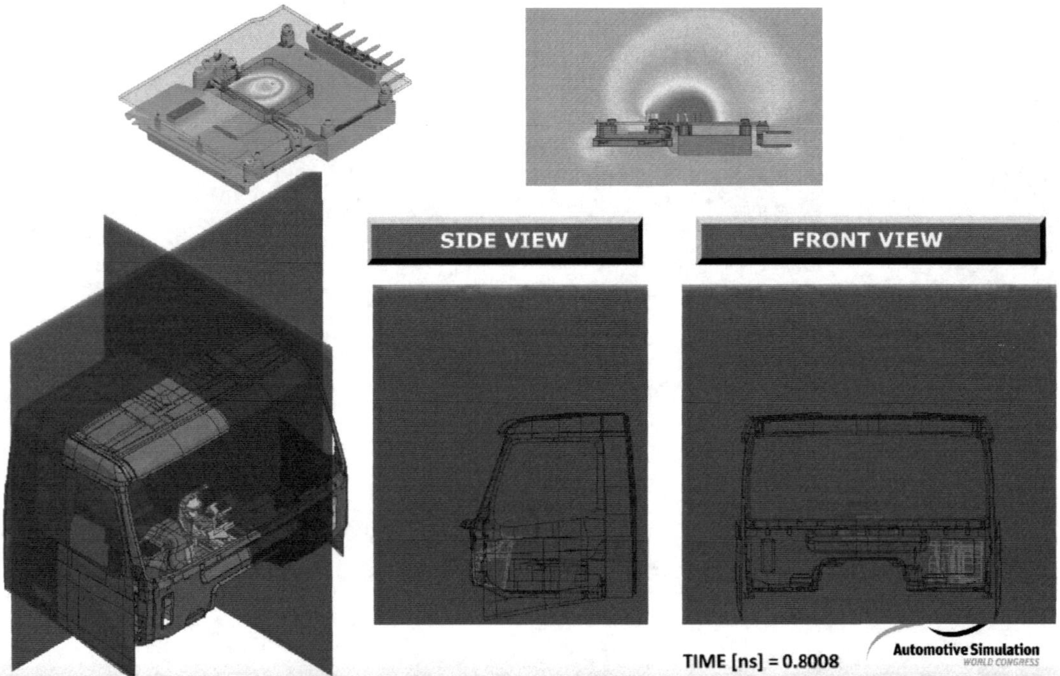

Automotive Simulation WORLD CONGRESS

Vehicle 3D Geometry Models

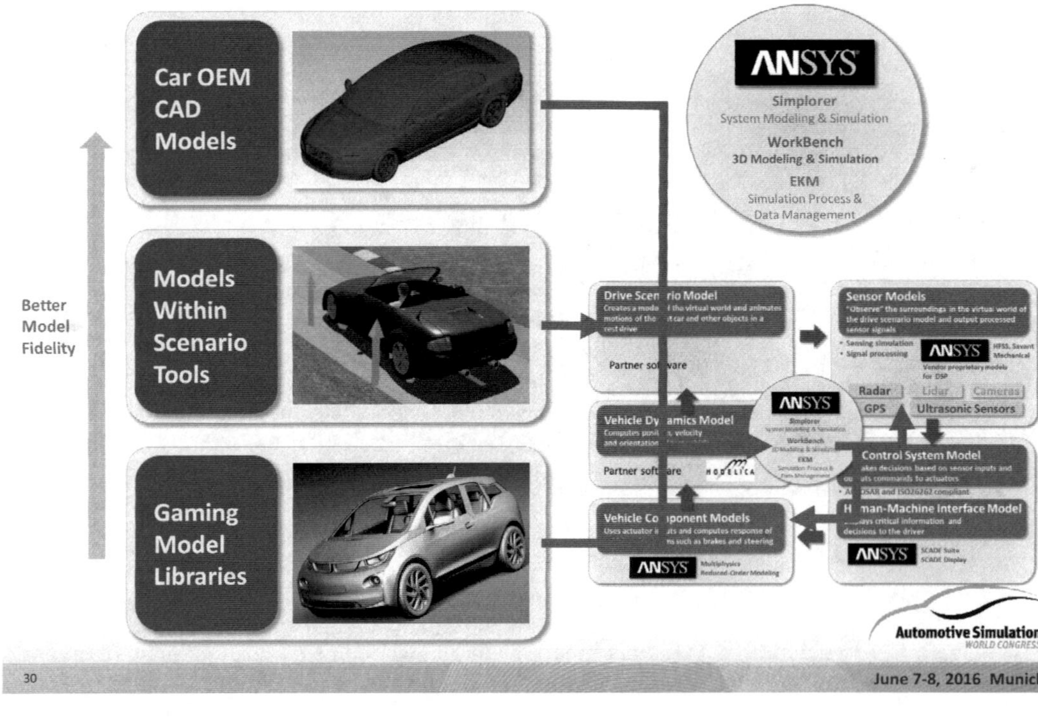

Automotive Simulation WORLD CONGRESS

Summary

- Simulation necessary to address complexity of ADAS loop
  - ANSYS capabilities + Partner Software completes the loop
- Also applicable to ADAS components – to ensure reliability of ADAS components:
  - Thermal-structural integrity
  - Power integrity
  - Signal integrity

Automotive Simulation WORLD CONGRESS

31 June 7-8, 2016 Munich

Automotive Simulation WORLD CONGRESS

June 7-8, 2016 Munich

Thank You!

Automotive Simulation WORLD CONGRESS

June 7-8, 2016 Munich

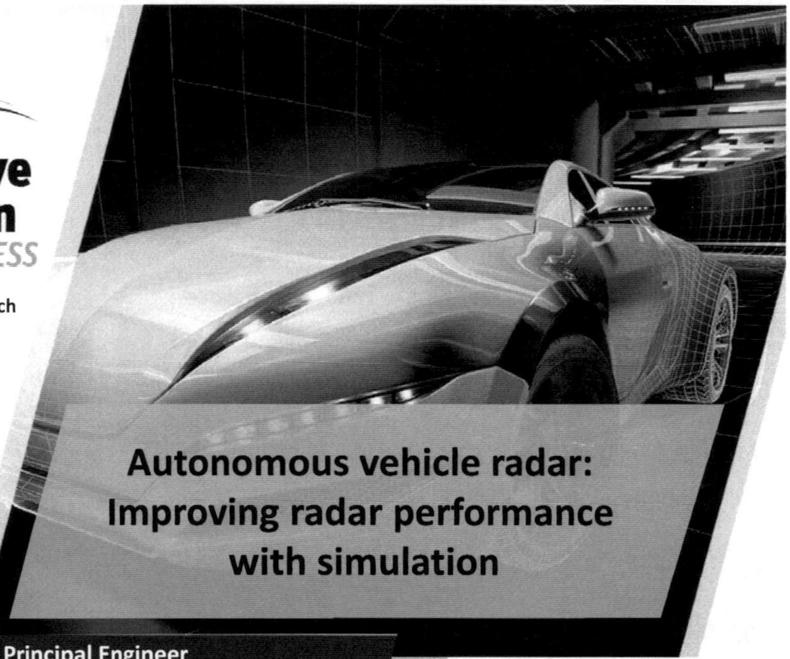

Autonomous vehicle radar: Improving radar performance with simulation

Markus Kopp, Principal Engineer
ANSYS Inc.

HFSS SBR+, Installed Fascia Antennas

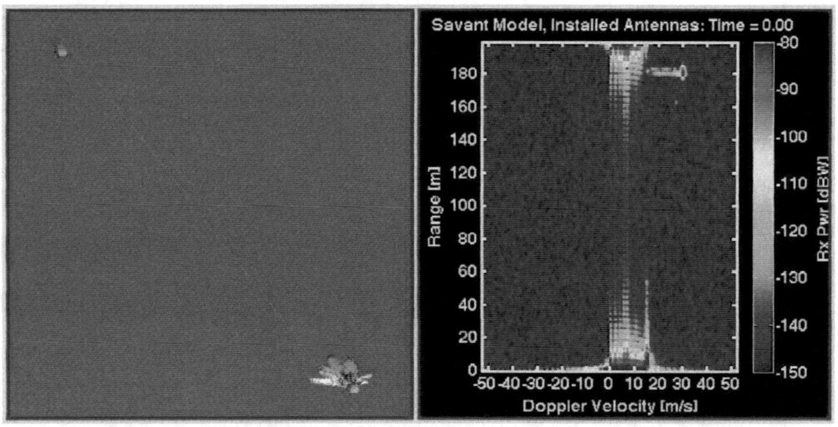

Value Proposition

- **Advantages of Simulations**
- Test and Measurement is very valuable but…
  - Expensive & time consuming
  - Difficult to measure all scenarios
  - Lacks insight to physics involved
- Finding problems too late
  - Delays in delivery = lost revenue
  - Potential for major increase in cost to resolve issues
  - In extreme scenarios, loss of life
- Simulation extremely powerful
  - Cost is low relative to test and measurements
  - Rapidly simulate many scenarios
  - Insight to physical mechanisms/root causes of problems
  - Complementary to T&M

Automotive Simulation
DELCROSS TECHNOLOGIES

Outline

- Automotive Radar Overview
- HFSS/HFSS SBR+ (Savant) Hybrid Approach
- Impact of Fascia Designs
- Range/Doppler Simulations
- Summary

Automotive Simulation WORLD CONGRESS

Automotive Radar Overview

- **Many Applications**
- Adaptive Cruise Control
- Collision Warning
- Mitigation & Brake Support
- Autonomous vehicles

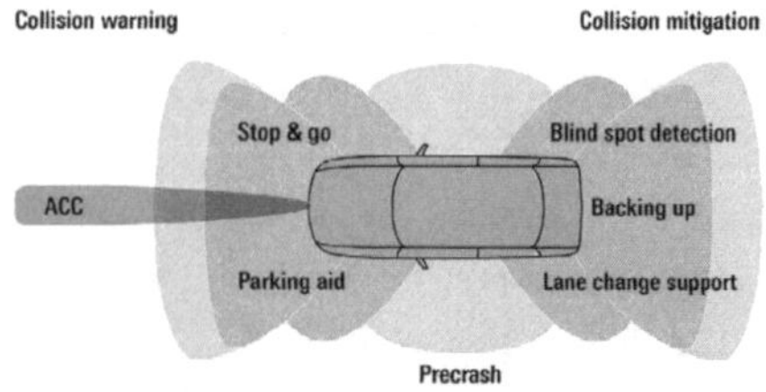

- **Design Considerations**
- Need to function in many different fascia designs
- Installed performance vs. free space performance
- Performance in variety of traffic conditions
- Environmental factors (rain, ice, mud, etc.)
- Target Identification

HFSS/HFSS SBR+ Hybrid Approach

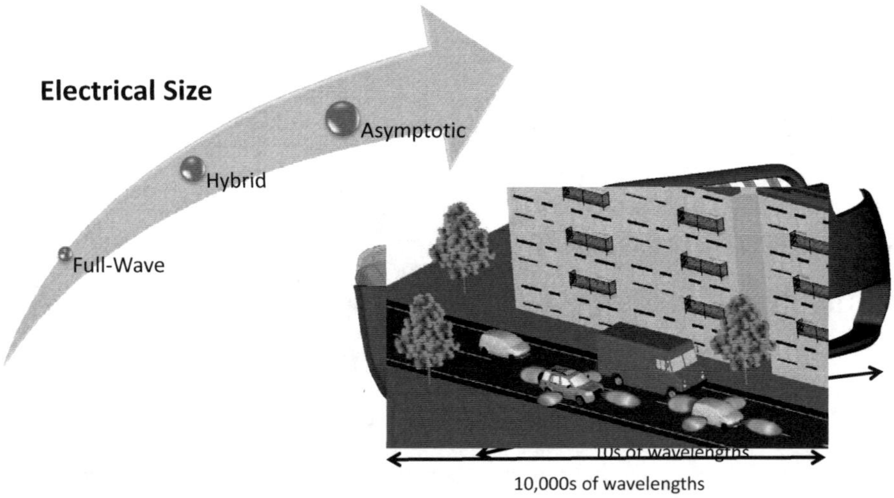

HFSS SBR+

- **What is SBR ?**
- SBR = Shooting and Bouncing Rays
  - Asymptotic technique
    - Complementary capability to full-wave solvers
    - Electrically large platforms (*i.e.*, many wavelengths in dimension)
  - Extends physical optics (PO) to multiple bounces with GO ray tracing
- **The HFSS (Delcross) SBR+ Technique**
- Build on SBR with additional physics
  - Physical Theory of Diffraction (PTD)
  - Creeping Wave
  - Uniform Theory of Diffraction (UTD)
  - Curved Surface Divergence Factor
- Acceleration
  - Multicore, GPU and MPI for massive acceleration
  - Algorithmic acceleration

Auto Radar Model

- **76.5 GHz Design**
- Transmit array
  - Rectangular, microstrip patch array
  - 23 dBi directivity
- Two receive arrays
  - Linear arrays
  - 20 dBi directivity
- Radome
  - 0.7 mm
  - $\varepsilon_r = 3.05 - j0.0183$

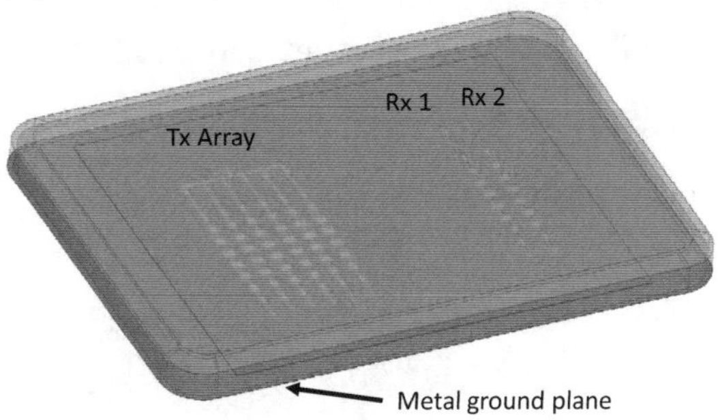

HFSS/HFSS SBR+ Hybrid Approach

- HFSS simulates isolated performance of the radar
- Electric/Magnetic fields captured over Huygens surface
- Convert field data into equivalent current sources
- Savant uses current sources to compute installed performance

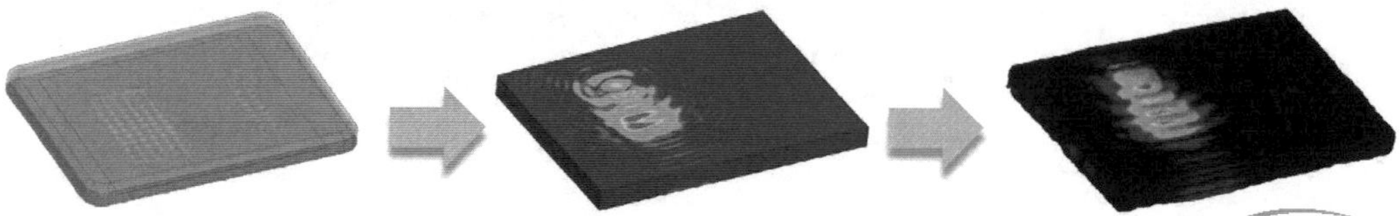

Automotive Simulation WORLD CONGRESS

Isolated Performance Validation Tests

- Compute isolated performance of Tx and Rx arrays
- HFSS: full-wave FEM solution
- HFSS SBR+: equivalent current sources from HFSS
- Compare solutions and confirm similar performance

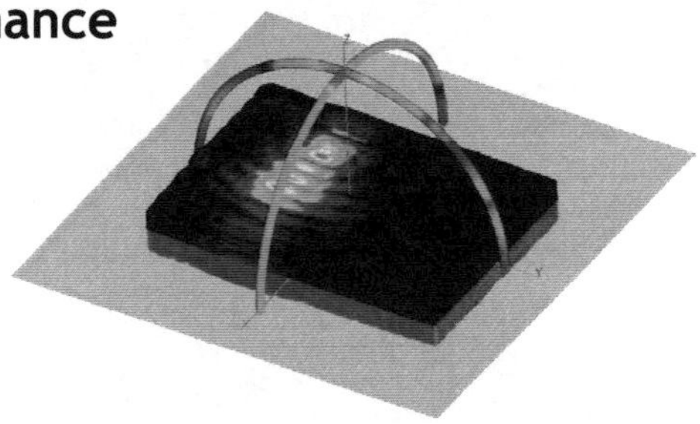

Automotive Simulation WORLD CONGRESS

Transmit Array Results

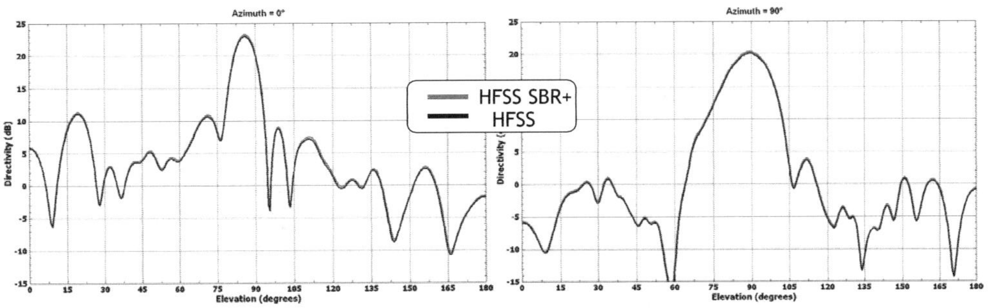

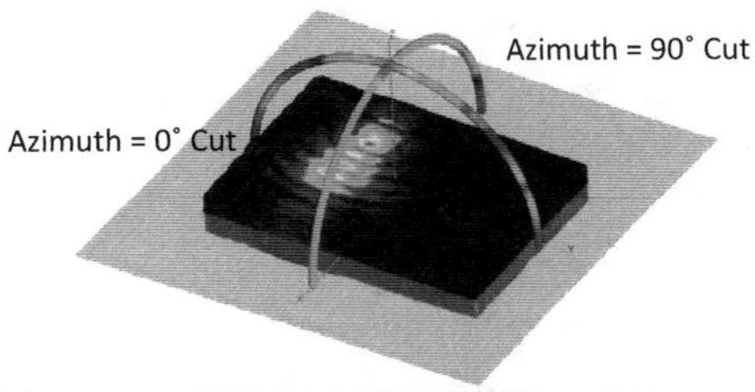

Receive Array #1 Results

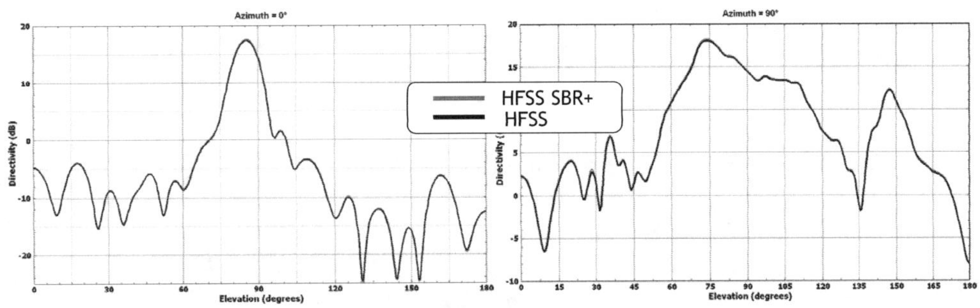

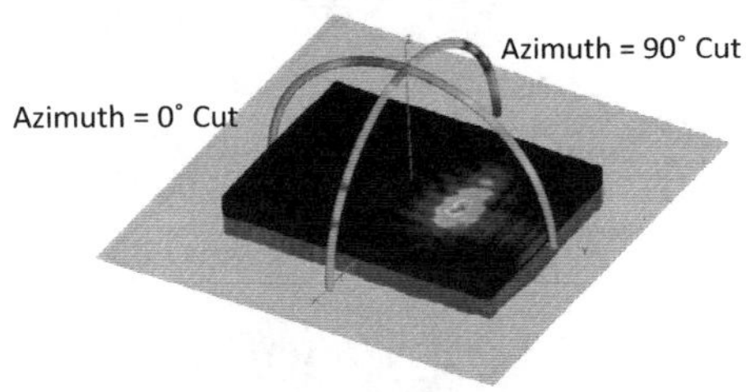

Receive Array #2 Results

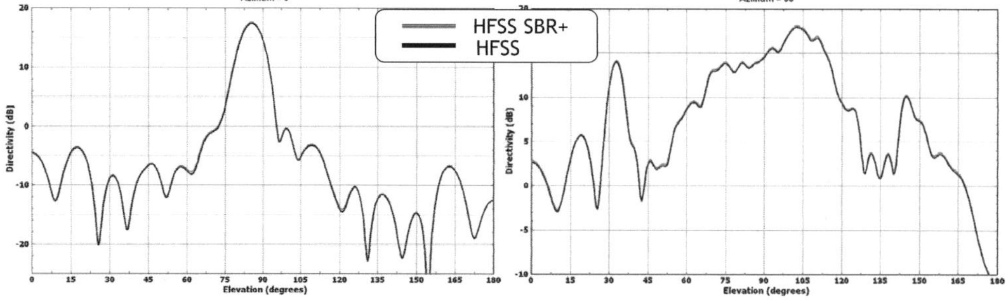

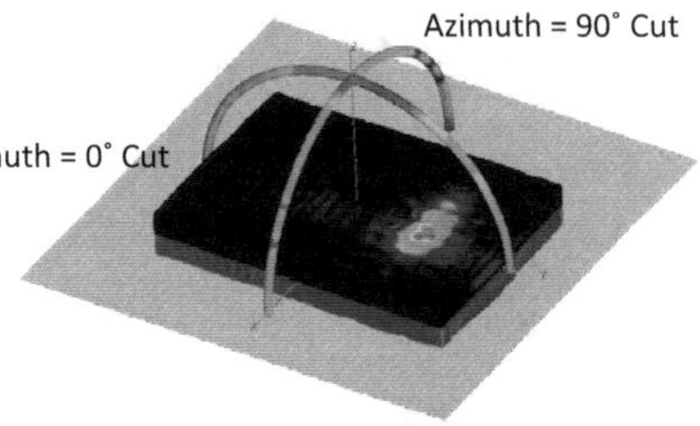

Installed Performance

- **Install radar in two locations in fascia CAD model**
- **Compute far-field patterns for Tx and Rx arrays**

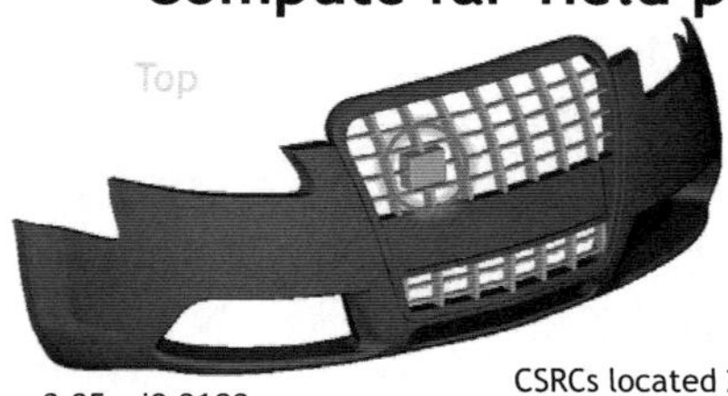

Grill
ε_r = 3.05 – j0.0183
t = 3.5 mm
Bumper
ε_r = 2.3 – j0.004
t = 10 mm

CSRCs located 2λ behind surface of enclosure in grill

Positioned above finite ground plane (red plate in image)

Ray Visualization in HFSS SBR+

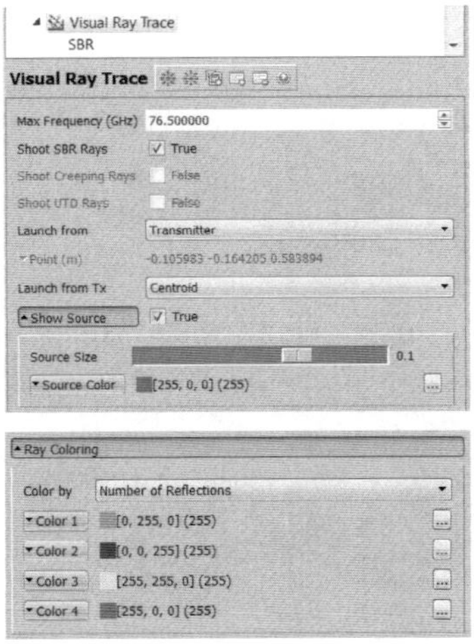

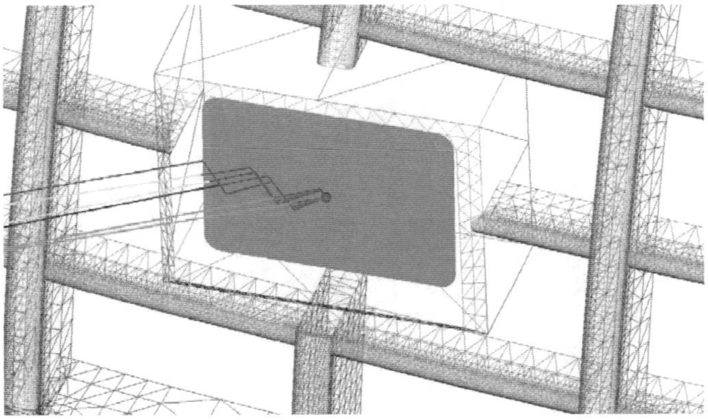

Multi-Scale Dimensions

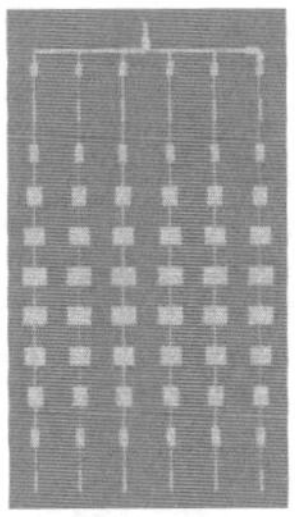

Sub-wavelength elements

13 λ x 18 λ x 1.2 λ

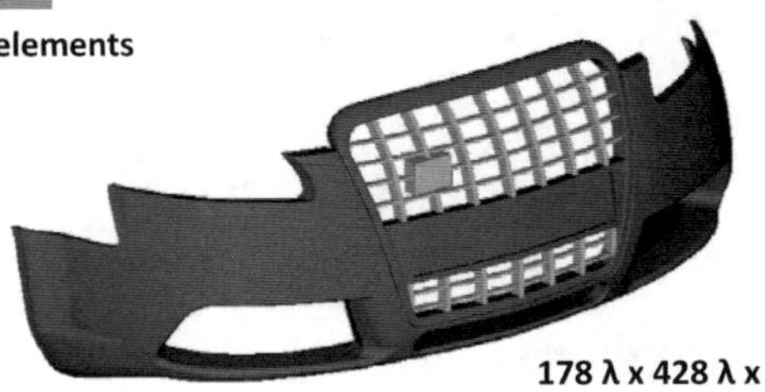

178 λ x 428 λ x 227 λ

Transmit Array Results

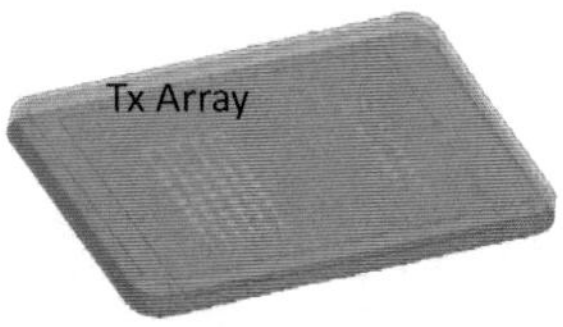

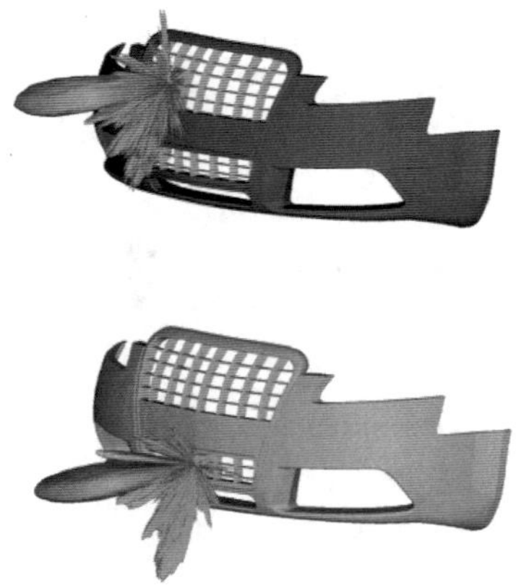

Runtime: 20 hours on 5 node clustε

Automotive Simulation WORLD CONGRESS

Transmit Array Results

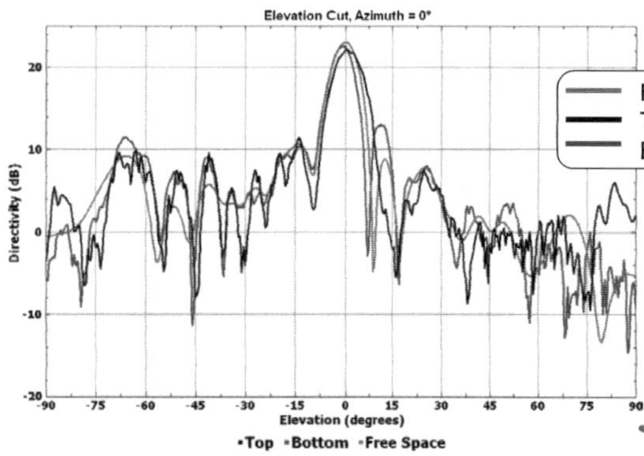

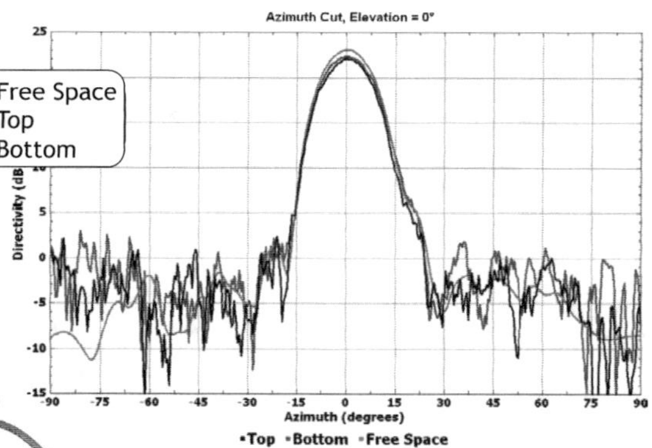

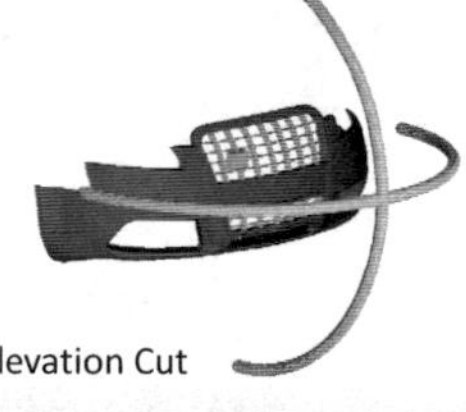

Automotive Simulation WORLD CONGRESS

Receive Array 1 Results

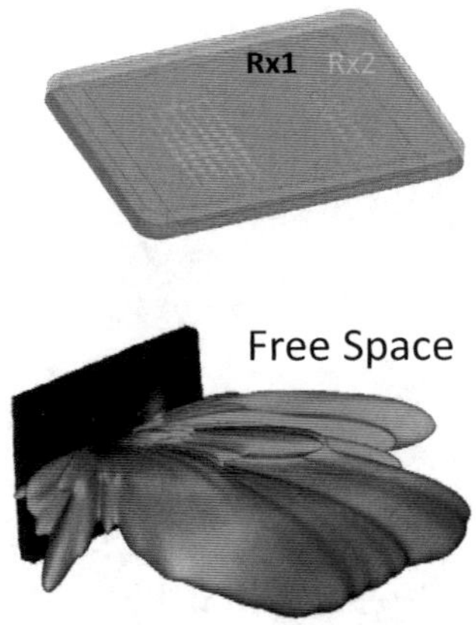

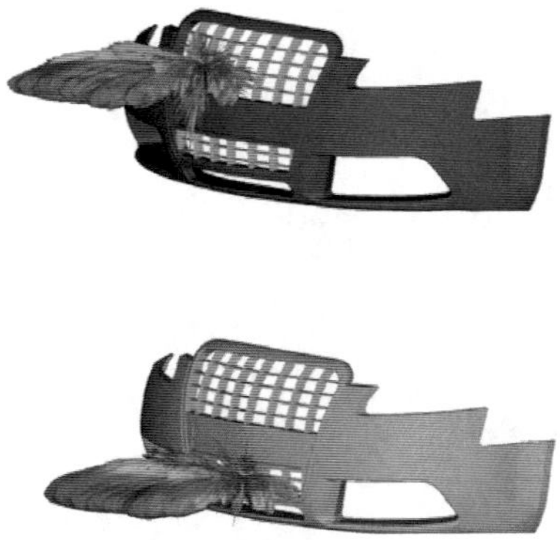

Runtime: 20 hours on 5 node cluste

Receive Array 1 Results

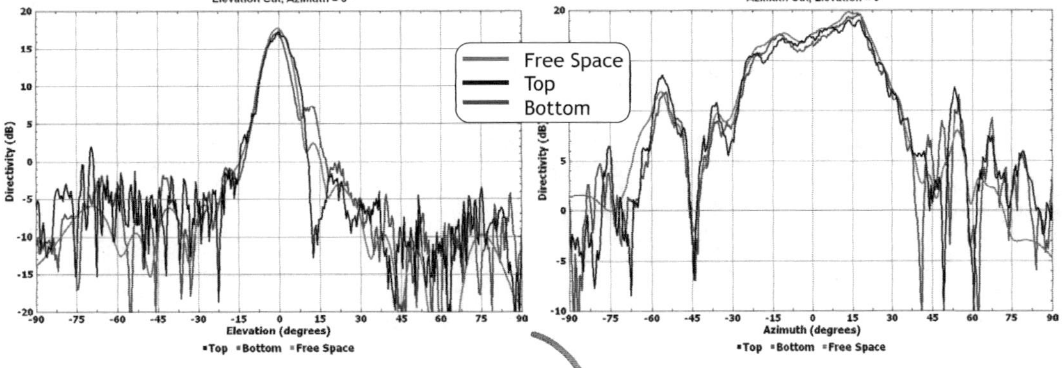

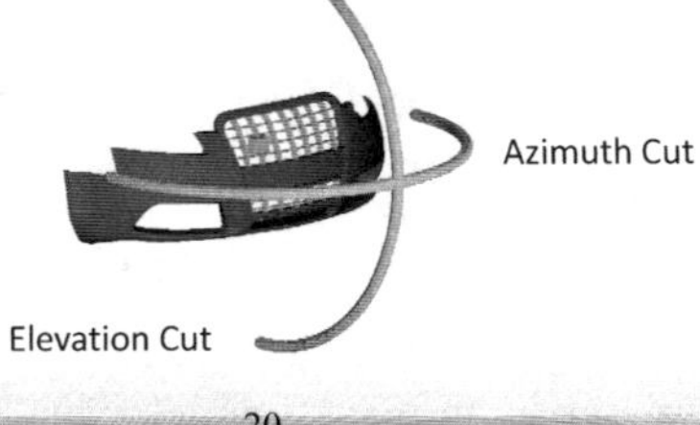

Receive Array 2 Results

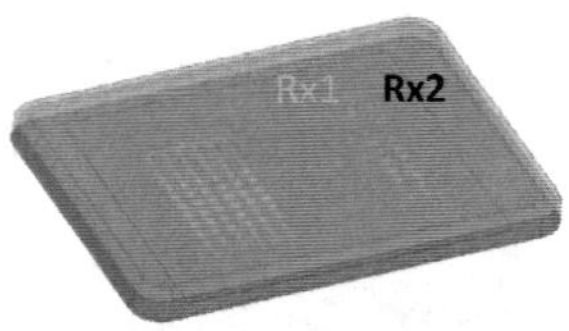

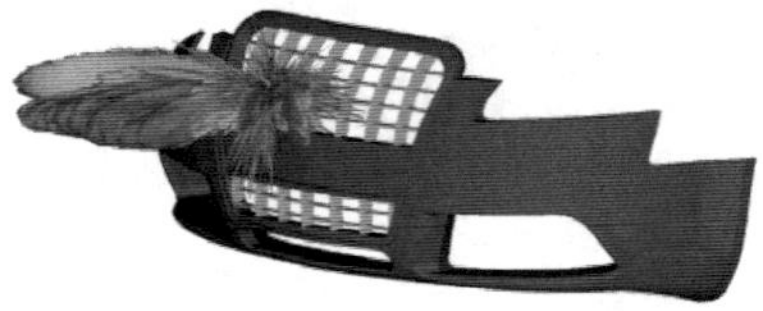

Free Space

Runtime: 20 hours on 5 node cluster

Receive Array 2 Results

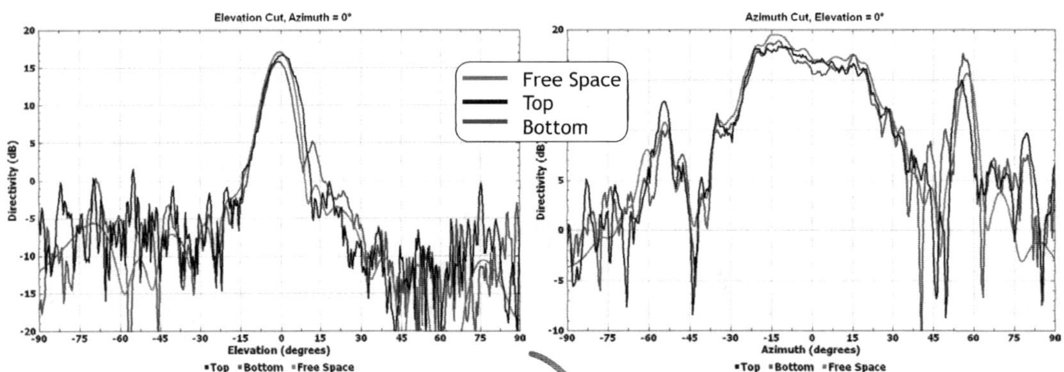

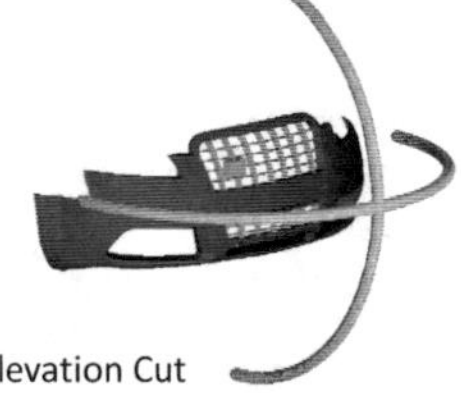

Radar Scenario Simulation with HFSS SBR+

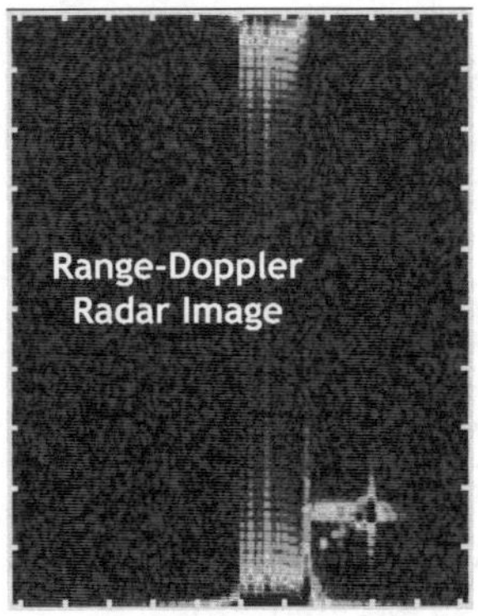

Pulse-Doppler Radar Background

- Issue series of pulses to measure target range and velocity

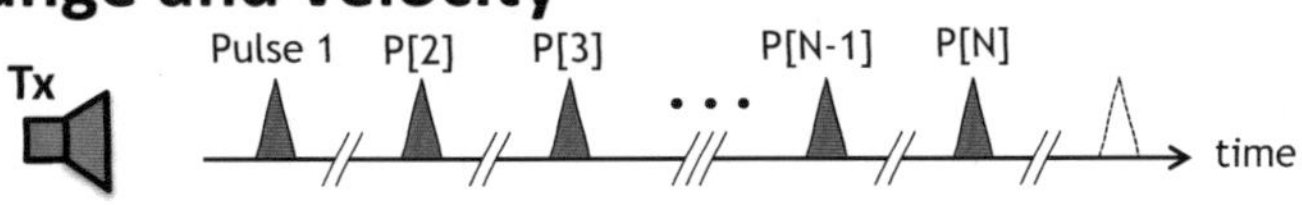

- Delay between Tx and Rx pulse determines range
- Frequency shift determines velocity
  - Velocity component perpendicular to radar

Range-Doppler Image Frame

(how things look to the radar at one moment of time)

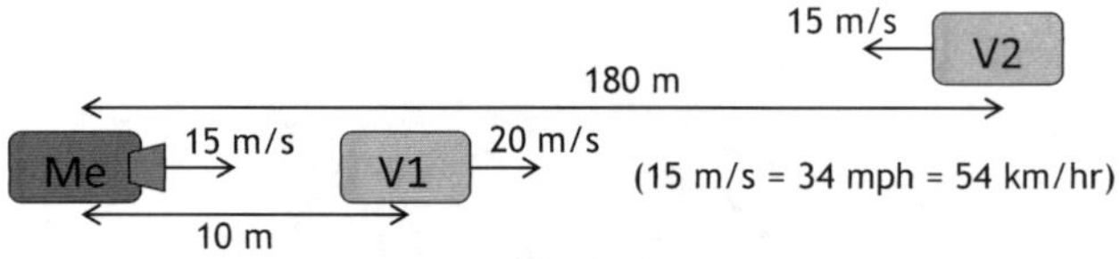

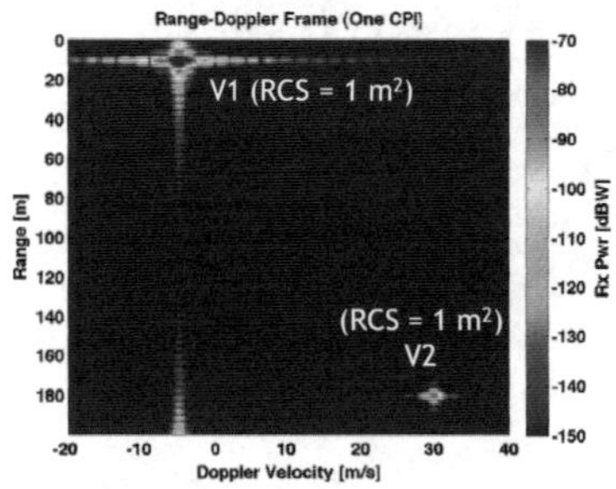

With pulse-Doppler radar, objects can be isolated and tracked in a 2-D space.

This range-Doppler image is idealized

- No radar system noise
- No ground clutter or bounce

Automotive Simulation WORLD CONGRESS

Coherent Processing Interval

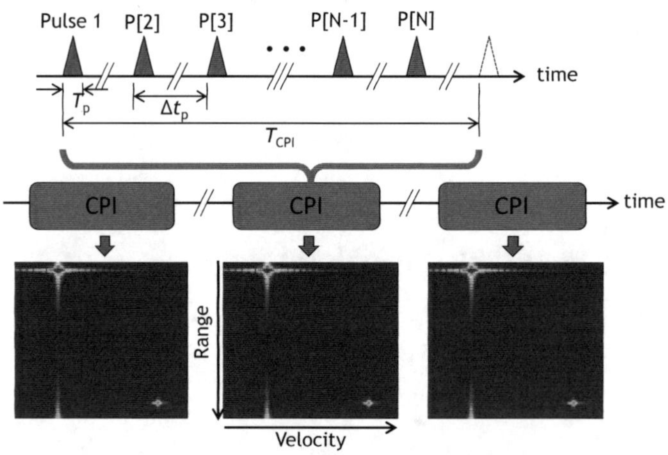

Automotive Simulation DELCROSS TECHNOLOGIES

Simulating Pulse-Doppler Radar in HFSS SBR+

- **HFSS SBR+ is a frequency-domain code**
  - Generates Tx-to-Rx signal vs. frequency
  - Synthesize pulse from N_f frequency samples
  - Fourier transform into "range profile"
- **Next pulse: advance scenario by Δt_p**
  - Move objects according to their velocities
  - Repeat Savant simulation
- **Repeat for *N* pulses**
  - Generates Rx signal data for a single CPI
- **Advance scenario to next CPI**

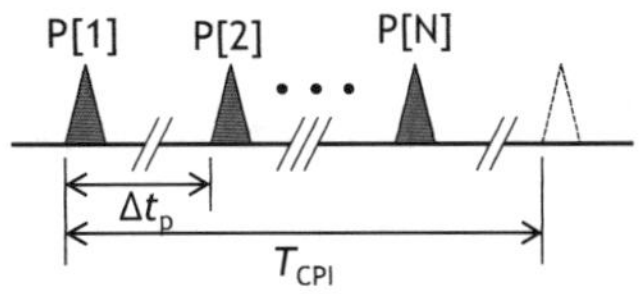

Automotive Simulation WORLD CONGRESS

Pulse-Doppler Radar Properties

- f_c = 76.5 GHz
- Range Resolution = 1 m → BW = 150 MHz
- Range Ambiguity = 200 m → N_f = 200
  - Ex: range-to-target = 15 m?, 215 m?, 415 m?
- Velocity Resolution = 0.5 m/s = 1.1 mph,
- Velocity Ambiguity = 105 m/s = 235 mph
- *N* = 210, Δt_p = 18.7 μs, T_{CPI} = 3.92 ms

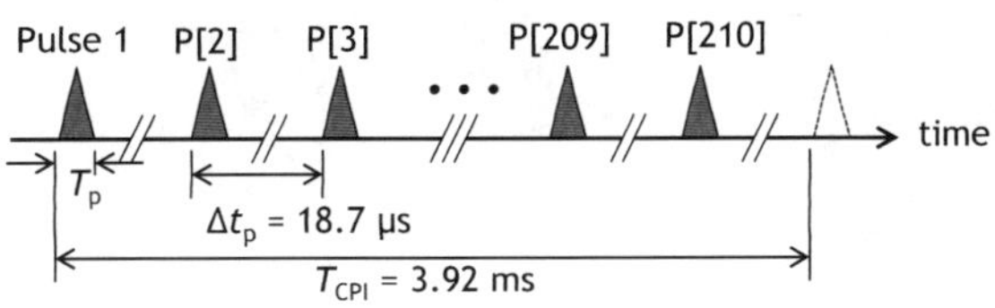

Automotive Simulation WORLD CONGRESS

What this means for HFSS SBR+ Simulations

- **Assume**
- Car speed = 15 m/s = 54.4 kph
- CPI frame rate = 10 fps
- Scenario time = 6 s

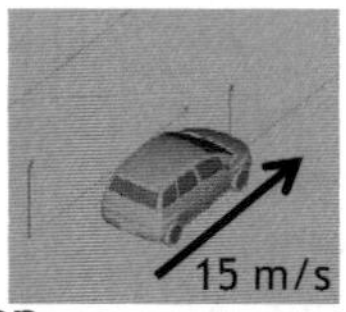

- **Car advances …**
- 0.28 mm with each Δt_p = 19 μs time step
- 5.9 cm over course of each T_{CPI} = 3.9 ms
- 1.44 m at end of CPI to start of next CPI

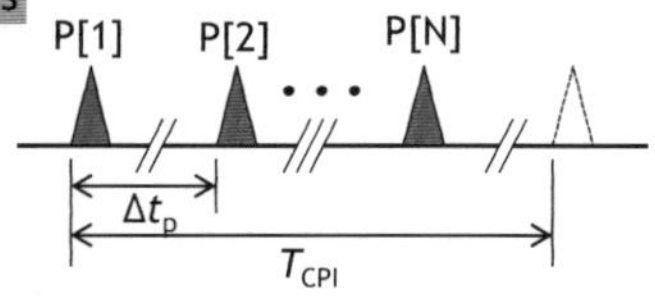

- **Entire simulation: 12,810 Savant runs**
- On scene measuring 51,000λ on a side
- Each with 200 freqs

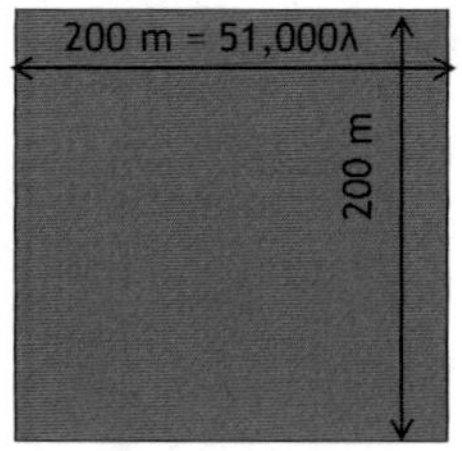

- **Speed-up**
- Use ray-extrapolation
- Only one HFSS SBR+ simulation per CPI
- Cuts effort by factor of N (210)
- Entire scenario runs in 90 min on laptop

System Noise

- **Real radar systems have system noise**
- Added during post-processing

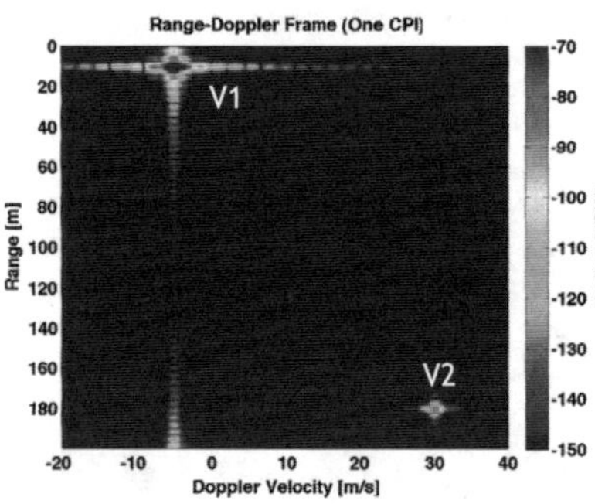

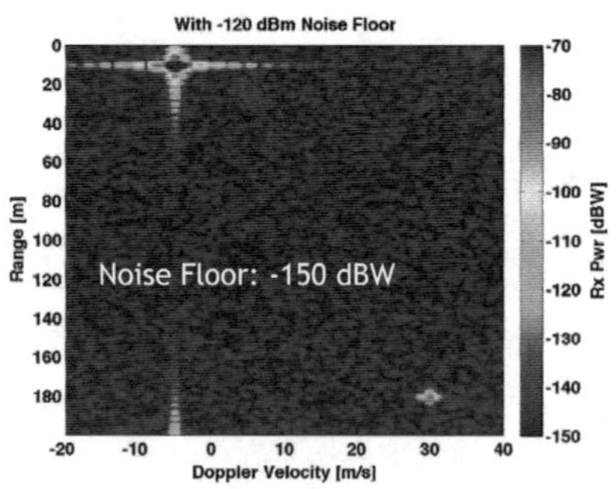

Environment: Rough Surface Scattering

- At 76.5 GHz, much of the world looks like a rough surface (wavelength is 3.9 mm)
- Rough: Road, terrains, trees
- Smooth: car panels, glass

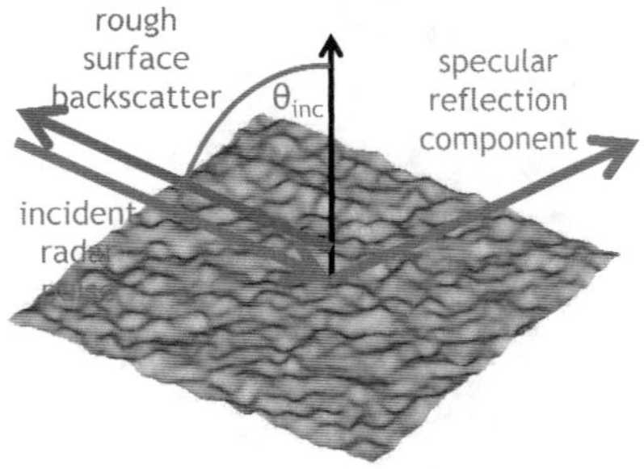

Road Return

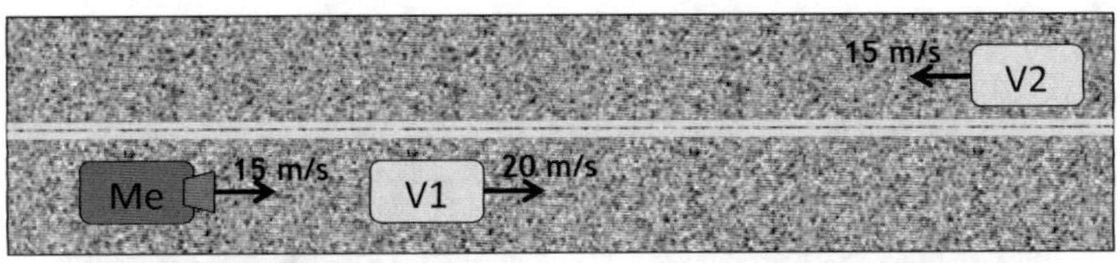

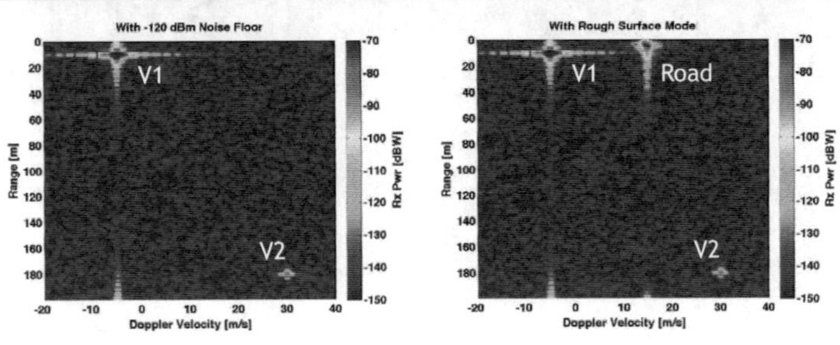

Scenario Setup

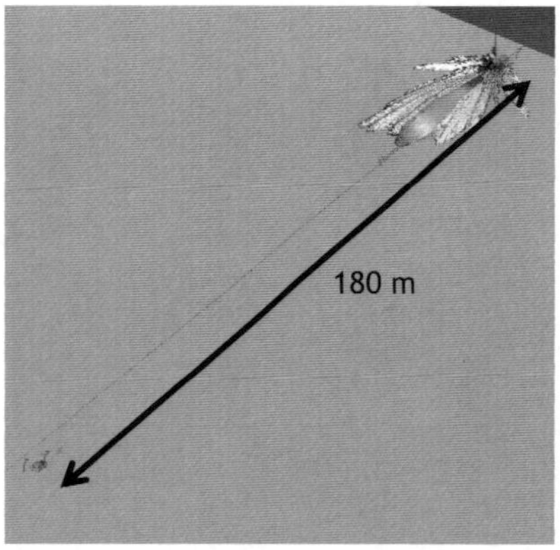

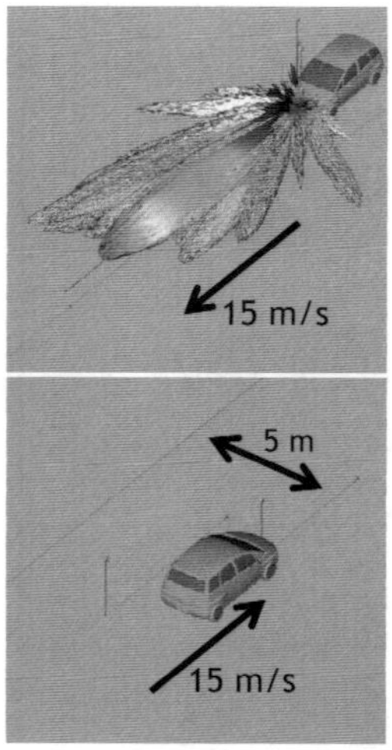

Point-Scatter Model, Parametric Beams

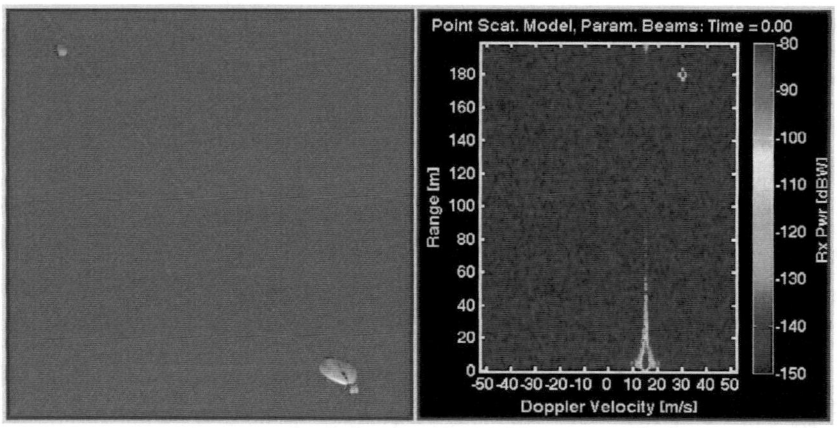

HFSS SBR+, Parametric Beams

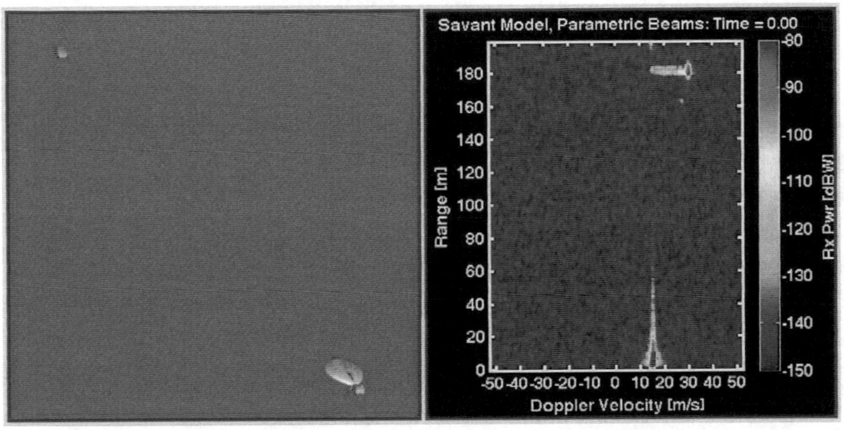

Automotive Simulation

HFSS SBR+, Installed Fascia Antennas

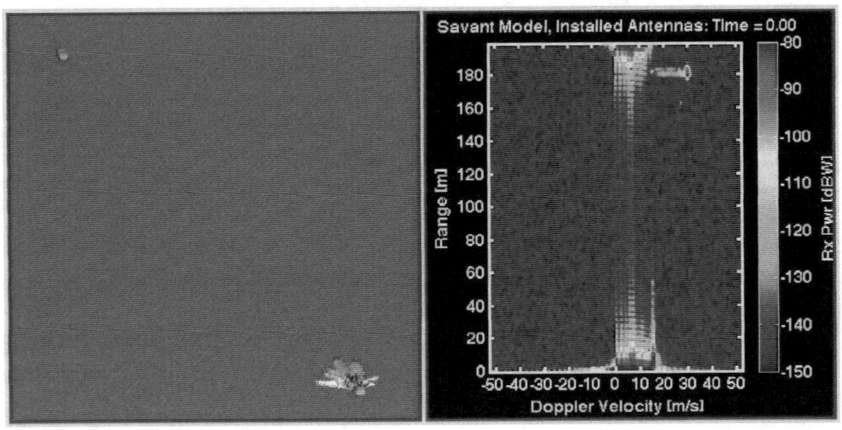

Automotive Simulation

Analysis

Installed antenna patterns have large impact

- Strong lobe points directly down

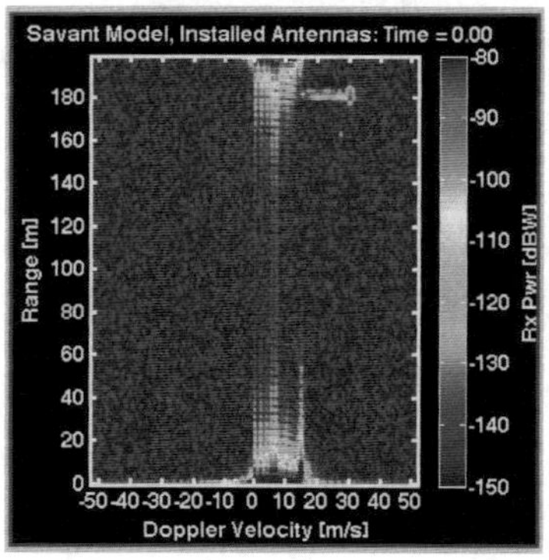

Automotive Simulation
DELCROSS TECHNOLOGIES

What's Next

- **More complex and realistic scenarios**
- Multiple vehicles
- Mixed ground clutter
  - Grass, Road Conditions, Terrain
- Environmental Features
  - Building, Trees, People
- **Other types of radars**
- FMCW

Automotive Simulation WORLD CONGRESS
June 7-8, 2016 Munich

Summary

- HFSS SBR+ allows you to simulate electrically large, realistic problems that you may have previously considered impossible
- For complex, multi-scale problems such as automotive radar, hybrid electromagnetic solutions are critical for practical and accurate solutions
- Radar design
- Fascia Impact
- Large scene range/Doppler simulations

Automotive Simulation WORLD CONGRESS

Why isn't the IoT more like the Internet? (and how to fix it)

ANSYS Electronics Simulation Expo

Ali Nicholl
Head of Sales and Marketing

7th June, 2016

Iotic Labs

ali.nicholl@iotic-labs.com
@ac_nicholl

Iotic Labs

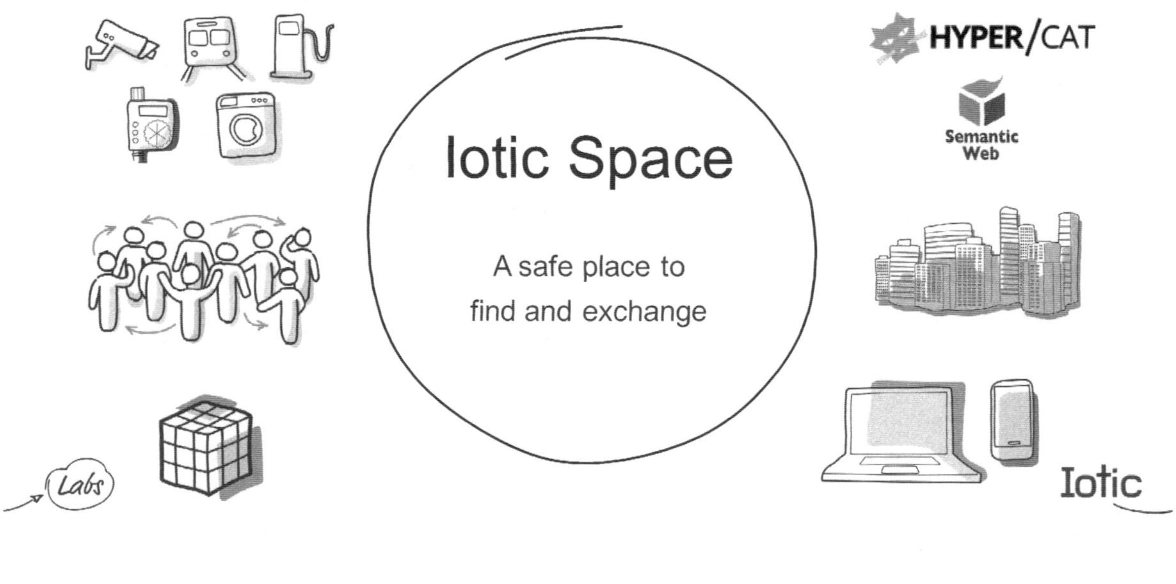

Just imagine…

We have been led astray by names and titles

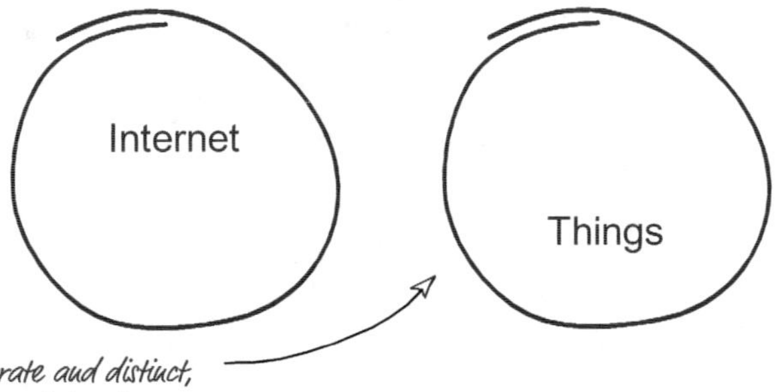

The idea that somehow Internet and IoT are separate and distinct, or at least one is the subset of the other

Labs

Iotic

Just imagine…

We have been led astray by names and titles

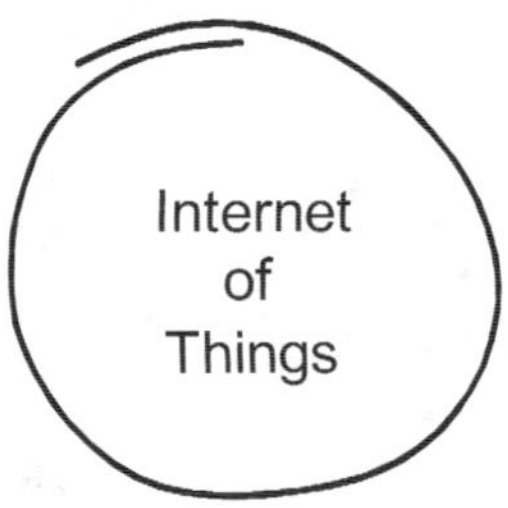

Not an Internet of Things, an Internet with Things in it

Labs

Iotic

No man is an island

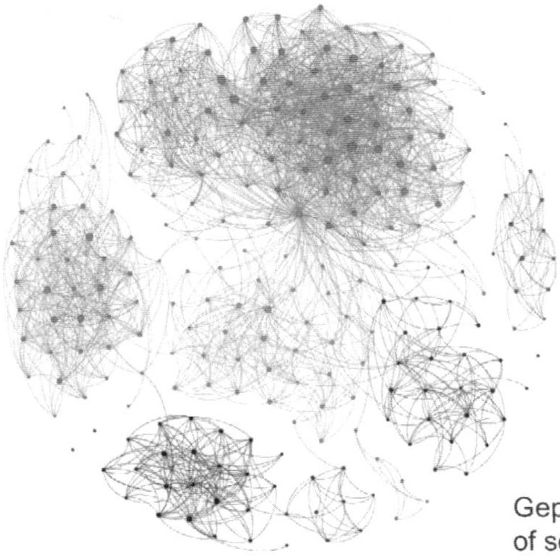

Gephi visualisation
of social network

Labs

Iotic

No industry is an Island

Labs

Iotic

6

Collaboration and creation in communities

Be and expert and share that expertise
Through collaboration we can innovate faster, do more

We can be more than the sum of our parts

Labs

Iotic

Internet of Transformation

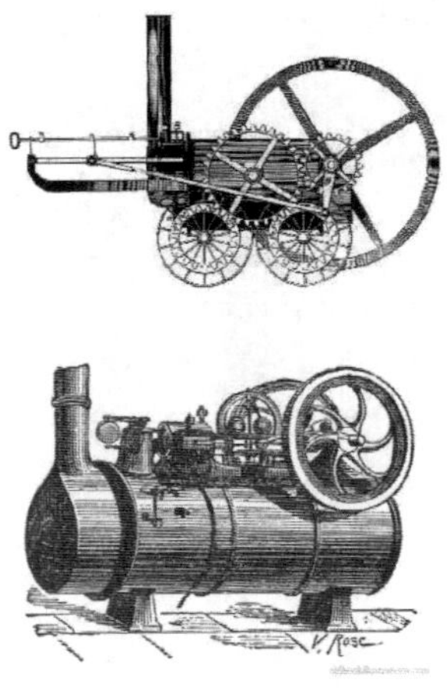

Where have we seen this before?

The industrial revolution.
Reusing repurposing, contributing, leveraging technology in previously unimagined ways.

Labs

Iotic

Enchanting and enhancing

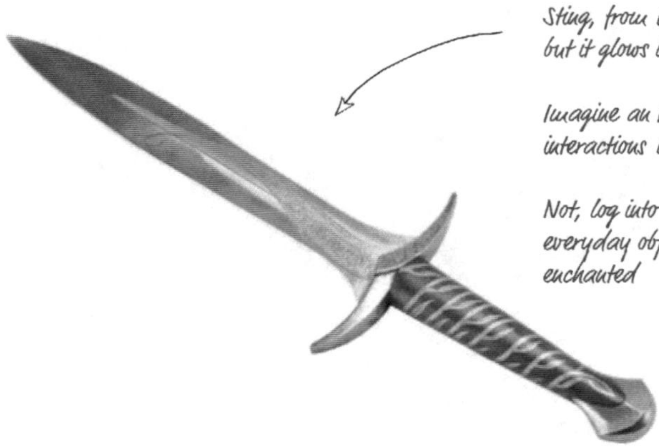

Enchanted objects that appear like magic.

Sting, from Lord of the Rings, is always a sword, but it glows in the presence of orcs and goblins.

Imagine an IoT world were the focus was on interactions not interfaces.

Not, log into our app to check for enemies, but the everyday objects and services I use are enhanced and enchanted

Labs

Iotic

Internet of...

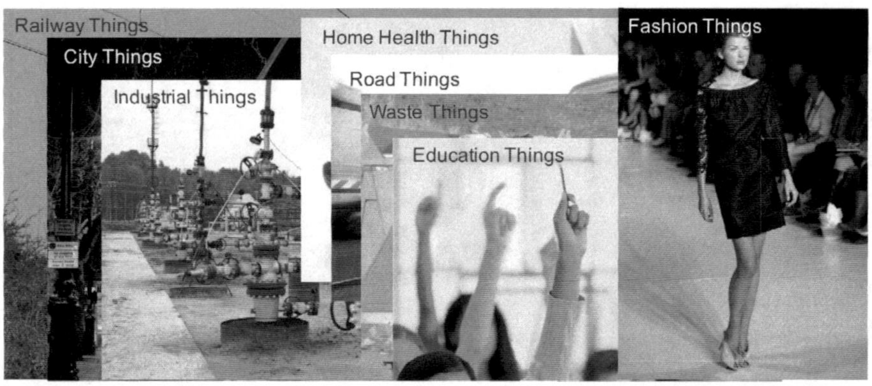

"We can begin to see the solution to problems we do not yet understand"

Labs

Iotic

User-centric

How do we fix the Internet of Things?

Make it more like the Internet
- ubiquitious and personal.

I control what I share, with whom and when. The user decides levels of engagement and personalises their experience. They chose what they are interest and create their own communities around those interests.

Iotic Space from Iotic Labs

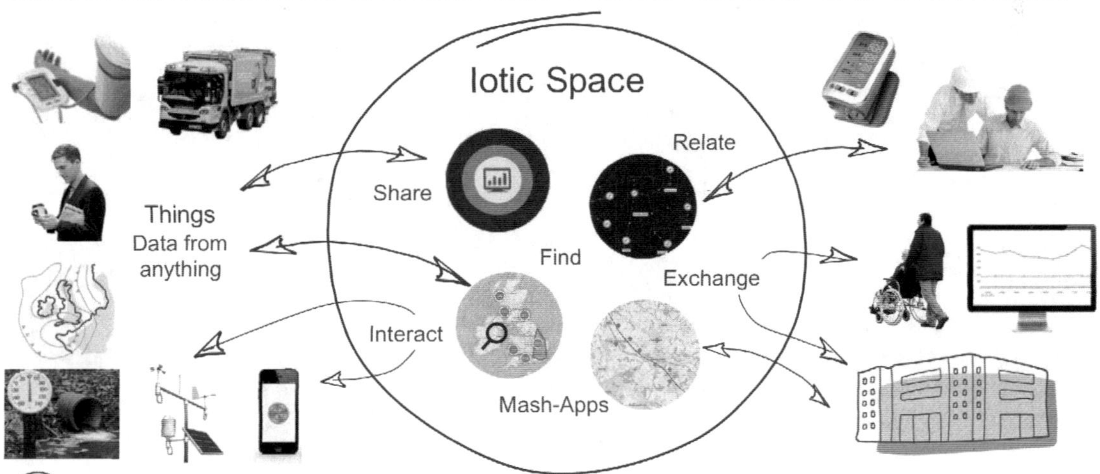

Share and combine data from anything with anything

What does this mean for development?

- How can we leverage what already exists?
- How can/will others leverage what we are developing?
- How can we maximise the potential uses?
- How does our adaptability compare to the competition?
- Can we benefit from, and benefit, a community of users?

Don't lock in users or organisations but connect them.

Not more interfaces but interoperability.

Labs

Iotic

All we need is the Internet

- Provide a service, do not sell solutions
- Context is king
- Explore rather than travel
- Be an expert and collaborate
- Just Start

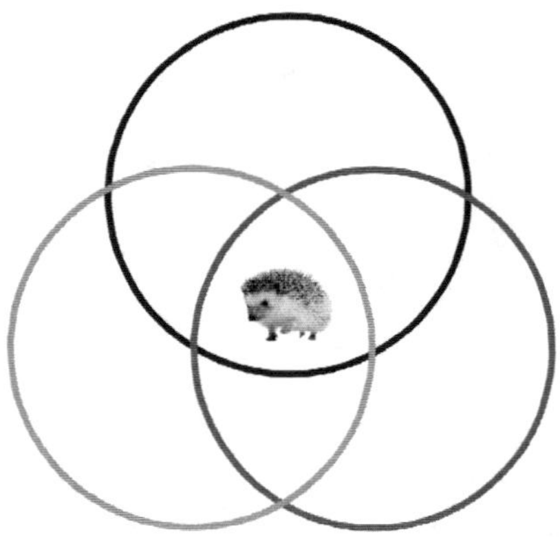

Create on the fly, connect the previously unrelated, generate enchantment through innovation and we can begin to see solutions to problems we have not yet defined nor understood.

Labs

Iotic

Thank you

Ali Nicholl
Iotic Labs

@ac_nicholl
@IoticLabs

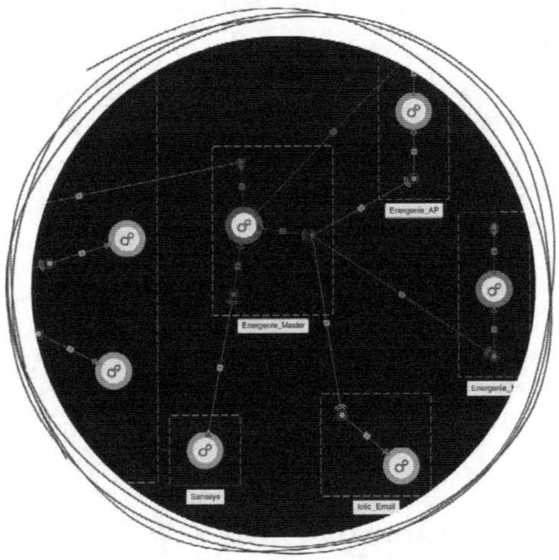

Explore Iotic Space at iotic-labs.com

Labs

Iotic

Electronics Simulation EXPO

June 7-8, 2016 Munich

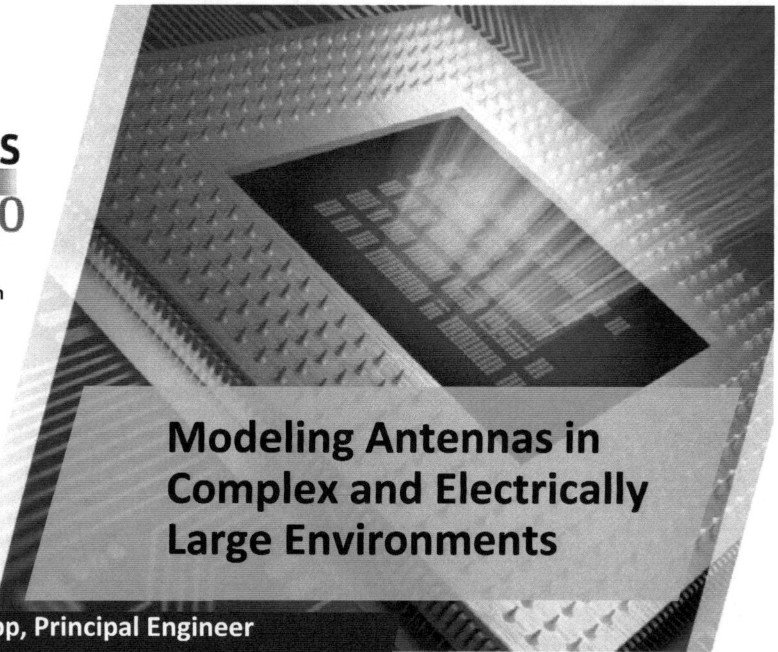

Modeling Antennas in Complex and Electrically Large Environments

Markus Kopp, Principal Engineer
ANSYS Inc.

Installed Antenna Analysis – Why?

- **Antennas are usually designed and tested in isolation or under ideal conditions**
- **Mounting antennas on realistic platforms or placing them in realistic environments changes their performance**

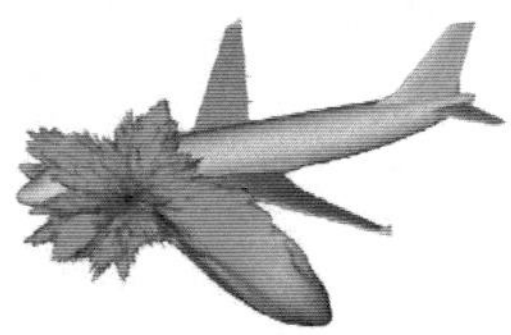

- **Installed antenna performance impacts RF system performance**
  - **RF Links and RF Cosite Interference**
  - **Explore the installed performance early in the design cycle!**

Video Presentation in Original e-media

Video Presentation in Original e-media

Electronics Simulation EXPO

ANSYS High Frequency Electromagnetic Solvers

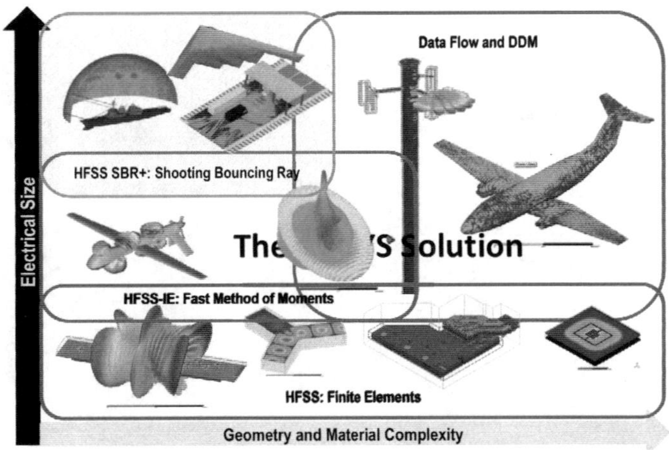

Electronics Simulation EXPO

The ANSYS Installed Antenna Work Flow

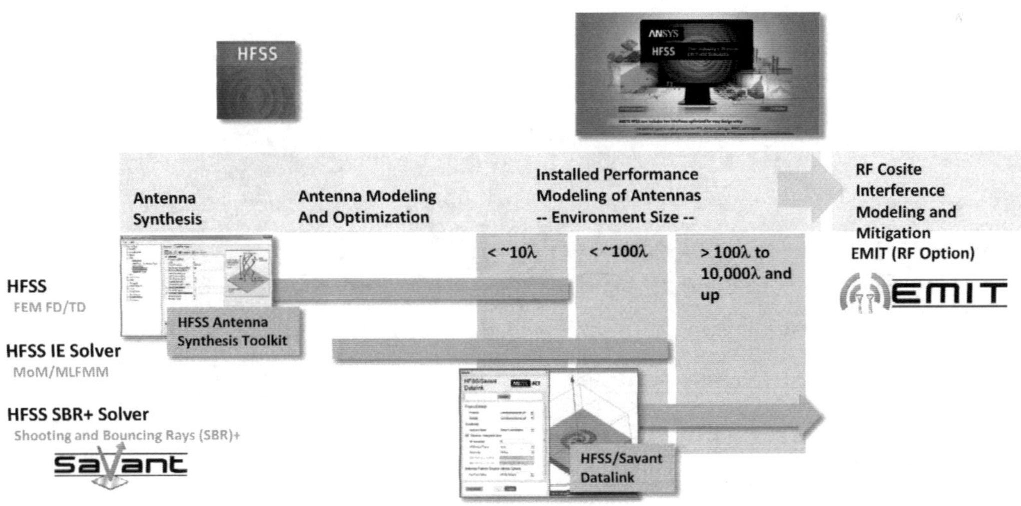

Electronics Simulation EXPO

Installed Performance of Automotive Radars

Antenna Model in HFSS

- **77 GHz Radar Array Design with HFSS**
- **Configure for Tx and Rx modes**
- **Tx high gain, narrow beam**
- **Rx broad azimuth**

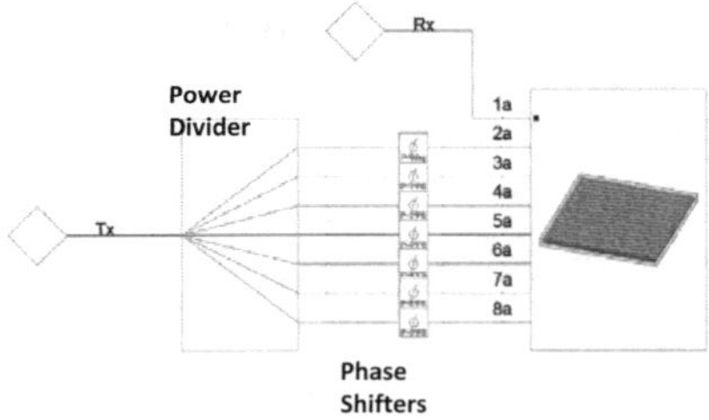

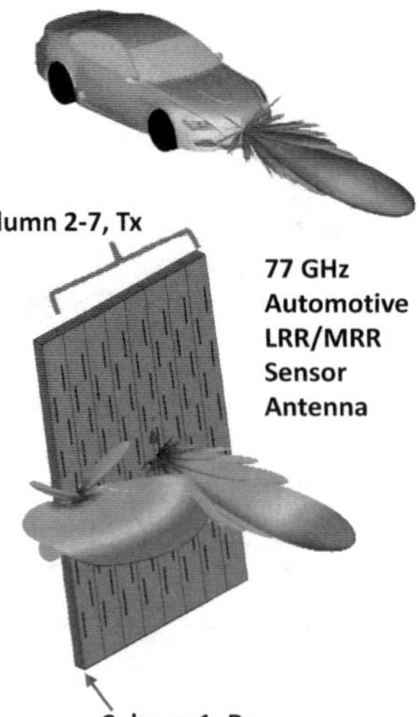

Installed Performance of Automotive Radars

- **Effect of fascia and bumper are needed**
- **HFSS solution for the array is placed behind the grille housing**
- **Problem size is 440λ at 78.5 GHz**

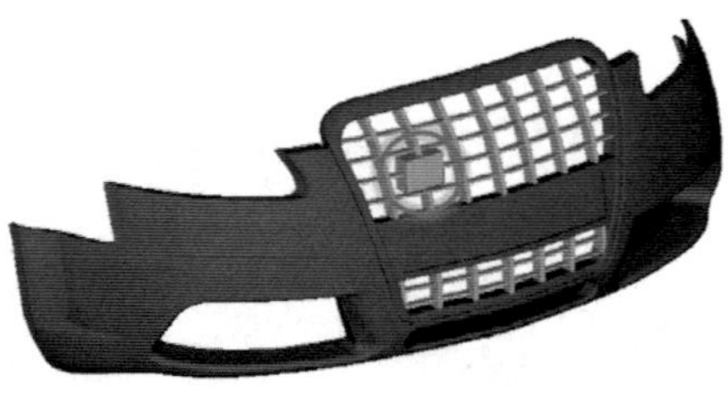

Grill
ε_r = 3.05 – j0.0183
t = 3.5 mm
Bumper
ε_r = 2.3 – j0.004
t = 10 mm

Installed Antenna Analysis In HFSS SBR+

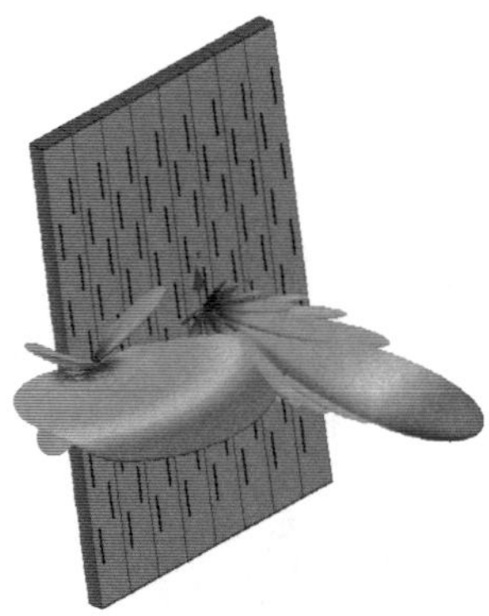

HFSS Antenna Model

Installed Performance of Automotive Radars

- SBR rays from one ray burst location
- Reflected and transmitted rays that coincide with exit at (0,80) (Az, El)
- Colors indicate bounce and reflection order
- 81,400 SBR ray sources (current sources) in this model

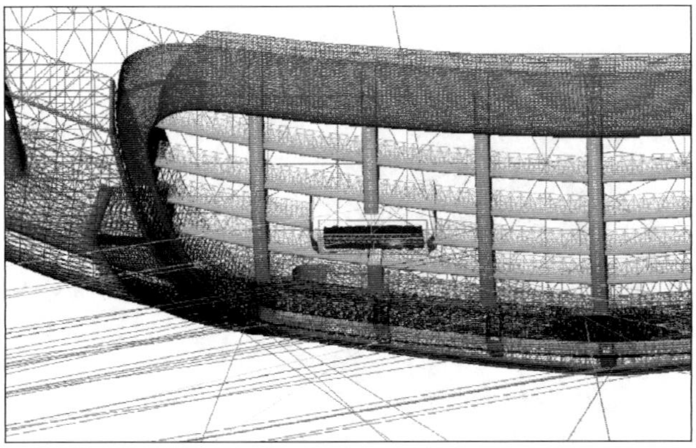

SBR Ray Shoot for a single ray source,
Rays exiting at Az = +80, El = 0 deg shown

Installed Performance of Automotive Radars

Tx Pattern – No Platform Interaction

129,600 Far-Field Obs Points

Tx Pattern – Installed Performance

Sim Time: 3.75 Hr 5-node HPC Cluster (30 total CPU cores, 10 GPUs)

Installed Performance of Automotive Radars

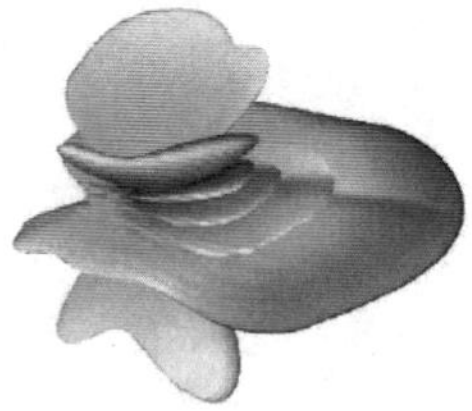

Rx Pattern – No Platform Interaction

129,600 Far-Field Obs Points

Rx Pattern – Installed Performance

Sim Time: 3.5 Hr 5-node HPC Cluster
(30 CPU total cores, 10 GPUs)

Electronics Simulation EXPO

Installed Performance of Automotive Radars

Tx Array Performance

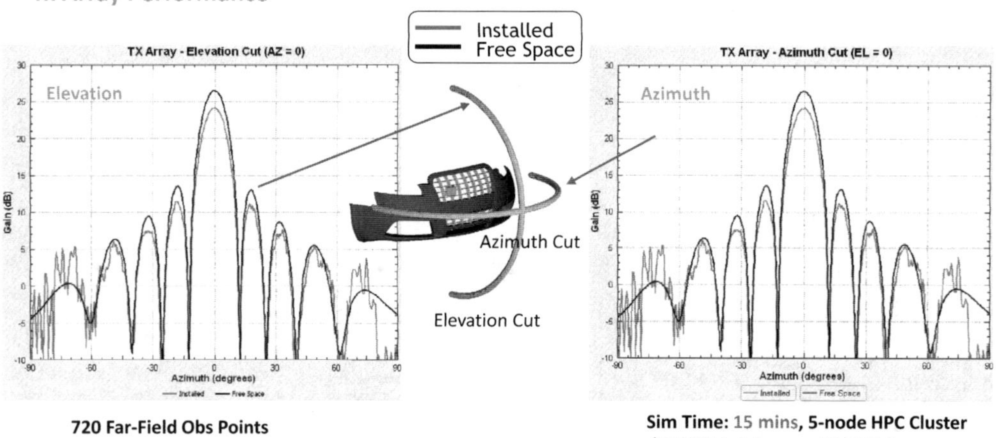

720 Far-Field Obs Points

Sim Time: 15 mins, 5-node HPC Cluster
(30 CPU total cores, 10 GPUs)

Electronics Simulation EXPO

Installed Performance of Automotive Radars

Rx Array Performance

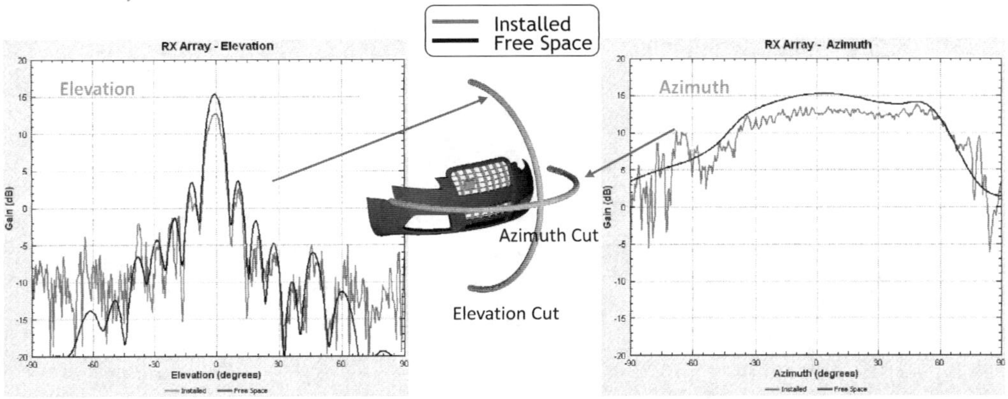

720 Far-Field Obs Points

Sim Time: 15 mins, 5-node HPC Cluster
(30 CPU total cores, 10 GPUs)

Electronics Simulation EXPO

Installed Performance of Satellite Antennas

- Hybrid analysis example:
  - HFSS models the feed horn
  - Savant/SBR+ models reflectors
  - Savant/SBR+ models reflectors, buss and solar panels
- Installed performance for the antenna system?
- Cross-pol radiation increase?

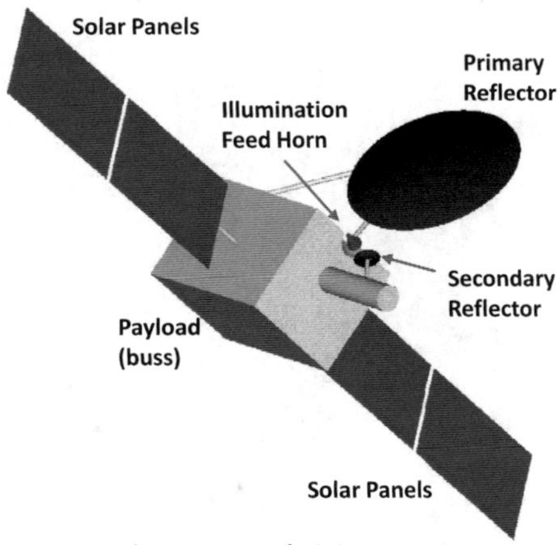

Max dimension: 155λ @ 3 GHz

Electronics Simulation EXPO

Installed Performance of Satellite Antennas (Far Fields)

Co-Pol, Elevation Cut

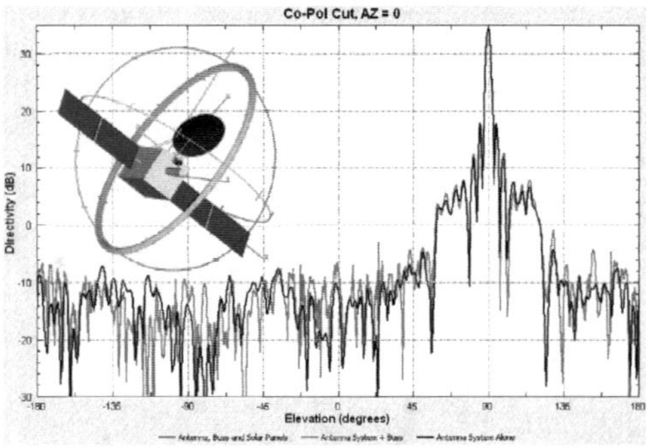

X-Pol, Elevation Cut

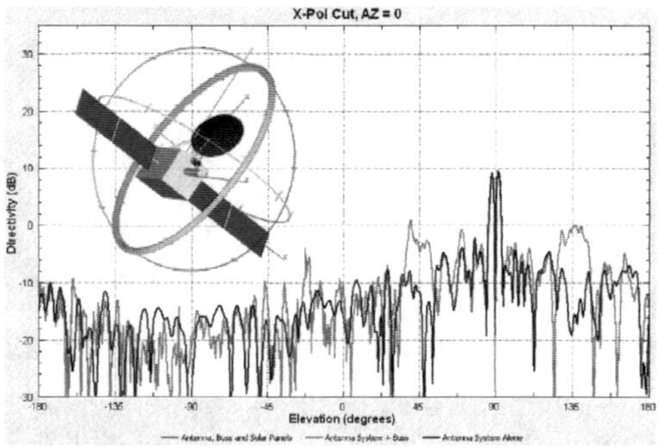

Antenna System (isolated)
Antenna + Payload/buss
Antenna + Payload + Solar Panels

Electronics Simulation EXPO

Installed Performance of Satellite Antennas (Near Fields)

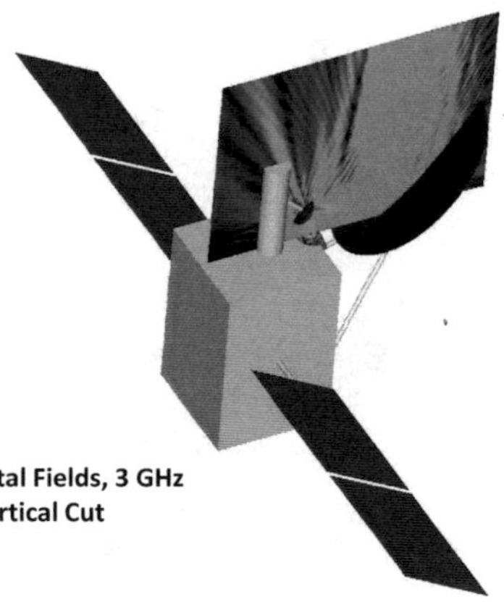

Total Fields, 3 GHz
Vertical Cut

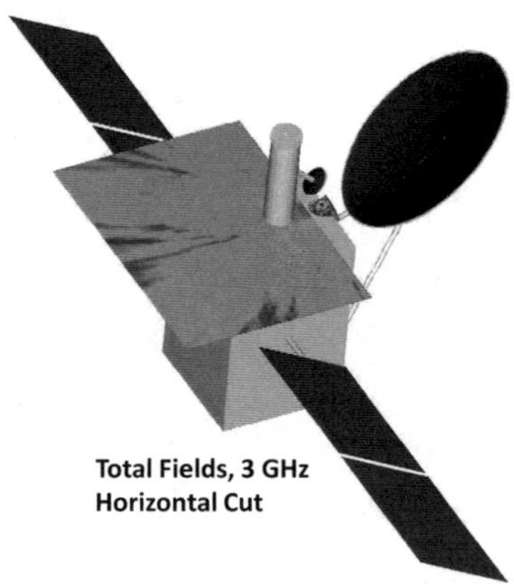

Total Fields, 3 GHz
Horizontal Cut

Electronics Simulation EXPO

Antennas Installed Under Electrically Large Radomes

- GlobalHawk UAV carries a Ku-band SATCOM dish under radome
- Objective: Model radome effects with respect to pointing direction
- Want to include both Radome and full airframe effects
- Problem size (14 GHz): $630\lambda \times 1{,}620\lambda \times 187\lambda$ (190 E6 λ^3)

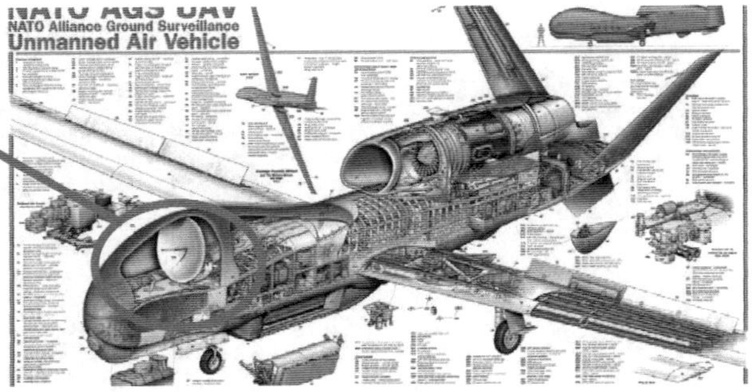

http://www.dmitryshulgin.com/2015/06/08/natos-first-hawk/

Radome:	Layer 1:	10 mm, ε_r = 3.3-0.23j
	Layer 2:	2 mm, ε_r = 5.2-0.11j
	Layer 3:	5 mm, ε_r = 2.6-0.17j

Electronics Simulation EXPO

Antennas Installed Under Electrically Large Radomes

Beam Scans in AZ,
Fixed at 20 deg EL
Freq = 14 GHz

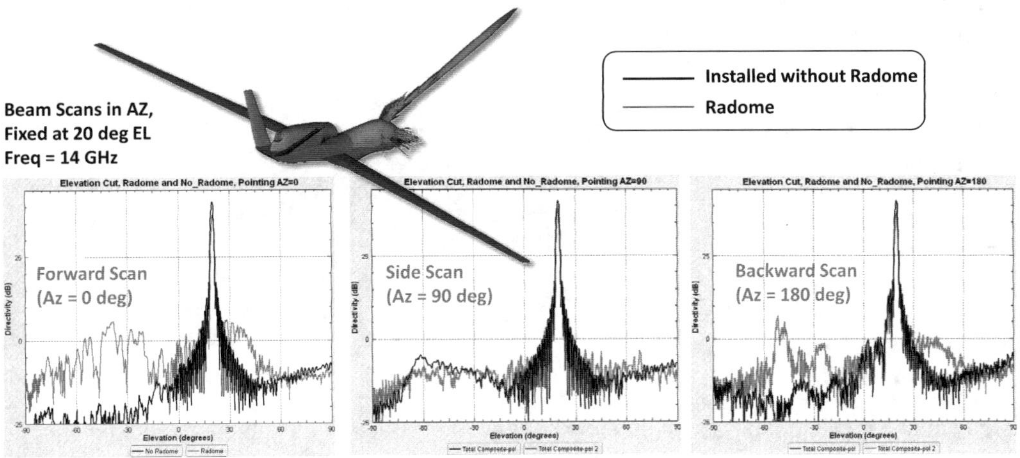

Electronics Simulation EXPO

Antennas Installed Under Electrically Large Radomes

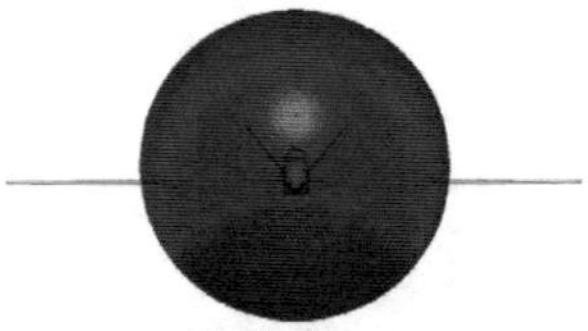

No Radome

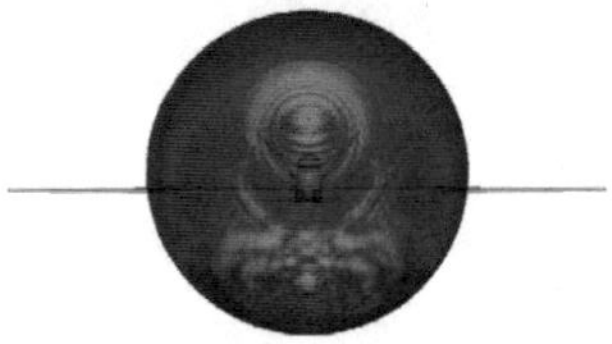

Radome

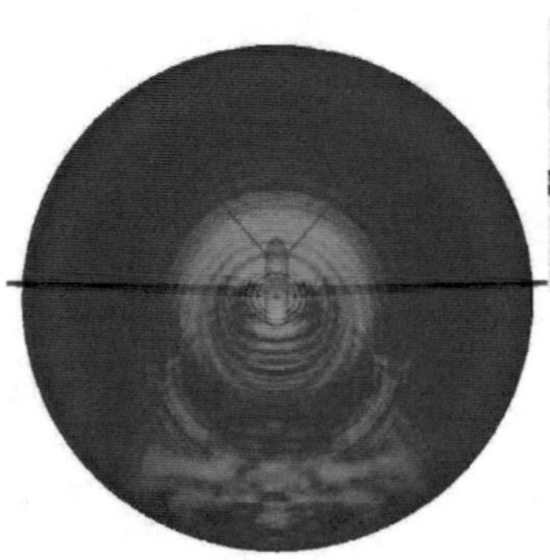

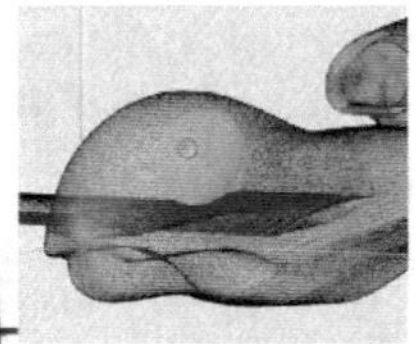

Beam Scans in AZ,
Fixed at 20 deg EL
Freq = 14 GHz

Installed Performance of Aircraft/Aerospace Antennas

- SBR+ can model installed phased array antennas
- Array models may come from HFSS or may be modeled within using canonical antenna types available in SBR+
- Example:
  - A320 with side-scanning conformal phased array at 1 GHz
  - 286 Horizontal Dipole elements modeled as distributed current sources, triangular lattice spacing
  - Cos2 magnitude tapering, edge taper to -40 dB
  - Size: 127λ x 113λ x 11λ (160 E03 λ^3)

Simulation Time: 75 min/beam position
(Dual-core laptop PC)

Installed Performance of Toll Tag Monitors (IoT)

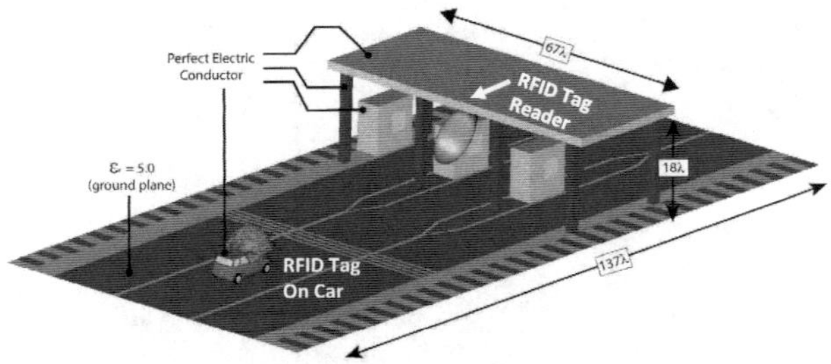

ISM Band Tag and Reader
(902 – 928 MHz)
Automobile with Tag; Gateway with Reader

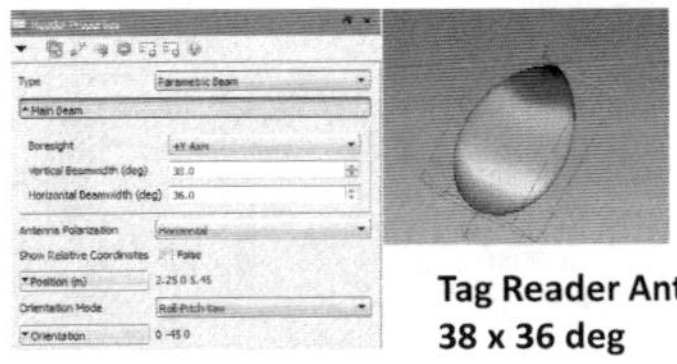

Tag Reader Ant
38 x 36 deg
HPol

Simple wire dipole
RFID Tag (915 MHz)
HPol

Installed pattern of RFID Tag (915 MHz)

Installed Performance of Toll Tag Monitors (IoT)

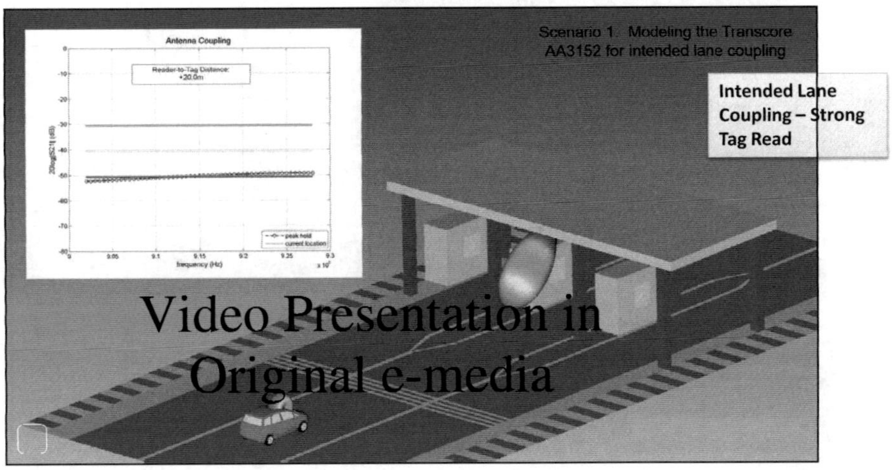

Installed Performance of Toll Tag Monitors (IoT)

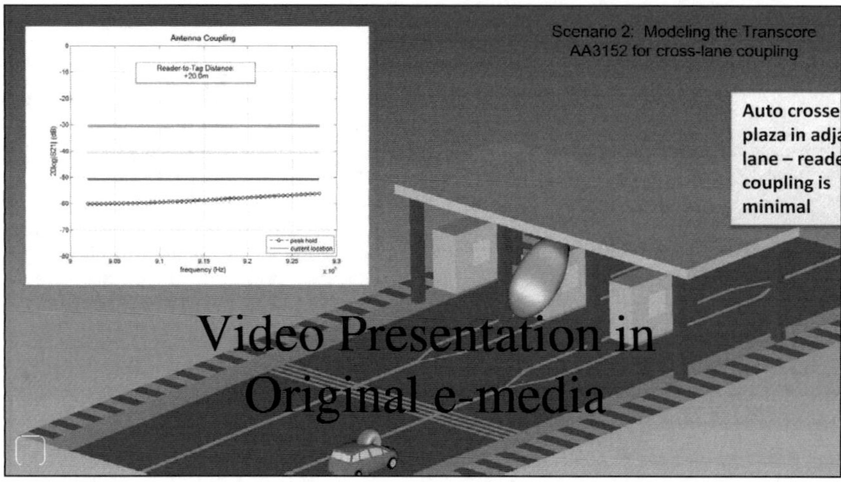

Auto crosses toll plaza in adjacent lane – reader coupling is minimal

Installed Performance of Toll Tag Monitors (IoT)

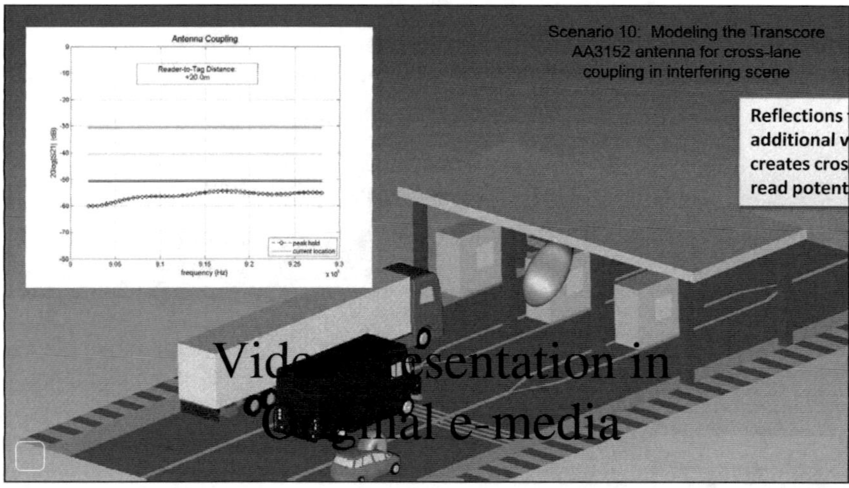

Reflections from additional vehicles creates cross-lane read potential

Installed Performance of Toll Tag Monitors (IoT)

E-Fields @ Tag Height, Top View (915 MHz)

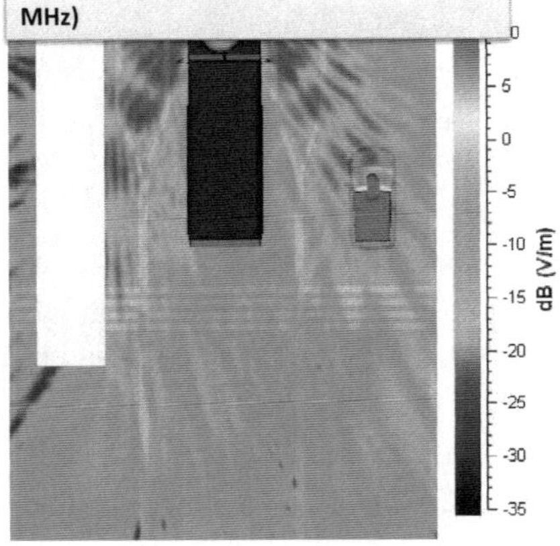

Interaction Ray Tracing Visualization (915 MHz)

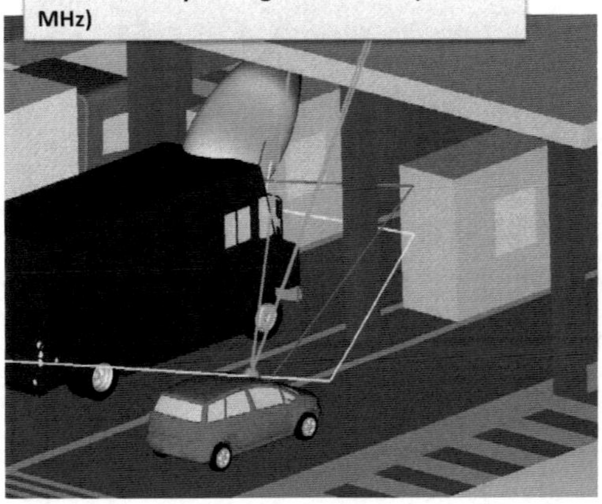

Electronics Simulation EXPO

Build With Confidence – Test Less

- HFSS SBR+ models interactions efficiently between antennas and electrically large hosts
- HFSS SBR+ models antenna-to-antenna coupling in complex and large environments
- ANSYS provides a complete antenna design and analysis flow, from nanometers to kilometers!

Electronics Simulation EXPO

For More Information...

- **Speaker:**
  - **Markus Kopp,**
  - ***Markus.Kopp@ansys.com***
- **Web:**
  - **www.ansys.com/Products/Electronics/Installed-Antenna-Performance**

Video Presentation in Original e-media

Thank you for your attention!

Electronics Simulation EXPO

25 June 7-8, 2016 Munich

CADFEM®

ANSYS
Competence Center FEM

Simulation ist mehr als Software®

Optimization of a Dual Band Antenna using `Statistics on Structures`

Marc Vidal, Sebastian Wolff, Christian Römelsberger

1

CADFEM®

ANSYS
Competence Center FEM

Fluids | Structures | Electronics | Software

© CADFEM 2015

Internet of Things and Antennas

- Number of connected devices growing rapidly
  - ca. 30,000,000,000 devices by 2020!
- Automotive, Households, Industrial, Consumer, Communications, …
- Wearable and mobile devices wirelessly connected

Unleash the IoT

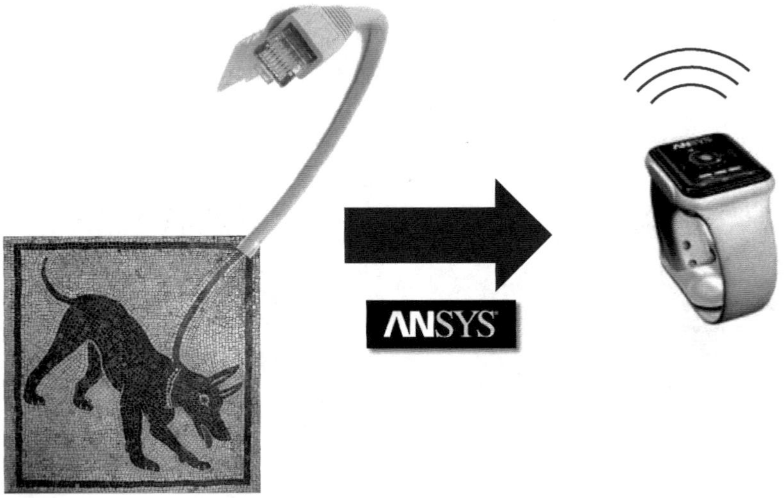

CADFEM

ANSYS Competence Center FEM

Challenges in Antenna Design

- Communication of wireless devices through various frequency bands using different technologies and standards
  - NFC
  - WLAN
  - Bluetooth
  - ZigBee
  - …
- Wearable and mobile devices should be small

→ Save space by using multiple band antennas

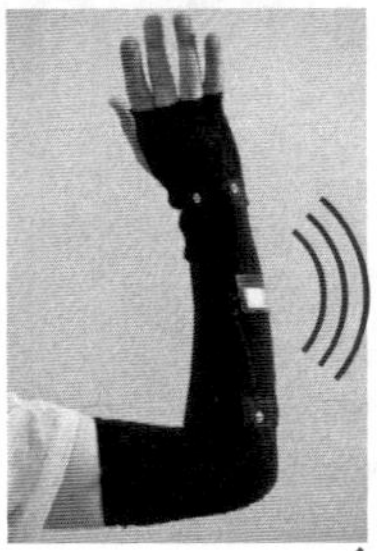

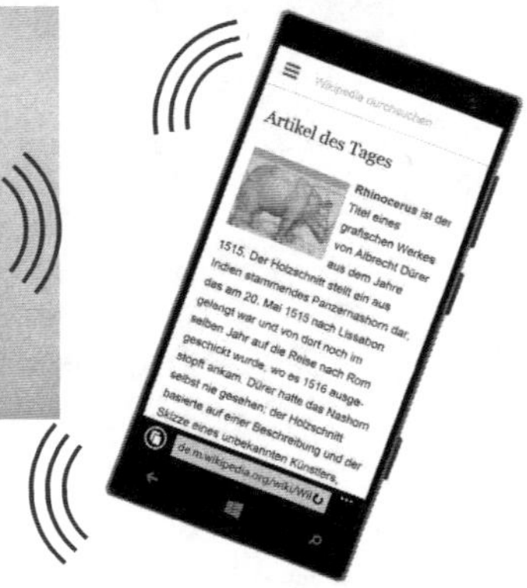

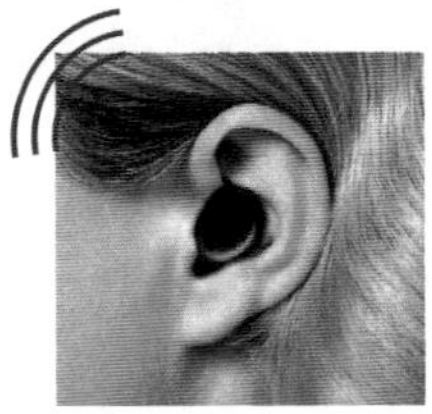

Source: Wikipedia:
Nicolas Sadoc, Nick Pitsas, CSIRO

5

CADFEM

ANSYS Competence Center FEM

Engineering Task:

Design antenna structure that works at the given frequency bands and fits into a given small space!

6

CADFEM

ANSYS Competence Center FEM

Simulation Model – Dual Band Slot Antenna

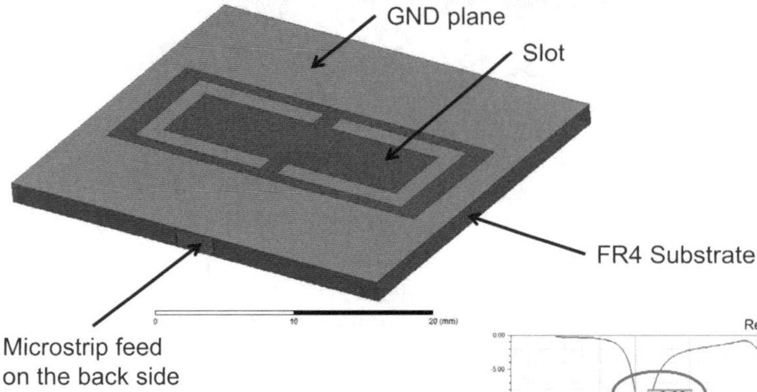

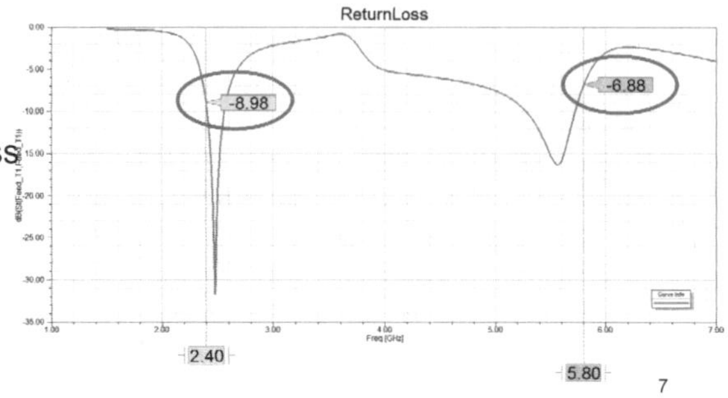

- Need to move minima of return loss to the right frequencies
- Make 5.8GHz minimum deeper

CADFEM

ANSYS Competence Center FEM

ANSYS HF Ecosystem

HFSS The Industry's Premier EM Field Simulator

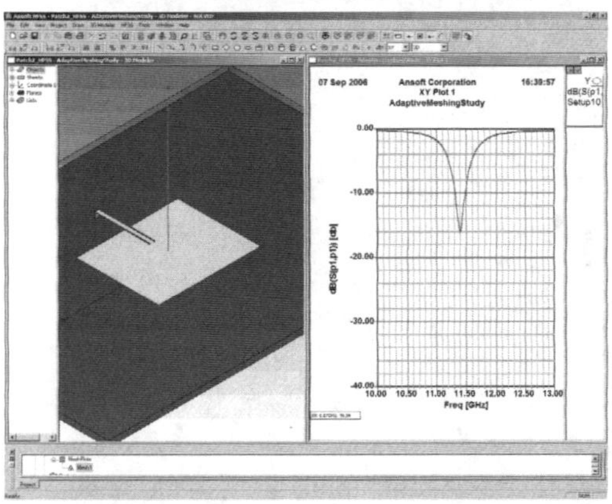

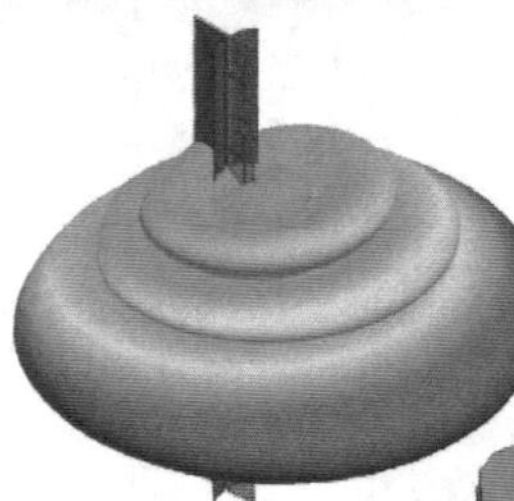

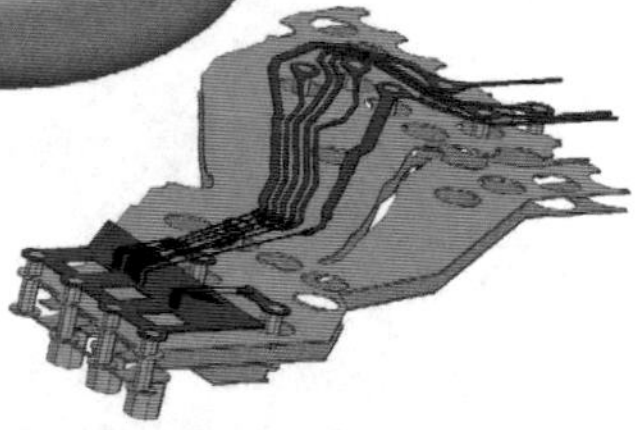

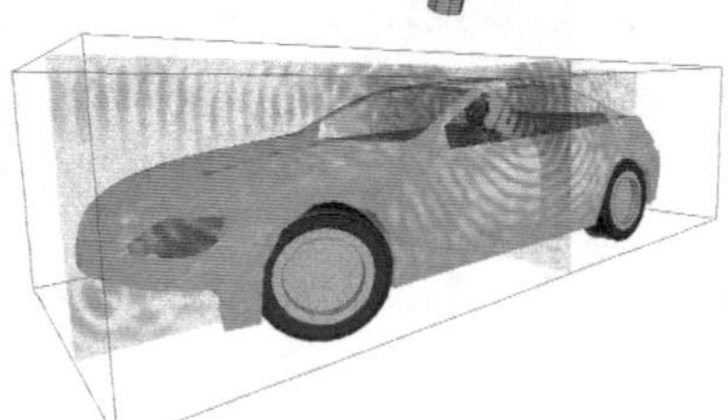

CADFEM

ANSYS Competence Center FEM

Simulation Model – Dual Band Slot Antenna

- Which parameter shall be taken for ~~a manual~~ variation?

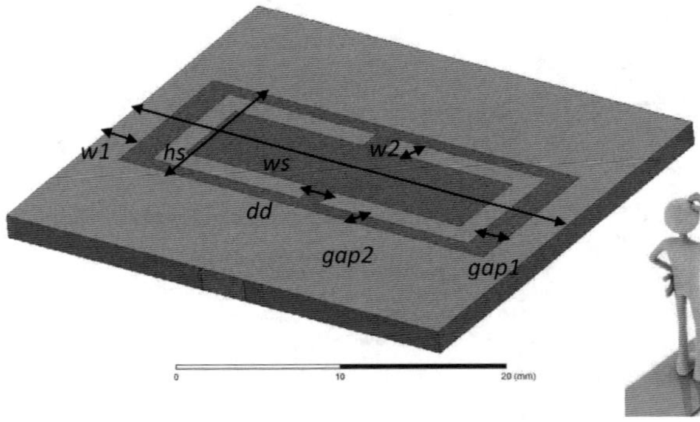

lfeed
wfeed
ws
hs
w1
w2
gap1
gap2
dd

CADFEM

ANSYS Competence Center FEM

First approach

- Scalar value: Return Loss at 2.4GHz and at 5.8GHz minimal

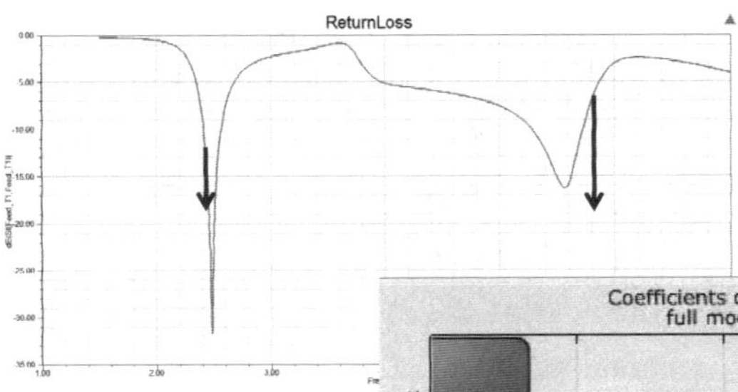

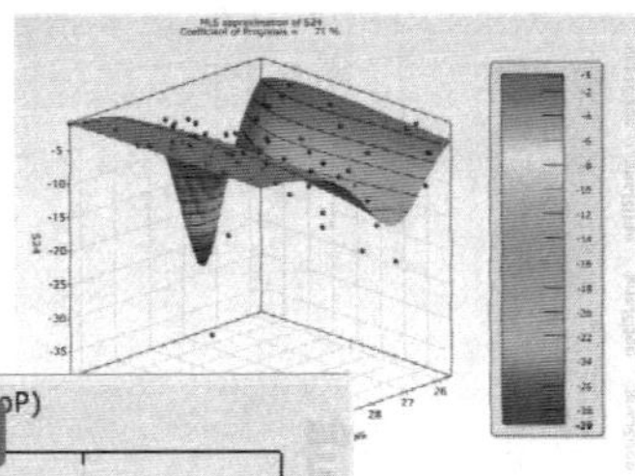

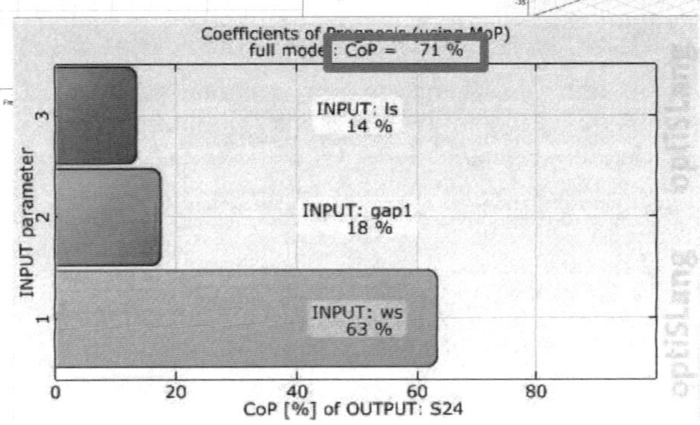

CADFEM

ANSYS Competence Center FEM

Use of Field Metamodel of Optimal Prognosis (F-MOP)

- **High flexibility** in definition of goals after DOE

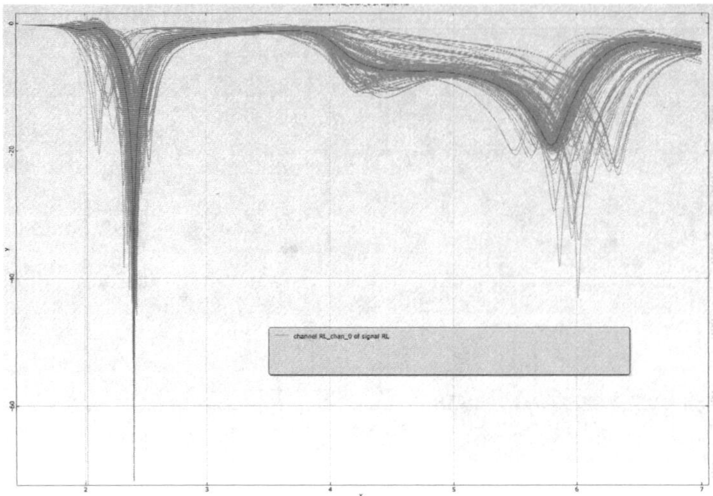

CADFEM

ANSYS Competence Center FEM

Field Meta Model

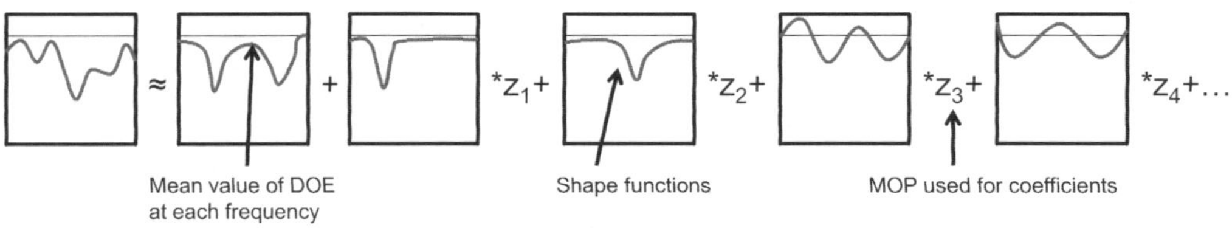

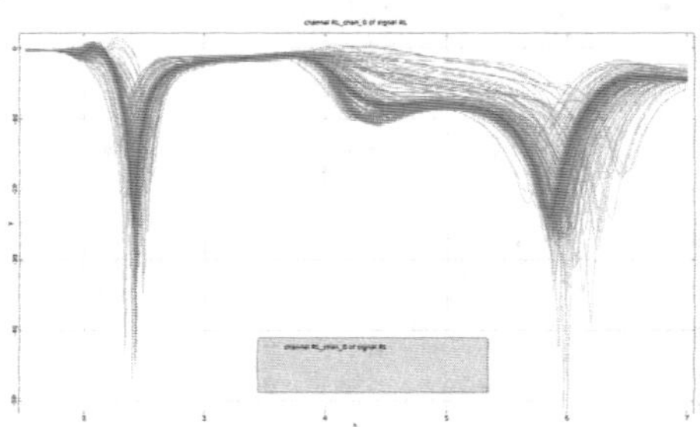

CADFEM

ANSYS Competence Center FEM

Field Meta Model Workflow

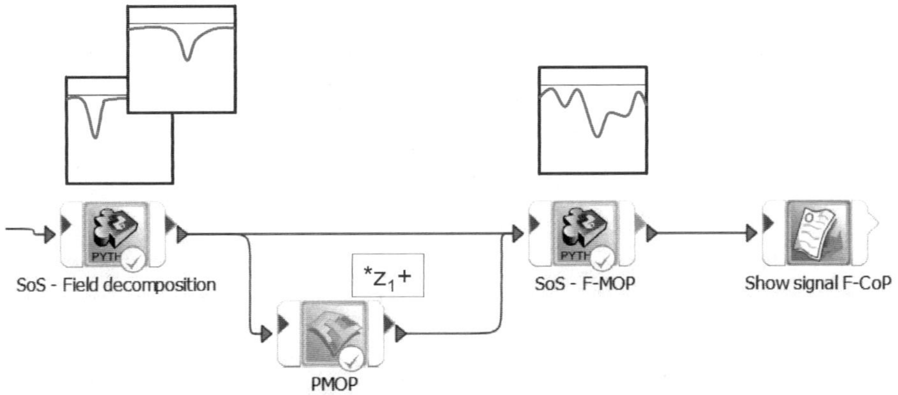

Determine 'shape functions'

Calculate MOP for coefficients

Reassemble signals

CADFEM

ANSYS Competence Center FEM

Plan / Workflow

DOE (optiSLang, HFSS) ⇒ **SoS** ⇒ **Optimization (optiSLang)** ⇒ **Validation (HFSS)**

- Fully automated sampling

- 'Shape Functions' for signals
- F-MOP
- F-CoP

- Use F-MOP as raw data
- Decide on optimization goal, algorithm

CADFEM

ANSYS Competence Center FEM

F-CoP for return loss

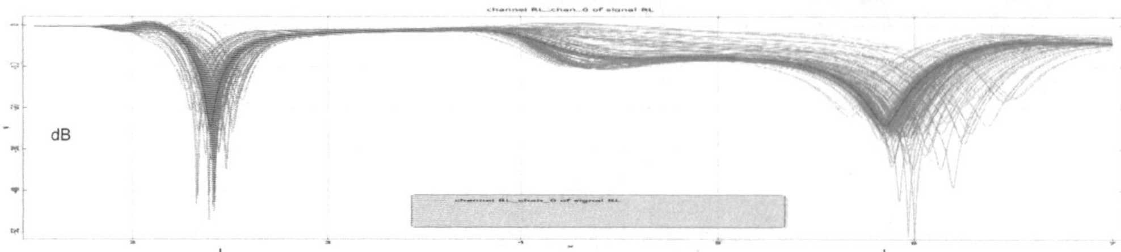

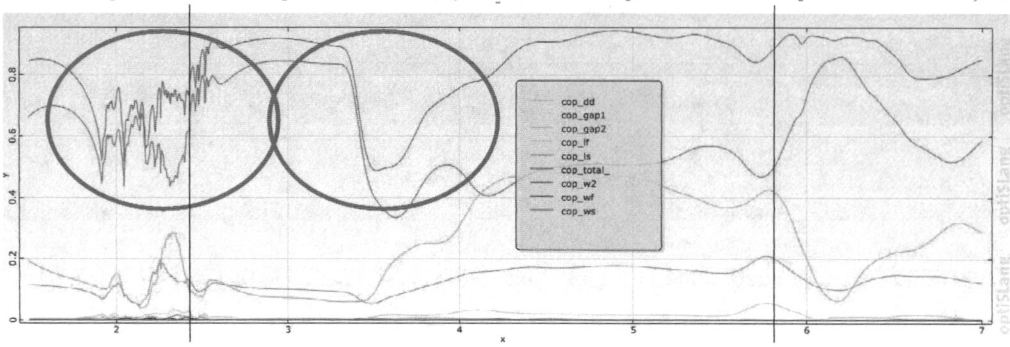

CADFEM

ANSYS Competence Center FEM

F-CoP for S Parameters

$$RL = 10\log_{10}\left(\Re(S_{11})^2 + \Im(S_{11})^2\right)$$

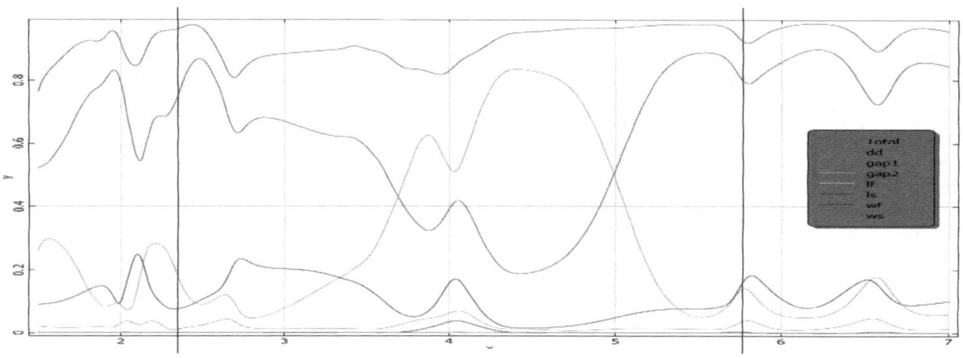

CADFEM

ANSYS Competence Center FEM

Reformulation of the Optimization Goal

- Extract positions and values of the minima of a given signal (from F-MOP)

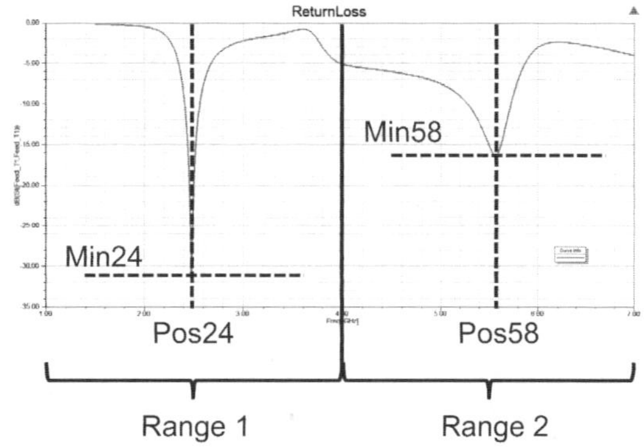

- Optimization Goal:
  - Minimize Max(Min24,Min58)
- Constraints:
  - Pos24=2.4
  - Pos58=5.8

CADFEM

ANSYS Competence Center FEM

Optimization using ARSM on F-MOP

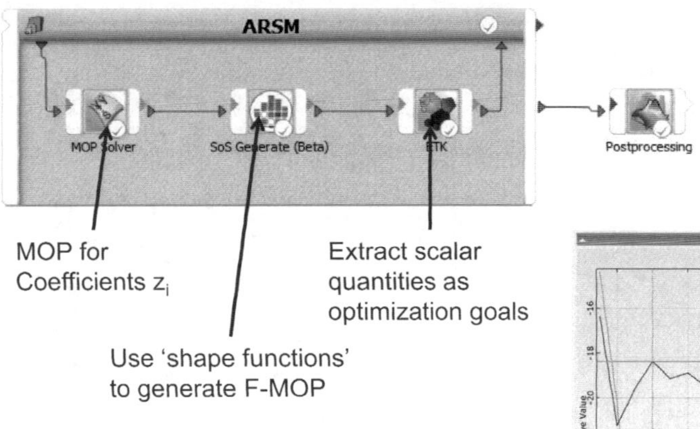

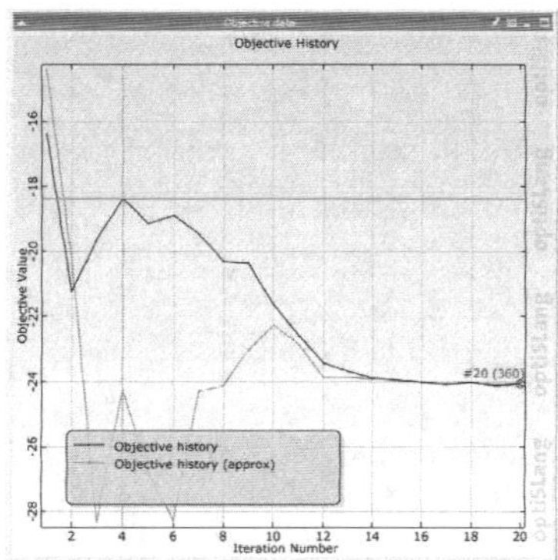

CADFEM

ANSYS
Competence Center FEM

Optimization

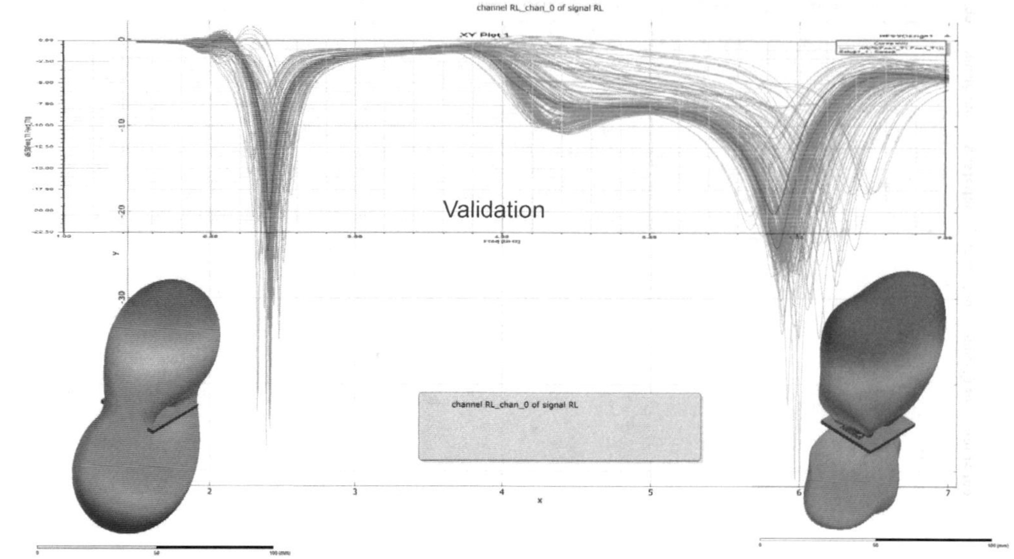

CADFEM

ANSYS
Competence Center FEM

Summary HF simulation and SoS

- Fast
- Accurate
- Full view
- High flexibility

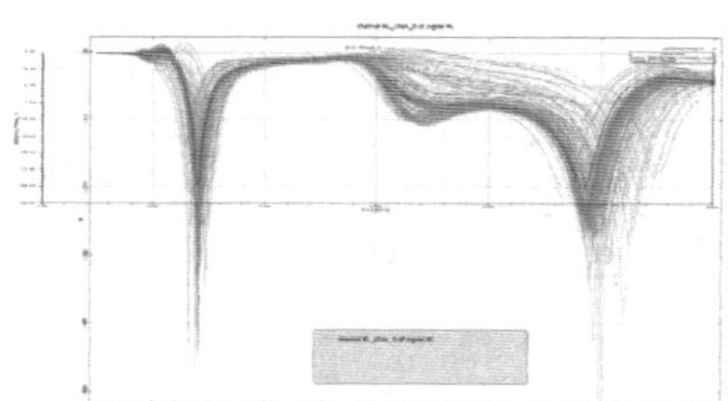

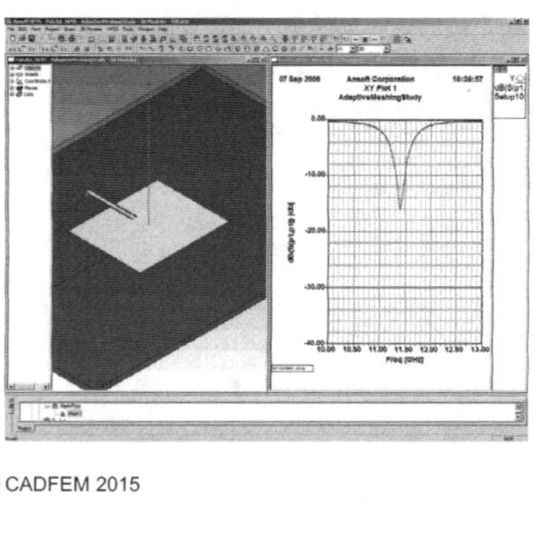

F-CoP

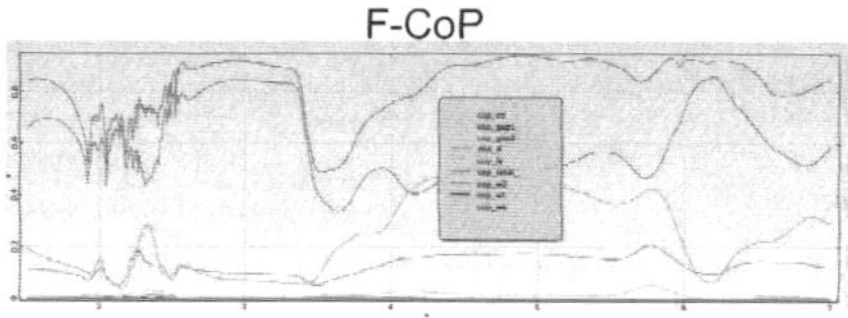

Electronics Simulation EXPO

June 7-8, 2016 Munich

Speeding Up Electromagnetic Simulations of Single Chips to Complete Module Laminate Assemblies at Qorvo®

Tony DeVarco, Director Manufacturing Solutions
SGI

Everyday Challenges Electronic in Design Automation (EDA)

- Time to market
- Time to tape out
- Must be right the first time – (no time for re-do's)
- Internal pressures to reduce costs
- Must gain/keep competitive edge

Increasing Productivity Is Difficult

- **Some EDA tools currently lag behind ASIC technology movement .**
- **As a result:**
  - Corners are cut
  - Models are simplified
  - There may be no time for re-evaluation of models / or redo of model analysis
- **Therefore:**
  - Accurate models that match those used by foundries are not used for analysis, due to
    - Model complexity
    - Compute resource limitations and
    - Long run time in standard cluster compute farms
- **One Solution: Running ANSYS® HFSS on SGI® UV™**
- **The following is an overview of a project with Qorvo, ANSYS and SGI**

Electronics Simulation EXPO

Qorvo Overview

qorvo all around you

- Merging TriQuint and RFMD technologies, collective experience and intellectual resources, Qorvo sets the pace as a global leader in scalable and dynamic Radio Frequency (RF) solutions for mobile, infrastructure and defense applications.
- Qorvo has a large number of design centers with a EM user base that ranges from novice, infrequent to advanced.

Goals of using the SGI UV Platform

- Simplified method of generating successful simulation results and the exportation of s-parameter data.
- Reduction of laminate simulation time to allow a greater number of simulations within one day.
- Processing different layout versions of a nominal design in a few hours rather than days.

Electronics Simulation EXPO

SGI® UV™ Platform

Suggested for EDA/CAE Workloads

- Up to 64 sockets (640 cores), 16 TB shared memory/rack
- Intel® Xeon® Processor E5-4627 v3 10-core (2.60 GHz)
- NVIDIA® Tesla® and Intel® Phi™ support
- NVIDIA® Quadro K5200 or M6000 graphics support
- SGI NUMALink™ 6 Interconnect
- SUSE or RedHat
- Altair PBS Professional Batch Scheduler
- SGI® VizServer™ available for remote visualization

Electronics Simulation EXPO

Goal - Simplify General Simulations

qorvo all around you

Identification of the 'sweet spot' for optimum performance' for general Qorvo EM simulations

- Reviewed past simulation resource consumption
- Using a wide range of designs review processing performance using 2 core increments

Results

- Creation of a submission script for PBS pro where the default execution satisfies greater than 80% of simulation requests. This interface executes the simulation, exportation of the Touchstone data, and generation of a memory usage report.

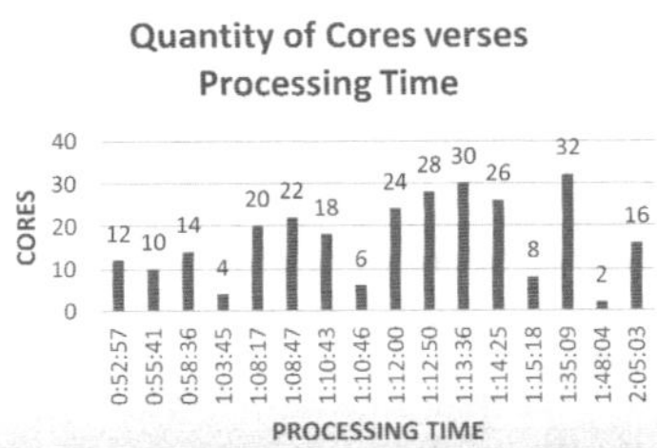

Electronics Simulation EXPO

Goal - Simplify General Simulations

qorvo. all around you

- **Greater than 80% of user simulation requests use the same default execution ' zhfss -l '**
- **300+ simulations are processed weekly**

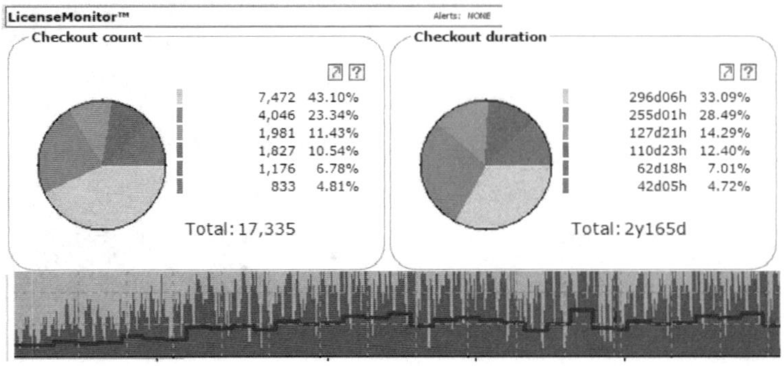

Electronics Simulation EXPO

Performance Analysis

qorvo. all around you

- SGI analysis was testing core usage above 32 cores for efficiency

<u>LEGEND</u> 2014=2014.0.7 Linux binaries 2015=2015.0.0 binaries SGI = 2014.0.7

CKT1_2014		CKT1_2015		CKT1_SGI		CKT3_2014		CKT3_2015		CKT3_SGI		CKT2_2014		CKT2_2015		CKT2_SGI	
14cores	0:52:39	12cores	0:52:57	2cores	1:59:20	16cores	0:44:23	8cores	0:48:35	1:50:33	2cores	12cores	2:05:48	8cores	2:18:38	6:20:58	2cores
10cores	1:05:36	10cores	0:55:41	4cores	1:18:44	12cores	0:46:07	10cores	0:49:21	1:29:58	4cores	10cores	2:09:10	14cores	2:27:02	6:20:40	4cores
8cores	1:05:50	14cores	0:58:36	6cores	1:34:59	10cores	0:47:50	22cores	0:57:37	1:23:03	6cores	16cores	2:09:58	4cores	2:34:57	3:06:04	6cores
20cores	1:05:53	4cores	1:03:45	8cores	1:09:33	8cores	0:48:42	20cores	0:58:15	1:00:36	8cores	8cores	2:10:06	12cores	2:50:56	3:04:15	8cores
22cores	1:07:49	20cores	1:08:17	10cores	3:27:16	6cores	0:50:41	18cores	1:00:05	2:30:27	10cores	6cores	2:15:03	6cores	2:58:02	2:59:26	10cores
6cores	1:09:52	22cores	1:08:47	12cores	2:17:12	4cores	0:56:43	24cores	1:00:31	1:25:36	12cores	4cores	2:21:41	20cores	3:03:56	8:08:06	12cores
18cores	1:10:17	18cores	1:10:43	14cores	2:45:44	24cores	1:00:35	14cores	1:00:43	3:28:56	14cores	18cores	2:43:59	18cores	3:07:53	6:19:11	14cores
16cores	1:11:57	6cores	1:10:46	16cores	1:12:04	18cores	1:00:56	28cores	1:00:53	1:01:57	16cores	20cores	2:46:42	22cores	3:10:51	3:36:41	16cores
24cores	1:12:44	24cores	1:12:00	18cores	2:22:40	20cores	1:01:02	26cores	1:01:13	2:35:13	18cores	22cores	2:53:23	24cores	3:14:52	3:28:19	18cores
26cores	1:14:53	28cores	1:12:50	20cores	1:12:09	22cores	1:02:13	30cores	1:02:15	1:30:06	20cores	28cores	2:55:01	28cores	3:20:18	0:44:07	20cores
30cores	1:16:00	30cores	1:13:36	22cores	2:21:07	26cores	1:02:47	12cores	1:06:35	1:07:07	22cores	26cores	2:55:24	26cores	3:26:19	6:40:16	22cores
28cores	1:16:04	26cores	1:14:25	24cores	2:25:10	28cores	1:05:35	4cores	1:08:54	1:12:52	24cores	24cores	2:56:36	30cores	3:40:35	3:12:20	24cores
12cores	1:18:54	8cores	1:15:18	26cores	2:47:58	30cores	1:18:07	2cores	1:15:32	2:01:56	26cores	30cores	3:01:14	10cores	4:47:19	3:19:19	26cores
4cores	1:21:20	32cores	1:35:09	28cores	1:13:05	32cores	1:40:14	6cores	1:15:44	2:13:51	28cores	32cores	4:56:47	32cores	5:13:49	5:05:33	28cores
2cores	1:40:32	2cores	1:48:04	30cores	1:14:34	2cores	1:42:31	16cores	1:23:29	1:51:17	30cores	14cores	5:45:21	16cores	6:28:54	3:31:43	30cores
32cores	1:40:44	16cores	2:05:03	32cores	1:14:10	14cores	2:03:15	32cores	1:24:50	1:02:49	32cores	2cores	0:00:00	2cores	0:00:00	3:21:08	32cores
				36cores	1:13:05					1:08:40	36cores					5:29:57	36cores
				40cores	1:13:08					1:54:14	40cores					3:37:23	40cores
				44cores	0:00:00					0:00:00	44cores					0:00:00	44cores
				48cores	2:42:42					4:05:50	48cores					4:58:25	48cores
				52cores	1:19:13					1:58:41	52cores					1:47:18	52cores
				56cores	1:29:16					5:02:21	56cores					4:15:01	56cores
				60cores	1:33:23					1:41:21	60cores					3:28:56	60cores
				64cores	1:39:53					1:57:53	64cores					3:31:21	64cores

Electronics Simulation EXPO

Goal - Simplify General Simulations

The standard production environment is LSF grid simulation servers where one condition for simulation failures exceeding the LSF job memory inputs/capabilities.

Benefits of the SGI® UV™ system

- **No user input for simulation memory needs and no simulation failures for insufficient memory**
- **The SGI UV is viewed as the silver-bullet resource: When the capabilities of the standard production LSF grid servers are exceeded – the UV2 can get the job done . . .**

Goal – Reduce Laminate Simulation Time

Typical Laminate Simulation

- Complex laminate projects can consume multiple LSF grid servers and take 8 hours or longer to complete.
- Simulations are conducted in two phases due to limitations of the EM software and LSF hardware. 1st phase meshing/adaptive solution; 2nd phase - distribute frequency sweep point solution across multiple systems but restricted by memory constraints.

Benefits of the SGI UV system

- Current laminate designs - the UV2 system allows the potential for a greater number of distributed frequency sweeps with no current memory constraints, resulting in faster results.
- Testing of performance using greater than 32 cores, up to 64 cores. Results - using a large number of cores is not the solution, using the optimum number of cores will yield faster results.

Goal – Simplify CORNERS SIMS

qorvo
all around you

Typical CORNERS PROCESSING

- Using a nominal design multiple layout designs meeting process corner testing are created. Multiple layouts ranging from 12 to 16 are sequentially processed usually taking days to complete all simulations.

Goals

- Remove manual creation of multiple layouts and provide interface to use standard corner values, or allow user inputs.
- Execute parallel processing of auto-created layouts.

Electronics Simulation EXPO

11 June 7-8, 2016 Munich

Goal – Simplify CORNERS SIMS

qorvo
all around you

Solution

- A joint development between ANSYS and Qorvo using ANSYS Designer that resulted in a programmable Optimetrics python script that auto-generates an optimetric definition of multiple layouts within 5-20 minutes.
- Implementation of ANSYS's DSO license feature for parallel optimetric simulations.
- Integration with current custom scripts creating PBSpro job requests.

Electronics Simulation EXPO

12 June 7-8, 2016 Munich

Test Results– Simplify CORNERS

qorvo all around you

Project V0104
- Nominal Design generation of Individual corner layouts* took 3-6 hours.
- Corners Sequential simulations took 20 hours.

* Not a complete assemble of corner layouts

Project V0104 using corners script, SGI and DSO
- Nominal Design generation of Individual corner layouts** took 4 minutes
- Corners simulations parallel processed took 3 hours 7 min.

** A complete assemble of corner layouts.

Electronics Simulation EXPO

13 June 7-8, 2016 Munich

General Summary - SGI UV Benefits

qorvo all around you

Standard Simulations

- **Silver-bullet Resource**: When the capabilities of the standard production LSF grid servers are exceeded – the SGI UV delivers fast results because of the availability of 2TB RAM 160 cores.
- (2/28 project) using standard 32 cores resources **15 hours, 40 minutes.**
  - Using the SGI and 64 cores it took ***8 hours and 17 minutes***
- No user input for simulation memory needs / no simulation failures for insufficient memory.

Corners Simulations

- Allows signification reduction of processing time for optimetric defined layout variations.

Electronics Simulation EXPO

14 June 7-8, 2016 Munich

Realize Your Product Promise®

"Using lessons learned from aerospace and railways in electronics, software and system development of autonomous vehicles"

Eric BANTEGNIE, VP & GM,

ANSYS Systems Business Unit

Example ANSYS A&D customers...

ADASI
ANTONOV
Alenia Aermacchi
AWE
AIRBUS
aselsan
Astronics
Avio Aero
BERIEV
Airship do Brasil
Bechtel Marine Propulsion Corporation
BAE SYSTEMS
BOEING
Bundeswehr
CETC
CRANE
中国航天
Esterline
CMC Electronics
COMAC
DGA
DASSAULT AVIATION
DIEHL Aerospace
DLR
Elbit Systems
EMBRAER
FIREFLY
IAI ELTA
ELV
ESG
GE Aviation
Honeywell
GENERAL ATOMICS
GENERAL DYNAMICS
AVIC
HAL
KAWASAKI HEAVY INDUSTRIES, LTD.
KIHOMAC
KAI KOREA AEROSPACE INDUSTRIES, LTD.
L3 communications
NASA
LIEBHERR
isro
LOCKHEED MARTIN
MEGGITT
Mectron
COBHAM
MILLENNIUM SPACE SYSTEMS
NORTHROP GRUMMAN
OAK
PIAGGIO AERO
Parker
ONERA
Pratt & Whitney
RAFAEL
Raytheon
Rockwell Collins
RHEINMETALL
Rolls-Royce
SAIC
SANDEL
上汽集团
SAFRAN
SAAB
SHIMADZU
SELEX
SKURKA
SPACEX
SUKHOI
TAI
THALES
United Technologies
Ultra Electronics
Williams International
xmobots
ZODIAC AEROSPACE
ANSYS

A wide range of UAV systems !

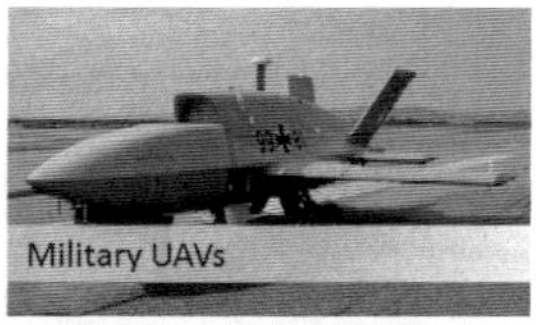

DASSAULT NEURON

AIRBUS BARRACUDA

GENERAL ATOMICS PREDATOR

UTACS WATCHKEEPER

A wide range of UAV systems !

XMOBOT NAURU 500

TAI ANKA

Piaggio P.1HH Hammerhead

...Most aircrafts operate autonomously more than 80% of their flight time...

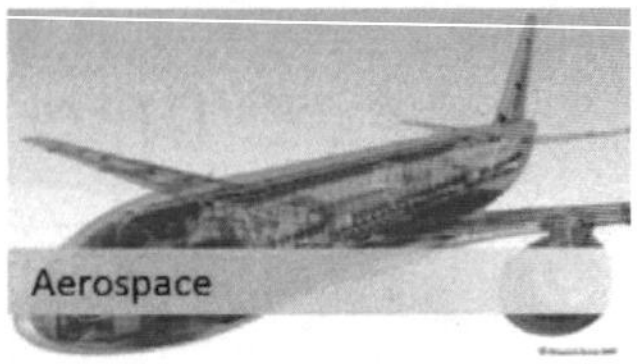

...and have become flying computers...

ANSYS

Example: Airbus A380

AIRBUS

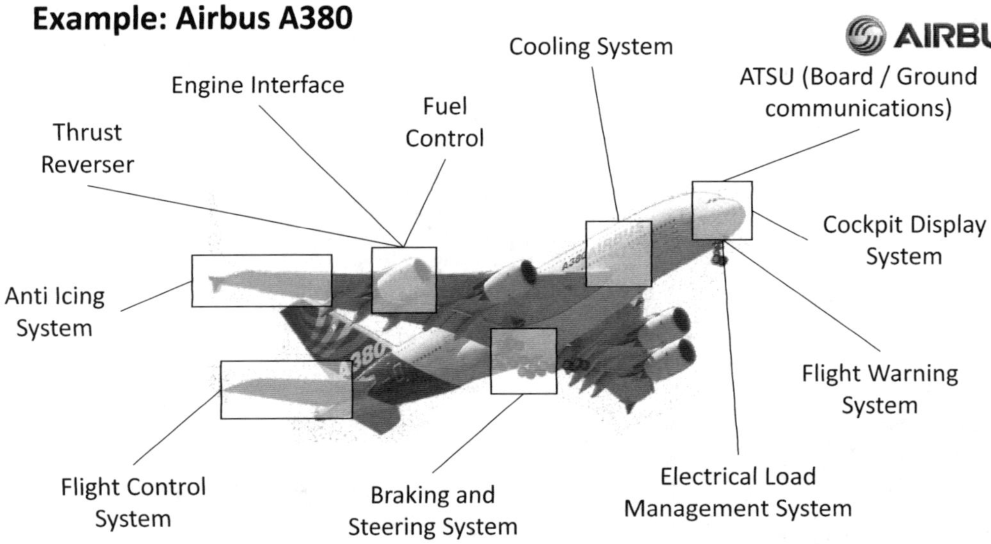

Zero bug in operations in10 years in 8 Mio LoCs (source: Airbus)

Embedded Software Controls

Safety & Security

ANSYS

The Airbus A380 Cockpit

Example impact when sensors are wrong...

As the error affected the redundant sensors as well as the primary ones, the rocket was left with no yaw control, which resulted in the failure

Example ANSYS Rail Transportation Customers

80% of new rail systems are controlled via V2X infrastructure

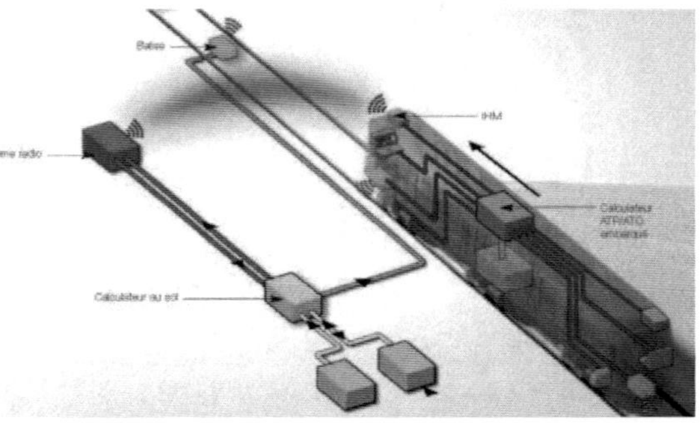

Communication & Connectivity

Example impact when automated train control software controls are turned off…

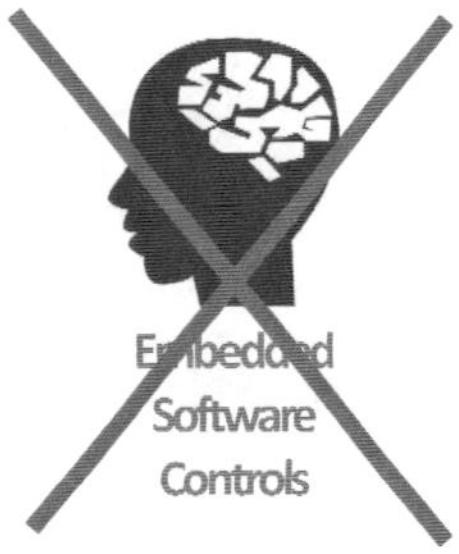

Key Design Challenges common to UAVs, Aircrafts and Autonomous Rail Systems….

Does this also apply to Automotive Autonomous Vehicles ?

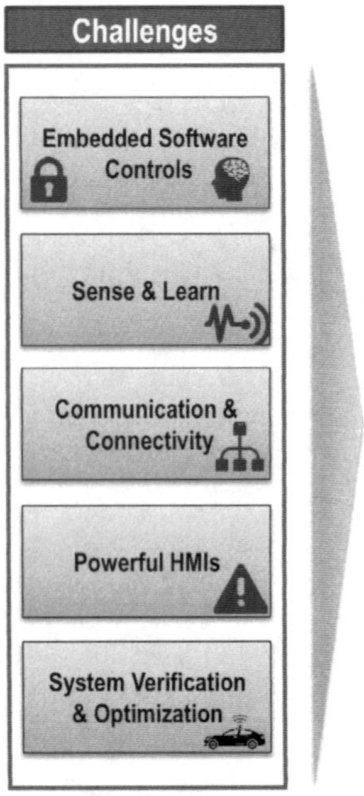

Safe, Secure & Certified Embedded Software Controls

100 MLOCs , MultiCore processors, ISO 26262 Certification, MILS Architecture...

Efficient Surround Sensing & Deep Learning

Radar, Lidar, Cameras, GPS, Speed Sensors, Video feeds, Image Processing, Sensor Voting & Fusion, Image Pattern Recognition

Networking and V2X

Reliable On-Board (CAN , Flexray...) and Ground Systems (V2V, V2I) communication

Safe, Secure, Certified and Powerful HMIs

Complex Driver Machine Interfaces, Human Factors

Overall System Safety, Security and Performance Verification and Optimization

Safety & Security Liabilities >> Company Valuations ?

Are Autonomous Cars testable physically at all ?...

RAND Corporation Study 2016 findings « Driving to Safety, How many miles of Driving Would It take to Demonstrate Autonomous Vehcile Reliability »

Nidhi Kalra, Susan M. Paddock

http://www.rand.org/pubs/research_reports/RR1478.html

« Autonomous vehicles would have to be driven hundreds of millions of miles and sometimes hundreds of billions of miles to demonstrate reliability in terms of fatalities and injuries »

« Even under aggressive testing assumptions, existing fleets would take tens and sometimes hundreds of years to drive these miles »

« Test driving alone cannot provide sufficient evidence for demonstrating autonomous vehicle safety »

Product Development Process Implications

Safe, Secure & Certified Embedded Software Controls

Traditionnal manual software development and testing costs and liabilities become unbearable: Need a Certified and Automated Development Environment

Efficient Surround Sensing & Deep Learning

Need to Simulate the entire Electronics to uncover Failure Modes

Networking and V2X

Complex and evolving antenna and communication chip/package/systems

Safe, Secure, Certified and Powerful HMIs

HMI designs become hugely complex and include safety and security critical components: Need a Certified and Automated Development Environment

Overall System Safety, Security and Performance Verification and Optimization

System Verification based on only physical testing becomes irrelevant

ANSYS Solutions

- Certified Embedded Software Development
- Antenna Design & Placement Simulation
- Chip-Package-System
- Sensors & MEMS Simulation
- Power Management Simulation
- Designing for Harsh Environments
- Virtual System Prototyping

ANSYS

Realize Your Product Promise®

ANSYS

ANSYS Solutions for Autonomous Vehicles

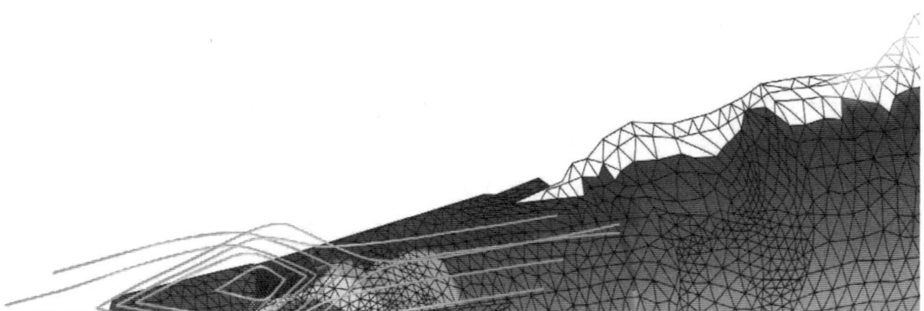

ANSYS R17: A Complete Simulation Platform

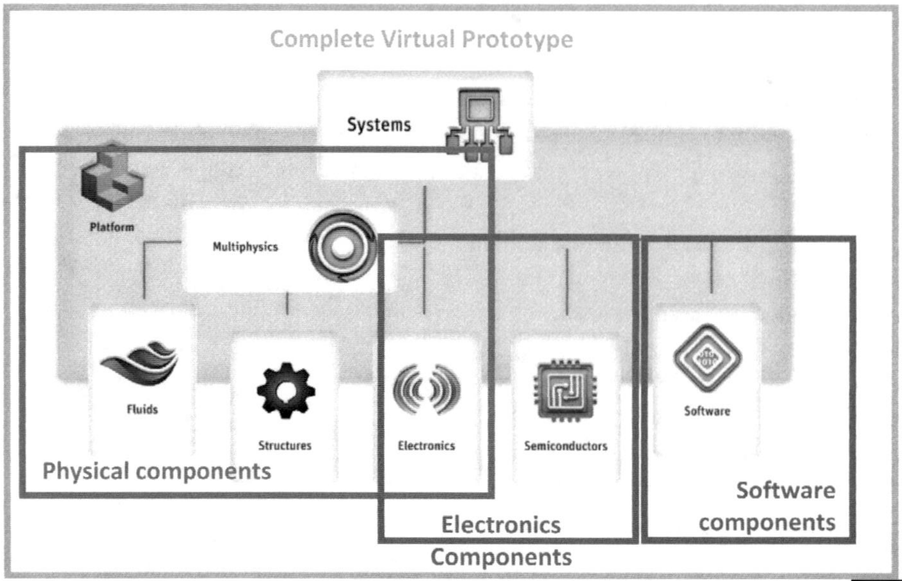

ANSYS

17 © 2015 ANSYS, Inc. June 23, 2016 ANSYS Confidential

ANSYS R17: A Complete Simulation Platform for Autonomous Vehicles

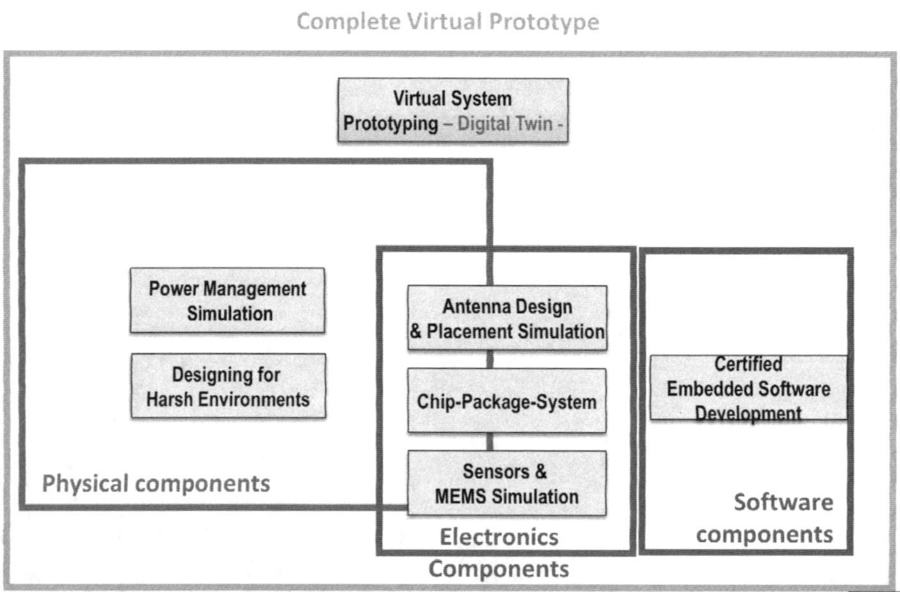

ANSYS

18 © 2015 ANSYS, Inc. June 23, 2016 ANSYS Confidential

ANSYS ensures ADAS/Infotainment Physical Reliability

Electrical/Thermal Efficiency

Coupled Electrical/Thermal analysis on PCB predicts hot-spots from Joule Heating effects

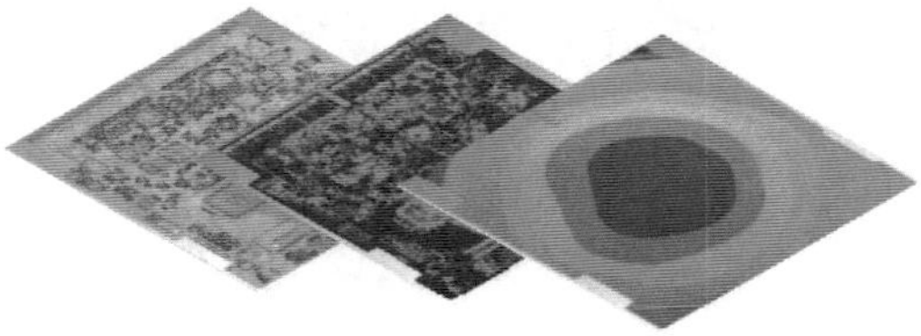

- ✓ Reducing board development times
- ✓ Reducing system cost
- ✓ Improving Reliability

Thermal & Structural Integrity

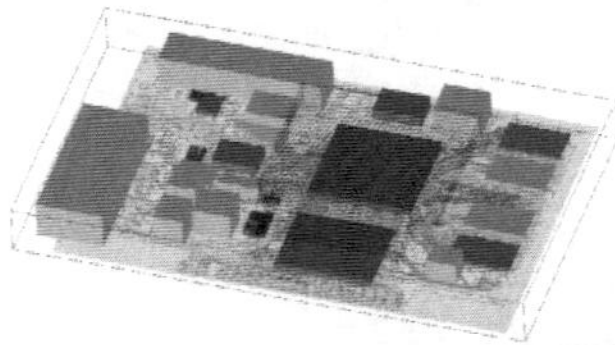

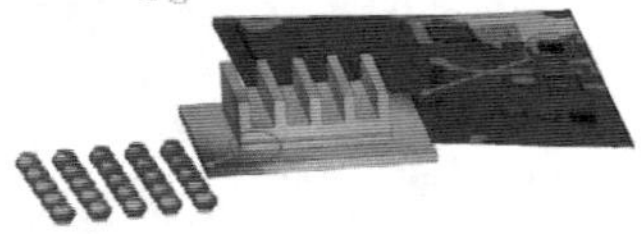

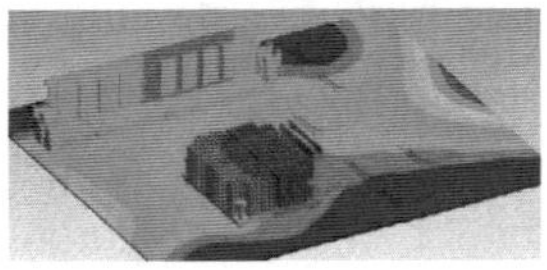

Failure Detection for :

- Stress
- Deformation
- Thermal Cycling
- Shock Analysis
- Vibration

- ✓ Improve component reliability
- ✓ Examine humidity impact
- ✓ Optimize cooling solution

ANSYS

ANSYS Enables Upfront Simulations of Key Electronic Systems

Drive Train

- Hybrid Drive Train
- Electric Motor
- Wireless Power Transfer
- Battery/Fuel Cell

Communication

- Vehicle to Vehicle
- Vehicle to Roadside
- Vehicle to Owner

Advanced Driver Assistance (ADAS)

- Collision Avoidance
- Location/Position
- Adaptive Cruise Control

Regulatory Compliance

- Susceptibility
- Radiated EMI/EMC
- Conducted EMI/EMC

ANSYS

Software in Cars is Safety and Security

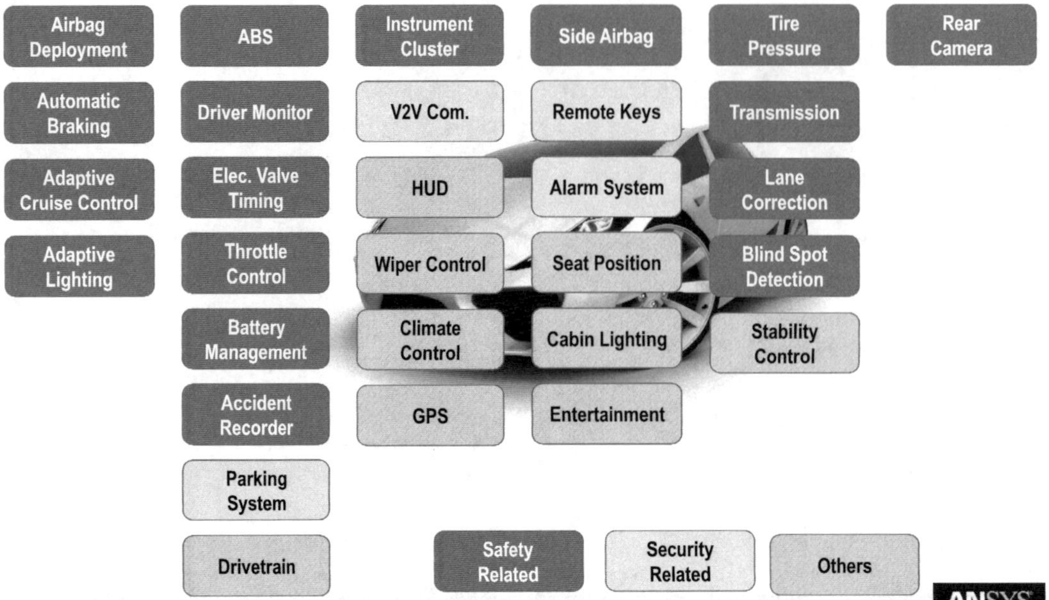

ANSYS

ANSYS SCADE enables automated and certified production of embedded software controls and HMIs

Embedded Control and HMI Software Design

Efficient modeling of controls, logic, HMIs and algorithm designs within a single environment

Integrated Suite for Prototyping, Modeling, Simulation, Verification, and Optimization

Efficient debugging and optimization of software models and code size, speed and performance

Certified Code Generation

*Automatic C and Ada certified code generators (DO-178B/C, EN 50128, **ISO 26262**, IEC 61508)*
Enables 80% embedded code production and testing cost reduction

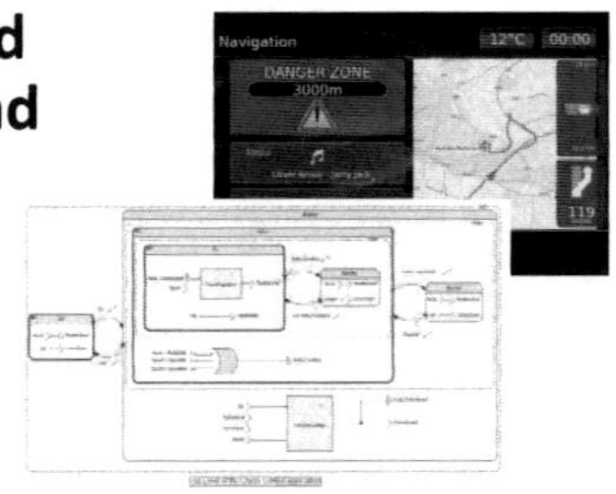


```
[...]
void Button_ABC_N(inC_Button_ABC_N *inC,
      outC_Button_ABC_N *outC)
{
  /* ABC_N::Button::SM1::SSM_SM1_dispatch_sel */
  SSM_Button_SM1_ST SSM_SM1_dispatch_sel;

  if (outC->init)
   {
    outC->init = kcg_false;
    SSM_SM1_dispatch_sel =
      SSM_SM1_Unselected__ABC_N;
   }
  else
   {
    SSM_SM1_dispatch_sel = outC->M_pre_;
   }
  switch (SSM_SM1_dispatch_sel) {
   case SSM_SM1_Locked__ABC_N :
    outC->foreground = white_ABC_N;
    outC->background = green_ABC_N;
    if (inC->Unlock)
     {
      outC->M_pre_ = SSM_SM1_Preselected__ABC_N;
     }
    else
     {
      outC->M_pre_ = SSM_SM1_Locked__ABC_N;
     }
    break;
   case SSM_SM1_WaitUnlock__ABC_N :
    outC->foreground = black_ABC_N;
    outC->background = grey_ABC_N;
    if (inC->Unlock)
     {
      outC->M_pre_ = SSM_SM1_Unselected__ABC_N;
     }
    else
     {
      outC->M_pre_ = SSM_SM1_WaitUnlock__ABC_N;
     }
    break;
[...]
```

ANSYS

ANSYS SCADE TÜV SÜD Qualification

CERTIFICATE
No. Z10 14 09 55460 007

Holder of Certificate: Esterel Technologies SAS

Factory(ies): 55460

Certification Mark:

Product: Software Tool for Safety Related Development

Model(s): Code Generator SCADE Suite KCG 6.4

Parameters: The code generator, classified as T3 offline support tool according to IEC 61508-4 and EN 50128 is qualified for the use in safety-related software development according to IEC 61508, EN 50128 and ISO 26262.

The report EE86102C is a mandatory part of this certificate.

Tested according to:

Test report no.: EE86102C

Date: 2014-09-23

Page 1 of 1

TÜV SÜD has qualified ANSYS SCADE Suite and SCADE Display KCG for ISO 26262-8:2011 (ASIL D)

ANSYS provides formal TÜV Certification Kits guaranteed certification credits:

- No need to verify generated code compliance against the software model
- MIL verification => SIL verification

ANSYS

With ANSYS SCADE, MIL and SIL can be done in one single step and code generation is automatic

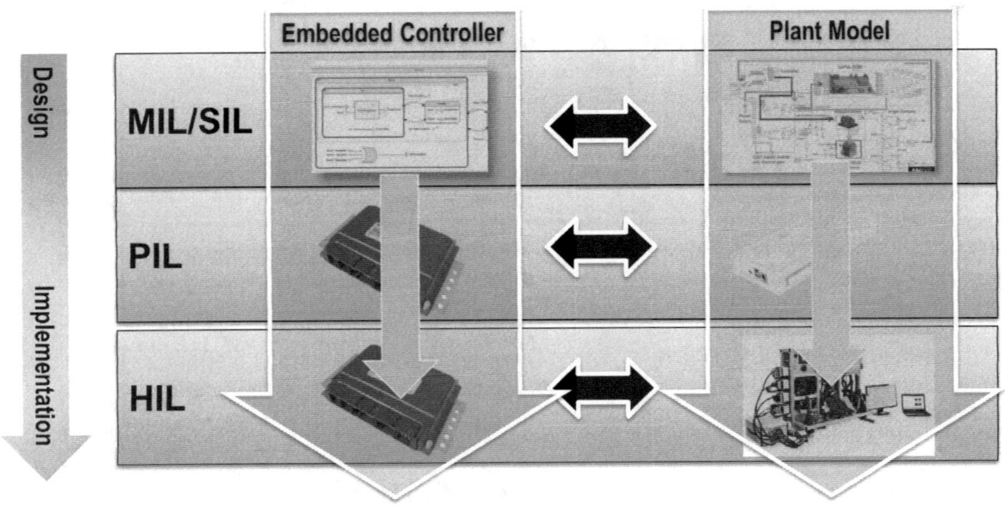

Streamlined Process from Design to Implementation !

ANSYS

ANSYS @ Piaggio Aerospace

Video Presentation in
Original e-media

Autonomous Vehicles, ADAS and IoT

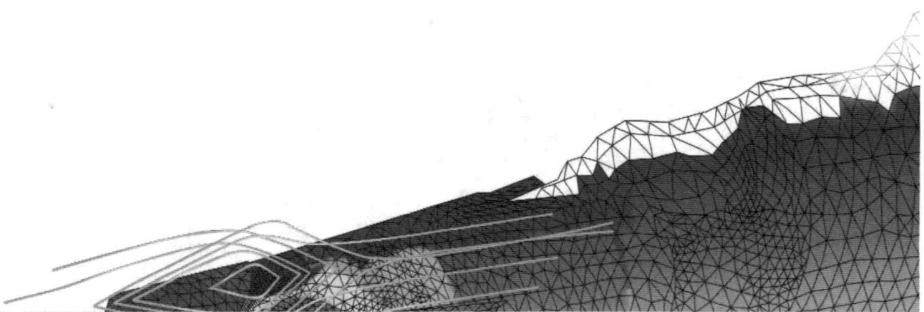

Simulation Driven ADAS Development with ANSYS

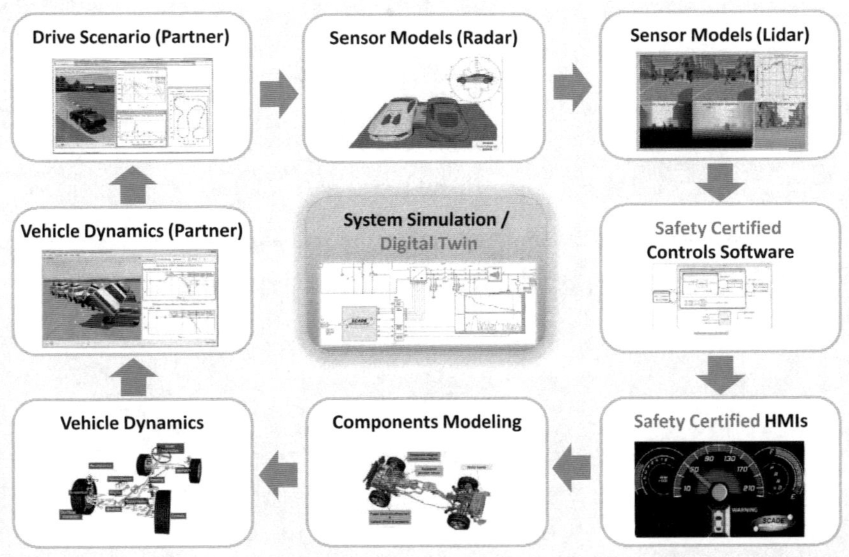

ANSYS

Simulation has been used so far to Engineer better products and improve Bottom line.

ANSYS

System Simulation coupled with IoT can now be used to Better Operate Products and improve Top Line.

ANSYS

ANSYS and GE Work Together To Enable a World of Smarter Machines

ANSYS And GE Work Together To Enable A World Of Smarter Industrial Machines

Released : 29 Sep 2015

PITTSBURGH, Sept. 29, 2015 /PRNewswire/ -- ANSYS (NASDAQ: ANSS) announced today that it joined GE's (NYSE: GE) Digital Predix early adopter program. This move combines the power of the leading engineering simulation platform with the first and only industrial cloud offering for industrial data and analytics, GE's Predix cloud platform. As part of the Predix early adopter program, the two companies will demonstrate a pilot integration of ANSYS flagship simulation platform with GE's Predix industrial internet platform today at GE's Minds + Machines event in San Francisco.

Simulation Platform for the Industrial Internet

ANSYS

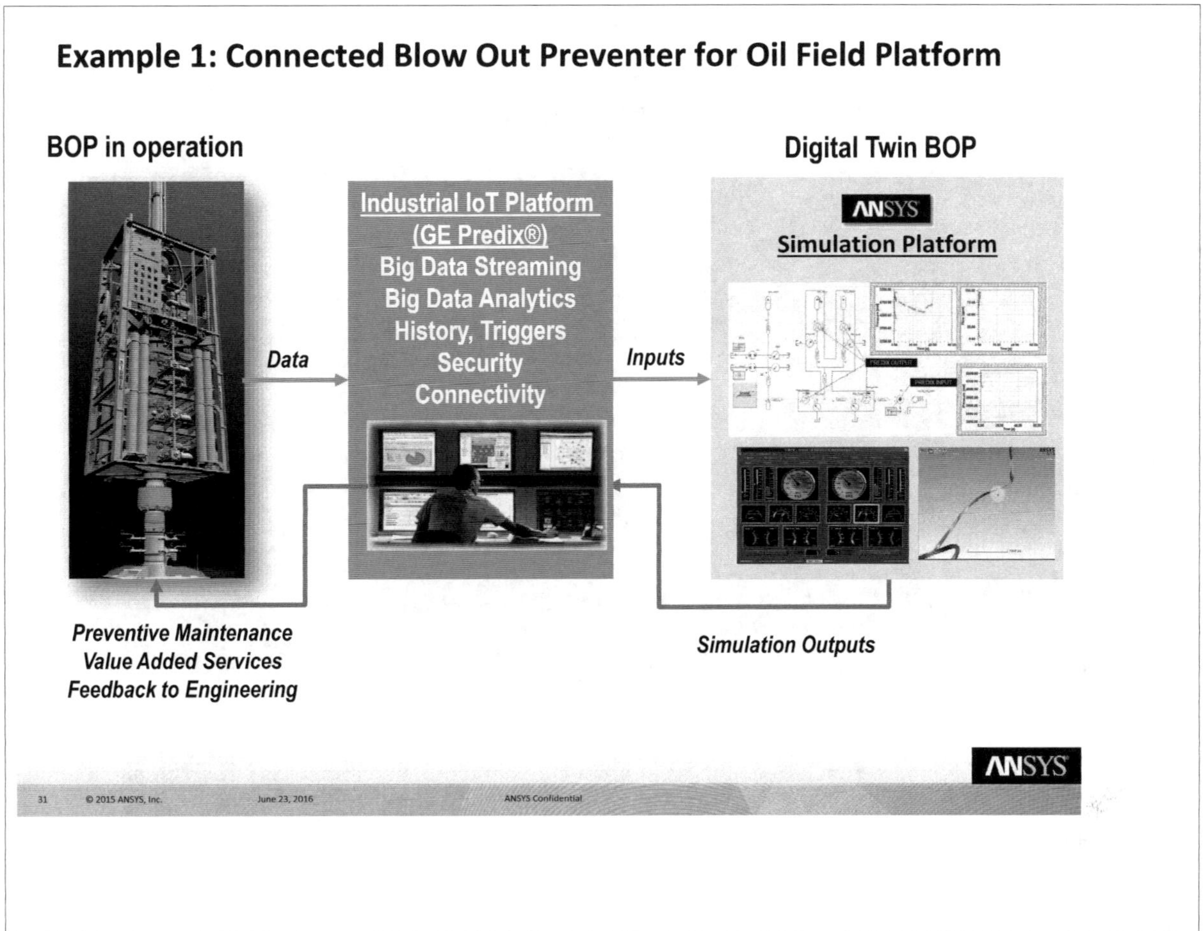
Example 1: Connected Blow Out Preventer for Oil Field Platform
BOP in operation
Digital Twin BOP
Industrial IoT Platform
(GE Predix®)
Big Data Streaming
Big Data Analytics
History, Triggers
Security
Connectivity
Data
Inputs
ANSYS
Simulation Platform
Preventive Maintenance
Value Added Services
Feedback to Engineering
Simulation Outputs
ANSYS
31 © 2015 ANSYS, Inc. June 23, 2016 ANSYS Confidential

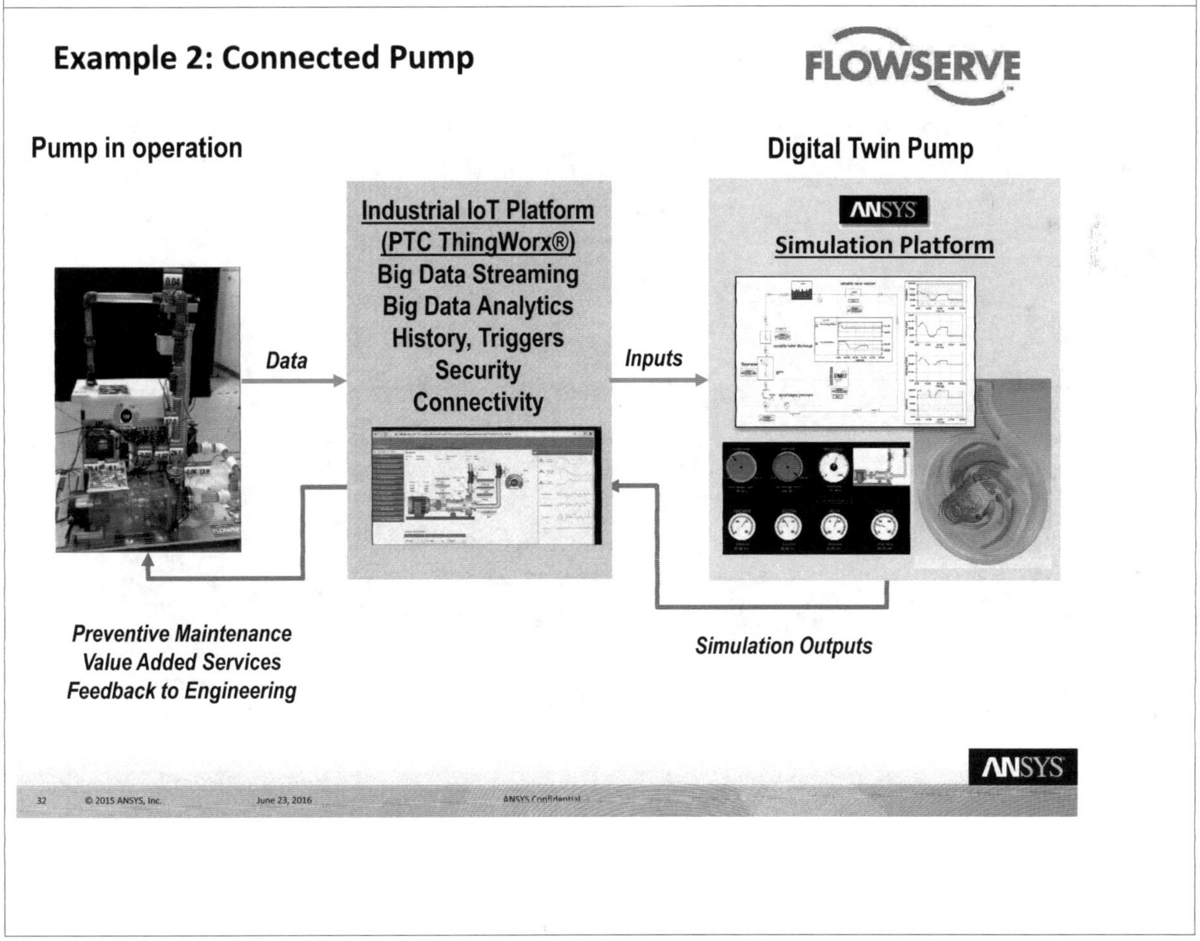
Example 2: Connected Pump
FLOWSERVE
Pump in operation
Digital Twin Pump
Industrial IoT Platform
(PTC ThingWorx®)
Big Data Streaming
Big Data Analytics
History, Triggers
Security
Connectivity
Data
Inputs
ANSYS
Simulation Platform
Preventive Maintenance
Value Added Services
Feedback to Engineering
Simulation Outputs
ANSYS
32 © 2015 ANSYS, Inc. June 23, 2016 ANSYS Confidential

Flowserve Leverages ANSYS to predict product performance for real-world operating conditions

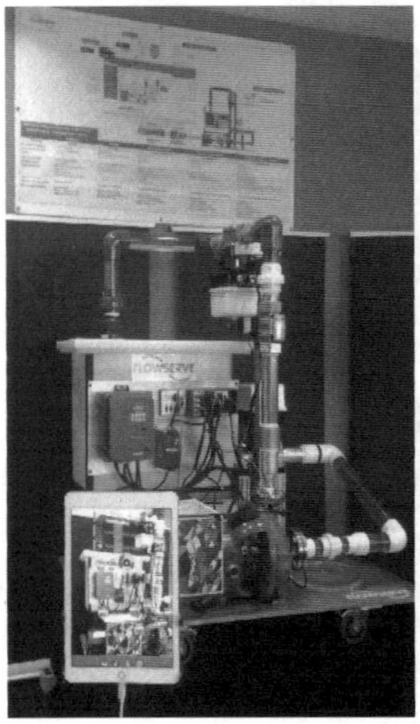

- Anomaly detection and alerts
- Failure Prediction
- Prescription
- Data Mining
- Realtime Monitoring
- Performance Optimization
- Remote Diagnostics
- Prognostics
- Next Gen Design

A Connected Smart Asset

Using Simulation in Operations will enable a massive increase of Customer Value.. and an improved feedback to engineering

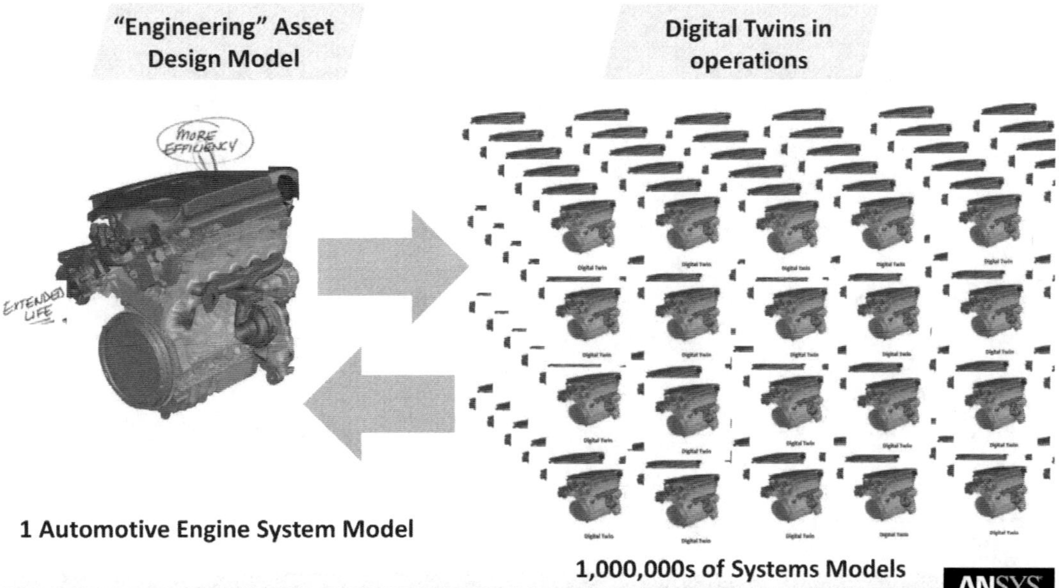

ANSYS

The next 20 years

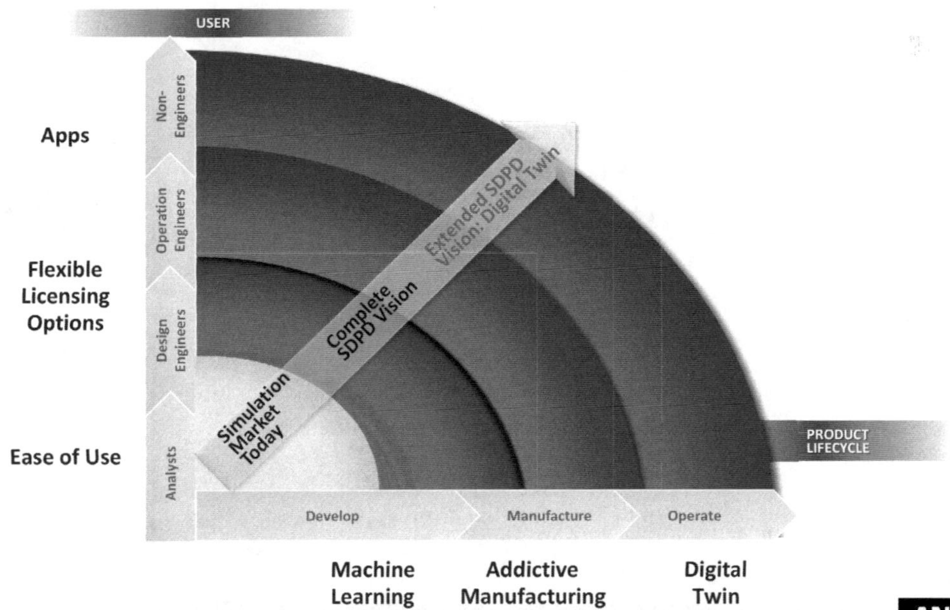

ANSYS

Physics, Electronics, Embedded Software

IoT-capable	Complete Systems	Simulation in Operations

ONLY ANSYS

ASWC-AESE 2016: Awards Ceremony

Thomas Zanzinger

Area Director Germany, Austria, Nordics

Lukas Schäfer (BMW)

"Lagrangian Spark Ignition Model – Accounting for Spark-Channel Dynamics, Diffusion and Detailed Chemical Kinetics"

Andreas Boemer & Philipp Morsch (Deutz)

"Efficient simulation of boiling effects at engine cooling jackets"

ANSYS

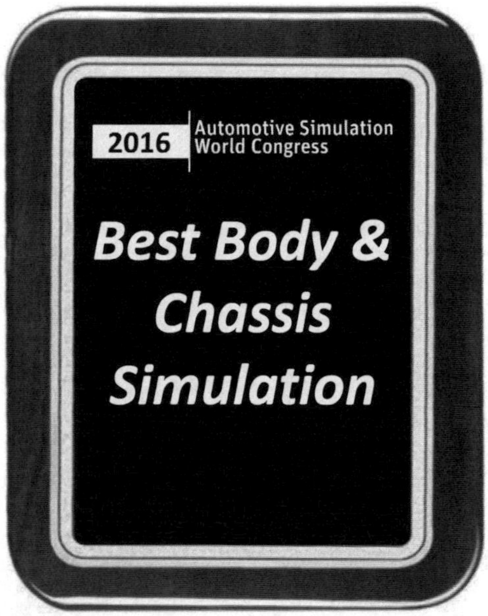

Johan Brunberg & Eysteinn Helgason (Volvo Cars)

"Optimization of Turbo Inlet Duct"

ANSYS

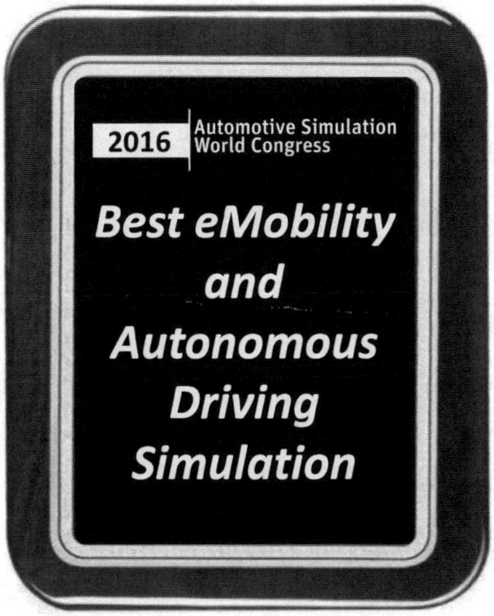

Marc Brück (EMMotive)

"Integral Approach for Simulation-Driven Development of Electric Traction Motors in Automotive Applications"

Jonathan Bayani (Tesla Motors)

"Automotive System Virtual Prototyping and Design for Reliability"

Daniel Krätschmer (Bosch)

"Simulation Concept for the Virtual Assessment of the Transient-Thermal Behaviour of PCB Based ECUs"

ANSYS

Gabriele Capello (U-Blox)

"Thermal Gradients in a Cellular Module for 2G/3G Cellular Communications"

ANSYS

Stefan Weiland (ZF-TRW)

"Efficient CAE-Tools and Processes to Support Brake-Squeal Countermeasure Development"

ANSYS

Thank you to all presenters!

ANSYS

THE BMW VISION AND STRATEGY IN ENGINE SIMULATION.

FROM SUPPORTER TO DRIVING FORCE.

Automotive Simulation World Congress 2016 I Dirk Linse

BMW GROUP THE NEXT 100 YEARS

IN-CYLINDER FLOW SIMULATION OF A BMW FOUR-CYLINDER GASOLINE ENGINE.

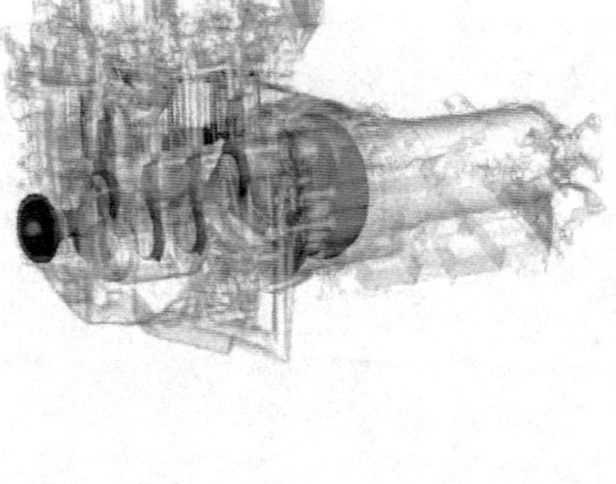

WORKFLOW TO MINIMIZE USER INTERACTION TIME AND TO OBTAIN USER INDEPENDENT RESULTS.

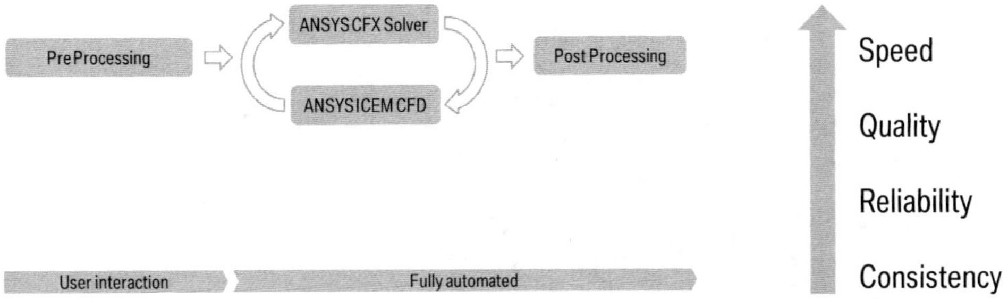

DEVELOPMENT OF MODELS TO PRODUCE PREDICTIVE AND ACCURATE RESULTS.

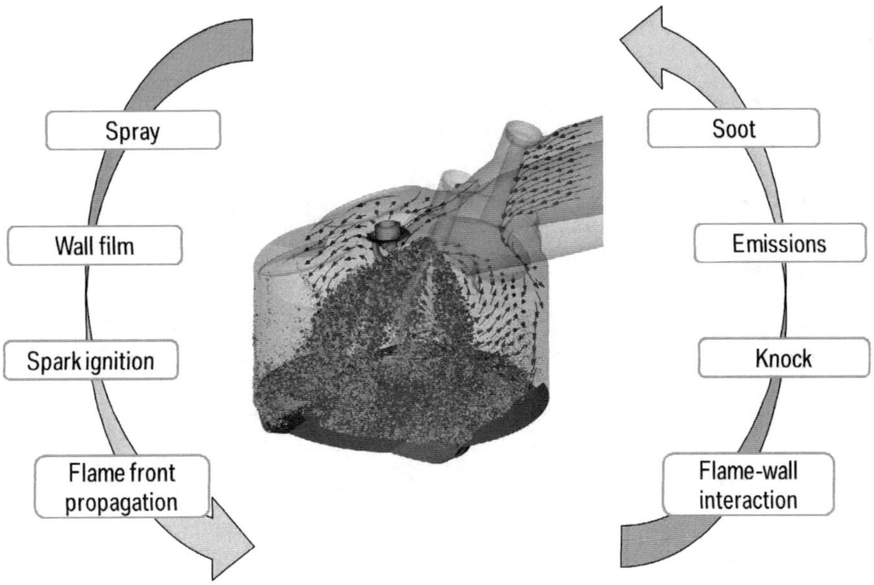

VALIDATION AND VERIFICATION TO PREDICT WITH CONFIDENCE. TRUST YOUR OWN SIMULATION RESULTS.

Laboratory Flames

Flow Bench

Engine

BMW HPC PRIVATE CLOUD ENABLES US TO CONSIDERABLY IMPROVE OUR SIMULATION METHODOLOGY.

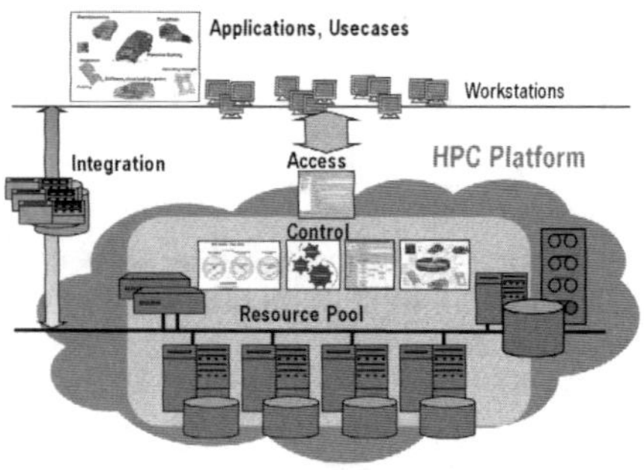

HPC Private Cloud

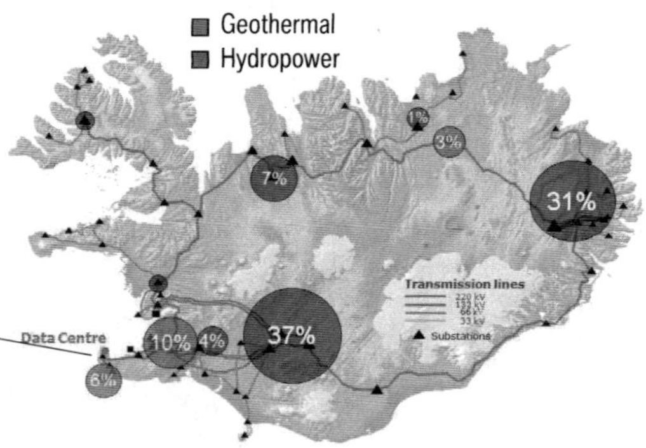

Iceland

INTAKE PORT EVALUATION USING THE STATIONARY FLOW BENCH.

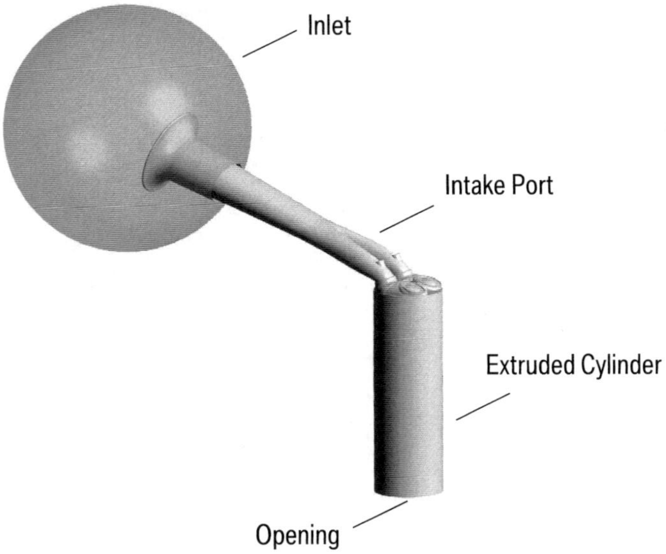

SCALE RESOLVING SIMULATION ESTABLISHED IN THE DEVELOPMENT PROCESS OF INTAKE PORTS.

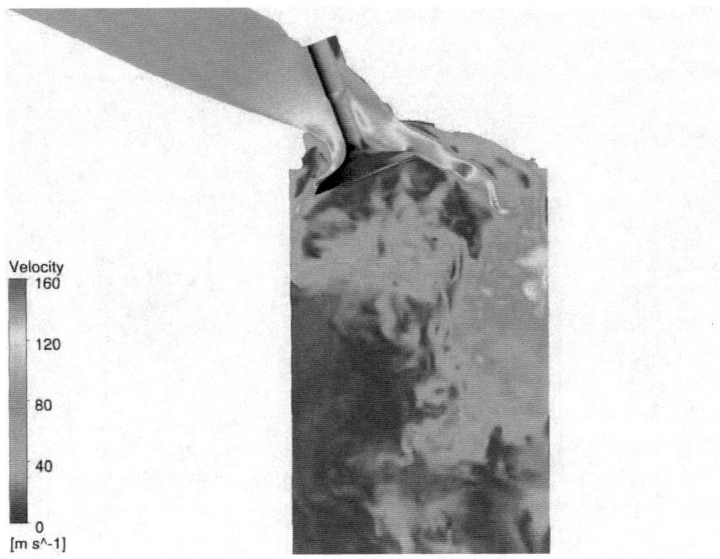

Stationary Flow Bench
40 million nodes
4 weeks @ 1000 cores

DERIVATION OF NEW BEST-PRACTICE SETTINGS TO OBTAIN USER AND MESH INDEPENDENT RESULTS.

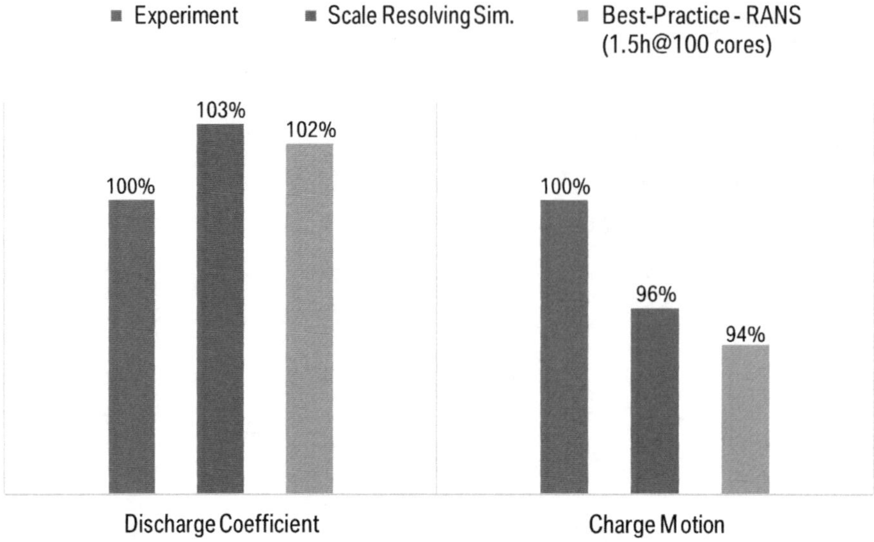

NEW PROCESS ACCOUNTING FOR TOLERANCES IN THE CASTING PROCESS OF INTAKE PORTS.

Core Offset

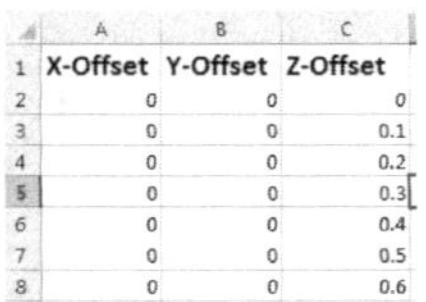

	A	B	C
1	X-Offset	Y-Offset	Z-Offset
2	0	0	0
3	0	0	0.1
4	0	0	0.2
5	0	0	0.3
6	0	0	0.4
7	0	0	0.5
8	0	0	0.6

Parametrized CAD

Meshing

Simulation

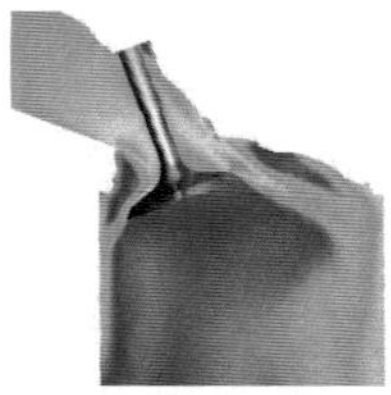

Fully automated

EVALUATION OF THE IMPACT OF TOLERANCES ON CHARGE MOTION AND DISCHARGE COEFFICIENT.

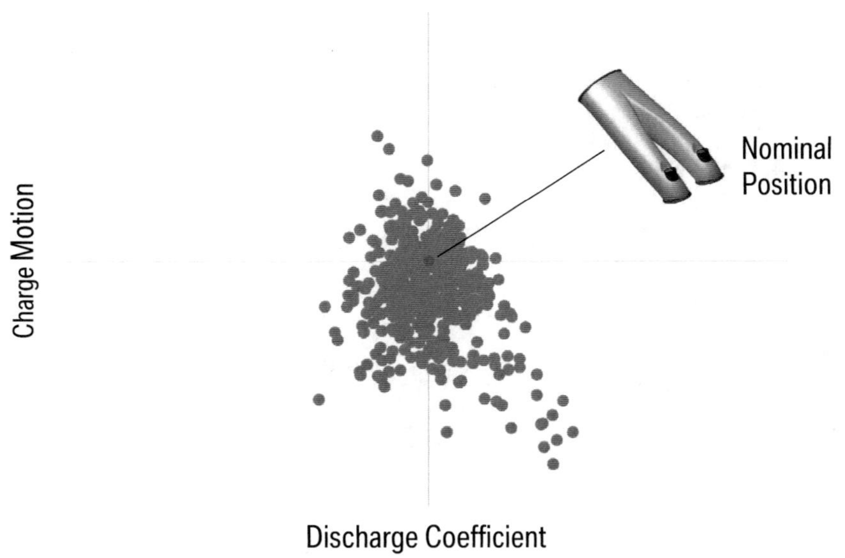

DETERMINATION OF THE STATISTICS OF FUNCTIONAL PARAMETERS IN THE PRODUCTION PROCESS.

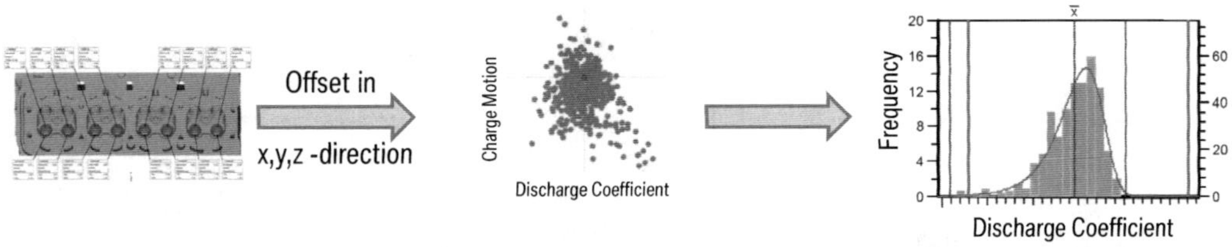

Tactile Measurement

Simulation Database

Statistics

WE NEED TO ENHANCE OUR METHODOLOGY TO STRENGTHEN SIMULATION IN THE EARLY PHASE OF DEVELOPMENT.

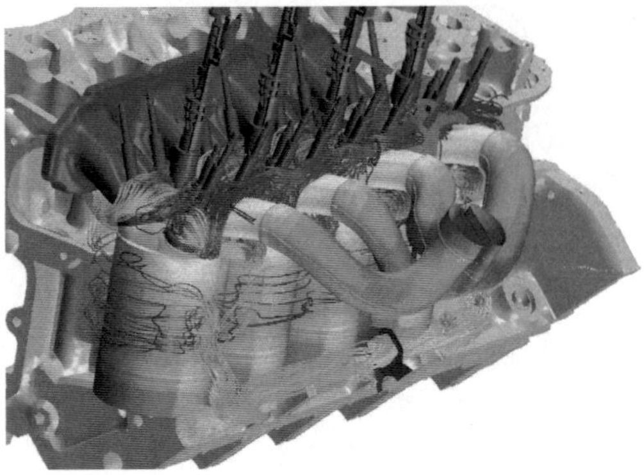

Multicylinder

Multiphysics

Multiscale

IT IS NECESSARY TO EMBRACE NEW SKILLS. FROM SUPPORTER TO DRIVING FORCE.

"Old" Skills	New Skills
IT	Management
Programming	Communication
Meshing	Stress Resistance

HOW DO WE UTILIZE SIMULATION IN THE FUTURE?

Routine-Based

Optimize

Improvements

Creative

Innovate

Competitive Advantage

AUTHOR INDEX